Testing of Materials and Elements in Civil Engineering

Testing of Materials and Elements in Civil Engineering

Volume 2

Editor

Krzysztof Schabowicz

MDPI • Basel • Beijing • Wuhan • Barcelona • Belgrade • Manchester • Tokyo • Cluj • Tianjin

Editor
Krzysztof Schabowicz
Wroclaw University of Science
and Technology
Poland

Editorial Office
MDPI
St. Alban-Anlage 66
4052 Basel, Switzerland

This is a reprint of articles from the Special Issue published online in the open access journal *Materials* (ISSN 1996-1944) (available at: https://www.mdpi.com/journal/materials/special_issues/testing_civil_engineering).

For citation purposes, cite each article independently as indicated on the article page online and as indicated below:

LastName, A.A.; LastName, B.B.; LastName, C.C. Article Title. *Journal Name* **Year**, *Volume Number*, Page Range.

Volume 2
ISBN 978-3-0365-1890-9 (Hbk)
ISBN 978-3-0365-1891-6 (PDF)

Volume 1-2
ISBN 978-3-0365-1577-9 (Hbk)
ISBN 978-3-0365-1578-6 (PDF)

Contents

About the Editor ix

Krzysztof Schabowicz
Testing of Materials and Elements in Civil Engineering
Reprinted from: *Materials* **2021**, *14*, 3412, doi:10.3390/ma14123412 1

Jeongjun Park and Gigwon Hong
Strength Characteristics of Controlled Low-Strength Materials with Waste Paper Sludge Ash
(WPSA) for Prevention of Sewage Pipe Damage
Reprinted from: *Materials* **2020**, *13*, 4238, doi:10.3390/ma13194238 21

Jarosław Michałek and Maciej Sobótka
Assessment of Internal Structure of Spun Concrete Using Image Analysis and
Physicochemical Methods
Reprinted from: *Materials* **2020**, *13*, 3987, doi:10.3390/ma13183987 43

Sung-Sik Park, Jung-Shin Lee and Dong-Eun Lee
Aggregate Roundness Classification Using a Wire Mesh Method
Reprinted from: *Materials* **2020**, *13*, 3682, doi:10.3390/ma13173682 69

Wiesław Trąmpczyński, Barbara Goszczyńska and Magdalena Bacharz
Acoustic Emission for Determining Early Age Concrete Damage as an Important Indicator of
Concrete Quality/Condition before Loading
Reprinted from: *Materials* **2020**, *13*, 3523, doi:10.3390/ma13163523 79

Małgorzata Wutke, Anna Lejzerowicz and Andrzej Garbacz
The Use of Wavelet Analysis to Improve the Accuracy of Pavement Layer Thickness Estimation
Based on Amplitudes of Electromagnetic Waves
Reprinted from: *Materials* **2020**, *13*, 3214, doi:10.3390/ma13143214 95

Adam Zieliński and Maria Kaszyńska
Calibration of Steel Rings for the Measurement of Strain and Shrinkage Stress for
Cement-Based Composites
Reprinted from: *Materials* **2020**, *13*, 2963, doi:10.3390/ma13132963 113

Anna Adamczak-Bugno and Aleksandra Krampikowska
The Acoustic Emission Method Implementation Proposition to Confirm the Presence and
Assessment of Reinforcement Quality and Strength of Fiber–Cement Composites
Reprinted from: *Materials* **2020**, *13*, 2966, doi:10.3390/ma13132966 129

Florin Dumbrava and Camelia Cerbu
Experimental Study on the Stiffness of Steel Beam-to-Upright Connections for Storage
Racking Systems
Reprinted from: *Materials* **2020**, *13*, 2949, doi:10.3390/ma13132949 151

Grzegorz Świt, Aleksandra Krampikowska, Tadeusz Pała, Sebastian Lipiec and Ihor Dzioba
Using AE Signals to Investigate the Fracture Process in an Al–Ti Laminate
Reprinted from: *Materials* **2020**, *13*, 2909, doi:10.3390/ma13132909 179

Jong Yil Park, Sung-Han Sim, Young Geun Yoon and Tae Keun Oh
Prediction of Static Modulus and Compressive Strength of Concrete from Dynamic Modulus
Associated with Wave Velocity and Resonance Frequency Using Machine Learning Techniques
Reprinted from: *Materials* **2020**, *13*, 2886, doi:10.3390/ma13132886 . . . . **203**

**Janusz Kozubal, Roman Wróblewski, Zbigniew Muszyński, Marek Wyjadłowski and
Joanna Stróżyk**
Non-Deterministic Assessment of Surface Roughness as Bond Strength Parameters between
Concrete Layers Cast at Different Ages
Reprinted from: *Materials* **2020**, *13*, 2542, doi:10.3390/ma13112542 . . . . **227**

Yingliang Tan, Bing Zhu, Le Qi, Tingyi Yan, Tong Wan and Wenwei Yang
Mechanical Behavior and Failure Mode of Steel–Concrete Connection Joints in a Hybrid Truss
Bridge: Experimental Investigation
Reprinted from: *Materials* **2020**, *13*, 2549, doi:10.3390/ma13112549 . . . . **249**

Michał Wróbel, Agnieszka Woszuk and Wojciech Franus
Laboratory Methods for Assessing the Influence of Improper Asphalt Mix Compaction on
Its Performance
Reprinted from: *Materials* **2020**, *13*, 2476, doi:10.3390/ma13112476 . . . . **265**

Dominik Logoń, Krzysztof Schabowicz and Krzysztof Wróblewski
Assessment of the Mechanical Properties of ESD Pseudoplastic Resins for Joints in Working
Elements of Concrete Structures
Reprinted from: *Materials* **2020**, *13*, 2426, doi:10.3390/ma13112426 . . . . **283**

**Zbigniew Ranachowski, Przemysław Ranachowski, Tomasz Dębowski, Adam Brodecki,
Mateusz Kopec, Maciej Roskosz, Krzysztof Fryczowski, Mateusz Szymków, Ewa Krawczyk
and Krzysztof Schabowicz**
Mechanical and Non-Destructive Testing of Plasterboards Subjected to a Hydration Process
Reprinted from: *Materials* **2020**, *13*, 2405, doi:10.3390/ma13102405 . . . . **301**

Krzysztof Schabowicz, Paweł Sulik and Łukasz Zawiślak
Identification of the Destruction Model of Ventilated Facade under the Influence of Fire
Reprinted from: *Materials* **2020**, *13*, 2387, doi:10.3390/ma13102387 . . . . **319**

Bohdan Stawiski and Tomasz Kania
Tests of Concrete Strength across the Thickness of Industrial Floor Using the Ultrasonic Method
with Exponential Spot Heads
Reprinted from: *Materials* **2020**, *13*, 118, doi:10.3390/ma13092118 . . . . **333**

Czesław Bywalski, Maciej Kaźmierowski, Mieczysław Kamiński and Michał Drzazga
Material Analysis of Steel Fibre Reinforced High-Strength Concrete in Terms of Flexural
Behaviour. Experimental and Numerical Investigation
Reprinted from: *Materials* **2020**, *13*, 1631, doi:10.3390/ma13071631 . . . . **355**

Yueqi Shi, Changhong Li and Dayu Long
Study of the Microstructure Characteristics of Three Different Fine-grained Tailings Sand
Samples during Penetration
Reprinted from: *Materials* **2020**, *13*, 1585, doi:10.3390/ma13071585 . . . . **375**

Marek Słoński, Krzysztof Schabowicz and Ewa Krawczyk
Detection of Flaws in Concrete Using Ultrasonic Tomography and Convolutional
Neural Networks
Reprinted from: *Materials* **2020**, *13*, 1557, doi:10.3390/ma13071557 . . . . **407**

Janusz Juraszek
Fiber Bragg Sensors on Strain Analysis of Power Transmission Lines
Reprinted from: *Materials* **2020**, *13*, 1559, doi:10.3390/ma13071559 . . . . **423**

Anna Karolak, Jerzy Jasieńko, Tomasz Nowak and Krzysztof Raszczuk
Experimental Investigations of Timber Beams with Stop-Splayed Scarf Carpentry Joints
Reprinted from: *Materials* **2020**, *13*, 1435, doi:10.3390/ma13061435 . . . . **435**

Tomasz Trapko and Michał Musiał
Effect of PBO–FRCM Reinforcement on Stiffness of Eccentrically Compressed Reinforced
Concrete Columns
Reprinted from: *Materials* **2020**, *13*, 1221, doi:10.3390/ma13051221 . . . . **451**

Czesław Bywalski, Michał Drzazga, Maciej Kaźmierowski and Mieczysław Kamiński
Shear Behavior of Concrete Beams Reinforced with a New Type of Glass Fiber Reinforced
Polymer Reinforcement: Experimental Study
Reprinted from: *Materials* **2020**, *13*, 1159, doi:10.3390/ma13051159 . . . . **469**

Lukasz Skotnicki, Jarosław Kuzniewski and Antoni Szydlo
Stiffness Identification of Foamed Asphalt Mixtures with Cement, Evaluated in Laboratory and
In Situ in Road Pavements
Reprinted from: *Materials* **2020**, *13*, 1128, doi:10.3390/ma13051128 . . . . **487**

Ewa Szewczak, Agnieszka Winkler-Skalna and Lech Czarnecki
Sustainable Test Methods for Construction Materials and Elements
Reprinted from: *Materials* **2020**, *13*, 606, doi:10.3390/ma13030606 . . . . **507**

Dariusz Bajno, Lukasz Bednarz, Zygmunt Matkowski and Krzysztof Raszczuk
Monitoring of Thermal and Moisture Processes in Various Types of External Historical Walls
Reprinted from: *Materials* **2020**, *13*, 505, doi:10.3390/ma13030505 . . . . **531**

Mario Bačić, Meho Saša Kovačević and Danijela Jurić Kaćunić
Non-Destructive Evaluation of Rock Bolt Grouting Quality by Analysis of Its
Natural Frequencies
Reprinted from: *Materials* **2020**, *13*, 282, doi:10.3390/ma13020282 . . . . **547**

About the Editor

Krzysztof Schabowicz

A specialist in the field of building engineering, building law, diagnostics and maintenance of engineering structures. He deals with ventilated fac̦ades, in particular production technology and testing of external fiber cement cladding in the field of detection, identification and classification of degradation and damage processes, as well as the methodology of these tests. He conducts scientific research and development works related to the implementation of non-destructive diagnostics devices and technologies in construction works, including the use of artificial intelligence. Author and co-author of 5 books, over 200 publications and 9 patents. Has more than 500 citations in Web of Science. He serves as an Editor of Materials (MDPI) and Editorial Board member of Civil Engineering and Architecture (HRPUB), and Nondestructive Testing and Diagnostics (SIMP). He developed more than 200 reviews of journal and conference articles.

Editorial

Testing of Materials and Elements in Civil Engineering

Krzysztof Schabowicz

Faculty of Civil Engineering, Wrocław University of Science and Technology, Wybrzeże Wyspiańskiego 27, 50-370 Wrocław, Poland; krzysztof.schabowicz@pwr.edu.pl

Abstract: This issue is proposed and organized as a means to present recent developments in the field of testing of materials in civil engineering. For this reason, the articles highlighted in this issue should relate to different aspects of testing of different materials in civil engineering, from building materials and elements to building structures. The current trend in the development of materials testing in civil engineering is mainly concerned with the detection of flaws and defects in elements and structures using destructive, semi-destructive, and nondestructive testing. The trend, as in medicine, is toward designing test equipment that allows one to obtain a picture of the inside of the tested element and materials. Very interesting results with significance for building practices of testing of materials and elements in civil engineering were obtained.

Keywords: testing; diagnostic; building materials; elements; civil engineering

Citation: Schabowicz, K. Testing of Materials and Elements in Civil Engineering. *Materials* **2021**, *14*, 3412. https://doi.org/10.3390/ma14123412

Received: 29 May 2021
Accepted: 18 June 2021
Published: 20 June 2021

1. Introduction

The field of testing of materials in civil engineering is very wide, and is interesting from an engineering and scientific point of view [1–3]. This issue is proposed and organized as a means to present recent developments in the field of testing of materials in civil engineering. For this reason, the articles highlighted in this issue should relate to different aspects of testing of different materials in civil engineering, from building materials and elements to building structures [4–6]. The current trend in the development of materials testing in civil engineering is mainly concerned with the detection of flaws and defects in elements and structures using destructive, semi-destructive, and nondestructive testing.

This issue mainly focuses on different novel testing approaches, the development of single and hybrid measurement techniques, and advanced signal analysis. The topics of interest include but are not limited to the testing of materials and elements in civil engineering, testing of structures made of novel materials [7–9], condition assessment of civil materials and elements, detecting defects invisible on the surface, damage detection and damage imaging, diagnostics of cultural heritage monuments, structural health monitoring systems, modeling and numerical analyses, nondestructive testing methods, and advanced signal processing for nondestructive testing [10,11].

2. Description of the Articles Presented in the Issue

Grouted rock bolts represent one of the most used elements for rock mass stabilization, as analyzed by [12], and reinforcement and the grouting quality have a crucial role in the load transfer mechanism. At the same time, the grouting quality, as well as the grouting procedures, are the least controlled in practice. This paper deals with the non-destructive investigation of grouting percentage through an analysis of the rock bolt's natural frequencies after applying an artificial longitudinal impulse to its head by using a soft-steel hammer as a generator. A series of laboratory models, with different positions and percentages of the grouted section, simulating grouting defects, were tested. A comprehensive statistical analysis was conducted and a high correlation between the grouting percentage and the first three natural frequencies of rock bolt models has been established. After validation of FEM numerical models based on experimentally obtained values, a further analysis

1

includes consideration of grout stiffness variation and its impact on the natural frequencies of rock bolts [12].

In their paper [13], in order to create and make available design guidelines, recommendations for energy audits, data for analysis and simulation of the condition of masonry walls susceptible to biological corrosion, deterioration of comfort parameters in rooms, or deterioration of thermal resistance, were given. The paper analyzes various types of masonry wall structures occurring in and commonly used in historical buildings over the last 200 years. The summary is a list of results of particular types of masonry walls and their mutual comparison. On this basis, a procedure path has been proposed which is useful for monitoring heat loss, monitoring the moisture content of building partitions, and improving the hygrothermal comfort of rooms. The durability of such constructions has also been estimated, and the impact on the condition of the buildings that have been preserved and are still in use today was assessed [13].

In [14], the laboratory testing of the construction materials and elements is a subset of activities inherent in sustainable building materials engineering. Two questions arise regarding the test methods used: the relation between test results and material behavior in actual conditions on the one hand, and the variability of results related to uncertainty on the other. The paper presents the analysis of the results and uncertainties of the two simple independent test examples (bond strength and tensile strength) in order to demonstrate discrepancies related to the ambiguous methods of estimating uncertainty, and the consequences of using test methods when method suitability for conformity assessment has not been properly verified. These examples are the basis for opening a discussion on the test methods development direction which makes it possible to consider them as "sustainable". The paper addresses the negative impact of the lack of complete test models, taking into account proceeding with the uncertainty regarding erroneous assessment risks. Adverse effects can be minimized by creating test methods appropriate for the test's purpose (e.g., initial or routine tests) and handling uncontrolled uncertainty components. Sustainable test methods should ensure a balance between widely defined tests and evaluation costs and the material's or building's safety, reliability, and stability [14].

The article by [15] presents the possibilities of using foamed asphalt in the recycling process to produce the base layer of road pavement constructions, in Polish conditions. Foamed asphalt was combined with reclaimed asphalt pavement (RAP) and hydraulic binder (cement). Foamed asphalt mixtures with cement (FAC) were made, based on these ingredients. To reduce stiffness and cracking in the base layer, foamed asphalt (FA) was additionally used in the analyzed mixes containing cement. The laboratory analyses allowed estimating the stiffness and fatigue durability of the conglomerate. In the experimental section, measurements of deflections are made, modules of pavement layers are calculated, and their fatigue durability is determined. As a result of the research, new fatigue criteria for FAC mixtures and the correlation factors of stiffness modulus and fatigue durability in situ with the results of laboratory tests were developed. It is anticipated that FAC recycling technology will provide durable and safe road pavements [15].

The article by [16] presents experimental tests of a new type of composite bar that has been used as shear reinforcement for concrete beams. In the case of shearing concrete beams reinforced with steel stirrups, according to the theory of plasticity, the plastic deformation of stirrups, and stress redistribution in stirrups cut by a diagonal crack, are permitted. Tensile composite reinforcement is characterized by linear-elastic behavior throughout the entire strength range. The most popular type of shear reinforcement is closed-frame stirrups, and this type of fiber-reinforced polymer (FRP) shear reinforcement was the subject of research by other authors. In the case of FRP stirrups, rupture occurs rapidly, without the shear reinforcement being able to redistribute stress. An attempt was made to introduce a quasi-plastic character into the mechanisms transferring shear by appropriately shaping the shear reinforcement. Experimental material tests covered the determination of the strength and deformability of straight glass fiber-reinforced polymer (GFRP) bars and GFRP headed bars. Experimental studies of shear-reinforced beams with GFRP stirrups and

GFRP headed bars were carried out. This allowed a direct comparison of the shear behavior of beams reinforced with standard GFRP stirrups and a new type of shear reinforcement: GFRP headed bars. Experimental studies demonstrated that GFRP headed bars could be used as shear reinforcement in concrete beams. Unlike GFRP stirrups, these bars allow stress redistribution in bars cut by a diagonal crack [16].

The paper by [17] examines the effect of PBO (P-phenylene benzobisoxazole)–FRCM (fabric-reinforced cementitious matrix) reinforcement on the stiffness of eccentrically compressed reinforced concrete columns. Reinforcement with FRCM consists of bonding composite meshes to the concrete substrate by means of mineral mortar. Longitudinal and/or transverse reinforcements made of PBO (P-phenylene benzobisoxazole) mesh were applied to the analyzed column specimens. When assessing the stiffness of the columns, the focus was on the effect of the composite reinforcement itself, the value and eccentricity of the longitudinal force, and the decrease in the modulus of elasticity of the concrete, with increasing stress intensity in the latter. Dependencies between the change in the elasticity modulus of the concrete and the change in the stiffness of the tested specimens were examined. The relevant standards, providing methods of calculating the stiffness of composite columns, were used in the analysis. Regarding columns, which were strengthened only transversely with PBO mesh, reinforcement increases their load capacity, and at the same time, the stiffness of the columns increases due to the confinement of the cross-section. The stiffness depends on the destruction of the concrete core inside its composite jacket. In the case of columns with transverse and longitudinal reinforcement, the presence of longitudinal reinforcement reduces longitudinal deformations. The columns failed at higher stiffness values in the whole range of the eccentricities [17].

The paper by [18] presents the results of an experimental investigation into stop-splayed scarf joints which was carried out as part of a research program at the Wroclaw University of Science and Technology. A brief description of the characteristics of scarf and splice joints appearing in historical buildings is provided, with special reference to stop-splayed scarf joints (so-called "bolt of lightning" joints) which were widely used, for example, in Italian Renaissance architecture. Analyses and studies of scarf and splice joints in bent elements as presented in the literature are reviewed, along with selected examples of analyses and research on tensile joints. It is worth noting that the authors in practically all the cited literature draw attention to the need for further research in this area. Next, the results of the authors' own research on beams with stop-splayed scarf joints, strengthened using various methods, e.g., by means of drawbolts (metal screws), steel clamps and steel clamps with wooden pegs, which were subjected to four-point bending tests, are presented. Load-deflection plots were obtained for load-bearing to bending of each beam in relation to the load-bearing of a continuous reference beam. A comparative analysis of the results obtained for each beam series is presented, along with conclusions and directions for further research [18].

Non-destructive testing of concrete for defects detection, using acoustic techniques, is currently performed mainly by human inspection of recorded images [19]. The images consist of the inside of the examined elements, obtained from testing devices such as the ultrasonic tomograph. However, such an automatic inspection is time-consuming, expensive, and prone to errors. To address some of these problems, this paper aims to evaluate a convolutional neural network (CNN) toward an automated detection of flaws in concrete elements using ultrasonic tomography. There are two main stages in the proposed methodology. In the first stage, an image of the inside of the examined structure is obtained and recorded by performing ultrasonic tomography-based testing. In the second stage, a convolutional neural network model is used for the automatic detection of defects and flaws in the recorded image. In this work, a large and pre-trained CNN is used. It was fine-tuned on a small set of images collected during laboratory tests. Lastly, the prepared model was applied for detecting flaws. The obtained model has proven to be able to accurately detect defects in examined concrete elements. The presented approach for automatic detection of flaws is being developed with the potential to not only detect defects of one type but also to classify various types of defects in concrete elements [19].

The reliability and safety of power transmission depend first and foremost on the state of the power grid, and mainly on the state of the high-voltage power line towers [20]. The steel structures of existing power line supports (towers) have been in use for many years. Their in-service time, the variability in structural, thermal and environmental loads, the state of foundations (displacement and degradation), the corrosion of supporting structures and lack of technical documentation are essential factors that have an impact on the operating safety of the towers. The tower state assessment used to date, consisting of finding the deviation in the supporting structure apex, is insufficient because it omits the other necessary condition, the stress criterion, which is not to exceed allowable stress values. Moreover, in difficult terrain conditions, the measurement of the tower deviation is very troublesome, and for this reason, it is often not performed. This paper presents a stress-and-strain analysis of the legs of 110 kV power line truss towers with a height of 32 m. They have been in use for over 70 years and are located in especially difficult geotechnical conditions—one of them is in a gravel mine on an island surrounded by water, and the other stands on a steep, wet slope. Purpose-designed fiber Bragg grating (FBG) sensors were proposed for strain measurements. Real values of stresses arising in the tower legs were observed and determined over a period of one year. Validation was also carried out based on geodetic measurements of the tower apex deviation, and a residual magnetic field (RMF) analysis was performed to assess the occurrence of cracks and stress concentration zones [20].

The paper by [21] explores the microstructural evolution characteristics of tailings sand samples from different types of infiltration failure during the infiltration failure process. A homemade small infiltration deformation instrument is used to test the infiltration failure characteristics of the tailings sand during the infiltration failure process. Evolutionary characteristics of the internal microstructure pores and particle distribution were also studied. Using CT (computerized tomography) technology to establish digital image information, the distribution of the microscopic characteristics of the particle distribution and pore structure after tailing sand infiltration were studied. Microscopic analysis was also performed to analyze the microscopic process of infiltration and destruction, as well as to see the microscopic structural characteristics of the infiltration and destruction of the total tailings. The test results show that there are obvious differences in the microstructure characterization of fluid soil and piping-type infiltration failures. Microstructure parameters have a certain functional relationship with macro factors. Combining the relationship between macrophysical and mechanical parameters and microstructural parameters, new ideas for future research and the prevention of tailings sand infiltration and failure mechanisms are provided [21].

The paper by [22] presents the results of tests for flexural tensile strength ($f_{ct,fl}$) and fracture energy (G_f) in a three-point bending test of prismatic beams with notches, which were made from steel fiber-reinforced high-strength concrete (SFRHSC). The registration of the conventional force–displacement (F–δ) relationship and unconventional force-crack tip opening displacement (CTOD) relationship was made. On the basis of the obtained test results, estimations of the parameters $f_{ct,fl}$ and G_f in the function of the fiber-reinforcement ratio were carried out. The obtained results were applied to building and validating a numerical model with the use of the finite element method (FEM). A non-linear concrete damaged plasticity model CDP was used for the description of the concrete. The obtained FEM results were compared with the experimental ones that were based on the assumed criteria. The usefulness of the flexural tensile strength and fracture energy parameters for defining the linear form of weakening of the SFRHSC material under tension was confirmed. The author's own equations for estimating the flexural tensile strength and fracture energy of SFRHSC, as well as for approximating deflections (δ) of SFRHSC beams as the function of crack tip opening displacement (CTOD) instead of crack mouth opening displacement (CMOD), were proposed [22].

The accepted methods for testing concrete are not favorable for determining its heterogeneity [23]. The interpretation of the compressive strength result as a product of destructive force and cross-section area is burdened with significant understatements. It is assumed erroneously that this is the lowest value of strength at the height of the tested sample. The top layer of concrete floors often crumble, and the strength tested using sclerometric methods does not confirm the concrete class determined using control samples. That is why it is important to test the distribution of compressive strength in a cross-section of concrete industrial floors with special attention to surface top layers. This study presents strength tests of borehole material taken from industrial floors using the ultrasonic method, with exponential spot heads with a contact surface area of 0.8 mm^2 and a frequency of 40 kHz. The presented research project anticipated the determination of strength for samples in various cross-sections at the height of elements and destructive strength in the strength testing machine. It was confirmed that for standard and big borehole samples, it is not possible to test the strength of concrete in the top layer of the floor by destructive methods. This can be achieved using the ultrasonic method. After the analysis, certain types of distributions of strength across concrete floor thickness were chosen from the completed research program. The gradient and anti-gradient of strength were proposed as new parameters for the evaluation of floor concrete quality [23].

Ventilated facades are becoming an increasingly popular solution for the external part of walls in buildings [24]. They may differ in many elements, among others, cladding (fiber cement boards, HPL plates, large-slab ceramic tiles, ACM panels, stone cladding), types of substructures, console supports, etc. The main element that characterizes ventilated facades is the use of an air cavity between the cladding and thermal insulation. Unfortunately, in some respects, they are not yet standardized and tested. Above all, the requirements for the falling-off of elements from ventilated facades during a fire are not precisely defined by, among other things, the lack of clearly specified requirements and testing. This is undoubtedly a major problem, as it significantly affects the safety of evacuation during a fire emergency. For the purposes of this article, experimental tests were carried out on a large-scale facade model, with two types of external facade cladding. The materials used as external cladding were fiber cement boards and large-slab ceramic tiles. The model of the large-scale test was 3.95 m × 3.95 m; the burning gas released from the burner was used as the source of fire. The facade model was equipped with thermocouples. The test lasted one hour, and the cladding materials showed different behavior during the test. Large-slab ceramic tiles seemed to be a safer form of external cladding for ventilated facades. Unfortunately, they were destroyed much faster, by about 6 min. Large-slab ceramic tiles were destroyed within the first dozen or so minutes, then their destruction did not proceed or was minimal. In the case of fiber cement boards, the destruction started from the eleventh minute and increased until the end of the test. The author referred the results of the large-scale test to testing on samples carried out by other authors. The results presented the convergence of the large-scale test with samples. External claddings were equipped with additional mechanical protection. The use of additional mechanical protection to maintain external cladding elements increases their safety but does not completely eliminate the problem of the falling-off of parts of the facade. As research on fiber cement boards and large-slab ceramic tiles has suggested, these claddings were a major hazard due to fall-off from the facade [24].

The aim of this study [25] was to investigate the effect of plasterboards' humidity absorption on their performance. The specimens' hydration procedure consisted of consecutive immersing in water and subsequent drying at room temperature. Such a procedure was performed to increase the moisture content within the material volume. The microstructural observations of five different plasterboard types were performed through optical and scanning electron microscopy. The deterioration of their properties was evaluated using a three-point bending test and a subsequent ultrasonic (ultrasound testing (UT)) longitudinal wave velocity measurement. Depending on the material porosity, a loss of UT wave velocity from 6% to 35% and a considerable decrease in material strength from 70% to 80% were observed. Four types of approximated formulae were proposed to describe the dependence of UT wave velocity on the board moisture content. It was found that the proposed UT method could be successfully used for the on-site monitoring of plasterboards' hydration processes [25].

Concrete structure joints are filled in mainly in the course of sealing works ensuring protection against the influence of water. This paper by [26] presents the methodology for testing the mechanical properties of ESD pseudoplastic resins (E—elastic deformation, S—strengthening control, D—deflection control) recommended for concrete structure joint fillers. The existing standards and papers concerning quasi-brittle cement composites do not provide an adequate point of reference for the tested resins. The lack of a standardized testing method hampers the development of materials universally used in expansion joint fillers in reinforced concrete structures, as well as the assessment of their properties and durability. An assessment of the obtained results by referring to the reference sample has been suggested in the article. A test stand and a method of assessing the mechanical properties results (including adhesion to the concrete surface) of pseudoplastic resins in the axial tensile test have been presented [26].

The compaction index is one of the most important technological parameters during asphalt pavement construction, which may be negatively affected by the wrong asphalt paving machine setting, weather conditions, or the mix temperature. Presented in [27], this laboratory study analyzes the asphalt mix properties in case of inappropriate compaction. The reference mix was designed for an AC 11 S wearing layer (asphalt concrete for a wearing layer with maximum grading of 11 mm). Asphalt mix samples used in the tests were prepared using a Marshall device with the compaction energy of 2×20, 2×35, 2×50, and 2×75 blows, as well as in a roller compactor where the slabs were compacted to various heights: 69.3 mm (+10% of nominal height), 66.2 mm (+5%), 63 mm (nominal), and 59.9 mm (-5%), which resulted in different compaction indexes. Afterward, the samples were cored from the slabs. Both Marshall samples and cores were tested for air void content, stiffness modulus in three temperatures, indirect tensile strength, and resistance to water and frost indicated by the ITSR value. It was found that either an insufficient or excessive level of compaction can cause a negative effect on the road surface performance [27].

The importance of surface roughness and its non-destructive examination has often been emphasized in structural rehabilitation. The innovative procedure presented in [28] enables the estimation of concrete-to-concrete strength, based on a combination of low-cost, area-limited tests and geostatistical methods. The new method removes the shortcomings of the existing one, i.e., it is neither qualitative nor subjective. The interface strength factors, cohesion and friction, can be estimated accurately based on the collected data on surface texture. The data acquisition needed to create digital models of the concrete surface can be performed by terrestrial close-range photogrammetry or other methods. In the presented procedure, limitations to the availability of concrete surfaces are overcome by the generation of subsequential Gaussian random fields (via height profiles) based on the semivariograms fitted to the digital surface models. In this way, the randomness of the surface texture is reproduced. The selected roughness parameters, such as mean valley depth and, most importantly, the geostatistical semivariogram parameter sill, were transformed into contact bond strength parameters based on the available strength tests. The proposed procedure estimates the interface bond strength based on the geostatistical methods applied to the numerical surface model and can be used in practical and theoretical applications [28].

The core part of a hybrid truss bridge is the connection joint that combines the concrete chord and steel truss-web members [29]. To study the mechanical behavior and failure mode of steel–concrete connection joints in a hybrid truss bridge, static model tests were carried out on two connection joints at the scale of 1:3, under a horizontal load that was provided by a loading jack mounted on the vertical reaction wall. The specimen design, experimental setup and testing procedure were introduced. In the experiment, the displacement, strain level, concrete crack and experimental phenomena were factually recorded. Compared with the previous study results, the experimental results in this study demonstrated that the connection joints had an excellent bearing capacity and deformability. The minimum ultimate load and displacement of the two connection joints were 5200 kN and 59.01 mm, respectively. Moreover, the connection joints exhibited multiple failure modes, including the fracture of gusset plates, the slippage of high-strength bolts, the local buckling of compressive splice plates, the fracture of tensile splice plates and concrete cracking. Additionally, the strain distribution of the steel–concrete connection joints followed certain rules. It is expected that the findings from this paper may provide a reference for the design and construction of steel–concrete connection joints in hybrid truss bridges [29].

The static elastic modulus (*Ec*) and compressive strength (*fc*) are critical properties of concrete [30]. When determining *Ec* and *fc*, concrete cores are collected and subjected to destructive tests. However, destructive tests require certain test permissions and large sample sizes. Hence, it is preferable to predict *Ec* using the dynamic elastic modulus (*Ed*), through non-destructive evaluations. A resonance frequency test performed according to ASTM C215-14, and a pressure wave (P-wave) measurement conducted according to ASTM C597M-16, are typically used to determine *Ed*. Recently, developments in transducers have enabled the measurement of shear wave (S-wave) velocities in concrete. Although various equations have been proposed for estimating *Ec* and *fc* from *Ed*, their results deviate from experimental values. Thus, it is necessary to obtain a reliable *Ed* value for accurately predicting *Ec* and *fc*. In this study, *Ed* values were experimentally obtained from P-wave and S-wave velocities in the longitudinal and transverse modes; *Ec* and *fc* values were predicted using these *Ed* values through four machine learning (ML) methods: support vector machine, artificial neural networks, ensembles, and linear regression. Using ML, the prediction accuracy of *Ec* and *fc* was improved by 2.5–5% and 7–9%, respectively, compared with the accuracy obtained using classical or normal-regression equations. By combining ML methods, the accuracy of the predicted *Ec* and *fc* was improved by 0.5% and 1.5%, respectively, compared with the optimal single variable results [30].

The paper by [31] describes tests conducted to identify the mechanisms occurring during the fracture of single-edge notches loaded in three-point bending (SENB) specimens made from an Al–Ti laminate. The experimental tests were complemented with microstructural analyses of the specimens' fracture surfaces and an in-depth analysis of acoustic emission (AE) signals. The paper presents the application of the AE method to identify fracture processes in the layered Al–Ti composite, using a non-hierarchical method for clustering AE signals (k-means) and analyses using waveform time domain, fast Fourier transform (FFT Real) and waveform continuous wavelet, based on the Morlet wavelet. These analyses made it possible to identify different fracture mechanisms in Al–Ti composites, which is very significant for the assessment of the safety of structures made from this material [31].

The aspects regarding the stiffness of the connections between the beams that support the storage pallets and the uprights are very important in the analysis of the displacements and stresses in the storage racking systems. The main purpose of the paper by [32] is to study the effects of both upright thickness and tab connector types on rotational stiffness and on the capable bending moment of the connection. For this purpose, 18 different groups of beam-connector-upright assemblies are prepared by combining three types of beams (different sizes of the box cross-section), three kinds of upright profiles (with a different thickness of the section walls), and two types of connectors (four-tab connectors and five-tab connectors). Flexural tests were carried out on 101 assemblies. For the assemblies containing the uprights with a thickness of 1.5 mm, the five-tab connector leads to a higher value of the capable moment and higher rotational stiffness than similar assemblies with four-tab connectors. A contrary phenomenon happens in the case of the assemblies containing those upright profiles having a thickness of 2.0 mm, regarding the capable design moment. It is shown how the safety coefficient of connection depends on both the rotational stiffness and capable bending moment [32].

Concrete shrinkage is a phenomenon that results in a decrease in volume in the composite material during the curing period. The method for determining the effects of restrained shrinkage is described in Standard ASTM C 1581/C 1581M–09a. This article [33] shows the calibration of measuring rings with respect to the theory of elasticity, and the analysis of the relationship of steel ring deformation to high-performance concrete tensile stress as a function of time. Steel rings equipped with strain gauges are used for the measurement of strain during the compression of the samples. The strain is caused by the shrinkage of the concrete ring specimen that tightens around steel rings. The method allows registering the changes to the shrinkage process over time and evaluating the susceptibility of concrete to cracking. However, the standard does not focus on the details of the mechanical design of the test bench. To acquire accurate measurements, the test bench needs to be calibrated. Measurement errors may be caused by an improper, uneven installation of strain gauges, imprecise geometry of the steel measuring rings, or incorrect equipment settings. The calibration method makes it possible to determine the stress in a concrete sample, leading to its cracking at the specific deformation of the steel ring [33].

The article by [34] proposes using the acoustic emission (AE) method to evaluate the degree of change in the mechanical parameters of fiber cement boards. The research was undertaken after a literature review, due to the lack of a methodology that would allow nondestructive assessment of the strength of cement–fiber elements. The tests covered the components cut out from a popular type of board available on the construction market. The samples were subjected to environmental (soaking in water, cyclic freezing–thawing) and exceptional (burning with fire and exposure to high temperature) factors, and then to three-point bending strength tests. The adopted conditions correspond to the actual working environment of the boards. When applying the external load, AE signals were generated which were then grouped into classes, and initially assigned to specific processes occurring in the material. The frequencies occurring over time for the tested samples were also analyzed, and microscopic observations were made to confirm the suppositions based on the first part of the tests. Comparing the results obtained from a group of samples subjected to environmental and exceptional actions, significant differences were noted between them, which included the types of recorded signal class, the frequency of events, and the construction of the microstructure. The degradation of the structure, associated with damage to the fibers or their complete destruction, results in the generation under load of AE signals that indicate the uncontrolled development of scratches, and a decrease in the frequency of these events. According to the authors, the methodology used allows the control of cement–fiber boards in use. The registration and analysis of active processes under the effect of payloads makes it possible to distinguish mechanisms occurring inside the structure of the elements, and to formulate a quick response to the situation when the signals indicate a decrease in the strength of the boards [34].

The article by [35] discusses one of the methods of dielectric constant determination in a continuous way, which is the determination of its value based on the amplitude of the wave reflected from the surface. Based on tests performed on model asphalt slabs, the research presented how the value of the dielectric constant changes, depending on the atmospheric conditions of the measured surface (dry, covered with water film, covered with ice, covered with snow, covered with de-icing salt). Coefficients correcting dielectric constants of hot mix asphalt (HMA) determined in various surface atmospheric conditions were introduced. It was proposed to determine the atmospheric conditions of the pavement with the use of wavelet analysis in order to choose the proper dielectric constant correction coefficient and, therefore, improve the accuracy of the pavement layer thickness estimation based on the ground-penetrating radar (GPR) method [35].

Phenomena occurring during the curing of concrete can decrease its mechanical properties, specifically its strength and serviceability, even before it is placed [36]. This is due to excessive stresses caused by temperature gradients, moisture changes, and chemical processes arising during the concreting and in hardened concrete. At stress concentration sites, microcracks form in the interfacial transition zones (ITZ) in the early phase and propagate deeper into the cement paste or to the surface of the element. Microcracks can contribute to the development of larger cracks, reduce the durability of structures, limit their serviceability, and, in rare cases, lead to their failure. It is thus important to search for a tool that allows objective assessment of damage initiation and development in concrete. The objectivity of the assessment lies in it being independent of the constituents and additives used in the concrete or of external influences. The acoustic emission-based method presented in this paper allows damage detection and identification in the early age of concrete (before loading) for different concrete compositions, curing conditions, temperature variations, and in reinforced concrete. As such, this method is an objective and effective tool for damage process detection [36].

The authors of [37] suggest a wire-mesh method to classify the particle shape of large amounts of aggregate. This method is controlled by the tilting angle and opening size of the wire mesh. The more rounded the aggregate particles, the more they roll on the tilted wire mesh. Three different sizes of aggregate, 11–15, 17–32, and 33–51 mm, were used for assessing their roundness after classification, using the sphericity index to sort them into rounded, sub-rounded/sub-angular, and angular. The aggregate particles with different sphericities were colored differently and then used for classification via the wire-mesh method. The opening sizes of the wire mesh were 6, 11, and 17 mm, and its frame was 0.5 m wide and 1.8 m long. The ratio of aggregate size to mesh-opening size was between 0.6 and 8.5. The wire mesh was inclined at various angles of 10°, 15°, 20°, 25°, and 30° to evaluate the rolling degree of the aggregates. The aggregates were rolled and remained on the wire mesh between 0.0–0.6, 0.6–1.2, and 1.2–1.8 m, depending on their sphericity. A tilting angle of 25° was the most suitable angle for classifying aggregate size ranging from 11–15 mm, while the most suitable angle for aggregate sizes of 17–32 and 33–51 mm was 20°. The best ratio for the average aggregate size to mesh-opening size for the aggregate roundness classification was 2 [37].

Taking into account the possibilities offered by two imaging methods, X-ray micro-computed tomography (µCT) and two-dimensional optical scanning, this article by [38] discusses the possibility of using these methods to assess the internal structure of spun concrete, particularly its composition after hardening (Michałek 2020). To demonstrate the performance of the approach based on imaging, laboratory techniques based on physical and chemical methods were used as verification. A comparison of the obtained results of applied research methods was carried out on samples of spun concrete, characterized by the layered structure of the annular cross-section. Samples were taken from the power pole E10.5/6c (Strunobet-Migacz, Lewin Brzeski, Poland) made by one of the Polish manufacturers of prestressed concrete E-poles precast in steel molds. The validation shows that optical scanning followed by appropriate image analysis is an effective method for evaluation of the spun concrete internal structure. In addition, such analysis can significantly complement the results of the laboratory methods used so far. In a fairly simple way, through the porosity image, it can reveal improperly selected parameters of concrete spinning, such as speed and time, and, through the distribution of cement content in the cross-section of the element, it can indicate compliance with the requirement for the corrosion durability of spun concrete. The research methodology presented in the paper can be used to improve the production process of poles made of spun concrete; it can be an effective tool for verifying concrete structure [38].

In the study by [39], the effects of the mixing conditions of waste-paper sludge ash (WPSA) on the strength and bearing capacity of controlled low-strength material (CLSM) were evaluated, and the optimal mixing conditions were used to evaluate the strength characteristics of CLSM with recyclable WPSA. The strength and bearing capacity of CLSM with WPSA were evaluated using unconfined compressive strength tests and plate bearing tests, respectively. The unconfined compressive strength test results show that the optimal mixing conditions for securing 0.8–1.2 MPa of target strength under 5% of cement content conditions can be obtained when both WPSA and fly ash are used. This is because WPSA and fly ash, which act as binders, have a significant impact on overall strength when the cement content is low. The bearing capacity of weathered soil increased from 550 to 575 kPa over time, and CLSM with WPSA increased significantly, from 560 to 730 kPa. This means that the bearing capacity of CLSM with WPSA was 2.0% higher than that of weathered soil immediately after construction; furthermore, it was 27% higher at 60 days of age. In addition, the allowable bearing capacity of CLSM corresponding to the optimal mixing conditions was evaluated, and it was found that this value increased by 30.4% until 60 days of age. This increase rate was 6.7 times larger than that of weathered soil (4.5%). Therefore, based on the allowable bearing capacity calculation results, CLSM with WPSA was applied as a sewage pipe backfill material. It was found that CLSM with WPSA performed better as backfill and was more stable than soil immediately after construction. The results of this study confirm that CLSM with WPSA can be utilized as sewage-pipe backfill material [39].

Non-destructive testing (NDT) methods are an important means to detect and assess rock damage [40]. To better understand the accuracy of NDT methods for measuring damage in sandstone, this study compared three NDT methods, including ultrasonic testing, electrical impedance spectroscopy (EIS) testing, computed tomography (CT) scan testing, and a destructive test method, elastic modulus testing. Sandstone specimens were subjected to different levels of damage through cyclic loading, and different damage variables derived from five different measured parameters—longitudinal wave (P-wave) velocity, first-wave amplitude attenuation, resistivity, effective bearing area and the elastic modulus—were compared. The results show that the NDT methods all reflect the damage levels for sandstone accurately. The damage variable derived from the P-wave velocity is more consistent with the other damage variables, and the amplitude attenuation is more sensitive to damage. The damage variable derived from the effective bearing area is smaller than that derived from the other NDT measurement parameters. Resistivity provides a more stable measure of damage, and damage derived from the acoustic parameters is less stable. By developing P-wave velocity-to-resistivity models based on theoretical and empirical relationships, it was found that differences between these two damage parameters can be explained by differences between the mechanisms through which they respond to porosity, since the resistivity reflects pore structure, while the P-wave velocity reflects the extent of the continuous medium within the sandstone [40].

The H spin-lattice relaxometry (T_1, longitudinal) of cement pastes with 0 to 0.18 wt % polycarboxylate superplasticizers (PCEs) at intervals of 0.06 wt % from 10 min to 1210 min was investigated in [41]. Results showed that the main peak in T_1 relaxometry of cement pastes was shorter and lower along with the hydration times. PCEs delayed and lowered this main peak in the T_1 relaxometry of cement pastes at 10 min, 605 min and 1210 min, which was highly correlated to its dosages. In contrast, PCEs increased the total signal intensity of T_1 of cement pastes at these three times, which still correlated to its dosages. Both changes of the main peak in T_1 relaxometry and the total signal intensity of T_1 revealed interferences in evaporable water during cement hydration by the dispersion mechanisms of PCEs. The time-dependent evolution of the weighted average T_1 of cement pastes with different PCEs between 10 min and 1210 min was found to be regular to the four-stage hydration mechanism of tricalcium silicate [41].

Control of technical parameters obtained by ready-mixed concrete may be carried out at different stages of the development of concrete properties, and by different participants involved in the construction investment process [42]. According to the European Standard EN 206 "Concrete—specification, performance, production and conformity", mandatory control of concrete conformity is conducted by the producer during production. As shown by the subject literature, statistical criteria set out in the standard, including the method for concrete quality assessment based on the concept of concrete family, continue to evoke discussions and raise doubts. This justifies seeking alternative methods for concrete quality assessment. This paper presents a novel approach to quality control and the classification of concrete based on combining statistical and fuzzy theories as a means of representation of two types of uncertainty: random uncertainty and information uncertainty. In concrete production, a typical situation when fuzzy uncertainty can be taken into consideration is the conformity control of concrete compressive strength, which is conducted to confirm the declared concrete class. The proposed procedure for quality assessment of a concrete batch is based on defining the membership function for the considered concrete classes and establishing the degree of belonging to the considered concrete class. It was found that concrete classification set out by the standard includes too many concrete classes of overlapping probability density distributions, and the proposed solution was to limit the scope of compressive strength to every second class so as to ensure the efficacy of conformity assessment conducted for concrete classes and concrete families. The proposed procedures can lead to two types of decisions: non-fuzzy (crisp) or fuzzy, which point to possible solutions and their corresponding preferences. The suggested procedure for quality assessment allows researchers to classify a concrete batch in a fuzzy way with the degree of certainty less than or equal to 1. The results obtained confirm the possibility of employing the proposed method for quality assessment in the production process of ready-mixed concrete [42].

In recent years, the application of fiber-reinforced plastics (FRPs) as structural members has been promoted [43]. Metallic bolts and rivets are often used for the connection of FRP structures, but there are some problems caused by corrosion and stress concentration at the bearing position. Fiber-reinforced thermoplastics (FRTPs) have attracted attention in composite material fields because they can be remolded by heating and manufactured at excellent speed compared with thermosetting plastics. In this paper, we propose and evaluate the connection method using rivets produced from FRTPs for FRP members. It was confirmed through material tests that an FRTP rivet provides stable tensile, shear, and bending strength. Then, it was clarified that a non-clearance connection could be achieved by the proposed connection method, so initial sliding was not observed, and connection strength linearly increased as the number of FRTP rivets increased through the double-lapped tensile shear tests. Furthermore, the joint strength of the beam using FRTP rivets could be calculated with high accuracy, using the method for bolt joints in steel structures through a four-point beam bending test [43].

Polymer pipes are used in the construction of underground gas, water, and sewage networks [44]. During exploitation, various external forces work on the pipeline which cause its deformation. In this paper, numerical analysis and experimental investigations of polyethylene pipe deformation at different external load values (500, 1000, 1500, and 2000 N) were performed. The authors measured the strains of the lower and upper surfaces of the pipe during its loading moment using resistance strain gauges, which were located on the pipe at equal intervals. The results obtained from computer simulation and experimental studies were comparable. An innovative element of the research presented in the article is the recognition of the impact of the proposed values of the load of polyethylene pipe on the change in its deformation [44].

The shear and particle-crushing characteristics of the failure plane (or shear surface) in catastrophic mass movements are examined with a ring shear apparatus, which is generally employed owing to its suitability for large deformations [45]. Based on the results of previous experiments on waste materials from abandoned mine deposits, we employed a simple numerical model based on ring shear testing using the particle flow code (PFC^{2D}). We examined drainage, normal stress, and the shear velocity-dependent shear characteristics of landslide materials. For shear velocities of 0.1 and 100 mm/s and normal stress (NS) of 25 kPa, the numerical results are in good agreement with those obtained from experimental results. The difference between the experimental and numerical results of the residual shear stress was approximately 0.4 kPa, for NS equal to 25 kPa and 0.9 kPa for NS equal to 100 kPa, for both drained and undrained conditions. In addition, we examined the particle-crushing effect during shearing using the frictional work concept in PFC. We calculated the work done by friction at both peak and residual shear stresses, and then used the results as crushing criteria in the numerical analysis. The frictional work at peak and the residual shear stresses ranged from 303 kPa·s to 2579 kPa·s for the given drainage and normal stress conditions. These results showed that clump particles were partially crushed at peak shear stress [45].

The main objective of the research presented in [46] was to develop a solution to the global problem of using steel waste obtained during rubber recovery during tire recycling. A detailed comparative analysis of the mechanical and physical features of concrete composite with the addition of recycled steel fibers (RSF) in relation to the steel fiber concrete commonly used for industrial floors was conducted. A study was carried out using micro-computed tomography and the scanning electron microscope to determine the fibers' characteristics, including the EDS spectrum. In order to designate full performance of the physical and mechanical features of the novel composite, a wide range of tests was performed, with particular emphasis on the determination of the tensile strength of the composite. This parameter, appointed by tensile strength testing for splitting, residual tensile strength testing (3-point test), and a wedge splitting test (WST), demonstrated the increase of tensile strength (vs. unmodified concrete) by 43%, 30%, and 70% respectively to the method. The indication of the reinforced composite's fracture characteristics using the digital image correlation (DIC) method allowed the researchers to illustrate the map of deformation of the samples during WST. The novel composite was tested in reference to the circular economy concept and showed 31.3% lower energy consumption and 30.8% lower CO_2 emissions than a commonly used fiber concrete [46].

Existing buildings, especially historical buildings, require periodic or situational diagnostic tests [47]. If a building is in use, advanced non-destructive or semi-destructive methods should be used. In the diagnosis of reinforced concrete structures, tests allowing assessment of the condition of the reinforcement and concrete cover are particularly important. The article presents the non-destructive and semi-destructive research methods that are used for such tests, as well as the results of tests performed for selected elements of a historic water tower structure. The assessment of the corrosion risk of the reinforcement was carried out with the use of a semi-destructive galvanostatic pulse method. The protective properties of the concrete cover were checked by the carbonation test and the phase analysis of the concrete, for which X-ray diffractometry and thermal analysis methods were used. In order to determine the position of the reinforcement and to estimate the concrete cover thickness distribution, a ferromagnetic detection system was used. The comprehensive application of several test methods allowed mutual verification of the results and the drawing of reliable conclusions. The results indicated a very poor state of reinforcement, loss in the depth of cover, and sulfate corrosion [47].

The paper by [48] presents the possibility of using low-module polypropylene dispersed reinforcement (E = 4.9 GPa) to influence the load-deflection correlation of cement composites. Problems have been indicated regarding the improvement of elastic range when using that type of fiber as compared with a composite without reinforcement. It was demonstrated that it was possible to increase the ability to carry stress in the Hooke's law proportionality range in the mortar and paste types of composites reinforced with low-module fibers, i.e., $V_f = 3\%$ (in contrast to concrete composites). The possibility of having good strengthening and deflection control in order to limit the catastrophic destruction process was confirmed. This paper identifies the problem of deformation assessment in composites with significant deformation capacity. Determining the effects of reinforcement based on a comparison with a composite without fibers is suggested as a reasonable approach, as it enables the comparison of results obtained by various universities under different research conditions [48].

Arcan shear tests with digital image correlation were used to evaluate the shear modulus and shear stress–strain diagrams in the plane defined by two principal axes of the material orthotropy [49]. Two different orientations of the grain direction, as compared to the direction of the shear force in specimens, were considered: perpendicular and parallel shear. Two different ways were used to obtain the elastic properties based on the digital image correlation (DIC) results from the full-field measurement and from the virtual strain gauges with the linear strains: perpendicular to each other and directed at an angle of $\pi/4$

to the shearing load. In addition, their own continuum structural model for the failure analysis in the experimental tests was used. The constitutive relationships of the model were established in the framework of the mathematical multi-surface elastoplasticity for the plane stress state. The numerical simulations performed by the finite element program after implementation of the model demonstrated the failure mechanisms from the experimental tests [49].

After obtaining the value of shear wave velocity (V_S) from the bender elements test (BET), the shear modulus of soils at small strains (G_{max}) can be estimated [50]. Shear wave velocity is an important parameter in the design of geo-structures subjected to static and dynamic loading. While bender elements are increasingly used in both academic and commercial laboratory test systems, there remains a lack of agreement when interpreting the shear wave travel time from these tests. Based on the test data of 12 Warsaw glacial quartz samples of sand, two different approaches were primarily examined for determining V_S. They are both related to the observation of the source and received *BE* signal, namely, the first time of arrival and the peak-to-peak method. These methods were performed through visual analysis of BET data by the authors, so that subjective travel time estimates were produced. Subsequently, automated analysis methods from the GDS Bender Element Analysis Tool (BEAT) were applied. Here, three techniques in the time-domain (TD) were selected, namely, the peak-to-peak, the zero-crossing, and the cross-correlation functions. Additionally, a cross-power spectrum calculation of the signals was completed, viewed as a frequency-domain (FD) method. Final comparisons between subjective observational analyses and automated interpretations of BET results showed good agreement. There is compatibility, especially between the two methods of the first time of arrival and cross-correlation, which the authors considered the best interpreting techniques for their soils. Moreover, the laboratory tests were performed on compact, medium, and well-grained sand samples with different curvature coefficients and mean grain sizes [50].

The reduction in natural resources and aspects of environmental protection necessitate alternative uses for waste materials in the area of construction [51]. Recycling is also observed in road construction, where mineral–cement emulsion (MCE) mixtures are applied. The MCE mix is a conglomerate that can be used to make the base layer in road pavement structures. MCE mixes contain reclaimed asphalt from old, degraded road surfaces, aggregate for improving the gradation, asphalt emulsion, and cement as a binder. The use of these ingredients, especially cement, can cause shrinkage and cracks in road layers. The article presents selected issues related to the problem of cracking in MCE mixtures. The authors of the study focused on reducing the cracking phenomenon in MCE mixes by using an innovative cement binder with recycled materials. The innovative cement binder, based on dusty by-products from cement plants, also contributes to the optimization of the recycling process in road surfaces. The research was carried out in the fields of stiffness, fatigue life, crack resistance, and shrinkage analysis of mineral–cement emulsion mixes. It was found that it was possible to reduce the stiffness and cracking in MCE mixes. The use of innovative binders will positively affect the durability of road pavements [51].

The windblown sand-induced degradation of glass panels influences the serviceability and safety of these panels. In this study, the degradation of glass panels subject to windblown sand at different impact velocities and impact angles was studied based on a sandblasting test simulating a sandstorm [52]. After the glass panels were degraded by windblown sand, the surface morphology of the damaged glass panels was observed using scanning electron microscopy, and three damage modes were found: a cutting mode, smash mode, and plastic deformation mode. The mass loss, visible light transmittance, and effective area ratio values of the glass samples were then measured to evaluate the effects of the windblown sand on the panels. The results indicate that, at high abrasive feed rates, the relative mass loss of the glass samples decreases initially and then remains steady with increases in impact time, whereas it increases first and then decreases with an increase in impact angle, such as that for ductile materials. Both the visible light transmittance and effective area ratio decrease with increases in the impact time and velocities. There exists a positive linear relationship between the visible light transmittance and effective area ratio [52].

Durability tests against fungal action for wood-plastic composites are carried out in accordance with European standard ENV 12038, but the authors of the manuscript try to prove that the assessment of the results performed according to these methods is imprecise and suffers from a significant error [53]. Fungal exposure is always accompanied by high humidity, so the result of tests made by such a method is always burdened with the influence of moisture, which can lead to a wrong assessment of the negative effects of the action of the fungus itself. The paper (Wiejak 2021) has shown a modification of such a method that separates the destructive effect of fungi from moisture accompanying the test's destructive effect. The functional properties selected to prove the proposed modification are changes in the mass and bending strength after subsequent environmental exposure. It was found that the intensive action of moisture measured in the culture chamber at about $(70 \pm 5)\%$, i.e., for 16 weeks, at (22 ± 2) °C, which was the fungi culture in the accompanying period, led to changes in the mass of the wood-plastic composites, amounting to 50% of the final result of the fungi resistance test, and changes in the bending strength amounting to 30–46% of the final test result. As a result of this research, the correction for assessing the durability of wood–polymer composites against biological corrosion has been proposed. The laboratory tests were compared with the products' test results following three years of exposure to the natural environment [53].

Reduced maintenance costs of concrete structures can be ensured by efficient and comprehensive condition assessment [54]. Ground-penetrating radar (GPR) has been widely used in the condition assessment of reinforced concrete structures and it provides completely non-destructive results in real-time. It is mainly used for locating reinforcement and determining concrete cover thickness. More recently, research has focused on the possibility of using GPR for reinforcement corrosion assessment. In this paper, an overview of the application of GPR in the corrosion assessment of concrete is presented. A literature search and study selection methodology were used to identify the relevant studies. First, the laboratory studies are shown. After that, the studies for application on real structures are presented. The results have shown that the laboratory studies have not fully illuminated the influence of the corrosion process on the GPR signal. In addition, no clear relationship was reported between the results of the laboratory studies and the on-site inspection. Although GPR has a long history in the condition assessment of structures, it needs more laboratory investigations to clarify the influence of the corrosion process on the GPR signal [54].

The paper by [55] attempts to compare three methods of testing floor slip resistance and the resulting classifications. Polished, flamed, brushed, and grained granite slabs were tested. The acceptance angle values (α_{ob}) obtained through the shod ramp test, slip resistance value (SRV), and sliding friction coefficient (μ) were compared in terms of the correlation between the series, the precision of each method, and the classification results assigned to each of the three obtained indices. It was found that the evaluation of a product for slip resistance was strongly related to the test method used and the resulting classification method. This influence was particularly pronounced for low-roughness slabs. This would result in risks associated with inadequate assessments that could affect the safe use of buildings and facilities [55].

Standard sensors for the measurement and monitoring of temperature in civil structures are liable to mechanical damage and electromagnetic interference [56]. A system of purpose-designed fiber-optic FBG sensors offers a more suitable and reliable solution—the sensors can be directly integrated with the load-bearing structure during construction, and it is possible to create a network of fiber-optic sensors to ensure not only temperature measurements but also measurements of strain and of the moisture content in the building envelope. The paper describes the results of temperature measurements of a building's two-layer wall using optical fiber Bragg grating (FBG) sensors, and of a three-layer wall using equivalent classical temperature sensors. The testing results can be transmitted remotely. In the first stage, the sensors were tested in a climatic test chamber to determine their characteristics. The paper describes the test results of temperature measurements carried out in the winter season for two multilayer external walls of a building, in relation to the environmental conditions recorded at that time, i.e., outdoor temperature, relative humidity, and wind speed. Cases are considered with the biggest difference in the level of the relative humidity of air recorded in the observation period. It was found that there is greater convergence between the theoretical and the real temperature distribution in the wall for high levels (~84%) of the outdoor air's relative humidity, whereas, at the humidity level of ~49%, the difference between theoretical and real temperature histories is substantial and totals up to 20%. A correction factor is proposed for the theoretical temperature distribution [56].

The paper by [57] deals with a complex analysis of acoustic emission signals that were recorded during freeze-thaw cycles in test specimens produced from air-entrained concrete. An assessment of the resistance of concrete to the effects of freezing and thawing was conducted on the basis of signal analysis. Since the experiment simulated the testing of concrete in a structure, a concrete block with a height of 2.4 m and width of 1.8 m was produced to represent a real structure. When the age of the concrete was two months, samples were obtained from the block by core drilling and were subsequently used to produce test specimens. Testing of the freeze-thaw resistance of concrete employed both destructive and non-destructive methods including the measurement of acoustic emission, which took place directly during the freeze-thaw cycles. The recorded acoustic emission signals were then meticulously analyzed. The aim of the conducted experiments was to verify whether measurement using the acoustic emission method during freeze-thaw (F-T) cycles is more sensitive to the degree of damage of concrete than the more commonly employed construction testing methods. The results clearly demonstrate that the acoustic emission method can reveal changes (e.g., minor cracks) in the internal structure of concrete, unlike other commonly used methods. The analysis of the acoustic emission signals using a fast Fourier transform revealed a significant shift of the dominant frequency toward lower values when the concrete was subjected to freeze-thaw cycling [57].

An original experimental method was used to investigate the influence of water and road salt with an anti-caking agent on the material used in pavement construction layers [58]. This method allowed the monitoring of material changes resulting from the influence of water and road salt with an anti-caking agent over time. The experiment used five different mineral road mixes, which were soaked separately in water and brine for two time intervals (2 days and 21 days). Then, each sample of the mix was subjected to tests of the complex module using the four-point bending (4PB-PR) method. The increase in mass of the soaked samples and the change in value of the stiffness modulus were analyzed. Exemplary tomographic (X-ray) imaging was performed to confirm the reaction of the road salt and anti-caking agent (lead agent) with the material. Based on the measurements of the stiffness modulus and absorption, the correlations of the mass change and the value of the stiffness modulus were determined, which may be useful in estimating the sensitivity of mixes to the use of winter maintenance agents, e.g., road salt with an anti-caking agent (sodium chloride). It was found that the greatest changes occur with mixes intended for base course layers (mineral cement mix with foamed asphalt (MCAS) and mineral-cement-emulsion mixes (MCE)), and that the smallest changes occur for mixes containing highly modified asphalt (HIMA) [58].

Cracking in non-load-bearing internal partition walls is a serious problem that frequently occurs in new buildings in the short term after putting them into service, or even before completion of construction [59]. Sometimes, it is so considerable that it cannot be accepted by the occupiers. The article presents tests of cracking in ceramic walls with a door opening connected in a rigid and flexible way along vertical edges. The first analyses were conducted using the finite element method (FEM), and afterward, the measurements of deformations and stresses in walls on deflecting floors were performed at full scale in the actual building structure. The measurements enabled the authors to determine floor deformations leading to the cracking of walls and to establish a dependency between the values of tensile stresses within the area of the door opening corners and their location along the length of walls, and the type of vertical connection with the structure [59].

The paper by [60] discusses the problems connected with the long-term exploitation of reinforced concrete post-tensioned girders. The scale of problems in the world related to the number of cable post-tensioned concrete girders built in the 1950s and still in operation is very large and possibly has very serious consequences. The paper presents an analysis and evaluation of the results of measurements of the deflection and strength and homogeneity of concrete in cable–concrete roof girders of selected industrial halls located in Poland, exploited for over 50 years. On the basis of the results of displacement monitoring in the years 2009 to 2020, the maximum increments of deflection of the analyzed girders were determined. Non-destructive, destructive, and indirect evaluation methods were used to determine the compressive strength of concrete. Within the framework of the indirect method recommended in standard PN-EN 13791, a procedure was proposed by the authors to modify the so-called base curve for determining compressive strength. Due to the age of the analyzed structural elements, a correction factor for the age of the concrete was taken into account in the strength assessment. The typical value of the characteristic compressive strength was within the range of 20.3–28.4 MPa. As a result of the conducted tests, the concrete class assumed in the design was not confirmed, and its classification depended on the applied test method. The analyzed girders, despite their long-term exploitation, can still be used for years, on the condition that regular periodical inspections of their technical condition are carried out. The authors emphasize the necessity for a permanent and cyclic diagnostic process and monitoring of the geometry of girders, as they are expected to operate much longer than was assumed by their designers [60].

The paper by [61] presents an implementation of purpose-designed optical fiber Bragg grating (FBG) sensors intended for the monitoring of real values of strain in reinforced road structures in areas of mining activity. Two field test stations are described. The first enables analysis of the geogrid on concrete and ground subgrades. The second models the situation of subsoil deformation due to mining activity at different external loads. The paper presents a system of optical fiber sensors registering strain and temperature dedicated to the investigated concrete mattress. Laboratory tests were performed to determine the strain characteristic of the FBG sensor–geogrid system with respect to standard load. As a result, it was possible to establish the dependence of the geogrid strain on the forces occurring within it. This may be the basis for an analysis of the mining activity effect on right-of-way structures during precise strain measurements of a geogrid using FBG sensors embedded within it. The analysis of the results of measurements in the aspect of forecasted and actual static and dynamic effects of mining on the stability of a reinforced road structure is of key importance for detailed management of road investment, and for the appropriate repair and modernization management of the road structure [61].

Geopolymer concrete (GPC) offers a potential solution for sustainable construction by utilizing waste materials [62]. However, the production and testing procedures for GPC are quite cumbersome and expensive, which can slow down the development of mix design and the implementation of GPC. The basic characteristics of GPC depend on numerous factors such as the type of precursor material, type of alkali activators and their concentration, and liquid to solid (precursor material) ratio. To optimize time and cost, artificial neural networks (ANN) can be a lucrative technique for exploring and predicting GPC characteristics. In this study, the compressive strength of fly-ash-based GPC, with bottom ash as a replacement for fine aggregates, as well as fly ash, is predicted using a machine learning-based ANN model. The data inputs are taken from the literature as well as in-house lab scale testing of GPC. The specifications of GPC specimens act as input features of the ANN model to predict compressive strength as the output, while minimizing error. Fourteen ANN models are designed which differ in the backpropagation training algorithm, the number of hidden layers, and the neurons in each layer. The performance analysis and comparison of these models in terms of mean squared error (MSE) and coefficient of correlation (R) resulted in a Bayesian regularized ANN (BRANN) model for the effective prediction of compressive strength of fly-ash and bottom-ash based geopolymer concrete [62].

The paper by [63] analyzes the issue of reduction of load capacity in fiber cement board during a fire. Fiber cement boards were put under the influence of fire by using a large-scale facade model. Such a model is a reliable source of knowledge regarding the behavior of facade cladding and the way fire spreads. One technical solution for external walls—a ventilated facade—is gaining popularity and is used more and more often. However, the problem of the destruction during a fire of a range of different materials used in external facade cladding is insufficiently recognized. For this study, the authors used fiber cement boards as the facade cladding. Fiber cement boards are fiber-reinforced composite materials, mainly used for facade cladding, but are also used as roof cladding, drywall, drywall ceilings and floorboards. This paper analyzes the effect of fire temperatures on facade cladding using a large-scale facade model. Samples were taken from external facade cladding materials that were mounted on the model at specific locations above the combustion chamber. Subsequently, three-point bending flexural tests were performed, and the effects of temperature and the integrals of temperature and time functions on the samples were evaluated. The three-point bending flexural test was chosen because it is a universal method for assessing fiber cement boards, as cited in Standard EN 12467. The test also allows easy reference to results in other literature [63].

The paper by [64] presents the possibilities of determining the range of stresses preceding the critical destruction process in cement composites, with the use of micro-events identified by means of a sound spectrum. The presented test results refer to the earlier papers in which micro-events (destruction processes) were identified, but without determining the stress level of their occurrence. This paper indicates a correlation of the stress level corresponding to the elastic range with the occurrence of micro-events in traditional and quasi-brittle composites. Tests were carried out on beams (with and without reinforcement) subjected to four-point bending. In summary, it is suggested that the conclusions can be extended to other test cases (e.g., compression strength), which should be confirmed by the appropriate tests. The paper also indicates a need for further research to identify micro-events. The correct recognition of micro-events is important for the safety and durability of traditional and quasi-brittle cement composites [64].

The paper by [65] contains the results of a newly developed residual-state creep test, performed to determine the behavior of a selected geomaterial in the context of reactivated landslides. Soil and rock creep is a time-dependent phenomenon in which deformation occurs under constant stress. Based on the examination results, it was found that the tested clayey material (from Kobe, Japan) shows tertiary creep behavior only under shear stress higher than the residual strength condition, and primary and secondary creep behavior under shear stress lower than or equal to the residual strength condition. Based on the data, a model for predicting the critical or failure time is introduced. The time until the occurrence of the conditions necessary for unlimited creep on the surface is estimated. As long-term precipitation and infiltrating water in the area of the landslides are identified as the key phenomena initiating collapse, the work focuses on the prediction of landslides, with identified surfaces of potential damage as a result of changes in the saturation state. The procedure outlined is applied to a case study, and considerations as to when the necessary safety work should be carried out are presented [65].

The article by [66] is focused on the medium-term negative effect of groundwater on the underground grout elements. This is the physical–mechanical effect of groundwater, which is known as erosion. We conducted a laboratory verification of the erosional resistance of grout mixtures. A new test apparatus was designed and developed, since there is no standardized method for testing at present. An erosion stability test of grout mixtures and the technical solutions of the apparatus for the test's implementation are described. This apparatus was subsequently used for the experimental evaluation of the erosional stability of silicate grout mixtures. Grout mixtures with activated and non-activated bentonite are tested. The stabilizing effect of cellulose relative to erosion stability has also been investigated. The specimens of grout mixtures were exposed to flowing water stress for a set period of time. The erosional stabilities of the grout mixtures were assessed on the basis of weight loss (WL) as a percentage of initial specimen weight. The lower the grout mixture weight loss, the higher its erosional stability, and vice versa [66].

3. Conclusions

As mentioned at the beginning, this issue was proposed and organized as a means of presenting recent developments in the field of non-destructive testing of materials in civil engineering. For this reason, the articles highlighted in this issue relate to different aspects of the testing of different materials in civil engineering, from building materials and elements to building structures. Interesting results, with significance for the materials, were obtained, and all of the papers have been precisely described.

Funding: This research received no external funding.

Institutional Review Board Statement: Not applicable.

Informed Consent Statement: Not applicable.

Conflicts of Interest: The authors declare no conflict of interest.

References

1. Schabowicz, K. Modern acoustic techniques for testing concrete structures accessible from one side only. *Arch. Civ. Mech. Eng.* **2015**, *15*, 1149–1159. [CrossRef]
2. Hoła, J.; Schabowicz, K. State-of-the-art non-destructive methods for diagnostic testing of building structures—Anticipated development trends. *Arch. Civ. Mech. Eng.* **2010**, *10*, 5–18. [CrossRef]
3. Hoła, J.; Schabowicz, K. Non-destructive diagnostics for building structures: Survey of selected state-of-the-art methods with application examples. In Proceedings of the 56th Scientific Conference of PANCivil Engineering Committee and PZITB Science Committee, Krynica, Poland, 19–24 September 2010. (In Polish).
4. Schabowicz, K.; Gorzelanczyk, T. Fabrication of fibre cement boards. In *The Fabrication, Testing and Application of Fibre Cement Boards*, 1st ed.; Ranachowski, Z., Schabowicz, K., Eds.; Cambridge Scholars Publishing: Newcastle upon Tyne, UK, 2018; pp. 7–39. ISBN 978-1-5276-6.
5. Drelich, R.; Gorzelanczyk, T.; Pakuła, M.; Schabowicz, K. Automated control of cellulose fiber cement boards with a non-contact ultrasound scanner. *Autom. Constr.* **2015**, *57*, 55–63. [CrossRef]
6. Chady, T.; Schabowicz, K.; Szymków, M. Automated multisource electromagnetic inspection of fibre-cement boards. *Autom. Constr.* **2018**, *94*, 383–394. [CrossRef]
7. Schabowicz, K.; Józwiak-Niedzwiedzka, D.; Ranachowski, Z.; Kudela, S.; Dvorak, T. Microstructural characterization of cellulose fibres in reinforced cement boards. *Arch. Civ. Mech. Eng.* **2018**, *4*, 1068–1078. [CrossRef]
8. Schabowicz, K.; Gorzelanczyk, T.; Szymków, M. Identification of the degree of fibre-cement boards degradation under the influence of high temperature. *Autom. Constr.* **2019**, *101*, 190–198. [CrossRef]
9. Schabowicz, K.; Gorzelanczyk, T. A non-destructive methodology for the testing of fibre cement boards by means of a non-contact ultrasound scanner. *Constr. Build. Mater.* **2016**, *102*, 200–207. [CrossRef]
10. Schabowicz, K.; Ranachowski, Z.; Józwiak-Niedzwiedzka, D.; Radzik, Ł.; Kudela, S.; Dvorak, T. Application of X-ray microtomography to quality assessment of fibre cement boards. *Constr. Build. Mater.* **2016**, *110*, 182–188. [CrossRef]
11. Ranachowski, Z.; Schabowicz, K. The contribution of fibre reinforcement system to the overall toughness of cellulose fibre concrete panels. *Constr. Build. Mater.* **2017**, *156*, 1028–1034. [CrossRef]
12. Bačić, M.; Kovačević, M.; Jurić Kaćunić, D. Non-Destructive Evaluation of Rock Bolt Grouting Quality by Analysis of Its Natural Frequencies. *Materials* **2020**, *13*, 282. [CrossRef]
13. Bajno, D.; Bednarz, L.; Matkowski, Z.; Raszczuk, K. Monitoring of Thermal and Moisture Processes in Various Types of External Historical Walls. *Materials* **2020**, *13*, 505. [CrossRef]
14. Szewczak, E.; Winkler-Skalna, A.; Czarnecki, L. Sustainable Test Methods for Construction Materials and Elements. *Materials* **2020**, *13*, 606. [CrossRef]
15. Skotnicki, L.; Kuźniewski, J.; Szydlo, A. Stiffness Identification of Foamed Asphalt Mixtures with Cement, Evaluated in Laboratory and In Situ in Road Pavements. *Materials* **2020**, *13*, 1128. [CrossRef] [PubMed]
16. Bywalski, C.; Drzazga, M.; Kaźmierowski, M.; Kamiński, M. Shear Behavior of Concrete Beams Reinforced with a New Type of Glass Fiber Reinforced Polymer Reinforcement: Experimental Study. *Materials* **2020**, *13*, 1159. [CrossRef]
17. Trapko, T.; Musiał, M. Effect of PBO–FRCM Reinforcement on Stiffness of Eccentrically Compressed Reinforced Concrete Columns. *Materials* **2020**, *13*, 1221. [CrossRef]

18. Karolak, A.; Jasieńko, J.; Nowak, T.; Raszczuk, K. Experimental Investigations of Timber Beams with Stop-Splayed Scarf Carpentry Joints. *Materials* **2020**, *13*, 1435. [CrossRef] [PubMed]
19. Słoński, M.; Schabowicz, K.; Krawczyk, E. Detection of Flaws in Concrete Using Ultrasonic Tomography and Convolutional Neural Networks. *Materials* **2020**, *13*, 1557. [CrossRef] [PubMed]
20. Juraszek, J. Fiber Bragg Sensors on Strain Analysis of Power Transmission Lines. *Materials* **2020**, *13*, 1559. [CrossRef]
21. Shi, Y.; Li, C.; Long, D. Study of the Microstructure Characteristics of Three Different Fine-grained Tailings Sand Samples during Penetration. *Materials* **2020**, *13*, 1585. [CrossRef]
22. Bywalski, C.; Kaźmierowski, M.; Kamiński, M.; Drzazga, M. Material Analysis of Steel Fibre Reinforced High-Strength Concrete in Terms of Flexural Behaviour. Experimental and Numerical Investigation. Experimental and Numerical Investigation. *Materials* **2020**, *13*, 1631. [CrossRef]
23. Stawiski, B.; Kania, T. Tests of Concrete Strength across the Thickness of Industrial Floor Using the Ultrasonic Method with Exponential Spot Heads. *Materials* **2020**, *13*, 2118. [CrossRef]
24. Schabowicz, K.; Sulik, P.; Zawiślak, Ł. Identification of the Destruction Model of Ventilated Facade under the Influence of Fire. *Materials* **2020**, *13*, 2387. [CrossRef] [PubMed]
25. Ranachowski, Z.; Ranachowski, P.; Dębowski, T.; Brodecki, A.; Kopec, M.; Roskosz, M.; Fryczowski, K.; Szymków, M.; Krawczyk, E.; Schabowicz, K. Mechanical and Non-Destructive Testing of Plasterboards Subjected to a Hydration Process. *Materials* **2020**, *13*, 2405. [CrossRef] [PubMed]
26. Logoń, D.; Schabowicz, K.; Wróblewski, K. Assessment of the Mechanical Properties of ESD Pseudoplastic Resins for Joints in Working Elements of Concrete Structures. *Materials* **2020**, *13*, 2426. [CrossRef] [PubMed]
27. Wróbel, M.; Woszuk, A.; Franus, W. Laboratory Methods for Assessing the Influence of Improper Asphalt Mix Compaction on Its Performance. *Materials* **2020**, *13*, 2476. [CrossRef] [PubMed]
28. Kozubal, J.; Wróblewski, R.; Muszyński, Z.; Wyjadłowski, M.; Stróżyk, J. Non-Deterministic Assessment of Surface Roughness as Bond Strength Parameters between Concrete Layers Cast at Different Ages. *Materials* **2020**, *13*, 2542. [CrossRef]
29. Tan, Y.; Zhu, B.; Qi, L.; Yan, T.; Wan, T.; Yang, W. Mechanical Behavior and Failure Mode of Steel–Concrete Connection Joints in a Hybrid Truss Bridge: Experimental Investigation. *Materials* **2020**, *13*, 2549. [CrossRef]
30. Park, J.; Sim, S.; Yoon, Y.; Oh, T. Prediction of Static Modulus and Compressive Strength of Concrete from Dynamic Modulus Associated with Wave Velocity and Resonance Frequency Using Machine Learning Techniques. *Materials* **2020**, *13*, 2886. [CrossRef]
31. Świt, G.; Krampikowska, A.; Pała, T.; Lipiec, S.; Dzioba, I. Using AE Signals to Investigate the Fracture Process in an Al–Ti Laminate. *Materials* **2020**, *13*, 2909. [CrossRef]
32. Dumbrava, F.; Cerbu, C. Experimental Study on the Stiffness of Steel Beam-to-Upright Connections for Storage Racking Systems. *Materials* **2020**, *13*, 2949. [CrossRef]
33. Zieliński, A.; Kaszyńska, M. Calibration of Steel Rings for the Measurement of Strain and Shrinkage Stress for Cement-Based Composites. *Materials* **2020**, *13*, 2963. [CrossRef]
34. Adamczak-Bugno, A.; Krampikowska, A. The Acoustic Emission Method Implementation Proposition to Confirm the Presence and Assessment of Reinforcement Quality and Strength of Fiber–Cement Composites. *Materials* **2020**, *13*, 2966. [CrossRef] [PubMed]
35. Wutke, M.; Lejzerowicz, A.; Garbacz, A. The Use of Wavelet Analysis to Improve the Accuracy of Pavement Layer Thickness Estimation Based on Amplitudes of Electromagnetic Waves. *Materials* **2020**, *13*, 3214. [CrossRef]
36. Trąmpczyński, W.; Goszczyńska, B.; Bacharz, M. Acoustic Emission for Determining Early Age Concrete Damage as an Important Indicator of Concrete Quality/Condition before Loading. *Materials* **2020**, *13*, 3523. [CrossRef]
37. Park, S.; Lee, J.; Lee, D. Aggregate Roundness Classification Using a Wire Mesh Method. *Materials* **2020**, *13*, 3682. [CrossRef] [PubMed]
38. Michałek, J.; Sobótka, M. Assessment of Internal Structure of Spun Concrete Using Image Analysis and Physicochemical Methods. *Materials* **2020**, *13*, 3987. [CrossRef] [PubMed]
39. Park, J.; Hong, G. Strength Characteristics of Controlled Low-Strength Materials with Waste Paper Sludge Ash (WPSA) for Prevention of Sewage Pipe Damage. *Materials* **2020**, *13*, 4238. [CrossRef]
40. Yin, D.; Xu, Q. Comparison of Sandstone Damage Measurements Based on Non-Destructive Testing. *Materials* **2020**, *13*, 5154. [CrossRef] [PubMed]
41. Pang, M.; Sun, Z.; Li, Q.; Ji, Y. 1H NMR Spin-Lattice Relaxometry of Cement Pastes with Polycarboxylate Superplasticizers. *Materials* **2020**, *13*, 5626. [CrossRef]
42. Skrzypczak, I.; Kokoszka, W.; Zięba, J.; Leśniak, A.; Bajno, D.; Bednarz, L. A Proposal of a Method for Ready-Mixed Concrete Quality Assessment Based on Statistical-Fuzzy Approach. *Materials* **2020**, *13*, 5674. [CrossRef]
43. Matsui, T.; Matsushita, Y.; Matsumoto, Y. Mechanical Behavior of GFRP Connection Using FRTP Rivets. *Materials* **2021**, *14*, 7. [CrossRef]
44. Gnatowski, A.; Kijo-Kleczkowska, A.; Chyra, M.; Kwiatkowski, D. Numerical–Experimental Analysis of Polyethylene Pipe Deformation at Different Load Values. *Materials* **2021**, *14*, 160. [CrossRef] [PubMed]
45. Jeong, S.; Kighuta, K.; Lee, D.; Park, S. Numerical Analysis of Shear and Particle Crushing Characteristics in Ring Shear System Using the PFC2D. *Materials* **2021**, *14*, 229. [CrossRef] [PubMed]

46. Pawelska-Mazur, M.; Kaszynska, M. Mechanical Performance and Environmental Assessment of Sustainable Concrete Reinforced with Recycled End-of-Life Tyre Fibres. *Materials* **2021**, *14*, 256. [CrossRef]
47. Tworzewski, P.; Raczkiewicz, W.; Czapik, P.; Tworzewska, J. Diagnostics of Concrete and Steel in Elements of an Historic Reinforced Concrete Structure. *Materials* **2021**, *14*, 306. [CrossRef] [PubMed]
48. Logoń, D.; Schabowicz, K.; Roskosz, M.; Fryczowski, K. The Increase in the Elastic Range and Strengthening Control of Quasi Brittle Cement Composites by Low-Module Dispersed Reinforcement: An Assessment of Reinforcement Effects. *Materials* **2021**, *14*, 341. [CrossRef]
49. Bilko, P.; Skoratko, A.; Rutkiewicz, A.; Małyszko, L. Determination of the Shear Modulus of Pine Wood with the Arcan Test and Digital Image Correlation. *Materials* **2021**, *14*, 468. [CrossRef] [PubMed]
50. Gabryś, K.; Soból, E.; Sas, W.; Šadzevičius, R.; Skominas, R. Warsaw Glacial Quartz Sand with Different Grain-Size Characteristics and Its Shear Wave Velocity from Various Interpretation Methods of BET. *Materials* **2021**, *14*, 544. [CrossRef] [PubMed]
51. Skotnicki, Ł.; Kuźniewski, J.; Szydło, A. Research on the Properties of mineral–Cement Emulsion Mixtures Using Recycled Road Pavement Materials. *Materials* **2021**, *14*, 563. [CrossRef]
52. Zhao, Y.; Liu, R.; Yan, F.; Zhang, D.; Liu, J. Windblown Sand-Induced Degradation of Glass Panels in Curtain Walls. *Materials* **2021**, *14*, 607. [CrossRef]
53. Wiejak, A.; Francke, B. Testing and Assessing Method for the Resistance of Wood-Plastic Composites to the Action of Destroying Fungi. *Materials* **2021**, *14*, 697. [CrossRef]
54. Tešić, K.; Baričević, A.; Serdar, M. Non-Destructive Corrosion Inspection of Reinforced Concrete Using Ground-Penetrating Radar: A Review. *Materials* **2021**, *14*, 975. [CrossRef] [PubMed]
55. Sudoł, E.; Szewczak, E.; Małek, M. Comparative Analysis of Slip Resistance Test Methods for Granite Floors. *Materials* **2021**, *14*, 1108. [CrossRef] [PubMed]
56. Juraszek, J.; Antonik-Popiołek, P. Fibre Optic FBG Sensors for Monitoring of the Temperature of the Building Envelope. *Materials* **2021**, *14*, 1207. [CrossRef] [PubMed]
57. Topolář, L.; Kocáb, D.; Pazdera, L.; Vymazal, T. Analysis of Acoustic Emission Signals Recorded during Freeze-Thaw Cycling of Concrete. *Materials* **2021**, *14*, 1230. [CrossRef]
58. Mackiewicz, P.; Mączka, E. The Impact of Water and Road Salt with Anti-Caking Agent on the Stiffness of Select Mixes Used for the Road Surface. *Materials* **2021**, *14*, 1345. [CrossRef]
59. Kania, T.; Derkach, V.; Nowak, R. Testing Crack Resistance of Non-Load-Bearing Ceramic Walls with Door Openings. *Materials* **2021**, *14*, 1379. [CrossRef] [PubMed]
60. Bednarz, L.; Bajno, D.; Matkowski, Z.; Skrzypczak, I.; Leśniak, A. Elements of Pathway for Quick and Reliable Health Monitoring of Concrete Behavior in Cable Post-Tensioned Concrete Girders. *Materials* **2021**, *14*, 1503. [CrossRef]
61. Juraszek, J.; Gwóźdź-Lasoń, M.; Logoń, D. FBG Strain Monitoring of a Road Structure Reinforced with a Geosynthetic Mattress in Cases of Subsoil Deformation in mining Activity Areas. *Materials* **2021**, *14*, 1709. [CrossRef]
62. Aneja, S.; Sharma, A.; Gupta, R.; Yoo, D. Bayesian Regularized Artificial Neural Network Model to Predict Strength Characteristics of Fly-Ash and Bottom-Ash Based Geopolymer Concrete. *Materials* **2021**, *14*, 1729. [CrossRef]
63. Schabowicz, K.; Sulik, P.; Zawiślak, Ł. Reduction of Load Capacity of Fiber Cement Board Facade Cladding under the Influence of Fire. *Materials* **2021**, *14*, 1769. [CrossRef] [PubMed]
64. Logoń, D.; Juraszek, J.; Keršner, Z.; Frantík, P. Identifying the Range of Micro-Events Preceding the Critical Point in the Destruction Process in Traditional and Quasi-Brittle Cement Composites with the Use of a Sound Spectrum. *Materials* **2021**, *14*, 1809. [CrossRef]
65. Bhat, D.; Kozubal, J.; Tankiewicz, M. Extended Residual-State Creep Test and Its Application for Landslide Stability Assessment. *Materials* **2021**, *14*, 1968. [CrossRef] [PubMed]
66. Boštík, J.; Miča, L.; Terzijski, I.; Džaferagić, M.; Leiter, A. Grouting below Subterranean Water: Erosional Stability Test. *Materials* **2021**, *14*, 2333. [CrossRef] [PubMed]

Article

Strength Characteristics of Controlled Low-Strength Materials with Waste Paper Sludge Ash (WPSA) for Prevention of Sewage Pipe Damage

Jeongjun Park [1] and Gigwon Hong [2],*

[1]   Incheon Disaster Prevention Research Center, Incheon National University, Incheon 22012, Korea; smearjun@hanmail.net
[2]   Institute of Technology Research and Development, Korea Engineering & Construction, Seoul 05661, Korea
*   Correspondence: gigwon_hong@kecgroup.kr

Received: 22 August 2020; Accepted: 21 September 2020; Published: 23 September 2020

Abstract: In this study, the effects of the mixing conditions of waste paper sludge ash (WPSA) on the strength and bearing capacity of controlled low-strength material (CLSM) were evaluated, and the optimal mixing conditions were used to evaluate the strength characteristics of CLSM with recyclable WPSA. The strength and bearing capacity of CLSM with WPSA were evaluated using unconfined compressive strength tests and plate bearing tests, respectively. The unconfined compressive strength test results show that the optimal mixing conditions for securing 0.8–1.2 MPa of target strength under 5% of cement content conditions can be obtained when both WPSA and fly ash are used. This is because WPSA and fly ash, which act as binders, have a significant impact on overall strength when the cement content is low. The bearing capacity of weathered soil increased from 550 to 575 kPa over time, and CLSM with WPSA increased significantly, from 560 to 730 kPa. This means that the bearing capacity of CLSM with WPSA was 2.0% higher than that of weathered soil immediately after construction; furthermore, it was 27% higher at 60 days of age. In addition, the allowable bearing capacity of CLSM corresponding to the optimal mixing conditions was evaluated, and it was found that this value increased by 30.4% until 60 days of age. This increase rate was 6.7 times larger than that of weathered soil (4.5%). Therefore, based on the allowable bearing capacity calculation results, CLSM with WPSA was applied as a sewage pipe backfill material. It was found that CLSM with WPSA performed better as backfill and was more stable than soil immediately after construction. The results of this study confirm that CLSM with WPSA can be utilized as sewage pipe backfill material.

Keywords: waste paper sludge ash (WPSA); controlled low-strength material (CLSM); unconfined compressive strength; bearing capacity; backfill material

---

1. Introduction

With the aging of buried pipes, incidents of ground subsidence around them have become increasingly frequent in South Korea. Underground cavities around sewage pipes in urban areas caused by aging and damage have been found to be the main cause of ground subsidence, accounting for approximately 82% of the total number of occurrences, as shown in Figure 1 [1,2].

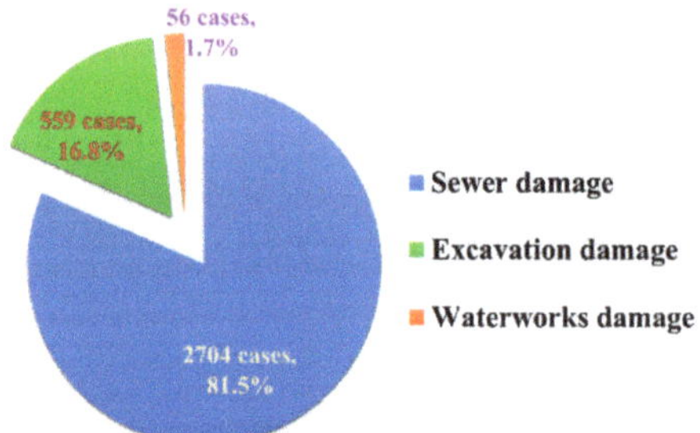

Figure 1. Causes of ground subsidence [1].

When aged sewage pipes are replaced, the ground is excavated and soil is backfilled after new sewage pipes are placed. The soil excavated on site is used as backfill in many cases, but there are also many cases in which the compaction of the ground around the pipes is not sufficient. The reason is that the compaction energy by equipment may affect the damage to sewer pipes when such compaction is performed. This problem can occur when the hydraulic filling of high-quality sand is difficult at the construction site. Specifically, it is a general restoration method that must be restored immediately after excavation construction, in order to secure the usability of the surface ground over sewage pipes, such as roads.

Poor compaction of the ground around sewage pipes causes stress concentration on the pipes when loaded from above, causing long-term damage as the pipes age. As this causes underground cavities that lead to ground subsidence around sewage pipes, it is necessary to use backfill materials that provide sufficient bearing capacity and stability of the ground around the pipes.

Studies on various backfill materials and methods have been conducted to determine how to prevent sewage pipe damage and ensure sufficient bearing capacity of the surrounding ground [3]. Controlled low-strength material (CLSM) was developed to address various problems caused by inadequate backfill materials and poor compaction [4].

CLSM is a construction material that applies the concept of low-strength concrete to geotechnical engineering. CLSM, which was developed in the United States in the mid-1900s, has been used as a backfill material for buried pipes, such as water and sewage pipes. The American Concrete Institute (ACI) has defined CLSM as material composed of sand, cement, admixture, fly ash, and water, and has suggested that the strength range be determined according to the method of excavation. For example, the suggested unconfined compressive strength is less than approximately 0.3 MPa when excavation is performed using manpower, 0.7–1.4 MPa when excavation is performed using machinery, and less than 8.3 MPa when excavation is not required. Unlike conventional soil backfill materials, CLSM is self-leveling, self-compacting, and flowable. In addition, it is possible to control its strength. Therefore, it can be used as a countermeasure against ground subsidence caused by the faulty installation of buried pipes and soil loss underground, as shown in Figure 2. It is also highly useful as an eco-friendly material because various recycled materials can be incorporated into it [5].

Figure 2. Backfilling of buried pipes using controlled low-strength material (CLSM) [6].

In recent years, various studies have been conducted on CLSM with coal ash as the main component to explore the feasibility of utilizing coal ash in CLSM on a large scale. Kim et al. [7] conducted unconfined compressive strength tests of CLSM with coal ash as the main component, in an effort to develop various construction materials and analyzed strength as a function of curing time and water content. Kong et al. [8] evaluated the strength characteristics of CLSM composed of fly ash, cement, water, and pond ash in place of sand, depending on the mixing ratio between pond ash and fly ash. Won and Lee [9] conducted research on the basic properties and strength characteristics of a CLSM mixture with bottom ash and a mixture with fly ash. Razak et al. [10] prepared a CLSM mixture using bottom ash and evaluated its performance under various curing conditions.

While several studies have been conducted on CLSM using coal ash as the main component, several others have been conducted on the development of new concrete mixtures using materials such as recycled concrete aggregate (RCA) and supplementary cementitious material (SCM) as replacements for Portland cement, a component of CLSM.

Hao et al. [11] evaluated the compressive and tensile strength of CLSM mixtures with fly ash, bottom ash, and paper sludge used as substitutes for cement and conducted research on methods for reducing carbon dioxide (CO_2) emissions. Kubissa et al. [12] measured compressive strength in the range of 0.52–4.29 MPa for CLSM containing fly ash and RCA. They also developed a mixture that achieved the same performance as conventional CLSM, using only a small amount of cement.

Naganathan et al. [13] prepared a CLSM mixture in which the amount of cement was minimized using RCA and fly ash, and they confirmed that the addition of fly ash improved the strength of the mixture. Ahmadi and Al-Khaja [14] compared the chemical properties of concrete mixtures with various paper sludge mixing ratios with those of existing concrete mixtures and confirmed that paper sludge can be used as a construction material. Frías et al. [15] and García et al. [16] developed a material that could be transformed into metakaolinite through the calcination of paper sludge, and used the material to evaluate the applicability of paper sludge as a cement substitute and its environmental impact through recycling. Monzó et al. [17], Horiguchi et al. [18], and Boni et al. [19] conducted research on the feasibility of replacing Portland cement with paper sludge, sewage sludge, and the waste obtained by incinerating them and derived management plans for industrial waste from an environmental perspective.

The amount of waste generated is increasing exponentially every year, due to the growing consumption caused by the growth of cities, and this increasing generation of waste is causing serious environmental problems. Therefore, it is urgent to identify ways to treat the ash emitted in large quantities by the incineration of waste and develop treatment technologies for such waste [20].

Fly ash and waste paper sludge ash (WPSA), which is obtained by incinerating paper sludge, are typical products of waste incineration. Various studies have been conducted on recycling WPSA. Heo et al. [21] reported that WPSA can be used as a lightweight embankment material because it has excellent engineering properties in comparison to fly ash and soil. They also noted that WPSA causes no problems in terms of environmental impact, because the proportion of lime (CaO) is nine times higher than that of fly ash and the concentrations of toxic heavy metals are lower than the established thresholds.

Lee et al. [22] conducted research on recycling WPSA as a construction material, by analyzing the quality characteristics of concrete and clay bricks fabricated by mixing with WPSA. Seo [23] evaluated the physical and engineering properties achieved when WPSA and fly ash were used as cement mixing materials. The possibility of stabilizing soil by improving the strength of clay using WPSA was evaluated in [24]. Bujulu et al. [25] evaluated the feasibility of using WPSA as a replacement for lime and cement in the stabilization of quick-clay and found that up to 50% of the lime and cement normally used can be replaced with WPSA.

Ahmad et al. [26] confirmed that WPSA is suitable for use as a concrete substitute by comparing the compressive and tensile strength of concrete, in which 5–20% cement was replaced with WPSA with those of ordinary concrete. Sani et al. [27] replaced ordinary Portland cement with 50–100% WPSA and compared the compressive strength obtained as a function of curing time. Ridzuan et al. [28] investigated the mixing characteristics of RCA and WPSA when used to replace Portland cement and evaluated the strength of CLSM based on the optimal content of WPSA (which yielded the maximum strength). Fauzi et al. [29] and Azmi et al. [30] evaluated the strength of WPSA and RCA mixtures. Bai et al. [31] and Mozaffari et al. [32] evaluated the applicability of WPSA and blast furnace slag through physical and chemical evaluation, using them as mixing materials and comparing the strength of mixes containing them with mixes containing ordinary Portland cement.

Concrete and grout materials such as CLSM have an important correlation with the environmental characteristics and material improvement belowground. Farzampour [33,34] analyzed the strength of concrete and grout materials according to various temperature and humidity conditions. The results confirmed that the behavior of concrete and grout material has a greater effect on temperature than humidity. In addition, it was found that the material does not reach its ultimate strength when the temperature is significantly low in the curing conditions. Mansouri et al. [35] analyzed the structural behavior of concrete, considering the steel fiber effect and curing time in order to improve its abrasion resistance. It was confirmed that abrasion resistance increased when the steel fiber ratio was high.

The results of various studies confirm that WPSA can be used to address various problems associated with waste treatment, the unavailability of natural aggregates, and environmental concerns. Studies have shown that CLSM with WPSA can be used as a backfill material to prevent sewage pipe damage, by ensuring sufficient bearing capacity. In this study, the strength and bearing capacity characteristics of CLSM with WPSA were analyzed. The strength characteristics were evaluated by conducting unconfined compressive strength tests of mixtures with various WPSA mixing ratios, and the bearing capacity was evaluated by conducting plate bearing tests.

2. Strength Evaluation of CLSM with WPSA

To evaluate the strength characteristics of CLSM with WPSA, a mix design process for determining the proportions of components (cement, sand, fly ash, and WPSA) was performed. In addition, flowability tests and unconfined compressive strength tests were conducted.

2.1. Materials

The CLSM used in this study was prepared by adding WPSA to conventional CLSM materials, resulting in mixtures of cement, sand, fly ash, and WPSA. Ordinary Portland cement was used as the cement, and Jumunjin standard sand was used as the sand. Table 1 lists the engineering properties of the standard sand.

Table 1. Engineering properties of Jumunjin standard sand in Korea.

Specific Gravity (Gs)	Maximum Dry Unit Weight ($\gamma_{d(max)}$; kN/m^3)	Optimum Moisture Content (ω_{opt}; %)	Internal Friction Angle (ϕ; $^\circ$)	USCS *
2.65	16.9	9.4	33.4	SP **

* Unified soil classification system, ** Poorly graded sands and gravelly sands, little or no fines

Fly ash was obtained from Ekons Co., Ltd. (Incheon, South Korea), and its main components were analyzed. The specific gravity of fly ash is 2.3, which is two-thirds that of ordinary cement, and it is known that the value increases as the iron content increases [36]. As Table 2 shows, SiO_2 and Al_2O_3 made up more than 90% of the fly ash used in this study, indicating that reactive oxides that can be used in the polymerization reaction were present in large quantities. SiO_2 can improve compressive strength in the long term, because it generates calcium silicate when it reacts with $Ca(OH)_2$, which is generated when cement undergoes hydration.

Table 2. Components of fly ash (FA).

Component	SiO$_2$	Al$_2$O$_3$	Fe$_2$O$_3$	CaO	K$_2$O	TiO$_2$	MgO	Na$_2$O	SO$_3$	Loss on Ignition
FA (%)	75.94	14.70	3.85	1.47	1.11	0.83	0.6	0.54	0.46	0.5

Scanning electron microscopy (SEM) imaging was performed using equipment (model; JSM-7001F) manufactured by JEOL Ltd., Tokyo, Japan, to analyze the structural characteristics of the fly ash. The results are shown in Figure 3. The particle size of the fly ash ranged from 1 to 100 μm, with an average particle diameter in the range of 20 to 30 μm, which is very similar to the particle size characteristics of cement. The fly ash particles were smooth spheres, with pores observed on the surfaces of relatively large particles. XRD (X-ray diffraction) analysis of fly ash was performed using equipment (model; smartlab) manufactured by Rigaku, Tokyo, Japan, and it showed that quartz (SiO_2) and mullite ($3Al_2O_3 \cdot 2SiO_2$) were present as crystalline substances, as shown in Figure 4. The general principle and operation of SEM and XRD can be confirmed in the research of Joseph et al. [37] and Borchert [38], respectively.

(a)   (b)

Figure 3. SEM images of fly ash: (a) 500×; (b) 2000×.

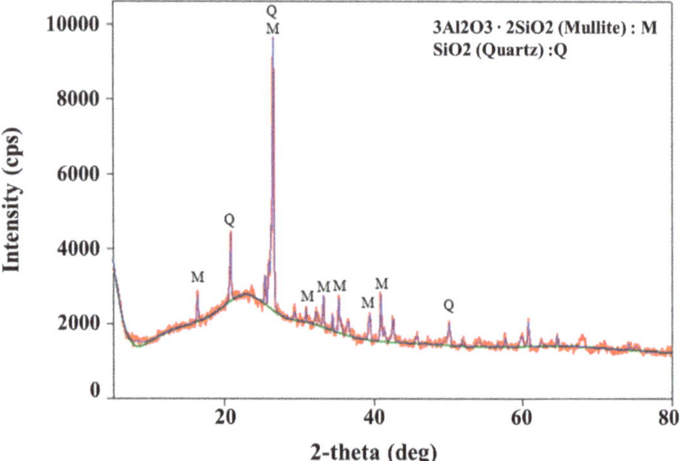

Figure 4. XRD results for fly ash.

The WPSA used in this study was obtained from Ekons Co., Ltd. (Incheon, South Korea). The results of the analysis of its main components are as follows. The specific gravity was 2.5, and CaO, SiO$_2$, and Al$_2$O$_3$ accounted for more than 80% of the main components (Table 3). CaO, which is a main component of cement, contributes to the excellent strength development of cement. Therefore, it is possible to adjust the strength development by adjusting the WPSA content. SEM imaging results for the WPSA showed that it was composed of ash particles and unburned fibers, with particle size ranging from to 2 to 100 μm. The WPSA included both spherical particles with smooth surfaces and plate-shaped particles (Figure 5). The results of XRD analysis of the WPSA confirmed that it was composed of CaO, CaCO$_3$, and C$_{12}$A$_7$, as shown in Figure 6.

Table 3. Components of waste paper sludge ash (WPSA).

Component	SiO$_2$	Al$_2$O$_3$	Fe$_2$O$_3$	CaO	K$_2$O	TiO$_2$	MgO	Na$_2$O	SO$_3$	Loss on Ignition
FA (%)	59.35	11.19	10.27	4.43	4.02	3.98	1.74	1.36	0.64	3.02

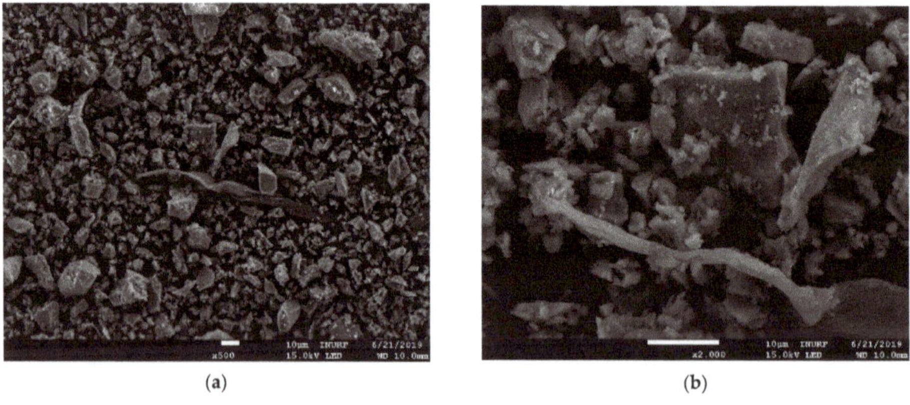

(a)   (b)

Figure 5. SEM images of WPSA: (a) 500×; (b) 2000×.

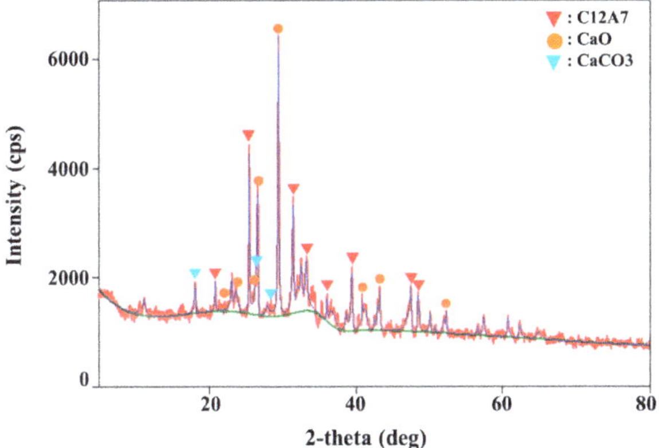

Figure 6. XRD results for WPSA.

2.2. Mix Design of CLSM

As the CLSM used in this study was intended for use as backfill material for sewage pipes, re-excavation in the future must be easy, and the time required for construction needs to be minimized. As mentioned previously, ACI recommends an unconfined compressive strength in the range of 0.7–1.4 MPa for CLSM when mechanical excavation is conducted. In this study, the criterion for unconfined compressive strength at an age of 28 days was determined to be 0.8–1.2 MPa, based on previous studies, because the CLSM developed in this study must allow re-excavation for the maintenance of sewage pipes.

Mix designs with cement proportions of 5% and 10% and sand proportions of 35, 40, 45, and 50 were developed to assess the effects of the proportions of cement and sand on the strength characteristics of CLSM. For each combination of cement and sand, mix designs with WPSA-to-fly-ash ratios of 1:0, 1:1, and 0:1 were developed. The mix design was determined for a relative quantitative comparison of WPSA and FA, considering the ratio of cement and sand. Table 4 summarizes the mix designs.

Table 4. Mix designs.

Case Number	1	2	3	4	5	6	7	8	9	10	11	12
Cement (%)						5						
Sand (%)		35			40			45			50	
WPSA (%)	60	30	0	55	27.5	0	50	25	0	45	22.5	0
Fly ash (%)	0	30	60	0	27.5	55	0	25	50	0	22.5	45
Case Number	**13**	**14**	**15**	**16**	**17**	**18**	**19**	**20**	**21**	**22**	**23**	**24**
Cement (%)						10						
Sand (%)		35			40			45			50	
WPSA (%)	55	27.5	0	50	25	0	45	22.5	0	40	20	0
Fly ash (%)	0	27.5	55	0	25	50	0	22.5	45	0	20	40

2.3. Experiment Details

ACI specifies that materials with excellent flowability should not exhibit noticeable material segregation, and their flowability value must be at least 200 mm. Flowability tests were conducted in accordance with ASTM (American Society for Testing and Materials) [39] to ensure water content conditions that would satisfy the flowability requirements for each mix design.

Unconfined compressive strength tests were conducted in accordance with ASTM [40]. In general, the unconfined compressive strength of a concrete material is based on a curing time of 28 days, but there are cases in which earlier strength information is required, depending on the use of the material.

CLSM, which was used in this study, must perform adequately immediately after construction, because it is used as backfill material for sewage pipes. Emery and Johnston [41] proposed a value of 0.1 MPa for the strength of CLSM at 1 day of age. NRMCA (National Ready Mixed Concrete Association) [42] and Crouch et al. [43] proposed a range of 0.1–0.5 MPa for the strength of CLSM at 3 days of age. In this study, the criteria set for strength at 1 and 28 days of age were 0.1 MPa and 0.8–1.2 MPa, respectively. Unconfined compressive strength was measured at curing times of 1, 7, 28, and 60 days, the latter to assess the long-term strength characteristics of CLSM.

The CLSM specimens used in the unconfined compression strength tests were 100 mm (D) × 200 mm (H) and were fabricated by producing a mix with appropriate water content for each mix design, based on the flowability test results. After initial curing, the specimens were subjected to water curing. Three specimens were tested for each curing time, and the average value was calculated. The specimens were subjected to unconfined compression at a rate of 1 mm/min. Photographs in Figure 7 illustrate the test procedure.

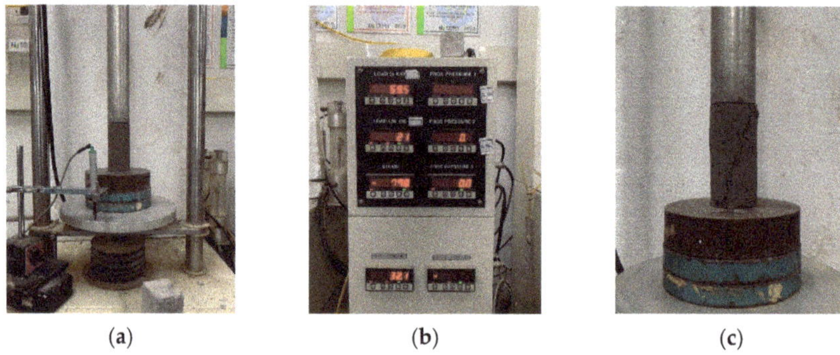

(a)   (b)   (c)

Figure 7. Procedure for unconfined compressive strength test: (**a**) specimen placement; (**b**) measurement of unconfined compressive strength; (**c**) specimen failure.

2.4. Flowability and Unconfined Compressive Strength

Figure 8 and Table 5 show the flowability test results for the various mix designs. The water content that satisfied the flowability criterion (200 mm) ranged from 24 to 32%. When the proportions of WPSA and fly ash were identical, the water content required to ensure flowability decreased as the amount of sand increased. In addition, the required water content increased as the WPSA content increased.

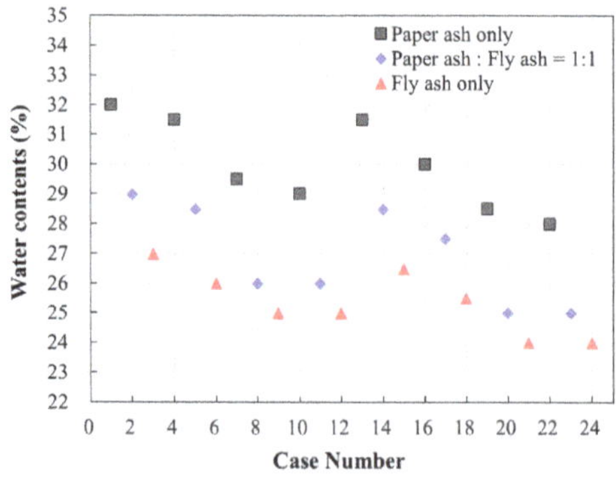

Figure 8. Water content required for adequate flowability.

Table 5. Results of flowability tests.

Case Number	Mix Design				Water Content Required For Adequate Flowability (%)
	Cement (%)	Sand (%)	WPSA (%)	FA (%)	
01			60	0	32.0
02		35	30	30	29.0
03			0	60	27.0
04			55	0	31.5
05		40	27.5	27.5	28.5
06	5		0	55	26.0
07			50	0	29.5
08		45	25	25	26.0
09			0	50	25.0
10			45	0	29.0
11		50	22.5	22.5	26.0
12			0	45	25.0
13			55	0	31.5
14		35	27.5	27.5	28.5
15			0	55	26.5
16			50	0	30.0
17		40	25	25	27.5
18	10		0	50	25.5
19			45	0	28.5
20		45	22.5	22.5	25.0
21			0	45	24.0
22			40	0	28.0
23		50	20	20	25.0
24			0	40	24.0

Table 6 presents the results of the unconfined compressive strength tests for the various mix designs, showing the average value of the results of three specimens, with minimal deviation for each curing time. The unconfined compressive strength tended to increase as the curing time increased, regardless of the mix design, and increased very little if at all after 28 days of age. The unconfined compressive strength increased as the proportion of cement increased, all other aspects of the mix design being equal. Based on these test results, the effects of various aspects of the mix designs on the unconfined compressive strength characteristics were analyzed.

Table 6. Unconfined compressive strength test results.

Case Number	Unconfined Compressive Strength (MPa)				Case No.	Unconfined Compressive Strength (MPa)			
	1 Day	7 Days	28 Days	60 Days		1 Day	7 Days	28 Days	60 Days
01	0.12	0.59	1.00	1.10	13	0.13	0.76	1.70	2.11
02	0.16	0.60	1.30	1.31	14	0.21	1.10	1.50	2.50
03	0.56	0.86	1.50	1.53	15	1.40	3.60	4.50	5.00
04	0.08	0.31	0.53	0.76	16	0.17	0.87	1.15	1.30
05	0.05	0.50	1.00	1.20	17	0.16	0.91	1.84	1.91
06	0.50	0.81	0.90	0.91	18	1.30	3.90	4.79	5.20
07	0.09	0.49	1.10	1.20	19	0.18	1.00	1.37	1.20
08	0.13	0.52	1.20	1.22	20	0.30	1.20	2.47	2.50
09	0.65	0.90	1.30	1.29	21	1.20	3.50	5.84	5.96
10	0.06	0.30	0.54	0.70	22	0.32	0.79	1.00	1.20
11	0.04	0.50	0.80	0.72	23	0.44	0.92	1.82	1.88
12	0.40	0.60	0.80	0.73	24	1.10	3.00	4.81	5.06

3. Results and Discussion

3.1. Unconfined Compressive Strength Versus Mixing Ratio between WPSA and Fly Ash

For cases in which only WPSA was used (i.e., WPSA/FA ratio of 1:0), the effect of the WPSA content on the unconfined compressive strength was assessed, as shown in Figure 9a,b. Regardless of the cement content (5% or 10%), the unconfined compressive strength increased steadily from 1 to 28 days of age, and no significant strength change occurred after 28 days. The strength increased as the WPSA content increased, and was thus lowest when the WPSA content was lowest (cases 10 and 22). When the cement content was high (10%), the strength was higher than when the cement content was low (5%), because the WPSA content was relatively lower.

When the mixing ratio between WPSA and fly ash was 1:1, the unconfined compressive strength of CLSM increased steadily from 1 to 28 days of age for all mix designs, but did not increase significantly beyond 28 days, as shown in Figure 9c,d. When the cement content was 5%, the strength decreased as the sand content increased. However, when the cement content was 10%, the strength characteristics differed depending on the sand content. This may be because both WPSA and fly ash, which act as binders, have a significant impact on the overall strength when the cement content is low (5%), whereas sand has a larger impact on the strength characteristics than ash materials when the cement content is high (10%).

Figure 9e,f show the unconfined compressive strength of CLSM containing only fly ash (i.e., WPSA/FA ratio of 0:1). When the cement content was 5%, the mix with the highest fly ash content exhibited the highest unconfined compressive strength. However, when the cement content was 10%, the mix with the lowest fly ash content exhibited the highest unconfined compressive strength. The strength increased significantly as the cement content increased, all other aspects of the mix design being equal. Especially, the result of case 21 shows that FA, cement, and sand are the mixing conditions with the maximum unconfined compressive strength when only FA is applied to CLSM. These results indicate that fly ash has a significant influence on the strength development of CLSM.

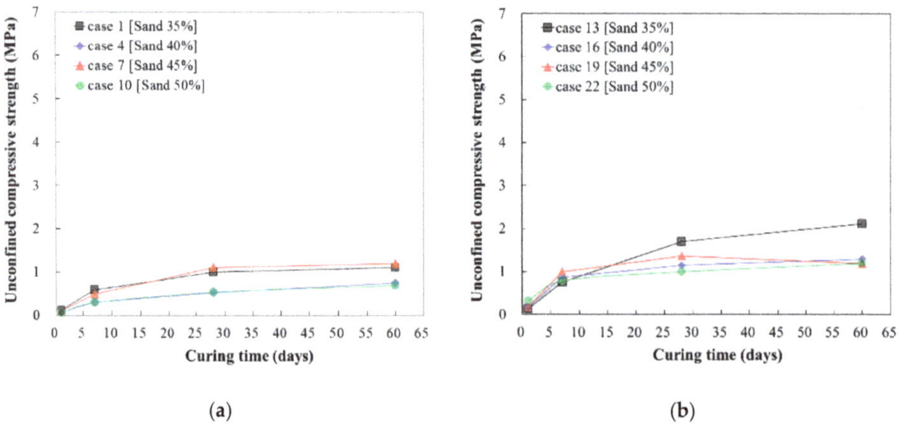

(a) (b)

Figure 9. *Cont.*

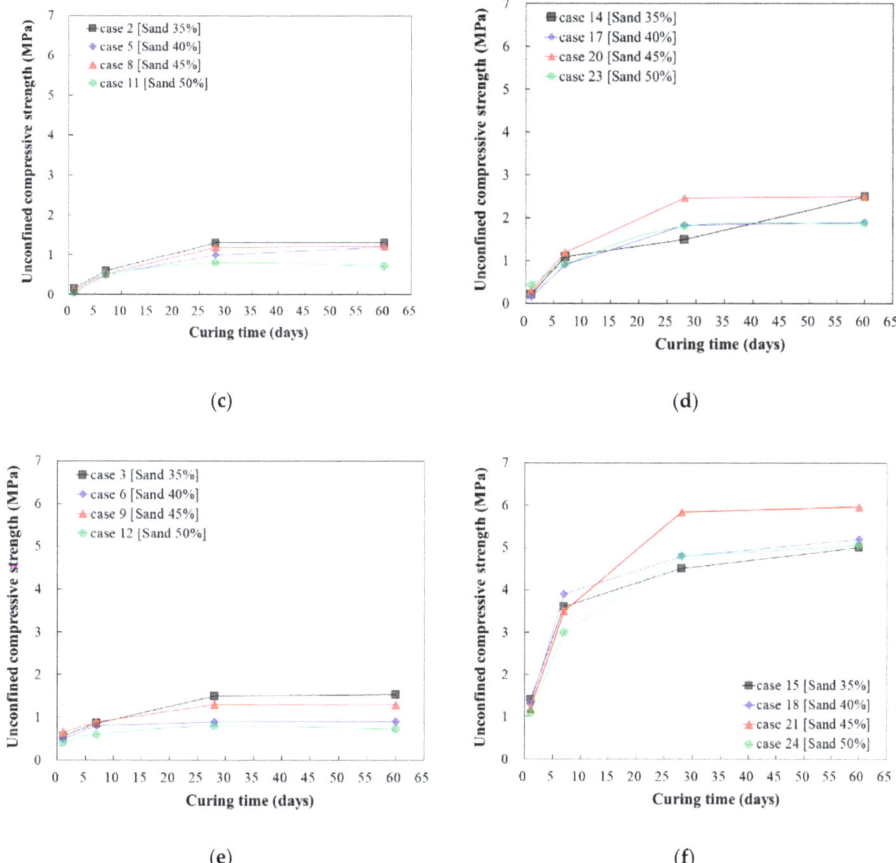

(c)

(d)

(e)

(f)

Figure 9. Unconfined compressive strength versus mixing ratio between WPSA and fly ash: (**a**) WPSA/FA = 1:0 (cement 5%); (**b**) WPSA/FA = 1:0 (cement 10%); (**c**) WPSA/FA = 1:1 (cement 5%); (**d**) WPSA/FA = 1:1 (cement 10%); (**e**) WPSA/FA = 0:1 (cement 5%); (**f**) WPSA/FA = 0:1 (cement 10%).

3.2. Unconfined Compressive Strength Versus Sand Content

The effect of the cement content on the unconfined compressive strength for a given sand content was evaluated. As shown in Figure 10, the rates of strength increase and strength achieved were significantly higher when the cement content was higher, regardless of the sand content and WPSA/fly ash mixing ratio. The lowest rate of increase in unconfined compressive strength was 18.3% when the sand content was 45%, and the highest was 593% when the sand content was 50%. As mentioned above, it was confirmed that, for the range of sand content considered, the rate of strength increase was highest when only fly ash was used.

Mix designs that minimized the cement and sand content were selected to evaluate the applicability of CLSM by recycling WPSA. For cement content of 5%, cases 1, 2, and 8 were found to yield appropriate mix characteristics. For cement content of 10%, cases 16 and 19 were found to yield appropriate mix characteristics. In this study, however, the strength criteria for 28 days (0.8–1.2 MPa) and 1 day (0.1 MPa) of age were determined based on findings from previous studies on how to ensure the desired excavation conditions. As such, case 8, which corresponded to stable strength characteristics, was determined to be the optimal mix design, based on short-term (1 day) and long-term (60 day) strength, as well as cement content.

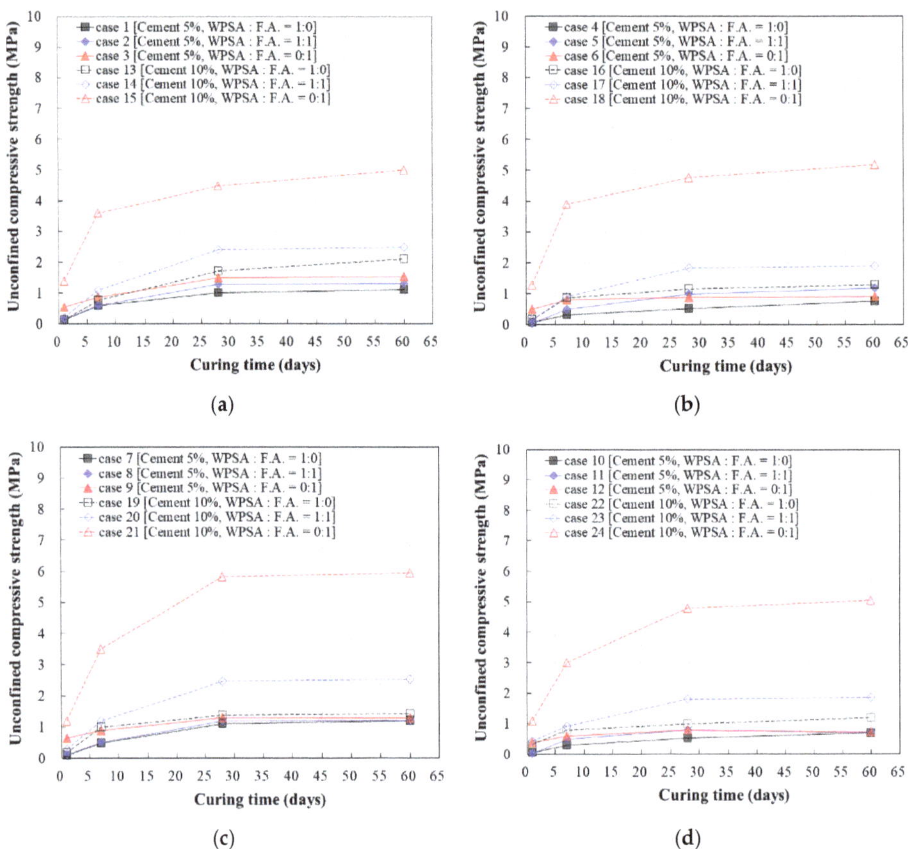

Figure 10. Unconfined compressive strength versus sand content: (**a**) sand 35%; (**b**) sand 40%; (**c**) sand 45%; (**d**) sand 50%.

4. Bearing Capacity Evaluation of CLSM Using Plate Bearing Test

A site was prepared for plate bearing tests of CLSM with WPSA. Plate bearing tests were conducted at a location where CLSM was produced according to the mix design for case 8, which was determined to be the optimal mix design, and was used as backfill material at a location where weathered granite soil, which is commonly used as backfill material, was used. Based on the test results, the bearing capacity of the CLSM backfill and weathered granite soil backfill was evaluated.

4.1. Materials

Various laboratory tests were conducted to assess the engineering properties of the weathered soil used as control backfill. The specific gravity was 2.66 and the USCS (Unified soil classification system) soil classification was SP (Figure 11a). A maximum dry unit weight of 20.2 kN/m^3 was observed at an optimal water content of 11.7% (Figure 11b).

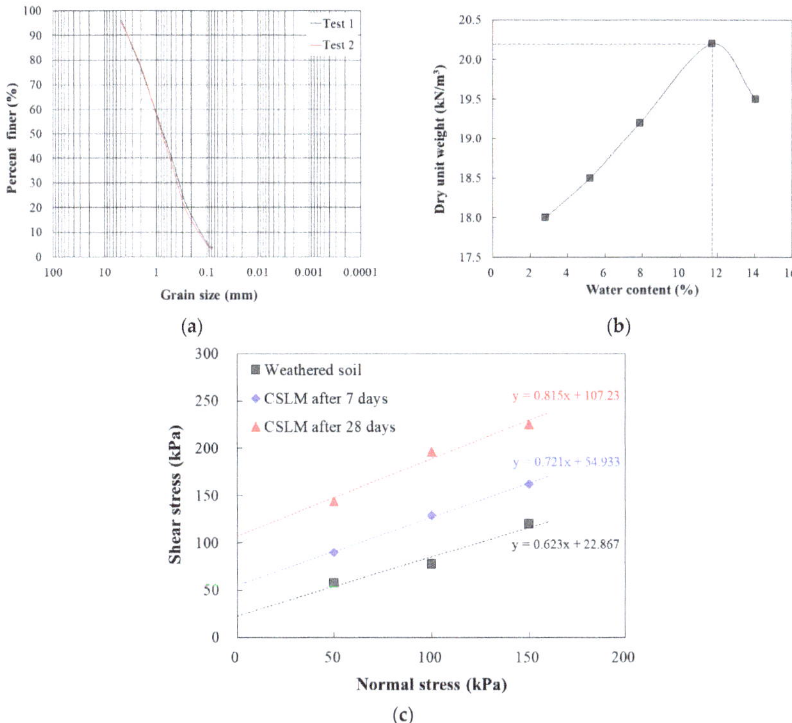

Figure 11. Engineering properties of weathered soil and CLSM: results of (**a**) sieve analysis of weathered soil; (**b**) compaction tests of weathered soil; (**c**) direct shear test.

The cohesion and internal friction angle of the weathered soil determined from direct shear testing were compared with those of CLSM. For this purpose, the CLSM test results at 7 and 28 days of age, when the rate of increase of unconfined compressive strength was high, were used. As shown in Figure 11c and Table 7, CLSM exhibited significantly higher shear strength than weathered soil. The cohesion and internal friction angle of CLSM after 28 days were 368% and 23% higher, respectively, than those of weathered soil.

Table 7. Comparison of shear strength between weathered soil and CLSM with WPSA.

Classification		Normal Stress (kPa)			Strength Parameters	
		50	100	150	Cohesion (c; kPa)	Internal Friction Angle (ϕ; °)
Weathered soil	Shear stress (kPa)	57.8	77.6	120.1	22.9	31.9
CSLM after 7 days		90.0	129.0	162.1	54.9	35.8
CSLM after 28 days		144	196.7	225.5	107.2	39.2

4.2. Plate Bearing Test Procedure

At the site, a large soil tank was installed in the ground. The tank was separated in the middle so that the plate bearing test could be conducted, depending on the application of CLSM. The load was applied in six steps, in each of which the load was less than 98 kPa, or one-sixth of the test target load. After the load was increased in each loading step, it was maintained for at least 15 min. Ground settlement was measured at 1, 2, 3, 5, 10, and 15 min from the time of loading to 15 min, while the load was held constant using an LVDT (Linear Variable Differential Transformer) with an accuracy of 0.01 mm. It was assumed that settlement had stopped when settlement was less than 0.01 mm after 15 min or less than 1% of cumulative settlement after 1 min. Plate bearing tests were conducted at 1, 7, 14, 28, and 60 days of age to assess the effect of curing time on unconfined compressive strength, and the yield strength was expressed by a P–S curve developed from the test results. Figure 12 shows photographs of the plate bearing test procedure.

(a)

(b)

(c)

(d)

(e)

(f)

Figure 12. Plate bearing test procedure: (**a**) ground excavation and soil box installation; (**b**) ground construction; (**c**) buried sewage pipe; (**d**) CLSM construction; (**e**) plate bearing test; (**f**) test measurement.

4.3. Evaluation Results

Figure 13 shows the plate bearing test results for the ground backfilled with weathered soil, for comparison with the bearing capacity of the ground where CLSM was applied. The P–S curve was used to calculate the load, corresponding to 10% settlement of the load plate diameter. A safety factor of 3 was applied to the allowable bearing capacity, based on the consideration that continuous cyclic loading could be applied to the ground in which sewage pipes were buried. It was found that the load corresponding to settlement equal to 10% of the load plate diameter increased from 550 to 575 kPa as the curing time increased. The bearing capacity of the weathered soil appeared to increase, because ground compaction was achieved by the tests conducted on selected curing days, but the rate of increase was judged to be insignificant.

Figure 14 shows the plate bearing test results for the ground where CLSM with WPSA was applied. When the results were analyzed in the same manner as those for weathered soil, it was found that the load increased from 560 to 730 kPa, as the curing time increased. When the rate of load increase was evaluated with respect to curing time, the load was found to increase by 20.5%, 24.1%, 28.6%, and 30.4% at 7, 14, 28, and 60 days of age, relative to the load at 1 day. These results show that the load increased continuously up to 28 days, but the rate of increase was not high after that. This trend is similar to that observed in the unconfined compression test results.

Figure 15 and Table 8 show the allowable bearing capacity of weathered soil and CLSM calculated from the plate bearing test results. As mentioned previously, the allowable bearing capacity was calculated by applying a safety factor of 3 to the load measured in the plate bearing test. The test results for weathered soil show that the allowable bearing capacity increased from 1 to 60 days of age by approximately 4%. In the case of ground where CLSM was applied, however, the allowable bearing capacity gradually increased as the curing time increased; it increased most significantly from 1 to 7 days of age, and then slowly after 28 days. The weathered soil and CLSM exhibited similar allowable bearing capacity immediately after construction.

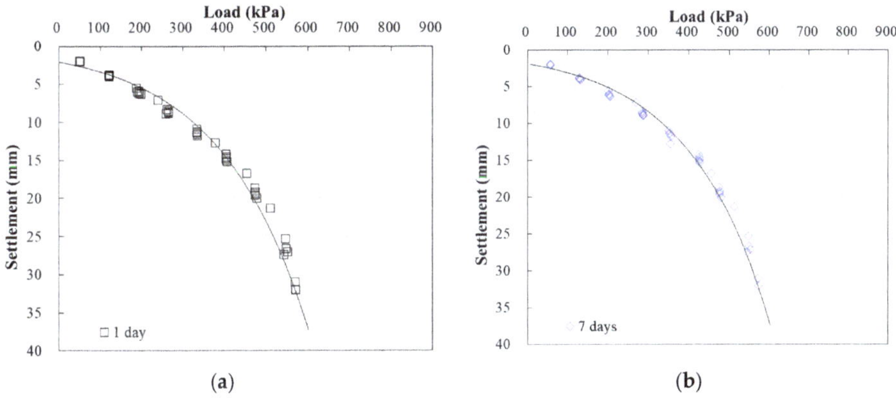

(a)   (b)

Figure 13. *Cont.*

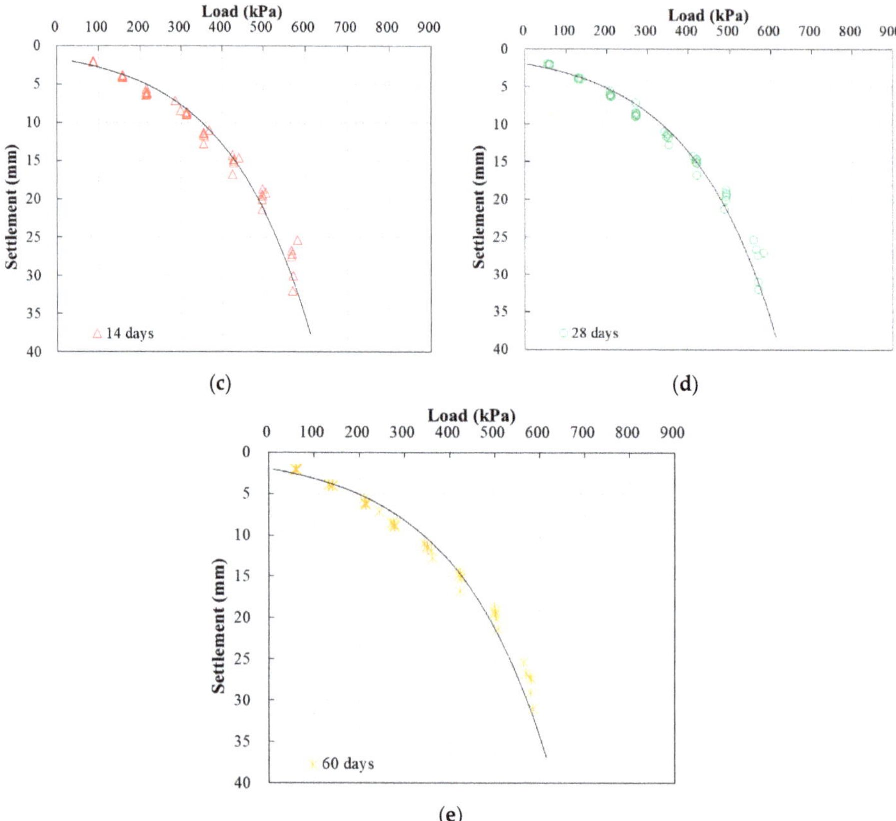

Figure 13. Plate bearing test results for weathered soil over time, with elapsed time of (**a**) 1 day; (**b**) 7 days; (**c**) 14 days; (**d**) 28 days; (**e**) 60 days.

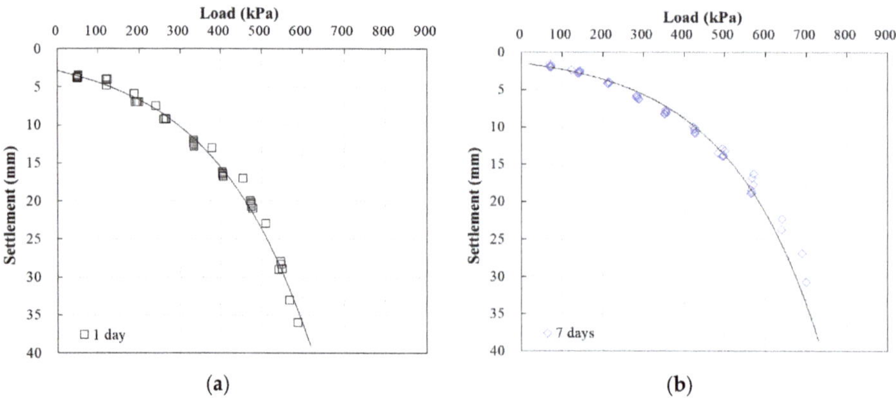

Figure 14. *Cont.*

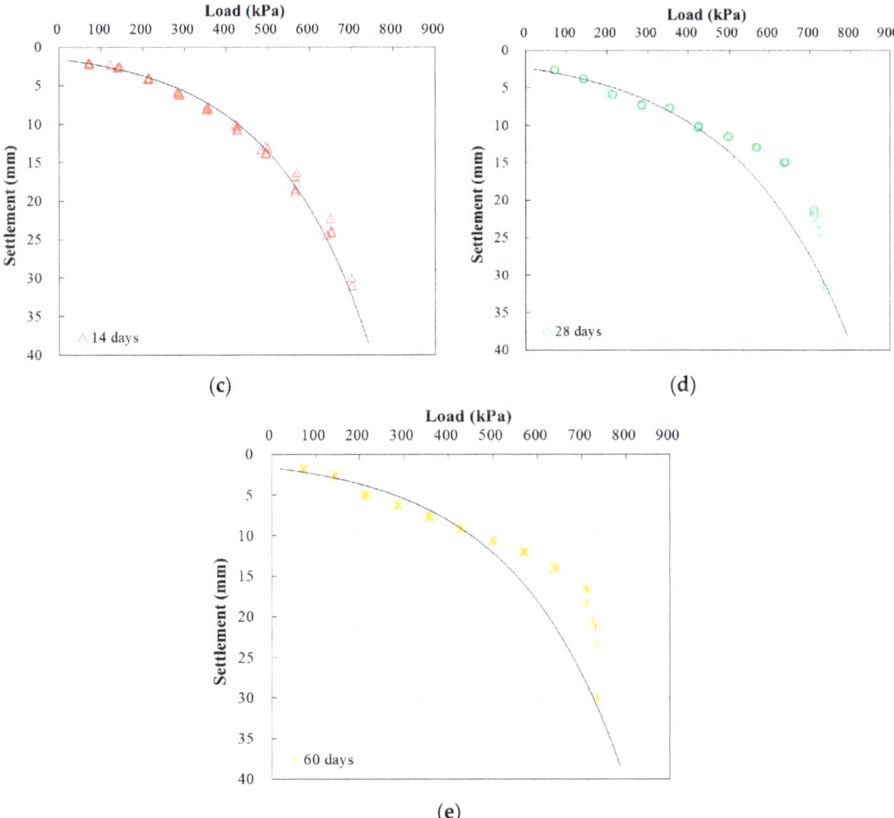

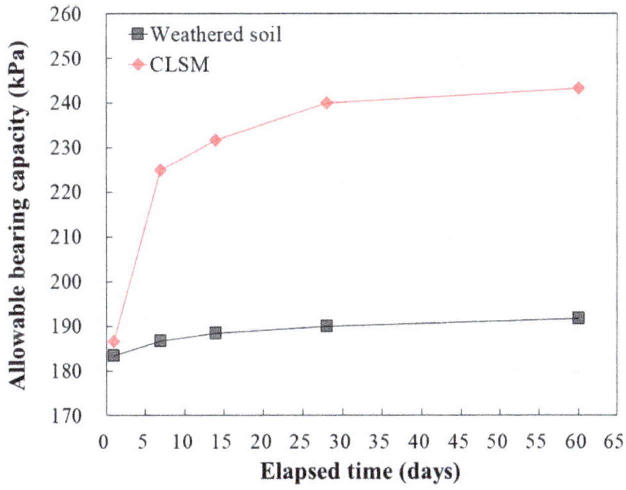

Figure 14. Plate bearing test results for CLSM over time, with elapsed time of (**a**) 1 day; (**b**) 7 days; (**c**) 14 days; (**d**) 28 days; (**e**) 60 days.

Figure 15. Comparison of allowable bearing capacity of weathered soil and CLSM.

Table 8. Allowable bearing capacity of weathered soil and CLSM.

Classification		Elapsed Time (Days)				
		1	7	14	28	60
Allowable bearing capacity (kPa)	Weathered soil	183.3	186.7	188.3	190.0	191.7
	CSLM	186.7	225.0	231.7	240.0	243.3

These results suggest that when CLSM with WPSA is used as sewage pipe backfill material, it is possible to achieve acceptable performance immediately after construction. In addition, it was confirmed that CLSM with WPSA can be used as a backfill material that ensures higher stability than conventional soil backfill, because its allowable bearing capacity increases over time.

5. Conclusions

In this study, the strength characteristics of controlled low-strength material (CLSM) made with recyclable waste paper sludge ash (WPSA) were evaluated, to assess its ability to prevent ground subsidence caused by poor compaction of the ground around sewage pipes and ensure adequate bearing capacity of the ground above the pipes. The unconfined compressive strength of WPSA was evaluated with respect to the mixing conditions, and the bearing capacity of CLSM produced with the optimal mixing conditions was evaluated. The results of this study are summarized as follows:

(1) When only WPSA was used, unconfined compressive strength increased steadily over time from 1 to 28 days, regardless of the cement content (5% or 10%). The strength did not change significantly after 28 days. Unconfined compressive strength increased as WPSA content increased.

(2) When the mixing ratio of WPSA to fly ash was 1:1, both materials, which act as binders, had a significant impact on strength when the cement content was low, but sand had a larger impact on the strength characteristics than ash material when the cement content was high.

(3) The strength and rate of strength increase were significantly higher when the cement content was high than when it was low, regardless of the sand content and WPSA–fly ash mixing ratio. The rate of strength increase was highest when only fly ash was used.

(4) The strength corresponding to 10% settlement of the loading plate diameter was calculated based on the load–settlement relationship of the plate bearing test results. The strength of weathered soil increased from 550 to 575 kPa with increased age, and the increase rate of strength decreased with increased age. Furthermore, CLSM with WPSA increased from 560 to 730 kPa. In addition, the increase rate of CLSM strength increased, and then decreased after 28 days of age; this tendency was the same in weathered soil.

(5) The allowable bearing capacity of weathered soil and CLSM was calculated from plate bearing test results. The allowable bearing capacity of weathered soil increased by approximately 4.5%, with aging from 1 to 60 days. In the case of the ground where CLSM was applied, the allowable bearing capacity gradually increased as the curing time increased. Weathered soil and CLSM exhibited similar allowable bearing capacity immediately after construction.

The results of this study confirm that CLSM with WPSA can be utilized as a sewage pipe backfill material, that can ensure higher stability than soil backfill. However, limited mixing conditions of WPSA and FA were applied in this study. Therefore, it is necessary to perform experiments and analyses on subdivided mixing conditions of WPSA and FA, in order to analyze the effect of WPSA on the strength of CLSM in detail.

Author Contributions: Conceptualization, J.P. and G.H.; Methodology, J.P.; Validation, G.H.; Formal analysis, J.P. and G.H.; Investigation, J.P.; Resources, G.H.; Writing—original draft preparation, G.H.; Writing—review and editing, J.P.; Visualization, J.P. and G.H.; Supervision, J.P.; Funding acquisition, G.H. All authors have read and agreed to the published version of the manuscript.

Funding: This research was funded by the Incheon Green Environment Center under the Korea Ministry of Environment (MOE) (grant number: 19-03-3-50-54).

Conflicts of Interest: The authors declare no conflict of interest.

References

1. Bae, Y.-S.; Kim, K.-T.; Lee, S.-Y. The Road Subsidence Status and Safety Improvement Plans. *J. Korea Acad. Coop. Soc.* **2017**, *18*, 545–552. [CrossRef]
2. Korea institute of geoscience and mineral resources. Research on causes and policy suggestions by sinkhole type. In *Research Report*; Korea Institute of Geoscience and Mineral Resources: Daejeon, Korea, 2014; pp. 18–39.
3. Lee, K.H.; Kim, J.D.; Hyun, S.C.; Song, Y.S.; Lee, B.S. Deformation behavior of underground pipe with controlled low strength materials with marine dredged soil. *J. Korea Soc. Hazard Mitig.* **2007**, *7*, 129–137.
4. Controlled Low Strength Materials. In *ACI Committee 229*; American Concrete Institute: Farmington Hill, MI, USA, 1999.
5. Ryu, Y.-S.; Han, J.-G.; Chae, W.-R.; Koo, J.-S.; Lee, D.-Y. Development of Rapid Hardening Backfill Material for Reducing Ground Subsidence. *J. Korean Geosynth. Soc.* **2015**, *14*, 13–20. [CrossRef]
6. Lee, D.; Kim, D.-M.; Ryu, Y.-S.; Han, J.-G. Development and Application of Backfill Material for Reducing Ground Subsidence. *J. Korean Geosynth. Soc.* **2015**, *14*, 147–158. [CrossRef]
7. Kim, J.H.; Cho, S.D.; Kong, J.Y.; Jung, H.S.; Chun, B.S. Curing characteristics of controlled low strength material made coal ashes. *J. Korean Geoenviron. Soc.* **2010**, *11*, 77–85.
8. Kong, J.Y.; Kang, H.N.; Chun, B.S. Characteristics of unconfined compressive strength and flow in controlled low strength materials made with coal ash. *J. Korean Geotech. Soc.* **2010**, *26*, 75–83.
9. Won, J.P.; Lee, Y.S. Properties of controlled low strength material containing bottom ash. *J. Korea Concr. Inst.* **2001**, *13*, 294–300.
10. Razak, H.A.; Naganathan, S.; Hamid, S.N.A. Performance appraisal of industrial waste incineration bottom ash as controlled low-strength material. *J. Hazard. Mater.* **2009**, *172*, 862–867. [CrossRef] [PubMed]
11. Wu, H.; Yin, J.; Bai, S. Experimental Investigation of Utilizing Industrial Waste and Byproduct Materials in Controlled Low Strength Materials (CLSM). *Adv. Mater. Res.* **2013**, *639*, 299–303. [CrossRef]
12. Kubissa, W.; Jaskulski, R.; Szpetulski, J.; Gabrjelska, A.; Tomaszewska, E. Utilization of Fine Recycled Aggregate and the Calcareous Fly Ash in CLSM Manufacturing. *Adv. Mater. Res.* **2014**, *1054*, 199–204. [CrossRef]
13. Naganathan, S.; Mustapha, K.N.; Omar, H. Use of recycled concrete aggregate in controlled low-strength material (CLSM). *Civ. Eng. Dimens.* **2012**, *14*, 13–18.
14. Ahmadi, B.; Al-Khaja, W. Utilization of paper waste sludge in the building construction industry. *Resour. Conserv. Recycl.* **2001**, *32*, 105–113. [CrossRef]
15. Frias, M.; Garcia, R.; Vigil, R.; Ferreiro, S. Calcination of art paper sludge waste for the use as a supplementary cementing material. *Appl. Clay Sci.* **2008**, *42*, 189–193. [CrossRef]
16. García, R.; de la Villa, R.V.; Vegas, I.; Frias, E.; de Rojas, M.S. The pozzolanic properties of paper sludge waste. *Constr. Build. Mater.* **2008**, *22*, 1484–1490. [CrossRef]
17. Monzó, J.; Payá, J.; Borrachero, M.V.; Morenilla, J.J.; Bonilla, M.; Calderón, P. Some strategies for reusing residues from waste water treatment plants: Preparation of binding materials. In Proceedings of the Conference on the Use of Recycled Material in Building and Structures, Barcelona, Spain, 8–11 November 2004; p. a289.
18. Horiguchi, T.; Fujita, R.; Shimura, K. Applicability of Controlled Low-Strength Materials with Incinerated Sewage Sludge Ash and Crushed-Stone Powder. *J. Mater. Civ. Eng.* **2011**, *23*, 767–771. [CrossRef]
19. Boni, M.R.; D'Aprile, L.; De Casa, G. Environmental quality of primary paper sludge. *J. Hazard. Mater.* **2004**, *108*, 125–128. [CrossRef]
20. Bratina, B.; Šorgo, A.; Kramberger, J.; Ajdnik, U.; Zemljič, L.F.; Ekart, J.; Šafarič, R. From municipal/industrial wastewater sludge and FOG to fertilizer: A proposal for economic sustainable sludge management. *J. Environ. Manag.* **2016**, *183*, 1009–1025. [CrossRef]
21. Heo, Y.; Lee, C.K.; Lee, M.W.; Ahn, K.K. A study on utilization method of paper ash in industrial waste. *J. Korean Soc. Saf.* **1999**, *14*, 135–141.
22. Lee, C.K.; Ahn, K.K.; Heo, Y. Development of the bricks using paper ash. *J. Korean Geoenviron. Soc.* **2003**, *4*, 47–56.

23. Seo, S.K. A Study on the Physicochemical Activation of Paper Sludge Ash & Fly Ash and Their Applications for Cement Admixture. Ph.D. Thesis, Hanyang University, Seoul, Korea, 2017.

24. Khalid, N.; Mukri, M.; Kamarudin, F.; Arshad, M.F. Clay soil stabilized using waste paper sludge ash (WPSA) mixtures. *Electron. J. Geotech. Eng.* **2012**, *17*, 1215–1225.

25. Bujulu, P.M.S.; Sorta, A.R.; Priol, G.; Emdal, A.J. Potential of waste paper sludge ash to replace cement in deep stabilization of quick clay. In Proceedings of the 2007 Annual Conference of the Transportation Association of Canada, Session on Characterization and Improvement of Soils and Materials, Saskatoon, SK, Canada, 14–17 October 2007.

26. Ahmad, S. Study of Concrete Involving Use of Waste Paper Sludge Ash as Partial Replacement of Cement. *IOSR J. Eng.* **2013**, *3*, 6–15. [CrossRef]

27. Sani, M.S.H.M.; Muftah, F.B.; Rahman, M.A. Properties of Waste Paper Sludge Ash (WPSA) as cement replacement in mortar to support green technology material. In Proceedings of the 3rd International Symposium & Exhibition in Sustainable Energy & Environment (ISESEE 2011), Melaka, Malaysia, 1–3 June 2011.

28. Ridzuan, A.R.M.; Fauzi, M.A.; Ghazali, E.; Arshad, M.F.; Fauzi, M.A.M. Strength assessment of controlled low strength materials (CLSM) utilizing recycled concrete aggregate and waste paper sludge ash. In Proceedings of the 2011 IEEE Colloquium on Humanities, Science and Engineering Research, Penang, Malaysia, 5–6 December 2011.

29. Fauzi, M.A.; Sulaiman, H.; Ridzuan, A.R.M.; Azmi, A.N. The effect of recycled aggregate concrete incorporating waste paper sludge ash as partial replacement of cement. *AIP Conf. Proc.* **2016**, *1774*, 030007.

30. Azmi, A.M.; Fauzi, M.A.; Nor, M.D.; Ridzuan, A.R.M.; Arshad, M.F. Production of controlled low strength material utilizing waste paper sludge ash and recycled aggregate concrete. In Proceedings of the 3rd International Conference on Civil and Environmental Engineering for Sustainability, MATEC Web of Conferences, Kuala Lumpur, Malaysia, 15–17 August 2016; p. 01011.

31. Bai, J.; Chaipanich, A.; Kinuthia, J.M.; O'Farrell, M.; Sabir, B.; Wild, S.; Lewis, M. Compressive strength and hydration of wastepaper sludge ash–ground granulated blastfurnace slag blended pastes. *Cem. Concr. Res.* **2003**, *33*, 1189–1202. [CrossRef]

32. Mozaffari, E.; O'Farrell, M.; Kinuthia, J.M.; Wild, S. Improving strength development of waste paper sludge ash by wet-milling. *Cem. Concr. Compos.* **2006**, *28*, 144–152. [CrossRef]

33. Farzampour, A. Temperature and humidity effects on behavior of grouts. *Adv. Concr. Constr.* **2017**, *5*, 659–669.

34. Farzampour, A. Compressive Behavior of Concrete under Environmental Effects. In *Compressive Strength of Concrete*; IntechOpen: London, UK, 2019. [CrossRef]

35. Mansouri, I.; Shahheidari, F.S.; Hashemi, S.M.A.; Farzampour, A. Investigation of steel fiber effects on concrete abrasion resistance. *Adv. Concr. Constr.* **2020**, *9*, 367–374.

36. Won, H. A laboratory Study on Anti-Acid Mortar Using Calcium-Aluminate Cement and Blast-Furnace Slag with Fly-Ash. Master's Thesis, Seoul National University of Technology, Seoul, Korea, 2010.

37. Joseph, I.G.; Dale, E.N.; Patrick, E.; David, C.J.; Charles, E.L.; Eric, L.; Linda, S.; Joseph, R.M. The SEM and Its Modes of Operation. In *Scanning Electron Microscopy and X-ray Microanalysis*; Springer: Boston, MA, USA, 2003; pp. 21–60.

38. Borchert, H. X-ray Diffraction. In *Solar Cells Based on Colloidal Nanocrystals*; Springer Series in Materials Science: Cham, Switzerland, 2014; pp. 79–94.

39. ASTM. *Standard Test Method for Flow Consistency of Controlled Low Strength Material (CLSM)*; ASTM International: West Conshohocken, PA, USA, 2004; p. D6103.

40. ASTM. *Standard Test for Preparation and Testing of Controlled Low Strength Material (CLSM) Test Cylinders*; ASTM International: West Conshohocken, PA, USA, 2004; p. D4832.

41. Emery, J.; Johnston, T. *Unshrinkable Fill for Utility Cut Restorations*; American Concrete Institute Special Publication: Farmington Hills, MI, US, 1986; pp. 187–212.

42. NRMCA. *Guide Specification for Controlled: Low Strength Materials (CLSM)*; National Ready Mixed Concrete Association: Alexandria, VA, USA, 1995; pp. 5–7.
43. Crouch, L.; Gamble, R.; Brogdon, J.; Tucker, C. Use of High-Fines Limestone Screenings as Aggregate for Controlled Low-Strength Material (CLSM). In *The Design and Application of Controlled Low-Strength Materials (Flowable Fill)*; ASTM International: West Conshohocken, PA, USA, 1998; p. 45.

 *materials*

Article

Assessment of Internal Structure of Spun Concrete Using Image Analysis and Physicochemical Methods

Jarosław Michałek and Maciej Sobótka *

Faculty of Civil Engineering, Wroclaw University of Science and Technology, 50-370 Wroclaw, Poland; jaroslaw.michalek@pwr.edu.pl
* Correspondence: maciej.sobotka@pwr.edu.pl; Tel.: +48-71-320-4127

Received: 23 July 2020; Accepted: 8 September 2020; Published: 9 September 2020

Abstract: Taking into account the possibilities offered by two imaging methods, X-ray microcomputed tomography (μCT) and two-dimensional optical scanning, this article discusses the possibility of using these methods to assess the internal structure of spun concrete, particularly its composition after hardening. To demonstrate the performance of the approach based on imaging, laboratory techniques based on physical and chemical methods were used as verification. Comparison of obtained results of applied research methods was carried out on samples of spun concrete, characterized by a layered structure of the annular cross-section. Samples were taken from the power pole E10.5/6c (Strunobet-Migacz, Lewin Brzeski, Poland) made by one of the Polish manufacturers of prestressed concrete E-poles precast in steel molds. The validation shows that optical scanning followed by appropriate image analysis is an effective method for evaluation of the spun concrete internal structure. In addition, such analysis can significantly complement the results of laboratory methods used so far. In a fairly simple way, through the porosity image, it can reveal improperly selected parameters of concrete spinning such as speed and time, and, through the distribution of cement content in the cross-section of the element, it can indicate compliance with the requirement for corrosion durability of spun concrete. The research methodology presented in the paper can be used to improve the production process of poles made of spun concrete; it can be an effective tool for verifying concrete structure.

Keywords: concrete centrifugation; morphology; image processing; porosity; aggregate; cement

1. Introduction

Concrete centrifugation is the process of forming and compacting concrete mix due to the normal (radial) force generated during the spinning of the mold around its longitudinal axis at a speed of 500–700 rpm. Due to its specificity, this method is applicable only to hollow elements. As a result of the centrifugation process, concrete with a heterogeneous and layered structure is obtained [1–3]. Thus, the structure is different from that of precast or monolithic concrete elements, which in practice can be considered homogeneous. The spun concrete is marked by the fact that components with a larger mass (coarse grains) come to the outside of the cross-section, while components with a smaller mass (cement slurry) tend to remain inside (Figure 1). Under these conditions, the outer layer can achieve high compactness and, after hardening, high strength and resistance to chemical and mechanical impacts. The inner layer, on the other hand, consists of highly compacted, very thick cement paste and may, after hardening, obtain special resistance to water permeability.

Figure 1. The structure of spun concrete in the wall thickness of a power pole.

One of the first to describe the structure of spun concrete used in pipe production was Marquardt [4]. Based on his observations, he found that a greater difference between the specific gravity of the concrete mixture components led to faster and deeper fractionation. He recommended that well-sorted, mixed aggregate grains with similar petrographic properties and a maximum diameter of 15 mm were preferred for concrete mixtures subjected to centrifugation. He also discussed the change in cement content as a function of the wall thickness of spun concrete elements. He found that cement at 83% of the wall thickness is quite evenly distributed, and an increase in its content is observed only in the inner layer of the section with a thickness of 2–5 mm. He proposed to use layered centrifugation at lower mold rotational speeds to reduce layering.

In [5], Kuranovas and Kvedaras presented the concrete centrifugation process as a four-phase process. The first phase is feeding the concrete mix inside the mold, with its even distribution over the length of the mold at a low centrifuge rotational speed. Afterward, the rotational speed of the mold increases, and the process enters the second phase, where the centrifugal force begins to compress the concrete mix and a layered wall of the concrete section is formed. The third phase is a further increase in the rotational speed of the mold, at which the wall thickness of the concrete section is compacted and stabilized with some of the mixing water squeezed inside. The fourth phase occurs at the maximum rotational speed of the mold, during which the concrete mix is further compacted and excess mixing water is squeezed out. The element's wall reaches its designed thickness.

Kliukas, Jaras, and Lukoševičienė in [6] found that each phase of the concrete centrifugation process has a specific rotational speed, which depends on the dimensions of the element and the time needed to obtain the selected speed. According to [6], the duration of the first phase should be 3–4 min, the second and third phase should be 1–2 min, and the fourth phase should be 10–15 min. It should be noted that the centrifugation time of an element in a steel form also depends on the initial water–cement ratio of the concrete mix and the pressure caused by centrifugal force, depending on the rotational speed and the diameter of the product.

Adesiyun et al. [7–9] used a computer image analysis method to describe the structure of spun concrete allowing extraction of quantitative and qualitative information contained in the image of an element wall cross-section. For a precise description of the structure, the wall thickness of the sample was divided into 20 strips with a width of 2.25–3.00 mm (depending on the overall wall thickness). The tests were carried out with different combinations of concrete mix parameters, such as sand point (25–50%), cement content (410–530 kg/m^3), amount of plasticizer (0–2%), spinning speed (400–700 rpm), and its time (5–11 min). Computer image analysis allowed for graphical representation of the aggregate, cement matrix, and air-void distribution as a function of the wall thickness in individual samples. It was found that the volume fraction of aggregate decreases from the outside of the centrifuged sample to the inside, while the cement matrix content changes in the opposite direction (in the inner zone, its content is almost 100%). The air content in the cross-section of the wall was higher for internal rather than external layers. Segregation of components was found in all tested samples.

Völgyi et al. [10,11] found that segregation of the concrete mix during centrifugation depends to a large extent on the excess of the cement paste and the degree of compaction. Properties such as strength, porosity, hardness, and composition vary across the wall of the spun concrete due to the segregation of components. The latter can be reduced by using less paste and optimal compaction energy. A relationship between the consistency and the advisable compaction energy was proposed. The formula for the optimal centrifuge settings, i.e., time and speed of spinning, was derived to obtain the best strength properties of the sample.

This paper focuses on methods of assessing the internal structure of spun concrete, which can be used in practice to verify the selection of spinning parameters. In the first part of the work, standard methods for determining the composition of hardened concrete were used to assess its variability across the element wall. Next, based on preliminary results presented in [12], imaging techniques followed by appropriate image analysis were utilized to determine spatial distribution of pores, aggregate, and cement matrix. Comparison of obtained results was carried out for the samples of spun concrete power pole E10.5/6c made by one of the Polish manufacturers of prestressed concrete E-poles. The capabilities and performance of the research methods used were discussed. Conclusions regarding their potential use in practice were formulated.

2. Preparation of Test Samples

The E10.5/6c pole used for testing was made in one of the Polish prefabrication plants and was a precast, partly prestressed element shaped as a truncated cone with hollow core. The length of the pole was equal to $L = 10.5$ m. The annular shape of the cross-section resulted from the adopted technology of manufacturing poles by the method of concrete centrifugation in longitudinally unopenable steel mold [13]. The outer diameter of the E10.5/6c column increased from $d_t = 173$ mm at the top to $d_b = 330.5$ mm at the base with a constant taper $t = 15$ mm/m. The equivalent characteristic peak resistance force, applied at a distance of 0.17 m from the top, was $P_k = 6$ kN. The E10.5/6c pole is designed to be produced from C40/50 spun concrete.

When designing power poles made of spun concrete, it was taken into account that they are exposed to direct environmental impacts, described by standard [14], class XC4 (cyclically wet and dry concrete surfaces exposed to contact with water), and for structures located near motorways, class XF2 (vertical concrete surfaces exposed to freezing and de-icing agents from air). The durability of spun prestressed concrete poles is ensured by appropriate thickness of the concrete cover of the reinforcement. Concrete that protects steel against corrosion is marked by limit values describing, among others, the quantity and quality of components (e.g., minimum cement content, water–cement ratio w/c, and classified aggregates), as well as the minimum compressive strength of concrete and its absorbability. For the XC4 and XF2 environmental exposure class, the standard [14] imposed a minimum cement content of 300 kg/m^3, maximum $w/c = 0.5$, and minimum concrete strength class C30/37. A concrete mix recipe was obtained from the manufacturer for the production of the E10.5/6c pole (Table 1).

Table 1. Recipe for concrete mix used in the production of E10.5/6c pole.

Component	Quantity (kg/m^3)
Sand 0–2 mm	750
Gravel 2–8 mm	350
Gravel 8–16 mm	400
Basalt grit 8–11	370
CEM I 42.5R cement	400
Plasticizer	0.5
Superplasticizer	1.6
Water	105
w/c ratio	0.26

For laboratory tests, from an E10.5/6c power pole, about a 200-mm-long fragment was cut off (Figure 2a) to obtain a sample in the shape of a hollow truncated cone. The wall thickness of the fragment was about 60 mm. The fragment of the pole was then cut along the generatrix into slices about 10–20 mm thick (Figure 2b), with the prestressing wires being avoided. One of the slices was intended for testing the concrete structure with the use of imaging techniques, while the other was intended for testing the composition of hardened concrete using laboratory methods.

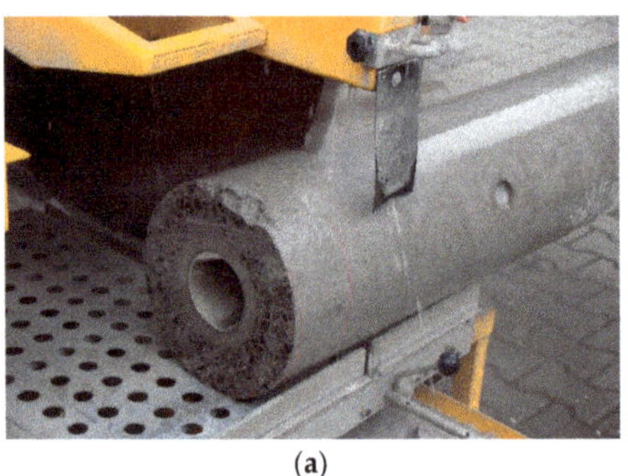

(**a**) (**b**)

Figure 2. Preparation of test samples: (**a**) cutting off the top of the pole; (**b**) a slice of concrete intended for laboratory testing.

3. Determining the Composition of Hardened Concrete

For testing the composition of hardened concrete, an estimation method was used in accordance with the instructions [15] based on the determination of apparent density of concrete, the content of parts insoluble in HCl, and the amount of components attached to the binder during its hardening. Since there is no direct method for determining the cement content of a concrete sample, the procedure consists of determining the content of soluble silica and calcium oxide and, on this basis, calculating the cement content. The method is based on the fact that the silicates in Portland cement are much more soluble than the silicate components normally contained in the aggregate. The same applies to the relative solubility of the calcium oxide components in cement and aggregate (except, however, limestone aggregate). The methods for determining the Portland cement content in concrete are described in standards [16,17], as well as in the already mentioned manual [15]. In addition, the extension of the research to determine water absorption and specific density according to [18,19] made it possible to calculate the total and open porosity. The composition of hardened concrete was determined independently for four layers of concrete pole section as shown in Figure 2b.

Due to the layered structure of spun concrete, for laboratory tests, samples were prepared by longitudinally cutting the concrete slices (Figure 2b) into four parts (as shown in Figure 3). Samples (layers) were marked with consecutive numbers 1–4, where sample 1 describes the inner layer of the annular cross-section, and sample 4 describes the outer layer. In addition, sample 5 was an entire slice of the concrete wall without splitting into layers. For aggregate with a grain diameter of up to 40 mm, the weight of the concrete sample should be at least 3 kg according to [15]. During the tests, due to the dimensions of separated centrifuged concrete samples, this condition was not met. Moreover, while the concrete slices were being cut, it was not possible to get distinct layers of precise dimensions (as in Figure 1); thus, the boundaries between the layers were in practice inexact, which was especially visible for the inner layer No. 1.

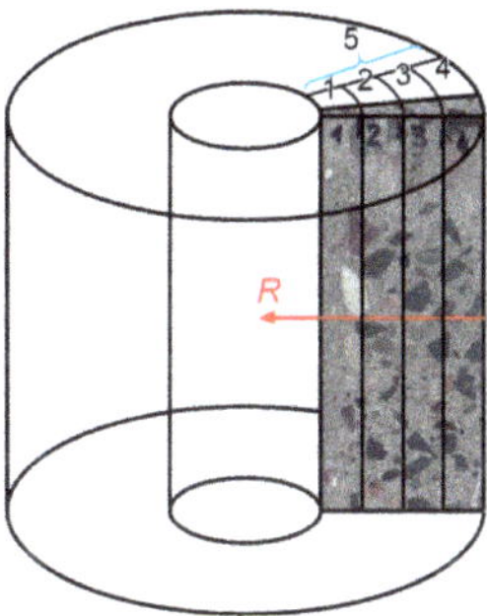

Figure 3. Sample shapes for laboratory tests.

The apparent density of concrete ρ_c was determined via the hydrostatic method due to the irregular shape of the samples. Determination was performed according to [18]. The samples were first dried at 105 °C to constant weight, i.e., until the changes in mass during 24 h were less than 0.2%. Dried samples were weighed after that. Then, the samples were saturated by immersion in water at a temperature of 20 °C until the changes in mass during 24 h were less than 0.2% and weighed. The volume of each sample was determined on the basis of its apparent mass in water and mass in air according to the following formula:

$$V = \frac{m_a - m_w}{\rho_w} \tag{1}$$

where m_a is the mass in air of saturated sample, m_w is the apparent mass in water of saturated sample, and $\rho_w = 998$ kg/m^3 is the density of water at 20 °C. When weighing the samples in water, the correction due to apparent mass of suspension wire and holder was made. The apparent density of dry concrete was calculated as

$$\rho_c = \frac{m_o}{V} \tag{2}$$

where m_o is the mass of dry sample. At the same time, water absorption of concrete can be calculated as

$$n_w = \frac{m_a - m_o}{m_o} \tag{3}$$

The procedure for determining the content of parts insoluble in HCl was as follows: after comminution, sieving through a 1-mm sieve, and drying to constant weight at 105 °C, the samples were sieved through a 0.2-mm sieve. Samples of about 2 g were tested using an aqueous solution (1:3) of hydrochloric acid (HCl) to determine the content of insoluble parts. Each sample was placed in a 250-mL beaker and 100 mL of HCl solution was added at room temperature. The contents of the beaker were mixed and the insoluble residue was triturated with a glass rod. After 15 min, the supernatant fluid was decanted. Next, 50 mL of HCl solution was poured onto the precipitate in a beaker and placed in a water bath at 90 °C for 15 min. Then, the contents of the beaker were washed twice with hot water and decanted. The precipitate remaining in the beaker was flooded with 50 mL of 5% Na$_2$CO$_3$ and placed in a water bath at 90 °C for 15 min, before being washed twice with hot water and decanted. The remaining precipitate was flooded with 50 mL of water, acidified with HCl (1:3) against methyl orange, while adding an excess of 3–4 drops of acid to the neutralized solution, after which the contents of the beaker were filtered. The filtered precipitate was washed six times with hot water until the reaction due to chlorides ceased. The washed precipitate was transferred to a weighed porcelain crucible and, after burning, the filter was calcined at 1000 °C to finally determine the mass of parts insoluble in HCl. Typically, it is assumed that the aggregate mass in the tested sample is equal to the mass of parts insoluble in HCl determined as a result of the analysis. However, when determining the aggregate content in this study, a preliminary analysis was performed to determine the aggregate behavior under the influence of hydrochloric acid. It turned out that basalt in the form of aggregate is resistant to HCl acid, while, after being pulverized, it dissolved partly in hydrochloric acid. Probably weathered basalt was used for concrete in some parts of the aggregate 8–11 mm (Table 1) [20]. Thus, when determining the aggregate content, account was taken of the presence of HCl-soluble parts in the aggregate, by introducing the appropriate correction. Finally, the aggregate mass percentage was calculated as

$$C_{agg,\%} = C_{ins,\%} \cdot corr = \frac{m_{ins}}{m_o} \cdot 100\% \cdot corr \tag{4}$$

where $C_{ins,\%}$ is the mass percentage of insoluble parts in concrete, m_{ins} is the mass of insoluble parts remaining from the sample of concrete, m_o is the initial mass of dry concrete sample, and $corr = 1/C_{ins,agg,\%}$ is the correction factor, in which $C_{ins,agg,\%}$ stands for mass percentage of HCl-insoluble parts in the aggregate. The aggregate content in kg/m^3 was calculated as

$$C_{agg} = C_{agg,\%} \cdot \rho_c \tag{5}$$

The content of compounds attached to cement during the setting and hardening of concrete (which are H_2O and CO_2) was determined on the basis of calcining losses. The samples were prepared according to the procedure for determining the aggregate content in concrete; they were additionally ground on a 0.06-mm sieve. Calcining losses were determined in an oxidizing atmosphere, while the samples were heated in the air at 950 °C until a constant sample mass was obtained. The constant mass was determined by successive heating for 15 min, followed by cooling and weighing. The percentage calcining loss was determined by the following formula:

$$C_{att,\%} = \frac{(m_o - m_{cal})}{m_o} \cdot 100\% \tag{6}$$

where m_o is the the mass of the sample tested, and m_{cal} is the the mass of the calcined sample. The obtained value of $C_{att,\%}$ was converted to the content of attached ingredients expressed in kg/m^3 according to the following formula:

$$C_{att} = \frac{C_{att,\%}}{100\%} \cdot \rho_c \tag{7}$$

Finally, the content of cement C_{cem} was calculated from the following condition:

$$C_{cem} = \rho_c - C_{agg} - C_{att} \tag{8}$$

The density of concrete ρ_{cr}, defined as the ratio of the mass of the dried sample to the volume of the solid part (without taking into account the volume of voids), was determined using the Le Chatelier flask method [19]. On the basis of the values of ρ_c and ρ_{cr}, total porosity was calculated as

$$\phi_{p,tot} = 1 - \frac{\rho_c}{\rho_{cr}} \tag{9}$$

Open porosity $\phi_{p,op}$ was estimated based on water absorption n_w as follows:

$$\phi_{p,op} = n_w \frac{\rho_c}{\rho_w} \tag{10}$$

The obtained parameters of concrete in the particular layers are presented in Table 2.

Table 2. Test results and calculations for individual samples 1–4 and average sample 5 (Figure 1).

Parameter	Sample No. 1	Sample No. 2	Sample No. 3	Sample No. 4	Sample No. 5
Apparent density ρ_c (kg/m^3)	2086	2170	2124	2175	2149
Specific density ρ_{cr} (kg/m^3)	2509	2597	2640	2592	2599
Water absorption n_w (%)	5.1	4.4	4.8	4.5	4.6
Open porosity $\phi_{p,op}$ (%)	10.7	9.5	10.3	9.7	10.0
Total porosity $\psi_{p,tot}$ (%)	16.9	16.4	19.6	16.1	17.3
Solid volume fraction s (%)	83.1	83.6	80.4	83.9	82.7
Aggregate mass percentage $C_{agg,\%}$ (%)	71.6	82.5	82.3	81.2	78.2
Aggregate content C_{agg} (kg/m^3)	1493.9	1790.6	1748.0	1766.4	1680.5
Attached ingredients mass percentage $C_{att,\%}$ (%)	5.5	4.1	3.8	4.6	3.7
Content of attached ingredients C_{att} (kg/m^3)	114.8	89.0	80.7	100.1	79.5
Cement mass percentage $C_{cem,\%}$ (%)	22.9	13.4	13.9	14.2	18.2
Cement content C_{cem} (kg/m^3)	477.8	290.8	295.2	308.9	391.1

4. Assessment of Morphology of Concrete Microstructure

4.1. Theoretical Assumptions

Research on the morphology of the spun concrete microstructure was divided into two parts due to the method of image acquisition: flat optical scanning and three-dimensional (3D) imaging in X-ray microcomputed tomography (μCT). Regardless of the method of obtaining the image, the analysis is aimed at characterizing selected morphology measures of the considered microstructure components: pores, aggregate, and cement matrix. In both cases, the following stages of research can be distinguished:

- obtaining an image;
- segmentation of the component under consideration;
- morphometric analysis.

In the conducted analyses, it was assumed that, due to the axial symmetry of the spun elements, the properties of the obtained material do not depend on the angular coordinate. Thus, at a fixed pole height (i.e., for a fixed coordinate value along the axis of rotation of the element), the parameters of the microstructure change only along the radius. The identified values were, therefore, referred to coordinate R in the radial direction (perpendicular to the axis of rotation of the spun concrete element). Coordinate R increases from $R = 0$ on the outer edge of the element toward the center of the cross-section in accordance with Figure 3.

The mathematical morphology methods were used to describe the distribution of concrete components (i.e., macropores, cement matrix, and aggregate) over the wall of the spun column. The theoretical foundations for these tools, based mainly on works [21–23], were described in more detail in the previous work of the authors [12].

On the basis of the analyses carried out, selected morphological statistical measures were determined: component volume fraction and so-called local thickness. The average volume fraction of the selected component is defined as the number of pixels in the component divided by the total number of pixels (in a selected area of the image). To describe the variability of the volume of the component as a function of R, a series of image subareas was selected, which are narrow circumferential bands of the concrete pole wall. The volume fraction $\phi(R)$ (dependent on coordinate R) is calculated in the band with the center line corresponding to coordinate R and the width ΔR. The method of selecting the aforementioned band is shown in Figure 4a. It is worth mentioning that, in the case of cross-section analysis as in [12], the shape of the band is annular (see Figure 4b).

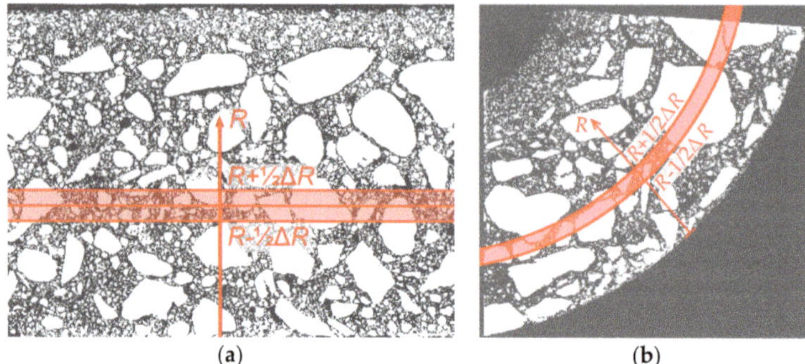

Figure 4. Selection of peripheral band: (**a**) in longitudinal section; (**b**) in cross-section [12].

Local structure thickness is a scalar field, defined only in the area occupied by the considered constituent. The procedure for determining the local element thickness generally consists of filling the area occupied by the considered constituent with circles of the largest possible diameter [24]. Then, the thickness of the element at a given point x is defined as the largest diameter of the circle, which entirely fits within the constituent under consideration and, at the same time, contains point x in its interior. This is schematically illustrated in Figure 5.

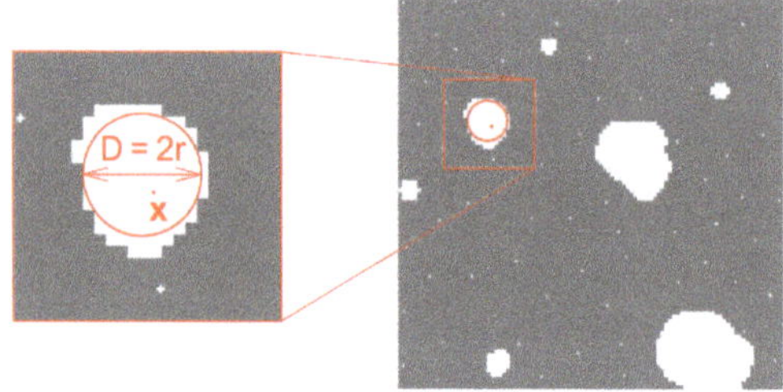

Figure 5. Element thickness at point x as the maximum diameter of the inscribed circle.

The calculations are performed first for the entire image, resulting in a map of the local thickness of the constituent under consideration. Then, to describe the variation of local thickness as a function of the wall thickness of the concrete element (i.e., relative to coordinate R), a procedure analogous to that for the volume fraction was used; specifically, the average in the appropriate circumferential band was determined as in Figure 4.

4.2. Two-Dimensional (2D) Imaging in an Optical Scanner

The research procedure used is based on the procedure given in [12]. It consists of the following consecutive actions:

1. cutting the sample;
2. surface preparation for scanning;
3. scanning;
4. segmentation of the considered constituent from the obtained image;
5. morphometric analysis of the considered component based on its binary image.

Staining the pores first and then the matrix (with a different color), after etching it with acid, allowed quantifying the morphology of the following components: air voids (macro-pores), cement matrix, and aggregate. Thus, the scope of analyses was expanded compared to the analysis presented in [12], where only aggregate morphology was described. To perform calculations as a part of the analysis of the morphology of individual components, the author's own procedure written in the Mathematica program was used. Selected image transformations were carried out in GIMP, ImageJ (Fiji distribution), and CTAn programs.

The first stage of sample preparation for scanning was to level the cut surface in such a way that it was as flat, even, and smooth as possible. For this purpose, a Struers LaboPol 5 grinding and polishing machine with an MD Piano grinding disc was used. The result of the optical scan of the surface of the test sample prepared in this way (before staining) is shown in Figure 6. The scan was performed using a standard office scanner with a resolution of 600 dpi, which, calculated using the pixel size, gives 42.33 μm/pix.

The pores were stained green by applying acrylic ink to the entire surface of the sample, which was then ground again. After this procedure, the sample was rescanned Figure 7.

Then, after etching the matrix with hydrochloric acid, it was stained yellow (Figure 8). The method of staining was analogous to that for pores; only the ink color changed.

To enable comparison of scans at individual stages of staining, all images were superimposed and matched to each other by appropriate rotation and shift. The pictures presented below show already "matching" images, for which the dimensions in pixels are the same and the position of the analyzed sample is the same. To exclude the background from the analysis, as well as inadequately polished fragments, edges jagged during cutting, and reinforcing bars, further analysis focused on the area of the image showing the correctly prepared surface of the concrete sample. This area is called the region of interest (ROI). The complement of ROI was not included in the analysis. For this part of the image, a white mask was applied. The image of the stained sample, limited by the aforementioned mask to ROI, is presented in Figure 9. By using ROI, most of the polishing marks, which were stained as pores, were eliminated from the analysis.

Figure 6. Raw scan of the tested sample.

4.2.1. Macropore Morphology

Pore segmentation was performed by subtracting images before and after staining, followed by global thresholding. The particular technique was described in more detail in previous work [12]. The segmentation result, i.e., the binary image of pores, is shown in Figure 10. Figure 11 shows a contour map of pore local thickness, as well as its histogram. Graphs of porosity and average pore thickness as a function of coordinate R are presented for the consideration sample in Figure 12. The brightest gray lines correspond to the width ΔR (as in Figure 4) equal to the pixel size. Darker lines correspond to greater values of adopted width ΔR. The same remark holds true throughout the remaining graphs in the paper.

Figure 7. Scan of tested sample after staining the pores.

Figure 8. Scan of the tested sample after staining the matrix.

Figure 9. Image of the tested sample after staining the matrix (limited to the region of interest (ROI)).

Figure 10. Binary image of pores (limited to ROI).

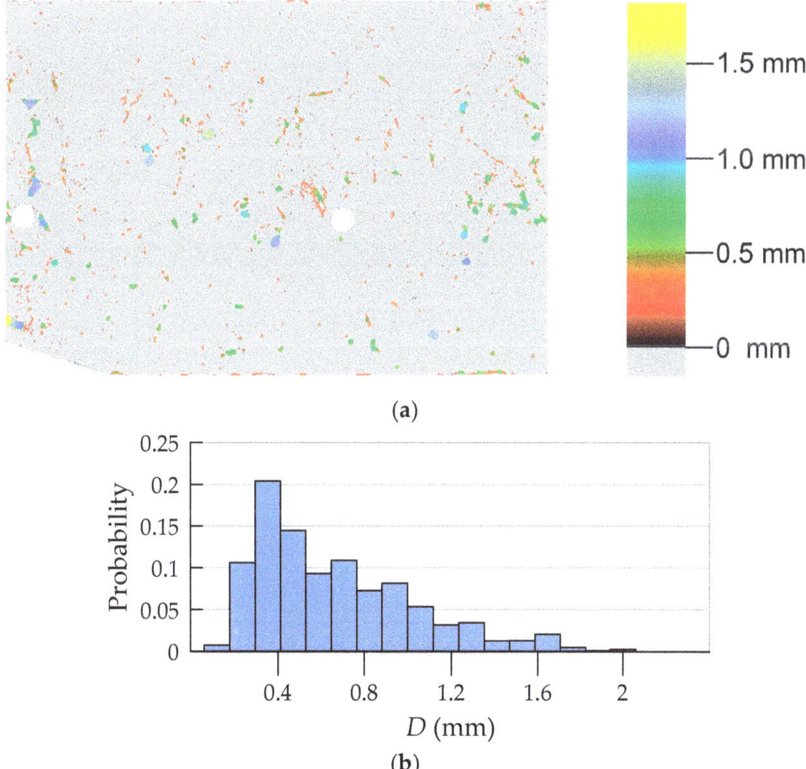

(**a**)

D (mm)

(**b**)

Figure 11. Pore local thickness: (**a**) contour map; (**b**) histogram.

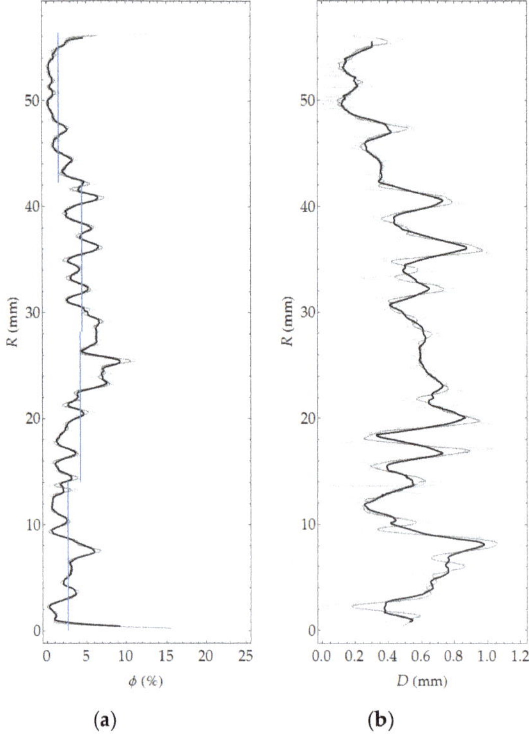

Figure 12. Porosity (**a**) and average local thickness of pores (**b**) as a function of R coordinate.

4.2.2. Cement Matrix Morphology

The matrix was segmented by extracting the appropriate color channels from the scan after staining the matrix, followed by thresholding. Next, the pores were removed and the calculations were carried out. Figure 13 presents a map of the local thickness of the cement matrix together with the histogram. Figure 14 shows the variability of the volume fraction and the average local thickness of the matrix as a function of coordinate R.

4.2.3. Aggregate Morphology

The aggregate was segmented as a complement of the area, being the union of the binary images of cement matrix and pores. The map of local aggregate thickness, as well as its histogram, is presented in Figure 15. Figure 16 shows the variability of the volume fraction and average aggregate thickness as a function of coordinate R.

4.3. 3D Imaging in Microcomputed Tomography

In order to describe the pore space network, a nondestructive technique, namely, μCT, was used. The sample was cut to a cuboid with dimensions allowing to obtain a test resolution of 21.40 μm/pix. Then, the prepared cuboid was fixed to the holder in the chamber of the device (Figure 17).

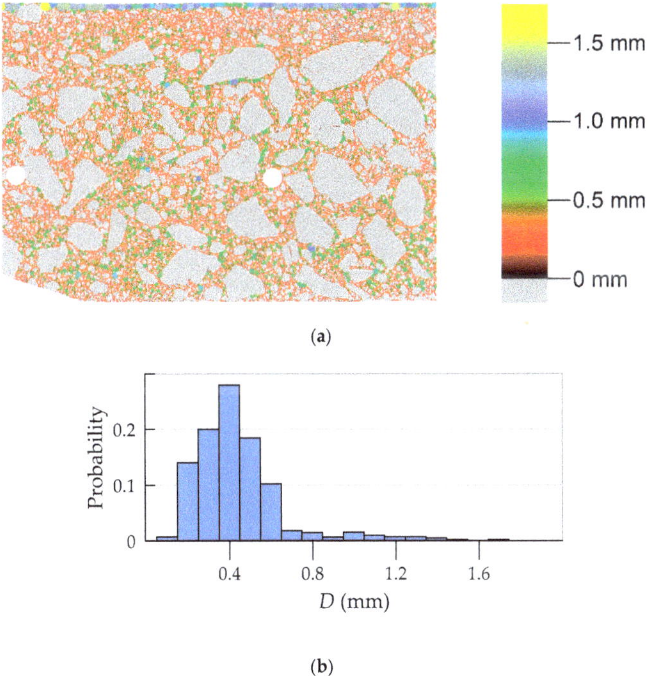

(**a**)

(**b**)

Figure 13. Local thickness of cement matrix: (**a**) contour map (limited to ROI); (**b**) histogram.

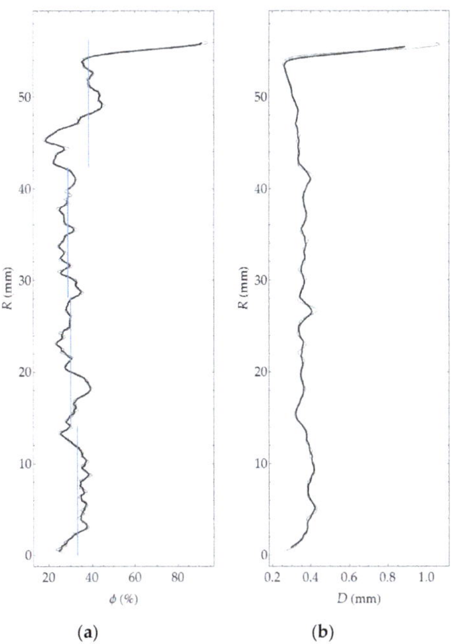

(**a**)   (**b**)

Figure 14. Volume fraction (**a**) and average local thickness (**b**) of cement matrix as a function of the element thickness.

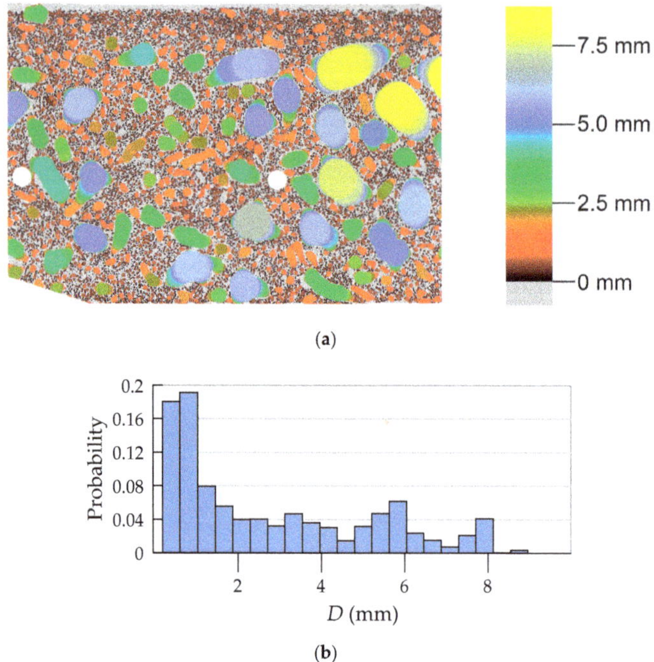

(a)

(b)

Figure 15. Local thickness of aggregate: (**a**) contour map (limited to ROI); (**b**) histogram.

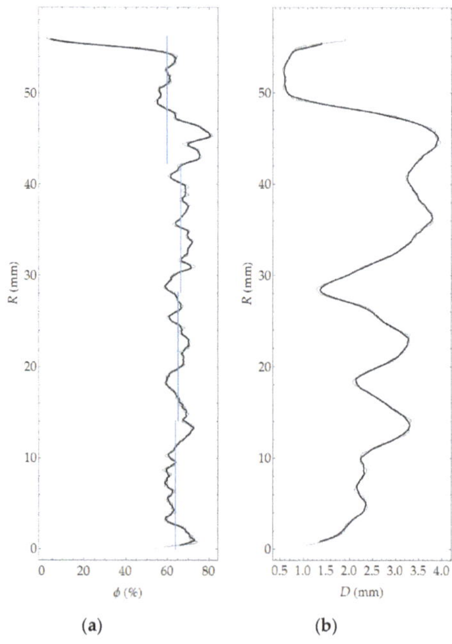

(a)   (b)

Figure 16. Volume fraction (**a**) and average aggregate thickness (**b**) as a function of coordinate *R*.

Figure 17. The sample attached to the holder and mounted in the chamber of X-ray scanner.

Scanning was performed in a Bruker Skyscan 1172 device (Bruker, Kontich, Belgium). The examination consisted of acquiring a series of X-ray projections, followed by the reconstruction of 3D image of the tested sample. The scanning parameters used are shown in Table 3.

Table 3. Selected, more important scanning parameters.

Parameter	Value
Source voltage (kV)	100
Source current (μA)	100
Image pixel size (μm)	21.38
Filter	Al 0.5 mm
Exposure (ms)	270
Rotation step ($^\circ$)	0.32
Frame averaging	On (8)
Random movement	On (10)
Use 360 rotation	Yes
Geometrical correction	On

Image reconstructions were made using the NRecon program based on the Feldkamp algorithm [25]. The set of reconstruction parameters is presented in Table 4.

Table 4. Selected, more important reconstruction parameters.

Parameter	Value
Pixel size (μm)	21.39861
Object bigger than field of view (FOV)	Off
Ring artefact correction	9
Beam hardening correction (%)	30
Threshold for defect pixel mask (%)	0
CS static rotation ($^\circ$)	-49.14
Minimum for CS to image conversion	0.000
Maximum for CS to image conversion	0.070

The reconstructed structure of the tested sample is shown in Figure 18a. In order to carry out a quantitative and qualitative assessment of the material, it was also necessary to specify in the reconstructed model the volume of interest (VOI). A quasi-cuboidal area was assumed, highlighted in green in Figure 18b.

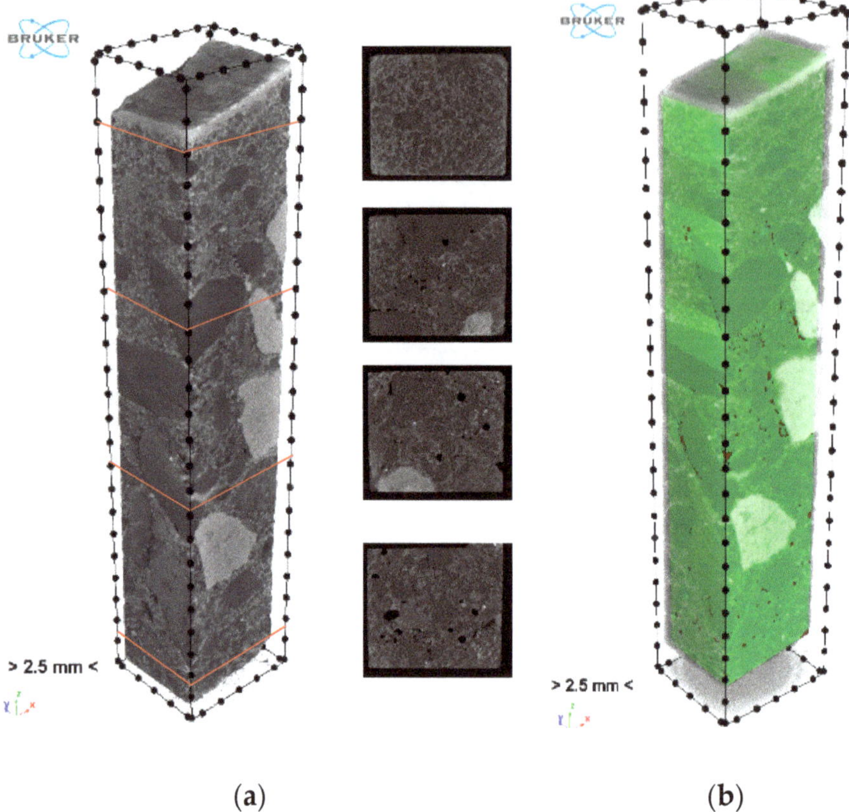

Figure 18. Reconstruction of the tested sample: (**a**) rendered three-dimensional (3D) view with exemplary cross-sections; (**b**) volume of interest (VOI) selection.

Macropore Morphology

The first stage of the analysis was pore segmentation using the threshold method preceded by the use of a smoothing filter. The spatial morphology of the pores extracted in this way is shown in the perspective view in Figure 19a, in which the pores are indicated in red and the reconstruction is shown in gray, for which a high level of transparency was set. Morphometric analysis was performed on the binary image of the pore space. In particular, porosity and local pore thickness were determined. The variability of these quantities as a function of coordinate R is shown in Figure 19b,c. Figure 19d presents the histogram of pore size. The variation with R was determined in a manner analogous to that used in 2D analyses. This time, however, subsequent layers (horizontal sections) of the 3D image were treated as circumferential bands.

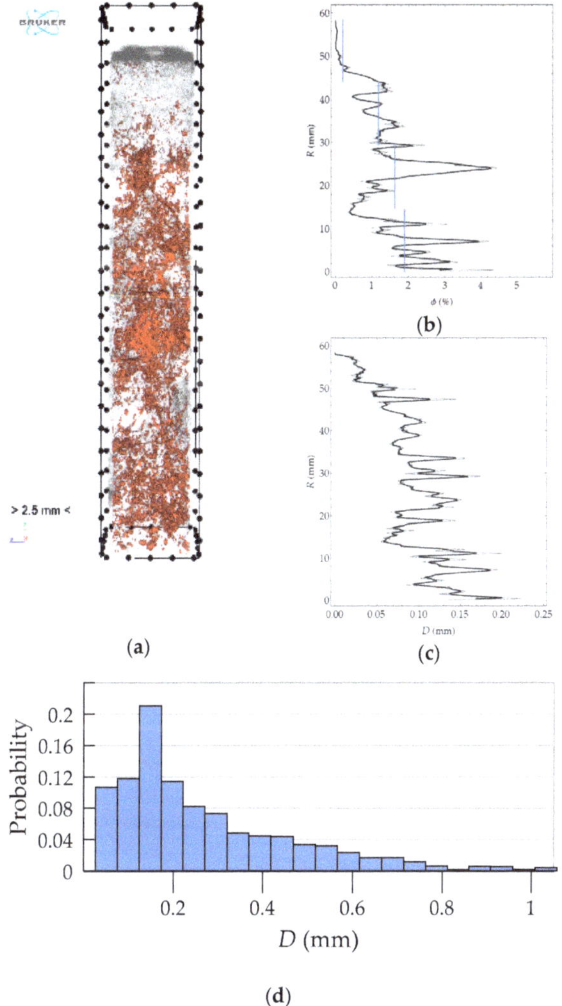

Figure 19. Results of morphometric analysis: (**a**) segmentation of pore space; (**b**) macroporosity as a function of coordinate *R*; (**c**) average local pore thickness as a function of coordinate *R*; (**d**) histogram of pore local thickness.

5. Test Results Obtained Using Different Methods

5.1. Pores

In general, all three methods used in the study allow determining porosity. The undoubted advantage of methods based on image analysis is the possibility of obtaining a virtually continuous function of porosity distribution, which in qualitative assessment (whether porosity changes as a function of the element thickness and what is the course of this variability) is very important. In addition, these methods make it possible to evaluate the morphology of pore space, in particular, to determine quantities such as pore width (see Figure 12) or shape parameters, e.g., sphericity. The laboratory method does not boast these advantages; in the field of assessment of porosity distribution, it allows in practice determining only the average porosity values in several arbitrarily determined layers of the element wall. In turn, the main limitation of the methods based on image analysis is linked to resolution. It is possible to obtain information only about pores larger than the pixel size (or voxel in 3D). With the test resolution used, on the order of tens of micrometers, such pores should be defined as structural macropores [26,27]. Obtaining any information about microporosity using methods based on image analysis requires extension of the methodology and the adoption of additional assumptions, e.g., [28,29]. Laboratory tests used in this work allow determining two types of porosity, i.e., total porosity and open porosity. The latter, due to the physical basis of the study (i.e., water absorption of concrete), should be primarily identified with capillary porosity. Apart from the limitations of the research methods used, they should be considered complementary to each other, i.e., giving detailed information on individual types of porosity: structural, total, and capillary (open).

Comparison of the results obtained with all of the methods used is only possible for average values in arbitrarily separated (four) layers of the wall of a spun pole, which corresponds to the division used in laboratory tests. For the results of optical scanning and tomography, the means were calculated on the basis of predetermined porosity distributions as a function of the wall thickness and referred to four layers equal in thickness. The average values of macroporosity in the layers calculated in this way are represented by blue lines in the graphs in Figures 12 and 19. The average values of the porosities in the layers according to all considered research methods are presented in Table 5.

Table 5. Comparison of results for different types of porosity.

Parameter	Layer 1	Layer 2	Layer 3	Layer 4
Total porosity from laboratory test $\phi_{p,tot}$ (%)	16.9	16.4	19.6	16.1
Capillary porosity (open) from laboratory test $\phi_{p,op}$ (%)	10.7	9.5	10.3	9.7
Macroporosity based on optical scanning $\phi_{p,m}$ (%)	1.5	4.5	4.3	2.7
Macroporosity from tomography $\phi_{p,m}$ (%)	0.2	1.2	1.6	1.9

Due to the differences in the research methodology, it is difficult to unequivocally compare the values or mutually verify the results of laboratory and image-based tests, as different types of porosity were determined. Of course, certain constraints should be met, e.g., both macroporosity and microporosity should be less than total porosity. This condition was met.

The comparison of macroporosity results obtained from optical scanning and tomography led to unexpected conclusions. Better resolution used in tomography (smaller pixel size) would suggest the possibility of extracting some portion of small pores, which cannot be observed by the second method with worse resolution. Then, the porosity of the tested sample should be lower in the case of optical scanning, in which the inferior resolution was used [30]. The presented results were the opposite. There can be several reasons for this surprising result. First, the image obtained through tomography results from mathematical reconstruction, and pixel/voxel size should not only be taken as a measure of accuracy. The resolution of the reconstructed image may be impaired due to inaccuracy of the projection images (causing blur) or noise. This may result in a lower porosity value. Secondly, the volume of material, due to the width of the sample subjected to scanning, may be too small to obtain representative results. It is usually assumed in concrete tests that the representative volume element (RVE) dimension should not be less than four aggregate diameters [26]. Fulfilling such a condition while maintaining adequate resolution is in practice very difficult. In the case of optical scanning, on the other hand, the possibility of "tearing out" fine grains when cutting and grinding the surface before scanning should be taken into account. Such an effect could possibly lead to increased porosity.

With regard to the assessment of porosity of the analyzed sample, thanks to the combined use of laboratory methods and imaging, the following findings were obtained:

- capillary porosity (open) is basically constant in all layers and its value is around 10%;
- total porosity is significantly higher only in layer 3 (by approximately 20%) compared to the other layers, where the value of total porosity is between 16% and 17%;

- structural macroporosity (as can be seen from both imaging methods used) shows the largest relative differences between individual layers and is even several times smaller in the middle part (i.e., in layer 1) compared to the other layers.

5.2. Aggregate

The aggregate content was determined using two methods: in laboratory tests and on the basis of plain optical scanning. General comments on the possibilities of the research techniques used are very similar to those given in Section 5.1. The surface scanning combined with image analysis gives the possibility of obtaining a virtually continuous function of the aggregate content distribution, as well as other morphology measures, characterizing, e.g., grain size (Figures 15 and 16). The laboratory method allowed determining the average values of aggregate content in four layers of the element wall. As in the case of pore analysis, the results of optical scanning had to be averaged within the four layers of equal thickness. Averaged values are presented in blue lines in the graph in Figure 16. In addition, to enable comparison of results from both methods, the volume fraction of aggregate was converted into its content as follows:

$$C_{agg} = \rho_{agg}\phi_{agg} \tag{11}$$

where C_{agg} is the aggregate content (kg/m^3), ϕ_{agg} is the volume of aggregate, and ρ_{agg} is the specific density of aggregate.

To apply Equation (11), it is necessary to know the specific density of aggregate ρ_{agg} in individual layers. Unfortunately, this value was not explicitly determined as part of the research. Therefore, two approaches were taken to estimate this value. In the first approach (*), a constant value equal to the average density of the aggregate used for the mixture was assumed, i.e., ρ_{agg} = 2710 kg/m^3; thus, its segregation (variability in wall thickness) due to centrifugation was not taken into account. In the second approach (**), it was assumed that the specific density of all solid particles is a sufficiently accurate estimation of aggregate density, i.e., $\rho_{agg} \approx \rho_{cr}$. This approach takes into account the variability due to centrifugation, and the estimation error should not be large, because aggregate represents the vast majority of the weight of the solids in the concrete mix (almost 80%); moreover, cement density is similar to aggregate density. Finally, regardless of the adopted estimation, good compliance of laboratory test results and imaging was obtained (Table 6). Relative differences in the results obtained using different methods did not exceed a few percent, while the values in individual layers differed by up to 20%.

Table 6. Comparison of results (aggregate content).

Parameter	Layer 1	Layer 2	Layer 3	Layer 4
Aggregate content based on laboratory determinations C_{agg} (kg/m^3)	1494	1791	1748	1766
Aggregate content calculated using optical scan results (*) C_{agg} (kg/m^3)	1505	1732	1729	1661
Aggregate content calculated using optical scan results (**) C_{agg} (kg/m^3)	1620	1801	1769	1731

While analyzing the results in Table 6, it should be borne in mind that, in the case of imaging, only the aggregate volume fraction ϕ_{agg} was determined intrinsically. Calculation of mass content in this case required additional data, which could be determined from the recipe for the mixture and physical characteristics of its components, or even on the basis of laboratory test results. However, the obtained compliance indicates that the results of both methods are consistent; the aggregate morphology determined from scanning corresponds to the average values of its content from tests based on physicochemical methods. The results of both test methods used can be considered complementary; they can be used together to analyze spun concrete structure. In particular, thanks to the use of known relationships (e.g., Equation (11)), by creating an appropriate correlation, the image analysis can be used to "extend" point information (in layers) to a continuous functional distribution as a function of the wall thickness.

The content of coarse aggregate decreases toward the inner cross-sectional area. The last inner layer is marked by almost zero aggregate content (Figure 16). It is interesting that the aggregate content as a function of the wall thickness is constant in all layers (Figure 16 and Table 6) except for the very inner layer (layer 1). The results of aggregate distribution as a function of the wall thickness of the cross-section confirm the authors' expectations and macroscopic observations, and they are consistent with the literature [9]. The aggregate content determined using both methods approximately corresponds to the amount of aggregate declared in the mixture recipe given by the manufacturer of spun concrete poles (Table 1).

5.3. Cement Content

From the optical scan, the volume fraction of the cement matrix can be determined. Similarly to the analysis of other concrete components, it is possible to assess the variability of the matrix volume fraction as a function of the wall thickness (see Figure 14). The laboratory test determines the average cement content expressed as the mass of cement used per volume unit of hardened concrete. To determine the relationship between these quantities, both of which in fact show the amount of cement, it is necessary to use some additional information. In order to compare the results obtained using different techniques, the content C_{cem} (kg/m^3) was referred to, because the requirements for the minimum value of this parameter are provided by the standard [14]. Cement content can be defined as

$$C_{cem} = \phi_{mat} \cdot \left(1 - \phi_{\mu p,mat}\right) \cdot C^* \tag{12}$$

where ϕ_{mat} is the volume share of cement matrix (including capillary micropores) according to the estimation made on the basis of optical scanning, $\phi_{\mu p,mat}$ is the cement matrix microporosity, and C^* is the mass of cement used per volume unit of solid parts of hardened cement matrix (including nonevaporable gel water); according to the relationships given in [31], $C^* = 1473$ kg/m^3 was assumed.

In Table 7, a comparison of obtained results is given. Due to the lack of detailed information on the microporosity of the matrix, an upper and lower estimation was made, assuming the following as the limit assumptions: (1) an even distribution of porosity within the matrix and aggregate and (2) full tightness of the aggregate, i.e., that all micropores are located in the cement matrix. The microporosity of the matrix corresponding to such assumptions $\phi_{\mu p,mat}$ can be estimated from the following inequalities:

$$\frac{\phi_{p,tot} - \phi_{p,m}}{\phi_{mat} + \phi_{agg}} \leq \phi_{\mu p,mat} \leq \frac{\phi_{p,tot} - \phi_{p,m}}{\phi_{mat}} \tag{13}$$

where $\phi_{p,tot}$ is the total porosity (from the laboratory test), $\phi_{p,m}$ is the macroporosity (based on scanning, average value in the layer), ϕ_{mat} is the volume fraction of cement matrix, and ϕ_{agg} is the volume fraction of the aggregate.

Table 7. Comparison of results (cement content).

Parameter	Layer 1	Layer 2	Layer 3	Layer 4
Cement content based on laboratory determinations C_{cem} (kg/m^3)	477.8	290.8	295.2	308.9
Upper cement content estimation calculated using optical scanning results $C_{cem,max}$ (kg/m^3)	478	371	374	422
Lower cement content estimate calculated using optical scanning results $C_{cem,min}$ (kg/m^3)	340	249	219	292

The specified ranges in Table 7 are quite wide; however, in terms of quality, cement content is "well rendered", i.e., the smallest values are in the middle layers (No. 2 and 3), and the highest value is in the inner one (Layer 1) (see Figure 20).

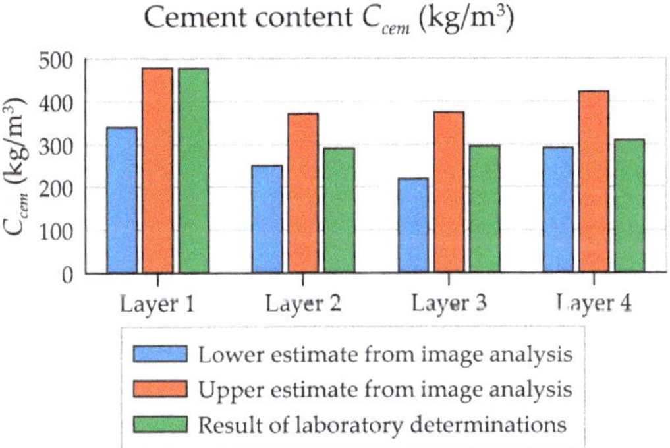

Figure 20. Estimation of cement content C_{cem} (kg/m^3) in particular layers of the element's wall.

The conclusion concerning the applied research methodology is similar to the previous analyses. The results are comparable and consistent; they complement each other. Imaging naturally results in geometrical relationships, i.e., a description of the cement matrix morphology, which is important information from the point of view of determining the effective parameters of concrete as composite using micromechanics tools or the theory of homogenization [21,32–34]. Information on the structure of concrete obtained from laboratory methods is limited to average values of cement content in arbitrarily separated layers. This information directly refers to the practical aspects of design and construction, related, for example, to the formulation of a recipe and control of the composition of concrete mix.

The description of the cement distribution as a function of the wall thickness of the cross-section obtained by means of both methods allows for straightforward qualitative description of this distribution. Optical scanning allowed determining the volume fraction of the cement matrix (Figure 12), with the cement matrix being segmented as a homogeneous component, although in fact it may contain a certain amount of the finest aggregate fraction. However, the laboratory study determined the average cement content expressed as the mass of cement used per volumetric unit of hardened concrete (Table 2). To determine the relationship between these quantities, some additional information was necessary. Due to the lack of detailed data on the microporosity of individual components (aggregate and matrix), a final estimation on the basis of the method of image analysis was made for the lower and upper cement content bounds in individual layers of the element wall (Table 7 and Figure 20). The obtained results show that the cement distribution was relatively constant as a function of the wall thickness, and an increase in its quantity was observed only in the inner layer of the cross-section (2 mm thick) (Figure 14 and Table 7). These research results coincide almost exactly with Marquardt's conclusions [4].

Cement content in individual layers determined by the laboratory method (Table 2) and by the image analysis method (lower estimate, Table 7) was lower than the amount of cement declared in the recipe by the manufacturer of spun concrete poles (Table 1). It can also be stated that, in the outer layer, the amount of cement, as a result of the centrifugation process, was reduced from 400 kg/m^3 (Table 1) to about 308 kg/m^3 (according to the laboratory method, Table 2) and to about 292 kg/m^3 (according to the lower estimation from the image analysis method, Table 7). Due to the durability of power poles made of spun concrete exposed to direct influence of atmospheric factors, described by standard [14] class XC4 (cyclically wet and dry concrete surfaces exposed to contact with water), and for structures located near motorways class XF2 (vertical concrete surfaces exposed to freezing and de-icing agents), the minimum cement content in the outer layer should not be less than 300 kg/m^3.

5.4. General Remarks

It was expected that, during concrete centrifugation, air and water would be squeezed out more from the outer, more compressed layers. Approaching the inner surface of the concrete section, less water would be squeezed out. From the inner layer of the concrete section, where the radial pressure is close to zero, water would not be removed at all. Following laboratory tests, it was observed that the total porosity of the inner layer (No. 3) increases (Table 2) compared to the porosity of other layers. This result confirmed the authors' expectations and previous research [5,7–9], in which it was found that the total porosity of concrete is greater for the inner layers than the outer layers. The other observation made in [7–9] was confirmed using image analysis methods. It was observed (Figure 21) that the pores (mostly in layer No. 3) were usually located next to the edges of large grains of aggregate on its outer side. Furthermore, the cement matrix had distinctly fewer aggregate inclusions in these locations compared to the other ones. Therefore, water and air were blocked there from being moved toward the inside of the cross-section as imposed by the centrifugal forces. Such a picture of the porosity and fine aggregate arrangement next to large aggregate grains indicates that the speed and spinning time of concrete in the pole were probably incorrectly selected, preventing the escape of excess air and water from the concrete mixture.

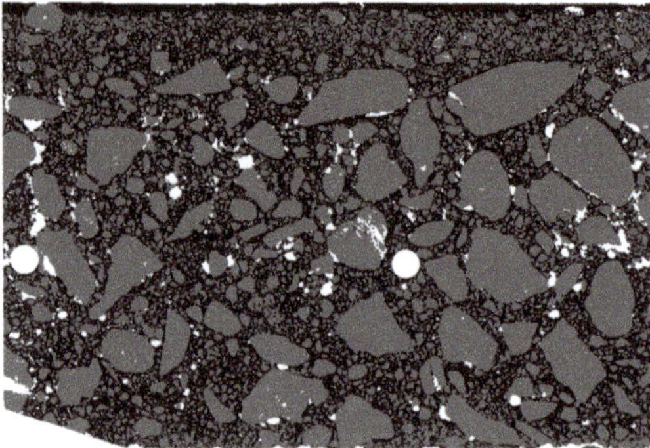

Figure 21. Arrangement of pores (white), aggregate (gray), and cement matrix (black) in the section of the element wall.

6. Summary and Conclusions

The aim of the article was to compare different methods for determining porosity, as well as cement and aggregate content, in the layered structure of spun concrete. The variability of these parameters considered as a function of the thickness of the concrete wall determines the macroscopic properties such as durability, load capacity, and rigidity of spun concrete poles. The article compares the results obtained using two methodologically different approaches: laboratory determinations based on chemical and/or physical methods and image analysis. In general, mass content per unit volume is obtained in laboratory tests, and the volume fraction of individual concrete components is obtained using image analysis methods. If the specific density of the component under consideration is known, conversion of one quantity to another becomes easy.

The main advantage of the methods based on image analysis is the possibility of obtaining a practically continuous function of the content of the considered concrete components as a function of the wall thickness of a spun pole. In this context, the main limitation of chemical/physical methods is the ability to determine only the average value for a few, arbitrarily separated layers of the concrete wall of a spun pole. In turn, the main limitation of image analysis methods is the pixel/voxel size and image quality, particularly due to sample preparation and image acquisition method.

In terms of concrete porosity, it should be stated that the attempt to validate the results obtained from imaging the structure of concrete using laboratory techniques (determining the composition of hardened concrete on the basis of physical and chemical methods) was only partially successful. The results obtained cannot be directly compared due to the fact that porosities are described on different scales. The methods based on image analysis allowed describing the structural porosity (i.e., macroporosity) of concrete, while the laboratory methods enabled describing total and open porosity (capillary microporosity). Quite importantly, imaging methods give the opportunity to analyze the macropore morphology. In particular, it was possible to visualize the location of air voids in relation to the other concrete components (Figure 21), which may have a direct impact on the assessment of the centrifugation process in terms of centrifugation parameters selection such as speed and time.

In terms of the description of aggregate distribution as a function of the wall thickness of the cross-section, very good compatibility of both methods was obtained. The image analysis method allowed determining the volume fraction of aggregate ϕ_{agg}. Calculation of aggregate mass content per unit volume in this case required additional information about the density of the aggregate, which could be determined either from the recipe for the concrete mixture and the physical characteristics of its components or from laboratory tests. The obtained consistency of the results from both methods shows that the methods are consistent, and the aggregate morphology determined by scanning corresponds to the average values of its content coming from research based on physicochemical methods.

Very good qualitative and quantitative compatibility of both methods in the field of aggregate morphology as a function of the wall thickness of the element was obtained. Thanks to the positive validation of laboratory methods and the method based on image analysis, it can be safely stated that optical scanning is a cheap, relatively fast, and effective method of assessing aggregate segregation as a function of the wall thickness of a spun concrete element.

The description of cement distribution as a function of the wall thickness of the element cross-section obtained using both methods allows for straightforward qualitative description of this distribution. However, full validation success was not achieved because the results based on optical scanning allow obtaining quantitative results only after taking into account additional calculations and information from laboratory tests. Thus, in this case, to get full information about the distribution of cement as a function of the wall thickness of the element, both methods must complement each other. Nevertheless, the information obtained as a result of testing with any of methods used allows assessing the cement content, which determines the requirement for durability of columns made of spun concrete.

An attempt to validate the imaging method with the use of laboratory techniques based on physical and chemical methods showed that optical scanning methods are relatively effective, and they can be a significant complement to the research methods used so far. The analyses also showed that there is a further need to conduct research in the area of spun concrete structure, with particular emphasis on the distribution of concrete porosity and cement content as a function of the wall thickness of the ring-shaped cross-section. The research methods presented in the work can be used to improve the production process of poles made of spun concrete as an effective tool for testing its structure.

Author Contributions: Conceptualization, J.M.; methodology, J.M. and M.S.; software, M.S.; validation, J.M. and M.S.; formal analysis, J.M. and M.S.; investigation, J.M. and M.S.; resources, J.M. and M.S.; data curation, J.M. and M.S.; writing—original draft preparation, J.M. and M.S.; writing—review and editing, J.M. and M.S.; visualization, J.M. and M.S.; supervision, J.M. and M.S.; project administration, J.M. and M.S.; funding acquisition, J.M. and M.S. All authors read and agreed to the published version of the manuscript.

Funding: This research received no external funding.

Conflicts of Interest: The authors declare no conflict of interest.

References

1. Michałek, J. Concrete lighting poles formerly and nowadays. *Mater. Bud.* **2016**, *5*, 82–83. (In Polish)
2. Michałek, J. Properties of spun concrete in the durability aspect of structures. In *Durability of Buildings and Protection Against Corrosion, Proceedings of the 13th Conference Kontra 2002, Zakopane, Poland, 22–25 May 2002*; Komitet Trwałości Budowli PZITB: Warszawa, Poland, 2002; pp. 175–184. (In Polish)
3. Trapko, T.; Michałek, J. Using of composite materials for strengthening of spun concrete electrical poles. *Przegląd Elektrotechniczny* **2012**, *88*, 267–273. (In Polish)
4. Marquardt, E. Geschleuderte Beton- und Eisenbetonrohre. *Die Bautech.* **1930**, *40*, 587–602.

5. Kuranovas, A.; Kvedaras, A.K. Centrifugally manufactured hollow concrete-filled steel tubular columns. *J. Civ. Eng. Manag.* **2007**, *XIII*, 297–306. [CrossRef]
6. Kliukas, R.; Jaras, A.; Lukoševičienė, O. Reinforced spun concrete poles—Case study of using chemical admixtures. *Materials* **2020**, *13*, 302. [CrossRef] [PubMed]
7. Adesiyun, A.; Kamiński, M.; Kubiak, J.; Łodo, A. Laboratory test on the properties of spun concrete. In Proceedings of the Third Interuniversity Research Conference, Szklarska Poręba, Poland, 22–26 November 1994; pp. 3–8.
8. Adesiyun, A.; Kamiński, M.; Kubiak, J.; Łodo, A. *Investigation of Spun-Cast Concrete Structure, School of Young Research Methodology of Concrete Structures*; Dolnośląskie Wydawnictwo Edukacyjne: Wrocław, Poland, 1996; pp. 13–19.
9. Kamiński, M.; Kubiak, J.; Łodo, A.; Adesiyun, A.; Kupski, J.; Michałek, J.; Oleszkiewicz, T. *Tests of Structural Elements with an Annular Cross-Section Made of Spun Concrete*; Dolnośląskie Wydawnictwo Edukacyjne: Wrocław, Poland, 1996. (In Polish)
10. Völgyi, I.; Farkas, G.; Nehme, S.G. Concrete strength tendency in the wall of cylindrical spun-cast concrete elements. *Period. Polytech. Civ. Eng.* **2010**, *54*, 23–30. [CrossRef]
11. Völgyi, I.; Farkas, G. Rebound testing of cylindrical spun-cast concrete elements. *Period. Polytech. Civ. Eng.* **2011**, *55*, 129–135. [CrossRef]
12. Michałek, J.; Pachnicz, M.; Sobótka, M. Application of nanoindentation and 2D and 3D imaging to characterize selected features of the internal microstructure of spun concrete. *Materials* **2019**, *12*, 1016. [CrossRef] [PubMed]
13. Kubiak, J.; Łodo, A.; Michałek, J. Production of power spun concrete poles in longitudinally non-openable forms. *Build. Mater.* **2015**, *6*, 38–39. (In Polish)
14. PN-EN 206+A1:2016-12. *Concrete. Specification, Performance, Production and Conformity*; Polish Committee for Standardization: Warsaw, Poland, 2017.
15. Jarmontowicz, A.; Krzywobłocka-Laurow, R. *Instruction No. 277 Determining the Composition of Hardened Concrete*; Instytut Techniki Budowlanej: Warszawa, Poland, 1986. (In Polish)
16. Standard Test Method for Portland-Cement Content of Hardened Hydraulic-Cement Concretete. ASTM C1084-19. 15 December 2019.
17. *BS 1881-124: 2015 Testing Concrete. Methods for Analysis of Hardened Concrete*; British Standards Institution: London, UK, 2015.
18. *PN-EN 12390-7: 2019-08 Testing Hardened Concrete. Part 7: Density of Hardened Concrete*; Polish Committee for Standardization: Warsaw, Poland, 2019.
19. *PN-EN 1936:2010 Natural Stone Test Methods. Determination of Real Density and Apparent Density, and of Total and Open Porosity*; Polish Committee for Standardization: Warsaw, Poland, 2010.
20. Kozłowski, S.; Parachoniak, W. Products of basalt weathering in the region of Luban in Lower Silesia. *Acta Geol. Pol.* **1960**, *X*, 287–326. (In Polish)
21. Torquato, S. Random Heterogeneous Materials. In *Microstructure and Macroscopic Properties*; Springer: New York, NY, USA, 2002.
22. Łydżba, D. Applications of asymptotic homogenization method in soil and rock mechanics. In *Scientific Papers of the Institute of Geotechnics and Hydrotechnics*; Wroclaw University of Technology: Wroclaw, Poland, 2002. (In Polish)
23. Różański, A. Random Composites: Representativity, Minimum RVE Size, Effective Transport Properties. Ph.D. Thesis, University Lille I, Lille, France, 2010.
24. Hildebrand, T.; Rüegsegger, P. A new method for the model-independent assessment of thickness in three-dimensional images. *J. Microsc.* **1997**, *185*, 67–75. [CrossRef]
25. Feldkamp, L.A.; Davis, L.C.; Kress, J.W. Practical cone-beam algorithm. *J. Opt. Soc. Am. A* **1984**, *1*, 612–619. [CrossRef]
26. Jamroży, Z. *Concrete and its Technology*; PWN: Warszawa, Poland, 2015. (In Polish)
27. Bednarek, Z.; Krzywobłocka-Laurów, R.; Drzmała, T. Effect of high temperature on the structure, phase composition and strength of concrete. *Zesz. Nauk. Sgsp/Szkoła Główna Służby Pożarniczej* **2009**, *38*, 5–27. (In Polish)
28. Bultreys, T.; Van Hoorebeke, L.; Cnudde, V. Multi-scale, micro-computed tomography-based pore network models to simulate drainage in heterogeneous rocks. *Adv. Water Resour.* **2015**, *78*, 36–49. [CrossRef]

29. Cid, H.E.; Carrasco-Núñez, G.; Manea, V.C. Improved method for effective rock microporosity estimation using X-ray microtomography. *Micron* **2017**, *97*, 11–21. [CrossRef] [PubMed]

30. Bossa, N.; Chaurand, P.; Vicente, J.; Borschneck, D.; Levard, C.; Aguerre-Chariol, O.; Rose, J. Micro-and nano-X-ray computed-tomography: A step forward in the characterization of the pore network of a leached cement paste. *Cem. Concr. Res.* **2015**, *67*, 138–147. [CrossRef]

31. Neville, A.M. *Properties of Concrete*; Polski Cement Sp. z o.o.: Kraków, Poland, 2000. (In Polish)

32. Łydżba, D.; Różański, A. Microstructure measures and the minimum size of a representative volume element: 2D numerical study. *Acta Geophys.* **2014**, *62*, 1060–1086. [CrossRef]

33. Łydżba, D.; Różański, A.; Stefaniuk, D. Equivalent microstructure problem: Mathematical formulation and numerical solution. *Int. J. Eng. Sci.* **2018**, *123*, 20–35. [CrossRef]

34. Różański, A.; Stefaniuk, D. Prediction of soil solid thermal conductivity from soil separates and organic matter content: Computational micromechanics approach. *Eur. J. Soil Sci.* **2016**, *67*, 551–563. [CrossRef]

Technical Note

Aggregate Roundness Classification Using a Wire Mesh Method

Sung-Sik Park [1], Jung-Shin Lee [1] and Dong-Eun Lee [2],*

[1] Department of Civil Engineering, Kyungpook National University, 80 Daehakro, Bukgu, Daegu 41566, Korea; sungpark@knu.ac.kr (S.-S.P.); jhjs14@knu.ac.kr (J.-S.L.)

[2] School of Architecture, Civil, Environment and Energy, Kyungpook National University, 80 Daehakro, Bukgu, Daegu 41566, Korea

* Correspondence: dolee@knu.ac.kr; Tel.: +82-53-950-7540

Received: 22 July 2020; Accepted: 18 August 2020; Published: 20 August 2020

Abstract: Herein, we suggest a wire mesh method to classify the particle shape of large amounts of aggregate. This method is controlled by the tilting angle and opening size of the wire mesh. The more rounded the aggregate particles, the more they roll on the tilted wire mesh. Three different sizes of aggregate: 11–15, 17–32, and 33–51 mm were used for assessing their roundness after classification using the sphericity index into rounded, sub-rounded/sub-angular, and angular. The aggregate particles with different sphericities were colored differently and then used for classification via the wire mesh method. The opening sizes of the wire mesh were 6, 11, and 17 mm and its frame was 0.5 m wide and 1.8 m long. The ratio of aggregate size to mesh-opening size was between 0.6 and 8.5. The wire mesh was inclined at various angles of 10°, 15°, 20°, 25°, and 30° to evaluate the rolling degree of the aggregates. The aggregates were rolled and remained on the wire mesh between 0.0–0.6, 0.6–1.2, and 1.2–1.8 m depending on their sphericity. A tilting angle of 25° was the most suitable angle for classifying aggregate size ranging from 11–15 mm, while the most suitable angle for aggregate sizes of 17–32 and 33–51 mm was 20°. The best ratio for the average aggregate size to mesh-opening size for aggregate roundness classification was 2.

Keywords: aggregate; classification; wire mesh; roundness; tilting angle; opening size

---

1. Introduction

Aggregates are necessary for manufacturing various construction materials such as concrete, mortar, brick, ballast, filter, backfill, and so forth. Natural aggregates with different shapes and sizes are obtained from rivers and seas, but they are becoming depleted. Currently, artificial aggregates are produced by crushing rock fragments from mountains. Both natural and artificial aggregates are widely used in many civil engineering structures, such as buildings, bridges, tunnels, roads, dams, among others. Moreover, a huge amount of aggregate is recycled from concrete structures and asphalt pavements, of which 85% is utilized in earthworks and road sub-bases in Korea [1]. The shape and size of aggregates contribute to their engineering characteristics such as interlocking, friction, and settlement. Rounded aggregates used in concrete give excellent performance and high strength whereas angular aggregates are utilized in road sub-bases and railway ballast for minimizing settlement and producing high interlocking [2]. Lee carried out a series of direct shear tests on recycled aggregates and showed a 4° higher friction angle for angular-shaped aggregates than rounded ones [3]. Therefore, the particle-shape classification for large amounts of aggregate is important because the aggregate's engineering characteristics depend on its shape. Even though the shape is more important than the size for engineering purposes, the aggregates are usually classified by size through sieve analysis in practice.

The measurement of particle shape is difficult due to varying edges and irregular surfaces. Various indexes such as the particle perimeter, convexity, sphericity, and shape factor have been suggested to classify the particle shape for mainly academic purposes [4–12]. A comparison of these indexes and their key parameters are introduced in Table 1. The determination of such indexes seems to be difficult for practical purposes, especially for the angularity factor. Krumbein [10] and Rittenhouse [12] both presented charts of aggregate shape for classifying them based on sphericity. The abrasion of particles can be quantitatively determined using the Krumbein chart and is in the range of 0.1–0.9. The roughness of angular particle edges is 0.1 before abrasion and 0.9 afterward. The Rittenhouse chart shows the degree to which the shape of a particle approaches a sphere. The sphericity of a particle is the ratio of the surface area of a sphere of the same volume as the particle to the actual surface area of the particle. The sphericity of a very irregularly shaped particle is 0.45 whereas one that is nearly spherical is 0.97 and one that is completely spherical is 1. However, the analysis of particle shapes using the aforementioned methods is limited as they depend on the judgment of the one doing the particle classification [13]. Digital image processing can be used to quantitatively analyze the characteristics of the shape and/or the shape coefficient of an aggregate particle [14–16]. Furthermore, stereoscopic image analysis can be performed using two cameras to determine the surface area characteristics of aggregates [17–19].

Table 1. Particle angularity index.

Index (Reference)	Equation	Key Parameter	Symbol
Roundness (Wadell [9])	$\dfrac{N}{\sum_{i=1}^{N}\frac{D}{d_i}}$	Corner diameter	N: number of D: diameter of largest inscribed circle d: diameter of corners
Sphericity (Krumbein [10])	$\Psi^3 = \dfrac{b \times c}{a^2}$	Diameter (a, b, c)	Ψ: Sphericity a: The longest diameter b: The longest diameter perpendicular to a c: Second longest diameter perpendicular to a
Angularity factor, AF (Sukumaran [14])	$AF = \dfrac{\sum_{i=1}^{N}\left(\beta_{i\,particle}-180\right)^2\left(\frac{360^2}{N}\right)}{3\times180^2-360^2/N}$	Number of sampling points and internal angle	N: Number of sampling points $\beta_{i\,particle}$: Internal angle
Volume ratio (Fei Xu [11])	$V_B = \frac{1}{6}\pi L_{max}^3$	External volume	V_B: Aggregatecircumscribed volume L: The longest length

The results of previous studies indicate that particle classification is usually performed on each particle rather than on the number of particles, and a practical classification method is still required in the field. Therefore, it is necessary to improve the way of evaluating the particle shape. In this study, a wire mesh method was investigated for classifying three different sizes of aggregates by controlling the tilt angle and using a suitable opening size of wire mesh. A classification chart for the wire mesh method is also proposed to simplify the classification of a large amount of aggregate.

2. Wire Mesh Method

2.1. Materials

Crushed aggregates used in more than 90% of Korean construction sites were used in this study. The aggregates were classified into three different sizes after performing a sieve analysis according to the KS F 2502 [20] standard. Particle size distribution curves are shown in Figure 1. Aggregate particles between 11 mm and 15 mm are termed small aggregates, those between 17 mm and 32 mm are called medium aggregates, and those between 33 mm and 51 mm are termed large aggregates. The small, medium, and large aggregates are commonly used in concrete making, road sub-base, and railroad ballast at the site, respectively [21,22].

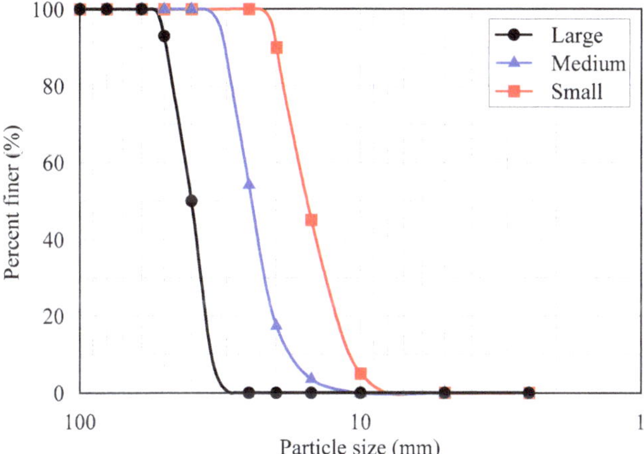

Figure 1. Particle size distribution curves of the three different sizes of aggregate.

The three types of the aggregates (small, medium, large) are illustrated in Figure 2. In addition, three wire mesh opening sizes were selected depending on the type of aggregate, as shown in Figure 3. The aggregate particles had a variety of shapes and their color was mostly blue. The specific gravity of the aggregates is 2.7 and detailed engineering properties of the aggregates are summarized in Table 2. A wear rate indicates the reduction in the size of aggregate due to friction. The bulk densities of the each size were obtained by the test according to ASTM C 29 [23]. The results of petrographic analysis from the scanning electron microscope (SEM) were shown in Figure 4, micro structure images (a, b) and mineral component spectrum from the energy dispersive X-ray spectroscopy (EDX) (c). Petrographic analysis indicated that the aggregate is a granite-based rock containing mainly silica and alumina.

Figure 2. Photographs of the three different sizes of aggregate: (**a**) Small, (**b**) Medium, (**c**) Large.

Figure 3. Photographs of the three different sizes of wire mesh: (**a**) 6 mm, (**b**) 11 mm, and (**c**) 17 mm.

Table 2. Physical properties of the used aggregates.

Bulk Density (kN/m^3)	Water Absorption (%)	Modified CBR * (%)	Wear Rate (%)	OMC ** (%)	Porosity (%)	Contamination Content (%)	Sand Equivalent (%)
15.20 (Small) 15.32 (Medium) 15.69 (Large)	6.16	58	36.5	11.3	2–3	0.11	45

* CBR: California bearing ratio, ** OMC: optimal moisture content.

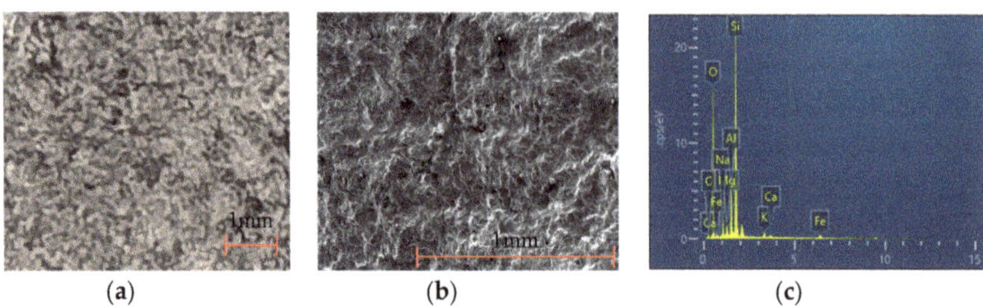

(a) (b) (c)

Figure 4. Petrographic analysis: (**a**) × 10 microscope, (**b**) × 50 microscope, (**c**) mineral components.

2.2. Wire Mesh Method

Herein, we suggest the particle shape classification of a large amount of aggregate using a wire mesh method. The concept of the particle shape classification using the wire mesh method is illustrated in Figure 5. The basic concept is that the greater the degree of particle roundness, the longer the moving distance while rolling on the inclined wire mesh plane. The aggregates were rolled from various tilting angles of the wire mesh that was determined by trial and error for proper classification. The aggregate particles with different sphericities were colored differently and then used for classification via the wire mesh method.

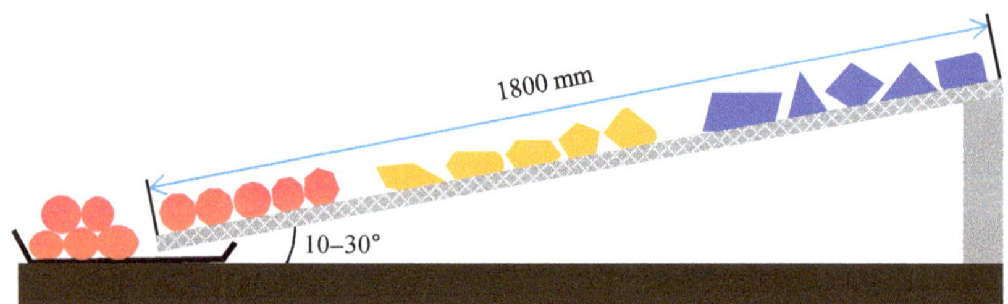

Figure 5. Illustration of particle-shape classification using the wire mesh method.

The frame containing the wire mesh was 1800 mm long and 500 mm wide. Three wire mesh plywood frames were connected with enough rolling distance depending on the sloping angle of the frame, which was a parameter considered during the experiment. When this technique is applied in the field, the size of the wire mesh frame can be modified to accommodate a suitable amount of aggregate. Suitable amounts of aggregate (1–5 kg) were poured onto the start position of the wire mesh, which was then tilted to the desired angle. The aggregates started to roll due to gravity depending on the particle shape. The wire mesh frame was inclined at various angles (10°, 15°, 20°, 25°, and 30°) to evaluate the rolling degree. The tilting angle was selected depending on the aggregate size and the wire mesh opening size. For the given aggregate and wire mesh used in this study, the majority of aggregates rolled down when the tilting angle was higher than 30°, and thus particle classification was not properly achieved. At a low tilting angle, it was difficult to classify the aggregate because only some of it moved. The aggregate that rolled and traveled the longest distance (from 1200–1800 mm) was classified as rounded, that which rolled and traveled to the middle of the wire mesh (from 600–1200 mm) was classified as sub-rounded/sub-angular, and that which remained on the inclined wire mesh without rolling or falling (from 0–600 mm) was classified as angular. These ranges were determined by trial and error.

3. Results and Discussion

3.1. Particle-Shape Classification Using a Sphericity Index

In this study, before applying a wire mesh method the sphericity proposed by Krumbein [10] was used for classifying aggregates. The sphericity Ψ is calculated as:

$$\Psi^3 = \frac{bc}{a^2} \tag{1}$$

where a, b, and c are the longest, intermediate, and shortest diameters of the aggregate particle, as shown in Figure 6. Sphericity is in the range of 0.1–0.9, and when it is close to 1, the particle shape is close to spherical, while ≤0.1 means that the particle shape is angular.

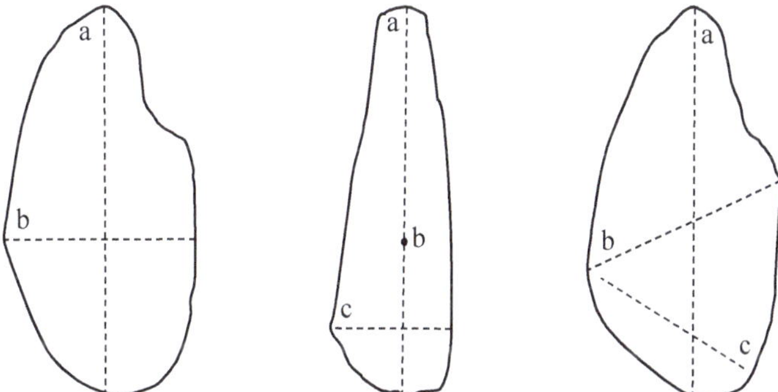

Figure 6. Particle dimension definition (**a**, **b**, and **c** are the longest, intermediate, and the shortest diameter).

The sphericities of three different sizes of aggregate particles (11–15, 17–32, and 33–51 mm) was measured, and based on these values, the aggregate particles were classified as rounded, sub-rounded/sub-angular, or angular. The corresponding sphericity indexes were 0.8–0.9, 0.5–0.7, and 0.3–0.4, respectively. As an example, the sphericity indexes and a comparison of 30 small-sized aggregates with different sphericities were included in Table 3.

3.2. Particle-Shape Classification Using the Wire Mesh Method

Three different aggregate-particle sizes of 11–15, 17–32, and 33–51 mm were used to classify the roundness of the aggregate. These aggregates had already been classified by the sphericity index to angular, sub-angular/sub-rounded, and rounded as described in Section 3.1. The wire mesh was inclined at various angles of 10°, 15°, 20°, 25°, and 30° to evaluate the rolling degree of the aggregates. When the tilting angle and opening size of the wire mesh had been optimally adjusted, the aggregates rolled and stopped on the tilted wire mesh depending on their roundness. The results showed that the angular aggregates remained on the wire mesh in the interval from 0.0–0.6 m from the top, the sub-angular/sub-rounded aggregates remained in the interval from 0.6–1.2 m, and the rounded aggregates remained in the interval between 1.2–1.8 m. For the small aggregate-particle size, some of the aggregate rolled down with a tilting angle of 20° but all the aggregates rolled down at 30°. The most suitable tilting angle for the small aggregate particles was 25°, as shown in Figure 7. For the middle- and large-sized aggregate particles, some of the aggregate rolled down with a tilting angle of 15° but all of it rolled down at 25°. The most suitable tilting angle for these aggregates was 20°. The ratio of the average aggregate size to mesh-opening size was approximately 2 for reasonable classification. When the ratio was less than 2, all of the aggregate rolled down. On the other hand, when the ratio was greater than 2, most of the aggregate particles rolled less far.

Table 3. Aggregate roundness classification.

Parameter	Sample (11–15 mm)									
	1	2	3	4	5	6	7	8	9	10
a (mm)	21.03	18.35	22.86	20.53	17.18	17.55	18.0	16.36	18.29	15.41
b (mm)	17.02	16.27	17.68	14.87	13.42	13.89	15.21	14.52	16.99	14.27
c (mm)	14.02	13.25	14.14	13.46	13.38	11.99	13.8	14.3	13.63	12.69
Ψ	0.8	0.9	0.8	0.8	0.9	0.8	0.9	0.9	0.9	0.9

Parameter	Sample (11–15 mm)									
	11	12	13	14	15	16	17	18	19	20
a (mm)	21.79	19.51	20.42	23.13	27.0	19.77	18.66	17.87	19.63	22.14
b (mm)	11.03	9.59	13.85	13.11	14.43	9.15	14.08	7.56	13.27	13.07
c (mm)	6.93	6.01	11.46	10.03	12.66	6.11	10.08	6.21	11.04	8.73
Ψ	0.5	0.5	0.7	0.6	0.6	0.5	0.7	0.5	0.7	0.6

Parameter	Sample (11–15 mm)									
	21	22	23	24	25	26	27	28	29	30
a (mm)	26.96	27.73	25.91	20.45	23.21	29.82	29.99	28.97	27.53	19.17
b (mm)	7.04	23.8	11.13	8.32	11.14	14.02	11.08	8.04	10.13	9.37
c (mm)	4.08	3.57	5.16	4.03	4.23	5.13	3.15	4.02	6.0	3.12
Ψ	0.3	0.4	0.4	0.4	0.4	0.4	0.3	0.3	0.4	0.4

a, the longest particle diameter; b, the intermediate particle diameter; c, the shortest particle diameter; Ψ, sphericity.

Figure 7. Aggregate shape classification via the wire mesh method (11–15 mm).

3.3. Comparison of Particle-Shape Classification Methods

The potential usage of the wire mesh method for classifying a large amount of aggregate was evaluated via comparison with the sphericity index. The relationship between the sphericity index (y) and the travel distance (x) from the top of small-sized aggregate particles is shown in Figure 8. The correlation showed a linear relationship (y = 0.42x + 0.22) and the correlation coefficient (R^2) is 0.97. The comparison between the sphericity and the wire mesh method showed a well-matched correlation. Similar relationships were observed for all three aggregate-particle sizes.

The tilting angle and opening size of wire mesh needs to be determined for accurately classifying aggregates using a wire mesh method. A classification chart for the opening size and tilting angle of the wire mesh is shown in Figure 9. It can be used to select a suitable tilting angle and size of mesh. Three classification levels are suggested: good, moderate, poor, and N.A. When the average lengths of the aggregate diameter were 34 and 24 mm, the most efficient tilting angle of the wire mesh was 20° whereas, for an average length of 14 mm, the tilting angle was 25°. The greater the aggregate-particle size, the greater its self-weight, so it is expected that even with a low tilting angle, large aggregate particles can roll down due to their self-weight. For the wire mesh method, the tilting angle of the wire mesh seems to decrease as the aggregate size increases. When the ratio of the average aggregate size to the opening size of the wire mesh was 2, all of the aggregate types were properly distinguished from one another. Hence, our method can be used for aggregate-particle shape verification and classification.

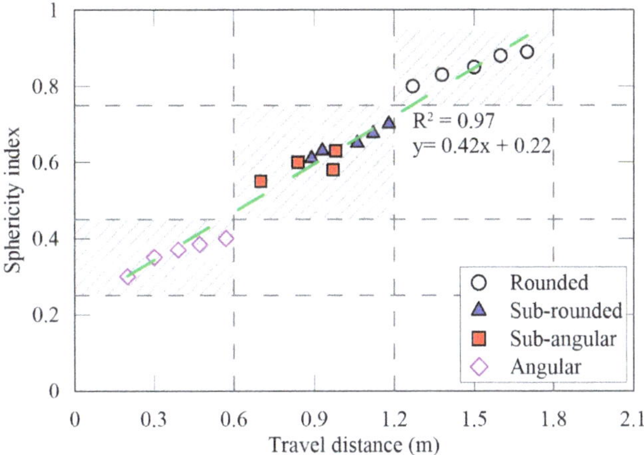

Figure 8. Relationship between the sphericity measurements and the wire mesh technique.

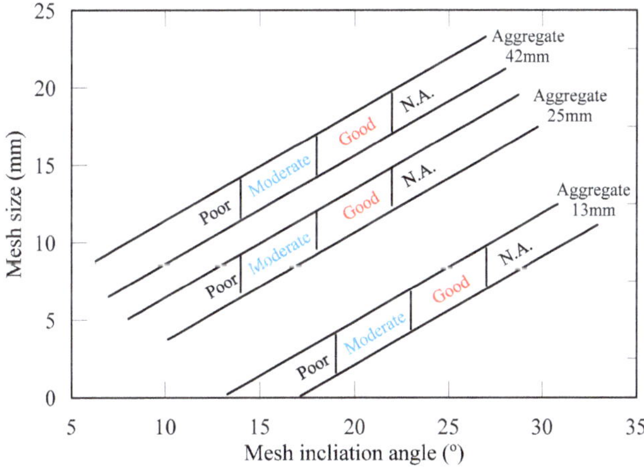

Figure 9. Aggregate shape classification with the wire mesh method.

4. Conclusions

In this study, the particle-shape classification of a large amount of aggregate was conducted using a wire mesh method. Three different types of aggregate that are usually used on construction sites were selected by size (11–15 mm, 17–32 mm, and 33–51 mm) to test the performance of the wire mesh classification. Based on the experimental results, the following conclusions can be drawn:

1. The sphericity of the three different sizes of aggregate was measured for classifying them as angular, sub-angular/sub-rounded, and rounded based on sphericity ranges of 0.3–0.4, 0.5–0.7, and 0.8–0.9, respectively.

2. The aggregates classified via sphericity were used to evaluate the wire mesh method. The opening size and tilting angle of the wire mesh were adjusted to properly classify the aggregate-particle shapes. When these factors are optimal, the angular aggregate remained on the wire frame in the interval from 0.0–0.6 m from the top, the sub-angular/sub-rounded aggregate remained in the interval from 0.6–1.2 m, and the rounded aggregate remained in the interval from 1.2–1.8 m.

3. As the tilting angle of the wire mesh increased, most of the aggregate rolled down and the classification became difficult. A suitable inclination angle of the wire mesh for the aggregate with a size range of 11–15 mm was 25°, while that for the aggregate with size ranges of 17–32 and 33–51 mm was 20°. When the opening size of the wire mesh was too small, all of the aggregate tended to roll down. On the other hand, when it was too large, none of aggregate rolled. The experimental results show that the ratio of the mesh-opening size to the average aggregate size is 1:2.

4. Classification via the wire mesh method showed similar results when using the sphericity index. Therefore, the wire mesh method can be used to classify a large amount of aggregate with different sizes and shapes at once by controlling the opening size and tilting angle of the wire mesh in practice.

Author Contributions: Conceptualization, S.-S.P. and J.-S.L.; methodology, S.-S.P. and J.-S.L.; validation, S.-S.P., J.-S.L., and D.-E.L.; formal analysis, S.-S.P., J.-S.L., and D.-E.L.; investigation, S.-S.P. and J.-S.L.; data curation, S.-S.P. and J.-S.L.; writing—original draft preparation, S.-S.P. and J.-S.L.; writing—review and editing, S.-S.P. and J.-S.L.; supervision, S.-S.P.; funding acquisition, S.-S.P. and D.-E.L. All authors have read and agreed to the published version of the manuscript.

Funding: This work was supported by a National Research Foundation of Korea (NRF) grant funded by the Korean government (MSIT) (No. NRF-2018R1A5A1025137).

Conflicts of Interest: The authors declare no conflict of interest.

References

1. Park, S.S.; Chen, K.; Lee, Y.J.; Moon, H.D. A Study on Crushing and Engineering Characteristics Caused by Compaction of Recycled Aggregates. *J. Korean Geotech. Soc.* **2017**, *33*, 35–44. [CrossRef]

2. Broek, D. Some Contributions of Electron Fractography to the Theory of Fracture. *Int. Metall. Rev.* **1974**, *19*, 135–182. [CrossRef]

3. Lee, S.H. Shear Characteristics of Recycled Aggregates. Master's Thesis, Kyungpook National University, Daegu, Korea, February 2017.

4. Bernhardt, I.C. *Particle Size Analysis: Classification and Sedimentation Methods (Vol. 5)*; Springer Science & Business Media: Berlin, Germany, 2012.

5. BS 812-105.1. *Methods for Determination of Particle Shape. Flakiness Index*; British Standards Institution: London, UK, 1989.

6. Li, L.; Chan, P.; Zollinger, D.G.; Lytton, R.L. Quantitative analysis of aggregate shape based on fractals. *Mat. J.* **1993**, *90*, 357–365.

7. Vallejo, L.E.; Zhou, Y. Fractal approach to measuring roughness of geomembranes. *J. Geotech. Eng.* **1995**, *121*, 442–446. [CrossRef]

8. Vallejo, L.E.; Zhou, Y. The relationship between the fractal dimension and Krumbein's roundness number. *Soils Found.* **1995**, *35*, 163–167. [CrossRef]

9. Wadell, H. Volume, shape and roundness of rock particles. *J. Geol.* **1932**, *40*, 443–451. [CrossRef]

10. Krumbein, W.C. Measurement and geological significance of shape and roundness of sedimentary particles. *J. Sediment. Res.* **1941**, *11*, 64–72. [CrossRef]

11. Fei, X. The Study on Evaluating method of grains shape of coarse aggregate. *J. Yangzhou Univ.* **2002**, *5*, 61–63. [CrossRef]

12. Rittenhouse, G. A visual method of estimating two-dimensional sphericity. *J. Sediment. Res.* **1943**, *13*, 79–81.

13. Sukumaran, B.; Ashmawy, A.K. Quantitative characterization of the geometry of discrete particles. *Geotechnique* **2001**, *51*, 171–179. [CrossRef]

14. Lee, J.I. Analysis on the Shape Properties of Granular Materials Using Digital Image Processing. Master's Thesis, University of Ulsan, Ulsan, Korea, February 2004.

15. Hwang, T.J. Study on the Particle Size Distribution Analysis for Large Aggregate Using Digital Image Processing. Ph.D. Thesis, Pusan National University, Pusan, Korea, February 2007.
16. Lee, H.; Chou, E.Y. Survey of image processing applications in civil engineering. In Proceedings of the Digital Image Processing: Techniques and Applications in Civil Engineering, Hawaii, HI, USA, 28 February–5 March 1993; Frost, J.D., Wright, J.R., Eds.; American Society of Civil Engineers: New York, NY, USA, 1993.
17. Mora, C.F.; Kwan, A.K. Sphericity shape factor, and convexity measurement of coarse aggregate for concrete using digitalimage processing, pergamon. *Cem. Concr. Res* **2000**, *30*, 351–358. [CrossRef]
18. Tang, X.; De Rooij, M.R.; Van Duynhoven, J.; Van Breugel, K. Dynamic volume change measurements of cereal materials by environmental scanning electron microscopy and videomicroscopy. *J. Microsc.* **2008**, *230*, 100–107. [CrossRef] [PubMed]
19. Hu, J.; Stroeven, P. Shape Characterization of Concrete Aggregate. *Image Anal. Stereol.* **2006**, *25*, 43–53. [CrossRef]
20. KS F 2502. *Standard Test Method for Sieve Analysis of Fine and Coarse Aggregates*; Korean Standards Association: Seoul, Korea, 2002.
21. KS F 2527. *Concrete Aggregate*; Korean Standards Association: Seoul, Korea, 2002.
22. KS F 2574. *Recycled Aggregate for Subbase*; Korean Standards Association: Seoul, Korea, 2004.
23. ASTM D2166. *Standard Test Method for Bulk Density ("Unit Weight") and Voids in Aggregate*; Annual Book of ASTM Standards: Montgomery, MD, USA, 2009.

 *materials*

Article

Acoustic Emission for Determining Early Age Concrete Damage as an Important Indicator of Concrete Quality/Condition before Loading

Wiesław Trąmpczyński *, Barbara Goszczyńska and Magdalena Bacharz

Faculty of Civil Engineering and Architecture, Kielce University of Technology, Al. Tysiąclecia Państwa Polskiego 7, 25-314 Kielce, Poland; bgoszczynska@tu.kielce.pl (B.G.); mbacharz@tu.kielce.pl (M.B.)
* Correspondence: wtramp@tu.kielce.pl; Tel.: +48-41-34-24-545

Received: 24 June 2020; Accepted: 5 August 2020; Published: 10 August 2020

Abstract: Phenomena occurring during the curing of concrete can decrease its mechanical properties, specifically strength, and serviceability, even before it is placed. This is due to excessive stresses caused by temperature gradients, moisture changes, and chemical processes arising during the concreting and in hardened concrete. At stress concentration sites, microcracks form in the interfacial transition zones (ITZ) in the early phase and propagate deeper into the cement paste or to the surface of the element. Microcracks can contribute to the development of larger cracks, reduce the durability of structures, limit their serviceability, and, in rare cases, lead to their failure. It is thus important to search for a tool that allows objective assessment of damage initiation and development in concrete. Objectivity of the assessment lies in it being independent of the constituents and additives used in the concrete or of external influences. The acoustic emission-based method presented in this paper allows damage detection and identification in the early age concrete (before loading) for different concrete compositions, curing conditions, temperature variations, and in reinforced concrete. As such, this method is an objective and effective tool for damage processes detection.

Keywords: early age concrete; acoustic emission method; damage processes detection before loading; strength of structures

1. Introduction

All engineering projects encounter a range of challenges associated with the most widely used building material, concrete. Being a major problem in current concrete construction, concrete cracking or damage requires a continuous search for new methods and improvement of the existing concrete assessment techniques. This is especially important for fresh concrete, which affects the behavior of concrete under load.

Due to the multilevel nature of concrete, with qualitatively distinct mechanisms taking place during the formation of the concrete, the interaction of various parameters must be considered and the ways to study these relationships and effects have to be found to detect damage. Concrete deterioration occurs primarily through technological cracks (microcracks) and different interfacial properties (cracks) formed at various structural levels, which propagate and initiate operational cracks affecting the usability and strength of concrete elements. The composition of the concrete largely affects its properties. Concrete has high compressive strength and is durable. It can be formed into virtually any shape. Weak points of this material include low tensile strength, shrinkage during the hardening process, and susceptibility to external influences, such as moisture [1], temperature, chemical influences, etc. [2,3].

Particularly important for concrete elements is the early period accompanied by a number of phenomena related to cement hydration [4–6].

Chemical reactions occurring in the cement paste during the hydration process, drying out (water evaporation), the cement paste properties themselves (e.g., bleeding in fresh concrete and temperature changes), as well as volumetric changes due to external factors (temperature and air humidity) cause swelling and chemical, plastic, autogenous, and drying shrinkage. These volume changes of hardening concrete generate natural stress, including "micro" stress. Stresses occur most often in the interfacial transition zones (ITZ) between the grains of aggregate and cement paste and decide on the mechanical properties of these zones and their microcracking. At stress concentrations exceeding the tensile strength of concrete, the microcracks may propagate into the deeper layer of the cement paste or to the surface of the element. Examples of damages in the concrete elements shown in Figure 1.

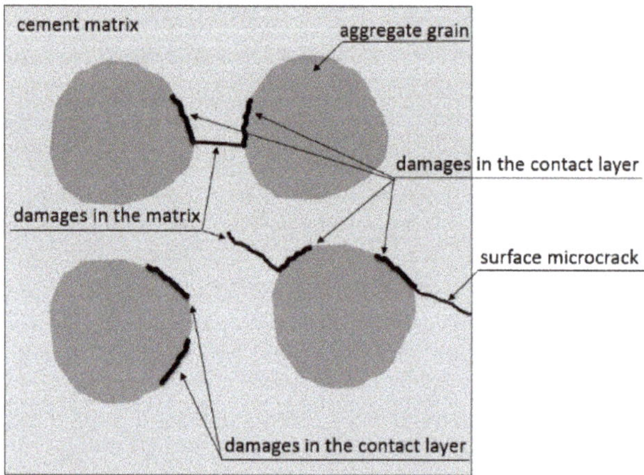

Figure 1. Examples of damages in the concrete microstructure, adapted from [6,7].

To mitigate microcracking in concrete, an addition of fly ash, an application of the blast-furnace slag cement or low density aggregate is good practice, as demonstrated in [8]. However, in the case of normal-weight concrete, under the influence of destructive external factors, such as high temperature, frost, and loading, these microcracks can develop into cracks, thereby reducing structural durability and serviceability and in rare cases lead to failures, e.g., walls in tanks [9], concrete slabs [10,11], precast elements [12], or other structural elements [13].

Objective assessment of damage formation and development in concrete, which is independent of the components, additives, and external impacts is essential.

Various non-destructive methods have been used for this purpose [14–16]. Acoustic emission (AE) is the technique capable of detecting, classifying [17], and locating [18,19] damage in concrete. Traditional use of acoustic emission methods in the building industry includes the monitoring of damage [20,21] and crack development under load [22–26], cement setting and curing [27–32], or an assessment of the ASR (alkali–silica reaction) in concrete [33,34].

The research analyzing stress in concrete is especially related to the acoustic emission phenomenon. The acoustic emission method enables the determination of basic parameters of fracture mechanics necessary to analyze the course of stress affecting concrete destruction. Depending on the grade of concrete tested, the criteria for the estimation of the level of stress were established [35,36]. Another approach aimed at estimating the correlation between acoustic emission and stress in compressed concrete is the technique that relies on the Gutenberg–Richter (GBR) law [37]. It was observed that the event frequency in concrete samples during compression corresponds to about 70% of the maximum stress.

Progressive damage of structural elements (Reinforced Concrete beams—RC beams) under bending is assessed using the Keiser effect by monitoring AE activity during cyclic loading [38].

The Keiser effect is used to estimate the stress to which the structural element was previously exposed. To estimate the Kaiser effect (according to NDIS-2421 by JSNDI—the Japanese Society for Non-Destructive Inspection) two ratios are calculated: the load ratio and calm ratio. Their values are the basis for damage qualification as intermediate, minor, or heavy. Digital image correlation (DIC) techniques supporting the AE method are applied for providing information about the level of damage in the RC beams [39]. The relaxation ratio may also be a good indicator of damage status. During initial stages of loading the deflections increase slightly (loading phase). The element is in a serviceable state up to 50% of the deflection limit. In a higher deflection range (50–85%) the structural element is no longer serviceable. Deflection higher than 85% represents the failure of the element. The acoustic emission method is also used for the observation of the crack mouth opening displacement (CMOD). The different nature of dissipated and emitted energy rates was observed in [38] during the loading process.

The methods performed on concrete subjected to compression and bending do not consider an influence of internal stress on concrete strength. At the initial stage of concrete setting, the cement paste shrinks and meets the resistance of aggregate grains that do not shrink. A self-balancing state of compressive and tensile stress arises. If the tensile stresses in the cement paste exceed the tensile strength, microdefects occur. These defects may form in the matrix and in the interfacial transition zone (ITZ) around the aggregate [4] (first destructive process). Internal microcracks interact with each other; they can join together in a damage network. This happens when the structure surrounding the internal microcracks in the cement paste is not able to transfer accumulated stresses. This is when the second destructive process arises. Furthermore, the heterogeneous increase in temperature in the cross-section of the element, as well as water evaporation from the surface layers causes the stretching in the outer zones and compression of the inner zones of the element. These non-stationary and non-linear temperature and humidity areas generate macrostress in the cross-section [36] that can lead to microcracks on the concrete surface (third destructive process) and then their propagation (fourth destructive process). These destructive processes result in discontinuities in the structure. Local structural defects initiate future destruction of the concrete and may reduce the strength of elements, causing their linear deformation and affecting serviceability functions [4,36,40].

There is no information about the assessment procedure of fresh concrete quality by acoustic emission before loading. In most cases analysis of non-loaded concrete is based on ring-down counting, which involves counting how many times the amplitude passes the fixed threshold or event-counting corresponding to number of AE waves recorded by a single sensor [41]. In these cases, acoustic emission signals the damage (crack formation) without being able to identify the underlying processes. Only some of the AE techniques, such as the methods described in [19,23,24], allow for effective identification and location of the destructive process.

The non-invasive acoustic emission method (modified IADP method—Identification of Active Destructive Processes method) presented in [42–44] has been shown to be suitable for investigating defect formation process at the early stage of hardening of young concrete.

The study presented in this paper demonstrates that this method is of a general nature and allows observation and identification of destruction processes regardless of the aggregate used, cement types, admixtures added, hardening conditions, temperature, or the presence of reinforcement. It also enables quantitative assessment of destructive processes, which can be important when assessing the strength properties of concrete.

The method can thus be applied to diagnosing elements made of reinforced concrete, controlling the concrete hardening stage, and supporting decision making (e.g., related to demolding), thereby ensuring the reliability of the structure.

2. Materials and Methods

A total of 30 samples (ten concrete series, W2, W3, W4, W5, W6, W7, W8, B2, B3, and B4, of three samples each—A, B, and C) were tested.

Twenty-one samples (W2–W8) were made of C30/37 concrete, six samples (B2, B3) were made of C40/50 concrete and three (W2) of C25/30 concrete. Except for sample W2 (100 mm × 100 mm × 500 mm), all samples had square cross-sections with 150 mm on each side and the length of 600 mm. Samples B2 were made with chemical admixtures (plasticizer and air entraining agent), other samples without admixtures. Samples denoted by "W" were made with limestone aggregate from the Trzuskawica quarry, while these marked with "B" with basalt aggregate from Górażdże quarry. All samples were made with cement CEMI 42,5N—MSR/NA from the Warta cement plant (Cementownia Warta S.A., Trębaczew, Poland) (except B4—CEMIII/A 42,5N—LH/HSR/NA from the cement plant in Małogoszcz, (Cementownia Lafarge Małogoszcz, Poland). The chemical compositions of the cements are compiled in Table 1. Mixture proportions of samples W2–W8 and B2–B4 are listed in Table 2.

Table 1. Cement composition (%).

Cement	CaO	MgO	SiO_2	Al_2O_3	Fe_2O_3	SO_3	Na_2O_{eq}	Cl^-
CEM I	66.03	0.79	21.23	3.66	3.21	2.63	0.43	0.076
CEM III						2.69	0.81	0.066

Table 2. Mixture proportions (kg/m^3).

Symbol	Aggregate 2–16	Sand 0–2	CEMI/CEMIII	Water	Air Entraining Agent	Plasticizer
W3–W8	1110 [1]	740	338 (CEMI)	169	0	0
W2	1073 [1]	777	290 (CEMI)	188	0	0
B2	1312 [2]	691	360 (CEMI)	150	0.36	1.98
B3, B4	1312 [2]	691	360 (CEMIII)	180	0	0

[1] limestone aggregate, [2] basalt aggregate.

Three samples B4 were made with basalt aggregate, blast furnace slag, and cement CEM III without any additions.

The W3 and W4 samples after fabrication were cured in water for 10 days and then tested for 58 days under cyclic temperature variations (Figure 2). Additionally, steel reinforcement was embedded in the W4 samples (Figure 3a).

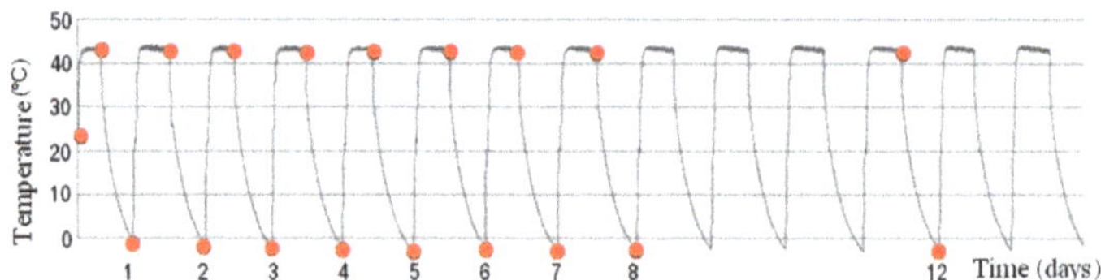

Figure 2. Temperature conditions during the first 14-days of test.

The samples in series W7 were tested for 58 days without water curing at a constant temperature (+22 ± 2 °C).

Before the test, the AE (Acoustic Emission) sensors were attached to one side of each sample (Figure 3b,c).

To provide appropriate conditions, the test stand was developed, comprising of a thermally and acoustically insulated chamber. A list of samples examined is shown in Table 3.

AE signals were recorded for 58 days in 12-h stages.

The proposed identification of active damage processes (IADP) method was presented in [19,21,24] and applied for damage identification and location in reinforced concrete beams under loading [22]. It relies on the study of AE signals produced by the process causing the deterioration of strength properties in structural elements. The results recorded in samples (AE signals) were compared with the reference signals obtained in the laboratory.

Then the modified version of this method was applied to detect damage in young concrete [23,42].

The IADP method outline is shown in Figure 4. This concept is based on the comparative analysis of waves generated by defects in concrete (detected by sensors) with a database of reference signals created earlier. Preamplifiers with a gain of 35 dB were used to amplify signals generated by defects. Then the signals were detected, transformed into electric signals, measured, recorded, analyzed, and assigned to the reference signals in the database using Noesis software and unsupervised learning methods.

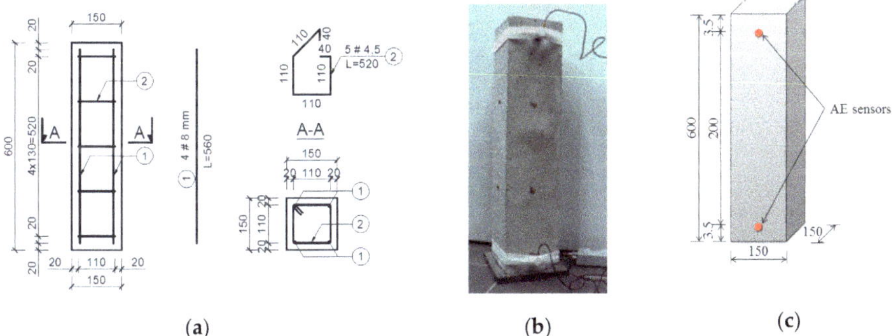

(a)   (b)   (c)

Figure 3. (a) Schematic diagram of reinforcement of W4 samples; (b) sample during the test; and (c) acoustic emission (AE) sensor arrangement (unit: mm).

Figure 3 shows the AE sensors arrangement on the test sample. Two piezoelectric sensors with a gain of 25–80 kHz allow not only detection of destructive processes (AE source) but also finding their linear location.

Table 3. Sample parameters and testing conditions.

Series	Water Curing (days)	Hardening Temperature	Cement Type	Concrete Class
W2	10	constant 22 ± 2 °C	CEMI	C25/30
W3	10	varied −5 to + 42 °C	CEMI	C30/37
W4 (reinforced)	10	varied −5 to + 42 °C	CEMI	C30/37
W5	10	constant 22 ± 2 °C	CEMI	C30/37
W6 (100 × 100 × 500)	10	constant 22 ± 2 °C	CEMI	C30/37
W7 (without curing)	none	constant 22 ± 2 °C	CEMI	C30/37
W8 (without curing)	none	varied −5 to +42 °C	CEMI	C30/37
B2 (with admixtures)	10	constant 22 ± 2° C	CEMI	C40/50
B3 different agregate type)	10	constant 22 ± 2 °C	CEMI	C40/50
B4	10	constant 22 ± 2 °C	CEM III	C30/37

The preliminary reference signal database was developed based on 12 parameters of the AE signal: counts, counts to the peak, amplitude signal duration, signal rise time, signal amplitude, signal energy, signal strength average, effective voltage, absolute energy, average frequency, reverberation frequency, and initiation frequency. There are four destructive processes, described in [2–4,7], which may be a source of AE in freshly made concrete before loading. In [42,43,45] damage processes were ascribed to four signal classes recorded in non-loaded concrete (Table 4).

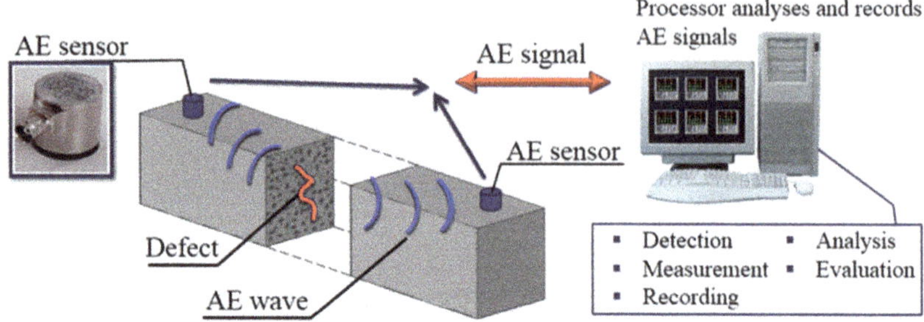

Figure 4. The concept of the method—IADP (Identification of Active Destructive Processes).

Table 4. Destructive processes by the IADP (Identification of Active Destructive Processes) method.

Denotation	AE (Acoustic Emission) Signal Class	Number of Destructive Process	The Source of the Destructive Process
●	Class 1	I	formation of internal microcracks
●	Class 2	II	propagation of internal microcracks
●	Class 3	III	formation of surface microcracks
●	Class 4	IV	propagation of surface cracks

3. Results

3.1. Test Results

The method of assessing the quality of early-age concrete must enable identification of internal defects, regardless of the components used for its manufacture or the conditions under which the structure of hardened concrete forms. The potential of the IADP method was analyzed in this context. Damage processes were identified based on the assessment of signal classes recorded during the test.

The AE signals were recorded for 12 h on days: 1–8, 12, 16, 20, 24, 28, 38, 46, and 57 using MISTRAS software. Then the proposed IADP method was used to analyze the signals (hits). The signals from the tests were compared by 12 AE parameters with signals from the database and assigned to particular classes. The reference database was first developed using K-means clustering and then verified in [43].

An analysis was performed on averaged results of the number of recorded signals (hits) with respect to destructive processes assigned to them. Concrete of each series was investigated by 6 sensors (two sensors were attached to one side of each of the three samples (A, B, and C) in a given series). All signals recorded by these 6 sensors capturing the processes were averaged for each series. The number of signals recorded on average by one sensor was analyzed.

The development of damage processes in series W3, W4, W6, and W7 and B2, B3, and B4 is shown in Figures 5, 7, 9, 10, 12, and 13. Corresponding images of side surface cracks (for samples A, B, and C) are presented in Figures 6, 8, and 11, except W6 and B2-B4 samples, because no cracks were observed on their surfaces.

3.1.1. Concrete W3—Curing in Water, Variable Hardening Temperature (+42 to −5 °C)

Figure 5 shows the values of destructive processes I–III captured during 56 days in samples W3. Throughout the test, 5195 signals (hits) were assigned to the initiation of internal microcracks (damage process I) and 94 AE signals were assigned to damage process II.

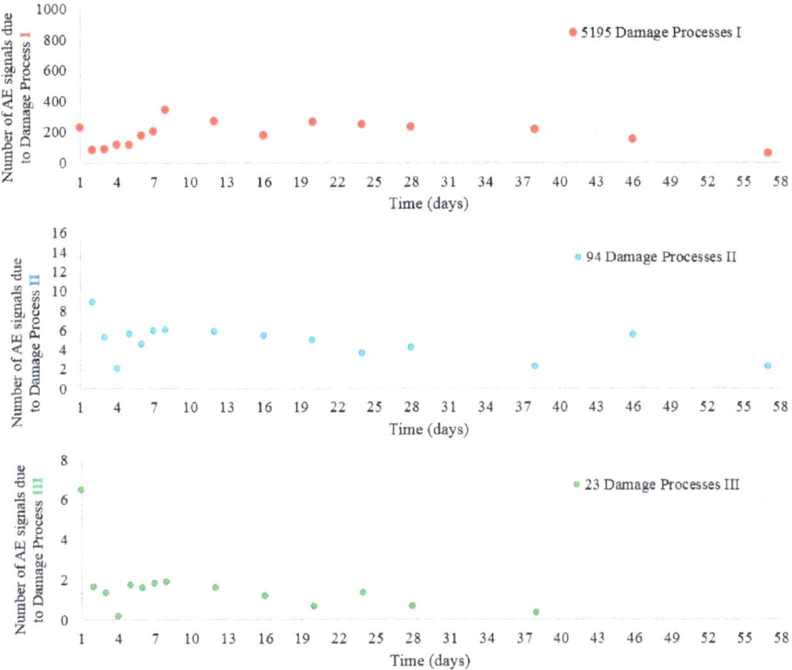

Figure 5. Number of damage processes I–III obtained from W3 samples.

Twenty-three hits assigned to surface microcrack formation (destructive process III) were captured in samples A, B, and C (Figure 6).

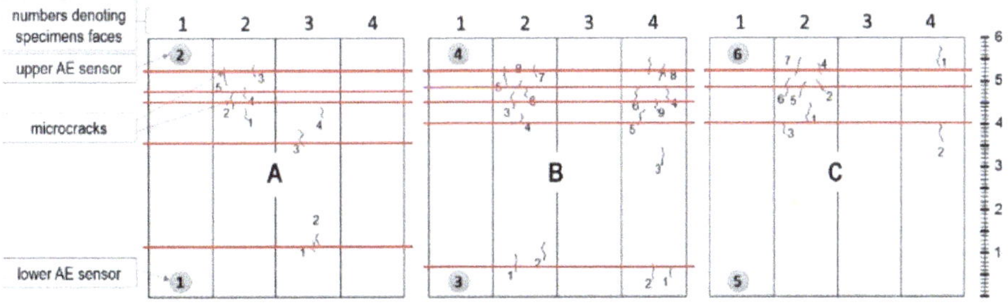

Figure 6. Surface microcrack distribution on sides of the W3 samples. Linear location of destructive processes III (AE signal class 3) in W3 samples obtained by the AE (Acoustic Emission) method is marked in red.

3.1.2. Reinforced Concrete W4—Curing in Water, Variable Hardening Temperature (+42 to −5 °C)

Figure 7 shows the values of destructive processes I–III captured during 56 days in samples W4. Most processes I and II were recorded during 20 days, later their number decreased. Throughout the test, 8608 hits were assigned to the initiation of internal microcracks (damage process I) and 90 AE signals were assigned to damage process II.

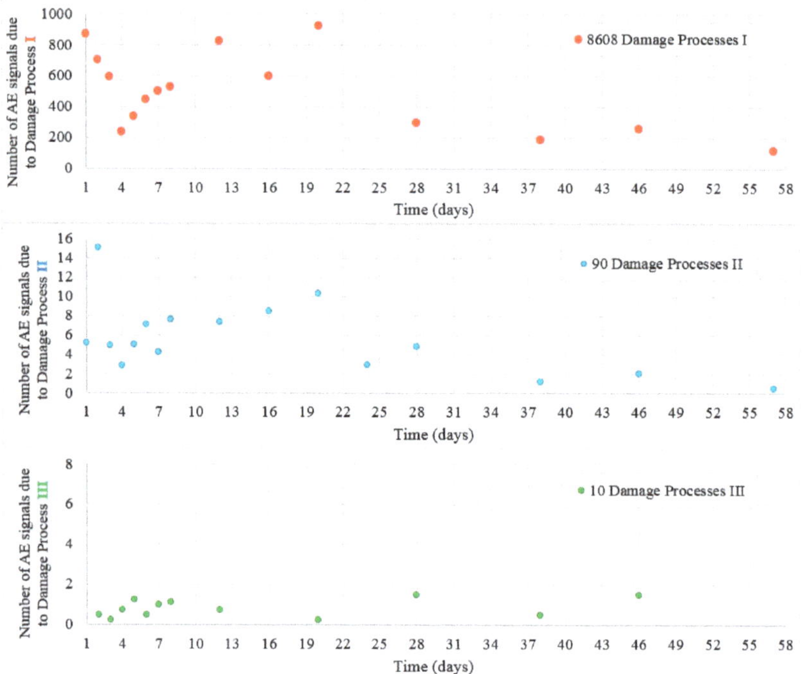

Figure 7. Number of damage processes I–III obtained from W4 samples.

Ten signals assigned to destructive process III (surface microcracks formation) were detected in samples A, B, and C (Figure 8) until day 46.

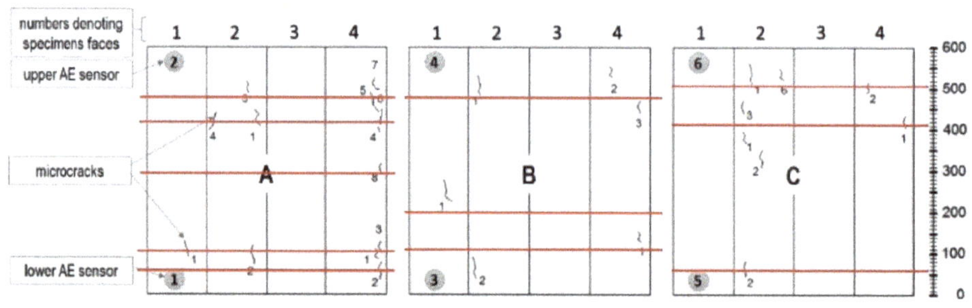

Figure 8. Surface microcracks distribution on sides of the W4 samples. Linear location of destructive processes III (AE signal class 3) in W4 samples obtained by the AE method is marked in red.

3.1.3. Concrete W6—Curing in Water, Constant Hardening Temperature of 22 °C

Figure 9 shows the values of destructive processes I and II recorded during 56 days in samples W6. Most processes were recorded during the first week of the test, then their number decreased. Throughout the test, 2560 hits were assigned to the initiation of internal microcracks (damage process I) and 94 AE signals were assigned to damage process II but not on every measuring day.

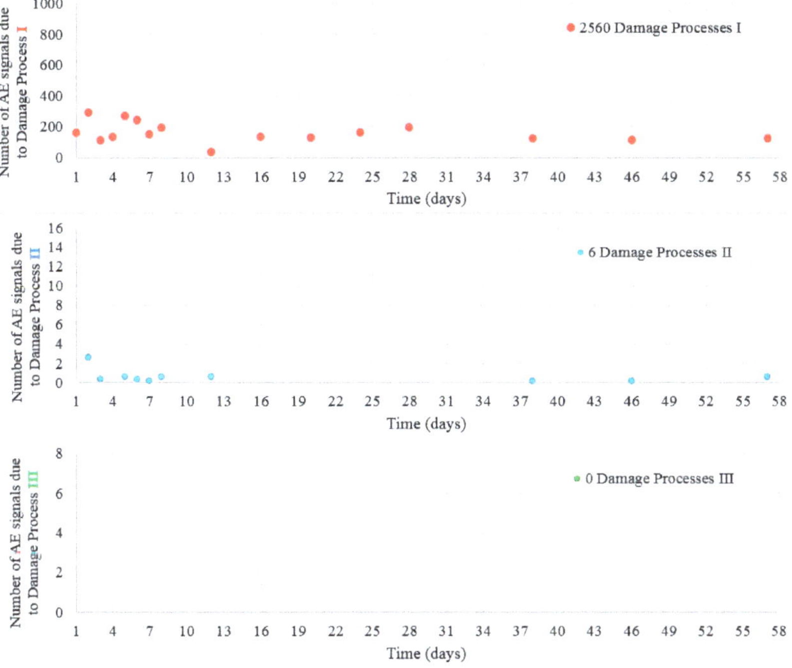

Figure 9. Number of damage processes I–III obtained from W6 samples.

Class III signals were not captured in the test and no microcracks at the surface of the samples were observed.

3.1.4. Concrete W7—Curing in Water, Constant Hardening Temperature of 22 °C

Figure 10 shows the values of destructive processes I-III captured during 56 days in samples W7. Most processes denoted as I were recorded during 20 days, later their number decreased. Throughout the test, 2795 hits were assigned to the initiation of internal microcracks (damage process I) and six AE signals were assigned to damage process II until day 24.

A few signals Class III assigned to the formation of surface microcracks were recorded in W7 samples. A single surface microcracks were detected on the sides of the samples (Figure 11).

3.1.5. Concrete B2 (with Admixtures)—Curing in Water, Constant Hardening Temperature of 22 °C

Figure 12 shows the values of destructive processes I and II recorded during 56 days in samples B2. Most processes were recorded during the initial days of the test, then their number decreased. Throughout the test, 4519 hits assigned to the initiation of internal microcracks (damage process I) were recorded together with 52 AE signals assigned to damage process II, which practically faded out after day 24 of the test.

Class III signals were not captured in the test and no microcracks at the surface of the samples were observed.

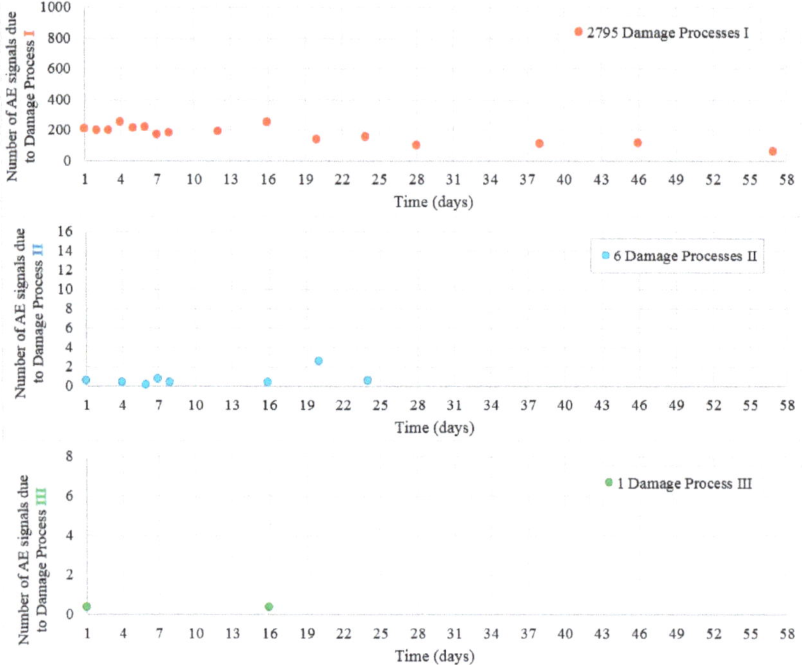

Figure 10. Number of damage processes I–III obtained from W7 samples.

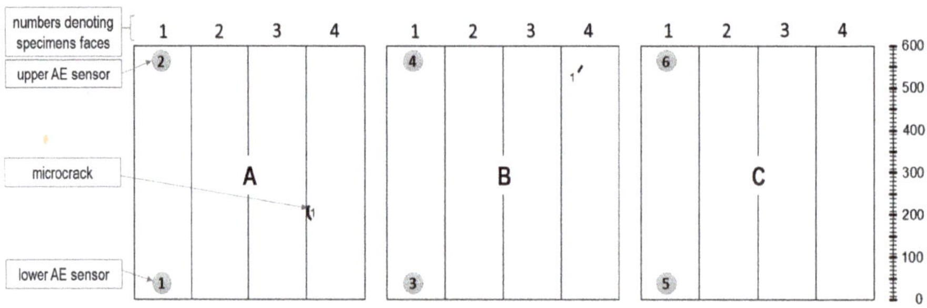

Figure 11. Surface microcracks distribution on sides of the W7 samples.

3.1.6. Concrete B4—Curing in Water, Constant Hardening Temperature of 22 °C

Figure 13 shows the values of damage processes I and II captured during 56 days in samples B4. Throughout the test, 2886 hits assigned to the initiation of internal microcracks (damage process I) and 19 AE signals assigned to damage process II were recorded.

Class III signals were not captured in the test and no microcracks at the surface of the samples were observed.

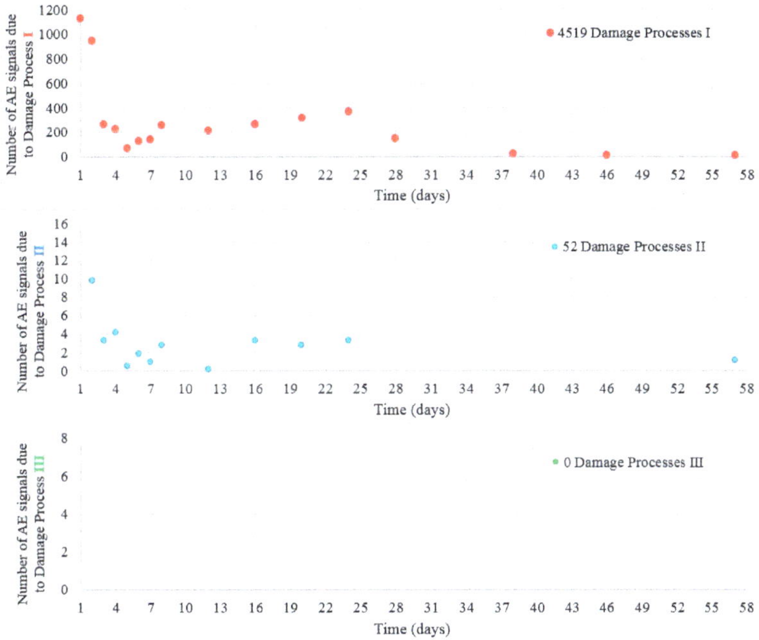

Figure 12. Number of damage processes I–III obtained from B2 samples.

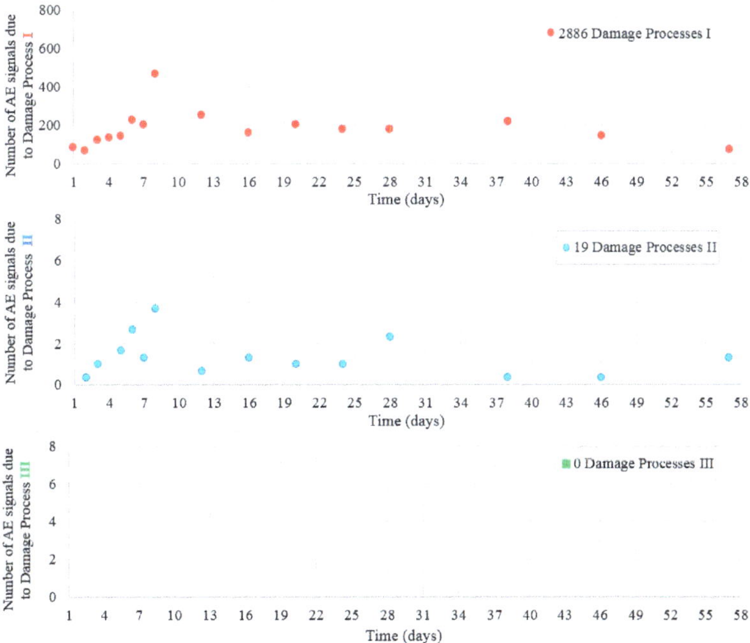

Figure 13. Number of damage processes I–III obtained from B4 samples.

3.2. Destructive Processes Analysis

The analysis of the destructive processes confirmed that the internal structure of the concrete was affected by a range of factors such as aggregate and cement type, curing conditions (especially moisture and ambient temperature) that influence cement hydration, dimensions of the tested element as well as reinforcing bars embedded in the samples.

The identified processes, number of signals and curing conditions are shown in Table 1.

Analysis of the results shows that Class 1 signals were recorded most often in the test. These signals correspond to internal microcrack formation. Most damage processes I were observed during the first week. Their number decreased over time but they did not fade out during 56 days.

The number of Class 2 signals assigned to damage process II (internal microcracks development) was almost an order of magnitude smaller.

Class 3 signals were not recorded in (W2, W6, and B2–B4) concrete samples cured after demolding during 10 days and then hardened at constant temperature (22 ± 2 °C). This indicates that destructive processes III (surface microcrack formation) did not occur, which was confirmed by the observation of the sample.

In addition to tracking the growth of individual destructive processes in time, the number of destructive processes recorded at any given time interval can be also analyzed. Several analyses based on this precise information about damage and damage development in early age concrete were performed.

The results selected for W3 and W4 concrete samples (Figure 14) show that in the case when reinforcement was used, the number of damage processes I (internal microcrack formation) increased (in the analyzed concrete samples by about 65%) due to the occurrence of additional interfacial transition zones (ITZ) between reinforcing bars and cement paste. Damage processes II (propagation of internal microcracks) in the analyzed samples slightly decreased and damage processes III (formation of microcracks on the surface of concrete) were limited by embedded reinforcement.

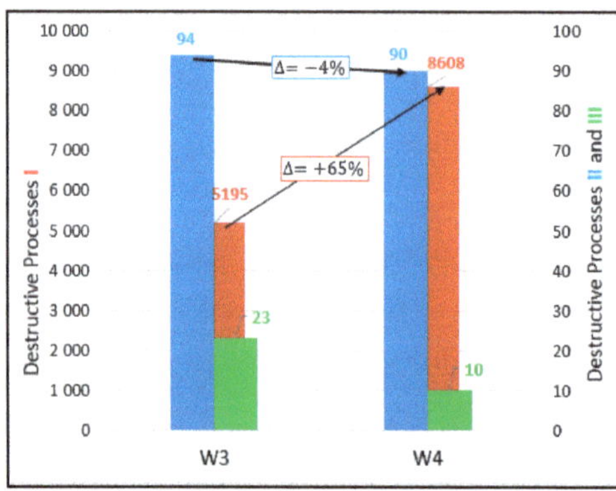

Figure 14. Number of hits accompanying processes I–III recorded in W3 and W4 samples (10 days curing, varied temperature, and with and without reinforcement).

In Table 5 shown the testing conditions of the samples with the results of number of AE signals and destructive processes in non-loaded concrete obtained by modified IADP acoustic emission method.

Figure 15 shows an influence of variable temperature on the number of damage processes in concrete hardening without initial curing. In the samples hardened at variable temperatures (−5 to +42 °C), far more destructive processes (I–III) were recorded compared to constant temperature conditions. This indicates a significant impact of the heating and cooling cycles on damage processes development in the early-age concrete, which may influence the strength of hardened concrete.

Table 5. Sample parameters, testing conditions, results, and detected damage.

Series	Curing Conditions (Days of Curing)	Hardening Temperature	Cement Type	Concrete Strength after 28 Days (MPa)	Process	No of Signals
W2	10	constant 22 ± 2 °C	CEMI	36.0	I	5980
					II	31
					III	0
W3	10	variable −5 to +42 °C	CEMI	44.5	I	5195
					II	94
					III	23
W4 (reinforced)	10	variable −5 to +42 °C	CEMI	44.5	I	8608
					II	90
					III	10
W5 [1]	10	constant 22 ± 2 °C	CEMI	40.1	I	2192
					II	9
					III	0
W6 (100 × 100 × 500)	10	constant 22 ± 2 °C	CEMI	40.1	I	2560
					II	6
					III	0
W7 (without curing)	none	constant 22 ± 2°C	CEMI	45.8	I	2795
					II	6
					III	1
W8 [1] (without curing)	none	variable −5 to +42°C	CEMI	41.2	I	5200
					II	42
					III	7
B2 [2] (with admixtures)	10	constant 22 ± 2 °C	CEMI	63.5	I	4519
					II	52
					III	0
B3 [2] different agregate type)	10	constant 22 ± 2 °C	CEMI	55.8	I	2984
					II	18
					III	0
B4	10	constant 22 ± 2 °C	CEM III	48.1	I	2886
					II	19

[1] The tests and analysis of W5 and W8 samples results were described in [45], [2] the test results of the B2 and B3 samples were described in [44].

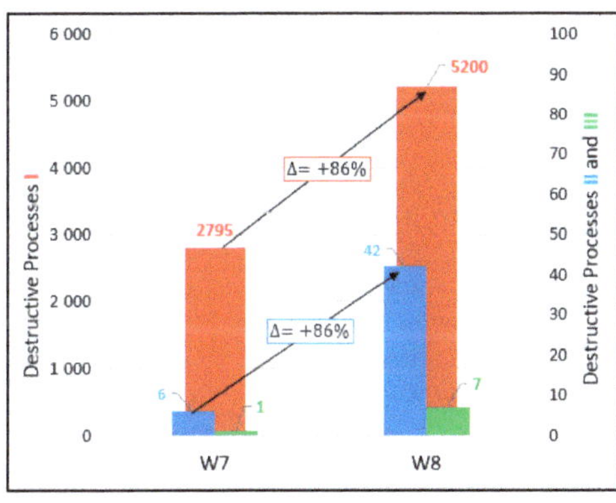

Figure 15. Number of hits accompanying processes I–III recorded in W7 and W8 samples (without curing and constant and varied temperature).

4. Discussion

The presented results for concrete tested under different curing conditions, hardening temperature, aggregate type, concrete strength, dimensions of the sample, presence of the admixtures, and presence of reinforcement show that the proposed AE method is the general method that allows for early age damage identification, damage tracking, and location.

In each of the analyzed cases, it was possible to select damage classes. The emergence of class 1 and class 2 signals does not represent direct effects on the strength level.

Evaluation of concrete strength based on the presented AE method has a qualitative nature. Therefore, it seems expedient to look for a correlation between the intensity of destructive processes and the strength of the concrete obtained.

The data obtained can also be used in several analyses of practical significance such as:

- In the case of hardening at varied temperature, class II damage (internal microcrack development) increases but class III (surface microcracks) decreases in reinforced concrete, which confirms that the reinforcement restricts most dangerous class III damage (Figure 14),
- All damage processes increase in the case of hardening at variable temperature (Figure 15).

Author Contributions: Conceptualization, B.G. and W.T.; Methodology, B.G. and W.T.; Software, M.B.; Validation, B.G., W.T. and M.B.; Formal Analysis, B.G.; Investigation, M.B.; Resources, B.G.; Data Curation, M.B.; Writing—Original Draft Preparation, W.T.; Writing—Review and Editing, W.T., B.G. and M.B.; Visualization, M.B.; Supervision, W.T.; Project Administration, B.G.; Funding Acquisition, B.G. and W.T. All authors have read and agreed to the published version of the manuscript.

Funding: This experiments were funded by Project POIG No.01.01.02-10-106/09-01

Acknowledgments: Research was carried out as part of the Project POIG No.01.01.02-10-106/09-01

Conflicts of Interest: The authors declare no conflict of interest.

References

1. Wu, Z.; Wong, H.S.; Buenfeld, N.R. Transport properties of concrete after drying-wetting regimes to elucidate the effects of moisture content hysteresis and microcracking. *Cem. Concr. Res.* **2017**, *98*, 136–154. [CrossRef]
2. Kurdowski, W. *Cement and Concrete Chemistry*; Springer: London, UK, 2014; pp. 1–677. ISBN 978-94-007-7944-0.
3. Neville, A.M. *Properties of Concrete*, 5th ed.; Prentice Hall: Lincoln, UK, 2011; pp. 1–872. ISBN 978-0-273-75580-7.
4. Bischop, J.; van Mier, J.G.M. How to study drying shrinkage microcracking in cement-based materials using optical and scanning electron microscopy? *Cem. Concr. Res.* **2002**, 279–287. [CrossRef]
5. Boshoff, W.; Combrinck, R. Modelling the severity of plastic shrinkage cracking in concrete. *Cem. Concr. Res.* **2013**, *48*, 34–39. [CrossRef]
6. Flaga, K. Impact of residual stress on stress destruction and strength parameters of concrete [Wpływ naprężeń własnych na destrukcję naprężeniową i parametry wytrzymałościowe betonu—in Polish]. *Inżynieria Bud.* **1995**, *6*, 315–322.
7. Golewski, G.; Sadowski, T. Analysis of brittle defects in concrete composites [Analiza kruchych uszkodzeń w kompozytach betonowych in Polish]. *Czas. Tech. (Bud.)* **2007**, *104*, 55–62.
8. Schindler, A.; Byard, B.; Tankasala, A. Mitigation of early-age cracking in concrete structures. *MATEC Web Conf.* **2019**, *284*, 07005. [CrossRef]
9. Flaga, K.; Furtak, K. Problem of thermal and shrinkage cracking in tanks vertical walls and retaining walls near their contact with solid foundation slabs. *Archit. Civil Eng. Environ.* **2009**, *2*, 23–30.
10. Linek, M.; Nita, P. Thermal Stress in Concrete Slab of the Airfield Pavement. *J. KONBiN* **2020**, *50*. [CrossRef]
11. Kwak, H.G.; Ha, S.J. Plastic shrinkage cracking in concrete slabs. *Part I Mag. Concr. Res.* **2006**, *58*, 505–516. [CrossRef]
12. Raczkiewicz, W.; Bacharz, K.; Bacharz, M.; Grzmil, W. Manufacturing Errors of Concrete Cover as a Reason of Reinforcement Corrosion in Precast Element—Case Study. *Coatings* **2019**, *9*, 702. [CrossRef]
13. Klemczak, B.; Knoppik-Wróbel, A. Early age thermal and shrinkage cracks in concrete structures—description of the problem. *Archit. Civil Eng. Environ.* **2011**, *2*, 35–47.

14. Ranachowski, Z. Application of acoustic emission to fault diagnosis in civil engineering. *Roads Bridges* **2012**, *2*, 65–87.

15. Ohtsu, M.; Isoda, T.; Tomoda, Y. Acoustic Emission Techniques Standardized for Concrete Structures. *J. Acoust. Emiss.* **2007**, *25*, 21–32.

16. Hoła, J.; Schabowicz, K. State-of-the-art. Non-destructive methods for diagnostics testing of building structures anticipated development trends. *Arch. Civil Mech. Eng.* **2010**, *10*, 5–18. [CrossRef]

17. Ohno, K.; Ohtsu, M. Crack classification in concrete based on acoustic emission. *Constr. Build. Mater.* **2010**, *24*, 2339–2346. [CrossRef]

18. Ohtsu, M.; Shigeishi, M.; Iwase, H.; Kyotanagit, W. Determination of crack location type and orientation in a concrete structures by acoustic emission. *Mag. Concr. Res.* **1991**, *43*, 127–134. [CrossRef]

19. Goszczyńska, B.; Świt, G.; Trąmpczyński, W.; Krampikowska, A.; Tworzewska, J.; Tworzewski, P. Experimental Validation of Concrete Crack Identification and Location with Acoustic Emission Method. *Arch. Civil Mech. Eng.* **2012**, *12*, 23–28. [CrossRef]

20. Nair, A.; Cai, C.S. Acoustic emission monitoring of bridges Review and case studies. *Eng. Struct.* **2010**, *32*, 1704–1714. [CrossRef]

21. Goszczyńska, B.; Świt, G.; Trąmpczyński, W. Monitoring of active destructive processes as a diagnostic tool for the structure technical state evaluation. *Bull. Polish Acad. Sci. Tech. Sci.* **2013**, *6*, 97–109. [CrossRef]

22. Goszczyńska, B.; Świt, G.; Trąmpczyński, W. Analysis of the microcracking process with the Acoustic Emission method with respect to the service life of reinforced concrete structures with the example of the RC beams. *Bull. Polish Acad. Sci. Tech. Sci.* **2015**, *63*, 55–63. [CrossRef]

23. Goszczyńska, B. Analysis of the process of crack initiation and evolution in concrete with acoustic emission testing. *Arch. Civil Mech. Eng.* **2014**, *14*, 134–143. [CrossRef]

24. Świt, G. *Predicting Failure Processes for Bridge—Type Structures made of Prestressed Concrete Beams Using the Acoustic Emission Method [Analiza Procesów Destrukcyjnych w Obiektach Mostowych z Belek Strunobetonowych z Wykorzystaniem Zjawiska Emisji Akustycznej in Polish]*; Kielce University of Technology: Kielce, Poland, 2011; pp. 1–179.

25. Karcili, M.; Alver, N.; Ohtsu, M. Application od AE rate-process analysis to damaged concrete structures due to earthquake. *Mater. Struct.* **2016**, *49*, 2117–2178. [CrossRef]

26. Wu, K.; Chen, B.; Yao, W. Study on the AE characteristics of fracture process of mortar concrete and steel-fiber-reinforced concrete beams. *Cem. Concr. Res.* **2000**, *30*, 1495–1500. [CrossRef]

27. Lura, P.; Couch, J.; Jensen, O.M.; Weiss, J. Early-age Acoustic emission measurements in hydrating cement paste: Evidence for cavitation during solidification due to self-desiccation. *Cem. Concr. Res.* **2009**, *39*, 426–432. [CrossRef]

28. Van, D.; Abeele, K.; Desadeleer, W.; De Schutter, G.; Wevers, M. Active and passive monitoring of the early hydration process in concrete using linear and non linear acoustics. *Cem. Concr. Res.* **2009**, *39*, 426–432.

29. Watanabe, T.; Ohno, T.; Hashimoto, C. Detection of Thermal Cracks in Early-Age Concrete by AE. *Concr. Res. Lett.* **2011**, *2*, 290–294.

30. Iliopoulos, S.N.; Khattabi, Y.E.; Angelis, D.G. Towards the establishment of a continuous nondestructive monitoring technique for fresh concrete. *J. Nondestruct. Eval.* **2016**, *35*, 37. [CrossRef]

31. Bischop, J.; van Mier, J.G.M. Drying shrinkage microcracking in cement-based materials. *Heron* **2002**, *47*, 163–184.

32. Chotard, T.J.; Smith, A.; Rotureau, D.; Fargot, D.; Gault, C. Acoustic emission characterization of calcium aluminate cement hydration at an early stage. *J. Eur. Ceram. Soc.* **2003**, *23*, 387–398. [CrossRef]

33. Sargolzahi, M.; Kodjo, S.A.; Rivard, P.; Rhazi, J. Effectiveness of nondestructive testing for the evaluation of alkali–silica reaction in concrete. *Constr. Build. Mater.* **2010**, *24*, 1398–1403. [CrossRef]

34. Świt, G.; Zapała-Sławeta, J. Application of acoustic emission to monitoring the course alkali-silica reaction. *Bull. Polish Acad. Sci. Tech. Sci.* **2020**, *68*, 169–178.

35. Newman, K.; Newman, I.B. Failure theories and design criteria for plain concrete. In *Proceedings of the Civil Engineering Materials Conference*; Te'eni, M., Ed.; Wiley-Interscience: London, UK, 1969; Part 2, pp. 963–996.

36. Hoła, J. Determination of initiating and critical stress levels in compressed plain and high-strength concrete by Acoustic Emission. *Arch. Acoust.* **2000**, *25*, 57–65.

37. Carni, D.L.; Scuro, C.; Lamonaca, F.; Sante Olivito, R. Damage analysis of concrete structures by means of acoustic emissions technique. *Compos. Part B* **2017**, *115*, 79–86. [CrossRef]

38. Sagar, R.V.; Prasad, B.K.R. Damage limit states of reinforced concrete beams subjected to incremental cyclic loading using ratio analysis of AE parameters. *Constr. Build. Mater.* **2012**, *35*, 139–148. [CrossRef]

39. Carpinteri, A.; Lacidogna, G.; Corrado, M.; Di Battista, E. Cracking and crackling in concrete-like materials: A dynamic energy balance. *Eng. Fract. Mech.* **2016**, *155*, 130–144. [CrossRef]

40. Kiernożycki, W. *Massive Caoncrete Structures [Betonowe Konstrukcje Masywne in Polish]*; Polski Cement: Cracow, Poland, 2003.

41. Miller, R.K. Acoustic Emission Testing. In *Nondestructive Testing Handbook*, 3rd ed.; Moore, P.O., Hill, E.v.K., Eds.; American Society for Nondestructive Testing: Columbus, OH, USA, 2005; Volume 6, pp. 1–446.

42. Bacharz, M. *Application of the Acoustic Emission Method to the Evaluation of Destructive Processes in Unloaded Concrete [Wykorzystanie Metody Emisji Akustycznej do Badania Procesów Destrukcyjnych w Betonie Nieobciążonym in Polish]*; Kielce University of Technology: Kielce, Poland, 2016.

43. Bacharz, M.; Trąmpczyński, W.; Goszczyńska, B. Acoustic emission method as a tool for the assessment of influence of the temperature variation on destructive processes created in early age concrete. *MATEC Web Conf.* **2019**, *284*, 1–12. [CrossRef]

44. Bacharz, M.; Goszczyńska, B.; Trąmpczyński, W. Analysis of destructive processes in non-loaded early-age concrete with acoustic emission method. *Procedia Eng.* **2015**, *108*, 245–253. [CrossRef]

45. Bacharz, M.; Trąmpczyński, W. Identification of active destructive processes in unloaded early-age concrete with the use of Acoustic Emission method. In Proceedings of the 2016 Prognostics and System Health Management Conference (PHM-Chengdu), Chengdu, China, 19–21 October 2016; pp. 1–6.

Article

The Use of Wavelet Analysis to Improve the Accuracy of Pavement Layer Thickness Estimation Based on Amplitudes of Electromagnetic Waves

Małgorzata Wutke [1],*, Anna Lejzerowicz [2] and Andrzej Garbacz [2]

[1] TPA Sp. z o.o., Parzniewska 8, 05-800 Pruszków, Poland
[2] Faculty of Civil Engineering, Warsaw University of Technology, Al. Armii Ludowej 16, 00-637 Warsaw, Poland; a.lejzerowicz@il.pw.edu.pl (A.L.); a.garbacz@il.pw.edu.pl (A.G.)
* Correspondence: malgorzata.wutke@tpaqi.com

Received: 17 June 2020; Accepted: 17 July 2020; Published: 19 July 2020

Abstract: The article discusses one of the methods of dielectric constant determination in a continuous way, which is the determination of its value based on the amplitude of the wave reflected from the surface. Based on tests performed on model asphalt slabs, it was presented how the value of the dielectric constant changes depending on the atmospheric conditions of the measured surface (dry, covered with water film, covered with ice, covered with snow, covered with de-icing salt). Coefficients correcting dielectric constants of hot mix asphalt (HMA) determined in various surface atmospheric conditions were introduced. It was proposed to determine the atmospheric conditions of the pavement with the use of wavelet analysis in order to choose the proper dielectric constant correction coefficient and therefore improve the accuracy of the pavement layer thickness estimation based on the ground penetrating radar (GPR) method.

Keywords: ground penetrating radar (GPR); HMA dielectric constant; wavelet analysis; road pavement thickness estimation

---

1. Introduction

The requirements for pavements, not only newly built, but also existing, maintained, and undergoing renovation are increasing along with new guidelines [1]. The proper recognition of the thickness of pavements layers [2,3] directly influences the quality assurance of assessing the current pavement load capacity and the correctness of the pavement repair technology being developed.

The measurement of road pavement thickness can be performed using non-destructive methods such as the ground penetrating radar (GPR) method [4–9], eddy currents (StratoTest device) [10], Impact Echo [11–14], and ultrasonic [15] or semi-destructive [16] method, which is drilling. However, the eddy current method is only suitable for measuring the thickness of new surfaces, since it requires placing an aluminum reflector on the bottom of the layer [10]. The Impact Echo method is used only to measure thick elements, and repaired surfaces often have a thickness smaller than 10 cm [12]. While the Impact Echo device has been developed that has many probes that can be implemented to study large areas [14], the device based on ultrasound allows only point measurements [15]. In addition, the high frequency of ultrasonic devices is associated with attenuation of the signal in shallow layers, which means that the signal may not reach the bottom of the layer and thickness measurement may not be possible. Therefore, it is concluded that GPR is the best device for the continuous measurement of road surface thickness.

Identifying the thickness of 10 km road pavement layers based on boreholes is almost twice as expensive as that based on GPR surveys. In addition to the price, the social burden and destructive interference in the pavement is increasing the likelihood of premature degradation;

however, when the structure is recognized based on GPR measurement, these negative consequences are much smaller. In addition, the detail of road construction information obtained from GPR measurement is incomparable to the information obtained from boreholes. A better diagnostic of road structures means, first, a greater safety of road users, and additionally, a possible reduction of costs of renovation works. Continuous recognition of the pavement structure enables an optimal and proper design of renovation treatments. By reducing the thickness of the layer by 1 cm on a 100 km stretch of road consisting of two lanes and a roadside, according to the present prices, we can save around PLN 600,000–800,000 on the asphalt wearing course and PLN 400,000–500,000 on the asphalt bonding and the asphalt base. This prompts an in-depth analysis of the issue of determining the thickness of road pavement layers using the GPR method.

The aim of this paper is an estimation of pavement thickness with GPR based on advanced signal analysis using wavelet transform.

2. Pavement Layers Thickness Estimation Based on GPR Method

2.1. Thickness Estimation Based on GPR Method

To calculate the layer thickness based on the GPR test, the following equation is used [5]:

$$d = \frac{c \cdot t}{2 \cdot \sqrt{\varepsilon_r}} \tag{1}$$

where:
d [cm]   layer thickness
c [cm/ns]   speed of light propagation in a vacuum, $c = 30$ cm/ns
t [ns]   two-way travel time between reflections from boundary surfaces
ε_r [−]   dielectric constant of the layer.

The unknown in Equation (1) is the dielectric constant of the medium, which depends on many factors, including among others the humidity of the medium, porosity, and mineral composition. The dielectric constant can be selected based on the values published in the literature (rough information, which is shown in Table 1 and Figure 1), which may be calculated based on a drilled core (accurate, but point information) or calculated based on the amplitude of the wave reflected from the surface (the value depends on the surface conditions).

2.2. Dielectric Constant of Hot Mix Asphalt (HMA) Published in Literature

The wide range of values of the dielectric constant of the pavement layers, which are taken for the thickness determinations, causes errors. Table 1 summarizes the dielectric constants of bitumen determined in microwave frequencies given in various sources. For example, according to [17], the range of dielectric constants of asphalt mixtures determined by the GPR is from 2 to 12, according to [18] from 2 to 4, according to [19] from 4 to 10, and according to [20] from 4 to 15.

Table 1. Dielectric constants of asphalt mixtures published in the literature.

Dielectric Constant [-]	Note
2–4	value given as "typical" for asphalt [18]
2–4	dry asphalt [17]
6–12	wet asphalt [17]
2.5–3.5	value given as "typical" for asphalt [21]
3–6	value given as "typical" for asphalt [22]
4.0–4.9	value determined based on SMA (stone mastic asphalt) field tests, dielectric constant calculated based on the reflected wave amplitudes [23]
4–8	value given as "typical" for asphalt [20]
8–15	slag asphalt [20]
4–10	value given as "typical" for asphalt [19]

As one can see then, the dielectric constants of materials published in the literature should be used only as an indication of the approximate value of the electrical properties of the tested medium. A way to accurately determine the dielectric constant is to drill a core and calculate it based on the thickness of the core. However, as a result, the dielectric constant value is obtained at one specific point, and its value can also change between drilling points, especially in the case of roads after repeated repairs. In addition, we strive to interfere as little as possible in a destructive way, which is drilling, into the conditions on the pavement and reduce its durability. To correctly determine the thickness of the medium by the GPR method, the dielectric constant should be determined continuously and in a completely non-destructive way.

2.3. Calculation of Dielectric Constants Based on the Amplitude of the Wave Reflected from the Surface

The method of determining the dielectric constants of a medium in a continuous way is to calculate it based on the amplitude of the wave reflected from the surface and the amplitude of the wave reflected from the ideal reflector, e.g., metal plate (reference amplitude). In this case, the following formula is used [5,6]:

$$\varepsilon_r = \left[\frac{1 + \frac{A_0}{A_m}}{1 - \frac{A_0}{A_m}} \right]^2 \tag{2}$$

where:

ε_r [−]   dielectric constant of the first layer of the medium
A_0 [−]   reflected wave amplitude on the border: air-tested surface
A_m [−]   reflected wave amplitude on the border: air-metal plate (reference amplitude).

However, as already mentioned, the result of determining the dielectric constant, as it is calculated based on the amplitude of the wave reflected from the surface, strictly depends on the conditions on the surface. In the presence of a water film on the surface, dielectric constants calculated based on the amplitude of the wave reflected from the surface increase, which was confirmed by research by the authors of this publication in the paper [24] (as a result of the presence of a water film on the asphalt pavement resulting from a small rainfall, an increase in dielectric constant from about 6, determined on a dry surface, to 9 determined on a wet surface was observed).

2.4. Impact of Incorrect Estimation of the Dielectric Constant on the Accuracy of Asphalt Pavement Thickness Determination

Figure 1 shows how an incorrect determination of the dielectric constant affects the error in measuring the asphalt pavement thickness using the GPR method. A surface made of HMA with a dielectric constant of 5 and thickness of 22 cm is assumed. Then, the two-way travel time is 3.3 ns (nanoseconds). Assuming such a time and taking the dielectric constants given in the literature from 2 to 15 [18,20] for calculations, the relative error in thickness estimation is determined as a commonly used formula [25]:

$$\Delta d = \frac{d - d_{ref}}{d_{ref}} \cdot 100\% \tag{3}$$

where:

Δd [%]   relative error in thickness determination using the GPR method
d [cm]   layer thickness determined by the GPR method
d_{ref} [cm]   reference thickness of the layer (thickness of the drilled core).

The relative error in thickness determination using the GPR method is from −58% (for a dielectric constant of 2) to +42% (for a dielectric constant of 15). It is noted that the assumption of a smaller dielectric constant than the real one has a stronger impact on the actual thickness value than using higher dielectric constant values.

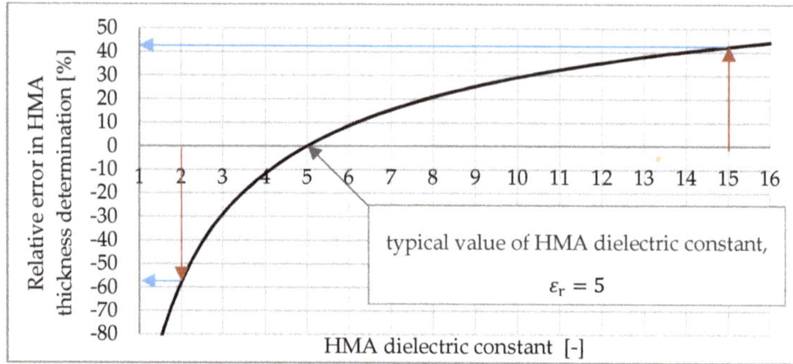

Figure 1. Impact of incorrect estimation of the dielectric constant of the asphalt mixture on the accuracy of thickness determination using the ground penetrating radar (GPR) method; blue arrows show the relative error in determining hot mix asphalt (HMA) thickness in situations when for HMA (22 cm thick and with an actual dielectric constant of 5), we incorrectly assume $\varepsilon_r = 2$ or $\varepsilon_r = 15$ (red arrows, the smallest/largest published HMA dielectric constant in the literature [18,20]).

The article proposes the use of one of the methods of advanced GPR signal analysis—wavelet analysis—as a support tool in determining the atmospheric conditions on the surface of the HMA during GPR tests (in determining whether the HMA surface was dry, wet, covered with ice, snow or de-icing salt). Depending on the surface conditions, the dielectric constants will be corrected accordingly.

3. Wavelet Analysis of GPR Signal

3.1. Theoretical Foundations of Wavelet Analysis

Wavelet analysis is based on wavelet functions—the equivalents of trigonometric functions found in the Fourier analysis [26]. Wavelet analysis splits the signal into components that are properly scaled and shifted (in time) by the basic wavelet (mother wavelet). Unlike trigonometric functions, wavelet functions are not periodic—they are irregular and asymmetrical, and their shape is similar to the shape of the pulse emitted by the radar antenna.

The idea of the wavelet transform is similar to that of the Fourier transform—they both rely on the decomposition of the examined signal into component functions, but instead of the harmonic components of the Fourier transform, wavelet functions of different scale and position are used. Scaling a wavelet means stretching or compressing it. Moving the wavelet, as the name implies, is a change in the position of the wavelet on the time axis. The results of the wavelet analysis are the wavelet coefficients, which are the sum relative to the time of the product of the signal and the scaled and shifted forms of the wavelet.

Wavelet coefficients describe how the wavelet function of a certain scale and position is similar to the signal fragment being considered. A smaller scale factor corresponds to a more 'compressed' wavelet. Larger scale values correspond to more 'stretched' wavelets. The more stretched the wavelet, the longer the signal range with which it is compared, and therefore the rougher the signal features represented by the wavelet coefficients that are obtained. The results are presented in the form of a wavelet scalogram showing the energy distribution of the signal in the coordinates time (shift)-scale [27]. It is the distribution of energy into individual wavelet coefficients as a function of scale and time. The scale allows determining what spectra the GPR pulse has at different time intervals (at different depths).

The results of the wavelet analysis depend on the type of the wavelet we choose as the wavelet mother. MatLab is a popular tool for wavelet analysis. The following wavelets can be used: Daubechies, Coiflet, Gaussian derivative, Haar, Symlets, Biortogonal, reverse Biortogonal, Meyer, discrete approximation of Meyer, Mexican hat, Morlet. The wavelet sets that can be used are Gaussian derivative, Shannon, and comprehensive Morlet. The wavelets that are the most often used in GPR signal analysis—Daubechies 4, 5 and 6, Bioorthogonal 3, 3.1, 3.5, and 7, and Symlet 6—will be discussed in more detail in the next section.

3.2. The Use of Wavelet Analysis in GPR Signal Interpretation

Reviewing the applications of wavelet analysis as a tool for processing signals emitted by non-destructive diagnostics devices mostly leads to monitoring the condition of engineering structures or its elements. Publication [28] presents the possibility of using wavelet analysis to monitor the condition of composite panels. It has been observed on wavelet spectrograms from various plates that the time–frequency structures differ in the dominant frequency bands depending on the moisture content of each panel. Publication [29] presents an application of continuous wavelet transform in vibration-based damage detection method for beams and plates. The promising application of a wavelet analysis of Impact Echo signals to assess the quality of concrete structure repairs is presented in publication [30].

The scope of applications of wavelet analysis of GPR signals in road infrastructure includes the use of wavelet analysis, among others, in the assessment of the backfill of the tunnel [31]. The tests were carried out using a 400 MHz ground-coupled antenna, and the GPR signal was compared to the Daubechies 4 wavelet. After analyzing the A-scan from a place where clays, silt, and water occur, faster suppression of the high-frequency component and slower suppression of the low-frequency component were observed.

Another application of wavelet analysis is the assessment of the condition of railway ballast in terms of the presence of impurities in it [32]. The standard deviations of the wavelet coefficients were considered to represent the signal scattering intensity. It turned out that the standard deviation decreases as the level of pollution increases. The measurements were made with an air-coupled 2.0 GHz antenna; 5 levels of signal decomposition with a Daubechies mother wave were used.

Wavelet analysis is also used to remove noise and interference, especially where the useful signal-to-noise ratio is so small that the correct signal is not visible. In measurements taken with 25, 50, 250, and 500 MHz ground-coupled antennas to distinguish geological layers, wavelets Bioorthogonal 3.1 (for 25 and 50 MHz antennas), Bioorthogonal 3.5, 3 (250 MHz), and 7 (500 MHz) were used. Noise reduction was based on 5-level decomposition of the recorded GPR signal [33]. In [34], it was checked which wavelets are best suited for removing noise from the GPR signal from measurement on a flexible surface using a 1 GHz antenna. Daubechies 6, Symlet 6, Biorthogonal, and Haar wavelets were tested (the Haar wavelet is significantly different from the GPR signal; this served to indicate the essence of the appropriate selection of the mother wavelet and to highlight the effect of this choice on the result); Daubechies 6 and Symlet 6 wavelets turned out to be the best for this application.

4. GPR Measurements of Asphalt Slabs in Various Atmospheric Conditions

4.1. Testing Area

Table 2 shows different constructions of tested model asphalt slabs with dimensions of 50 cm × 50 cm × 22 cm. Each slab consists of three layers with a total thickness of 22 cm. The slabs differ in their wearing course type, which is: MA—mastic asphalt being the most tight asphalt mix, SMA—stone mastic asphalt, AC—asphalt concrete, BBTM (*fr. beton bitumineuse trés mince*)—asphalt concrete for thin layers, and PA—porous asphalt, being the most porous asphalt mix. The slabs were placed on a metal plate to intensify wave reflection from the bottom of the slab (see Figure 2). The metal plate has a dielectric constant higher than asphalt; hence, during propagation by asphalt media, the wave phase does not change. This is important from the point of view of marking the bottom level of the slab.

Table 3 shows the atmospheric conditions when performing different GPR tests. The reference measurement was made at 28 °C (wI). Other measurements were made at a temperature below zero in the following order: on the dry surface of the slab (wII), after pouring water onto the slab (wIII), after ice formation on the slab surface (wIV), in the presence of a thin layer of fluffy snow on the slab (wV), and in the presence of de-icing salt on the surface of the slab (wVI). Measurements were carried out using a GSSI (Geophysical Survey Systems, Inc., Nashua, NH, USA) air-coupled antenna with a central frequency of 1.0 GHz.

Table 2. Constructions of model asphalt slabs.

Slab No.	Wearing Course (4 cm)	Bonding Layer (8 cm)	Base Layer (10 cm)
m1	MA 8		
m2	SMA 8		
m3	AC 8 S	AC 16 W	AC 22 P
m4	BBTM 8		
m5	PA 8 S		

Figure 2. Tested model asphalt slabs.

Table 3. Atmospheric conditions during GPR surveys.

Condition No.	Atmospheric Condition Description
wI	temperature 28 °C, dry slab surface
wII	temperature −5 °C, dry slab surface
wIII	temperature −5 °C, water film on the slab surface as a result of pouring water (about 5 L per slab)
wIV	temperature −5 °C, thin layer of ice on the slab surface
wV	temperature −2 °C, thin layer of fresh snow on the slab surface
wVI	temperature −2 °C, de-icing salt on the slab surface

4.2. A-Scans from m1–m5 Slabs Measurements in Various Atmospheric Conditions

In Figures 3–5, A-scans from measurements of slabs m1–m5 in conditions wI–wVI are shown. As a zero level, the minimum amplitude of the wave reflected from the slab surface was assumed. The bottom level is the minimum amplitude of the wave reflected from the bottom of the metal plate under the asphalt slab. Based on the A-scans, it is visible that both the propagation time through the slab and the reflection amplitude from the surface vary depending on the weather conditions.

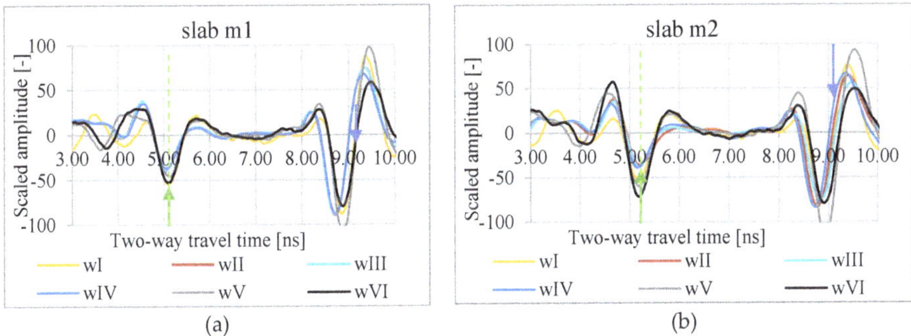

(a)   (b)

Figure 3. A-scans of m1 (**a**) and m2 (**b**) slabs in atmospheric conditions wI, wII, wIII, wIV, wV, and wVI; green arrow—assumed zero level, blue arrow—assumed bottom of the slab.

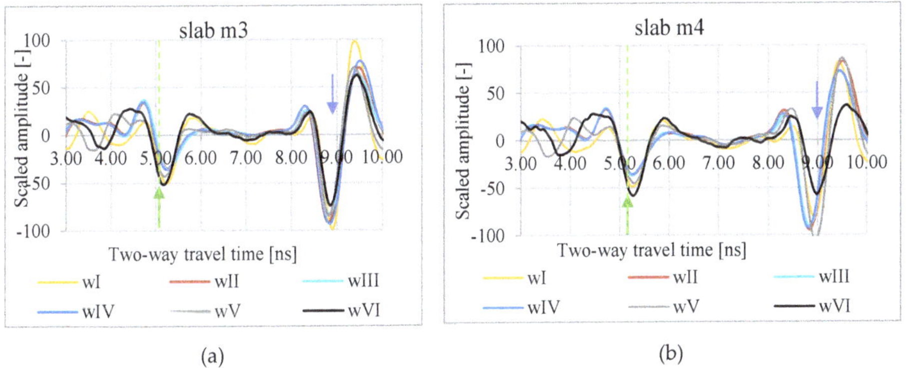

(a)   (b)

Figure 4. A-scans of m3 (**a**) and m4 (**b**) slabs in atmospheric conditions wI, wII, wIII, wIV, wV, and wVI; green arrow—assumed zero level, blue arrow—assumed bottom of the slab.

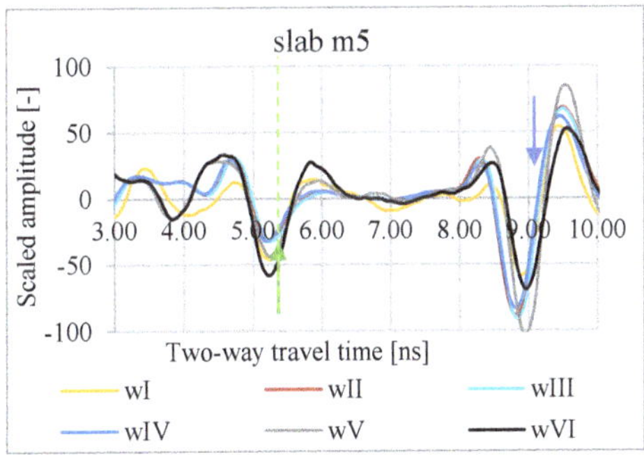

Figure 5. A-scans of m5 slab in atmospheric conditions wI, wII, wIII, wIV, wV, and wVI; green arrow—assumed zero level, blue arrow—assumed bottom of the slab.

4.3. Dielectric Constants of HMA

4.3.1. Dielectric Constants Calculated based on Propagation Time through the Slab of Known Thickness, Hereinafter Named the "B" Method

Table 4 shows two-way travel time through the slabs m1–m5 read directly from the GPR measurement. Knowing that the thickness of slabs m1–m5 is equal to 22 cm, and transforming Equation (1), their dielectric constants (ε_{r_B}) were calculated. The results are summarized in Table 5.

Table 4. Two-way travel time through the slabs.

t [ns]	m1	m2	m3	m4	m5
wI	3.78	3.71	3.66	3.63	3.61
wII	3.64	3.56	3.53	3.53	3.53
wIII	3.67	3.67	3.64	3.64	3.56
wIV	3.64	3.49	3.64	3.56	3.53
wV	3.75	3.75	3.59	3.67	3.67
wVI	3.79	3.75	3.63	3.71	3.75

Table 5. Dielectric constants calculated based on propagation time through the slabs ('B' method).

ε_{r_B} [−]	m1	m2	m3	m4	m5	Average ε_{r_B} [−]
wI	6.64	6.40	6.23	6.13	6.06	6.29
wII	6.16	5.89	5.79	5.79	5.79	5.88
wIII	6.26	6.26	6.16	6.16	5.89	6.15
wIV	6.16	5.66	6.16	5.89	5.79	5.93
wV	6.54	6.54	5.99	6.26	6.26	6.32
wVI	6.68	6.54	6.13	6.40	6.54	6.46

The differences in dielectric constants depending on the atmospheric conditions during the measurement calculated by the "B" method are not large (Figure 6), but they cannot be ignored. The following trends of apparent increases and decreases of the dielectric constant of the HMA are noted: temperature below zero (wII) causes an apparent decrease of the dielectric constant determined by the "B" method. Pouring the slab with water (wIII) causes an apparent increase in the dielectric constant. Freezing of the formed water film (wIV) causes the apparent decrease of the dielectric constant determined by the "B" method. The presence of snow (wV) on the surface causes an apparent increase in dielectric constant (except slab m3). The presence of salt on the slab surface (wVI) also causes an apparent increase in HMA dielectric constant marked by method "B".

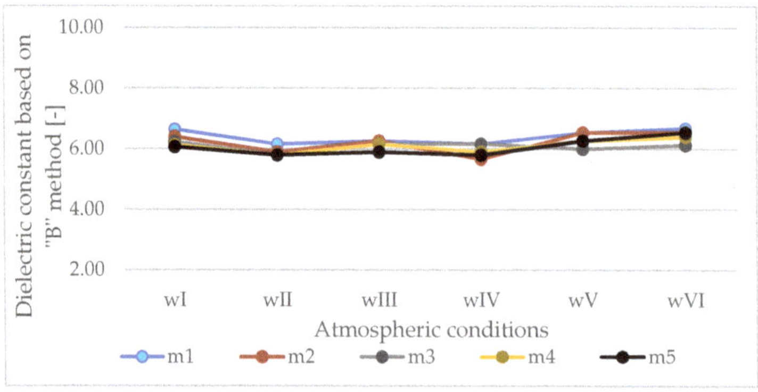

Figure 6. Dielectric constants calculated based on propagation time through the slabs ("B" method).

For calculations of dielectric constants using the "B" method, knowledge of thickness is required, which in road practice translates into drilling and taking cores from roads for being measured. The following are dielectric constants calculated based on measurements of the same slabs, under the same atmospheric conditions, but based on the amplitude of the wave reflected from the surface, i.e., by a method that does not require taking cores.

4.3.2. Dielectric Constants Calculated based on the Amplitude of the Wave Reflected from the Surface, Hereinafter Named the "A" Method

Calculations were made based on Equation (2). Table 6 shows the ratio of the wave amplitude reflected from the surface of the slab to the wave amplitude reflected from the metal plate (reference amplitude). Table 7 shows dielectric constants calculated based on amplitudes.

Table 6. Ratio of the wave amplitude reflected from the surface of the slab to the wave amplitude reflected from the metal plate.

A_0/A_m [−]	m1	m2	m3	m4	m5
wI	0.48	0.48	0.46	0.44	0.42
wII	0.47	0.45	0.42	0.42	0.38
wIII	0.48	0.45	0.43	0.45	0.39
wIV	0.44	0.44	0.43	0.42	0.38
wV	0.40	gross error	0.41	0.41	0.50
wVI	0.57	gross error	0.51	0.56	0.63

Table 7. Dielectric constants calculated based on the amplitudes ("A" method').

ε_{r_A} [−]	m1	m2	m3	m4	m5	Average ε_{r_A} [−]
wI	8.28	7.93	7.13	6.51	6.51	7.27
wII	7.85	6.07	6.07	6.06	4.98	6.21
wIII	7.91	6.98	6.98	6.84	5.15	6.77
wIV	6.62	6.55	6.33	6.02	4.98	6.10
wV	5.45	gross error	5.59	5.66	8.85	6.39
wVI	13.10	gross error	9.40	12.69	12.69	11.97

Based on the summaries in Table 7 and Figure 7, it is noted that the dielectric constants calculated based on the reflected wave amplitudes are significantly different from those calculated based on propagation time and known slab thickness. The temperature below zero (wII) of the HMA causes a significant decrease in the apparent value of its dielectric constant. Pouring the slab with water and the presence of a water film on the surface (wIII) causes an increase in dielectric constants calculated with the "A" method. Freezing of water on the surface (wIV) causes a decrease in dielectric constant again. The presence of snow (wV) decreases the dielectric constant (except for slab m5). The presence of salt on the surface (wVI) causes a significant increase in dielectric constant determined based on the amplitude of the wave reflected from the surface.

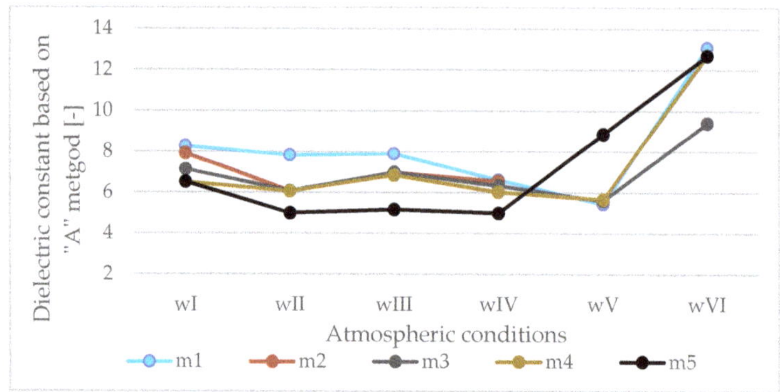

Figure 7. Dielectric constants calculated based on the amplitudes ("A" method).

4.3.3. Calculated Slabs Thicknesses Based on the Wave Amplitude Reflected from the Surface

Table 8 summarizes the thicknesses calculated based on dielectric constants determined based on the amplitudes of waves reflected from the surface. Table 9 shows the relative thickness determination error caused by calculating the dielectric constant based on amplitudes.

Table 8. Calculated slabs thicknesses based on the wave amplitude reflected from the surface.

d [cm]	m1	m2	m3	m4	m5
wI	19.70	19.76	20.56	21.34	21.22
wII	19.49	21.67	21.49	21.51	23.73
wIII	19.57	20.84	20.67	20.88	23.53
wIV	21.22	20.45	21.70	21.76	23.73
wV	24.09	gross error	22.78	23.14	18.50
wVI	15.71	gross error	17.76	15.62	15.79

Table 9. Relative thickness determinations error caused by calculating the dielectric constant based on amplitudes.

Δd [%]	m1	m2	m3	m4	m5	Mean of the Absolute Value
wI	−10	−10	−7	−3	−4	7
wII	−11	−1	−2	−2	8	5
wIII	−11	−5	−6	−5	7	7
wIV	−4	−7	−1	−1	8	4
wV	10	gross error	4	5	−16	7
wVI	−29	gross error	−19	−29	−28	21

The measurement error in the conditions adopted as a reference is up to −10%, and it is the accuracy of determining the thickness of layers by the method based on the amplitudes of the wave reflected from the surface. Measurements in temperature below zero cause the calculated thickness to be smaller than the actual thickness up to −11%, except for slab m5. The presence of water film on the slab's surface caused the calculated thickness to be up to −11% smaller than the actual thickness, except for slab m5. An icy slab surface caused the calculated thickness to be up to −7% smaller than the actual thickness, except for slab m5. The presence of snow on the slab surface caused the calculated thickness to be greater than the actual thickness of the slab (up to 10% greater), except for slab m5 (thickness −16% smaller). As a result of the decrease in freezing temperature during the presence of de-icing salt on the HMA surface of the asphalt mixture, apparently smaller thicknesses are obtained (up to −29%).

4.3.4. Correction Coefficients for Dielectric Constants Determined based on the Amplitudes Reflected from the Surface

Based on Equation (4), the correction coefficients for dielectric constants determined based on amplitudes to dielectric constants determined based on propagation time through the slab of known thickness were calculated. Their values are summarized in Table 10. The reference conditions were the wI conditions (28 °C, dry slab surface). The dielectric constant determined based on amplitudes in all conditions other than wI should be multiplied by the correction coefficient k.

$$k = \frac{\varepsilon_{r_B}}{\varepsilon_{r_A}}.$$

(4)

where:

$k\ [-]$ correction coefficients for dielectric constants determined based on the amplitudes

$\varepsilon_{r_B}\ [-]$ dielectric constants calculated based on propagation time through the slabs ("B" method)

$\varepsilon_{r_A}\ [-]$ dielectric constants calculated based on the amplitudes ("A" method).

Table 10. Correction coefficients for dielectric constants determined based on the wave amplitude reflected from the surface.

k [−]	m1	m2	m3	m4	m5	Average k [−]
wI	0.80	0.81	0.87	0.94	0.93	0.87
wII	0.78	0.97	0.95	0.96	1.16	0.97
wIII	0.79	0.90	0.88	0.90	1.14	0.92
wIV	0.93	0.86	0.97	0.98	1.16	0.98
wV	1.20	gross error	1.07	1.11	0.71	1.02
wVI	0.51	gross error	0.65	0.50	0.52	0.55

During determinations of the dielectric constant based on the amplitudes in a reference condition, a dielectric constant correction factor k = 0.87 should be adopted. During determinations of the dielectric constant based on the amplitudes at the temperature of −5 °C on the dry slab surface, a dielectric constant correction factor k = 0.97 should be adopted. During measurements at −5 °C and with water film presence on the slab surface, a dielectric constant correction coefficient k = 0.92 should be used. During measurements at −5 °C and with ice presence on the slab surface, a dielectric constant correction coefficient k = 0.98 should be used. During measurements at −2 °C and snow on the slab surface, a dielectric constant k = 1.02 correction coefficient should be adopted. During measurements at −2 °C and de-icing salt on the slab surface, a dielectric constant correction factor k = 0.55 should be used.

The thicknesses calculated based on dielectric constants corrected by the mentioned coefficients and the relative thickness measurement error are presented in Tables 11 and 12, respectively.

Table 11. Calculated slabs thicknesses based on the corrected dielectric constant.

d_{cor} [cm]	m1	m2	m3	m4	m5
wI	21.1	21.2	22.0	22.9	22.7
wII	19.8	22.1	21.9	21.9	24.1
wIII	20.4	21.7	21.5	21.7	24.5
wIV	21.4	20.6	21.9	22.0	23.9
wV	23.8	gross error	22.5	22.9	18.3
wVI	21.3	gross error	24.1	21.2	21.4

Table 12. Relative thickness determinations error after correction of dielectric constant.

Δd [%]	m1	m2	m3	m4	m5	Mean of the Absolute Value
wI	−4	−4	0	4	3	3
wII	−10	0	−1	−1	10	4
wIII	−7	−1	−2	−1	11	5
wIV	−3	−6	0	0	9	4
wV	8	gross error	2	4	−17	6
wVI	−3	gross error	9	−4	−3	4

It is noted that the thickness measurements error has been significantly reduced in the case of the slab covered with de-icing salt. Attention is drawn to the fact that the correction coefficients have been calculated as the average of the coefficients for 5 slabs with different HMA wearing courses for the determined atmospheric conditions. To significantly reduce the error, a factor calculated for a specific HMA wearing course should be applied.

The coefficients presented above are a proposal for specific slab geometry (m1–m5), specific measurement conditions (wI–wVI), and selected frequency and antenna design (1.0 GHz air-coupled). The observed trends thrive to expand the base of dielectric constants in controlled atmospheric conditions with a constant change in temperature, amount and type of precipitation, and the amount of de-icing salt used.

When measuring using GPR on large sections, the atmospheric conditions of the surface usually are not known; it is not known if and in which sections the surface was covered with water film, ice, snow, or de-icing salt. Of course, video cameras are helpful in recording the surface atmospheric condition; while it will be possible to record where the water film was on the surface during the measurements, the presence of salt on the surface will not be recognized based on the video image.

In determining the surface conditions, wavelet analysis may be a useful tool to correctly select the proposed correction coefficient.

4.4. Wavelet Analysis of the GPR Signal

Wavelet analysis was performed using the Db6 wavelet. The type of wavelet was chosen based on a literature review and the selection of the group of wavelets that are most useful in GPR signal analysis as well as the thorough initial empirical analysis of the authors.

Wavelet analysis was performed on the unscaled signal from m3 slab measurements. Figures 8–10 compare the distribution of energy into individual wavelet coefficients as a function of scale and time. It was noticed that in some cases, the scale coefficients of signals obtained in different atmospheric conditions at a slab surface differ in magnitude. The largest magnitudes have signal scale factors from the measurement of a slab covered with de-icing salt, which is a positive result, because it is the presence of salt on the surface that strongly affects the value of the dielectric constant determined based on the amplitude, and, as previously noted, the presence of salt is not detectable by video recording. The results of the conducted analysis thrive at defining the atmospheric conditions on the surface (dry, covered with water film, ice, snow, and salt), which was obtained during the GPR survey.

There is a difference between the scalograms from the measurement signals taken at temperatures above zero (Figure 8a) and below zero (Figure 8b). Scale factors 9–60 at a distance of 180–250 samples have higher magnitudes when the measurement is performed at a temperature above zero. Scale factors 9–60 at a distance of 200–400 samples have higher magnitudes when the measurement is performed at a temperature below zero.

The scalograms from the measurement signals taken at temperatures below zero on a dry slab surface (Figure 8b), temperatures below zero and water film on the slab surface (Figure 9a), and temperatures below zero and a thin layer of ice on the slab surface (Figure 9b) do not differ significantly.

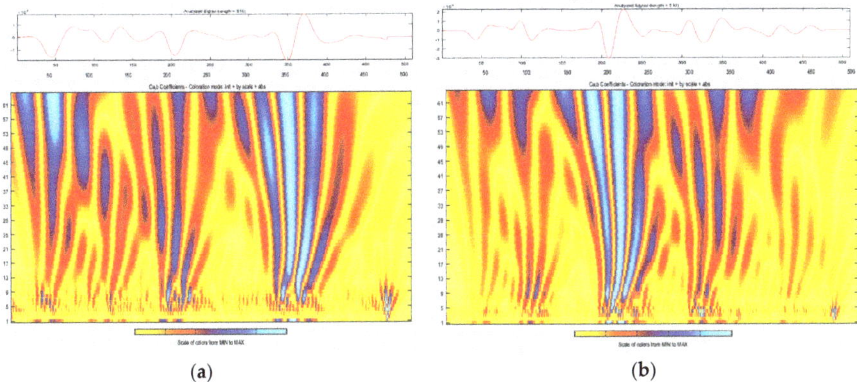

Figure 8. Wavelet analysis of A-scan from measurements in conditions (**a**) wI (temp. 28 °C, dry slab surface); (**b**) wII (temp. −5 °C, dry slab surface).

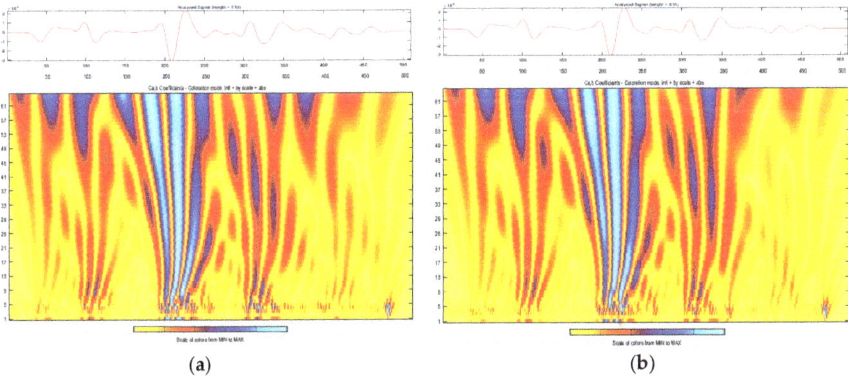

Figure 9. Wavelet analysis of A-scan from measurements in conditions (**a**) wIII (temp. −5 °C, water film on the slab surface); (**b**) wIV (temp. −5 °C, thin layer of ice on the slab surface).

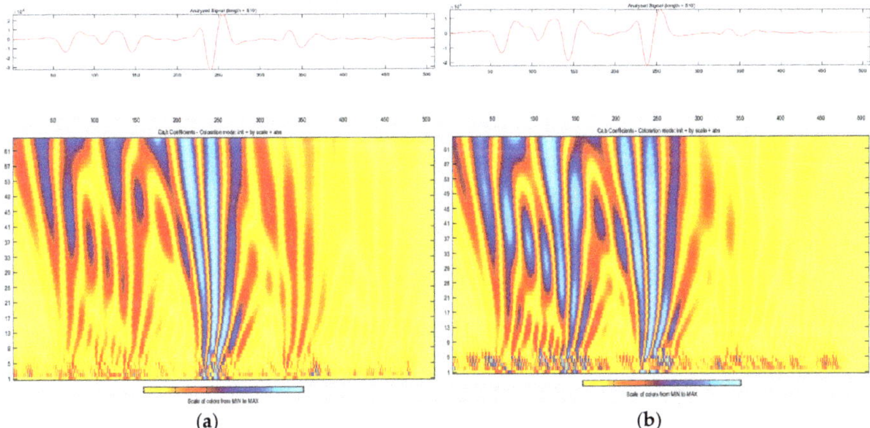

Figure 10. Wavelet analysis of A-scan from measurements in conditions (**a**) wV (temp. −2 °C, thin layer of fresh snow on the slab surface); (**b**) wVI (temp. −2 °C, de-icing salt on the slab surface).

Snow on the pavement at a temperature below zero (Figure 10a) results in the fact that the majority of energy falls on scale factors 9–60 at a distance of 200–300 samples. The effect of salt (Figure 10b) on the surface is that the scale factors 9–60 at a distance of 0–150 samples have very large amplitudes: the largest of those obtained during all other surface conditions.

5. Discussion

Based on the research and analysis, it was found that the value of the dielectric constant determined continuously based on the amplitude of the wave reflected from the surface is affected by the atmospheric conditions on the surface—its temperature, moisture, ice or snow, and presence of de-icing salt.

When the temperature is high, HMA dielectric constants values are higher than at temperatures below zero. The observed tendency of the apparent increase in HMA dielectric constants with the increase in its temperature results from the more energetic movement of molecules—this enables an easier formation and orientation of dipoles and requires continuing testing on a larger number of samples and/or test sites in a continuous temperature range. As a result of the presence of a water film on the surface and partly in the pores of the HMA (water itself is a dipole), after the application of an external electric field, as expected, an increase in the polarization capacity of the medium was observed. The freezing of water on the surface and partly in the pores caused a slight decrease in the dielectric constant. This is due to the formation of hydrogen bonds and retention of the dipole polarization mechanism of water. An increase in dielectric constant, as it was observed, was due to the presence of snow on the surface. The apparent increase in HMA dielectric constant due to the presence of salt can be combined with the presence of Na^- and Cl^+ ions as well as ionic polarization and increase in the material's ability to compensate the electric field; on the other hand, the electric field between the Na^- and Cl^+ ions and water molecules whose presence results from melting snow and ice on the surface of the slab should level out. This issue as well as the issue related to the presence of snow on the surface should be further investigated.

The measurement error in calculating the thickness based on reflection amplitude in the conditions adopted as a reference is up to −10%. The surface atmospheric conditions affect less or more the thickness value determined by the GPR method—and as a result of the thickness determination based on the dielectric constant calculated using the amplitude of the reflected wave from the surface below zero temperature, the thickness measurement error was up to $-11 \div +8\%$; in the case of a wave amplitude reflected from the surface covered with water film, it was up to $-11 \div +7\%$; in the case of a wave amplitude reflected from icy surfaces, it was up to $-7 \div 8\%$; in the case of a wave amplitude reflected from the surfaces covered with a thin layer of powdery snow, it was up to $-16 \div +10\%$, and in the case of a wave amplitude reflected from the surfaces covered with de-icing salt, it was up to −29% (see Table 9).

It was proposed to use advanced wavelet analysis to determine which atmospheric conditions on the surface of the pavement result in high/low dielectric constant values determined based on the amplitude. Promisingly, high differences in magnitudes of scale factors in relation to the reference conditions were obtained based on a signal from the measurements on the surface covered with de-icing salt, which is a very positive result, because the presence of salt on the surface is not observable on video records performed alongside GPR measurements.

For each of the analyzed atmospheric conditions, a correction factor for the dielectric constant calculated based on the wave amplitudes reflected from surfaces with different atmospheric conditions was proposed, whose value could be chosen based on wavelet analysis and exactly reflects the appearance of the wavelet scalogram (see Figures 8–10). During determinations of the dielectric constant based on the amplitudes in a reference condition, a dielectric constant correction factor $k = 0.87$ should be adopted. During determinations of the dielectric constant based on the amplitudes at the temperature of −5 °C on the dry slab surface, a dielectric constant correction factor $k = 0.97$ should be adopted. During measurements at −5 °C and with a water film presence on the slab surface, a dielectric constant correction coefficient k = 0.92 should be used. During measurements at −5 °C and with ice presence on the slab surface, a dielectric constant correction coefficient k = 0.98 should be used. During measurements at −2 °C and with snow on the slab surface, a dielectric constant k = 1.02 correction coefficient should be adopted. During measurements at −2 °C and with de-icing salt on the slab surface, a dielectric constant correction factor k = 0.55 should be used. As a result of applied correction factors, measurement errors of thickness determinations have been reduced, especially if there is de-icing salt on the surface: up to −4 ÷ +9% in the case of the surfaces covered with de-icing salt (see Table 12). In other cases, after applying correction factors, the thickness measurement error was up to ±4% in reference conditions, up to ±10% in the case of the surface below zero temperature, up to −7 ÷ +11% in the case of the surface covered with water film, up to −6 ÷ +9% in the case of icy surfaces, and up to −17 ÷ +8% in the case of the surfaces covered with a thin layer of snow. It is noted that the coefficients have been calculated as the average of the coefficients for 5 slabs with different HMA wearing courses. To significantly reduce the error, a factor calculated for a specific HMA wearing course should be used.

Such coefficients have been proposed for specific slab geometry (m1–m5), specific measurement conditions (wI–wVI), and selected antenna (1.0 GHz air-coupled). It is proposed to continue the research in order to build a base of dielectric constant values of various asphalt mixtures determined at different temperature and humidity conditions based on the amplitude of the wave reflected from the asphalt layer and the corresponding correction factors, thus increasing the accuracy of the thickness determination of the layer by the GPR method and reducing the number of cores necessary to achieve the desired accuracy of thickness determination.

6. Conclusions

Based on the research and analysis carried out, the following was concluded:

1. The value of dielectric constant that is necessary to determine the layer thickness based on the GPR method can be determined either based on the thickness of the cores ("B" method) or based on the amplitude of the wave reflected from the surface ("A" method). The "B" method is accurate, and a possibility of coring is usually limited; therefore, this method can be used for testing a homogeneous section in terms of the layer's material. The "A" method allows recognizing the thickness over a large distance without drilling. Its accuracy depends on the surface conditions that result from atmospheric conditions that are present during the test, and they require a proper correction of dielectric constant.

2. The value of the dielectric constant was calculated based on the amplitudes estimated in different atmospheric conditions: (changes in the temperature, moisture, presence of ice/snow, presence of de-icing salt). In the conducted test, the error of thickness measurement was:

    (a) up to 11% when measuring in negative ambient temperature
    (b) up to 11% during measurements of the asphalt's slab surface covered with water film
    (c) up to 8% during measurements of the ice-covered asphalt's slab surface
    (d) up to 16% when measuring asphalt slab with a surface covered with a thin layer of fresh snow
    (e) even 29% during measurements of asphalt slab with a surface covered with a layer of de-icing salt

3. The purpose of this work was to improve the accuracy of determining the thickness of layers based on dielectric constants calculated based on the wave amplitudes reflected from the surface. It was proposed to achieve it using wavelet analysis and based on the wavelet scalogram to determine the effect of surface conditions during the measurements. Then, the correction factor for the dielectric constant was calculated based on the amplitude of the wave reflected from the surface, and the thickness was calculated based on the corrected dielectric constant value.

After applying correction factors for dielectric constants, the error in determining the thickness was reduced (especially the error in measuring the thickness of the surface covered with de-icing salt). The average thickness determination errors were reduced:

(a)   when measuring in negative ambient temperature, to on average 4%

(b)   when measuring the asphalt's slab surface covered with water film, to on average 5%

(c)   when measuring the ice-covered asphalt's slab surface to on average 4%

(d)   when measuring the asphalt slab with surface covered with a thin layer of fresh snow, to on average 6%

(e)   when measuring the asphalt slab with a surface covered with a layer of de-icing salt, to on average 4%

It should be noted that the correction coefficients have been calculated as the average of the coefficients for 5 slabs with different HMA wearing courses for specific atmospheric conditions. To significantly reduce the error, a dielectric constant correction factor calculated for a specific HMA wearing course should be applied.

4.   The usefulness of wavelet analysis for determining the presence of de-icing salt on the tested surface has been demonstrated—its presence is clearly visible on the wavelet scalogram. This is particularly useful, because the salt significantly affects the value of the dielectric constant and causes a thickness measurement error 21% (in average)—and it is not possible to recognize the presence of de-icing salt on the surface neither based on the visual assessment of the surface, nor based on basic radargram analysis. No significant changes were observed on the wavelet scalograms from the signal recorded in the different surface conditions.

5.   Until the time of satisfactory development of the correction coefficients for dielectric constants, it is recommended to use the hybrid method for determining the thickness of layers using the GPR method—calculating the dielectric constant based on the amplitude of the wave reflected from the surface and each time determining the correction coefficient due to the surface condition based on drilled cores. In addition, selective wavelet analysis is recommended for the possible detection of the presence of de-icing salt when measurements are taken in the autumn and winter.

Author Contributions: Conceptualization, M.W., A.L. and A.G.; methodology, M.W., A.L.; software, M.W.; validation, M.W.; formal analysis, A.L. and A.G.; investigation, M.W.; resources, M.W., A.L. and A.G.; data curation, M.W.; writing—original draft preparation, M.W.; writing—review and editing, A.L. and A.G.; visualization, M.W.; supervision, A.L. and A.G.; project administration, M.W., A.L. and A.G.; funding acquisition, M.W., A.L., and A.G. All authors have read and agreed to the published version of the manuscript.

Funding: This paper was co-financed under the research grant of the Warsaw University of Technology no. 504/04512/1080/43.070046 supporting the scientific activity in the discipline of Civil Engineering and Transport.

Acknowledgments: The paper was prepared in the framework of the agreement between the TPA Sp. z o. o. research laboratory and a Faculty of Civil Engineering, Warsaw University of Technology on a cooperation in the field of research, education and human resources.

Conflicts of Interest: The authors declare no conflict of interest.

References

1.   Yi, L.; Zou, L.; Sato, M. Practical Approach for High-Resolution Airport Pavement Inspection with the Yakumo Multistatic Array Ground-Penetrating Radar System. *Sensors* **2018**, *18*, 2684. [CrossRef] [PubMed]

2.   Miechowski, T.; Sudyka, J.; Harasim, P.; Kowalski, A.; Kusiak, J.; Borucki, R.; Grączewski, A. Impact of Accuracy of Pavement Structure Identification on Road Reparation Projects, Research work of the Road and Bridge Research Institute *(IBDiM)*. 2006; pp. 1–76, (In Polish). Available online: https://www.gddkia.gov.pl/userfiles/articles/p/prace-naukowo-badawcze-zrealizow_3435//documents/wplyw-dokladnosci.pdf (accessed on 31 March 2020).

3.   Uniwersał, P. The impact of selected factors on the measured deflections and on the results of the identification of stiffness modules using the FWD (In Polish). *Nowocz. Bud. Inż.* **2016**, *6*, 62–65.

4. Benedetto, A.; Pajewski, L. *Civil Engineering Applications of Ground Penetrating Radar*; Springer Transactions in Civil and Environmental Engineering: Rome, Italy, 2015; ISBN 978-3-319-04812-3.
5. Lalagüe, A. Use of Ground Penetrating Radar for Transportation Infrastructure Maintenance. Ph.D. Thesis, Norwegian University of Science and Technology, Trondheim, Norway, June 2015.
6. Li, W.; Wen, J.; Xiao, Z.; Xu, S. Application of Ground-Penetrating Radar for Detecting Internal Anomalies in Tree Trunks with Irregular Contours. *Sensors* **2018**, *18*, 649. [CrossRef] [PubMed]
7. Hamrouche, R.; Saarenketo, T. Improvement of coreless method to calculate the average dielectric value of the whole asphalt layer of a road pavement. In Proceedings of the 15th International Conference on Ground Penetrating Radar, Brussels, Belgium, 30 June–4 July 2014; pp. 1–4.
8. Al-Qadi, I.L.; Lahouar, S.; Loulizi, A. Successful application of GPR for quality assurance/quality control of new pavements, [w:] transportation research record journal of the transportation research board. In Proceedings of the 82th Annual Meeting, Washington, DC, USA, 12–16 January 2003; pp. 1–24.
9. Zhang, J.; Ye, S.; Yi, L.; Lin, Y.; Liu, H.; Fang, G. A Hybrid Method Applied to Improve the Efficiency of Full-Waveform Inversion for Pavement Characterization. *Sensors* **2018**, *18*, 2916. [CrossRef] [PubMed]
10. Surface Testing Instruments. Available online: https://www.elektrophysik.com/pl/produkty/pomiar-grubosci-jezdni/stratotest-4100 (accessed on 18 February 2020).
11. Hoła, J.; Bień, J.; Schabowicz, K. Non-destructive and semi-destructive diagnostics of concrete structures in assessment of their durability, Bulletin of the Polish Academy of Sciences. *Tech. Sci.* **2015**, *63*, 87–96.
12. Adamczewski, G.; Garbacz, A.; Piotrowski, T.; Załęgowski, K. Application of combined NDT methods for assessment of concrete structures (In Polish). *Mater. Bud.* **2013**, *9*, 2–5.
13. Schabowicz, K.; Jawor, D.; Radzik, Ł. Analysis of non-destructive acoustic methods for testing renovated concrete structures (In Polish). *Mater. Bud.* **2015**, *11*, 103–105.
14. U.S. Department of Transportation, Federal Highway Administration. Available online: https://fhwaapps.fhwa.dot.gov (accessed on 18 February 2020).
15. Hoegh, K.; Khazanovich, L.; Yu, T.H. Ultrasonic Tomography for Evaluation of Concrete Pavement. *Transp. Res. Board* **2011**, *2232*, 85–94. [CrossRef]
16. Schabowicz, K. Non-Destructive Testing of Materials in Civil Engineering. *Materials* **2019**, *12*, 3237. [CrossRef] [PubMed]
17. Daniels, D.J. *Ground Penetrating Radar*, 2nd ed.; The Institution of Electrical Engineers: London, UK, 2004; pp. 1–761, ISBN 0-86341-360-9.
18. *D4748-10 Standard Test Method for Determining the Thickness of Bound Pavement Layers Using Short-Pulse Radar*; ASTM International: West Conshohocken, PA, USA, 2012; pp. 1–7.
19. *DMRB Design Manual for Roads and Bridges, HD 29/08 Data for Pavement Assessment: Ground Penetrating Radar for Pavement Assessment, Part 2, Volume 7, Chapter 6, Section 3*; Highways Agency: London, UK, 2008; pp. 1–94.
20. Saaranketo, T. Electrical Properties of Road Materials and Subgrade Soils and the Use of Ground Penetrating Radar in Traffic Infrastructure Surveys. Ph.D. Thesis, Faculty of Science of the University of Oulu, Oulu, Finland, November 2006; pp. 1–127.
21. Karczewski, J.; Ortyl, Ł.; Pasternak, M. *Outline of the GPR Method (In Polish)*, 2nd ed.; AGH Publishing: Kraków, Poland, 2011; pp. 1–345, ISBN 978-83-7464-422-8.
22. Willet, D.A.; Rister, B. *Ground Penetrating Radar—Pavement Layer Thickness Evaluation*; Kentucky Transportation Center: Lexington, Kentucky, 2002; pp. 1–36. Available online: https://uknowledge.uky.edu/cgi/viewcontent.cgi?article=1249&context=ktc_researchreports (accessed on 31 March 2020).
23. Al-Qadi, I.L.; Lahouar, S.; Loulizi, A. *Ground-Penetrating Radar Calibration at the Virginia Smart Road and Signal Analysis to Improve Prediction of Flexible Pavement Layer Thicknesses*; Virginia Tech Transportation Institute: Blacksburg, Virginia, 2005; pp. 1–66. Available online: https://vtechworks.lib.vt.edu/bitstream/handle/10919/46647/05-cr7.pdf?sequence=1&isAllowed=y (accessed on 31 March 2020).
24. Wutke, M.; Lejzerowicz, A.; Jackiewicz-Rek, W.; Garbacz, A. Influence of variability of water content in different states on electromagnetic waves parameters affecting accuracy of GPR measurements of asphalt and concrete pavements. In Proceedings of the 64 Scientific Conference of the Committee for Civil Engineering of the Polish Academy of Sciences and the Science Committee of the Polish Association of Civil Engineers (PZITB) (KRYNICA 2018), Krynica, Poland, 16–20 September 2018; Volume 262, pp. 1–9.
25. Iyengar, B. *A Textbook of Engineering Mathematics*; Laxmi Publications: New Delhi, India, 2004; p. 1425, ISBN 10: 8170083656.

26. Wavelet Transform (In Polish), Lectures at the AGH University of Science and Technology. Available online: http://home.agh.edu.pl/~{}jsw/oso/FT.pdf (accessed on 15 June 2017).

27. Batko, W.; Mikulski, A. Wavelet transform in the diagnosis of hoisting devices. *Diagnostyka* **2004**, *30*, 61–68. (In Polish)

28. Adamczak-Bugno, A.; Świt, G.; Krampikowska, A. Assessment of destruction processes in fibre-cement composites using the acoustic emission method and wavelet analysis. *Mater. Sci. Eng.* **2019**, *471*, 1–10. [CrossRef]

29. Rucka, M.; Wilde, K. Application of continuous wavelet transform in vibration based damage detection method for beams and plates. *J. Sound Vib.* **2006**, *297*, 536–550. [CrossRef]

30. Garbacz, A.; Piotrowski, T.; Courard, L.; Kwaśniewski, L. On the evaluation of interface quality in concrete repair system by means of impact-echo signal analysis. *Constr. Build. Mater.* **2017**, *134*, 311–323. [CrossRef]

31. Zhang, S.; Li, Y.; Fu, G.; He, W.; Hu, D.; Cai, X. Wavelet Packet Analysis of Ground–Penetrating Radar Simulated Signal for Tunnel Cavity Fillings. *J. Eng. Sci. Technol.* **2018**, *11*, 62–69. [CrossRef]

32. Shangguan, P.; Al-Quadi, I.L.; Leng, Z. Development of Wavelet Technique to Interpret Ground-Penetrating Radar Data for Quantifying Railroad Ballast Conditions. *J. Transp. Res. Board Natl. Acad.* **2012**, *2289*, 95–102. [CrossRef]

33. Javadi, M.; Ghasemzadeh, H. Wavelet analysis for ground penetrating radar applications: A case study. *J. Geophys. Eng.* **2017**, *14*, 1189–1202. [CrossRef]

34. Baili, J.; Lahouar, S.; Hergli, M.; Al-Qadi, I.L.; Besbes, K. GPR signal de-noising by discrete wavelet transform. *NDT E Int.* **2009**, *42*, 696–703. [CrossRef]

 *materials*

Article

Calibration of Steel Rings for the Measurement of Strain and Shrinkage Stress for Cement-Based Composites

Adam Zieliński * and Maria Kaszyńska

Faculty of Civil Engineering and Architecture, West Pomeranian University of Technology in Szczecin, al. Piastów 50, 70-311 Szczecin, Poland; maria.kaszynska@zut.edu.pl
* Correspondence: adam.zielinski@zut.edu.pl

Received: 18 June 2020; Accepted: 30 June 2020; Published: 2 July 2020

Abstract: Concrete shrinkage is a phenomenon that results in a decrease of volume in the composite material during the curing period. The method for determining the effects of restrained shrinkage is described in Standard ASTM C 1581/C 1581M–09a. This article shows the calibration of measuring rings with respect to the theory of elasticity and the analysis of the relationship of steel ring deformation to high-performance concrete tensile stress as a function of time. Steel rings equipped with strain gauges are used for measurement of the strain during the compression of the samples. The strain is caused by the shrinkage of the concrete ring specimen that tightens around steel rings. The method allows registering the changes to the shrinkage process in time and evaluating the susceptibility of concrete to cracking. However, the standard does not focus on the details of the mechanical design of the test bench. To acquire accurate measurements, the test bench needs to be calibrated. Measurement errors may be caused by an improper, uneven installation of strain gauges, imprecise geometry of the steel measuring rings, or incorrect equipment settings. The calibration method makes it possible to determine the stress in a concrete sample leading to its cracking at specific deformation of the steel ring.

Keywords: restrained ring test; autogenous shrinkage cracking; concrete cracking test; concrete shrinkage cracking test; restrined ring calibration

---

1. Introduction

The shrinkage of composite materials is a phenomenon where the material reduces its volume as a result of drying, carbonation, and autogenous processes [1–5]. If an element is not restrained and can freely change its volume, the structure remains intact. However, when the shrinkage is restrained, the lack of free strain results in the development of internal stresses that lead to cracking.

One of the basic research methods for the controlled reduction of concrete shrinkage deformations is the use of ring methods. Presumably, the first tests of this type were carried out by Carlson and Reading [6] in the 1940s, where the result of the research was the age of cracking of concrete ring samples. The geometry and cross-section of the concrete ring can be selected based on the size of the aggregate. The degree of limitation depends on the modulus of elasticity and width of the two rings: the concrete ring and the rigid steel ring limiting the free deformability of the composite. However, height is a generally accepted parameter. Different geometries of limiting rings [7–9] and annular concrete samples [7,10–12] were developed. Two steel measuring rings were used: external and internal, where an additional external ring was used to limit deformations caused by autogenous swelling and the thermal expansion of concrete [13]. Studies on elliptical rings have been implemented to achieve earlier concrete cracking [14,15]. Two standards for ring tests have been developed in the USA: the AASHTO bridge standard T 334-08 and ASTM 1581M–09a.

Established in the standard ASTM C 1581/C 1581M–09a "Determining Age at Cracking and Induced Tensile Stress Characteristics of Mortar and Concrete under Restrained Shrinkage" dimensions of the steel and concrete rings mean that the tensile stresses due to the constraints are similar to the tensile stresses due to the drying of the outer surface of concrete samples. This configuration of boundary stress causes uniform straining of the concrete section. A similar value of the edge tensile stress determines the fracture of the concrete sample as a result of exceeding its tensile strength [16]. The rings method uses strain gauge measurement of the steel ring strain caused by the shrinkage of concrete. The significant advantage of this method is that the recording of strains starts right after the sample is formed.

In modern concretes with low water/cement ratios, the overall shrinkage is significantly affected by the autogenous shrinkage, which occurs in the first stage of hardening. High-performance concretes undergo autogenous shrinkage even up to 200 μm/m after the first day of maturing. In the case of traditional concretes with a water/cement ratio of 0.5, the value of autogenic shrinkage after 28 days reaches 100 μm/m and in practical conditions is negligible [1]. Cracking caused by the shrinkage increases the penetration depth of water and aggressive substances that cause the corrosion of rebar, concrete leaching, and as a result, the deterioration of concrete's durability and structural failure. So far, a lot of research has been done to improve durability and minimize concrete susceptibility to cracking. The studies analyzed the impact of changing climatic conditions affecting the fracture rate of concrete samples [10,17,18] and the rate at which drying begins [19,20]. The effect of concrete composition on cracking susceptibility was also investigated [7,9,21–23]. The research also included the effect of internal curing soaked aggregate [24,25], fibers [7–9,26,27], admixtures reducing shrinkage [28,29]. Numerical simulation tests were also performed in predicting concrete susceptibility to cracking based on ring methods [30–32].

Tests performed in accordance with the ASTM C 1581/C 1581M–09a standard allow determining concrete sample cracking time as a result of restrained shrinkage exceeding concrete tensile strength. However, it is not possible to determine the exact value of the shrinkage; instead, the strain of the steel ring needs to be measured. Before test measurements can be used in further analysis, the steel measuring rings must be calibrated. The calibration process eliminates measurement errors caused by strain gauge installation, which could give different results than those calculated with theoretical equations. Those errors can significantly affect or even disrupt the mesurements entirely. Tests performed on calibrated steel rings using the restrained ring method allow accurately measuring strains in steel rings and make it possible to determine tensile stresses in concrete ring samples.

The article presents the calibration process of three steel measuring rings. Using calibrated restrained rings, the testing procedure was carried out for two self-compacting high-performance concretes with light and natural aggregate. Obtained steel ring deformation values and developed tensile stresses in annular concrete samples were analyzed for two maturation conditions: deformation due to autogenous and drying shrinkage—the side formwork removed after 24 hours of concreting—and deformation due to autogenous shrinkage only without side surface drying effects. The use of various test modes has made it possible to check the measurement precision and stability of strain development during short- and long-term tests.

2. Research Problem

The aim of the study was to calibrate three steel measuring rings for deformation registration in accordance with values resulting from the theory of elasticity. A novelty of this test is the calibration stand and procedure dedicated to measuring steel rings strain according to ASTM C 158/C 1581M–09a, which obtained a patent for an invention.

3. Methods and Experiment Program

3.1. Description of the Test Bench

The basic scheme of the calibration test bench is shown in Figure 1. The steel measuring ring equipped with strain gauges installed circumferentially on the internal surface must be set in the center of the outer shielding ring and fixed to the bottom plate. To apply external pressure for calibration, a rubber inflatable collar should be placed in between the measuring ring and the outer shielding. Then, the rings should be covered with a rigid top plate. Bottom and top plates should be made of a non-deformable material such as steel and fixed to each other with bolts. The outer ring should be 5 mm higher than the measurement ring to allow free deformation. Such a design of the test bench allows for the application of compressive stresses on the inner measuring ring from fixed outer shielding and fixed horizontal plates.

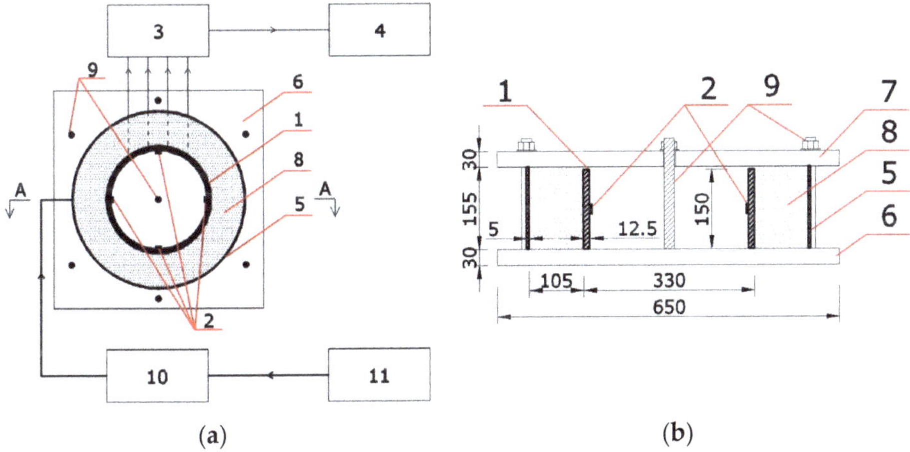

(a)   (b)

1–steel measuring ring, 2–circumferential strain gauges, 3–strain gauge bridge, 4–registration equipment (computer), 5–outer formwork ring, 6–bottom plate, 7–top plate, 8–rubber inflatable collar, 9–bolt fastening, 10–digital manometer, 11–air compressor

Figure 1. Block diagram of the calibration system for steel measuring rings: (**a**) top view; (**b**) section A-A.

The rubber collar should be connected through a digital manometer to the air compressor for simultaneous recording of its pressure and the deformation of the measuring ring. The steel measuring ring is connected with cables to strain gauge bridge and measurement equipment. The calibration system shown in Figure 1 uses a strain gauge bridge with internal temperature compensation.

The system used in the lab ulitizes a strain gauge bridge without the internal temperature compensation, which requires connecting strain gauges with Wheatstone half- or full-bridge circuits. Each measuring point consisted of a pair of strain gauges, which were vertically and annularly glued to the inner surface of the steel ring. The temperature compensation was provided by the strain gauges placed in the vertical axis, which are a part of a circuit of another measuring ring. The setup is shown in Figure 2 and a block diagram is presented in Figure 3. Calibration was carried out for three measuring rings, with four pairs of strain gauges spaced every 90 degrees. Strain gauges were installed in the circumferential direction, halfway up the inner surface of the steel rings. To compensate the temperature impact, recordings were taken from strain gauges installed in an additional measuring ring that was not actively involved in calibration, as shown in Figure 4a.

(a)   (b)

Figure 2. Calibration test bench: (**a**) placement of measuring ring, rubber collar, and shielding ring; (**b**) isolated with the top plate, bottom plate, and outer shielding ring.

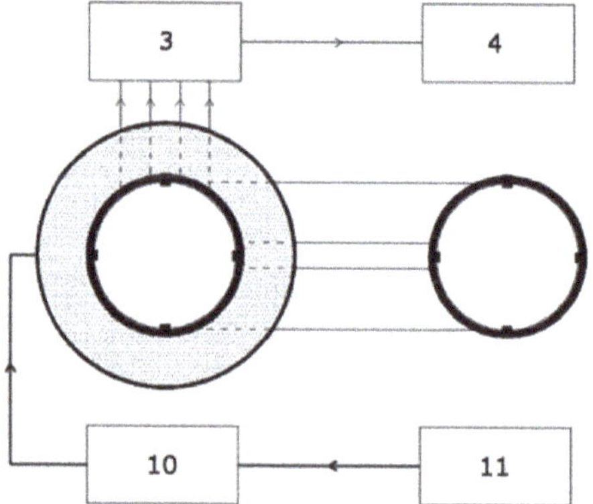

3–strain gauge bridge, 4–database, 10–digital manometer, 11–air compressor

Figure 3. Calibration system utilizing a strain gauge bridge without internal temperature compensation: 4 pairs of strain gauges.

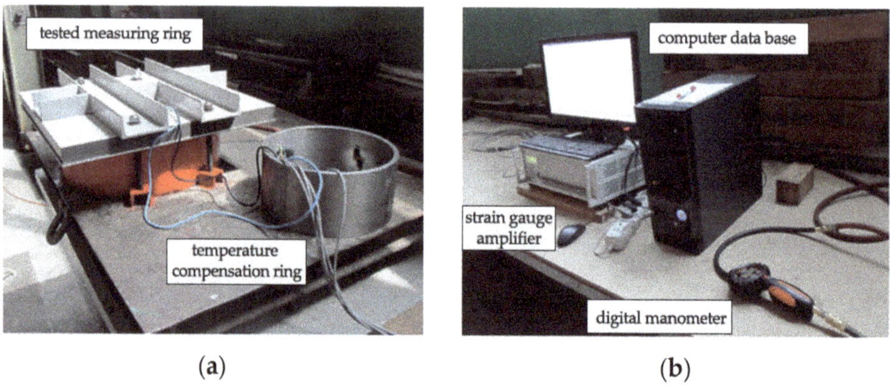

(a)   (b)

Figure 4. Components of the calibration system: (**a**) measuring rings during the test; (**b**) registration of the strain and air pressure.

3.2. Experimental Procedure

First, the passive stage of the calibration begins with placing the measuring ring on the test bench, connecting it tightly to the plates, and connecting the measuring equipment and the compressor. The active calibration process starts in the second stage, as shown in Figure 3. Air pumped by the compressor passed through the hose with a digital manometer to the collar. Once the space between the ring and shield plate fills, the collar starts to impose even radial pressure on the surrounding surfaces, including the external surface of the steel ring. The strain gauges register the change in resistance and send the impulse to the gauge bridge responsible for calculating the strain of the steel ring. From the gauge bridge, the signal is sent to the computer, which shows the measurements as a continuous graph of ring strain function. Figure 4 shows the test bench during the ring calibration process.

Additionally, to minimize friction between the expanding collar and the measuring ring, the outer surfaces of the measuring ring, collar, and inner surface of the outer ring were covered with synthetic oil before the test. The friction of expanding torus on the outer surface of the measuring ring can cause discrepancies and uneven strain. This results from Poisson's ratio for steel and can induce cumulative measurement error for each calibration of measuring rings.

The measurements allow acquiring a time function of pressure and strain. The result is visible as a linear dependency between circumferential strain and radial stress. A comparison of functions, which were both acquired from the measurements and calculated from theoretical equations, allows determining the calibration coefficient for the tested ring. The calibration allows comparing the results of measured strains for three independent rings.

Calibration analysis was made individually for three steel rings, using a theoretical function [6]:

$$\sigma_R = -\varepsilon_\theta \cdot E_s \cdot \frac{r_{os}^2 - r_{is}^2}{2r_{os}^2} \tag{1}$$

where σ_R represents the external pressure imposed on the steel ring (MPa), ε_θ represents the circumferential strain of the steel ring (m/m·10^{-3}), E_s represents the elasticity modulus of the steel ring (GPa), r_{os} represents the outer radius of the steel ring (mm), and r_{is} represents the inner radius of the steel ring (mm).

Based on the calibrated relationship of circumferential deformation of measuring rings ε_θ, to the value of radial pressure σ_R, the peripheral stress course in concrete ring samples is determined. The largest value of peripheral stresses in the concrete sample is recorded in the nearest zone of the radial stress of the steel ring—on the inner surface of the concrete sample [6]:

$$\sigma_{\theta max,c} = \sigma_R \cdot \left(\frac{r_{oc}^2}{r_{ic}^2} + 1\right) / \left(\frac{r_{oc}^2}{r_{ic}^2} - 1\right) \tag{2}$$

where $\sigma_{\theta max,c}$ represents the maximum circumferential stress in concrete specimen (MPa), r_{oc} represents the outer radius of the concrete specimen (mm), and r_{ic} represents the inner radius of the concrete specimen (mm).

4. Results

4.1. Calibration Test Results

The steel ring deformations were recorded individually for each of the four circumferential strain gauges as a function of time and depending on the acting pressure. To eliminate potential measurement errors and to increase the precision of the calibration, the measurement of the pressure acting on each ring and the measurement of strain at each strain gauge was taken 6 times. This allowed incorporating three measuring cycles, turning the steel ring around the rubber collar each time, with two measurements per cycle. Then, the mean value of the steel ring strain could be calculated.

The influence of air pressure that ranged from 0 to 5.5 bars on the strain function in time was consistent and repeatable for each tested ring, as shown in Figure 5.

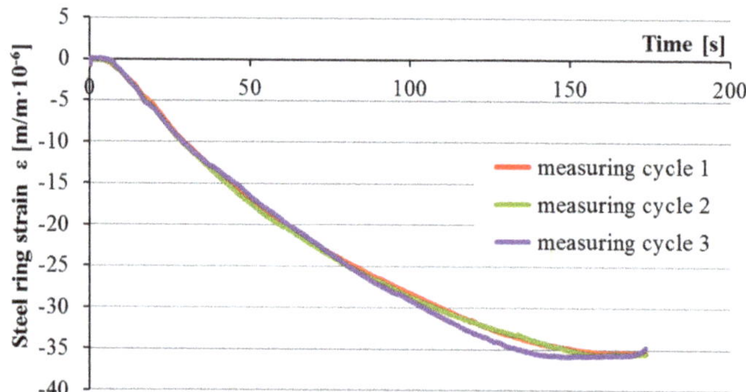

Figure 5. Steel ring B strains in relation to external pressure ranging from 0 to 5.5 bar.

The measured values of steel ring strains per each gague and for each trial test are shown in Table 1. The table also shows the average strains per each gauge from all trials and the average strain for the whole ring per each trial.

Based on the deviations shown in Table 1, it can be observed that the strain gauges of rings A and B were installed properly and the geometry of the ring is within 2%. It is assumed that deviation of up to 5% accounts for fabrication imprefections, and its impact is negligible. Deviations above 5% and up to 15% require the application of a calibration factor, which is derived and applied to an individual strain gauge or to the whole ring. Larger strain deviation requires elimination of the measuring ring from the tests. In such a situation, it is necessary to remove faulty strain gauges and verify ring geometry.

Table 1 also shows that for ring C, the measured values differed by 6% from the theoretical ring model. Circumferential strain gauges No. 1, 2, and 4 on rings A, B, and C record similar strain values, while strain gauge No. 3 on ring C shows value lower than the values for the corresponding strain gauge on rings A and B. This indicates a poor installation of the third circumferential strain gauge and a correct geometry of the steel ring. As mentioned above, for ring C, a calibration coefficient has to be applied due to the measured strain divergance off the theoretical values within 15%. Tolerance ranges between ±5% and ±15% were analyzed for each circumferential strain gauge and for the mean ring deformation value relative to the theoretical value.

When strain gauges record differenciated strain values at a constant pressure level, this indicates their incorrect or non-parallel installation on the inner surface of the ring. However, if all recorded deformation values are similar and lower or higher than the theoretical value, then most likely, the measuring ring geometry differs.

Figure 6 shows the measurement accuracy of the tested rings relative to the theoretical strain values. Rings A and B show strain values close to the ones calculated from Equation (1), whereas ring C had an extensive measurement error.

Figure 7 shows the measured function of circumferential strain–radial stresses for steel rings and the theoretical curve. The strain function for rings A and B develops in accordance to the theoretical relationship. Based on this, it can be stated that rings A and B are calibrated properly, and there is no need for additional amplification through a calibration factor. The strains of ring C differ significantly from theoretical calculations. To properly calibrate ring C, it is necessary to change the slope coefficient of the circumferential strain–radial stress function.

Table 1. Measured circumferential strains for individual strain gauges under constant pressure (5.5 bars).

Measuring Cycle (MC)		Repetition	1: Starting Position		2: 90° Turn		3: 180° Turn		Gauge Mean	Ring Mean/ Theoretical Ring	Deviation
			1 time	2 time	1 time	2 time	1 time	2 time			
Ring A	Strain per gauge $[m/m \cdot 10^{-6}]$	1	−37.26	−36.98	−35.54	−37.01	−36.67	−36.21	−36.61	101.9%	1.9%
		2	−36.13	−36.11	−36.67	−36.23	−36.54	−36.04	−36.29	101.0%	1.0%
		3	−35.47	−36.14	−35.54	−36.16	−37.24	−36.54	−36.18	100.7%	0.7%
		4	−36.55	−36.07	−36.33	−36.31	−36.18	−35.73	−36.20	100.7%	0.7%
	Ring Mean per MC		−35.35	−36.33	−36.02	−36.43	−36.66	−36.13	**−36.32**	**101.1%**	**1.1%**
Ring B	Strain per gauge $[m/m \cdot 10^{-6}]$	1	−35.82	−36.05	−35.25	−35.99	−35.71	−35.89	−35.79	99.6%	−0.4%
		2	−35.43	−35.64	−35.97	−35.16	−35.58	−35.16	−35.49	98.8%	−1.2%
		3	−35.36	−35.64	−35.92	−35.74	−34.84	−35.86	−35.56	99.0%	−1.0%
		4	−35.03	−36.04	−35.14	−35.81	−35.81	−35.11	−35.49	98.8%	−1.2%
	Ring Mean per MC		−35.41	−35.84	−35.57	−35.68	−35.49	−35.51	**−35.58**	**99.0%**	**−1.0%**
Ring C	Strain per gauge $[m/m \cdot 10^{-6}]$	1	−34.98	−35.11	−35.57	−35.27	−35.78	−35.47	−35.36	98.4%	−1.6%
		2	−35.51	−34.14	−34.05	−34.52	−34.54	−34.68	−34.57	96.2%	−3.8%
		3	−30.57	−30.87	−31.21	−30.94	−30.56	−30.68	−30.81	85.7%	−14.3%
		4	−34.89	−34.71	−34.94	−34.48	−34.5	−34.64	−34.69	96.6%	−3.4%
	Ring Mean Per MC		−33.99	−33.71	−33.94	−33.8	−33.85	−33.87	**−33.86**	**94.2%**	**−5.8%**
	Theoretical ring								**−35.93**	**100.0%**	**0.0%**

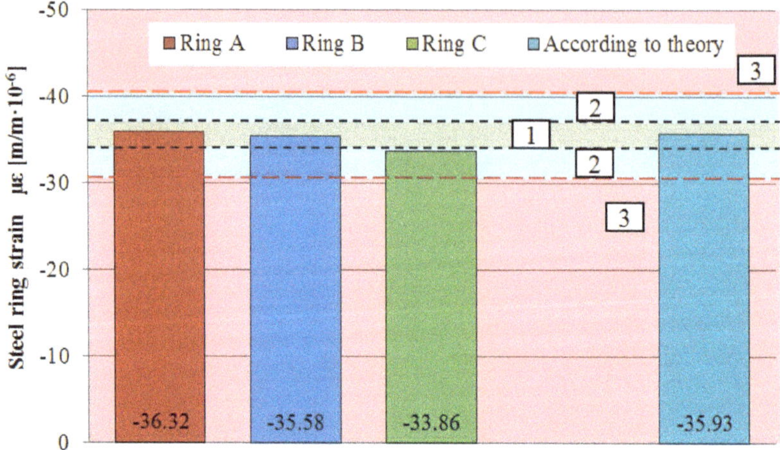

Figure 6. Circumferential deformation of tested steel rings under 5.5 bar pressure with allotment to measuring correctness zones: 1—deformation of the ring within ±5% tolerance, 2—deformation of the ring requires the use of a calibration factor within ±15%, 3—incorrect registration of ring deformation.

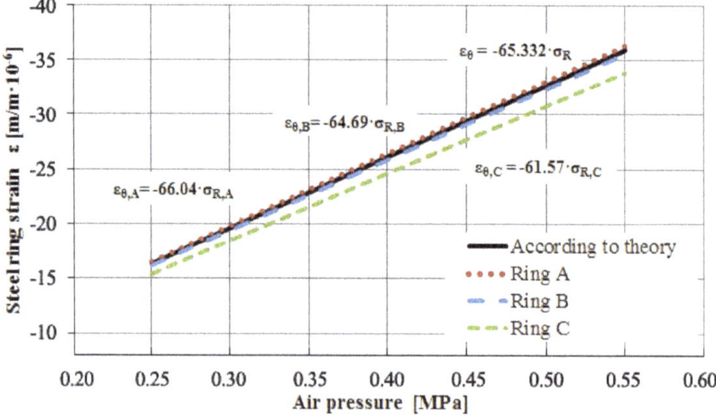

Figure 7. Determination of calibration coefficient for individual ring.

4.2. Calibration Coefficient for Individual Ring

The result of calibration process is an individually determined ring calibration coefficient (3) that adjusts the slope coefficient of the measured value plot to the theoretical plot. This coefficient accounts for ring geometrical imperfections and faulty strain gauge installation. The calibration coefficients for the three measuring rings considered are shown in Table 2.

$$\gamma_c = \frac{\varepsilon_{\theta.t}}{\varepsilon_{\theta.m}} \tag{3}$$

where $\varepsilon_{\theta.t}$ represents the theoretical circumferential strain of the steel ring at given pressure (m/m·10⁻⁶), $\varepsilon_{\theta.m}$ represents the measured circumferential strain of the steel ring at given pressure (m/m·10⁻⁶), and γ_c represents the calibration coefficient.

Table 2. Error tolerance and calibration coefficients.

Ring	Strain at 5.5 Bars [m/m·10^{-6}]	Deviation [%]	Calibration Coefficient [-]
A	−36.32	1.1	1.000
B	−35.58	−1.0	1.000
C	−33.86	−5.8	1.061
Theory	−35.93	-	-

The recorded strains of rings A and B are within the lower bound tolerance of 5%, so they do not need to be calibrated, and they can be directly used in further analyses. The measured strains for ring C must be computed, including the calibration coefficient, accordingly to the equation:

$$\varepsilon_{\theta.n} = \gamma_c \cdot \varepsilon_{\theta.n.m} \tag{4}$$

where $\varepsilon_{\theta.n}$ represents the measured circumferential strain of the steel ring "n" (m/m·10^{-6}), and $\varepsilon_{\theta.n.m}$ represents the recorded circumferential strain of the steel ring "n".

The calibration coefficient can also be used to rectify the concrete cracking time, as shown in Equation (5). In the case of a uniform deviation of recorded strains from all the strain gauges of a given ring, this clearly indicates stiffness that deviates from the stiffness of the theoretical ring. In such a situation, when the deformation deviation is in the range of 5% to 15%, it is reasonable to modify the recorded concrete cracking time with a calibration factor. Based on the results in Table 1, only one C-ring strain gauge read values significantly below the theoretical value, which clearly indicates the mounting error of this strain gauge and no reason to modify the cracking time for this ring.

$$t_{crack.n} = \frac{t_{crack,n.m}}{\gamma_c} \tag{5}$$

where $t_{crack.n}$ represents the measured cracking time of the steel ring "n" after the calibration (days), and $t_{crack.n.m}$ represents the recorded cracking time of the steel ring "n" (days).

The use of such calibration is necessary for each measuring ring, which was prepared for susceptibility to cracking tests in accordance with the Standard ASTM C 1581/C 1581M–09a.

4.3. σ-ε Relation

The use of calibration coefficients for each measuring ring allows for a common interpretation of results, averaging the deformation values, determining of the average cracking time as a mean of the cracking times for individual samples, and determining the function of circumferential deformation of the measuring ring ε_θ to maximum values of circumferential stresses in concrete ring samples $\sigma_{\theta max,c}$. Figure 8 presents the linear relationship of the discussed parameters.

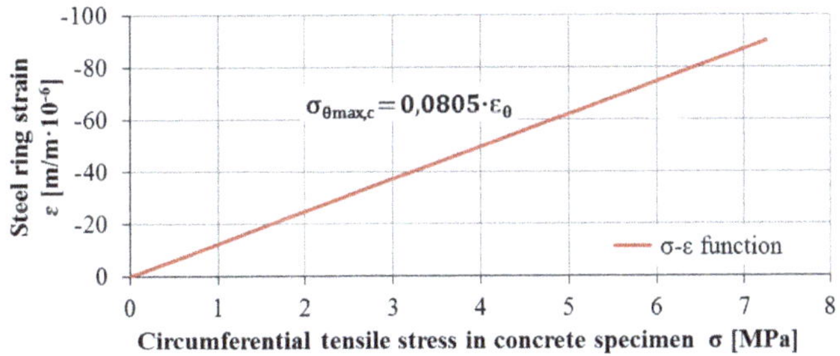

Figure 8. Theoretical relationship between steel ring strain and concrete ring sample tensile stress.

5. Experimental Research

An analysis of the impact of the steel measuring ring calibration was carried out for two self-compacting concretes: concrete C-1 with fine and coarse natural aggregate, and concrete C-2 with pre-soaked fine and coarse lightweight aggregate. Two types of concrete shrinkage tests were performed for both concretes analyzed; the first was based on concrete deformation after 24 hours from concreting, while the second did not involve sample deformation.

The composition of concrete mixes under consideration is shown in Table 3. Annular concrete samples were formed around the steel measuring rings, and their geometry was in agreement with the requirements of ASTM C 1581/C 1581M–09a. The measuring stands were placed in a climatic chamber where tests were carried out at a constant temperature T = 20 ± 2 °C and relative humidity RH = 50 ± 3%. The designed concretes were to have a high susceptibility to cracking under the influence of total shrinkage.

Table 3. Composition and notification of concrete mixes.

Concrete	Cement 42,5R [kg/m^3]	Fly Ash [kg/m^3]	Silica Fume [kg/m^3]	Water [kg/m^3]	SP [kg/m^3]	Aggregate [kg/m^3]			
						Natural		Lightweight	
						0–2	2–8	0–4	4–8
C1/450/NA	450	72	38	155	11	624	1072	-	-
C2/450/NA-LWA	450	72	38	155	7.65	-	-	310	540

Deformation tests were carried out simultaneously on three calibrated measuring rings, as shown in Figure 9.

(a)   (b)

Figure 9. Restrained shrinkage test of concrete: (**a**) concrete samples insulated and subjected to autogenous shrinkage; (**b**) side formwork removal after 24 h of concreting and measurement of the impact of the drying shrinkage.

Figures 10 and 11 present the results of type 1 steel ring deformation tests and the development of tensile stresses on the inner surface of concrete samples from the moment of their formation, followed by deformation after 24 h, and until their cracking as a result of progressive drying shrinkage.

The performed deformation tests allowed to conduct two separate analyses. The first analysis concerned the deformations of the measuring ring C before and after calibration, taking into account the determined calibration factor. On its basis, it can be concluded that calibration validates the C ring relative to rings A and B. Therefore, the deformation values are characterized by a low standard deviation and allow for determination of the average deformation development affecting the correct interpretation of the results.

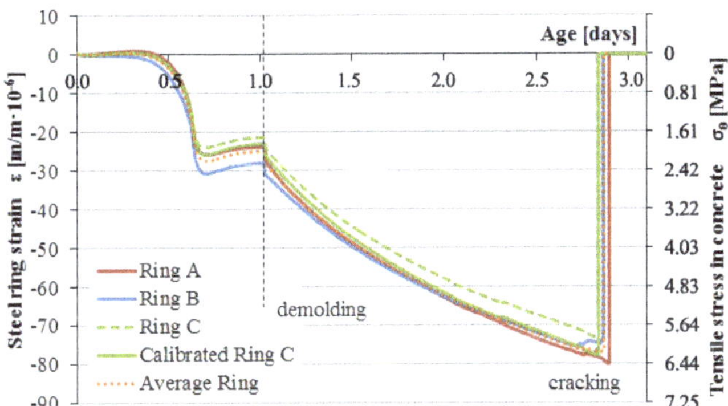

Figure 10. Strain development in the steel rings and stress progress in concrete C-1 induced by the total shrinkage.

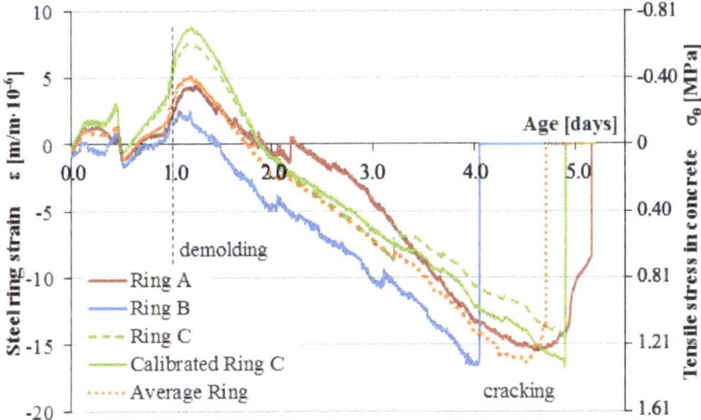

Figure 11. Strain development in the steel rings and stress progress in concrete C-2 induced by the total shrinkage.

The second analysis referred to the interpretation of the material properties of concrete based on the relationship of the steel ring deformation to tensile stress on the inner surface of the annular concrete samples as a function of time. The natural aggregate used in C-1 concrete resulted in higher strength parameters as well as a more airtight and homogeneous structure compared to C-2 concrete with lightweight aggregate. Yet, C-1 concrete cracked in the third day after concreting at the average deformation value of the steel ring of −76.8 µm/m and a mean tensile stress of 6.2 MPa at the inner surface of concrete samples. The dynamic development of autogenous shrinkage in the first day and the additional impact of drying shrinkage after one day resulted in a rapid loss of strength due to the cracking of C-1 concrete samples. In the case of C-2 concrete, no autogenous shrinkage was observed in the first day and there was moderate development of the shrinkage from drying out after sample deforming. Light soaked aggregate led to internal care, which caused a slower development of shrinkage and stress. The use of lightweight aggregate extended the cracking time to about 5 days and reduced the strength of the concrete. C-2 concrete cracking occurred at the average deformation value of the steel ring of −16.3 µm/m, causing an average inner surface tensile stress of 1.3 MPa. Figure 12 shows the morphology of concrete sample cracks after the loss of strength due to autogenous and drying shrinkage. The development of deformation of the measuring rings reflects the homogeneity of the material structure. Hence, it is observed that for concrete C-2, the deformation course was more irregular.

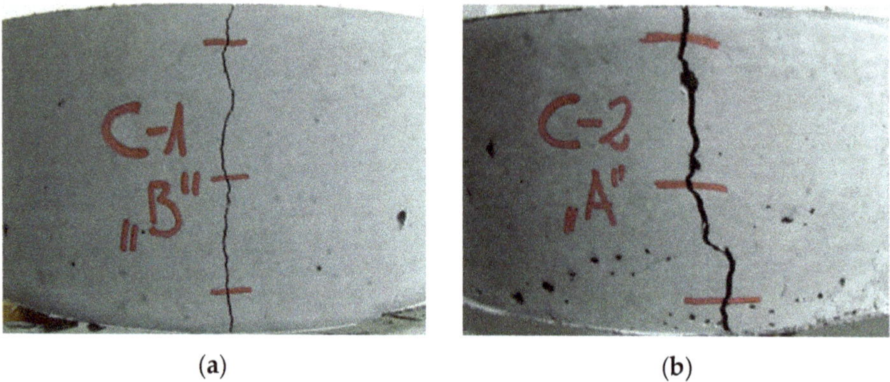

(a)
(b)

Figure 12. Cracked concrete ring specimens: (**a**) high-performance concrete with coarse natural aggregate 2-8, crack width = 0.9 mm; (**b**) high-performance concrete with coarse lightweight aggregate 4-8, crack width = 2.4 mm.

In the following type 2 restrained concrete tests, the impact of steel rings calibration on the correctness of measurements over a longer period of time was analyzed. Ring samples of C-1 and C-2 concretes were not deformed after 1 day but remained insulated for 28 days. At that time, only autogenous shrinkage developed, and its impact on steel ring deformations was analyzed. The test results are shown in Figures 13 and 14.

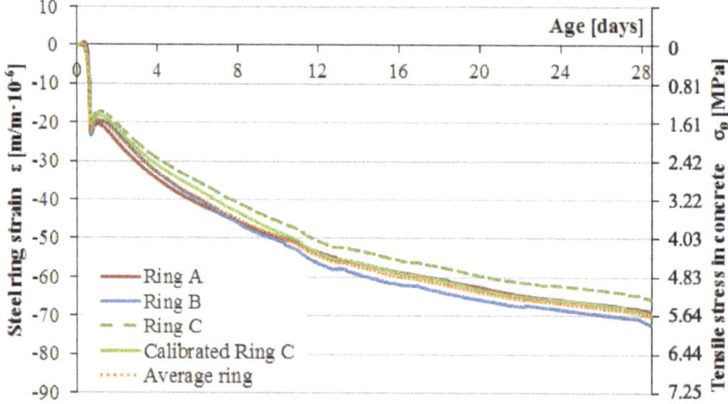

Figure 13. Strain development in the steel rings and stress progress in concrete C-1 induced by the autogenous shrinkage.

The measurement of deformation of steel rings under the influence of autogenous shrinkage, especially for concrete C-1, showed the correctness of the calibration procedure in the range of 28 days. For the C-ring, the results before and after calibration are presented. The application of the calibration factor for the C-ring deformation course allowed for correct analysis of the results and the determination of concrete susceptibility to cracking in both short and long measurement periods.

Based on the analysis of the development of parameters of C-1 concrete with natural aggregate, a monotonic increase in the deformation of the measuring rings can be noticed as a result of the continuous development of autogenous shrinkage of concrete. Within 28 days, concrete does not show susceptibility to cracking at a given limitation level. On the other hand, the nature of the increase and the value of the average tensile stress at the inner surface of the concrete samples at the level of 5.5 MPa may indicate the development of micro-cracks in the structure and fracture of the samples at a later time. A lack of sample cracking within 28 days is caused by the increase in concrete strength during the test and by the absence of drying shrinkage.

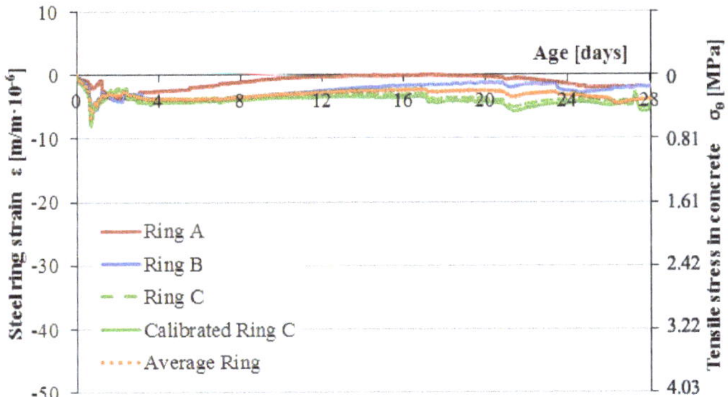

Figure 14. Strain development in the steel rings and stress progress in concrete C-2 induced by the autogenous shrinkage.

Analysis of the deformation progress of measuring rings for C-2 concrete with lightweight aggregate showed no impact of autogenous shrinkage. In the whole measuring range, the soaked lightweight aggregate showed curing properties, as a result of which autogenous shrinkage did not develop in concrete ring samples. The registration of steel ring deformations throughout the entire measuring range was between 0 and -10 μm/m, generating minimal tensile stress in concrete samples.

Figure 13 proves the necessity of steel ring C calibration, where the calibrated strain values converge with strains for rings A and B. The uncalibrated, recorded strain of ring C was plotted as well and shows an approximate deviation of concrete tensile stresses after 28 days of about 0.4 MPa, which translates to underestimation of about 7% relative to mean stress from all the samples.

6. Conclusions

In the calibration test, the pressure applied to the measuring ring by the rubber collar imitates the load caused by the shrinkage of concrete. Measurement of the air pressure with the digital manometer determines the graph of the circumferential strain–radial stress function. The calibration test additionally eliminates the error caused by the geometry and elastic modulus of the material. Pneumatic calibration allows compensating for errors due to improper strain gauge installation by the application of a test-determined calibration coefficient, which translates registered strains into calibrated circumferential strains closely aligned to theoretical values.

The calibration procedure allowed for simuntanous strain measurements under given stress for all rings of the test bench. Acquired calibration functions are used to calculate the mean values of the results, which can be used in further studies. Calibrated strains help determine the stresses occurring at the moment of cracking of concrete ring samples using a standard rigid measuring ring. This enables the classification of concrete's susceptibility to cracking.

Short-term and long-term tests confirm the effectiveness of calibration to correctly interpret the test results on concrete susceptibility to cracking using restrictive rings. The applied calibration method extends the scope of tests with the correct analysis of the average deformability of steel rings and determination of the value of tensile stress in concrete at a given level of steel ring deformation.

In the future, it is planned to carry out research on the impact of the percentage of mineral additives on the time of cracking of concrete ring samples caused by the effect of autogenous shrinkage.

7. Patents

No. PL225785: Method for the calibration of measuring rings used for measuring their deformability in result of the shrinking strain of poured-in materials and the system for the calibration of measuring rings used for measuring their deformability as a result of the shrinking strain of poured-in materials.

Author Contributions: Conceptualization, A.Z. and M.K.; methodology, A.Z.; software, A.Z.; validation, A.Z.; formal analysis, A.Z.; investigation, A.Z., M.K.; writing—original draft preparation, A.Z., M.K.; writing—review and editing, A.Z., M.K.; supervision, A.Z., M.K. All authors have read and agreed to the published version of the manuscript.

Funding: This research received no external funding.

Acknowledgments: The authors would like to recognize the Faculty of Civil Engineering and Architecture of West Pomeranian University of Technology Szczecin and thank for enabling the research described in this article. We thank our technician Ryszard Wojtaszewski and reviewer Dr. Patryk J. Wolert whose suggestions greatly improved this manuscript.

Conflicts of Interest: The authors declare no conflict of interest.

References

1. Aïtcin, P.C. *High Performance Concrete*, 1st ed.; CRC Press: London, UK, 1998. [CrossRef]
2. Tazawa, E. *Autogenous Shrinkage of Concrete*, 1st ed.; A & FN Spon Press: London, UK, 1999. [CrossRef]
3. Xie, T.; Fang, C.; Mohamad Ali, M.S.; Visintin, P. Characterizations of autogenous and drying shrinkage of ultra-high performance concrete (UHPC): An experimental study. *Cem. Concr. Compos.* **2018**, *91*, 156–173. [CrossRef]
4. Yang, Y.; Sato, R.; Kawai, K. Autogenous shrinkage of high-strength concrete containing silica fume under drying at early ages. *Cem. Concr. Res.* **2005**, *35*, 449–456. [CrossRef]
5. Dueramae, S.; Tangchirapat, W.; Chindaprasirt, P.; Jaturapitakkul, C.; Sukontasukkul, P. Autogenous and drying shrinkages of mortars and pore structure of pastes made with activated binder of calcium carbide residue and fly ash. *Constr. Build. Mater.* **2020**, *230*, 116962. [CrossRef]
6. Carlson, R.W.; Reading, T.J. Cracking of Concrete. *J. Boston Soc. Civ. Eng.* **1942**, *29*, 98–109.
7. Hogancamp, J.; Grasley, Z. The use of microfine cement to enhance the efficacy of carbon nanofibers with respect to drying shrinkage crack resistance of Portland cement mortars. *Cem. Concr. Compos.* **2017**, *83*, 405–414. [CrossRef]
8. Yoo, D.-H.; Banthia, N.; Yoon, Y.-S. Ultra-high performance fiber-reinforced concrete: Shrinkage strain development at early ages and potential for cracking. *J. ASTM Int.* **2017**, *45*, 2061–2070. [CrossRef]
9. Briffaut, M.; Bendoudjema, F.; D'Aloia, L. Effect of fibres on early age cracking of concrete tunnel lining. Part I: Laboratory ring test. *Tunn. Undergr. Space Technol.* **2016**, *59*, 215–220. [CrossRef]
10. Carlson, R.W.; Reading, T.J. Model study of shrinkage cracking in concrete building walls. *ACI Struct. J.* **1988**, *85*, 395–404.
11. Hossain, A.B.; Weiss, J.W. The role of specimen geometry and boundary conditions on stress development and cracking in the restrained ring test. *Cem. Concr. Compos.* **2006**, *26*, 189–199. [CrossRef]
12. Grzybowski, M.; Shah, S.P. Shrinkage cracking of fiber reinforced concrete. *ACI Mater. J.* **1990**, *87*, 138–148.
13. Dean, S.W.; Schlitter, J.L.; Senter, A.H.; Bentz, D.P.; Nantung, T.; Weiss, W.J. A Dual Concentric Ring Test for Evaluating Residual Stress Development due to Restrained Volume Change. *J. ASTM Int.* **2010**, *7*, 1–13. [CrossRef]
14. Dong, W.; Zhou, X.; Wu, Z. A fracture mechanics-based method for prediction of cracking of circular and elliptical concrete rings under restrained shrinkage. *Eng. Fract. Mech.* **2014**, *131*, 687–701. [CrossRef]
15. Dong, W.; Zhou, X.; Wu, Z.; Kastiukas, G. Effects of specimen size on assessment of shrinkage cracking of concrete via elliptical rings: Thin vs. thick. *Comput. Struct.* **2016**, *174*, 66–78. [CrossRef]
16. Bentur, A.; Kovler, K. Evaluation of early age cracking characteristics in cementitious systems. *Mater. Struct.* **2003**, *36*, 183–190. [CrossRef]
17. Li, Z.; Qi, M.; Li, Z.; Ma, B. Crack width of high-performance concrete due to restrained shrinkage. *J. Mater Civ. Eng.* **1999**, *11*, 214–223. [CrossRef]

18. Shah, H.R.; Weiss, J.W. Quantifying shrinkage cracking in fiber reinforced concrete using the ring test. *Mater. Struct.* **2006**, *39*, 887–899. [CrossRef]
19. Beushausen, H.; Chilwesa, M. Assessment and prediction of drying shrinkage cracking in bonded mortar overlays. *Cem. Concr. Res.* **2013**, *53*, 256–266. [CrossRef]
20. Kovler, K.; Sikuler, J.; Bentur, A. Restrained shrinkage tests of fibre-reinforced concrete ring specimens: Effect of core thermal expansion. *Mater. Struct.* **1993**, *26*, 231–237. [CrossRef]
21. Yu, Y.; Zhu, H. Effects of Rubber Size on the Cracking Resistance of Rubberized Mortars. *Materials* **2019**, *12*, 3132. [CrossRef]
22. Huang, L.; Hua, J.; Kang, M.; Luo, Q.; Zhou, F. Influence of Steel Plates and Studs on Shrinkage Behavior and Cracking Potential of High-Performance Concrete. *Materials* **2019**, *12*, 342. [CrossRef]
23. Ma, L.; Zhao, Y.; Gong, J. Restrained early-age shrinkage cracking properties of high-performance concrete containing fly ash and ground granulated blast-furnace slag. *Constr. Build. Mater.* **2018**, *191*, 1–12. [CrossRef]
24. Kaszyńska, M.; Zieliński, A. Influence of mixture composition on shrinkage cracking of lightweight self-consolidating concrete. *Brittle Matrix Compos.* **2012**, *10*, 265–274. [CrossRef]
25. Kaszyńska, M.; Zieliński, A. Effect of lightweight aggregate on minimizing autogenous shrinkage in Self-Consolidating Concrete. *Procedia Eng.* **2015**, *108*, 608–615. [CrossRef]
26. Wu, X.; Zhou, J.; Kang, T.; Wang, F.; Ding, X.; Wang, S. Laboratory Investigation on the Shrinkage Cracking of Waste Fiber-Reinforced Recycled Aggregate Concrete. *Materials* **2019**, *12*, 1196. [CrossRef] [PubMed]
27. Saradar, A.; Tahmouresi, B.; Mohseni, E.; Shadmani, A. Restrained Shrinkage Cracking of Fiber-Reinforced High-Strength Concrete. *Fibers* **2018**, *6*, 12. [CrossRef]
28. Shah, S.P.; Karaguler, M.E.; Sarigaphuti, M. Effects of shrinkage-reducing admixtures on restrained shrinkage cracking of concrete. *ACI Mater. J.* **1992**, *89*, 291–295.
29. Weiss, J.; Lura, P.; Rajabipour, F.; Sant, G. Performance of shrinkage-reducing admixtures at different humidities at and early ages. *ACI Mater. J.* **2008**, *105*, 478–486. [CrossRef]
30. Al-musawi, H.; Huang, H.; Guadagnini, M.; Pilakoutas, K. A numerical study on the effect of restrained shrinkage on rapid hardening plain and recycled clean steel fibre concrete overlays. *Constr. Build. Mater.* **2020**, *244*, 117723. [CrossRef]
31. Briffaut, M.; Benboudjema, F.; Torrenti, J.M.; Nahas, G. Numerical analysis of the thermal active restrained shrinkage ring test to study the early age behavior of massive concrete structures. *Eng. Struct.* **2011**, *33*, 1390–1401. [CrossRef]
32. Radlińska, A.; Kaszyńska, M.; Zieliński, A.; Ye, H. Early-Age Cracking of Self-Consolidating Concrete with Lightweight and Normal Aggregates. *J. Mater. Civ. Eng.* **2018**, *30*, 04018242. [CrossRef]

Article

The Acoustic Emission Method Implementation Proposition to Confirm the Presence and Assessment of Reinforcement Quality and Strength of Fiber–Cement Composites

Anna Adamczak-Bugno * and Aleksandra Krampikowska

Faculty of Civil Engineering and Architecture, Kielce University of Technology, Aleja Tysiąclecia Państwa Polskiego 7, 25-314 Kielce, Poland; akramp@tu.kielce.pl
* Correspondence: aadamczak@tu.kielce.pl; Tel.: +48-723-790-363

Received: 18 June 2020; Accepted: 30 June 2020; Published: 2 July 2020

Abstract: This article proposes to use the acoustic emission (AE) method to evaluate the degree of change in the mechanical parameters of fiber–cement boards. The research was undertaken after a literature review, due to the lack of a methodology that would allow nondestructive assessment of the strength of cement–fiber elements. The tests covered the components cut out from a popular type of board available on the construction market. The samples were subjected to environmental (soaking in water, cyclic freezing–thawing) and exceptional (burning with fire and exposure to high temperature) factors, and then to three-point bending strength tests. The adopted conditions correspond to the actual working environment of the boards. When applying the external load, AE signals were generated, which were then grouped into classes, and initially assigned to specific processes occurring in the material. The frequencies occurring over time for the tested samples were also analysed, and microscopic observations were made to confirm the suppositions based on the first part of the tests. Comparing the results obtained from a group of samples subjected to environmental and exceptional actions, significant differences were noted between them, which included the types of recorded signal class, the frequency of events, and the construction of the microstructure. The degradation of the structure, associated with damage to the fibers or their complete destruction, results in the generation under load of AE signals that indicate the uncontrolled development of scratches, and a decrease in the frequency of these events. According to the authors, the methodology used allows the control of cement–fiber boards in use. The registration and analysis of active processes under the effect of payloads makes it possible to distinguish mechanisms occurring inside the structure of the elements, and to formulate a quick response to the situation when the signals indicate a decrease in the strength of the boards.

Keywords: cement–fiber boards; acoustic emission method; k-means algorithm; wavelet analysis; fiber composites

---

1. Introduction

Fiber–cement is a natural material that is widely used in construction. Its creator was Czech engineer Ludwig Hatschek. At the beginning of the last century, to produce such elements, asbestos was used. In the 1970s and 1980s, asbestos cement sheets were popularly called eternit [1–5].

Since then, the method of fiber–cement production has undergone many transformations. First of all, asbestos has been eliminated and replaced with completely harmless materials. The appearance of fiber–cement products has also changed significantly [6–8].

The relatively high strength parameters of fiber–cement boards are the result of their production process. The components are made of Portland cement (90%) and rayon staple (10%). Cement binds the material and determines its final strength. Cellulose provides the right amount of water in the cement setting process, fills the gaps and increases the density of the product. It also serves as the reinforcement. In the production process, mineral materials are also used as fillers to improve the flexibility and appearance of the boards [9–13].

The fiber–cement production technology consists of applying successive thin layers of a mixture of materials, which are pressed well together before the slow hardening process. The applied technological process allows fiber–cement to achieve high strength parameters [14–16].

As mentioned above, boards can be used both inside and outside buildings. Installation of the components in an outdoor environment involves exposure of the boards to weather conditions. Therefore, it is considered appropriate to perform tests on fiber–cement components taking into account the conditions to which they will be exposed during use. Among the typical interactions, the authors indicate atmospheric precipitation and the recurring process of freezing and defrosting the boards. Certain factors, specified in the standard guidelines, should be taken into account when testing the physical and mechanical parameters. Attempts to evaluate the condition of fiber–cement components must also take into account the likelihood of special factors, which include high temperatures and direct fire [17–20].

The paper presents the application of the acoustic emission (AE) method, based on unsupervised pattern recognition (k-means algorithm), to evaluate the change of the mechanical parameters of fiber–cement components. Three-point bending tests were performed on material samples in an air-dry state, which were then subjected to environmental (soaking in water, and recurring freezing and defrosting) and special (flaming with a gas burner and exposing to a high temperature—230 °C for 180 min) operating factors. According to the authors, the environmental factors listed above correspond to the typical operating conditions of external cladding boards, which must always be taken into account in the test procedures. The flame action reflected the fire conditions that impact directly on the partition covered with cement–fiber boards. The high temperature was also intended to reflect fire conditions (range of the high temperature zone) and heat accumulation in the material. The recorded AE signals (generated by active developing destructive processes) were divided into four classes. Each of them reflects certain processes taking place in the structure of components under an external load. The frequencies occurring over time in each of the tested samples were also analysed.

After the three-point bending strength test, fragments of fractures were extracted from the tested components for microstructure tests, performed with a scanning electron microscope. Parallel to the observations of the fracture surfaces, an analysis of the elemental composition of the material was performed using the EDS (Energy Dispersive X-Ray Spectroscopy) microprobe.

The authors undertook the indicated research topics because they believe that the use of the AE method may enable the assessment of the condition of the cement–fiber boards in operation. So far, most of the research on cement–fiber boards has been devoted to the influence of operational factors [21,22] and the action of high temperatures, examined by testing some of the physicochemical parameters of the boards, especially their bending strength (Modulus of Rupture MOR)). The nondestructive testing of fiber–cement boards was mainly limited to detecting imperfections arising at the production stage. Articles by Drelich et al. [23] and Schabowicz and Gorzelańczyk [24] present the possibility of using Lamb waves in a non-contact ultrasonic scanner to detect such defects. In the literature on the subject, there is little information on the use of other nondestructive testing methods for fiber–cement boards. The research described in the works of Chady et al. [25] and Chady and Schabowicz [26] showed that the terahertz (T-Ray) method is suitable for testing fiber–cement boards. Terahertz signals are very similar to ultrasonic signals, but their interpretation is more complicated. In [27], the microtomography method was used to identify delaminations and low-density areas in fiber–cement boards. Test results indicate that this method clearly reveals differences in the microstructures of the elements. Therefore, the microtomography method can be a useful tool for testing the structure of fiber–cement boards, in which defects may arise due to manufacturing errors. However, this method can only be used for small size boards.

It should be noted that, so far, few cases of testing fiber–cement boards using acoustic emissions have been reported. Ranachowski and Schabowicz et al. [27] conducted pilot studies on fiber–cement boards manufactured by extrusion, including exposing boards to 230 °C for 2 h. In this research, an acoustic emission method was used to determine the impact of cellulose fibers on boards' strength, and they attempted to distinguish between the AE events emitted by the fibers and the cement matrix. The results of these tests confirmed the usefulness of this method for testing fiber–cement boards. In [28], by Gorzelańczyk et al., the use of the acoustic emission method was proposed to examine the effect of high temperatures on fiber–cement boards. It should be noted that the effect of high temperatures on concrete, as well as the interrelationships associated with this process, have been extensively described using the acoustic emission method; an example is Ranachowski's work [5]. AE was used to assess the state of concrete-like material and consider the effects of extreme conditions on it, such as fire or frost (papers [29–32]). It should also be mentioned that the acoustic emission method is often used to test thin materials, e.g., steel and polymer composites, and even fragile food products [15,33]. Nondestructive tests have also been used to show decay, and the correlation between static and energy performances [34].

Analysing the results of the conducted works [35–39], it can be stated that for cement matrix panels reinforced with fibers, the biggest threat is the damage or degradation of the reinforcing fibers, as well as any decrease in the degree of binding between the matrix and reinforcement, as these decrease the mechanical parameters of the composites. In turn, taking into account the fact that these panels are mounted on facades of public buildings classified as high, a decrease in the strength of cladding elements may be associated with a real threat to human health and life. On the other hand, the wide applicability of the acoustic emission method suggests that it will also give positive results when assessing the condition of fiber–cement boards.

To sum up, it has been found that there is no methodology in the existing literature for assessing the condition of fiber–cement boards. For this reason, the authors conducted research using the acoustic emission method, the results of which clearly show that the signals recorded for samples with high mechanical parameters differ significantly from those for elements with a degraded structure and reduced strength. The differences consist in the number of recorded signals, belonging to classes corresponding to the destructive processes occurring in the material, as well as the frequencies emitted.

2. Materials and Methods

The tests were performed on rectangular samples cut out from a 3100 × 1250-mm fiber–cement board. Information about the components intended for the three point bending strength test is presented in Table 1.

Table 1. The conditioning and testing plan.

Sample Determination	Type of Conditioning	Sample Length [mm]	Sample Width [mm]	Sample Thickness [mm]	Number of Samples in the Group
BS	Air-dry state	300	50	8	5
BW	Soaking in water	300	50	8	5
BM	50 freezing–defrosting cycles	300	50	8	5
BP	A direct flame of a gas burner for 10 min	300	50	8	5
BC	Firing in a laboratory oven at 230 °C for 3 h	300	50	8	5

The tests were performed using a Zwick Roell universal testing machine. The bending speed for each sample was 0.1 mm/min (each test was carried out with a constant increase in deflection). The scheme of the test stand is shown in Figure 1. The axial distance between the support points ls was 200 cm, and the radius of the support points and the loading beam r was 10 mm. Based on the measurement data, the *MOR* bending strength of the elements was calculated in accordance with EN 12467 Fibre-cement flat sheets - Product specification and test methods.

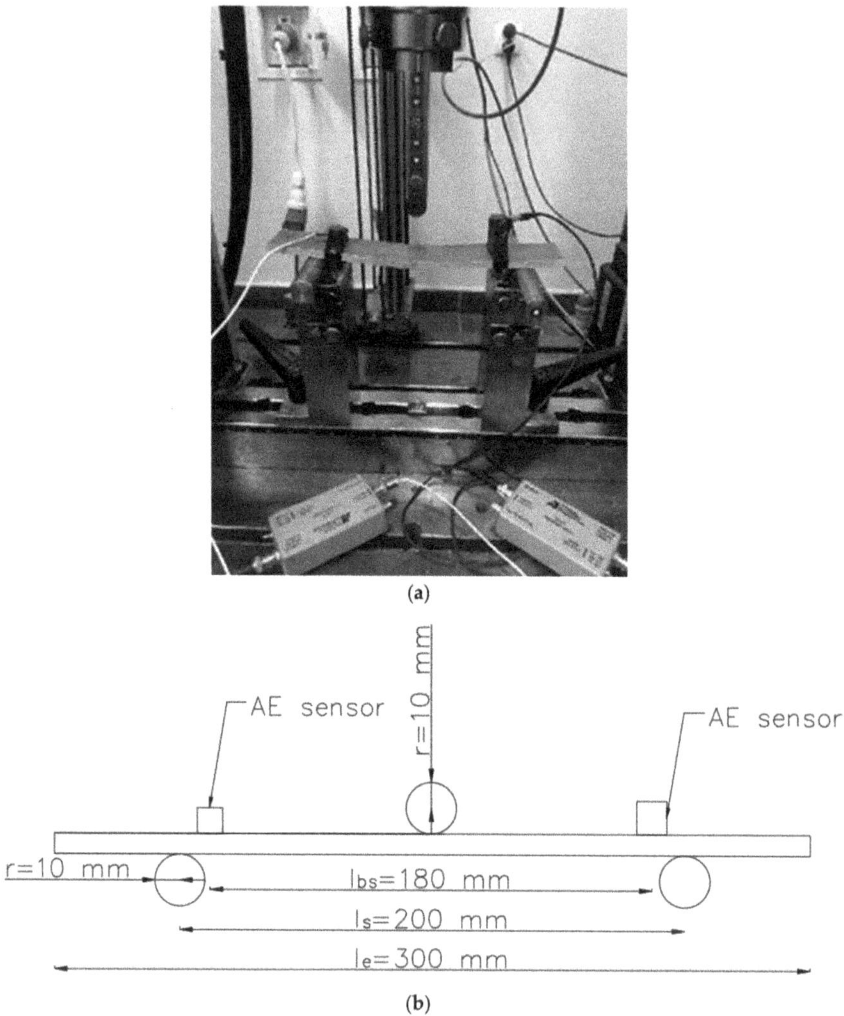

Figure 1. The fiber–cement element in three-point bending with installed AE sensors: (**a**) photograph of the specimen; (**b**) scheme of loading and AE sensors.

During each test, acoustic emission signals were recorded. The eight-channel Vallen AE processor board was used for this purpose. The acquisition was carried out using two sensors with built-in preamplifiers—VS30-SIC (25–80 kHz range) and VS150-RIC (100–450 kHz range). During the tests, 13 typical AE parameters (AE signal duration, AE rise time, mean effective voltage, number of counts, number of counts to maximum signal amplitude, amplitude of AE signal, signal energy, average frequency, reverberation frequency, initiation frequency, absolute energy of the AE signal, signal strength and average signal level), and the values of strength increase and deformation of the samples, were recorded. The sensors were attached with a clamp near the supports on the inside.

Analysing the literature, it can be assumed that the course of the destruction of ordinary concrete (mortar, cement matrices) under short-term loading is three-stationary. These stages are stable microcracks initiation, stable microcracks development and propagation, and unstable microcracks propagation.

The stage of stable crack initiation is characterised by microcracks, that were initiated already at the stage of this material's formation, appearing at isolated points of the concrete material, in the form of micro-gaps, pores and local concentrations of tensile stresses. The formation of these microcracks alleviates existing stress concentrations, leading to the restoration of the balance of internal forces. It is characteristic that at this stage of destruction, the existing microcracks do not develop, while the phenomenon of their increase occurs.

The increase in load causes the destruction of concrete to enter the second stage, in which two simultaneous processes occur: the phenomenon of crack propagation created in the first stage, and the further formation of stable microcracks. The cracks multiply and spread in a stable manner, in the sense that if the external load increase is stopped, the development of the cracks will also cease.

The third, final stage occurs when, as a result of a further increase in load, the crack system develops to such an extent that it becomes unstable. Under the influence of the released energy of deformation, the cracks spread automatically until the structure is completely destroyed. Destruction at this stage can occur even without further increase in external load.

The works of Kanji Ono, Othsu and Fowler [40–43] state that the sources of the AE signals in elements made of a concrete (cement) composite can be:

- The creation of microcracks;
- crack formation and propagation;
- cracks closing (friction at the concrete–concrete border);
- friction at the concrete–reinforcement border;
- corrosion, erosion;
- plasticisation and destroying of the reinforcement.

The recorded signals were divided into classes using the k-means algorithm, using Vallen software. In the course of the procedure, a standard sample was selected, in which, during the bending process, all the processes characteristic of the fiber–cement material were generated. The standard file was used to divide the signals recorded for the remaining samples.

The test components came from the type of fiber–cement board that is available on the construction market, which has a broad application range (installed inside and outside buildings). As declared by the manufacturer, the samples contained Portland cement, mineral binders, natural organic reinforcing fibers and synthetic organic reinforcing fibers. Elemental composition was confirmed by performing analyses using an EDS micro probe during microscopic observations (Figure 2). The comparison of the data provided by the manufacturer with the EDS analysis performed is presented in Table 2. The analysis of the elemental composition of the matrix and fiber allowed us to assess the type of binder used. In the previous research on fibrous cement materials carried out by the authors, it was found that cement matrices are made using only Portland cement, or Portland cement with the addition of other mineral binders. Performing such an analysis in this case gave the information that Portland cement and a silicate mineral binder were used in the production of the matrix. Technical data from the tested boards are presented in Table 3.

Microscopic observations were performed after the strength tests, and were combined with the AE signal generation. A 10 × 10-mm piece was cut out from the fracture surface of each sample. The components placed in the microscope chamber had not been sprayed before. The observations were performed at 5 kV with an LFD (Low vacuum Secondary Electron) detector designed for operation in low vacuum.

Table 2. Composition data of the tested board.

Manufacturer Information about Composition of Boards	Result of Observation and EDS Analysis
Portland cement	Presence confirmed
Mineral binders	Silicate binder
Natural organic reinforcing fibers	Cellulose fibers
Synthetic organic reinforcing fibers	PVA (Polyvinyl Alcohol) fibers

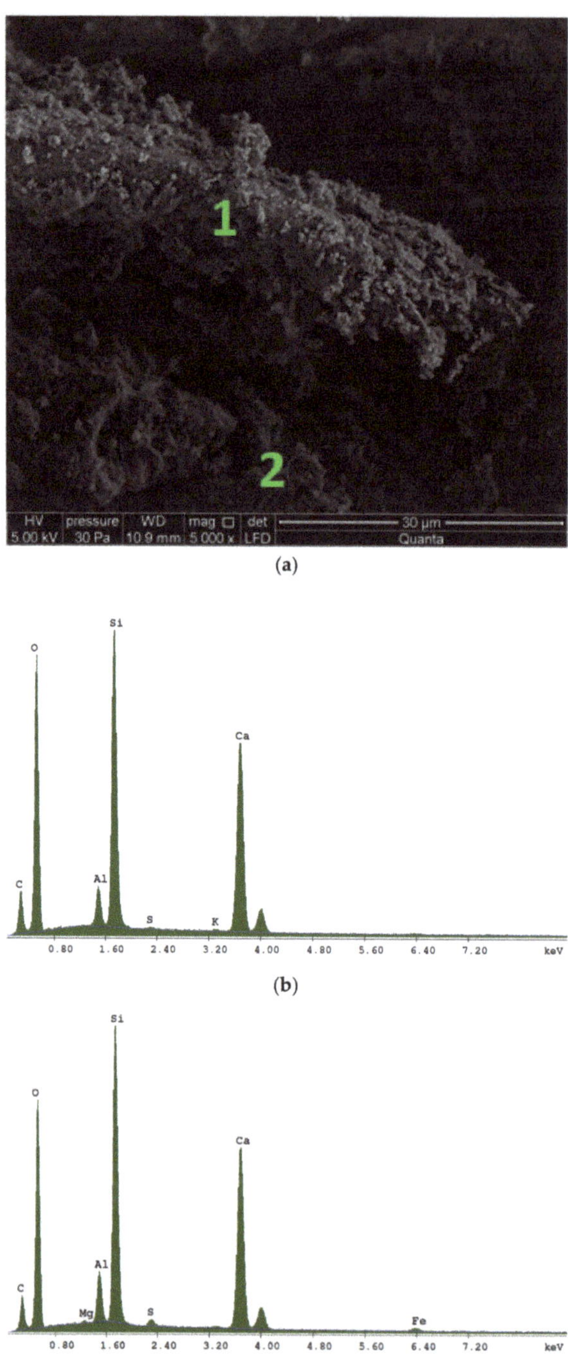

Figure 2. EDS analysis results for a mock sample: (**a**) distribution of measuring points; (**b**) analysis results from point 1—fiber; (**c**) Analysis results from point 2—matrix.

Table 3. Technical data of the tested board.

Parameter	Conditions	Value	Unit
Density	Dry condition	>1.65	g/cm^3
Flexural strength	Perpendicular	>23	N/mm^2
	Parallel	>18.5	N/mm^2
Modulus of elasticity		12,000	N/mm^2
Stretching at humidity	30–95%	1.0	mm/m
Porosity	0–100%	>18	%

2.1. K-Means Algorithm in the AE Method

To build the base of reference signals, used to assess the change of mechanical parameters in cement–fiber composites, the pattern recognition method was used, specifically, the version with arbitrary division into classes (unsupervised)—USPR. Arbitrary pattern analysis is mainly used when creating a database of reference signals if the number of classes is unknown. The k-means grouping method was used to divide the signals into classes corresponding to the destructive processes in cement composites.

K-means is a standard cluster analysis algorithm, in which the value of parameters determining the number of groups to be extracted from a data set is initially determined. Representatives are randomly selected, so it is important that they are as far apart as possible [44–47]. The selected components are the seedbed of the groups (prototypes). In the next step, each component of the set is assigned to the nearest group. Initial groups are designated at this stage. In the next step, a centre is calculated for each group, based on the arithmetic mean of the coordinates of the objects assigned to a group. Then, all the objects are considered and reallocated to the nearest (with respect to their distance from individual centroids) group. New group centres are designated until the migration of objects between clusters ceases. According to the same principle, the assignment correctness of particular objects to particular groups is checked. If in the next two runs of the algorithm there is no change in the division made (at which point it is said that the stabilisation is achieved), the processing is finished [48–50]. In this method, the number of groups is constant and consistent with the k parameter; only the group object assignment can be changed. In the k-means method, the search for the optimal division corresponds to the designation of the prototypes of groups that minimises the following criteria function [51,52]:

$$J_{(v,B)} = \sum_{i=1}^{k} \sum_{k=1}^{N} b_{ik} d^2(v_i, v_k), \tag{1}$$

In this function, $d(v,x)$ is the distance of the element represented by the vector x from the group designated by the prototype (centroid, centre of the group) v, N is the number of the set O, B is the division matrix, and the other parameters have the same meaning as stated above. The principle of the method can be described as follows:

1. Preliminary division of the set into k clusters;
2. Calculation of an individual centroid for each cluster (centre of the group);
3. Assigning of each element of the set to the nearest group (in this case, the distance from the group is the same as the distance from the centroid);
4. Repetition of the previous two steps until the changes related to the object assignment to clusters ceases.

Creating a base of reference signals involves several stages. These are:

- generating signals in the laboratory while destroying specially designed samples in a certain way;
- comparison of signals received from samples with signals generated during the destruction of model beams subjected to various destruction processes;
- verification of reference signals on the basis of monitoring results of various types and lengths of composite panels loaded for destruction;

- final verification on the elements during their normal operation (this stage will be carried out in subsequent tests).

The characteristics of the signals contained in the database include the geometric, energy and frequency parameters of the signals. In addition, the database contains typical noise signals.

It should be emphasised that the main purpose of the research was to register the AE signals generated by destructive processes. Therefore, no additional tests, e.g., deformation of concrete under compression, or measurement of Young's modulus, were carried out, and no statistical calculations were carried out regarding the spread of parameters. This approach was derived from the fact that the actual element lacks information about the state of the concrete at the test site, and the purpose of the research was to collect the most real AE signals generated during the destruction of the tested samples.

A total of 13 AE parameters was used to create the base of reference signals for the destructive processes that take place in cement–fiber boards:

1. AE signal duration (duration);
2. AE rise time (rise time);
3. Mean effective voltage (RMS);
4. Number of counts;
5. Number of counts to maximum signal amplitude (counts to peak);
6. Amplitude AE signal (amplitude);
7. Signal energy (energy);
8. Average frequency AE (average frequency);
9. Reverberation frequency;
10. Initiation frequency;
11. Absolute energy of the AE signal (absolute energy);
12. Signal strength;
13. Average signal level AE (ASL).

The second step was the adoption of the basic parameters necessary to create a base of reference signals, namely:

- expected number of destructive processes (based on literature and our own research);
- measures of distance between clusters—in this case, the Euclidean distance with time distribution;
- the number of iterations needed to find the optimal number of classes (minimum 1,000,000).

The standard file, with four classes obtained in this way, was then checked with real samples in laboratory tests of individually destructive processes.

The result of the research was the obtaining of a base of reference signals intended for the assessment of the technical condition of the boards. Using the base of reference signals, individual AE parameters (individual graphs) are only an illustrations of the processes taking place, not a source of analysis. That is why it is so important to use BIG DATA analysis to create "BLACK BOX", which can be used by persons without any academic knowledge of AE to analyse the technical condition of the structure.

In the presented research, the issue of signal localisation was not addressed. This issue will be addressed in further studies on full-size elements. So far, the focus has been on identifying active destructive processes.

2.2. Scanning Electron Microscopy with Elemental Composition Analysis

Scanning electron microscopy (SEM) allows the recording of a surface image of samples at high magnification, by means of secondary or backscattered electron recording. Unlike optical microscopy, it allows for much higher magnifications, with incomparably higher resolution. The electron beam surface scanning of samples is a simple and quick way to obtain images that reflect differences in the elemental composition of the sample [53–55]. Accessories, such as the EDS microprobe analyser, allow for quick elementary composition determination at points and areas of different sizes. The element identification is based on the recording of the X-ray energy spectrum emitted by the sample atoms induced by the electron beam. The software automatically determines the elementary composition of the sample based on the characteristic radiation. The combination of electron microscopy with elementary analysis allows us to record high-resolution images at very high magnifications, determining the elementary composition of very small objects (even less than 1 mm), and creating change profiles for the composition of components and two-dimensional colour maps of element distribution on samples' surfaces. The method is considered to be non-invasive as the destructive effect of the electron beam on samples is very rare, and, in addition, such effect occurs (if it occurs) on a microscopic scale. The scanning electron microscope does not require the spraying of conductive layers on the surface of the materials to be tested, thanks to a special measurement mode at low pressure (called low vacuum measurement) [56].

3. Results

3.1. Results of AE Signal Analysis

3.1.1. Individual AE Signal Class Distribution Analysis

After dividing the recorded AE signals into four classes, using k-means algorithms, two AE parameters were analysed (as an illustration of ongoing processes)—signal strength and signal duration. The number of classes was input by the authors into the software used to analyse the AE signals. The value adopted corresponds to the nature of the material's work, and allows us to link individual classes with specific processes. Another criterion explaining the imposed number of classes was the level of individual signals matched to the appropriate classes, which in this case was over 90%. The following individual AE signal classes were assigned to processes occurring in the structure of the tested material:

- class 1 (green)—formation of microcracks;
- class 2 (red colour)—crack development;
- class 3 (yellow)—cracking of reinforcing fibers, delamination of the structure, detaching the fibers from the matrix;
- class 4 (colour blue)—breakdown of reinforcing fibers, sample destruction.

The paper presents graphs of the descriptors over time, including their division into classes, for mock samples from each of the tested series. The results obtained for samples within the group were similar.

Analysing Figures 3–7, it can be seen that for samples in the air-dry state exposed to environmental factors (BS, BW and BM), the signals of 1–3 classes can be seen from the beginning of the external load action. Shortly after applying the force, Class 3 signals begin to appear, indicating that the fiber destruction process has begun. As the cracks in the tensile zone progress and the cracks deepens, Class 4 signals of gradual fiber breakage start to appear at the bottom of the sample. Exceeding the stress limits in the material results in a rapid increase in the recorded descriptors.

In the case of samples subjected to direct fire (BP) and high temperature (BC), two classes of signals were observed in the recorded time waveforms—Class 1 and Class 2. The authors related this fact to the possibility of the damage or degradation of the reinforcing fibers under the influence of temperature, which was associated with a change in the way the components are destroyed. It was found that dry samples and samples exposed to environmental factors, due to the presence of reinforcing fibers, were destroyed by exceedingly high bending stresses. On the other hand, the samples exposed to temperature cracked due to too much shear stress, likely because of the degradation of the fibers. To confirm these assumptions, a microscopic analysis of the fractures extracted from the tested components was conducted. During the analysis of the characteristics of the recorded descriptors, it was also found that during the bending of flamed and fired samples, the occurrence of most AE events was associated with a lower strength of signals than those found in the cases of samples from other groups. According to the authors, this fact confirms previous assumptions, which in their opinion validates the use of the AE method for assessing the degree of change of the mechanical parameters of the fiber–cement boards.

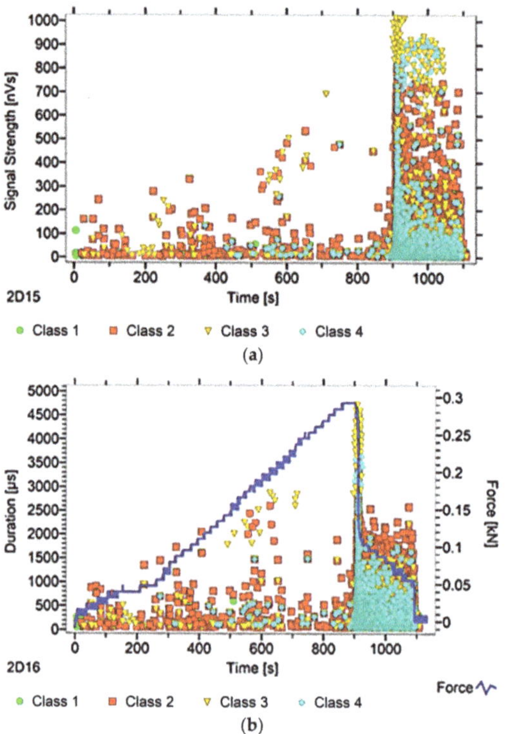

Figure 3. AE signal graphs for the mock BS sample: (**a**) signal strength distribution in time; (**b**) signal duration distribution in time, with plotted force increment curve.

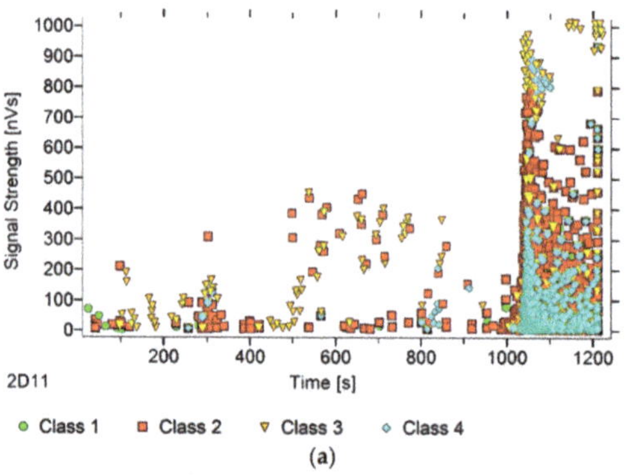

Figure 4. *Cont.*

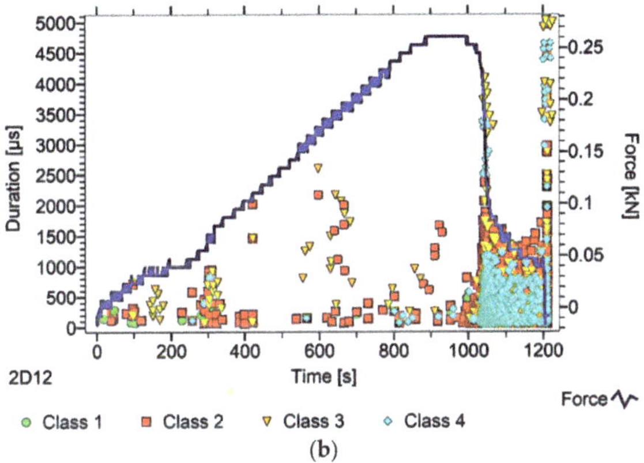

Figure 4. AE signal graphs for the mock BW sample: (**a**) signal strength distribution in time; (**b**) signal duration distribution in time, with plotted force increment curve.

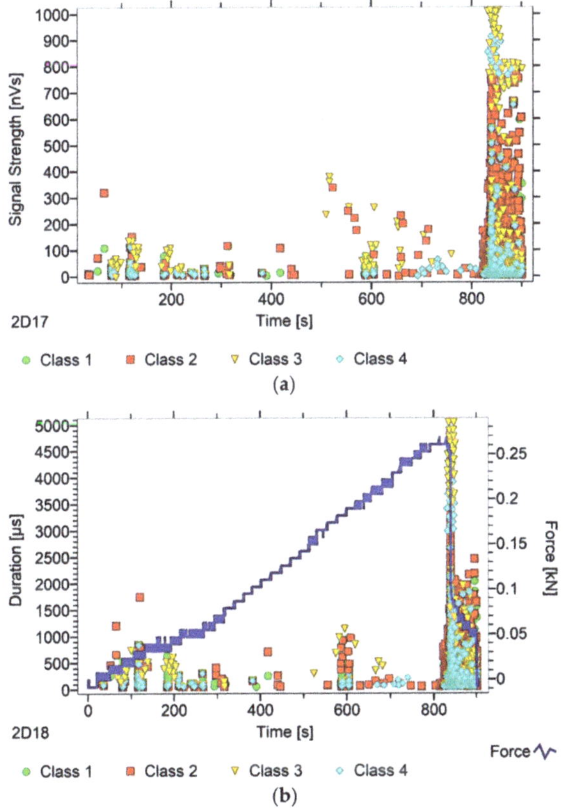

Figure 5. AE signal graphs for the mock BM sample: (**a**) signal strength distribution in time; (**b**) signal duration distribution in time, with plotted force increment curve.

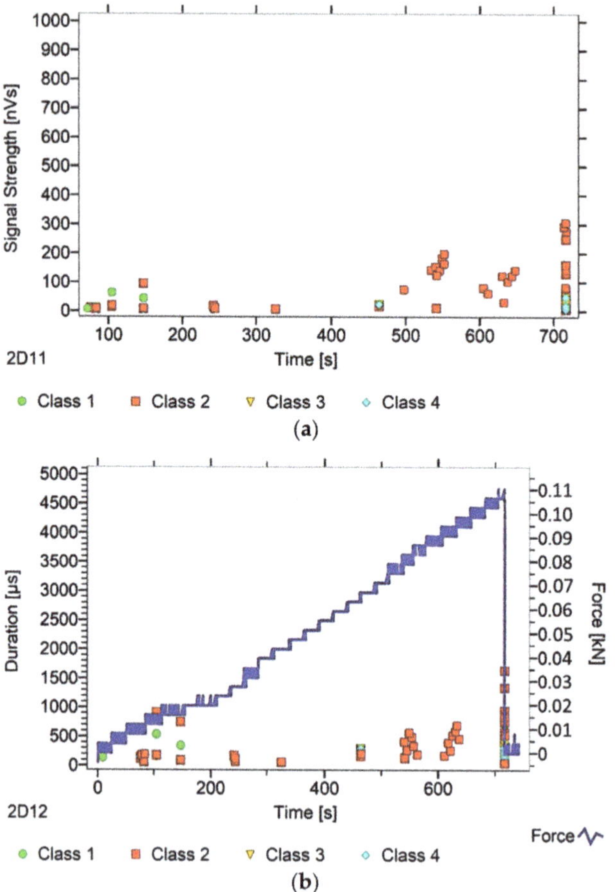

Figure 6. AE signal graphs for the mock BP sample: (**a**) signal strength distribution in time; (**b**) signal duration distribution in time, with plotted force increment curve.

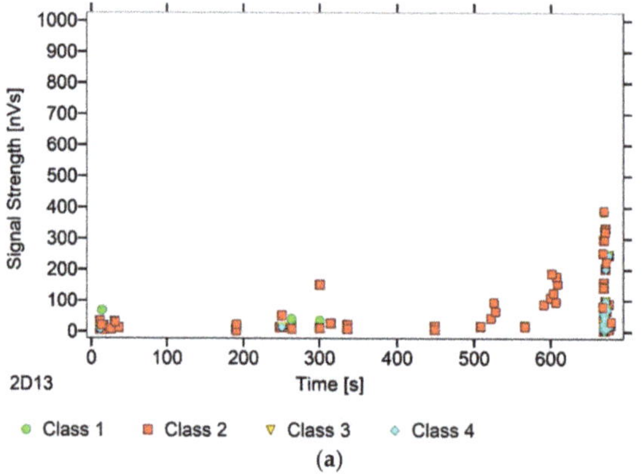

Figure 7. *Cont.*

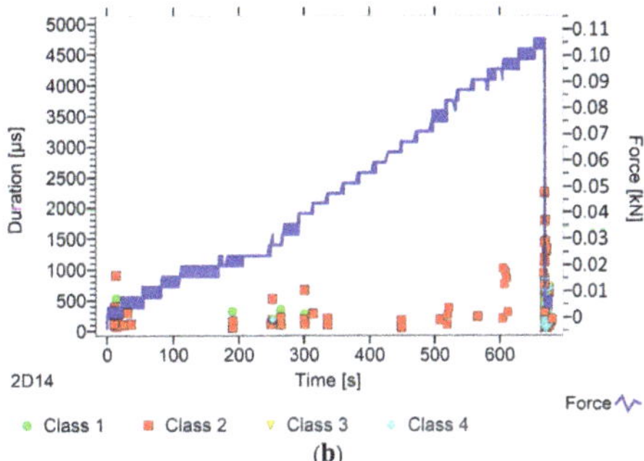

(b)

Figure 7. AE signal graphs for the mock BC sample: (**a**) signal strength distribution in time; (**b**) signal duration distribution in time, with plotted force increment curve.

Analysing the classes of AE signals occurring in the analysed patterns for samples BP and BC, it was found that the absence of Class 3 and 4 signals in the recorded events clearly indicates damage to the reinforcing fibers, or the possibility of delamination and voids in the material structure. It should also be noted that there is a close correlation between the type of AE signal classes recorded, and the destructive force value. This gives rise to the conclusion that the use of the AE method may allow the analysis of fiber–cement components in terms of the occurrence of material defects already formed at the production stage (e.g., discontinuities, uneven distribution of reinforcing fibers).

In summary, it was found that in the case of samples in the air-dry state, soaked in water and cyclically frozen and defrosted, the process of sample destruction was ductile. The result of the bending force was a deepening and widening of the crack in the element's tension zone. After reaching maximum strength, the samples were gradually unloaded. On the other hand, samples exposed to flames and high temperatures broke in a fragile manner. Along with reaching the maximum value of force, the element rapidly split into two parts. The differences in destruction mechanisms clearly influenced the types of recorded signal classes.

3.1.2. Analysis of Frequencies Accompanying Changes in Mechanical Parameters

Using the options available in the Vallen software, the frequencies accompanying the event emissions were extracted for the recorded data. The analysis results are presented in Figures 8–12. For a more readable representation of the frequency ranges emitted by the material, two frequency distribution graphs are provided for each of the mock sample series. The first one is used to present the frequencies from the whole tested run, and the second one details the frequencies of signals recorded before the destruction moment.

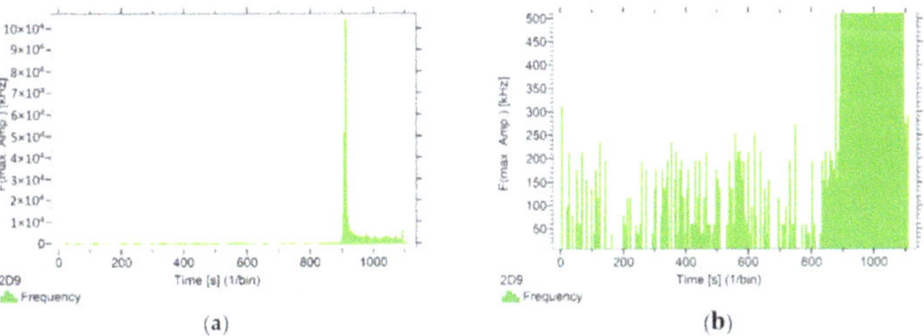

(a)   **(b)**

Figure 8. Frequency distribution graph during the test of the mock BS sample: (**a**) taking into account the entire frequency range; (**b**) specifying the frequency range before destruction.

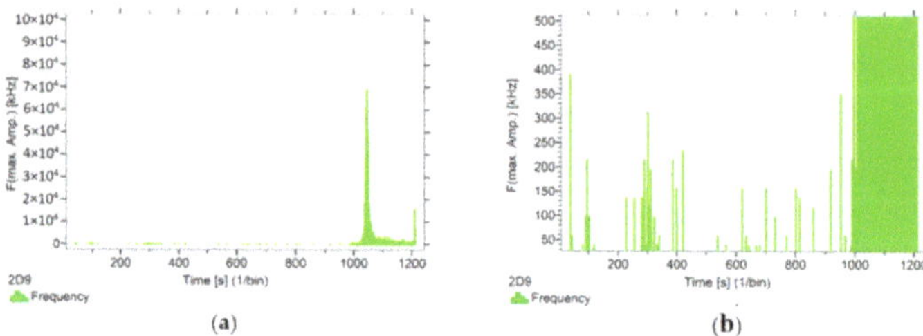

Figure 9. Frequency distribution graph during the test of the mock BW sample: (**a**) taking into account the entire frequency range; (**b**) specifying the frequency range before destruction.

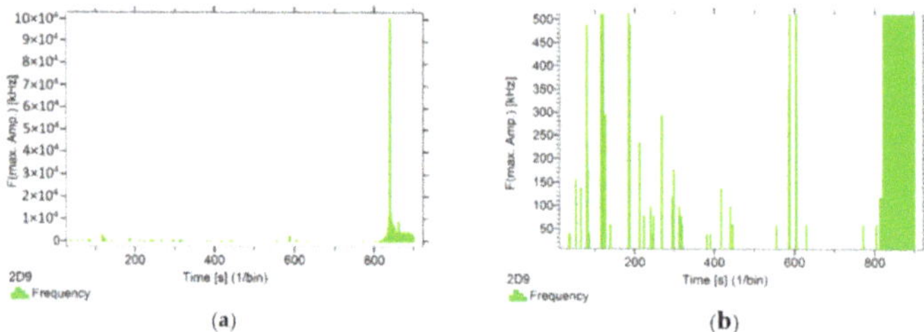

Figure 10. Frequency distribution graph during the test for the mock BM sample: (**a**) taking into account the entire frequency range; (**b**) specifying the frequency range before destruction.

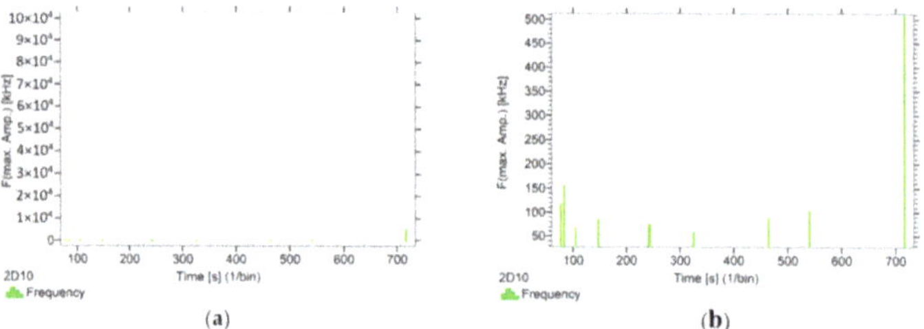

Figure 11. Frequency distribution graph during the test of the mock BP sample: (**a**) taking into account the entire frequency range; (**b**) specifying the frequency range before destruction.

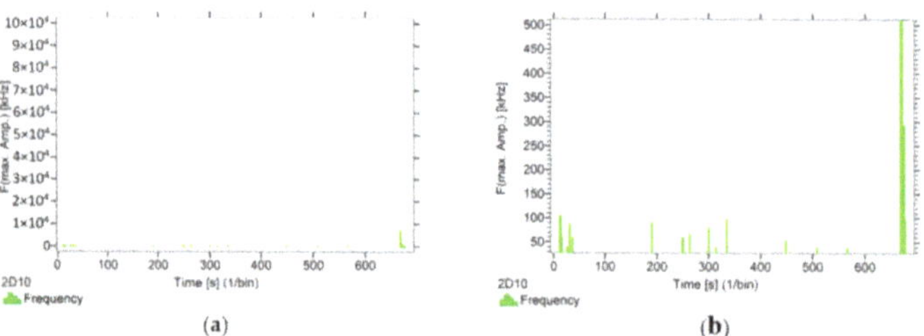

Figure 12. Frequency distribution graph during the test of the mock BC sample: (**a**) taking into account the entire frequency range; (**b**) specifying the frequency range before destruction.

Analysing the graphs presented above, it can be concluded that the results from subjecting the tested components to two groups of operating factors (environmental and exceptional) illustrated significant differences in the emitted frequency ranges. Changes in the mechanical parameters of the samples in the air-dry state, water-soaked and cyclically frozen then defrosted under an external load, are associated with low and high frequency signals. Most of the recorded frequencies exceed the

200 kHz threshold, and some events produce sounds of 300–500 kHz. The situation is different for flamed and fired samples. The bending of components exposed to high temperatures caused events at much lower frequencies, only some of which exceed 100 kHz.

3.2. Results of Microscopic Analyses

Microstructure images of the extracted sample fractures are shown in Figure 13. All images show a 250-fold magnification of the surface.

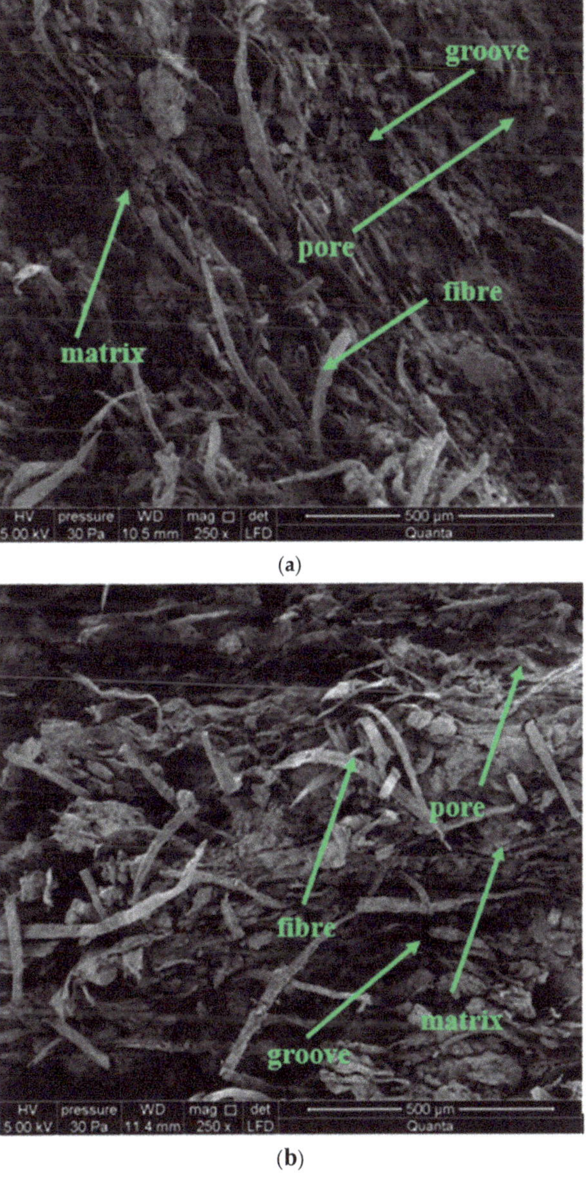

(**a**)

(**b**)

Figure 13. *Cont.*

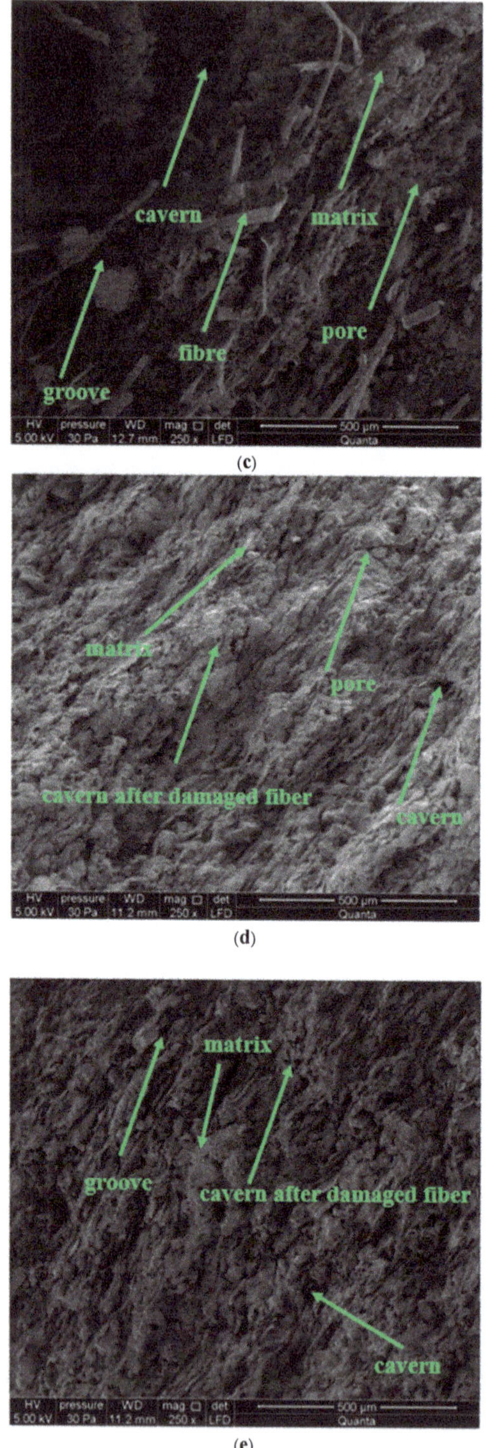

Figure 13. Image of the mock microstructure: (**a**) of BS sample, (**b**) of BW sample, (**c**) of BM sample, (**d**) of BP sample, and (**e**) of BC sample.

Based on the analysis of the results shown in Figure 13, it should be concluded that the macrostructure of each of the analysed board samples before microscopic examination was determined, visually, as compact. Microscopic observations determined that the structure of the samples were fine-porous, with pore size up to 50 μm. Deep grooves of up to 500 μm in width were found on the fracture surfaces. A high density of irregularly distributed cellulose fibers was observed at the tested fractures (Figure 13a–c), except for the flamed sample and the fired sample. In the burned sample, it was observed that most of the fibers are fired or blended into the matrix. Flaming causes gradual burning of fibers, and the degradation of their structure, depending on the flame range (Figure 13d,e). Caverns and grooves from the torn fibers were observed. The matrix structure is defined as granular with numerous delaminations. No space was observed between the fibers and the matrix, indicating a strong bond between them. In the analysis of the flamed sample (Figure 13d) a locally altered matrix structure was observed.

According to the authors, the microscopic observations performed confirmed previous assumptions, which justify the necessity and legitimacy of the AE method for evaluating the degree of change in the mechanical parameters of fiber–cement composites.

4. Discussion

Tables 4–9 present a summary of the results of the tests. Analysing the data contained in them, it can be seen that the type of factor to which fiber–cement boards are exposed has a significant impact on the processes taking place inside its structure. According to the authors, this fact can be the basis for creating a professional system for assessing the condition of used boards.

Table 4. Summary of research results for BS samples.

Sample	Number of Class 1 Signals	Number of Class 2 Signals	Number of Class 3 Signals	Number of Class 4 Signals	Max. Signal Energy [eu]	Max. Force [kN]	MOR [MPa]	Max. Frequency before Destruction [kHz]
BS_1	1747	8644	2040	2648	5049	0.29	27.19	316
BS_2	2042	7256	2165	2543	5126	0.30	28.13	403
BS_3	1635	7269	1847	2714	5098	0.29	27.19	359
BS_4	1798	9842	2264	2796	4863	0.30	28.13	426
BS_5	2249	11365	2073	2379	5194	0.30	28.13	511
Arithmetic average	1894.2	8875.2	2077.8	2616	5066	0.296	27.75	403
Standard deviation	221.74	1574.10	139.28	144.59	111.81	0.005	0.46	65.88

Table 5. Summary of research results for BW samples.

Sample	Number of Class 1 Signals	Number of Class 2 Signals	Number of Class 3 Signals	Number of Class 4 Signals	Max. Signal Energy [eu]	Max. Force [kN]	MOR [MPa]	Max. Frequency before Destruction [kHz]
BW_1	2365	7424	2084	2215	5055	0.26	24.38	398
BW_2	1823	7982	2152	2196	5139	0.28	26.25	452
BW_3	1964	8214	1836	2364	5024	0.29	27.19	374
BW_4	2159	6920	2174	1974	5097	0.28	26.25	512
BW_5	2144	7563	2296	2259	4936	0.28	26.25	367
Arithmetic average	2091	7620.6	2108.4	2201.6	5050.2	0.278	26.06	420.6
Standard deviation	184.64	450.68	152.44	127.81	69.03	0.01	0.92	54.58

Table 6. Summary of research results for BM samples.

Sample	Number of Class 1 Signals	Number of Class 2 Signals	Number of Class 3 Signals	Number of Class 4 Signals	Max. Signal Energy [eu]	Max. Force [kN]	MOR [MPa]	Max. Frequency before Destruction [kHz]
BM_1	1736	6422	1759	2168	5023	0.26	24.38	1023
BM_2	2054	6948	2052	2232	5102	0.27	25.31	415
BM_3	2148	6536	1866	2085	5074	0.26	24.38	504
BM_4	1947	7049	2173	2266	5130	0.28	26.25	362
BM_5	2072	7197	1945	1987	5069	0.26	24.38	388
Arithmetic average	1991.4	6830.4	1959	2147.6	5079.6	0.266	24.94	538.4
Standard deviation	142.93	299.82	143.73	101.26	35.75	0.008	0.75	246.98

Table 7. Summary of research results for BP samples.

Sample	Number of Class 1 Signals	Number of Class 2 Signals	Number of Class 3 Signals	Number of Class 4 Signals	Max. Signal Energy [eu]	Max. Force [kN]	MOR [MPa]	Max. Frequency before Destruction [kHz]
BP_1	17	67	4	4	1823	0.11	10.31	113
BP_2	25	84	3	8	1526	0.09	8.44	109
BP_3	12	66	4	10	1465	0.10	9.38	126
BP_4	10	72	7	9	1506	0.11	10.31	94
BP_5	20	79	3	12	1764	0.9	8.44	116
Arithmetic average	16.8	73.6	4.2	8.6	1616.8	0.10	9.38	111.6
Standard deviation	5.42	6.95	1.47	2.65	146.80	0.009	0.84	10.44

Table 8. Summary of research results for BC samples.

Sample	Number of Class 1 Signals	Number of Class 2 Signals	Number of Class 3 Signals	Number of Class 4 Signals	Max. Signal Energy [eu]	Max. Force [kN]	MOR [MPa]	Max. Frequency before Destruction [kHz]
BC_1	20	70	4	4	1885	0.15	14.06	113
BC_2	14	84	2	10	1624	0.10	9.38	135
BC_3	10	112	1	11	1894	0.09	8.44	92
BC_4	23	62	4	6	1733	0.11	10.31	84
BC_5	11	97	2	8	2014	0.11	10.31	141
Arithmetic average	15.6	85	1.44	6.56	1830	0.11	10.5	113
Standard deviation	5.08	18.04	1.2	2.56	136.25	0.02	1.91	22.58

Table 9. Summary of research results.

Sample	Average Number of Class 1 Signals	Average Number of Class 2 Signals	Average Number of Class 3 Signals	Average Number of Class 4 Signals	Average Max. Signal Energy [eu]	Average Max. Force [kN]	Average MOR [MPa]	Average Max. Frequency before Destruction [kHz]
BS	1894.2	8875.2	2077.8	2616	5066	0.296	27.75	403
BW	2091	7620.6	2108.4	2201.6	5050.2	0.278	26.06	420.6
BM	1991.4	6830.4	1959	2147.6	5079.6	0.266	24.94	538.4
BP	16.8	73.6	4.2	8.6	1616.8	0.10	9.38	111.6
BC	15.6	85	1.44	6.56	1830	0.11	10.5	113

An important element in the aforementioned system of intelligent supervision of the safety of the use of structures using fiber–cement elements will be the AE method. There is no information in the literature on the construction of such a system. Of course, there are scientific papers devoted to the use of the AE method as a research tool, but only to determine the phenomena occurring in a cement–fiber composite. In previous publications [57,58], the AE method has been used to assess the degree of destruction of boards that were previously affected by high temperatures. However, the analysis of registered descriptors without grouping the signals did not allow the direct connection of the AE signals with the endurance parameters of the plates, or the linking of them to specific destructive processes occurring in the structure. Therefore, these publications carried out additional analyses of registered AE events using artificial neural networks, which allowed for the initial division of events into signals corresponding to matrix cracking and fiber cracking. However, the results presented in this article show the possibility of creating an engineering system based on the pattern recognition method, i.e., creating a black box that can be operated by any engineer, not just a well-educated person. The use of the AE method allows one to monitor in real time, and intervene when the need arises, and not when a failure occurs.

5. Conclusions

This article presents a proposal for using the AE method to assess the condition of fiber–cement boards. The research was carried out on five groups of elements exposed to environmental and exceptional factors. During the three-point bending tests, AE signals were recorded, which were then grouped into four classes using the k-means algorithm. Based on the presented results, it was found that the indicated method can be successfully used to confirm or deny the presence, and to assess the condition of, reinforcing fibers in boards, which directly affect the strength parameters of these elements. The frequencies recorded before destruction were also traced for each sample. It has been noted that the frequency values are significantly different for samples with reinforcements, compared to elements with fired or degraded fibers. Initial hypotheses were confirmed during microscopic studies using a scanning electron microscope.

The following specific conclusions were drawn on the basis of the tests performed:

1. The implementation of the k-means grouping method based on the analysis of the AE signal parameters gives positive results in the classification of destructive processes taking place in the structure of fiber–cement boards.

2. The application of the k-means grouping method allows the distinguishing of the processes taking place in the structure of fiber–cement boards subjected to an external load.

3. Tracking events assigned to specific signal classes allows for evaluating the changes in the mechanical parameters of the material.

4. The presence of reinforcing fibers significantly affects the distribution and the number of AE signals of individual classes, and the strength of the boards.

5. The frequencies emitted by changes in the fiber–cement structure are closely linked to the presence of fiber reinforcement.

6. It was found that the application of the AE method enables the effective detection and monitoring of the initiation of changes in the structure, as well as the separation and identification of AE signals corresponding to different types of processes affecting the change of the mechanical parameters of fiber–cement boards.

7. The developed reference signal base provides a theoretical basis for the application of the AE technology, based on the standard for the detection and monitoring of cracks and delamination propagation in full-size fiber–cement components.

Author Contributions: Conceptualization, A.A.-B. and A.K.; methodology, A.K.; software, A.A.-B.; validation, A.A.-B. and A.K.; formal analysis, A.A.-B.; investigation, A.A.-B.; resources, A.K.; data curation, A.A.-B.; writing—original draft preparation, A.A.-B.; writing—review and editing, A.K.; visualization, A.A.-B.; supervision, A.K.; project administration, A.A.-B.; funding acquisition, A.K. All authors have read and agreed to the published version of the manuscript.

Funding: The tests were implemented thanks to the financial support of the Kielce University of Technology within the framework of the statutory work No. 02.0.06.00/2.01.01.00.0000 SUBB. BKWB. 20.001.

Conflicts of Interest: The authors declare no conflict of interest.

References

1. Biagiotti, J.; Puglia, D.; Kenny, J.M. A review on natural fibre-based composites-part I. *J. Nat. Fibers* **2004**, *1*, 37–41. [CrossRef]

2. John, M.; Thomas, S. Biofibres and biocomposites. *Carbohydr. Polym.* **2008**, *71*, 343–364. [CrossRef]

3. Akhavan, A.; Catchmark, J.; Rajabipour, F. Ductility enhancement of autoclaved cellulose fibre reinforced cement boards manufactured using a laboratory method simulating the Hatschek process. *Constr. Build. Mater.* **2017**, *135*, 251–259. [CrossRef]

4. Ardanuy, M.; Claramunt, J.; Toledo Filho, R.D. Cellulosic fibre reinforced cement-based composites: A review of recent research. *Constr. Build. Mater.* **2015**, *79*, 115–128. [CrossRef]

5. Ranachowski, Z.; Jóźwiak-Niedźwiedzka, D.; Brandt, A.M.; Debowski, T. Application of acoustic emission method to determine critical stress in fibre reinforced mortar beams. *Arch. Acoust.* **2012**, *37*, 261–268. [CrossRef]

6. Fernández-Carrasco, L.; Claramunt, J.; Ardanuy, M. Autoclaved cellulose fibre reinforced cement: Effects of silica fume. *Constr. Build. Mater.* **2014**, *66*, 138–145. [CrossRef]

7. Soroushian, P.; Elzafraney, M.; Nossoni, A.; Chowdhury, H. Evaluation of normal-weight and light-weight fillers in extruded cellulose fibre cement products. *Cem. Concr. Compos.* **2006**, *28*, 69–76. [CrossRef]

8. Savastano, H.; Warden, P.G.; Coutts, R.S.P. Microstructure and mechanical properties of waste fibre–cement composites. *Cem. Concr. Compos.* **2005**, *27*, 583–592. [CrossRef]

9. Neithalath, N.; Weiss, J.; Olek, J. Acoustic performance and damping behaviour of cellulose-cement composites. *Cem. Concr. Compos.* **2004**, *26*, 359–370. [CrossRef]

10. Horikoshi, T.; Ogawa, A.; Saito, T.; Hoshiro, H.; Fischer, G.; Li, V. Properties of polyvinyl alcohol fiber as reinforcing materials for cementitious composites. In Proceedings of the International RILEM Workshop on High Performance Fiber Reinforced Cementitious Composites in Structural Applications, Aachen, Germany, 5–7 September 2006; pp. 145–153.

11. Liu, J.; Li, C.; Liu, J.; Cui, G.; Yang, Z. Study on 3D spatial distribution of steel fibres in fibre reinforced cementitious composites through micro-CT technique. *Constr. Build. Mater.* **2013**, *48*, 656–661. [CrossRef]

12. Roma, L.C.; Martello, L.S.; Savastano, H., Jr. Evaluation of mechanical, physical and thermal performance of cement-based tiles reinforced with vegetable fibers. *Constr. Build. Mater.* **2008**, *22*, 668–674. [CrossRef]

13. Ardanuy, M.; Claramunt, J.; García-Hortal, J.A.; Barra, M. Fiber-matrix interactions in cement mortar composites reinforced with cellulosic fibers. *Cellulose* **2011**, *18*, 281–289. [CrossRef]

14. Savastano, H.; Warden, P.G.; Coutts, R.S.P. Brazilian waste fibres as reinforcement for cement-based composites. *Cem. Concr. Compos.* **2000**, *22*, 379–384. [CrossRef]

15. Jawaid, M.; Khalil, H.A. Cellulosic/synthetic fibre reinforced polymer hybrid composites: A review. *Carbohydr. Polym.* **2011**, *86*, 1–18. [CrossRef]

16. Tonoli, G.H.D.; Fuente, E.; Monte, C.; Savastano, H. Effect of fibre morphology on flocculation of fibre–cement suspensions. *Cem. Concr. Res.* **2009**, *39*, 1017–1022. [CrossRef]

17. Claramunt, J.; Ventura, H.; Parés, F.; Ardanuy, M. Natural fibre nonwovens as reinforcement for cement mortar composites. In Proceedings of the B. Abstr. 1st Int. Conf. Nat. Fibers Sustain. Mater. Adv. Appl., Guimarães, Portugal, 9–11 June 2013; Fangueiro, R., Ed.; Universidade do Minho: Braga, Portugal, 2013; pp. 191–192.

18. Fidelis, M.E.A.; Silva, F.D.; Filho, R.D. The influence of fiber treatment on the mechanical behavior of jute textile reinforced concrete. *Key Eng. Mater.* **2014**, *600*, 469–474. [CrossRef]

19. Agopyan, V.; Savastano, H.; John, V.M.; Cincotto, M. Developments on vegetable fibre–cement based materials in São Paulo, Brazil: An overview. *Cem. Concr. Compos.* **2005**, *27*, 527–536. [CrossRef]

20. Ramakrishna, G.; Sundararajan, T. Studies on the durability of natural fibres and the effect of corroded fibres on the strength of mortar. *Cem. Concr. Compos.* **2005**, *27*, 575–582. [CrossRef]

21. Claramunt, J.; Ardanuy, M.; García-Hortal, J.A. Effect of drying and rewetting cycles on the structure and physicochemical characteristics of softwood fibres for reinforcement of cementitious composites. *Carbohydr. Polym.* **2010**, *79*, 200–205. [CrossRef]

22. Pizzol, V.D.; Mendes, L.M.; Savastano, H.; Frías, M.; Davila, F.J.; Cincotto, M.A.; John, V.M.; Tonoli, G.H.D. Mineralogical and microstructural changes promoted by accelerated carbonation and ageing cycles of hybrid fibre–cement composites. *Constr. Build. Mater.* **2014**, *68*, 750–756. [CrossRef]

23. Drelich, R.; Gorzelańczyk, T.; Pakuła, M.; Schabowicz, K. Automated control of cellulose fibre cement boards with a non-contact ultrasound scanner. *Autom. Constr.* **2015**, *57*, 55–63. [CrossRef]

24. Schabowicz, K.; Gorzelańczyk, T. A non-destructive methodology for the testing of fibre cement boards by means of a non-contact ultrasound scanner. *Constr. Build. Mater.* **2016**, *102*, 200–207. [CrossRef]

25. Chady, T.; Schabowicz, K.; Szymków, M. Automated multisource electromagnetic inspection of fibre-cement boards. *Autom. Constr.* **2018**, *94*, 383–394. [CrossRef]

26. Chady, T.; Schabowicz, K. Non-destructive testing of fibre-cement boards, using terahertz spectroscopy in time domain. *Badania Nieniszczące i Diagnostyka* **2016**, *1*, 62–66. (In Polish)

27. Schabowicz, K.; Ranachowski, Z.; Józwiak-Niedźwiedzka, D.; Radzik, Ł.; Kudela, S.; Dvorak, T. Application of X-ray microtomography to quality assessment of fibre cement boards. *Constr. Build. Mater.* **2016**, *110*, 182–188. [CrossRef]

28. Gorzelańczyk, T.; Schabowicz, K.; Szymków, M. Non-destructive testing of fibre-cement boards, using acoustic emission. *Przegląd Spawalnictwa* **2016**, *88*, 35–38. (In Polish)

29. Pazdera, L.; Topolar, L. Application acoustic emission method during concrete frost resistance. *Russ. J. Nondestruct. Test.* **2014**, *50*, 127–131. [CrossRef]

30. Shields, Y.; Garboczi, E.; Weiss, J.; Farnam, Y. Freeze-thaw crack determination in cementitious materials using 3D X-ray computed tomography and acoustic emission. *Cem. Concr. Compos.* **2018**, *89*, 120–129. [CrossRef]

31. Topolar, L.; Kucharczykowa, B.; Kocab, D.; Pazdera, L. The Acoustic Emission Parameters Obtained during Three-point Bending Test on Thermal-stressed Concrete Specimens. *Procedia Eng.* **2017**, *190*, 111–117. [CrossRef]

32. Mpalaskas, A.; Matikas, T.; Aggelis, D. Acoustic emission of fire damaged fiber reinforced concrete. In *Smart Materials and Nondestructive Evaluation for Energy Systems 2016, Proceedings of the International Society for Optics and Photonics, Las Vegas, NV, USA, 1 April 2016*; SPIE: Las Vegas, NV, USA, 2016; Volume 9806, p. 980618.

33. Marzec, A.; Lewicki, P.; Ranachowski, Z.; Debowski, T. The influence of moisture content on spectral characteristic of acoustic signals emitted by flat bread samples. In Proceedings of the AMAS Course on Nondestructive Testing of Materials and Structures, Centre of Excellence for Advanced Materials and Structures, Warszawa, Poland, 20–22 May 2002; pp. 127–135.

34. Lucchi, E. Non-invasive method for investigating energy and environmental performances in existing buildings. In Proceedings of the PLEA Conference on Passive and Low Energy Architecture, Louvain-la-Neuve, Belgium, 13–15 July 2011; pp. 571–576.

35. Dębowski, T.; Lewandowski, M.; Mackiewicz, S.; Ranachowski, Z.; Schabowicz, K. Ultrasonic tests of fibre-cement boards. *Przegląd Spawalnictwa* **2016**, *10*, 69–71. (In Polish)

36. Schabowicz, K.; Józwiak-Niedźwiedzka, D.; Ranachowski, Z.; Kudela, S.; Dvorak, T. Microstructural characterization of cellulose fibres in reinforced cement boards. *Arch. Civ. Mech. Eng.* **2018**, *4*, 1068–1078. [CrossRef]

37. Adamczak-Bugno, A.; Gorzelańczyk, T.; Krampikowska, A.; Szymków, M. Non-destructive testing of the structure of fibre-cement materials by means of a scanning electron microscope. *Badania Nieniszczące i Diagnostyka* **2017**, *3*, 20–23. (In Polish)

38. Adamczak-Bugno, A.; Krampikowska, A. The basics of a system for evaluation of fiber-cement materials based on acoustic emission and time-frequency analysis. *Math. Biosci. Eng.* **2020**, *17*, 2218–2235. [CrossRef] [PubMed]

39. Adamczak-Bugno, A.; Świt, G.; Krampikowska, A. Assessment of destruction processes in fiber-cement composites using the acoustic emission method and wavelet analysis. *IOP Conf. Ser. Mater. Sci. Eng.* **2018**, *214*, 1–9.

40. Tinkey, B.V.; Fowler, T.J.; Klingner, R.E. *Nondestructive Testing of Prestressed Bridge Girders with Distributed Damage, Research Raport 1857-2*; Center for Transportation Research University of Texas: Austin, TX, USA, 2002.

41. Ono, K.; Ohtsu, M. Crack classification in concrete based on acoustic emission. *Constr. Build. Mater.* **2010**, *24*, 2339–2346. [CrossRef]

42. Golaski, L.; Gębski, P.; Ono, K. Diagnostics of reinforced concrete bridges by acoustic emission. *J. Acoust. Emiss.* **2002**, *20*, 83–98.

43. Suzuki, H.; Kinjo, T.; Hayashi, Y.; Takemoto, M.; Ono, K. Wavelet Transform of Acoustic Emission Signals. *J. Acoust. Emiss.* **1996**, *14*, 69–84.

44. Goszczyńska, B.; Swit, G.; Trąmpczyński, W. Application of the IADP acoustic emission method to automatic control of traffic on reinforced concrete bridges to ensure their safe operation. *Arch. Civ. Mech. Eng.* **2016**, *16*, 867–875. [CrossRef]

45. Goszczyńska, B.; Swit, G.; Trąmpczyński, W. Analysis of the microcracking process with the Acoustic Emission method with respect to the service life of reinforced concrete structures with the example of the RC beams. *Bull. Pol. Acad. Sci. Tech. Sci.* **2015**, *63*, 55–65. [CrossRef]

46. Krampikowska, A.; Pała, R.; Dzioba, I.; Swit, G. The Use of the Acoustic Emission Method to Identify Crack Growth in 40CrMo Steel. *Materials* **2019**, *12*, 2140. [CrossRef]

47. Krampikowska, A.; Adamczak-Bugno, A. Assessment of the technical condition of prefabricated elements using the acoustic emission method. *Sci. Rev. Eng. Environ. Sci.* **2019**, *28*, 356–365. [CrossRef]

48. Goszczyńska, B. Analysis of the process of crack initiation and evolution in concrete with acoustic emission testing. *Arch. Civ. Mech. Eng.* **2014**, *14*, 134–143. [CrossRef]

49. Goszczyńska, B.; Świt, G.; Trąmpczyński, W.; Krampikowska, A. Application of the acoustic emission (AE) method to bridge testing and diagnostics comparison of procedures. In Proceedings of the IEEE Prognostics and System Health Management Conference, Beijing, China, 23–25 May 2012; pp. 1–10.

50. Swit, G. Diagnostics of prestressed concrete structures by means of acoustic emission. In Proceedings of the 2009 8th International Conference on Reliability, Maintainability and Safety, Chengdu, China, 20 July 2009; Volume 2, pp. 958–962.

51. Proverbio, E.; Venturi, V. Reliability of Nondestructive Tests for on Site Concrete Strength Assessment. In Proceedings of the 10DBMC, Lyon, France, 17–20 April 2005.

52. Olaszek, P.; Casas, J.R.; Swit, G. On-site assessment of bridges supported by acoustic emission. In Proceedings of the Institution of Civil Engineers-Bridge Engineering, London, UK, 2 June 2016; Volume 169, pp. 81–92.

53. Adamczak-Bugno, A.; Świt, G.; Krampikowska, A. Scanning electron microscopy in the tests of fibre-cement boards. In Proceedings of the MATEC Web of Conferences, 2018, 3rd Scientific Conference Environmental Challenges in Civil Engineering (ECCE), Opole, Poland, 23–25 April 2018; Volume 174, p. 02015.

54. Coutts, R.S.P.; Kightly, P. Microstructure of autoclaved refined wood-fibre cement mortars. *J. Mater. Sci.* **1982**, *17*, 1801–1806. [CrossRef]

55. Mohr, B.J.; Biernacki, J.J.; Kurtis, K.E. Microstructural and chemical effects of wet/dry cycling on pulp fiber–cement composites. *Cem. Concr. Res.* **2006**, *36*, 1240–1251. [CrossRef]

56. Tonoli, G.H.D.; Santos, S.F.; Savastano, H.; Delvasto, S.; De Gutiérrez, R.M.; de Murphy, M.D.M.L. Effects of natural weathering on microstructure and mineral composition of cementitious roofing tiles reinforced with fique fibre. *Cem. Concr. Compos.* **2011**, *33*, 225–232. [CrossRef]

57. Szymków, M.; Schabowicz, K.; Gorzelańczyk, T. Non-destructive assessment of the effect of high temperature on the destruction of the structure of the fiber-cement board. *Non-Destr. Test. Diagn.* **2018**, *4*, 44–45. (In Polish)

58. Schabowicz, K.; Gorzelańczyk, T.; Szymków, M. Identification of the Degree of Degradation of Fibre-Cement Boards Exposed to Fire by Means of the Acoustic Emission Method and Artificial Neural Networks. *Materials* **2019**, *4*, 656. [CrossRef]

Article

Experimental Study on the Stiffness of Steel Beam-to-Upright Connections for Storage Racking Systems

Florin Dumbrava [1,2] and Camelia Cerbu [1,*]

[1] Department of Mechanical Engineering, Faculty of Mechanical Engineering, Transilvania University of Brasov, B-dul Eroilor, No. 29, 500036 Brasov, Romania; florin.dumbrava@unitbv.ro

[2] Product Development Department, S.C. Dexion Storage Solutions S.R.L, Str. Campului Nr. 1A, 505400 Rasnov, Romania

* Correspondence: cerbu@unitbv.ro; Tel.: +40-722-491-398

Received: 5 June 2020; Accepted: 28 June 2020; Published: 1 July 2020

Abstract: The aspects regarding the stiffness of the connections between the beams that support the storage pallets and the uprights is very important in the analysis of the displacements and stresses in the storage racking systems. The main purpose of this paper is to study the effects of both upright thickness and tab connector type on the rotational stiffness and on the capable bending moment of the connection. For this purpose, a number of 18 different groups of beam-connector-upright assemblies are prepared by combining three types of beams (different sizes of the box cross section), three kinds of uprights profiles (with a different thickness of the section walls), and two types of connectors (four-tab connectors and five-tab connectors). Flexural tests were carried out on 101 assemblies. For the assemblies containing the uprights having the thickness of 1.5 mm, the five-tab connector leads to a higher value of the capable moment and higher rotational stiffness than similar assemblies with four-tab connectors. A contrary phenomenon happens in case of the assemblies containing the upright profiles having a thickness of 2.0 mm regarding the capable design moment. It is shown how the safety coefficient of connection depends on both the rotational stiffness and capable bending moment.

Keywords: storage systems; stiffness; tab connector; flexural test; capable design moment

1. Introduction

In recent years, there has been a significant increase in consumption, which has led to an increase of producing the storage racking systems. The products are often stored in pallet raking systems in super-markets or in warehouses of products or raw materials within factories.

Pallet racking systems are self-supporting structures that need to support considerable vertical weights. They are generally made of two main types of components: frames and beams. The frames are made of thin-walled upright profiles that have perforations in their length through which the joints with the beams are usually realized by using metallic connectors with tabs [1,2]. Such connectors are welded to both ends of the beams. Horizontal and diagonal braces welded between the uprights ensure the stability of the frame in a cross-aisle direction. The beams provide the stability in down-aisle direction and the stiffness of the connections is an important characteristic of the storage racking systems from this point of view.

Besides metallic connectors with tabs, there are other types of connections used in constructions, which include metallic connectors with screws or bolts [3], wood-steel-wood connections whose external parts are made of wood and the internal part is made of steel [4], and connectors with tabs combined with additionally fixing using bolts [5,6]. The last type of connections are commonly known as speed lock connections [7].

In a simple way, the connections used between beams and uprights are either hinged (pin-connected) or rigid connections. In practice, for almost all connections, even for those considered as rigid connections, there may be certain rotations that influence the internal forces, especially the bending moments. Not all the connections that are considered as being rigid connections are completely rigid in reality. On the other hand, the connections considered as being hinged are not perfectly hinged because, in practice, frictions can take place and the friction influences the rotation of the connection. Therefore, estimating the stiffness of the connections becomes a necessity, especially for the connections used for the storage racking structures.

Following the emergence of EN 1993-1-3, Eurocode 3 [8], a classification of the connections used between beams and uprights was made, based on their corresponding stiffness. The connections that are investigated within this research fall into the category of those considered as being semi-rigid connections. The stiffness of such connections is experimentally determined, according to the EN 15512 standard [9]. It is essential to understand that is crucial to determine the rotational stiffness of the connections and the capable bending moment because, otherwise, the racking storage systems do not work safety.

The literature is devoid of publications on the behavior in bending tests, of the beam-to-upright connections used in storage racking systems, and results related to the effect of the geometry of different components (beam, upright profile, and beam-end connector) on the connection stiffness and capable moment. Some recent studies [2,10] on the beam-to-upright connections reported the effects of both depth of the beam and type of the tab connection on the connection stiffness and on the failure modes of the connections with tabs in bending tests. It was shown that the common failure modes are tearing of the uprights near the slots for tabs and cracking of the tabs of the connectors [2]. No remarks were made about the failure modes of the beams in bending tests of the beam-to-upright connections. These papers did not make a comparison of the failure modes depending on the number of the tabs for each type of the connector involved. The authors showed the effects of the cross section shapes of the upright profiles on the connections' stiffness.

From a theoretical point-of-view, some researchers proposed in recent published works [11–13] the mechanical model, which includes five basic deformable components (tab in bending, wall of the upright profile in bearing and bending, tab connector in bending and shear) of the beam-to-upright connections in order to predict the initial rotational stiffness of the connections.

Regarding the beams used for storage racking systems, because these are cold-formed steel structural elements with thin-walls, distortional buckling failures may occur in bending and the effects of the shearing force must be accurately evaluated in bending, especially in case of the thin-walled elements having opened a cross section [14,15]. An interesting study [16] presents the results concerning the elastic shear buckling loads and ultimate strength by using numerical simulations of the stresses and strains in case of the finite element models of cold-formed steel channels with slotted webs subjected to shear.

Another critical issue of the metallic storage structures is related to the fire design methodologies so that all thin-walled structural elements to keep the load-bearing capacity as long as possible in the event of the accidental fire. In practice, in order to improve the load-bearing capacity in fire conditions, there are some types of passive or active protection systems [17]. Passive protection systems generally refer to special materials, manufacturing, or surface coating technologies in order to delay the spread of fire [4]. However, in the case of storage systems for supermarkets or warehouses, it usually uses active fire protection systems, which include automatic fire detection and extinguishing equipment (for example, a network of water pipes that are equipped with discharge nozzles called sprinklers) [17]. Regarding the cold-formed steel beams with opened and closed cross sections, a unique fire design methodology validated by experimental tests, was developed in Reference [18]. Experimental and numerical results concerning the mechanical behavior of the cold-formed steel columns subjected in fire conditions are presented in Reference [19]. Regarding the connections, there are numerical models used to predict the load-bearing capacity of the connections with and without passive protection. This is similar to the numerical model for wood-steel-wood connections [4].

Since the main target of the researchers from the industry of the storage racking systems is to increase the rotational stiffness of the beam-to-upright connections, some innovative beam-end connectors with tabs having one additional bolt fixed in the side part of the upright [1] or two additional bolts fixed in the front part of the upright [3,6] were proposed and investigated in monotonic and cyclic bending tests. However, it is known that the increase of the rotational stiffness of the beam-to-upright connections leads to the rise of the bending moment at the beam ends when the beams are mechanically loaded in storage systems. It follows that the beam-to-uprights connections should be evaluated by taking into account the ratio between the capable bending moment and the real value of the bending moment. Moreover, the external force is a distributed force applied on the length of the beam in a real case of loading while the force is applied to the free end of the beam in the bending tests used in research on beam-to-upright connections, according to EN 15512 [9] and MH16.1 [20].

In this context, in the present research, the stiffness and capable moment is experimentally determined in bending tests carried-out on 18 different groups of beam-connector-upright assemblies prepared by combining three types of beams (different sizes of the box cross section), three kinds of uprights profiles (with different thickness of the section wall), and two types of connectors (four-tab connector and five-tab connector). This work is a part of the report research on the mechanical behavior of the thin-walled metal parts used in the racking storage systems subjected to bending and buckling [21].

The main purpose of this research to analyze the effects of the connector type with tabs and the effects of the dimensions of the assembled elements (beams and uprights) used in the racking storage systems on both the rotational stiffness and the capable bending moment at failure of that connection in order to predict the safety coefficients of the beam-to-upright connections. For this purpose, the following objectives are established: (i) testing of different groups of assemblies between the beams and the uprights by using two types of connectors (with four or five tabs), (ii) comparison of the stiffness of the connections in function of the type of the connector, (iii) analysis of the effects of the wall thickness of the upright profile and of the size of a beam cross section on the rotational stiffness and on the capable moment at failure for all beam-to-upright connections involved, (iv) comparison of estimated safety coefficients for all beam-to-upright connections tested in order to predict the best type of connection for each group containing beams of the same size for the case of the beam loaded by the distributed weight of the pallets, (v) comparison of the estimated maximum deflections for beams with the same type of semi-rigid connections at both ends by taking into account the stiffness experimentally obtained.

The authors also aim to use the results obtained experimentally by focusing on the stiffness of the connections in order to compare the maximum deflections of the beams having such semi-rigid connectors at both ends with the maximum deflections computed for the beams pin-connected or with rigid connectors at both ends. Bending moments developed at the midpoint of the beam and at the ends of the beam are also evaluated for different boundary conditions.

2. Materials and Methods

2.1. Beam-to-Upright Connections Tested

The main structural elements of such a semi-rigid joint used for the pallet racking structures are shown in Figure 1.

The upright (Figure 1c,d) is a cold-formed omega-profile made of a thin steel sheet (Figure 2). The upright profile has perforations along its length that facilitate the connection with the beam end connectors welded at the beam end. The main dimensions of the cross sections of the uprights involved in this research are shown in Table 1.

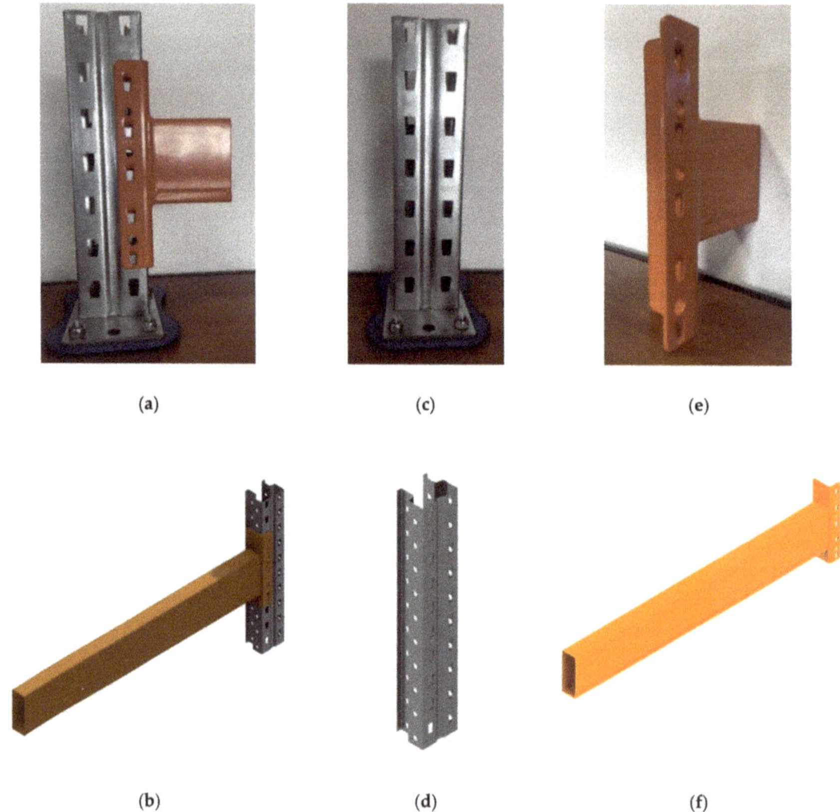

(a)   (c)   (e)

(b)   (d)   (f)

Figure 1. Main elements of the connections tested: (**a**,**b**) upright-connector-beam assembly, (**c**,**d**) upright, (**e**,**f**) beam with welded beam end connector.

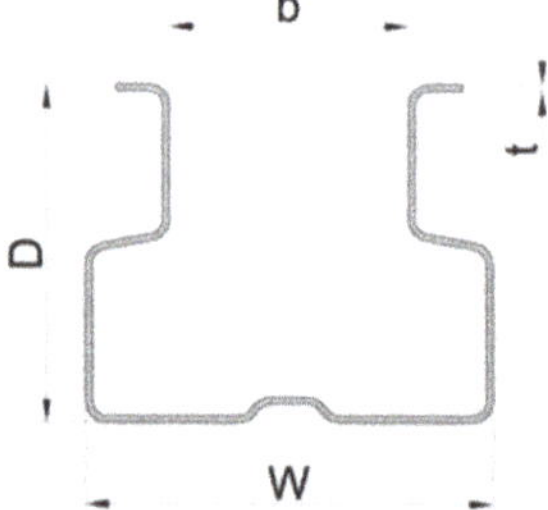

Figure 2. Cross section of the upright profile (dimensions D, W, b, t are given in Table 1).

Table 1. Dimensions of the cross sections of the uprights.

Type of the Upright	Code of the Upright	D * (mm)	W * (mm)	B * (mm)	T * (mm)
90 × 1.50	I				1.50
90 × 1.75	II	70	90	50	1.75
90 × 2.00	III				2.00

* Dimensions of the uprights are shown in Figure 2.

Beams having a box cross section (Figure 1e,f) are manufactured by using two C-profiles assembled together in a rectangular shape (Figure 3). The connectors are welded at both ends of the beam and those connectors are used to hook the beam by perforations of the upright. The main dimensions of the cross sections of the beams involved in this study are shown in Table 2.

In Table 4, all upright-connector-beam assemblies tested in bending are shown in this research in order to determine the stiffness of those connections. The corresponding identification codes are also shown in Table 4. The main components of each assembly are: upright, beam, and beam and connector. Three types of upright profiles were used, which have the same geometry of the cross-section (Figure 2) but have different thicknesses of the profile wall (Table 1). Three types of box beams (Figure 3) are used whose dimensions are given in Table 2. For each type of upright used in an assembly, two types of connectors were considered, which include a four-tab connector and a five-tab connector. Lastly, six different types of upright-connector-beam assemblies were tested for each beam type (Table 4). Multiple individual tests were performed for each type of assembly (last column of Table 4). Lastly, 101 individual tests were carried out.

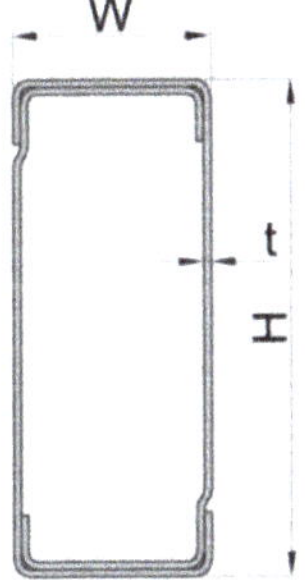

Figure 3. Shape and dimensions of the cross-sections of the beams (dimensions H, W, t are given in Table 2).

Table 2. Dimensions of the beams' cross sections involved.

Beam Type	Beam Code	H^* (mm)	W^* (mm)	t^* (mm)
BOX 90	A	90	40	1.50
BOX 100	B	100	40	1.50
BOX 110	C	110	40	1.50

* Dimensions of the uprights are shown in Figure 3.

In Table 3, the overall dimensions (Figure 4) are shown as well as the designation codes corresponding to both types of beam end connectors used in this research.

Table 3. Dimensions of the beam end connectors.

Connector Type	Connector Code	No. of Tabs	H^* (mm)	D^* (mm)	W^* (mm)	Thickness (mm)
5-tab connector	5L	5	238	63	41	4.0
4-tab connector	4L	4	190	63	41	4.0

* Dimensions of the beam end connectors are shown in Figure 4.

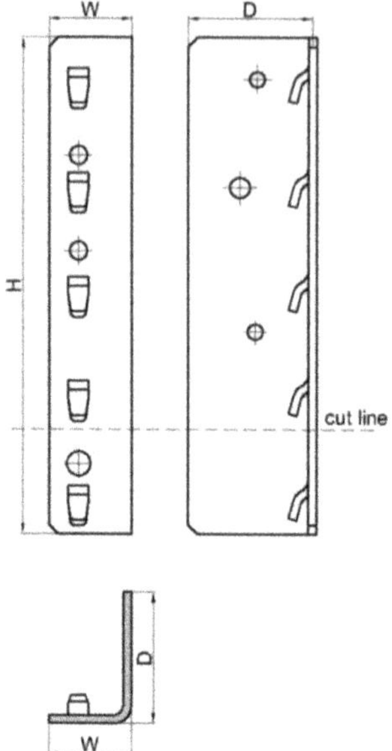

Figure 4. Shape and dimensions of the beam end connectors involved in the research (dimensions H, W, D are given in Table 3).

Table 4. Upright-connector-beam assemblies tested.

Assembly Identification Code	Upright	Beam	Connector Type	Number of Tests for Assembly
A-I-4L	90 × 1.50		4L	5
A-I-5L			5L	6
A-II-4L	90 × 1.75	A (BOX 90)	4L	6
A-II-5L			5L	6
A-III-4L	90 × 2.00		4L	6
A-III-5L			5L	6
B-I-4L	90 × 1.50		4L	5
B-I-5L			5L	6
B-II-4L	90 × 1.75	B (BOX 100)	4L	6
B-II-5L			5L	6
B-III-4L	90 × 2.00		4L	3
B-III-5L			5L	5
C-I-4L	90 × 1.50		4L	6
C-I-5L			5L	6
C-II-4L	90 × 1.75	C (BOX 110)	4L	6
C-II-5L			5L	6
C-III-4L	90 × 2.00		4L	5
C-III-5L			5L	6

2.2. Work Methods

2.2.1. Tensile Tests

The component parts (upright, beam, and beam end connector) are made of steel. Tensile tests were carried out for the tensile specimens cut from each component part of the assemblies involved in order to experimentally obtain the following characteristics of the materials: yield stress, ultimate stress, and elongation until maximum force. The material properties are used in data processing obtained in bending tests of the connections in accordance with EN 15512 standard [9].

Tensile specimens were cut from each component part (upright, beam end connector, and beam), as shown in Figure 5.

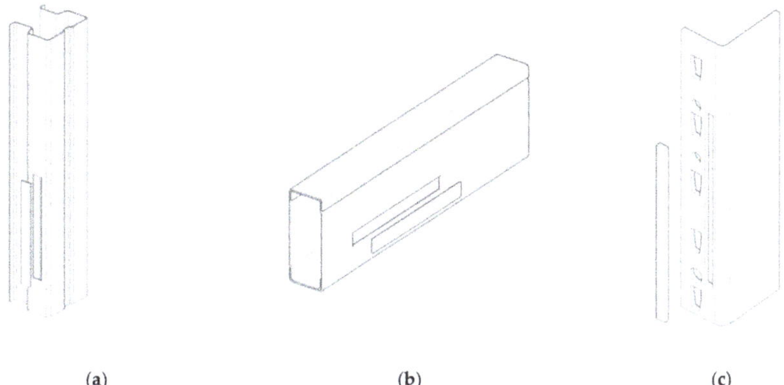

(a)   (b)   (c)

Figure 5. Locations from where we have cut the samples to make tensile tests: (**a**) from upright, (**b**) from beam, and (**c**) from beam end connector.

Universal testing machine INSTRON 3369 (INSTRON, Norwood, MA, USA) with digital controls and a maximum force of 50 kN was used in tensile tests. The tensile testing machine pulls the sample clamped at both ends and records the tensile force related to the elongation of the tensile specimen until it ruptures. By using the force-elongation curve, the stress-strain curve is obtained and we get the yield stress and ultimate stress. In tensile tests, the strain rate was equal to 0.00025 per second until the strain $\varepsilon = 0.003$ and then the strain rate equaled 0.0067 per second until to failure in accordance with the standard EN ISO 6892 [22].

2.2.2. Bending Test of the Connections

Tests have been carried out according to the standard EN 15512 [9]. The scheme of loading used in the bending test of the connection and the photo of the experimental setup are shown in Figure 6.

In accordance with Figure 6 and EN 15512 [9], a short upright cut between two sets of perforations was connected to a very stiff testing frame in two points apart at the distance g = 470 mm from each other. A short beam with a length of 650 mm is connected to the upright region by means of the connector. Sideways movement and twisting of the beam were prevented. The beam is able to move freely in the vertical direction of the load by guiding the free end vertically because a steel plate welded at the free end of the beam is sliding between the bearings. The vertical force was applied at distance b = 400 mm from the face of the upright region by using the force transducer of type S-E-G Instruments (S-E-G Instruments AB, Bromma, Sweden) whose accuracy is 0.001 kN and maximum force is 12.5 kN.

Rotation angle θ of the beam end at the connection with the upright was measured by using the displacement transducers whose rods are in permanent contact with the plate fixed to the beam close to the connector so that the distance is equal to 50 mm (Figure 6a). The displacement transducers of type WA-100 (manufactured by HBM–Hottinger Baldwin Messtechnik Gmbh, Darmstadt, Germany) may record displacements with a size less than 100 mm having the accuracy of 0.001 mm.

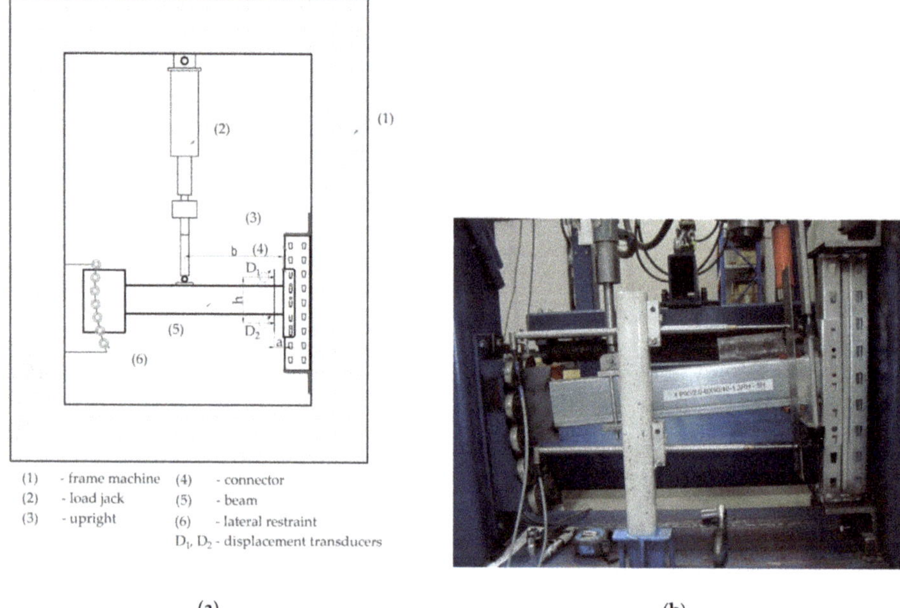

Figure 6. Bending test of the connection according to EN 15512 [9]: (**a**) scheme of the test stand and (**b**) photo of the experimental setup.

The force transducer and both displacement transducers are connected to the QuantumX MX840 Universal Measuring Amplifier (manufactured by Hottinger HBM, Darmstadt, Germany) that transmits the data to Easy Catman software (version 3.1, Hottinger Baldwin Messtechnick Gmbh, Darmstadt, Germany) on the computer. The data acquisition device QuantumX MX840 (Hottinger Baldwin Messtechnick Gmbh, Darmstadt, Germany) works at a frequency of 19.2 kHz.

In the beginning of the bending test, an initial force of approximately 10% of the maximum failure load is applied and, then, the vertical force is increased gradually until failure occurs. Tests were repeated identically for all the upright-connector-beam assemblies involved in this study (Table 4). A minimum of five tests were made for each kind of upright-connector-beam assembly shown in Table 4. In this way, the scattering rate of the results is also analyzed.

The value of the external force F acquired by the force transducer is used to compute the bending moment M by using Equation (1).

$$M = Fb, \tag{1}$$

in which b is the dimension shown in Figure 6a.

The deflections D_1 and D_2 acquired by the displacement transducers (Figure 6a) are used to compute the rotation angle θ of the beam end at the connection by using Equation (2).

$$\theta = (D_2 - D_1)/h \tag{2}$$

in which h is the distance between the rods of the displacement transducers (Figure 6a). In Equation (2), the deflection D_1 has a positive value while the deflection D_2 is a negative value. The rotation angle θ of the beam end computed with Equation (2) is expressed in radians.

The failure mode of the component parts (beam end connector, upright) was also noted in the testing report in the case of each bending test carried out.

The bending moment M and rotation angle θ experimentally obtained were corrected in accordance with the standard EN 15512 [9]. Therefore, the correction procedure is described below.

The corrections are required because there are variations of the yield stress of the material corresponding to each component part (upright, beam, connector) and variations of the thickness corresponding to those components. As a result, the correction factor C_m was computed by using Equation (3) and C_m must be less than or equal to 1, according to EN 15512 [9].

$$C_m = \left[\left(\frac{f_y}{f_t} \right)^{\alpha} \frac{t}{t_t} \right]_{max} \text{ and } C_m \leq 1, \tag{3}$$

where f_t is the yield stress obtained by testing the tensile specimens, f_y is the nominal yield stress and the values are given in standard EN 10346 [23], t_t is the measured thickness for the tensile specimen, t is the design thickness, $\alpha = 0$ when $f_y \geq f_t$, and $\alpha = 1$ when $f_y < f_t$.

To make the correction for a moment-rotation curve (M–θ), some steps should be covered in accordance with the EN 15512 standard [9]. First, for each bending test of a connection, the moment-rotation curve (M–θ) is plotted and the slope of the curve K_0 is measured at the origin. Then, the elastic rotation M/K_0 is subtracted from the measured rotation θ to obtain the plastic rotation θ_p by using Equation (4), according to EN 15512 [9].

$$\theta_p = \theta - M/K_0. \tag{4}$$

The corrected moment M_n is computed by using Equation (5).

$$M_n = M \cdot C, \tag{5}$$

where C is another correction factor related to C_m whose value is computed by using Equation (6) and must be less than or equal to 1, according to EN 15512 [9].

$$C = 0.15 + C_m \leq 1. \tag{6}$$

The corrected value of the elastic rotation θ_e is computed by using Equation (7).

$$\theta_e = M_n/K_0 \tag{7}$$

and the corrected rotation angle θ_n is computed by using Equation (8), according to EN 15512 [9].

$$\theta_n = \theta_p + \theta_e = \theta_p + M_n/K_0. \tag{8}$$

The adjusted moment-rotation curve (M_n–θ_n) is plotted and its initial slope K_0 is the same for the initial moment-rotation curve (M–θ).

The failure moment M_{ni} is considered to be the maximum corrected moment from the adjusted moment-rotation curve (M_n–θ_n) shown in Figure 7, which is plotted for each upright-connector-beam assembly. The subscript i of the failure moment M_{ni} represents the number of the test with i = 1,n, where n is the number of tests corresponding to each upright-connector-beam assembly. The moment M_m is the mean value of failure moments M_{ni} and it is computed with Equation (9), according to EN 15512 [9].

$$M_m = \frac{1}{n} \sum_{i=1}^{n} M_{ni}. \tag{9}$$

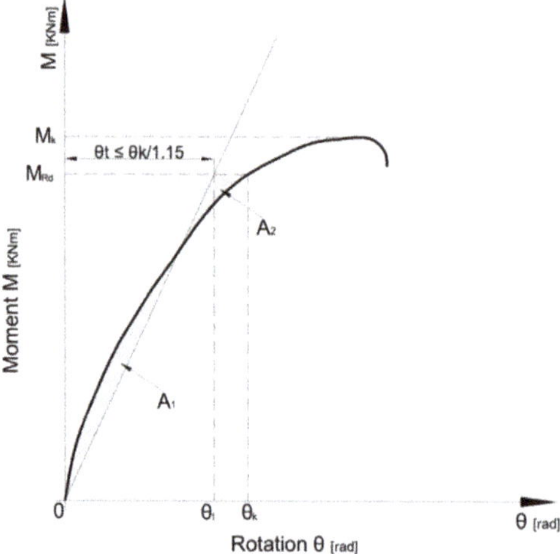

Figure 7. Computing of the rotational stiffness k_{ni}, according to EN15512 [9].

The standard deviation of the adjusted test results denoted with s is computed with Equation (10), according to EN 15512 [9].

$$s = \sqrt{\frac{1}{n-1}\sum_{i=1}^{n}(M_{ni} - M_m)^2}. \tag{10}$$

For each upright and connector assembly, the characteristic failure moment M_k is the characteristic failure moment computed by Equation (11), according to EN 15512 [9].

$$M_k = M_m - sK_{s,}. \tag{11}$$

where K_s is the coefficient based on a 95% fractile at a confidence level of 75% in accordance with EN 15512 [9].

The design moment for the connection is denoted by M_{Rd} and is given by Equation (12) [9].

$$M_{Rd} = \eta\frac{M_k}{\gamma_M}, \tag{12}$$

where γ_M is a partial safety factor for the connection and $\gamma_M = 1.1$, according to EN 15512 [9]. η is the variable moment reduction factor selected by the designer so that $\eta \leq 1$.

The rotational stiffness k_{ni} ($i = 1,n$) of the connection is obtained as the slope of a line through the origin, which isolates equal areas (A_1 and A_2 in Figure 7) located between that line and the experimental curve below the design moment M_{Rd}. The rotational stiffness k_{ni} is computed for each test using Equation (13), according to EN 15512 [9].

$$k_{ni} = \frac{1.15M_{Rd}}{\theta_{ki}}, \tag{13}$$

where θ_{ki} ($i = 1,n$) is the rotation corresponding to the design moment M_{Rd} on the adjusted moment-rotation curve (M_n–θ_n) shown in Figure 7.

The design rotational stiffness k_m is computed by Equation (14) [9].

$$k_m = \frac{1}{n}\sum_{i=1}^{n}k_{ni}. \tag{14}$$

2.2.3. Theoretical Aspects Regarding the Effects of the Stiffness of Connection on Deflections

In strength calculus of the beams of the racking storage systems, the beam is considered to be subjected to uniformly distributed force q given by the specific weight of the goods on the pallets stored on the shelf. It is considered that the modulus of rigidity EI in bending is considered to be constant along the axis of the beam with the length L. By considering the traditional methods to analyze the beams in a racking storage system for which the connections between beam and upright are considered as pin-connections (hinged), the rotation θ is equal to $qL^3/24EI$ at the beam end at that connection while the bending moment M_b developed at the pin-connection is equal to zero. In case the end connections considered as rigid-like is the fixed support, the rotation θ is equal to zero at that connection while the bending moment M_b is equal to $qL^2/12EI$.

On the other hand, for the semi-rigid connections between the beam and upright region, the values of the rotation angle θ and of the bending moment M_b are between the values corresponding to the pin-connection and fixed support. Depending on the rigidity of the connection, in Figure 8, a classification of the connections is presented, according to EN 1993-1-3, Eurocode 3 [8].

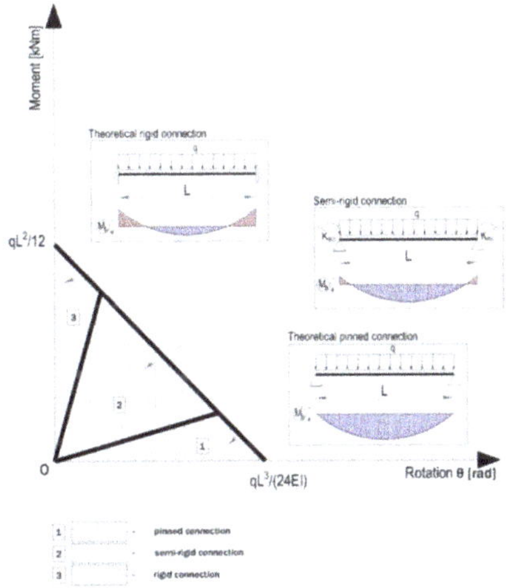

Figure 8. Classes of the beam-end-connections related to the rigidity, according to EN 1993-1-3 [8].

For the beam connected with the uprights of the storage system, by using semi-rigid connections whose rigidity k_m was experimentally determined, the bending moment M_{end} developed at the ends of the beam subjected to the uniformly distributed force q, is computed by using Equation (15).

$$M_{end} = \frac{qL^3}{24EI\left(\frac{1}{k_m} + \frac{L}{2EI}\right)}. \tag{15}$$

The bending moment M_{mid} developed at the middle cross section of the beam is computed using Equation (16).

$$M_{mid} = \frac{qL^2}{8} - M_{end}. \tag{16}$$

In order to analyze the capacity of the semi-rigid connections of the beam, it is computed by the safety coefficient c of the connection as the ratio between the design moment M_{Rd} computed by using Equation (12), based on experimental results, and the bending moment M_{end} developed at the end of the beam with semi-rigid connections at both ends and loaded with the uniformly distributed force q. The safety coefficient c of the connection is computed by Equation (17).

$$c = \frac{M_{Rd}}{M_{end}}.$$ (17)

The value of the maximum deflection v_{max} having two semi-rigid connections at both ends, whose rigidity k_m was experimentally determined, is computed by using Equation (18).

$$v_{max} = \frac{qL^4(10EI + Lk_m)}{384EI(2EI + Lk_m)}.$$ (18)

Practically, customizing Equation (18) for the case of the beam pin-connected (hinged) at both ends for which the rigidity of the connection k_m is equal to zero, leads to the maximum deflection v_{max} equal with $5qL^4/384EI$. On the other hand, customizing Equation (18) for the case of the beam fixed at both ends (embedded ends) for which the rigidity of the connection k_m tends to infinity, leads to the maximum deflection v_{max} equal to $qL^4/384EI$.

3. Results

3.1. Tensile Properties of the Materials

In Table 5, the tensile properties corresponding to the tensile specimens cut from each component are shown. The tensile test results for the steel corresponding to the four-tab connector are identical with the ones corresponding to the five-tab connector for the same beam height because we used connectors with five tabs for all beams and we had cut the connector to obtain four-tab connectors. The decision regarding the cutting of the five-tab connector was taken in order to have the same material for both kinds of connectors.

Table 5. Material properties for tested components.

Component Corresponding to the Tensile Specimen	Measured Thickness	Yield Stress Rp,02 f_t *	Ultimate Stress f_u *	Strain *
	(mm)	(MPa)	(MPa)	(%)
Upright I (90 × 1.50)	1.48	447.9	528.9	19.2
Upright II (90 × 1.75)	1.80	496.2	536.2	15.3
Upright III (90 × 2.00)	2.09	535.1	590.2	18.5
Beam A (box 90)	1.25	360.6	459.5	38.8
Beam B (box 100)	1.23	379.8	429.5	29.7
Beam C (box 110)	1.27	344.2	411.6	38.1
4-tab connector Box 90	4.04	409.9	469.8	25.6
5-tab connector Box 90	4.04	409.	469.8	25.6
4-tab connector Box 100	4.07	395.3	464	32.6
5-tab connector Box 100	4.07	395.3	464	32.6
4-tab connector Box 110	4.05	444.7	503.5	24.5
5-tab connector Box 110	4.05	444.7	503.5	24.5

* The values were obtained in tensile tests within this research carried-out by INSTRON 3369 machine.

3.2. Results of Bending Tests

In this research, all upright-connector-beam assemblies presented in Table 4 were tested according to the testing procedure presented in EN 15512 [9].

First, the moment-rotation curves (M–θ) are represented by using the data acquired with the force and displacement transducers and by using Equations (1) and (2). Then, the corrected moment-rotation curves $(M_n–θ_n)$ are plotted for each upright-connector-beam assembly, after the corrections caused by the different thickness and yield strength were made by using Equations (5) and (8). The moment-rotation curves (M_n–θ_n) corresponding to all sets of the assemblies are plotted in Appendix A. There are a total of 101 curves plotted. In Figure A1, it is shown that the curves for the beam-connector-upright assemblies containing a type A beam. In the same manner, in Figures A2 and A3, the curves for the beam-connector-upright assemblies containing a type B beam and a type C beam, respectively. In order to comparatively analyze the effects of the connector type on the mechanical behavior of the beam-upright connections in bending the moment-rotation mean curves (M_n–θ_n) are plotted in Figures 9–11. It may be observed that the behavior of the connections is nonlinear and the connectors with five tabs are always stiffer than the connectors with four tabs for all assemblies tested.

The shapes of the moment-rotation (M_n–θ_n) curves shown in Figures A1-A3 added in the Appendix A, show a good repeatability recorded for each set of beam-connector-upright assemblies tested, concerning both the initial slopes of these curves (which gives the rotational stiffness) and the maximum values of the capable bending moments.

All values for the design bending moment $M_{Rd,i}$ and for the rotational stiffness k_{ni} ($i = 1,n$) are determined by bending tests of the upright-connector-beam assemblies and by using Equations (12) and (13), respectively. In order to make the comparisons for the test results, the stiffness k_{ni} was evaluated for the maximum bending moment $M_{Rd,i}$ when $\eta = 1$. All results are presented in Table 6.

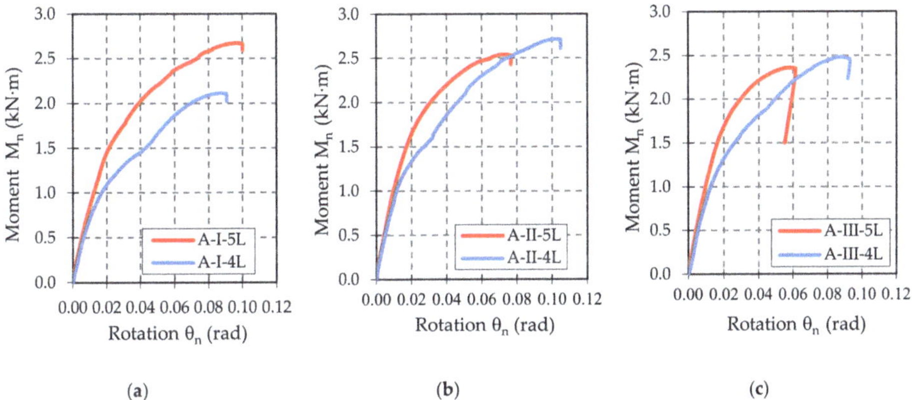

Figure 9. Comparison of the moment-rotation curves recorded for: (**a**) A-I-4L and A-I-5L assemblies, (**b**) A-II-4L and A-II-5L assemblies, and (**c**) A-III-4L and A-III-5L assemblies.

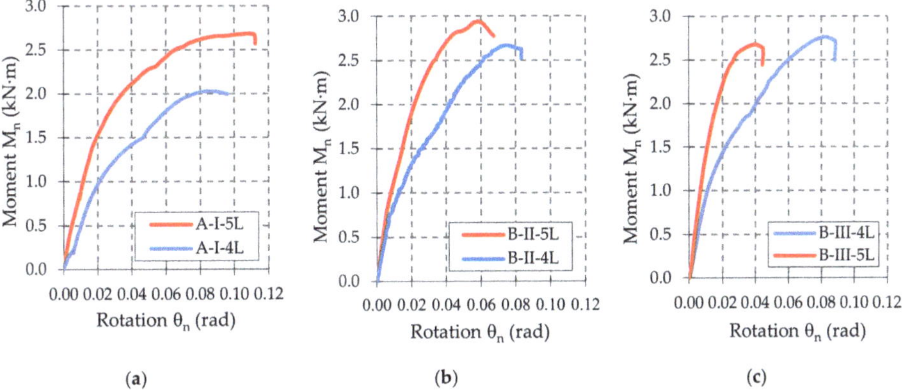

Figure 10. Comparison of the moment-rotation curves recorded for: (**a**) B-I-4L and B-I-5L assemblies, (**b**) B-II-4L and B-II-5L assemblies, and (**c**) B-III-4L and B-III-5L assemblies.

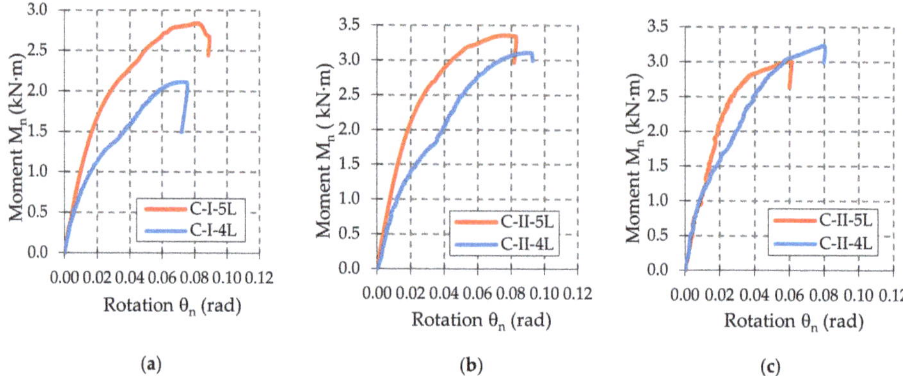

(a)   (b)   (c)

Figure 11. Comparison of the moment-rotation curves recorded for: (**a**) C-I-4L and C-I-5L assemblies, (**b**) C-II-4L and C-II-5L assemblies, and (**c**) C-III-4L and C-III-5L assemblies.

The average values and standard deviations are also shown in Table 6 for the design bending moment M_{Rd} and for the rotational stiffness k_m corresponding to each group of beam-connector-upright assemblies tested in bending. The value of the standard deviation corresponding to the design bending moment M_{Rd} for each group of assemblies, divided by the average value M_{Rd}, is generally less than 3% and this means a high level of confidence of the experimental tests. Similar remarks may also be made regarding the values of the standard deviation corresponding to the rotational stiffness k_m. The low values of stdev compared to the average values show that the degree of scattering of the results is small. The differences between the results obtained for assemblies of the same group are justified by the geometric imperfections of the cross sections, which are characteristic of the cold formed steel profiles, especially in the area of the slots of the upright profile.

The results regarding both the corrected moment M_{ni} and the rotational stiffness k_{ni} are comparatively shown in Figures 12–14 for each type of beam used in the assembly, i.e., for beam A, beam B, or beam C, respectively. It notes that the degree of scattering of the results is small from the point-of-view of both the corrected moment M_{ni} and rotational stiffness k_{ni} corresponding to each group of assemblies tested (Figures 12–14). Comparing the results corresponding to the assemblies composed by the same type of beam and upright, we noticed that the design rotational stiffness k_m increases as the upright thickness increases. It is also evident that the design rotational stiffness k_m is greater for the assembly containing the five-tab connector than the value corresponding to the assembly containing the four-tab connector (Table 6, Figures 12b, 13b and 14b).

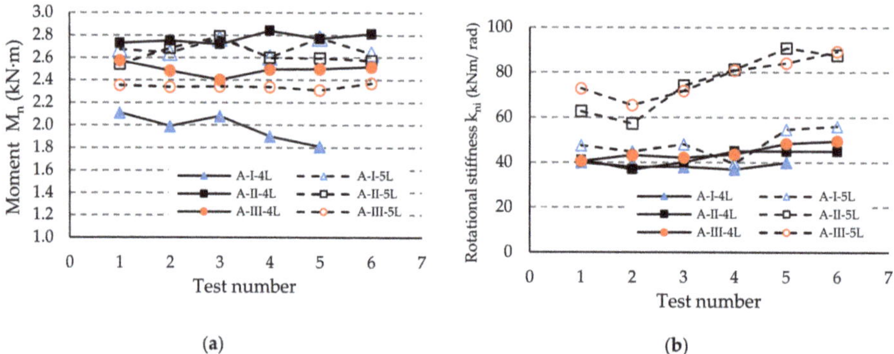

(a)   (b)

Figure 12. Comparison of the results obtained for the assemblies with the beam of type A concerning: (**a**) capable corrected moment M_n, and (**b**) rotational stiffness k_{ni} of the connections.

Table 6. Test results in bending tests for all assemblies tested.

Group A

Assembly code	Design Moment			Rotational Stiffness		
	M_{ni} (kNm)	M_{Rd} (kNm)	Stdev (kNm)	k_{ni} (kNm/rad)	k_m (kNm/rad)	Stdev (kNm/rad)
A-I-4L	2.11 1.99 2.08 1.90 1.81	1.54	0.12	40 38 38 37 40	39.0	1.34
A-I-5L	2.68 2.65 2.77 2.60 2.78 2.64	2.3	0.07	47 45 48 40 55 56	48.0	6.18
A-II-4L	2.73 2.75 2.72 2.84 2.77 2.81	2.43	0.05	41 37 40 45 45 45	42.0	3.37
A-II-5L	2.54 2.69 2.79 2.60 2.59 2.57	2.21	0.09	63 57 74 81 91 88	76.0	13.5
A-III-4L	2.57 2.48 2.40 2.49 2.50 2.52	2.16	0.06	41 43 42 44 48 50	45.0	3.57
A-III-5L	2.36 2.34 2.34 2.34 2.31 2.37	2.09	0.02	73 65 72 81 84 89	77.0	8.90

Group B

Assembly code	Design Moment			Rotational Stiffness		
	M_{ni} (kNm)	M_{Rd} (kNm)	Stdev (kNm)	k_{ni} (kNm/rad)	k_m (kNm/rad)	Stdev (kNm/rad)
B-I-4L	2.03 2.06 2.08 2.07 2.07	1.84	0.02	34 36 35 38 37	35.9	1.38
B-I-5L	2.69 2.75 2.62 2.58 2.76 2.92	2.24	0.12	55 59 46 61 58 68	57.8	7.29
B-II-4L	3.04 2.67 2.87	2.03	0.18	56.1 54.3 57.6	56.0	1.65
B-II-5L	2.66 2.78 2.75 2.85 2.87 2.94	2.36	0.10	65 77 76 100 105 94	86.1	15.7
B-III-4L	2.76 2.74 2.78 2.75 2.70 2.71	2.43	0.03	50 50 47 53 55 52	51.2	2.97
B-III-5L	2.52 2.49 2.55 2.55 2.68	2.18	0.07	80 86 98 114 131	102	20.9

Group C

Assembly code	Design Moment			Rotational Stiffness		
	M_{ni} (kNm)	M_{Rd} (kNm)	Stdev (kNm)	k_{ni} (kNm/rad)	k_m (kNm/rad)	Stdev (kNm/rad)
C-I-4L	2.12 2.13 2.13 1.95 2.14 1.96	1.70	0.08	44 44 45 41 44 40	43.7	1.26
C-I-5L	2.84 3.05 3.01 2.60 2.61 2.69	2.15	0.20	72 73 73 80 81 86	77.4	5.71
C-II-4L	3.08 3.03 3.07 3.11 3.08	2.74	0.03	50.1 49.3 52.8 53.1 56.8	52.4	2.97
C-II-5L	3.29 3.39 3.36 3.35 3.47 3.35	2.95	0.06	79 93 81 80 93 77	83.7	7.25
C-III-4L	3.28 3.27 3.24 3.20 3.27 3.24	2.89	0.03	63 67 64 64 61 58	63.0	2.93
C-III-5L	2.92 3.01 2.99 2.94 3.00 3.08	2.61	0.06	107 101 116 121 119 128	115	9.76

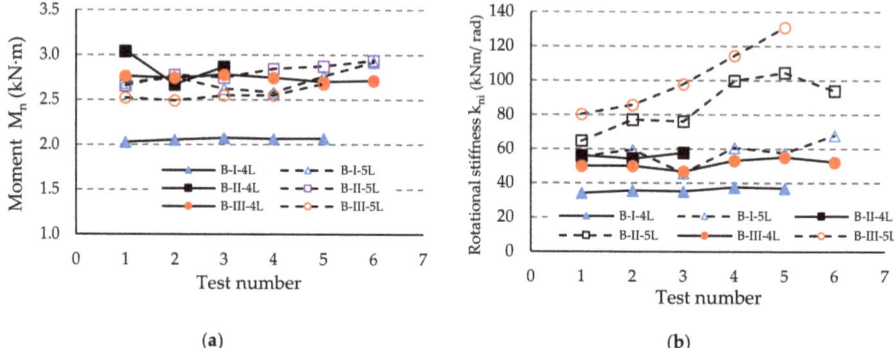

Figure 13. Comparison of the results obtained for the assemblies with the beam of type B concerning: (**a**) capable corrected moment M_n and (**b**) rotational stiffness k_{ni} of the connections.

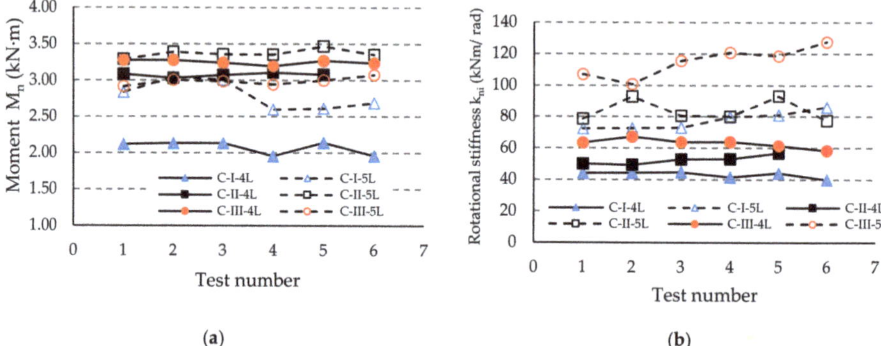

Figure 14. Comparison of the results obtained for the assemblies with the beam of type C concerning: (**a**) capable corrected moment M_n, and (**b**) rotational stiffness k_{ni} of the connections.

Regarding the capable load, the design moment M_{Rd} corresponding to the assembly containing the upright II having a thickness of 1.75 mm is greater than the design moment M_{Rd} corresponding to the assembly containing the upright I with a thickness of 1.5 mm (Table 6, Figures 12a, 13a and 14a). However, the design moment M_{Rd} corresponding to the assembly containing the upright III with a thickness of 1.5 mm and the five-tab connector is less than the design moment M_{Rd} for the assembly with the same upright region and with four connectors, even if the stiffness is bigger for the assembly with a five-tab connector comparatively with the stiffness for the assembly with a four-tab connector. Moreover, the design moment M_{Rd} for the assembly containing upright III and a four-tab connector is much closer to the design moment M_{Rd} for the assembly containing upright II and a five-tab connector in the case of all types of beams (A, B, or C).

3.3. Failure Modes in Bending Tests

The failure mode was changed from assembly to another assembly tested. Upright is the main failure mode, especially when the upright thickness is 1.50 mm (Figure 15). This happens because, when the beam is loaded and the connector rotates, the first two tabs are tearing the upright's slots because the upper part of the beam is subjected to tensile stresses (Figure 15c).

It was observed that, when the thickness of the upright increases, for the upright having a thickness of 1.75 mm, the failure mode was changed. All the elements (upright, beam, and beam end connector) start to be a part of the failure mode (Figure 16). In this case, the rotation angle θ deceases (Figure 16a), especially as the beam height increases and the bottom part of the beam is compressed and it starts to locally buckle (Figure 16b). The connector is also bending (Figure 16a).

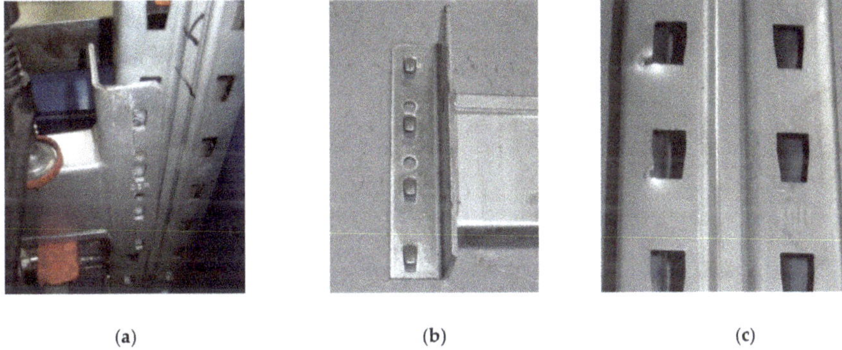

(a) (b) (c)

Figure 15. Failure modes of the upright region with a thickness equal to 1.50 mm: (**a,b**) failure mode of the connector and (**c**) failure mode of the upright region.

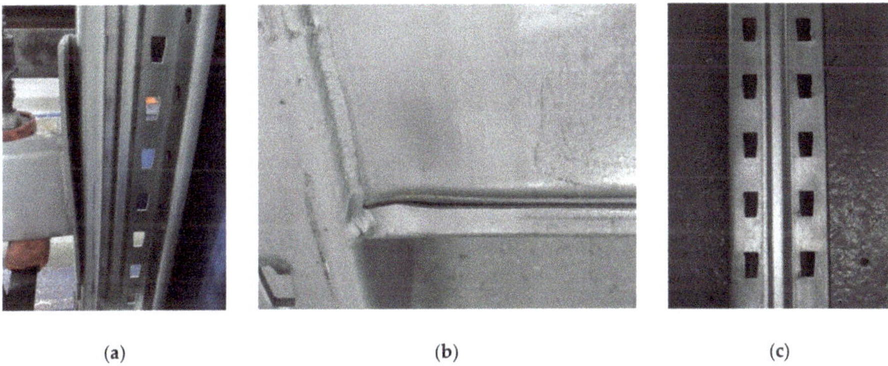

(a) (b) (c)

Figure 16. Failure modes in case of the assembly with an upright region having a thickness equal to 1.75 mm: (**a**) failure mode of the connector, (**b**) failure mode of the beam, and (**c**) failure mode of the upright region.

For uprights having a thickness of 2.00 mm, the effects of the action of the connector tabs on the upright region is very small, especially for the five-tab connector (Figure 17a,c). The rotation of the connection is even smaller (Figure 17a), which means that the biggest rotational stiffness corresponds to this situation. The main failure modes were bending of the connector and the local buckle of the beam near the connector (Figure 17b). For assembly with a four-tab connector (Figure 18) because the beam rotates more than in a situation of the assembly with a five-tab connector, the first top tab still leaves a mark on the upright slots (Figure 18c).

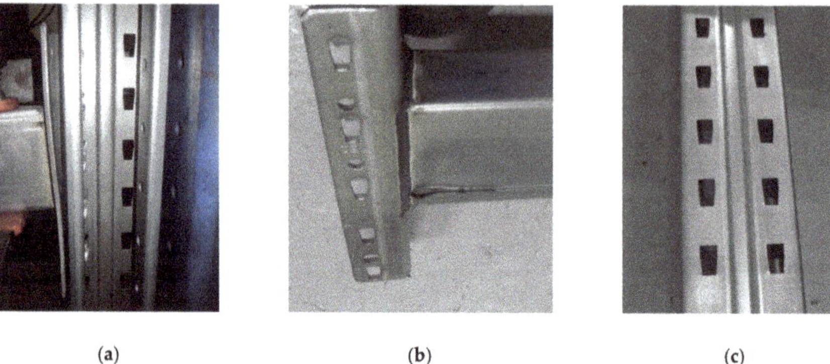

(a)   (b)   (c)

Figure 17. Failure modes in case of the assembly with the upright region having a thickness equal to 2.00 mm and a connector with five tabs: (**a**) failure mode of the connector, (**b**) failure mode of the beam, and (**c**) failure mode of the upright region.

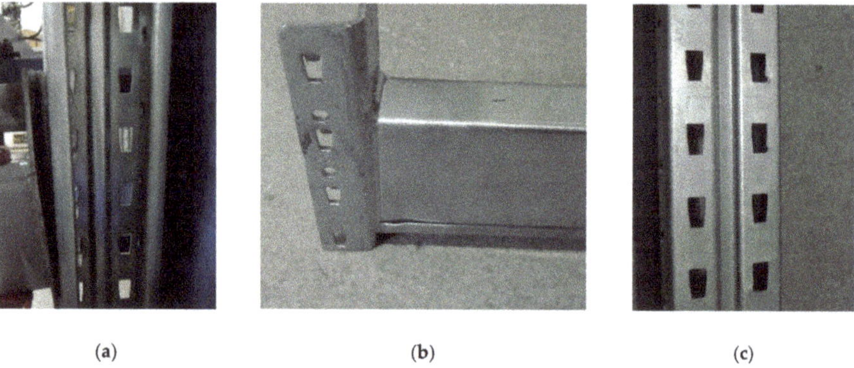

(a)   (b)   (c)

Figure 18. Failure modes in case of the assembly with an upright region having a thickness equal to 2.00 mm and a connector with four tabs: (**a**) failure mode of the connector, (**b**) failure mode of the beam, and (**c**) failure mode of the upright region.

4. Discussion

Analyzing the experimental results summarized in Table 6 and moment-rotation curves (M–θ) shown in Figures 9–11, we note the following important findings.

- For the assemblies containing the uprights of type I with a thickness of 1.50 mm, the five-tab connector leads to a higher value of the design moment M_{Rd} and higher rotational stiffness k_m than in the assemblies with four-tab connectors. This behavior was noticed for all three types of beams (A, B, C) with different heights of the beam being cross sectional (Table 6).
- When the upright thickness increases (upright III), a change in the failure modes was observed because the five-tab connector rotates less than the four-tab connector (Figures 17 and 18). The rotational stiffness k_m of the connection is higher for the assemblies with five-tab connectors than for the assemblies with four-tab connectors. On the contrary, the capable design moment M_{Rd} decreases for a five-tab connector, which is more visible in the bending tests applied to the assembly that contain the uprights of type III having a thickness of 2.00 mm (Table 6).
- As the upright thickness increases, the portion of the moment-rotation curves corresponding to the plastic behavior for the assemblies containing four-tab connectors is greater than the curve portion in plastic corresponding to the assemblies containing the five-tab connectors (Figures 9–11). The mechanical behavior of the material is characteristic of elastic-plastic metallic materials [10,24].

- For the assemblies containing B or C beams with a section height of 100 mm or 110 mm, respectively, and containing a five-tab connector, the design moment M_{Rd} and rotational stiffness k_m of the joints increase due to the upright sections having a thickness of 1.5 mm or 1.75 mm. For the assemblies containing the upright sections with a thickness of 2.00 mm, the rotational stiffness k_m still increases, but design moment M_{Rd} of the connection decreases. In the assemblies with a 5-tab connector, the capable design moment M_{Rd} corresponding to the upright III having the thickness of 2.00 mm, is lower than the capable moment M_{Rd} corresponding to the upright I with a thickness of 1.75 mm upright regions in case of all types of beams.

In practice, the values of the design moment M_{Rd} and rotational stiffness of the connection experimentally obtained and given in Table 6 are very useful in order to check if the beam-connector-upright assembly works safely from both strength and stiffness points-of-view.

We considered a typical case for storage pallet racking systems for which two beams having the length of 2.7 m must support the weight of three wooden pallets (Figure 19a). The maximum total mass for one pallet loaded with goods is 500 kg per pallet, which means that the total weight is approximately 15,000 N for three pallets. This weight applied to one shelf is supported by two beams, which means approximately 7500N applied to the beam length of 2700 mm, which leads to a distributed force of 2.78 N/mm by assuming that the weight is uniformly distributed on the beam (Figure 19b). We considered that this beam is connected with upright regions by one type of semi-rigid connections tested for which the experimental results regarding the design moment M_{Rd} and rotational stiffness of the connection are synthesized in Table 6.

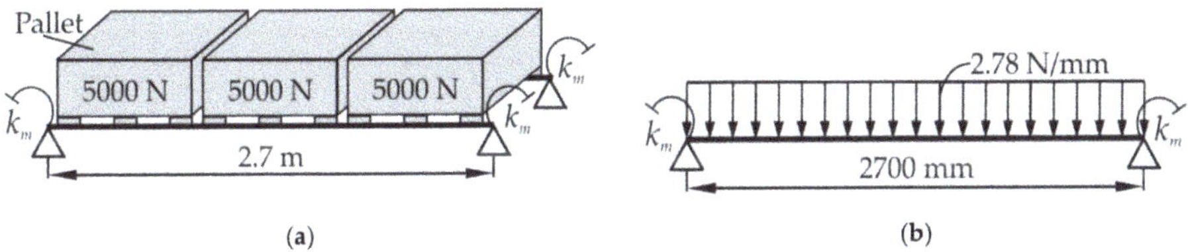

Figure 19. Scheme of loading: (**a**) loading for two beams, and (**b**) equivalent scheme of loading for calculus (k_m represents the rotational stiffness of the connections).

We aim to comparatively analyze the safety coefficient c for all beam-connector-upright assemblies tested to show in this way the reserve of each connection so that it works in the safety range. For this purpose, the moment developed at the end of the beam (at the level of beam end connection) is computed, which is denoted by M_{end} by using Equation (15) for the uniformly distributed force q of 2.78 N/mm applied to the beam with the length L of 2700 mm.

The safety coefficient c corresponding to each beam-connector-upright assembly involved in this study is computed by using Equation (17). Lastly, the results are synthesized and analyzed comparatively in Table 7 for all assemblies involved in this research. In Table 7, for each class of assemblies corresponding to one type of beam, the smallest and the highest value of the safety coefficient c is highlighted by using superscript symbols for those values (see the footnotes of Table 7).

In Table 7, we may remark that, for the assemblies containing types A or C beams, the best assemblies are based on the upright of type II and beam-end connector with four tabs (A-II-4L and C-II-4L in Table 7). Just in case of the beam of type B, the highest value of the safety coefficient c is for the assembly containing the upright of type III, which has the greatest thickness of 2 mm, and the beam-end connector with four tabs (B-III-4L in Table 7).

For each class of assemblies corresponding to a certain type of beam, the smallest safety coefficient c corresponds to the assembly for which the beam is connected with the upright of type III by using the beam-end connector with five tabs (A-III-5L, B-III-5L, and C-III-5L in Table 7).

Table 7. Results for the safety coefficient c of the connection for the maximum deflection in case of the assemblies tested for the case of the real loading in the pallet racking storage system.

Assembly Code	Rotational Stiffness k_m (kN·m/rad)	Design Moment M_{Rd} (kN·m)	Moment at End M_{end} (kN·m)	Safety Coefficient c	Maximum Deflection v_{max} (mm)	Ratio v_{all} [***]/ v_{max} (%)
A-I-4L	39.0	1.54	0.52	2.94	12.35	1.09
A-I-5L	48.0	2.30	0.60	3.82	11.74	1.15
A-II-4L	42.0	2.43	0.55	4.41 [*]	12.14	1.11
A-II-5L	76.0	2.21	0.79	2.80	10.29	1.31
A-III-4L	45.0	2.16	0.58	3.75	11.94	1.13
A-III-5L	77.0	2.09	0.79	2.63 [**]	10.25	1.32
B-I-4L	35.9	1.84	0.41	4.47	10.31	1.31
B-I-5L	57.8	2.24	0.58	3.88	9.31	1.45
B-II-4L	56.0	2.03	0.57	3.59	9.38	1.44
B-II-5L	86.1	2.36	0.74	3.20	8.34	1.62
B-III-4L	51.2	2.43	0.53	4.57 [*]	9.58	1.41
B-III-5L	102.0	2.18	0.81	2.70 [**]	7.91	1.71
C-I-4L	43.7	1.70	0.40	4.22	8.27	1.63
C-I-5L	77.4	2.15	0.60	3.56	7.30	1.85
C-II-4L	52.4	2.74	0.46	5.94 [*]	7.99	1.69
C-II-5L	83.7	2.95	0.63	4.65	7.15	1.89
C-III-4L	63.0	2.89	0.53	5.50	7.68	1.76
C-III-5L	115.0	2.61	0.76	3.42 [**]	6.52	2.07

[*] The greatest safety coefficient of the connection for the group assemblies corresponding to one type of beam. [**] The smallest safety coefficient of the connection for the group assemblies corresponding to one type of beam. [***] The allowable deflection v_{all} is 13.5 mm for the beam length of 2700 mm, according to EN15512 [9].

It was very interesting to remark that the ratio between the highest safety coefficient and the smallest safety coefficient is equal to 1.68 (i.e., 4.41/ 2.63) for the assemblies containing the beam of type A, 1.69 (i.e., 4.57/ 2.70) for the assemblies containing the beam of type B, and 1.73 (i.e., 5.94/ 3.42) for the assemblies containing the beam of type C. The conclusion is that, for any type of beam involved in this research, there is a reserve of approximately 70% regarding the safety coefficient, which depends on the type of the upright and end-beam connector used.

On the other hand, the beam used in racking storage systems must obey another requirement: the allowable deflection v_{all} is equal to a maximum of 5% from the beam length, according to EN15512 [9]. This is called a stiffness condition. This means that the allowable deflection v_{all} must be 13.5 mm for the beam length of 2700 mm.

For the practical example considered, the maximum deflection v_{max} of the beam having two semi-rigid connections at both ends is computed by using Equation (18) for all assemblies involved in this research and the results are shown in Table 7. In order to check the stiffness condition, the ratio between the allowable deflection v_{all} of 13.5 mm and the maximum deflection v_{max} for the beam with the same semi-rigid connection at both ends is given for each type of beam-connector-upright assembly involved. It is remarked that all values computed for the v_{all}/v_{max} ratio are greater than 1 and this means that the stiffness condition is obeyed for all assemblies. As expected, the ratio v_{all}/v_{max} increases as the thickness for the beam or for the upright increases. In Table 7, it is observed that the best assemblies from the stiffness point-of-view (i.e., the greatest values of the ratio v_{all}/v_{max}) do not correspond to the best values for the safety coefficient c.

For better understanding the importance of the experimental determination of the connection's rotational stiffness, a comparative study is presented for the same practical study case corresponding to the beam of type B for the following beam models.

(i) beam B shown in Figure 20a, for which the rotational stiffness of the connections is equal to 0 that corresponds to hinged beam ends,

(ii) beam B shown in Figure 20b, for which the rotational stiffness k_m of both end connections is equal to 35.9 kNm/rad corresponding to the semi-rigid connection,

(iii) beam B shown in Figure 20c, for which the rotational stiffness k_m of both end connections is equal to 102 kNm/rad corresponding to the semi-rigid connection,

(iv) beam B shown in Figure 20d, for which the rotational stiffness k_m of the connections is infinite and corresponds to the rigid connections (fixed beam ends).

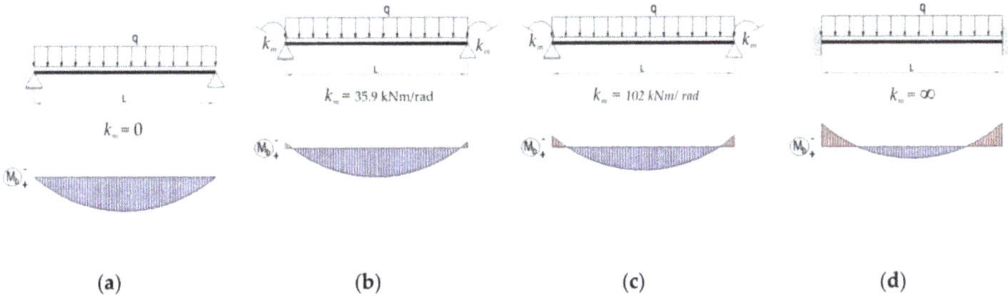

Figure 20. Bending moment diagram in case of different boundary conditions for beam B: (**a**) hinged beam ends, (**b**) semi-rigid connector with four tabs, (**c**) semi-rigid connector with five tabs, and (**d**) rigid connection (k_m represents the rotational stiffness of both end connections).

Furthermore, using RFEM software (Dlubal software Gmbh, Tiefenbach, Germany), it investigates the bending moment M_{mid} developed at the level of the cross section of the beam located at the midpoint of the beam, the bending moment M_{end} developed at the level of the connection, and the maximum deflection. Beam finite elements with two nodes were used in a numerical model of the beam loaded, as shown in Figure 19b. For finite element analysis (FEA) by using RFEM software, the beam corresponding to each case (Figure 20) was divided in 10 segments. Tridimensional support conditions were used in FEA at the beam ends in order to define the following boundary conditions for all cases: (i) beam hinged at both ends – all degrees of freedom are set to zero at supports, except the rotation with respect to the axis perpendicular to the loading plane (Figure 20a), (ii) beam with 4-tab connections at both ends – all degree of freedom are set to zero and the rotational stiffness k_m is set to 35.9 kN·m/rad at both end supports (Figure 20b), (iii) beam with 5-tab connections at both ends—all degrees of freedom are set to zero and the rotational stiffness k_m is set to 35.9 kN·m/rad at both end supports (Figure 20c), (iv) rigid connections (embedded) at both ends—all degrees of freedom are set to zero (Figure 20d). It is mentioned that all results are reported in case of the assemblies corresponding to the type B beam. In case of the beam models with semi-rigid connections at both ends (Figure 20b,c), only the cases corresponding to the extreme values of the rotational stiffness k_m are taken into account.

For better visual comparisons, the bending moment diagrams and deflection diagrams are graphically shown in Figure 21 and in Figure 22, respectively, for all cases shown in Figure 20. In the same manner, the distribution of the equivalent stress by Von Mises failure theory is comparatively plotted in Figure 23.

The results shown in Figures 21–23, show that it is important to determine the stiffness of such semi-rigid connections because the deflection, the bending moment, and stresses are very much influenced by the boundary conditions.

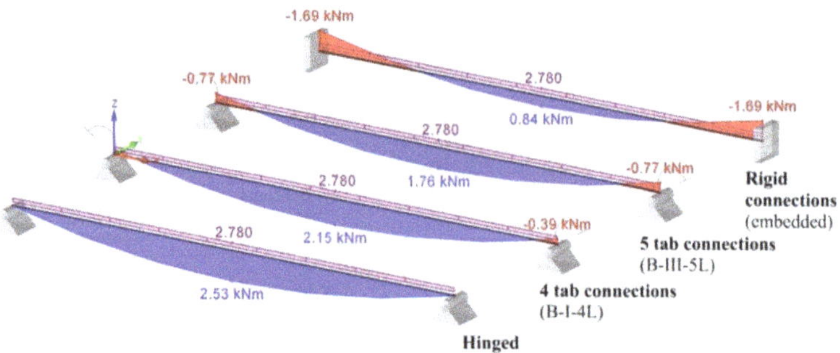

Figure 21. Comparison of the bending moment diagrams for different boundary conditions in case of the type B beam.

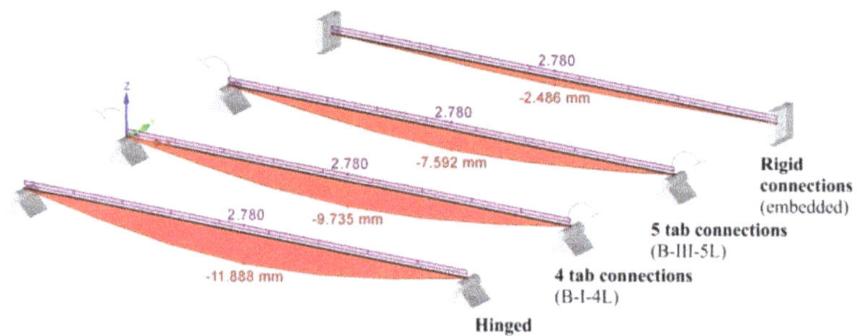

Figure 22. Comparison of the maximum deflection v_{max} for different boundary conditions in case of the type B beam.

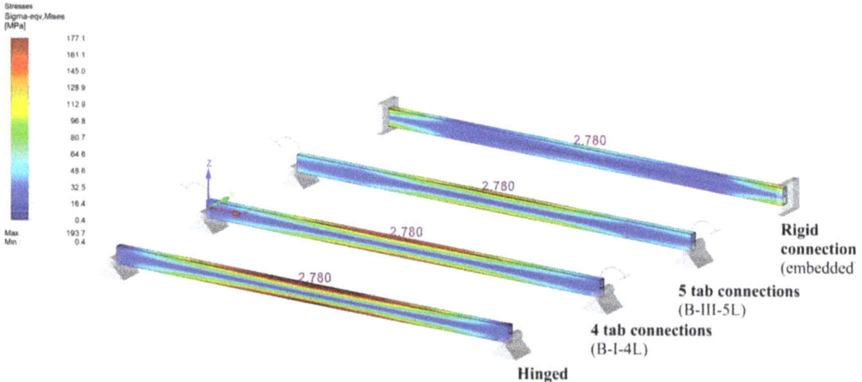

Figure 23. Comparison of the distribution of the equivalent stresses (von Mises) for different boundary conditions in case of the type B beam.

5. Conclusions

Semi-rigid connections are often used in the industry of the pallet racking storage systems. This experimental study presents the mechanical behavior of the beam-end connectors used to assembly the beam with the upright regions. The main objective of this research was to investigate the influence of the type of the beam-end connectors (four-tab connector and five-tab connector) and also the effects of the thickness of the upright section wall on the capacity of the connections when the beam of the storage system is mechanically loaded. For all upright-connector-beam assemblies analyzed, the moment-rotation curves were plotted and the capable moment and the rotational stiffness of the connections are compared.

The research proved that, for the assemblies containing the uprights of type I having a thickness of 1.50 mm, the five-tab connector leads to a higher value of the design moment M_{Rd} and higher rotational stiffness k_m than in the assemblies with four-tab connectors. The rotational stiffness k_m is greater by approximately 23.1%, with 61% and with 77.1% for the assemblies containing five-tab connectors than for the assemblies containing five-tab connectors in case of type A, B, and C beams, respectively. The increase of the design moment M_{Rd} is approximately equal to 49.3%, 21.7%, and 26.5% for the assemblies containing five tab connectors than for the assemblies containing five-tab connectors in case of type A, B, and C beams, respectively.

On the contrary, for the assemblies containing the uprights of type III with a thickness of 2.00 mm, the capable design moment M_{Rd} decreases for the assemblies with five-tab connectors with respect to the values recorded for the four-tab connector. The decrease of the design moment M_{Rd} is indeed small, it is approximately equal to 3.2%, 10.3%, and 9.7% for the assemblies containing five-tab connectors than for the assemblies containing five-tab connectors in case of type A, B, and C beams, respectively.

It was shown that, for each class of assemblies corresponding to a certain type of beam, the highest value recorded for the rotational stiffness k_m obtained in bending tests of the connections, does not lead to the highest value of the safety coefficient c for that connection. Moreover, for a beam of type A and C, the best assemblies (A-II-4L and C-II-4L) from a safety coefficient point-of-view lead to the reduction of the mass of the racking storage system due to the thickness of the upright region and due to using the beam-end connector with four tabs. Additionally, this involves reducing material costs.

For the practical study case of the beam having the length of 2.7 m for which the same semi-rigid connector is used at both ends, the stiffness condition is obeyed, according to EN 15512 standard [9], for all types of beam-connector-upright assemblies involved in this research.

The experimental results concerning the rotational stiffness and the capable moment, obtained in this research are very important in modeling and simulation of the stresses and strain states in racking storage systems as long as the rotational stiffness of the beam-end connector is one of the input data in finite element analysis. In this context, the experimental results and test methods shown in this study can be used by the researchers who work in the field of the racking storage systems in order to obtain improved numerical models for such mechanical structures and good results by finite element analysis.

Some limitations of the research presented in this paper are related to the following aspects: (i) just one type of cross-section shape was considered for the upright profile, (ii) results obtained for both the rotational stiffness and design bending moment corresponding to the connections involved in this research, are valid for room temperature and are not valid in fire situations.

Taking into account the above limitations, there are some further directions of research identified. One of these research directions is to repeat the tests for similar groups of assemblies containing another type of thin-walled upright profile regarding the shape of the cross-section and, then, comparing the results with the ones presented in this paper in order to check if the effects of the upright thickness are the same. Another study could be made to make a numerical model or an analytical model in order to predict the rotational stiffness and the load-bearing capacity of the connections involved in this study for the accidental fire situations and to predict the time interval for maintaining the load-bearing capacity from the beginning of the fire.

Author Contributions: Conceptualization, F.D. and C.C. Formal analysis, C.C. Investigation, F.D. and C.C. Methodology, F.D. Supervision, C.C. Validation, F.D. and C.C. Visualization, F.D. Writing of the original draft, C.C. and F.D. Writing—review and editing, C.C. All authors have read and agreed to the published version of the manuscript.

Funding: This research received no external funding.

Acknowledgments: The authors would like to thank and are grateful to SC Dexion Storage Solutions SRL (Dexion) for good collaboration for providing the materials tested and for the technical support. This paper is published with the kind permission of the General Manager Mr. Brian Howson. The support provided by Transilvania University of Brasov is also greatly acknowledged.

Conflicts of Interest: The authors declare no conflict of interest.

Appendix A

In Figure A1: the moment-rotation (M_n–θ_n) curves recorded for the beam-connector-upright assemblies containing type A beams in all tests carried out.

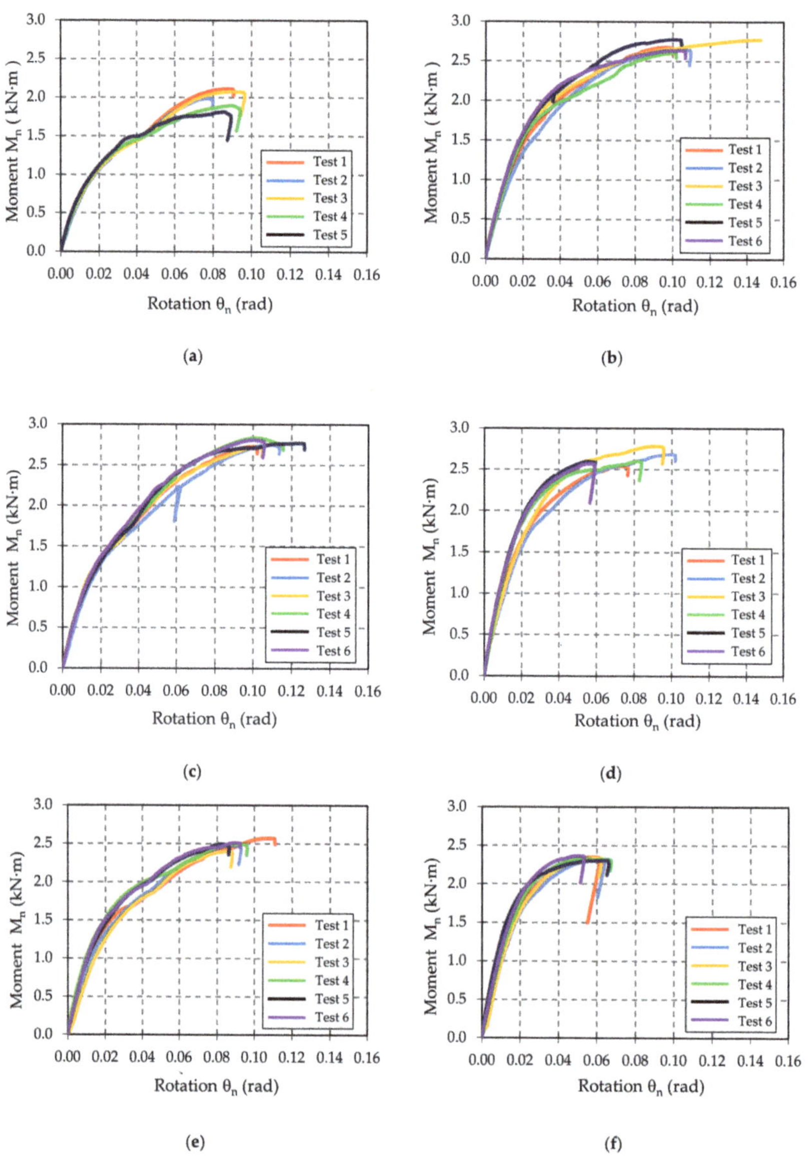

Figure A1. Moment-rotation (M_n–θ_n) curves recorded for the beam-connector-upright assemblies: (**a**) A-I-4L, (**b**) A-I-5L, (**c**) A-II-4L, (**d**)A-II-5L, (**e**) A-III-4L, and (**f**) A-III-5L.

In Figure A2, the moment-rotation (M_n–θ_n) curves recorded for the beam-connector-upright assemblies are shown, which contain type B beams in all tests carried out.

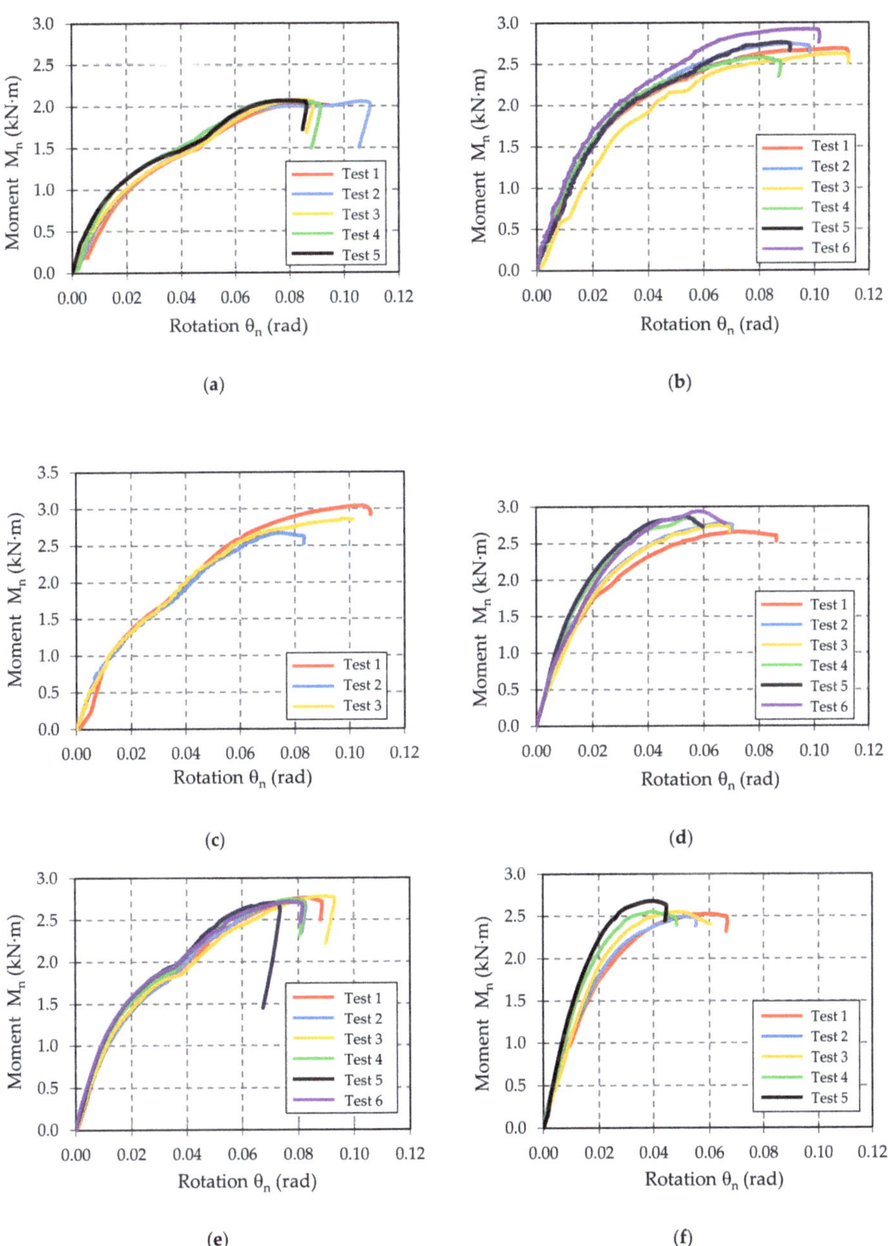

Figure A2. Moment-rotation (M_n–θ_n) curves recorded for the beam-connector-upright assemblies: (**a**) B-I-4L, (**b**) B-I-5L, (**c**) B-II-4L, (**d**) B-II-5L, (**e**) B-III-4L, and (**f**) B-III-5L.

In Figure A3, the moment-rotation (M_n–θ_n) curves recorded for the beam-connector-upright assemblies are shown, which contain type C beams in all tests carried out.

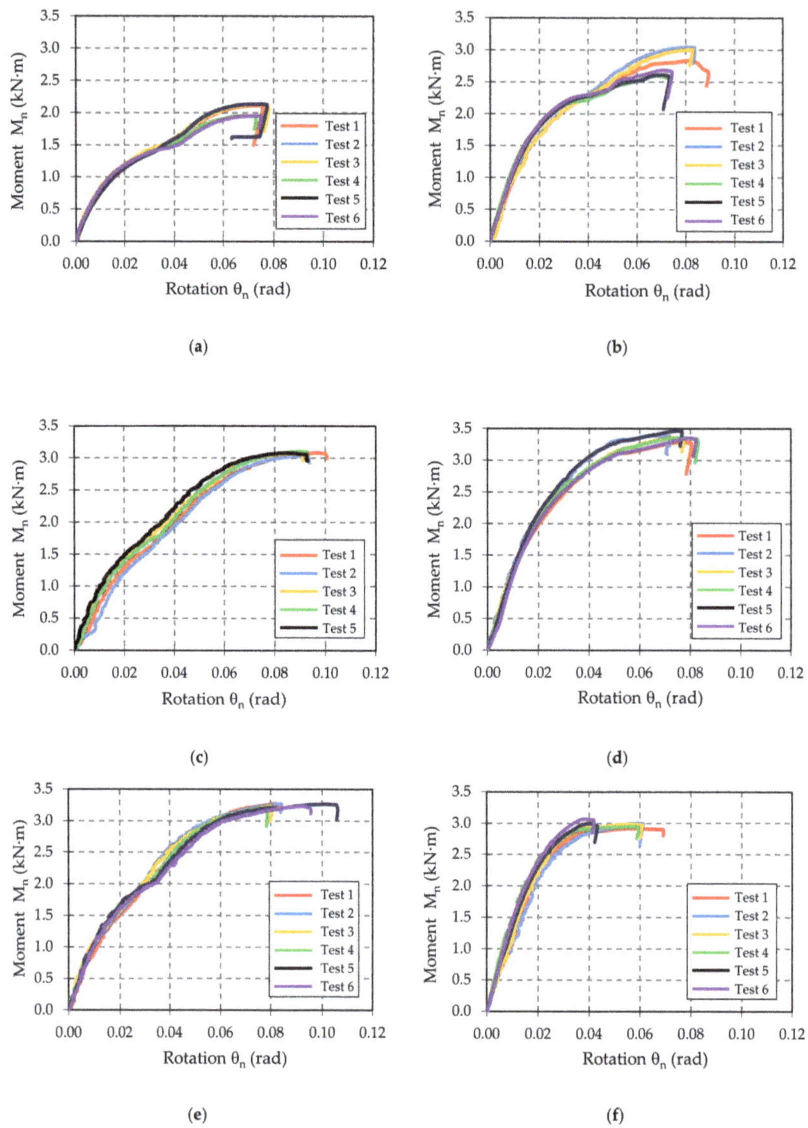

Figure A3. Moment-rotation (M_n–θ_n) curves recorded for the beam-connector-upright assemblies: (**a**) C-I-4L, (**b**) C-I-5L, (**c**) C-II-4L, (**d**) C-II-5L, (**e**) C-III-4L, and (**f**) C-III-5L.

References

1. Shariati, M.; Tahir, M.M.; Wee, T.C.; Shah, S.N.R.; Jalali, A.; Abdullahi, M.M.; Khorami, M. Experimental investigations on monotonic and cyclic behavior of steel pallet rack connections. *Eng. Fail. Anal.* **2018**, *85*, 149–166. [CrossRef]
2. Zhao, X.Z.; Wang, T.; Chen, Y.Y.; Sivakumaran, K.S. Flexural behavior of steel storage rack beam-to-upright connections. *J. Constr. Steel Res.* **2014**, *99*, 161–175. [CrossRef]
3. Dai, L.S.; Zhao, X.Z.; Rasmussen, K.J.R. Flexural behaviour of steel storage rack beam-to-upright bolted connections. *Thin-Walled Struct.* **2018**, *124*, 202–217. [CrossRef]

4. Elza, M.M.; Fonseca, L.S.; Pedro, A.S. Numerical model to predict the effect of wood density in wood–steel–wood connections with and without passive protection under fire. *J. Fire Sci.* **2020**, *38*, 122–135. [CrossRef]

5. Gusella, F.; Lavacchini, G.; Orlando, M. Monotonic and cyclic tests on beam-column joints of industrial pallet racks. *J. Constr. Steel Res.* **2018**, *140*, 92–107. [CrossRef]

6. Dai, L.S.; Zhao, X.Z.; Rasmussen, K.J.R. Cyclic performance of steel storage rack beam-to-upright bolted connections. *J. Constr. Steel Res.* **2018**, *148*, 28–48. [CrossRef]

7. Yin, L.F.; Tang, G.; Zhang, M.; Wang, B.J.; Feng, B. Monotonic and cyclic response of speed-lock connections with bolts in storage racks. *Eng. Struct.* **2016**, *116*, 40–55. [CrossRef]

8. Eurocode 3: Design of steel structures: Part 1-3: General rules. Supplementary rules for cold-formed members and sheeting. In *Design of Joints*; EN 1993-1-3; European Committe for Standardization: Brussels, Belgium, 2006; p. 12.

9. Standard, B. *Steel Static Storage Systems—Adjustable Pallet Racking Systems—Principles for Structural Design*; EN 15512; European Committe for Standardization: Brussels, Belgium, 2009.

10. Mohan, V.; Prabha, P.; Rajasankar, J.; Iyer, N.R.; Raviswaran, N.; Nagendiran, V.; Kamalakannan, S.S. Cold-formed steel pallet rack connection: An experimental study. *Int. J. Adv. Struct. Eng. (IJASE)* **2015**, *7*, 55–68. [CrossRef]

11. Zhao, X.Z.; Dai, L.S.; Wang, T.; Sivakumaran, K.S.; Chen, Y.Y. A theoretical model for the rotational stiffness of storage rack beam-to-upright connections. *J. Constr. Steel Res.* **2017**, *133*, 269–281. [CrossRef]

12. Gusella, F.; Orlando, M.; Thiele, K. Evaluation of rack connection mechanical properties by means of the Component Method. *J. Constr. Steel Res.* **2018**, *149*, 207–224. [CrossRef]

13. Gusella, F.; Arwade, S.R.; Orlando, M.; Peterman, K.D. Influence of mechanical and geometric uncertainty on rack connection structural response. *J. Constr. Steel Res.* **2019**, *153*, 343–355. [CrossRef]

14. Degtyareva, N.V.; Degtyarev, V.V. Experimental investigation of cold-formed steel channels with slotted webs in shear. *Thin-Walled Struct.* **2016**, *102*, 30–42. [CrossRef]

15. Degtyareva, N.V. Review of Experimental Studies of Cold-Formed Steel Channels with Slotted Webs under Bending. *Procedia Eng.* **2017**, *206*, 875–880. [CrossRef]

16. Degtyarev, V.V.; Degtyareva, N.V. Finite element modeling of cold-formed steel channels with solid and slotted webs in shear. *Thin-Walled Struct.* **2016**, *103*, 183–198. [CrossRef]

17. Garrido, C. MECALUX-Warehouse Storage Solutions. Fire Protection Measures for Metal Racks and Warehouses. Available online: https://www.mecalux.com/logistics-items/fire-protection-measures-for-metal-racks-and-warehouses (accessed on 23 June 2020).

18. Laim, L.; Rodrigues, J.P.C. Fire design methodologies for cold-formed steel beams made with open and closed cross-sections. *Eng. Struct.* **2018**, *171*, 759–778. [CrossRef]

19. Craveiro, H.; Rodrigues, J.P.C.; Laim, L. Cold-formed steel columns at both ambient and fire conditions. *J. Struct. Fire Eng.* **2018**, *9*, 189–202. [CrossRef]

20. RMI. Specification for the design, testing and utilization of industrial steel storage racks. In *Pallet Beam-to-Column Connection Tests*; Rack Manufactures Institute: Charlotte, NC, USA, 2008; p. 4.

21. Dumbrava, F. *Research Report: Experimental Research on Stress States and Deformations of Elements (Columns, Beams, etc.) in the Construction of Thin-Walled Metal Structures*; Transilvania University of Brasov: Brașov, Romania, 2020.

22. European Committe for Standardization. *Metallic materials. Tensile Testing. Part 1: Method of Test at Room Temperature*; BS EN ISO 6892-1; European Committe for Standardization: Brussels, Belgium, 2009.

23. European Committe for Standardization. *Continuously Hot-Dip Coated Steel Flat Products—Technical Delivery Conditions*; BS EN 10346; European Committe for Standardization: Brussels, Belgium, 2009.

24. Cerbu, C.; Teodorescu-Draghicescu, H. Aspects on modeling the mechanical behavior of aluminum alloys with different heat treatments. *J. Comput. Appl. Mech.* **2017**, *12*, 13. [CrossRef]

 *materials*

Article

Using AE Signals to Investigate the Fracture Process in an Al–Ti Laminate

Grzegorz Świt [1,*], **Aleksandra Krampikowska** [1], **Tadeusz Pała** [2], **Sebastian Lipiec** [2] and **Ihor Dzioba** [2]

[1]   Department of Strength of Materials, Concrete Structures and Bridges, Faculty of Civil Engineering and Architecture, Kielce University of Technology, Al. 1000-lecia PP 7, 25-314 Kielce, Poland; akramp@tu.kielce.pl

[2]   Department of Machine Design, Faculty of Mechatronics and Mechanical Engineering, Kielce University of Technology, Al. 1000-lecia PP 7, 25-314 Kielce, Poland; tpala@tu.kielce.pl (T.P.); slipiec@tu.kielce.pl (S.L.); pkmid@tu.kielce.pl (I.D.)

*   Correspondence: gswit@tu.kielce.pl

Received: 7 June 2020; Accepted: 26 June 2020; Published: 29 June 2020

Abstract: The paper describes tests conducted to identify the mechanisms occurring during the fracture of single-edge notches loaded in three-point bending (SENB) specimens made from an Al–Ti laminate. The experimental tests were complemented with microstructural analyses of the specimens' fracture surfaces and an in-depth analysis of acoustic emission (AE) signals. The paper presents the application of the AE method to identify fracture processes in the layered Al–Ti composite using a non-hierarchical method for clustering AE signals (k-means) and analyses using waveform time domain, waveform time domain (autocorrelation), fast Fourier transform (FFT Real) and waveform continuous wavelet based on the Morlet wavelet. These analyses made it possible to identify different fracture mechanisms in Al–Ti composites which is very significant to the assessment of the safety of structures made of this material.

Keywords: Al–Ti laminate; fracture; acoustic emission diagnostic; pattern recognition; clustering AE signal

1. Introduction

Design requirements imposed on contemporary structural members frequently make it necessary to use materials that combine different strengths and mechanical properties. Examples of such materials include composites, particularly laminates, which may consist of several layers of materials with different strengths and mechanical properties, selected depending on the specific needs of the user. Usually, one of the layers of laminate is responsible for structural strength, while the remaining layers may have special properties—corrosion protection, thermal insulation and sealing or damping of high-energy impact loads. The object under test described in this paper is an Al–Ti laminate consisting of three layers: outer lamella from an Al alloy (AA2519), inner lamella from an Al alloy (AA1050) and an outer lamella from a Ti alloy (Ti6Al4V).

Various techniques and technologies are used during production to join different metallic materials in the laminates—adhesive bonding, high-temperature welding, friction welding [1–4] and, increasingly, explosion welding [5–9]. This technology uses the explosive energy for mutual penetration of the bonded materials. The technology of explosion bonding is fairly new, and it is used to make various types of items consisting of different materials; in each specific case, the parameters of the process are selected based on experience with laboratory specimens. That is why, in order to use elements made of layered materials created by explosion welding, it is necessary to know their strength and mechanical properties. The least investigated and, at the same time, very important area, which frequently determines the strength of the layered material, is the interface of the bonded materials—lamellas. There are no theoretical studies concerning explosion bonding of different materials and mechanical characteristics of the created laminates, as a result of which the most reliable method is to assess their strength properties using experimental means [10–15]. On the other hand, the commonly applied methods of determining material characteristics have been prepared for homogeneous materials, which is why it is difficult to apply them directly to laminates, and, consequently, the determined physical and mechanical properties cannot be regarded as the actual material characteristics of the laminates.

The AE technique is commonly used to detect and monitor damage and development of such damage in various structures, and it is currently recognised as one of the most reliable and well-established methods of non-destructive testing (NDT) [16]. Acoustic emission is a very efficient and effective method of detecting cracking and fatigue of metals, glass fibre, wood, composite materials, ceramics, concrete and plastics [17–19]. It can also be used to detect faults and pressure leaks in tanks or pipes or monitor the progress of corrosion in welded joints [20].

Unlike other techniques, which can only detect geometric discontinuities, AE methods can detect fibre tearing, delamination of adjacent layers in laminated composite plates, matrix cracking and fibre pull-out [21]. Most AE signals are caused by friction or by friction among the damaged components of the composite. Potential application of the AE technique to the assessment of damage to composite materials was discussed in References [22,23], with the mechanisms behind the cracking of composite materials being described in Reference [23]. Studies of local damage to composite materials based on an analysis of the acoustic emission signal were carried out by Marec et al. [24]. The results of the tensile tests clearly identified damage mechanisms in various composite materials: cross-ply composites and sheet moulding compound (SMC). The tests indicated an evolution of damage in these materials over time until global failure, and they identified the most critical damage mechanisms. It was also found that the generated AE signals were subject to scattering and attenuation due to the elasticity of the waves. Most studies carried out in order to monitor AE signals in metals were focused on samples with the shape of thin plates [16,25]. However, the propagation of waves in thin plates is dispersive by definition [26] due to the different phase velocities at which different frequencies are propagated. Dang Hoang et al. [27] showed a relationship between the duration and energy of the AE signal and failures in aluminium plates joined together. In a similar study, the relationship between AE count rates and crack propagation rates in welded steel samples during fatigue was described in Reference [28]. Aggelis et al. [17] established that certain parameters of AE signals, such as the rise angle (RA), duration and rise time, are very sensitive to crack propagation rate, and can be useful in the characterisation of damage if the AE signals from rise time (RT) samples subject to fatigue are accordingly interpreted. The tests described in Reference [29] regarding the fatigue properties of steel and welds and fractographic and microstructural observations show that AE can be used as a tool to monitor damage caused by fatigue of the structure due to the fact of its sensitivity to changes within the crack.

That is why using AE in the study made it possible to identify certain peculiarities of the fracture process in the tested layered composite. Also, the AE signals were clustered depending on the processes that generated them using the iterative *k*-means method, which clusters AE parameters in a Euclidean space [30,31], and analysed them using waveform time domain, waveform time domain (autocorrelation), fast Fourier transform (FFT Real) and waveform continuous wavelet based on the Morlet wavelet [32].

Understanding the processes occurring in the contact areas of different materials has an important role in assessing laminate strength. Actually, these areas of contact of different layers are the most important in composites strength analysis. Theoretical and experimental studies on the effect on bonding strength are presented in References [33–35].

This paper presents the results of studies of the fracture process in Al–Ti laminate. To implement different fracture mechanisms the tests were carried out at two temperatures (i.e., $T_1 = 20$ °C and $T_2 = -50$ °C) on specimens with single-edge notches loaded in three-point bending (SENB). The force, specimens' deflection, crack opening and acoustic emission (AE) signals were recorded during the loading. Based on mechanical sensors signals the fracture toughness characteristics were obtained and differences in loading diagrams were established. Observation of the specimens' fracture surfaces using scanning electron microscope (SEM) showed some peculiarities of the cracking process of Al–Ti laminate in tested temperatures. Great attention was given to the analysis of AE signals to identify fracture mechanisms of Al–Ti laminate.

2. Material and Testing Methods

The object subject to laboratory tests was an Al–Ti laminate consisting of three layers: outer lamella from an Al alloy (AA2519), with a thickness of approximately 4.6 mm, inner lamella from an Al alloy (AA1050) with a thickness of approximately 0.2 mm, and an outer lamella from a Ti alloy (Ti6Al4V) with a thickness of approximately 4.6 mm. After the individual lamellas were bonded by explosion welding, the laminate was subject to heat treatment by soaking at 550 °C for 2 h, cooling to 165 °C and soaking at that temperature for 10 h. The authors of this paper analysed the fracture process in the laminate using specimens that had already been subject to welding and heat treatment. Information about the standard properties of base materials, welding process and heat treatment of the examined laminate was taken from References [12,15].

Experimental tests were carried out by the research team using SENB specimens (B = 10; W = 20; S = 80 mm) in three-point bending with a single-edge notch passing across all layers to a depth of 0.5W (Figure 1) at two temperatures: T_1 = 20 °C and T_2 = −50 °C. The strength characteristics of the component materials alloy at different temperature are given in Table 1. The specimens were prepared and loaded in accordance with ASTM E1820-09 [36]. During loading, the specimens were partially unloaded in order to determine the change in the compliance and calculate the growth of the crack. Tests at lowered temperature T_2 = −50 °C were conducted in a thermal chamber in the environment of nitrogen vapours. During the tests, temperature variations did not exceed ΔT = ±1 °C. In order to determine fracture toughness characteristics, force P, deflection of the specimens (sensor 3) and crack opening displacement (*COD*) (sensor 4) were recorded while the load was applied to the sample. Acoustic emission signals were recorded using a 24 channel μSAMOS acoustic emission processor with the AEwin and NOESIS 12.0 software developed by the Physical Acoustics Corporation (PAC) from the Princeton, NJ, USA. Two low-frequency sensors with a flat response curve in the 30–80 kHz range (VS30-SIC-40dB, manufactured by Vallen GmbH) (sensor 5) and two broadband sensors with a frequency range of 100–1200 kHz (WD 100–1200 kHz, manufactured by PAC) (sensor 6) were used to record AE signals in a broad frequency range (Figure 1b).

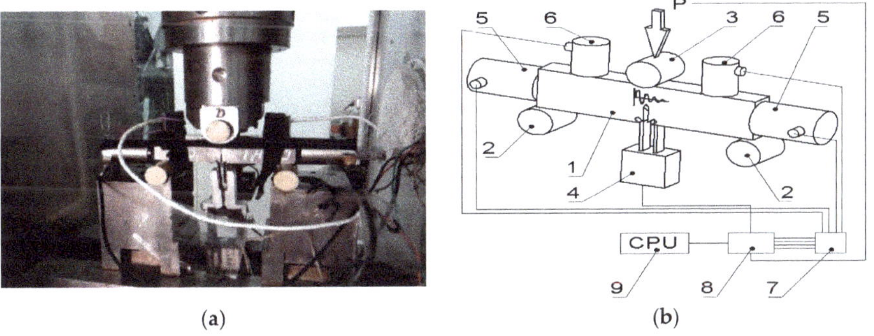

(a)   (b)

Figure 1. The SENB specimen in three-point bending with installed AE sensors: (**a**) photograph of the specimen in the thermal chamber; (**b**) scheme of loading and installation of mechanical and AE sensors.

Table 1. Mean values of the mechanical properties of AA2519 aluminium alloy and Ti6Al4V titanium alloy at different temperature.

Material	Temperature	σ_y	σ_u	E	A_5
	(°C)	(MPa)	(MPa)	(GPa)	(%)
AA2519	20	301	560	67.8	16.3
	−50	320	607	72.8	16.9
AA1050	20	150	194	69.2	18.2
	−50	161	208	74.3	18.5
Ti6Al4V	20	859	908	111.7	13.6
	−50	937	1140	117.8	13.8

AE parameters were clustered with the k-means method in a Euclidean space. This method involves the iterative search for the set of reference elements representing the individual clusters. Successive approximations of the reference elements are sought in each iteration through calculations using the indicated methods. Depending on the adopted assumptions, the reference element may be one of the elements of the X population or an element of a specific set $U \supseteq X$. In metric spaces, the reference element may be calculated as an arithmetic mean, and it accordingly represents the centre of gravity of the cluster.

In general, the input for the k-clustering algorithm is the set of objects X and the expected number of clusters k and the output is the division into subsets $\{C_1, C_2, \ldots, C_k\}$. Usually, k-clustering algorithms belong to the category of optimisation algorithms. Optimisation algorithms assume that there is a loss function k: $\{x|X \subseteq S\} \rightarrow R^+$ defined for every S subset. The purpose is to find the group with a minimal sum of the losses described by the following Formula (1):

$$E_q = \sum_{i=1}^{k} k(C_i) \tag{1}$$

In order to use the iterative algorithm, it is necessary to determine the dissimilarity measure used in the clustering process. In this case, it will be a point that is the resultant point for the particular cluster (representative element) calculated, for instance, as the geometric or arithmetic mean, described with the following Formula (2):

$$k(C_i) = \sum_{r=1}^{|C_i|} d\left(\bar{x}^i, x_r^i\right) \tag{2}$$

where: $\bar{x}^i$- arithmetic (or geometric) mean of the cluster.

The most important advantage of this method is the speed of data processing and analysis.

The research also used other grouping methods (Forgy, Normalized Normal Constraint (NNC), Fuzzy C-Means (FCM), Gaussian Mixture Decomposition (GMD), Gustafson-Kessel (GK), Hidden Markov Model (HMM), Autoregressive Hidden Markov Model (ARHMM)) to isolate destructive subprocesses based on a full statistical-mathematical approach to divide into groups [37–41].

Analyses using waveform time domain, waveform time domain (autocorrelation), fast Fourier transform (FFT Real) and waveform continuous wavelet (CWT) based on the Morlet wavelet were also conducted in order to verify the divisions made using the k-means method.

The waveform time domain chart shows a change of amplitude expressed in units of voltage V within a specific period. It is also useful in determining the rise time or duration of the acoustic signals for the indicated AE signal classes. The waveform time domain (autocorrelation) chart, in turn, is significant to technical diagnostics and to signal processing and transmission theory. The autocorrelation function is used to determine the rate of signal change and to detect periodic signals in "noisy" measurement signals—which is very important to the processing of diagnostic signals. The autocorrelation function $\varphi_x(\tau)$ of signal $x(t)$ with limited energy is described by the following Formula (3):

$$\varphi_X(\tau) = \int_{-\infty}^{\infty} x(t)x^0(t-\tau)dt \tag{3}$$

The value of autocorrelation function $\varphi_x(\tau)$ at point $\tau = 0$ is real, and it is equal to the energy of the $x(t)$ signal described with the following Formula (4):

$$\varphi_X(\tau) = \int_{-\infty}^{\infty} x|x(t)|^2 dt = E_x \tag{4}$$

This property can be derived directly from Formula (3) after substituting $\tau = 0$.

The charts of the fast Fourier transform (FFT Real) can be used for an analysis in the frequency domain to identify the key frequencies in the entire data set instead of examining every change in the time domain. The chart in the frequency domain shows the phase shift or strength of the signal at each frequency on which it exists. It shows how much signal is contained in the particular frequency band within a specific frequency range.

Important frequency and energy-related information of AE signals can be extracted with the waveform continuous wavelet chart based on the Morlet wavelet. Waveform continuous wavelet (CWT) in the time domain is described by the following Formula (5):

$$CWT(t,a) = \frac{1}{\sqrt{|a|}} \int_{-\infty}^{\infty} x(\tau)\gamma\left(\frac{\tau-t}{a}\right)d\tau \tag{5}$$

where t—time instant of the tested signal; $x(\tau)$—analysed signal; $\gamma(t)$—base filter function, so-called mother function; a—scale of the mother function.

The features extracted from this wavelet transform from can be significant to the detection of structural damage. By using a suitable family of the base signal filter function (so-called wavelet mother), it is possible to identify a temporary variation of the analysed signal. Another advantage of using *CWT* is the detection of the variation of the analysed value and/or—depending on the domain of the analysed response—identification of the location or time of damage. The Morlet wavelet was selected as the mother function for the analysis of the response signal obtained through the simulation of damage. The Morlet wavelet is currently used in various applications including successful use in damage diagnostics.

3. Results of Experimental Tests

The critical values of fracture toughness were calculated as *J*-integral at the point of reaching the maximum force using the Rice formula: $J_C = 2A_C/B(W-a_0)$, where A_C is the energy absorbed by plastic deformation and crack growth in the sample. For room temperature, $T_1 = 20\,°C$, the integral in accordance with the Rice formula reached $J_C = 46$ kN/m, and as the test temperature lowered, $T_2 = -50\,°C$, the critical value of *J*-integral decreased slightly, reaching $J_C = 40$ kN/m, which was consistent with expectations. For metals, particularly ferritic steels, a decrease of the test temperature will reduce critical fracture toughness characteristics [42–46].

However, careful observation and comparison of the *P–COD* load curves with the recorded AE signals indicates certain peculiarities of the fracture process in Al–Ti laminate specimens tested at two temperatures: $T_1 = 20\,°C$ and $T_2 = -50\,°C$ (Figure 2). For the lowered temperature, the *P–COD* curve has slightly lower values than for room temperature. It should be emphasised that in samples made from homogeneous materials, Al (AA2519) and Ti (Ti6Al4V) alloys [15], and in specimens from steel [15,47], an opposite trend could be observed—reducing the temperature increased the strength characteristics and Young's modulus, and it made the rise of the chart in the loading section accordingly steeper. Also, in the rising sections of the load curves, we can observe a deviation from linearity occurring when the force is approximately 4 kN for the specimens tested at temperature $T_2 = -50\,°C$ and approximately 6 kN for tests at temperature $T_1 = 20\,°C$.

The distributions of AE signals in both samples (Figure 2) clearly show that signals with a high strength (above 5.0×10^7 pV·s) appear when the force is decreasing, directly after reaching the maximum force levels which corresponds to the start of the main subcritical crack. Also, the AE signals with high strength for the sample tested at $T_2 = -50\,°C$ were located primarily in short time sections just after the maximum force level, which indicates a rapid growth of the subcritical crack. For the sample tested at $T_1 = 20\,°C$, AE signals with high strength also appeared while the load was being applied which indicates a continuous development of the subcritical crack.

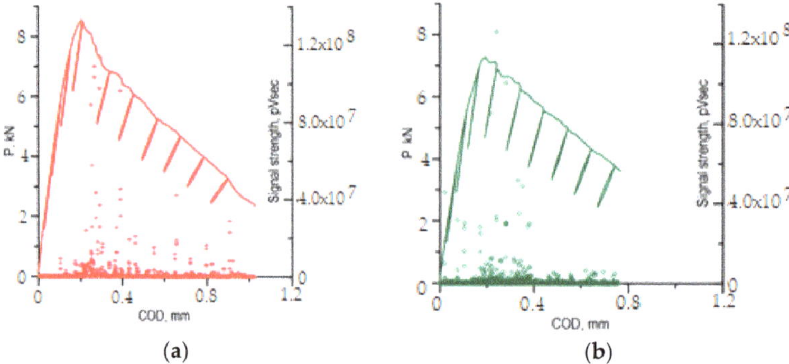

Figure 2. Load curves of the specimens with AE signals. (**a**) $T_1 = 20\,°C$; (**b**) $T_2 = -50\,°C$.

AE signals for which signal strength (SS) was within the range of 1.0×10^7–5.0×10^7 pV·s were very significant to the assessment of the generated destructive processes in the tested samples. These signals appeared in various phases of the loading of the samples tested at $T_2 = -50\,°C$ and $T_1 = 20\,°C$. At room temperature, most of these signals were recorded along the falling part of the *P–COD* load curve, representing the increase of subcritical crack. However, during the test at lowered temperature ($T_2 = -50\,°C$), the great majority of signals from that range were recorded in the specimen loading phase before the maximum force was reached, i.e., when the subcritical crack did not start yet. The observed difference in the appearance of signals from that range may indicate that the Al–Ti laminate fracture process develops differently at different temperatures. This assumption is also supported by the peculiarities observed in the *P–COD* curves as indicated above.

Fracture surfaces of the samples were observed with a scanning electron microscope (SEM), and the recorded AE signals were analysed using a non-hierarchical method for AE signal clustering based on *k*-means [30,31] and analysis using waveform time domain, waveform time domain (autocorrelation), fast Fourier transform (FFT Real) and waveform continuous wavelet based on the Morlet wavelet [32] in order to explain the peculiarities of the *P–COD* curves at different temperatures and identify the mechanisms generating the AE signals.

4. Tests of Fracture Surfaces and Microstructure Using SEM

Observations of the fracture surfaces of the tested specimens using an optical microscope and SEM clearly indicated the formation and development of two mutually perpendicular cracks—the main crack, which developed from the pre-crack in the same plane, and the delamination crack, which developed in a perpendicular plane between the component layers of Ti and Al alloys (Figure 3). The presence of delamination cracks was observed in all samples tested at both temperatures. In most cases, the delamination crack was formed on the Ti6Al4V alloy side, but there were also situations where the crack was formed at the AA2519 alloy layer, or even in parallel on both sides next to the base layers (Figure 3b).

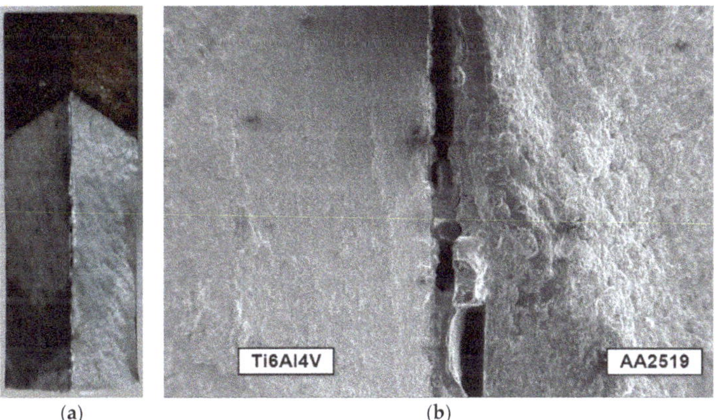

(a)   (b)

Figure 3. Delamination crack: (**a**) total view of fracture surface of the SENB specimen; (**b**) development of the delamination crack.

The nature of the delamination crack clearly indicates the presence of two mechanisms in the development of the fracture—brittle phase fracture and shear in the AA1050 aluminium alloy (Figure 4). Most likely, a brittle fracture may be initiated by the brittle fracture of the particles of Al and Ti intermetallic compounds [45] or oxides formed in the transition zone (TZ) between the Al alloy (AA1050) and Ti alloy (Ti6Al4V) during explosion welding (Figure 5). The presence of elements of metals and oxygen was confirmed through EDS analysis in the transition zone (Figure 6, Table 2). In the photographs (Figure 4), the areas of the fractured brittle phases are clearly visible as the light areas from which shear develops in the Al alloy (AA1050). In samples tested at temperature $T_1 = 20\ °C$, the growth of the delamination crack occurred primarily by shear (Figure 4a), and at the reduced temperature, $T_2 = -50\ °C$, there were large areas with brittle fracture (Figure 4b).

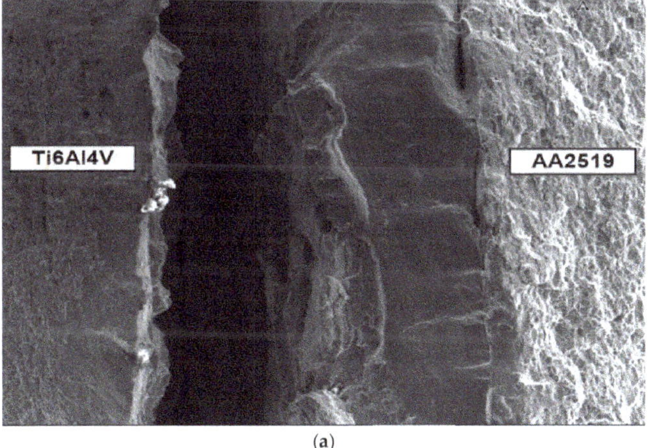

(a)

Figure 4. *Cont.*

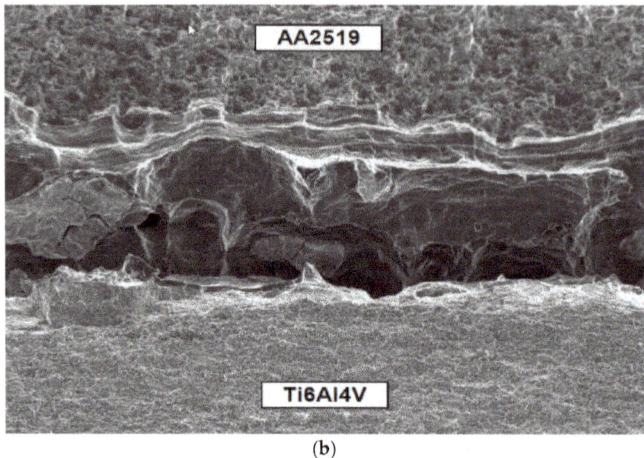

(**b**)

Figure 4. Nature of the delamination crack. (**a**) tests at $T_1 = 20\,^\circ$C; (**b**) tests at $T_2 = -50\,^\circ$C.

Observations of metallographic photographs of the connecting zone of the base materials (Figure 5) indicated that this narrow strip with a width of up to 40 μm (transition zone—TZ) between the material of the inner layer, AA1050 alloy, and the Ti6Al4V alloy (Figure 5b), was the weakest link in the Al–Ti laminate. This is precisely the strip formed by the mutual penetration of materials during explosion bonding. The appearance of brittle fracture in that zone is caused by different types of particles and material discontinuities present in the TZ.

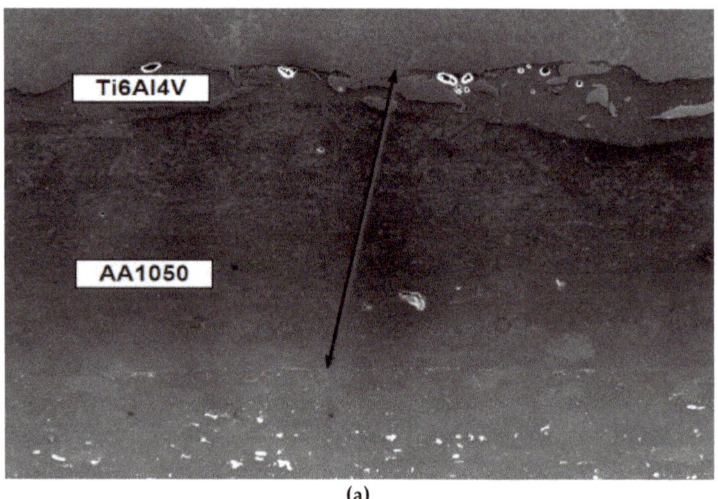

(**a**)

Figure 5. *Cont.*

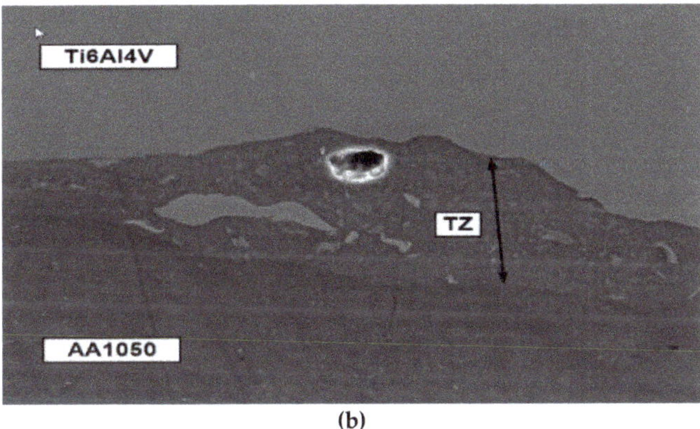

(b)

Figure 5. (**a**) inner layer and connecting zone of the base materials; (**b**) transition zone (TZ) between the AA1050 alloy and the Ti6Al4V alloy.

The complementary tests conducted using the EDS analysis showed that the TZ contained elements of metals from the base layers (Figure 6a), which formed Al and Ti intermetallic compounds [47], and oxygen (Figure 6b), which indicated the existence of metal oxides (Table 2).

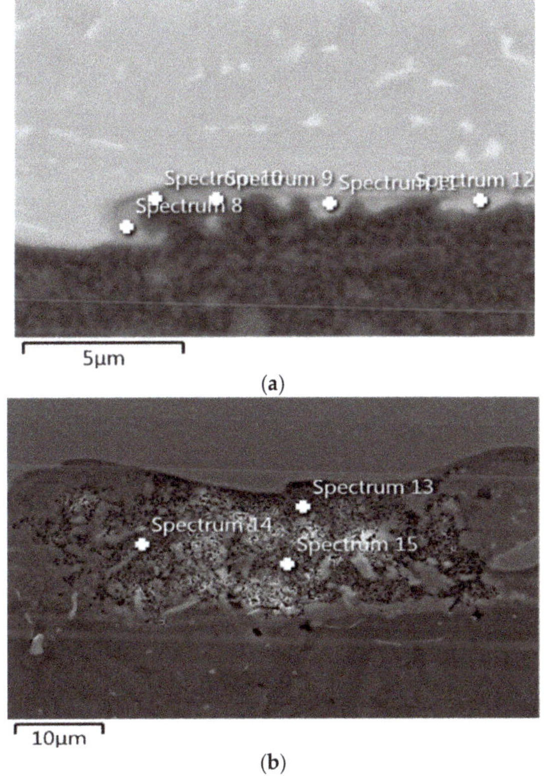

(a)

(b)

Figure 6. The point of EDS analysis in TZ in aria 1 (**a**) and aria 2 (**b**).

Table 2. The results of EDS analysis.

Result Type	Weight %					
Spectrum N	O	Al	Si	Ti	V	Total
Spectrum 8	-	32.90		64.53	2.57	100.00
Spectrum 9	-	33.22	0.38	63.57	2.83	100.00
Spectrum 10	-	32.35	0.32	64.66	2.67	100.00
Spectrum 11	-	30.80	0.33	66.19	2.69	100.00
Spectrum 12	-	26.15	0.30	70.47	3.08	100.00
Spectrum 13	13.42	48.56	7.63	28.93	1.46	100.00
Spectrum 14	14.44	51.46	8.07	24.90	1.14	100.00
Spectrum 15	15.97	46.82	10.27	25.89	1.04	100.00

The development of the main crack, which grew from the pre-crack in the same plane, occurred in both alloys: Al (AA2519) and Ti (Ti6Al4V), in accordance with the ductile mechanism through the formation and connection of voids. However, there are certain noticeable differences. In the Al alloy (AA2519), voids with a size of 10–20 μm are formed around fairly large particles with a size of 2–5 μm (Figure 7a). The morphology of the fracture surface of the Ti (Ti6Al4V) alloy, in turn, consists of smaller fragments of voids with a size of 2–7 μm (Figure 7b). The nature of the propagation of the main crack is similar in the samples tested at both temperatures.

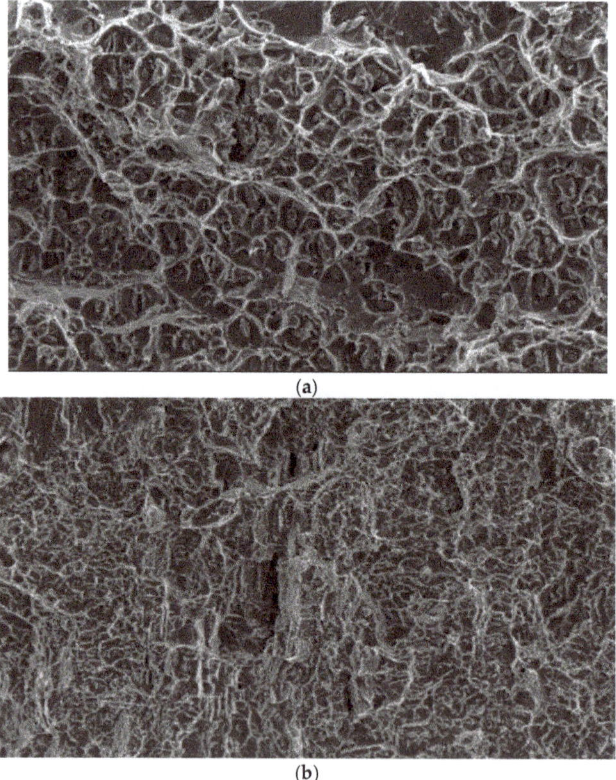

(a)

(b)

Figure 7. Fracture mechanism in the base materials. (**a**) Al alloy (AA2519); (**b**) Ti alloy (Ti6Al4V).

5. Analysis of AE Signals

The AE signals recorded during loading of the specimens at $T_1 = 20\,^\circ\text{C}$ and at $T_2 = -50\,^\circ\text{C}$ (Figure 2) were preliminarily clustered using the non-hierarchical method for AE signal clustering—k-means. As a result of the clustering, the signals were divided into five classes shown on point charts of AE signal strength (SS) pV·s, over time (s) (Figure 8).

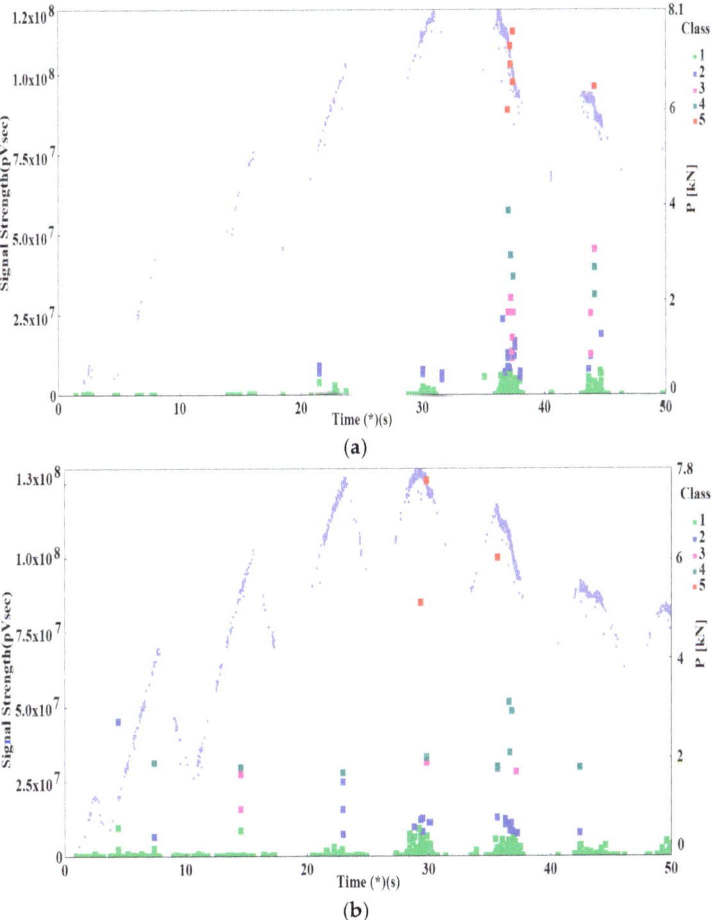

Figure 8. Point charts of signal strength (SS) over time with breakdown into classes describing the characteristic fracture mechanisms. (**a**) for $T_1 = 20\,^\circ\text{C}$; (**b**) for $T_2 = -50\,^\circ\text{C}$.

The charts also show the force acting on the sample, P (kN), which was recorded with a set of acoustic instruments only when the AE signals were present. Since the sample was loaded and partially unloaded, the AE signals intensify when the load is rising and diminish as the load is decreasing.

It is to notice that in the sample tested at $T_1 = 20\,^\circ\text{C}$ (Figure 8a), the great majority of AE signals from classes 1–5 occurred just after the maximum force was reached and later, i.e. after the main crack has propagated. When the load was rising, only AE signals with low signal strength (SS) were recorded—for class 1, the SS reached 8.0×10^6 pV·s, and, for class 2, the SS values were within the range of 4.8×10^6 to 2.4×10^7 pV·s.

During tests at temperature $T_2 = -50\ °C$ (Figure 8b), in the rising part of the load curve, up to the point at which maximum force was reached, numerous signals of classes 1 to 4 were recorded, which indicates an intense growth of the crack during that period. Just after the maximum force was reached, AE signals of classes 1 to 5 were recorded, with no signals of classes 2–5 being recorded after the force dropped below 0.68 P_{max} (less than 5.3 kN).

The preliminary clustering of AE signals using the k-means method was followed by an in-depth analysis of AE signals using the following charts: waveform time domain (WTD) (Figure 9), fast Fourier transform (FFT Real) and waveform continuous wavelet using the Morlet wavelet (Figures 10–14) and waveform time domain (autocorrelation)—complex Fourier series (Figure 15).

The preliminary clustering of AE signals using the *k*-means method was followed by an in-depth analysis of AE signals using the following charts: (WTD) (Figure 9), waveform frequency domain (Real)—FFT Real and waveform continuous wavelet using the Morlet wavelet (Figures 10–14) and waveform time domain (autocorrelation)—complex Fourier series (Figure 15). The charts shown in Figures 9–15 can be used to verify and adjust the preliminary clustering of AE signals. In Table 3 are presented ranges and maximum levels of AE signal characteristics obtained by this analysis.

Table 3. Ranges and maximum levels of AE signal characteristics.

Class	1	2	3	4	5
Signal Strength (SS) (pV·s)	1.8×10^4 ÷ 8.0×10^6	4.8×10^6 ÷ 2.4×10^7	1.0×10^7 ÷ 3.0×10^7	3.0×10^7 ÷ 5.8×10^7	8.8×10^7 ÷ 1.2×10^8
Amplitude (V)	0.43 (4%)	7.5 (75%)	8.8 (80%)	8.8 (88%)	8.81 (88%)
FFT Real (V)	±(20 ÷ 38)	±350	±1000	±(1000 ÷ 2000)	±(750 ÷ 1500)
Frequency (kHz)	100	250 ÷ 270	250 ÷ 270	60; 250 ÷ 300	60; 75
Duration (µs)	400	520	520	1000	4000
WTD (Auto Correlation) Energy (eu)	10	3000	11500	18000	12000

Class-5 signals (red points) are generated in connection with the ductile development of the main crack by the growth of voids, which is driven by the dislocation motion. They are characterised by a high amplitude of up to 8.81 V and low frequency level in the two dominant bands (Figure 14a,b): 60 and 70 kHz. The 60 kHz frequency is characteristic to aluminium alloys, and the frequency of 75 kHz characterises titanium alloys. These processes related to the growth of voids have long durations of up to 4000 µs (Figure 9e), high energy values of up to 12,000 eu (Figure 15e) and signal strength from 8.8×10^7 to 1.2×10^8 pV·s (Table 3).

Class-4 signals (Bottle Green points) also include two characteristic frequency ranges: approx. 60 kHz and 250–300 kHz. The structure of the class-4 signal indicates that it is generated in the shear process, plastic deformations appear first with a frequency of approximately 60 kHz, followed by shear fracture characterised by a frequency of 250–270 kHz (Figure 13a,b). These signals are accompanied by a high amplitude, reaching 8.8 V and duration of up to 1000 µs (Figure 9d). The signal strength for class-4 signal ranges from 3.0×10^7 to 5.8×10^7 pV·s, and the signal energy calculated from the WTD reaches up to 18,000 eu (Figure 15d).

Class-2 (dark blue) and class-3 (pink) AE signals are characterised by a similar frequency in the range of 250–270 kHz and duration of approximately 520 µs, but they differ in amplitude, signal strength and energy (Table 3).

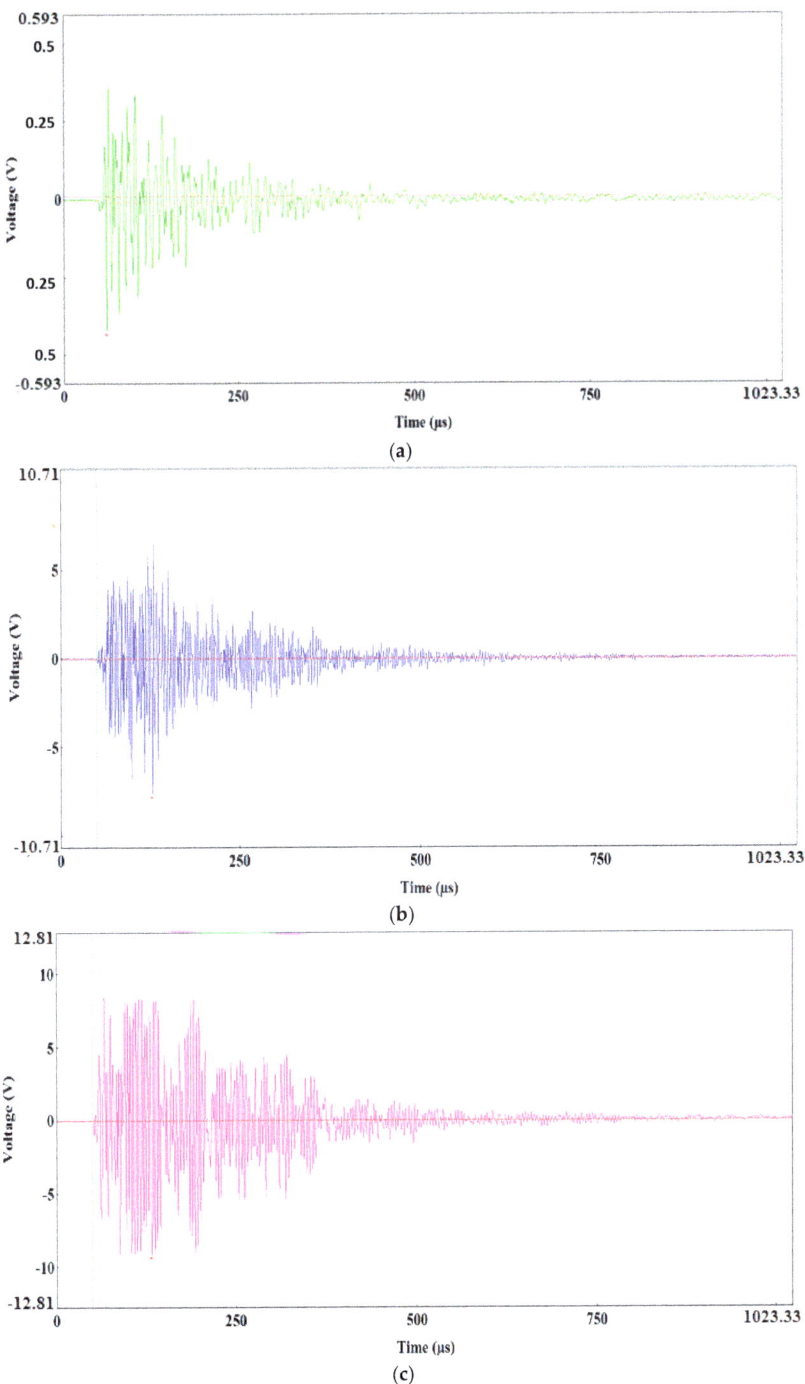

Figure 9. *Cont.*

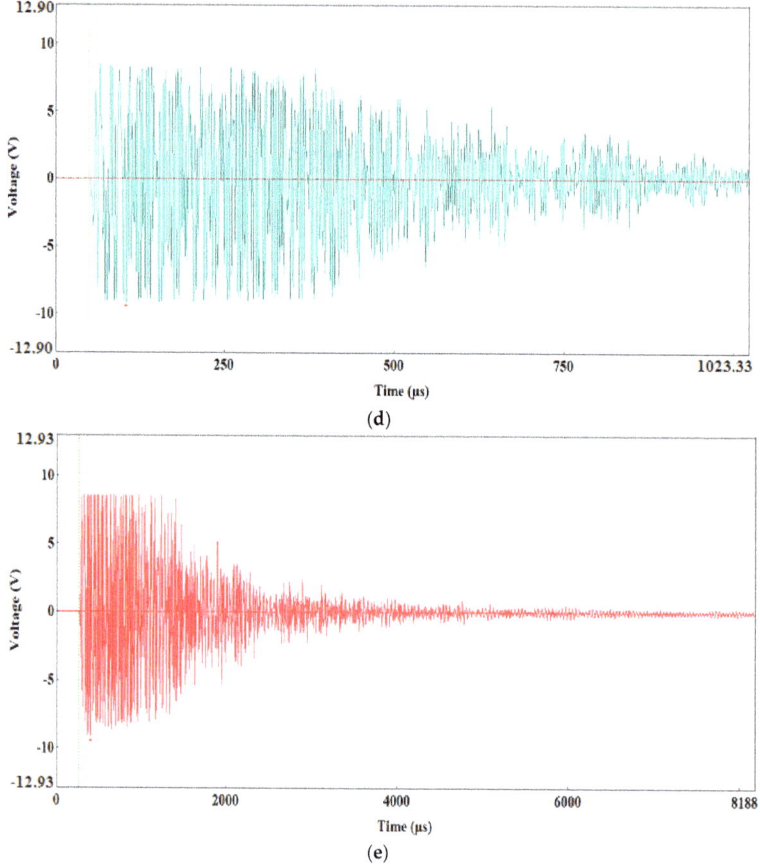

(d)

(e)

Figure 9. Waveform time domain charts for the individual classes of AE signals. (**a**) class 1; (**b**) class 2; (**c**) class 3; (**d**) class 4; (**e**) class 5.

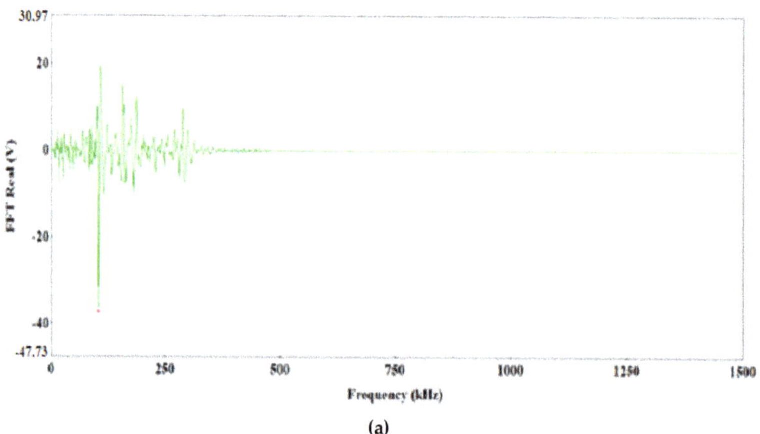

(a)

Figure 10. *Cont.*

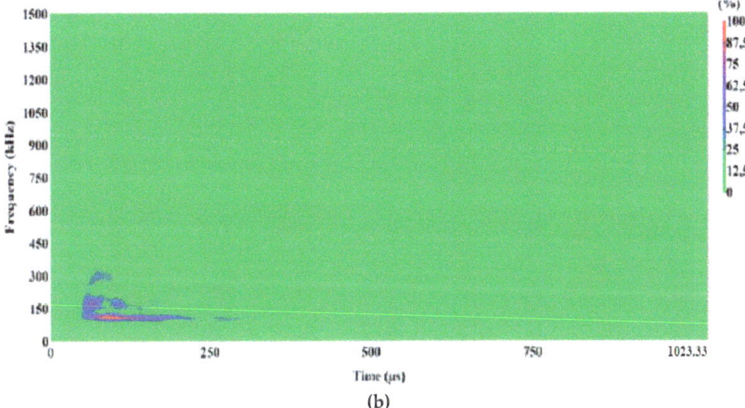

Figure 10. Waveform frequency domain (Real) (**a**) and waveform continuous wavelet (Morlet) (**b**) charts for the individual classes of AE signals (class 1).

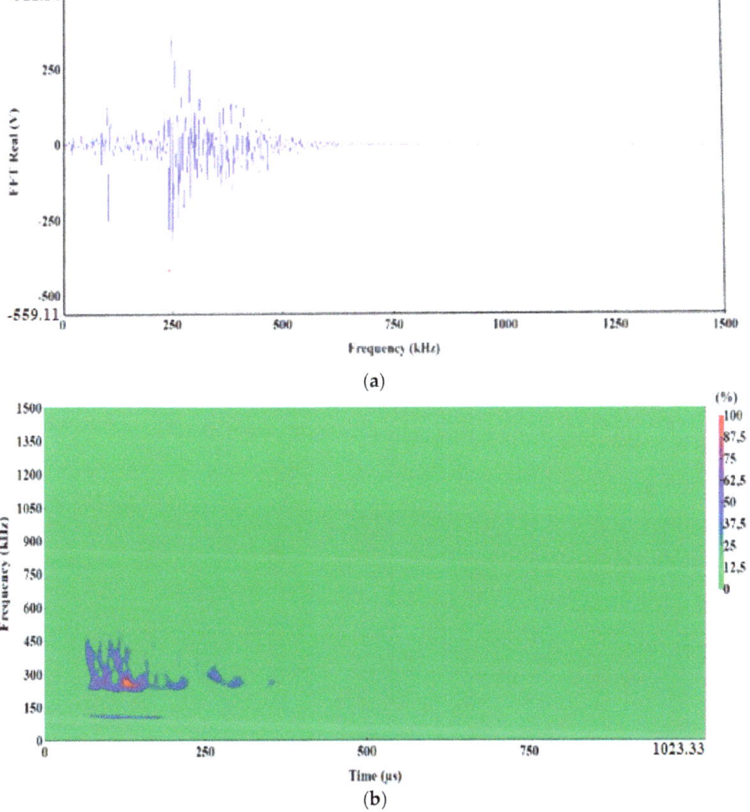

Figure 11. Waveform frequency domain (Real) (**a**) and waveform continuous wavelet (Morlet) (**b**) charts for the individual classes of AE signals (class 2).

A higher amplitude, reaching up to 8.8 V could be observed with class-3 signals. These signals reached signal strength in the range from 1.1×10^7 to 4.6×10^7 pV·s, and their maximum energy calculated from the WTD was 11,500 eu (Figure 9c, Figure 12a,b and Figure 15c).

Class-2 signals were characterised by lower signal strength, in the range from 4.8×10^6 to 2.4×10^7 pV·s and 3.5 times lower energy, not exceeding 3000 eu (Figure 9b, Figure 11a,b and Figure 15b), determined through the WTD analysis. Both types of signals were generated by the brittle fracture mechanism. Most likely, they characterise brittle phase fracture in the transition zone: class-3 signals—fracture of Al–Ti intermetallic compounds, whereas class-2 signals indicate oxide fracture.

Class-1 (Light green) AE signals with a frequency of approximately 100 kHz had a very low amplitude, not exceeding 0.4 V, and they were characterised by signal strength in the range from 1.8×10^4 to 8.0×10^6 pV·s and energy of up to 10 eu determined through the WTD analysis (Table 3, Figure 9a, Figure 10a,b and Figure 15a).

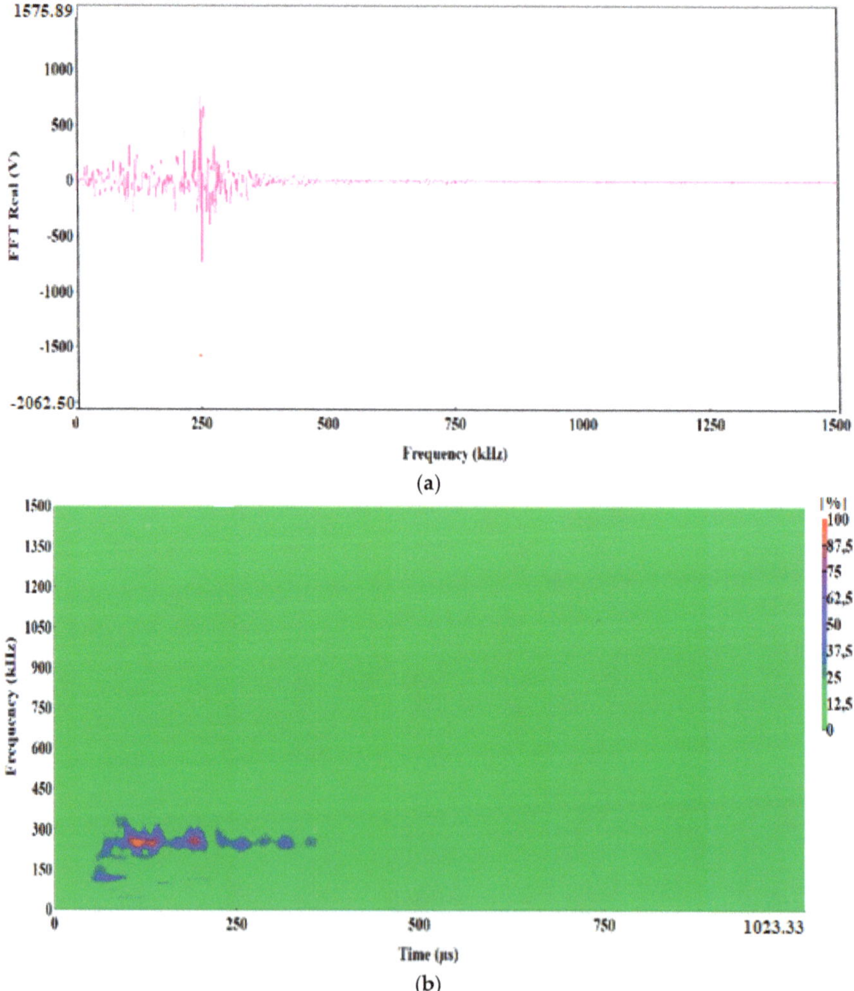

Figure 12. Waveform drequency domain (Real) (**a**) and waveform continuous wavelet (Morlet) (**b**) charts for the individual classes of AE signals (class 3).

These signals were generated by "noise", and they were usually caused by the friction of the rollers against the sample or by the operation of the strength testing machine. Class-1 signals will not be considered in further analysis of the fracture process.

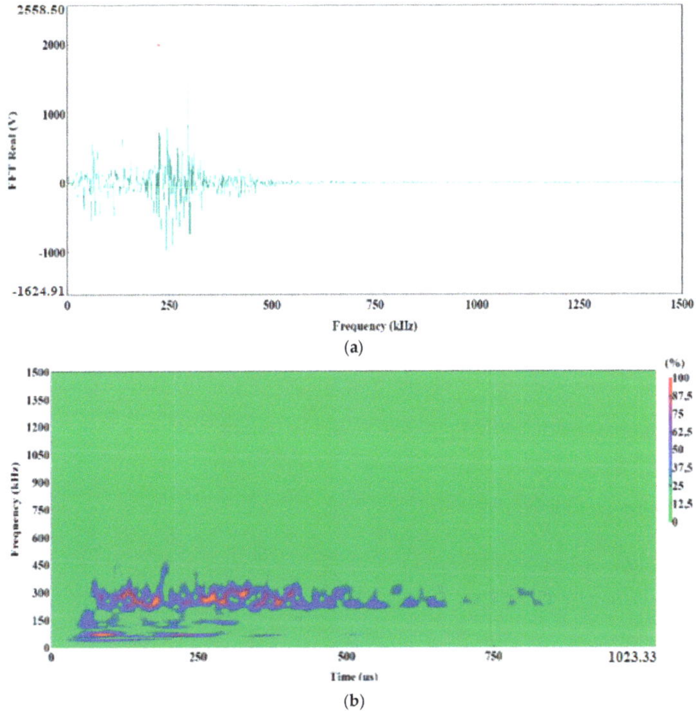

(a)

(b)

Figure 13. Waveform frequency domain (Real) (**a**) and waveform continuous wavelet (Morlet) (**b**) charts for the individual classes of AE signals (class 4).

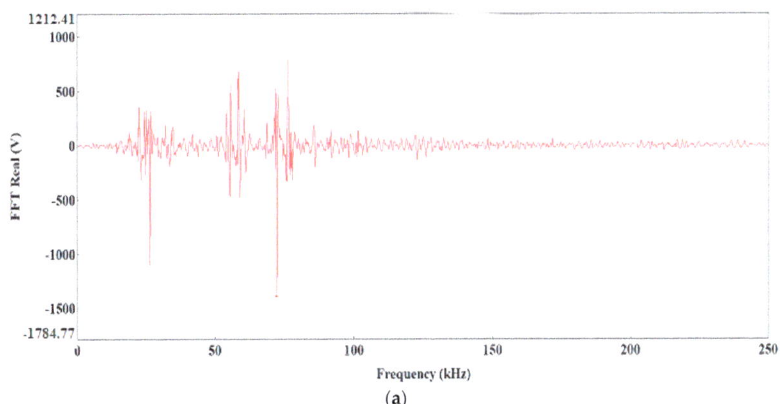

(a)

Figure 14. *Cont.*

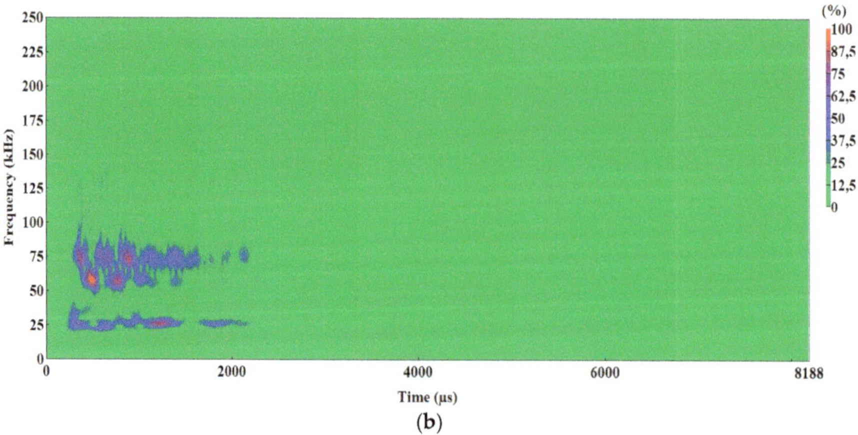

Figure 14. Waveform frequency domain (Real) (**a**) and waveform continuous wavelet (Morlet) (**b**) charts for the individual classes of AE signals (class 5).

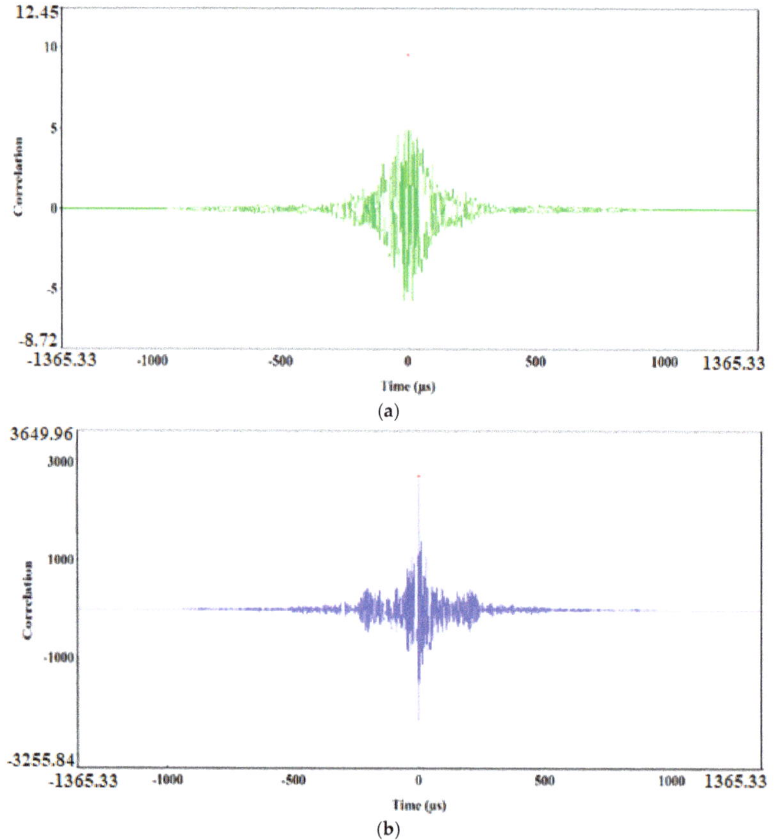

Figure 15. *Cont.*

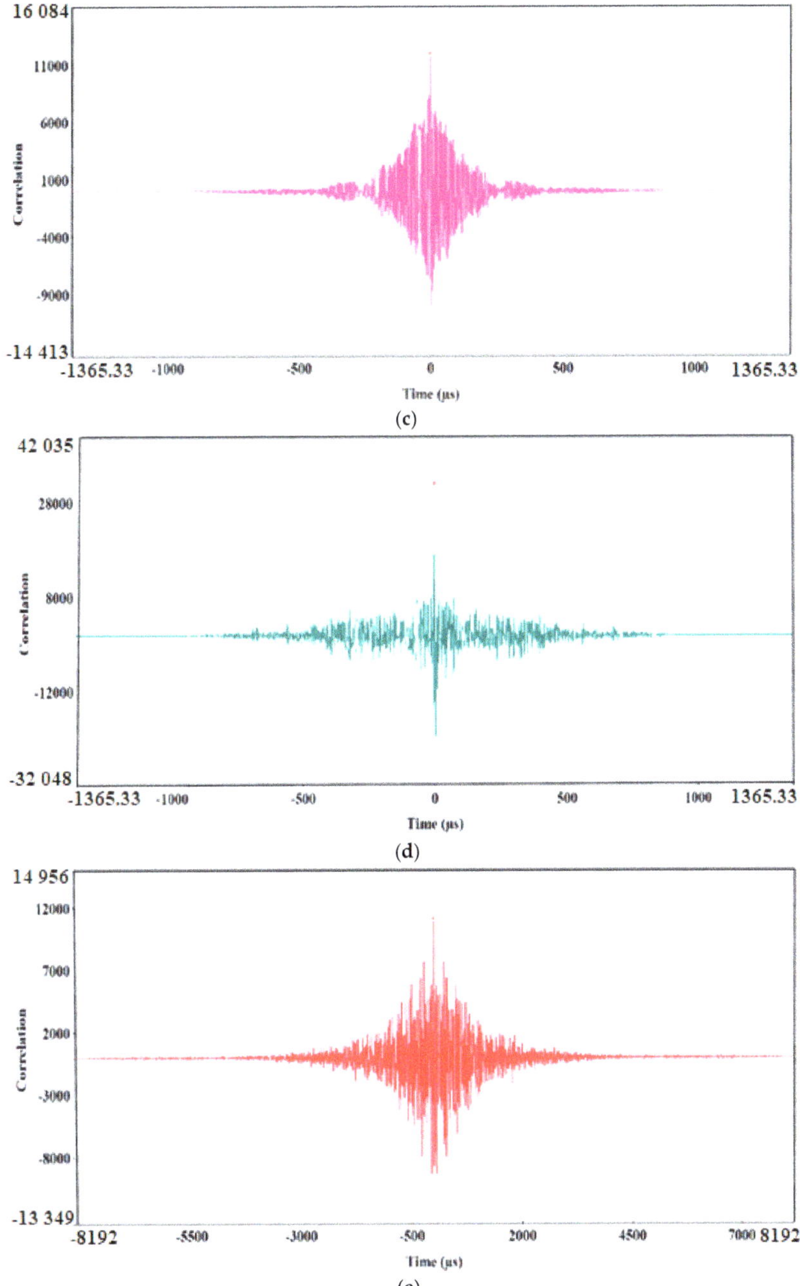

Figure 15. Waveform time domain (autocorrelation) chart for the individual classes of AE signals. (**a**) class 1; (**b**) class 2; (**c**) class 3; (**d**) class 4; (**e**) class 5.

6. Discussion

Based on the test results discussed above in this paper, we will now try to describe the fracture process in the Al–Ti laminate. During the test at temperature $T_1 = 20\,°C$, the recorded signals of AE indicated that when loading the specimens up to P_{max} only brittle phase fracture, most likely fracture of oxides, occurred in the transition zone between the AA1050 alloy and the Ti6Al4V alloy, which were well reflects by class-2 AE signals. At the time when value of force P_{max} was reached, fracture occurred in various components of the Al–Ti laminate according to different mechanisms. Brittle fracture of Al and Ti intermetallic particles occurs in the transition zone. This process was accurately illustrated by class-3 signals. Also, this was the time of the appearance of shear fracture, i.e. delamination crack, characterised by class-4 signals. Almost simultaneously to this process, in the base layers of the AA2519 and Ti6Al4V alloys, the main crack also developed due to the growth of voids, which was described by the class-5 signals. The growth of the delamination crack and the main crack decreases the force because the sample is loaded by displacement.

According to our results, the fracture process was slightly different in samples tested at $T_2 = -50\,°C$. AE signals of classes 2, 3 and 4 could be observed already at early stages of sample loading, before value P_{max} is reached. The signals were generated because the brittle fracture toughness of the materials was reduced along with the decrease in temperature [42–46]. That was why at $T_2 = -50\,°C$ brittle phase fracture in the transition zone occurred at a lower load (i.e., classes 2 and 3) which, in turn, led to the fracture caused by shear in the connecting layer of the AA1050 alloy (class 4). This means that the delamination crack was present already when the main crack started to appear. This conclusion was confirmed by the results of mechanical tests, which indicated that the compliance of the specimen has decreased. When the specimen reaches its maximum force, P_{max}, the main crack was propagated by the ductile mechanism of the growth of voids (class 5) and followed by subsequent stages of fracture by other mechanisms.

In a sense, the results presented in this paper are confirmed by Reference [48], where the authors described the evolution of the waves of AE signals along with the increase of the fatigue crack. It was found that the change of the shape of the AE wave was closely related to the physical condition of fracture loading and to the mechanism behind the growth of fatigue cracks. It was demonstrated that the load level could be associated with the acoustic emission signals present during the growth of fatigue crack.

The results presented in this paper are consistent with the results of fracture tests in metallic layered composites prepared by adhesive bonding [49–51], which indicate that different stages of the fracture process are represented by AE signals with different characteristic parameters.

Comprehensive and in-depth analysis of AE signals confirmed occurring various mechanisms of failure in the cracking process of the Al–Ti composite, which tested in the range of ambient to cryogenic temperatures [52].

7. Conclusions

This paper presents a comprehensive study of the fracture process in an Al–Ti laminate prepared by explosion welding. The testing methods used are modern and the results obtained are novelty. Mechanical load curves of SENB samples, *P–COD*, were examined in order to accurately interpret the fracture process; the fracture surfaces of the samples were thoroughly tested using SEM, and AE signals were recorded and subjected to various methods of analysis of wave packets of AE signals.

Based on the loading tests of the SENB specimens at $T_1 = 20\,°C$ and $T_2 = -50\,°C$ temperatures and the load curves, it was found that the *P–COD* curves had an atypical form, particularly at lowered temperatures.

Tests of the fracture surfaces of the specimens using optical and SEM microscopes found a delamination crack, which is usually initiated by the brittle phase fracture in the transition zone between the inner layer of the AA1050 alloy and the layer of the Ti6Al4V alloy. Then, the delamination crack develops primarily through the shear fracture mechanism. The propagation of the main crack in the base layers of the AA2519 and Ti6Al4V alloys occurs through the ductile mechanism of the growth of voids for two test temperatures.

The recording of AE signals and using different methods for the analysis of these signals, including non-hierarchical clustering methods (*k*-means) and analyses using Waveform Time Domain, Fast Fourier Transform (FFT Real), Waveform Continuous Wavelet using the Morlet wavelet and Waveform Time Domain (Autocorrelation), enabled the identification of four classes of signals and characterisation of the primary mechanisms of the fracture processes in the tested layered Al–Ti composite:

1. Class 2—AE signals of the brittle phase fracture of Al and Ti oxides;
2. Class 3—AE signals of the brittle phase fracture of Al and Ti intermetallic compounds;
3. Class 4—AE signals generated during the formation of the delamination crack through shear in the Al (AA1050) layer;
4. Class 5—AE signals generated during the development of the main crack in the base material layers, Al (AA2519) and Ti (Ti6Al4V) alloys.

The recorded AE signals were used to determine the primary differences in the fracture process of the Al–Ti laminate. At temperature $T_1 = 20\,°C$, the growth of the delamination crack and the main crack occur almost simultaneously. However, at temperature $T_2 = -50\,°C$, the delamination crack precedes the main crack, and the development of cracks in the base materials occurs without interaction between them. This explains the non-characteristic and illogical behaviours observe in the specimen load curves (*P–COD*) at different test temperatures.

These comprehensive tests indicated that the methods of analysing AE signals can be effectively used to identify the development of cracks in structural members. They can identify characteristic mechanisms of the formation and development of defects, also at very early stages of damage evolution, which cannot be achieved with other NDT methods. This fact can be used to create an automatic diagnostic system capable of determining the types and mechanisms of the development of potential damage at every stage of material use and assessing the reliability of the structure.

Author Contributions: Conceptualization, G.Ś., I.D. and A.K.; Methodology, G.Ś., D.I. and A.K.; Software, S.L. and A.K.; Validation, S.L. and T.P.; Formal Analysis, G.Ś., I.D. and A.K.; Investigation, T.P. and A.K.; Resources, G.Ś. and I.D.; Data Curation, A.K. and T.P.; Writing—Original Draft Preparation, G.Ś., I.D. and A.K.; Writing—Review and Editing, G.Ś., I.D. and A.K.; Visualization, A.K. and S.L.; Supervision, G.Ś. and I.D.; Project Administration, G.Ś.; Funding Acquisition, G.Ś. and I.D. All authors have read and agreed to the published version of the manuscript.

Funding: This research was funded by the National Science Centre, Poland (No. 2017/25/N/ST8/00179) and the Ministry of Science and Higher Education of Poland (No. 01.0.08.00/2.01.01.00.0000 MKPK 20.001 and 02.0.06.00/2.01.01.00.0000 SUBB. BKWB. 20.001.

Conflicts of Interest: The authors declare no conflict of interest and funders had no role in the design of the study; in the collection, analyses, or interpretation of data; in the writing of the manuscript, and in the decision to publish the results.

References

1. Mamalis, N.M.; Vaxevanidis, A.; Szalay, A.; Prohaszka, J. Fabrication of aluminium/copper bimetallics by explosive cladding and rolling. *J. Mater. Process. Technol.* **1994**, *44*, 99–117. [CrossRef]
2. Pozuelo, M.; Carreno, F.; Ruano, O.A. Delamination effect on the impact toughness of an ultrahigh carbon-mild steel laminate composite. *Compos. Sci. Technol.* **2006**, *66*, 2671–2676. [CrossRef]
3. Pozuelo, M.; Carreno, F.; Capeda-Jimenez, C.M.; Ruano, O.A. Effect of hot rolling on bonding harakterystics and impact behaviour of a laminated composite material based on UHCS-1.35pct C. *Metall. Mater. Trans. A* **2008**, *39*, 666–671. [CrossRef]
4. Groza, J.R.; Shackelford, J.F.; Lavernia, E.J.; Powers, M.T. *Materials Processing Handbook*; CRC Press: Boca Raton, FL, USA, 2007.
5. Balasubramanian, V.; Rathinasabapathi, M.; Raghukandan, K. Modelling of process parameters in explosive cladding of mildsteel and aluminium. *J. Mater. Process. Technol.* **1997**, *63*, 83–88. [CrossRef]

6. Gerland, M.; Presles, H.; Guin, J.; Bertheau, D. Explosive cladding of a thin Ni-film to an aluminium alloy. *Mater. Sci. Eng. A* **2000**, *280*, 311–319. [CrossRef]

7. Findik, F. Recent developments in explosive welding. *Mater. Des.* **2011**, *32*, 1081–1093. [CrossRef]

8. Gałka, A. Application of explosive metal cladding in manufacturing new advanced layered materials on the example of titanium Ti6Al4V – aluminum AA2519 bond. *High-Energetic Mater.* **2015**, *7*, 73–79.

9. Habib, M.A.; Keno, H.; Uchida, R.; Mori, A.; Hokamoto, K. Cladding of titanium and magnesium alloy plates using energy-controlled underwater three layer explosive welding. *J. Mater. Process. Technol.* **2015**, *217*, 310–316. [CrossRef]

10. Sun, X.; Tao, J.; Guo, X. Bonding properties of interface in Fe/Al clad tube prepared by explosive welding. *Trans. Nonferrous Met. Soc. China.* **2011**, *21*, 2175–2180. [CrossRef]

11. Aizawa, Y.; Nishiwaki, J.; Harada, Y.; Muraishi, S.; Kumai, S. Experimental and numerical analysis of the formation behavior of intermediate layers at explosive welded Al/Fe joint interfaces. *J. Manuf. Process.* **2016**, *24*, 100–106. [CrossRef]

12. Fronczek, D.M.; Wojewoda-Budka, J.; Chulist, R.; Sypien, A.; Korneva, A.; Szulc, Z.; Schell, N.; Zieba, P. Structural properties of Ti/Al clads manufactured by explosive welding and annealing. *Mater. Des.* **2016**, *91*, 80–89. [CrossRef]

13. Carvalho, G.H.S.F.L.; Mendes, R.; Leal, R.M.; Galvão, I.; Loureiro, A. Effect of the flyer material on the interface phenomena in aluminium and copper explosive welds. *Mater. Des.* **2017**, *122*, 172–183. [CrossRef]

14. Chu, Q.; Zhang, M.; Li, J.; Yan, C. Experimental and numerical investigation of microstructure and mechanical behavior of titanium/steel interfaces prepared by explosive welding. *Mater. Sci. Eng. A* **2017**, *689*, 323–331. [CrossRef]

15. Boroński, D.; Kotyk, M.; Maćkowiak, P.; Śnieżek, L. Mechanical properties of explosively welded AA2519-AA1050-Ti6Al4V layered material at ambient and cryogenic conditions. *Mater. Des.* **2017**, *133*, 390–403. [CrossRef]

16. Aggelis, D.; Matikas, T. Effect of plate wave dispersion on the acoustic emission parameters in metals. *Comput. Struct.* **2012**, *98*, 17–22. [CrossRef]

17. Aggelis, D.; Kordatos, E.; Matikas, T. Acoustic emission for fatigue damage characterization in metal plates. *Mech. Res. Commun.* **2011**, *38*, 106–110. [CrossRef]

18. Aicher, S.; Höfflin, L.; Dill-Langer, G. Damage evolution and acoustic emission of wood at tension perpendicular to fiber. *Eur. J. Wood Wood Prod.* **2001**, *59*, 104–116. [CrossRef]

19. Arul, S.; Vijayaraghavan, L.; Malhotra, S.K. Online monitoring of acoustic emission for quality control in drilling of polymeric composites. *J. Mater. Proc. Technol.* **2007**, *185*, 184–190. [CrossRef]

20. Ai, Q.; Liu, C.C.; Chen, X.R.; He, P.; Wang, Y. Acoustic emission of fatigue crack in pressure pipe under cyclic pressure. *Nucl. Eng. Des.* **2010**, *240*, 3616–3620. [CrossRef]

21. Diamanti, K.; Soutis, C. Structural health monitoring techniques for aircraft composite structures. *Pr. Aerosp. Sci.* **2010**, *46*, 342–352. [CrossRef]

22. Aymerich, F.; Staszewski, W. Experimental study of impact-damage detection in composite laminates using a cross-modulation vibro-acoustic technique. *Struct. Health Monit.* **2010**. [CrossRef]

23. Pappas, Y.Z.; Kostopoulos, V. Toughness characterization and acoustic emission monitoring of a 2-D carbon/carbon composite. *Eng. Fract. Mech.* **2001**, *68*, 1557–1573. [CrossRef]

24. Marec, A.; Thomas, J.H.; El Guerjouma, R. Damage characterization of polymer-based composite materials: Multivariable analysis and wavelet transform for clustering acoustic emission data. *Mech. Syst. Signal Pr.* **2008**, *22*, 1441–1464. [CrossRef]

25. Sedlak, P.; Hirose, Y.; Khan, S.A.; Enoki, M.; Sikula, J. New automatic localization technique of acoustic emission signals in thin metal plates. *Ultrasonics* **2009**, *49*, 254–262. [CrossRef] [PubMed]

26. Rose, J.L. *Ultrasonic Waves in Solid Media*; Cambridge University Press: Cambridge, UK, 2004.

27. Dang Hoang, T.; Herbelot, C.; Imad, A. Rupture and damage mechanism analysis of a bolted assembly using coupling techniques between AE and DIC. *Eng. Struct.* **2010**, *32*, 2793–2803. [CrossRef]

28. Roberts, T.; Talebzadeh, M. Acoustic emission monitoring of fatigue crack propagation. *J. Construct. Steel Res.* **2003**, *59*, 695–712. [CrossRef]

29. Han, Z.; Luo, H.; Cao, J.; Wang, H. Acoustic emission during fatigue crack propagation in a micro-alloyed steel and welds. *Mater. Sci. Eng. A* **2011**, *528*, 7751–7756. [CrossRef]

30. Krampikowska, A.; Dzioba, I.; Pała, R.; Swit, G. The Use of the Acoustic Emission Method to Identify Crack Growth in 40CrMo Steel. *Materials* **2019**, *12*, 2140. [CrossRef]

31. Swit, G. Acoustic Emission Method for Locating and Identifying Active Destructive Processes in Operating Facilities. *Appl. Sci.* **2018**, *8*, 1295. [CrossRef]

32. NOESIS 12.0. *Advanced Acoustic Emission Data Analysis Pattern Recognition & Neural Networks Software*; MISTRAS HELLAS A.B.E.E.: Princeton, NJ, USA, 2018.

33. Kong, J.; Ruan-Wu, C.; Luo, Y.X.; Zhang, C.L.; Zhang, C. Magnetoelectric effects in multiferroic laminated plates with imperfect interfaces. *Theor. Appl. Mech. Lett.* **2017**, *7*, 93–99. [CrossRef]

34. Li, D.; Liu, Y. Three-dimensional semi-analytical model for the static response and sensitivity analysis of the composite stiffened laminated plate with interfacial imperfection. *Compos. Struct.* **2012**, *94*, 1943–1958. [CrossRef]

35. Kim, J.S.; Oh, J.; Cho, M. Efficient analysis of laminated composite and sandwich plates with interfacial imperfections. *Compos. Part B Eng.* **2011**, *42*, 1066–1075. [CrossRef]

36. ASTM E1820-09. Standard Test Method for Measurement of Fracture Toughness. In *Annual Book of ASTM Standards*; ASTM International: West Conshohocken, PA, USA, 2011; Volume 03.01, pp. 1070–1118.

37. Omkar, S.N.; Suresh, S.; Raghavendra, T.R.; Mani, V. Acoustic emission signal classification using fuzzy c-means clustering. In Proceedings of the 9th International Conference: Neural Information Processing, Singapore, 18–22 November 2002.

38. Sawan, G.A.; Walter, M.E.; Marquette, B. Unsupervised learning for classification of acoustic emission events from tensile and bending experiments with open-hole carbon fiber composite samples. *Compos. Sci. Technol.* **2015**, *107*, 89–97. [CrossRef]

39. Ramasso, E.; Palcet, V.; Boubakar, M.L. Unsupervised Consensus Clustering of Acoustic Emission Time-Series for Robust Damage Sequence Estimation in Composites. *IEEE Trans. Instrum. Meas.* **2015**, *64*, 3297–3307. [CrossRef]

40. Warren Liao, T.; Hua, G.; Qu, J.; Blau, P.J. Grinding wheel condition monitoring with hidden Markov model-based clustering methods. *Mach. Sci. Technol.* **2006**, *10*, 511–538. [CrossRef]

41. Ramasso, E.; Butaud, P.; Jeannin, T.; Sarasini, F.; Palcet, V.; Godin, N.; Tirillo, J.; Gabrion, X. Learning the representation of raw acoustic emission signals by direct generative modelling and its use in chronology-based clusters identification. *Eng. Appl. Artif. Intell.* **2020**, *90*, 1–13. [CrossRef]

42. Anderson, T.L. *Fracture Mechanics: Fundamentals and Applications*; CRC Press Taylor & Francis Group: Milton Park, Abingdon, UK, 2005.

43. Khan, A.S.; Liu, H. Strain rate and temperature dependent fracture criteria for isotropic and anisotropic metals. *Int. J. Plast.* **2012**, *37*, 1–15. [CrossRef]

44. Neimitz, A.; Dzioba, I. The influence of the out-of- and in-plane constraint on fracture toughness of high strength steel in the ductile to brittle transition temperature range. *Eng. Fract. Mech.* **2015**, *147*, 431–448. [CrossRef]

45. Dzioba, I.; Pała, R. Strength and Fracture Toughness of Hardox-400 Steel. *Metals* **2019**, *9*, 508. [CrossRef]

46. Dzioba, I.; Lipiec, S. Fracture Mechanisms of S355 Steel—Experimental Research, FEM Simulation and SEM Observation. *Materials* **2019**, *12*, 3959. [CrossRef]

47. Bazarnik, P.; Adamczyk-Cieślak, B.; Gałka, A.; Płonka, B.; Snieżek, L.; Cantoni, M.; Lewandowska, M. Mechanical and microstructural characteristics of Ti6Al4V/AA2519 and Ti6Al4V/AA1050/AA2519 laminates manufactured by explosive welding. *Mater. Des.* **2016**, *111*, 146–157. [CrossRef]

48. Bhuiyan, M.Y.; Lin, B.; Giurgiutu, V. Acoustic emission sensor effect and waveform evolution during fatigue crack growth in thin metallic plate. *J. Intell. Mater. Syst. Struct.* **2017**, *29*, 1275–1284. [CrossRef]

49. Crawford, A.; Droubi, M.G.; Faisal, N.H. Analysis of Acoustic Emission Propagation in Metal-to-Metal Adhesively Bonded Joints. *J. Nondestruct. Eval.* **2018**, *37*, 1–19. [CrossRef]

50. Droubi, M.G.; Stuart, A.; Mowat, J.; Noble, C.; Prathuru, A.K.; Faisal, N.H. Acoustic emission method to study fracture (Mode-I, II) and residual strength characteristics in composite-to-metal and metal-to-metal adhesively bonded joints. *J. Adhes.* **2017**, *94*, 347–386. [CrossRef]

51. Pashmforoush, F.; Khamedi, R.; Fotouhi, M.; Hajikhani, M.; Ahmadi, M. Damage classification of sandwich composites using acoustic emission technique and k-means genetic algorithm. *J. Nondestr. Eval.* **2014**, *33*, 481–484. [CrossRef]
52. Boroński, D.; Dzioba, I.; Kotyk, M.; Krampikowska, A.; Pala, R. Investigation of the Fracture Process of Explosively Welded AA2519–AA1050–Ti6Al4V Layered Material. *Materials* **2020**, *13*, 2226. [CrossRef]

Article

Prediction of Static Modulus and Compressive Strength of Concrete from Dynamic Modulus Associated with Wave Velocity and Resonance Frequency Using Machine Learning Techniques

Jong Yil Park [1], Sung-Han Sim [2], Young Geun Yoon [3,*] and Tae Keun Oh [3,4,*]

[1] Department of Safety Engineering, Seoul National University of Science and Technology, Seoul 01811, Korea; jip111@seoultech.ac.kr
[2] School of Civil, Architectural Engineering and Landscape Architecture, Sungkyunkwan University, Suwon 16419, Korea; ssim@skku.edu
[3] Department of Safety Engineering, Incheon National University, Incheon 22012, Korea
[4] Research Institute for Engineering and Technology, Incheon National University, Incheon 22012, Korea
* Correspondence: yyg900@inu.ac.kr (Y.G.Y.); tkoh@inu.ac.kr (T.K.O.); Tel.: +82-032-835-8294 (T.K.O.)

Received: 27 May 2020; Accepted: 24 June 2020; Published: 27 June 2020

Abstract: The static elastic modulus (Ec) and compressive strength (fc) are critical properties of concrete. When determining Ec and fc, concrete cores are collected and subjected to destructive tests. However, destructive tests require certain test permissions and large sample sizes. Hence, it is preferable to predict Ec using the dynamic elastic modulus (Ed), through nondestructive evaluations. A resonance frequency test performed according to ASTM C215-14 and a pressure wave (P-wave) measurement conducted according to ASTM C597M-16 are typically used to determine Ed. Recently, developments in transducers have enabled the measurement of a shear wave (S-wave) velocities in concrete. Although various equations have been proposed for estimating Ec and fc from Ed, their results deviate from experimental values. Thus, it is necessary to obtain a reliable Ed value for accurately predicting Ec and fc. In this study, Ed values were experimentally obtained from P-wave and S-wave velocities in the longitudinal and transverse modes; Ec and fc values were predicted using these Ed values through four machine learning (ML) methods: support vector machine, artificial neural networks, ensembles, and linear regression. Using ML, the prediction accuracy of Ec and fc was improved by 2.5–5% and 7–9%, respectively, compared with the accuracy obtained using classical or normal-regression equations. By combining ML methods, the accuracy of the predicted Ec and fc was improved by 0.5% and 1.5%, respectively, compared with the optimal single variable results.

Keywords: concrete; static elastic modulus; dynamic elastic modulus; compressive strength; machine learning; P-wave; S-wave; resonance frequency test; nondestructive method

---

1. Introduction

During the design, construction, and maintenance of concrete structures, static elastic modulus (Ec) and compressive strength (fc) are critical properties for analyzing structural stability parameters such as member force, stress, deflection, and displacement [1,2]. In addition, these properties are indicators of concrete deterioration. Therefore, Ec and fc have been used to evaluate the conditions of structures such as pavements and bridge decks. The specimen cores of existing concrete structures are typically extracted and tested to determine Ec and fc, according to the recommended ASTM C469/C469M-14 guidelines. Ec is typically predicted from fc; however, this yields a relatively large error [3]. The standard testing methods for determining Ec and fc cannot be used to evaluate the entire area, owing to the need for test permissions and several specimens. Furthermore, the values of Ec and *fc* vary at different locations in the concrete structure, and these differences may increase with aging. Therefore, it is necessary to determine *Ec* and *fc* using a nondestructive evaluation (NDE) method that can be applied at several locations in a structure without damaging the structure.

The elastic modulus measured using the NDE method is generally called the dynamic elastic modulus (*Ed*). Ultrasonic pulse velocity (UPV) measurements based on the ASTM C597/ASTM C597M-16 guidelines and resonance frequency tests conducted in accordance with the ASTM C215-14 specifications are used to determine *Ed* [4,5]. As for ASTM C215, Subramaniam et al. have improved the standard equation using the second resonance frequency as well as the first. Although it has the advantage of finding the dynamic Poisson ratio, its accuracy is insignificant [6]. The *Ed* determined with the NDE method is generally higher than the *Ec* obtained by following the ASTM C469/C469M-14 recommendations [7]. The *Ed* determined with the P-wave measurement was the largest, and the *Ed* determined with the resonant frequency test was located in the middle compared to *Ec* [8,9]. The *Ed* value with the S-wave measurement was more correlated with *Ec* and *fc* than the *Ed* using the P-wave measurement and resonant frequency [10]. However, the S-wave is difficult to measure, and the dispersion of the measured values may be large at a low age. Additionally, it is more difficult to predict the properties of early aged and low-strength concrete (which include moisture and voids) and high-strength concrete. It has been known that the static modulus (secant modulus of elasticity) determined by the destructive test according to ASTM C469 and the dynamic modulus determined by non-destructive tests such as ASTM C597 and ASTM C215 have a nonlinear proportional relationship, and the static modulus is 10–20% less than the dynamic modulus [11]. Since the dynamic modulus is determined by the slope at a very low stress range, it is less statistically stable than the elastic modulus, but it has the advantage of being simple to test without the damaging of the specimen [12]. Therefore, it is essential to develop a method that can be used to more accurately predict *Ec* from *Ed* because it is difficult to measure all the specimens with the destructive test.

Various equations have been proposed for predicting *Ec* based on *Ed*. For example, Lydon and Balendran, BS 8110 Part 2, and Popovics proposed empirical equations for predicting *Ec* from *Ed*, using the NDE method [11–13]. Among previously proposed equations, the BS 8110 Part 2 equation yields similar values to the *Ed* value measured using the UPV through P-waves, and the Popovics equation provides results that are similar to *Ed* obtained through resonance frequency tests using longitudinal and transverse modes [10]. In addition, the Lydon and Balendran results are lower than those obtained using other equations [9]. Most of the previous studies attempted to relate Ec values with Ed values using statistical regression analysis such as linear or nonlinear regression [8]. However, these equations may lead to a significant error in predicting *Ec* if the Ed value obtained from another test method is used. As described above, because the deviation and the degree of change in the *Ed* value differs significantly for each NDE method, it is difficult to apply these equations for the accurate prediction of *Ec*.

Additionally, correlation equations have been proposed by CEB-FIP, the ACI 318 Committee, and the ACI 363 Committee as methods for predicting *fc* using *Ed* [14–16]. However, when predicting *fc* using *Ed*, the prediction error may increase in a step-by-step relational expression because the procedures for predicting *Ec* in *Ed* and *fc* in *Ec* are employed. Therefore, it is necessary to apply a method that can solve the problems of the existing prediction of linear regression and predict *Ec* and fc more accurately by utilizing various *Ed* that can be obtained through experiments. Recently, Chavhan and Vyawahare [17] proposed a study to predict fc from *Ed* with P-wave velocity tests but did not obtain a good relationship, even with only 22 specimens. Thus, for the relationship between *Ec* or fc and *Ed*, the classical or regression equations could not give an accurate estimation for *Ec* from *Ed* because the *Ed* value differs according to the test method [18]. Additionally, there exists considerable nonlinearity between *Ec* or *fc* and *Ed* [19].

In this regard, the machine learning (ML) technology has been applied and improved in many areas of concrete, but it has been used to predict *Ec* and *fc* limitedly with various components [20], to establish a correlation between *Ec* and fc [21], and to estimate fc from ultrasonic velocity and rebound hardness [22]. Furthermore, the estimation of *Ec* or fc from *Ed* has been little studied due to the difficulties in testing enough specimens. Thus, these studies have focused on obtaining more accurate predictions of concrete strength and conditions using (ML) algorithms such as support vector machine (SVM), ensemble, and artificial neural networks (ANNs) [21–24].

In most studies, ML methods contributed to predicting fc with higher accuracy than conventional linear regression methods. In addition, it was used to predict the characteristics of concrete using various variables as follows. Erdal et al. used an ANN and ensemble for comparing the accuracy of strength prediction for high-performance concrete, and the method of applying the ensemble was found to be slightly better [25]. Yuvaraj et al. confirmed the applicability of several variables to the SVM model by predicting the fracture characteristics of concrete beams according to the ratio of concrete mixtures from the SVM model [26]. Yan and Shi used four theoretical equations, linear regression, and ANN and SVM models to confirm the accuracy of *Ec* prediction with *fc*, and the ANN and SVM methods were found to be the best methods [27]. Cihan employed SVM, ANN, and ensemble to predict the compressive strength and slump value of concrete, and reported that ANN and ensemble were better than SVM [28]. Young et al. employed SVM, ANN, and linear regression (LR) to estimate 28-day intensity from 10,000 data with varying mixing ratios and reported that ANN and SVM were more accurate than LR [29]. Among the many ML methods, the SVM, ANN, ensemble, and LR methods contributed to the prediction of concrete quality using various variables. However, previous studies only considered cases such as the prediction of strength and the modulus of elasticity using different combinations of materials, ultrasonic pulse velocity, and *fc*. Very few studies have attempted to predict *Ec* and *fc* using two or more *Eds* and ML.

In this study, various combinations were used to overcome the differences in the characteristics of the eigen-data of *Eds* through the ML methods and to confirm the possibility of predicting actual *Ec* and *fc* values more accurately than conventional linear equations. Various *Ed* values were obtained with UPV measurements and resonance frequency tests. Additionally, by determining the S-wave velocity (*Vs*), according to the recommended guidelines of ASTM C215-14 and ASTM C597M-16 and by determining *Ed* using *Vs*, the differences with respect to existing *Ed* values were verified and then utilized to improve the accuracy of the prediction. Four ML methods—SVM, ANN, ensemble, and LR—were trained on four *Ed* datasets, and a five-fold validation was performed to prevent the overfitting of the prediction results. Subsequently, the predicted and measured values for *Ec*, *fc*, and Ed were compared using the mean squared error (MSE) and mean absolute percentage error (MAPE). Thus, this study aimed to confirm how much accuracy can be improved compared to that obtained with the classical and regression equation by predicting *Ec* and *fc* directly from the Ed value, which is a representative property value in the non-destructive testing method, associated with ML. An attempt was made to predict exactly how much and how the *Ec* and *fc* values will be affected not only by the four ML methods that can predict Ed but also by a combination thereof.

2. Materials and Methods

2.1. Materials and Preparation of Specimens

The concrete specimens were composed of Type I Portland cement, river sand, crushed granite with a size of up to 25 mm, and supplementary cementitious materials (SCMs), i.e., fly ash and slag cement. The concrete of this study replaces about 50% of the cement with fly ash (FA) and Granulated Blast Furnace Slag (GBFS), and is composed of a mixture that causes the development of the initial strength to be slow at a low age but the long-term strength to be high. It is mainly used in mass concrete that generates less heat of hydration and requires long-term strength. Two water/binder (W/B) ratios of 0.45 and 0.35 were used, which were expected to achieve 28-day target compressive strength values of 20 MPa and 40 MPa, respectively. Two concrete mixture groups—Mix 1 and Mix 2—were prepared; the proportions of these mixture groups are presented in Table 1. Two hundred and ninety-five concrete specimens were cast using 150 mm × 300 mm plastic molds, in accordance with the ASTM C31/C31M-12 specifications [30]. The specimens were removed from the molds after 24 h and then cured in water. *Ed, Ec,* and *fc* tests were performed at the different curing ages of 4, 7, 14, and 28 d. The 28-day average compressive strengths for Mixes 1 and 2 were 19.19 MPa and 43.94 MPa, respectively.

Table 1. Proportions of the concrete mixture groups [1].

ID	Cement Type	W/B	S/A	W	C	S	G	Unit Quantity (kg/m^3) Mineral Admixture		Chemical Admixture	
								FA	GBFS	AE (Binder%)	SP (Binder%)
Mix 1 (20 MPa)	Type I	0.45	0.46	259	121	777	934	58	69	0.9	-
Mix 2 (40 MPa)		0.35	0.47	308	166	761	886	81	85	-	1

[1] SCMs: Supplementary cementitious materials, W: water, C: cement, S: sand, G: crushed cobblestone, FA: fly ash, SC: slag cement, AE: air-entraining agent, and SP: superplasticizer.

2.2. Static Tests for Elastic Modulus and Compressive Strength

As a pretest preparation procedure, both ends of each specimen were polished and kept perpendicular to the measurement by removing protrusions on the specimen surface. The *Ec* and *fc* values of each specimen were determined using a universal testing machine (UTM, Instron, Norwood MA, USA) with a capacity of 1000 kN, according to the ASTM C469/C469M-14 and ASTM C39/C39M-14a specifications (Figure 1) [7,31]. The UTM was operated at a loading speed of approximately 0.28 MPa/s.

Figure 1. Experimental setup for determining static elastic modulus and compressive strength of concrete specimens.

2.3. P- and S-Wave Measurements for Calculating Dynamic Elastic Modulus

The P-wave velocity (*Vp*) was determined according to the ASTM C597M-16/C597M-16 recommendations, using a pair of P-wave transducers (MK 954, MKC Korea, Seoul, Korea) connected to a pulser-receiver (Ultracon 170, MKC Korea, Seoul, Korea) [4]. Figure 2a depicts the method of determining *Vp* using a receiving transducer. A longitudinal pulse of 52 kHz was transmitted through the specimens' interior, and the *Vp* values of the concrete specimens were determined. The measured signal was digitized to a sampling frequency of 10 MHz using an oscilloscope (NI-PXIe 6366, National Instruments, Austin, TX, USA). The *Ed* based on *Vp* is expressed as Equation (1):

$$Ed.Vp = \frac{(1+v)(1-2v)}{(1-v)}\rho V_p^2 \tag{1}$$

where *Ed.Vp* is the *Ed* of the specimen for the experimentally measured *Vp*, and ρ is the mass density of the concrete; it is equal to m/2πrL. *m*, *r*, and *L* are the mass, radius, and length of the specimens, respectively. A Poisson's ratio (ν) of 0.2 was used, which is the typical value for ordinary concrete.

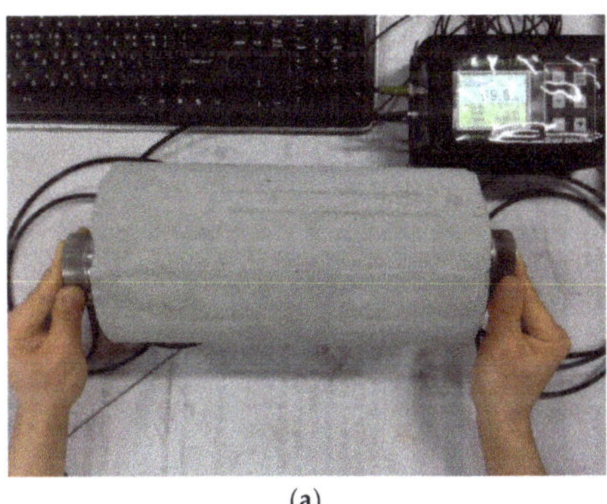
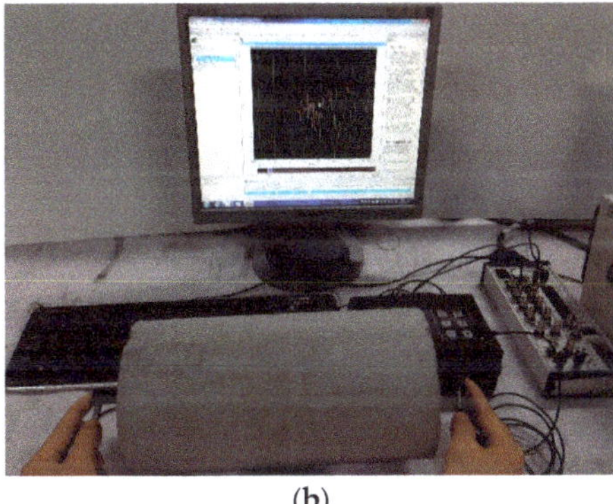

(a)   (b)

Figure 2. Methods for measuring (**a**) ultrasonic pulse velocity using a P-wave transducer (MK 954) and (**b**) ultrasonic shear-wave velocity using an S-wave transducer (ACS T1802).

The S-wave velocity *(Vs)* was determined using the measurement procedure for *Vp*. The measurement of *Vs* was similar to that of *Vp*, except that a pair of S-wave transducers (ACS T1802, ACS Group, Saarbrücken, Germany) were used. The signal driven using the pulsar was converted to a pulse in the transverse mode by the transducer and passed through the specimen. The signal measured using the receiving transducer was digitized using the oscilloscope. Figure 2b presents the method of measuring *Vs* and the ultrasonic signal measured using the S-wave receiver. The arrival of a stress wave through the specimen was determined based on the signals measured using a modified threshold method [32]. In this method, the approximate arrival time was first obtained using a threshold method [33]. Thereafter, the exact arrival time required to fit the line to a single datum was calculated. The S-wave travel time was defined at the intersection between the calculated zero signal level and the fitting line. Finally, *Vs* was calculated by dividing the length of the specimen (*L*) by the travel time (*t*) (*Vs* = *L*/*t*). In this study, the *Ed* based on *Vs* is expressed as Equation (2):

$$Ed.Vs = 2(1+\nu)\rho V_s^2 \tag{2}$$

where *Ed.Vs* is the *Ed* of the concrete specimen for the measured *Vs*, and ν is the Poisson's ratio.

2.4. Resonance Frequency Tests for Calculating Dynamic Elastic Modulus

The fundamental longitudinal and transverse resonance frequencies of the concrete specimens were determined to calculate the *Ed* value according to the ASTM C215-14 guidelines (Figure 3) [5]. A steel ball hammer with a diameter of 10 mm was used to generate stress waves in the concrete specimens through impact. The use of the steel ball hammer was effective in generating significantly low to 20 kHz frequency signals during the resonance frequency tests. The dynamic responses of the concrete specimens were measured using an accelerometer (PCB 353B16, PCB, Erie County, NY, USA) with a resonance frequency of approximately 70 kHz and an error of 5%, within the range of 1 to 10 kHz. The signals measured using the accelerometer were stabilized using a signal conditioner (PCB 482C16, PCB, Erie County, NY, USA) and digitized at a sampling frequency of 1 MHz using an oscilloscope (NI-PXIe 6366). The time signal was transformed into the frequency domain through a fast Fourier transform algorithm. The resonance frequency had the largest amplitude in the amplitude spectrum, and the frequency corresponding to the largest amplitude was considered as the fundamental resonance frequency in the longitudinal and transverse modes. The *Ed* based on this resonance frequency is defined using Equations (3) and (4):

$$Ed.LT = \alpha_{LT} m f_{LT}^2 (Pa) \tag{3}$$

$$Ed.TR = \alpha_{TR} m f_{TR}^2 (Pa) \tag{4}$$

where *Ed.LT* and *Ed.TR* are calculated using f_{LT} and f_{TR}, respectively, as the fundamental resonance frequencies. α_{LT} has a constant value depending on the dimensions of the specimen (5.093 (L/d^2)), where L is the length and d is the diameter. α_{TR} has a constant value, and its value depends on the +Poisson's ratio and specimen dimensions (1.6067 ($L^3 T/d^4$)), where T is the correction factor of the Poisson's ratio according to ASTM C215-14 [5], and m is the mass of the specimen in kg.

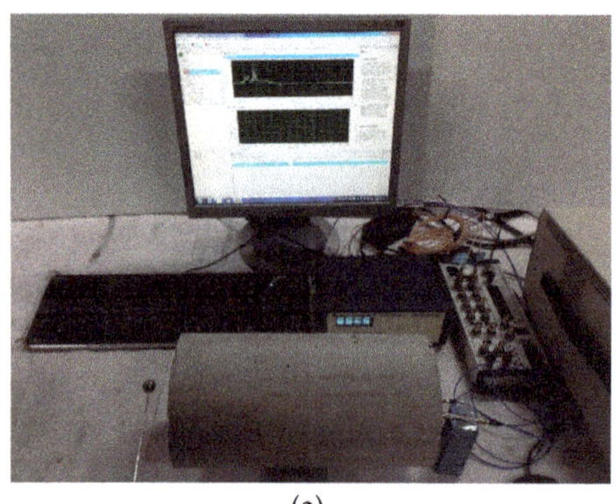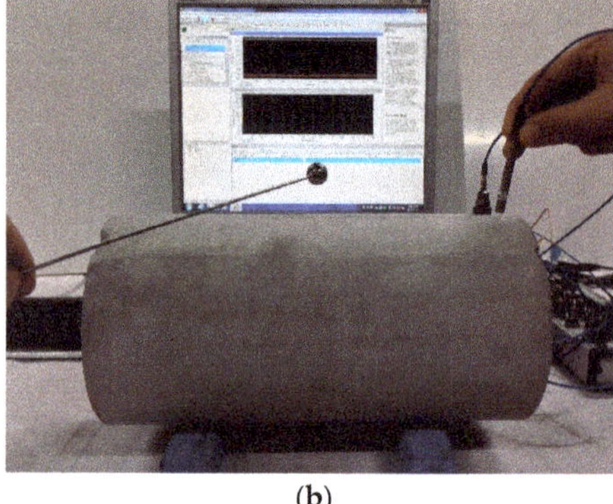

(a)   (b)

Figure 3. Resonance frequency test for (**a**) longitudinal mode and (**b**) transverse mode.

3. ML Techniques

3.1. SVM

The SVM analytical method, proposed by Vapnik in 1992, is a widely used ML tool for classification and regression [34]. SVM regression is considered a nonparametric technique because it is based on kernel functions, where a linear epsilon neglected during SVM (ε-SVM) regression is implemented; it is also known as L1 loss. In ε-SVM regression, the training data set contains values of both predictor variables and observed response variables. For each training point x, the SVM deviates from yn (the response variable value) by a value not higher than ε, as shown in Figure 4. Simultaneously, the objective is to determine the maximum flattened function f(x), as expressed in Equation (5):

$$f(x) = x'\beta + b \tag{5}$$

where β is the linear predictor coefficient, and b is the bias. In this study, the SVM regression model was implemented in the MATLAB R2019b (MathWorks, Natick, MA, USA) environment using "fitrsvm", which employs linear learning.

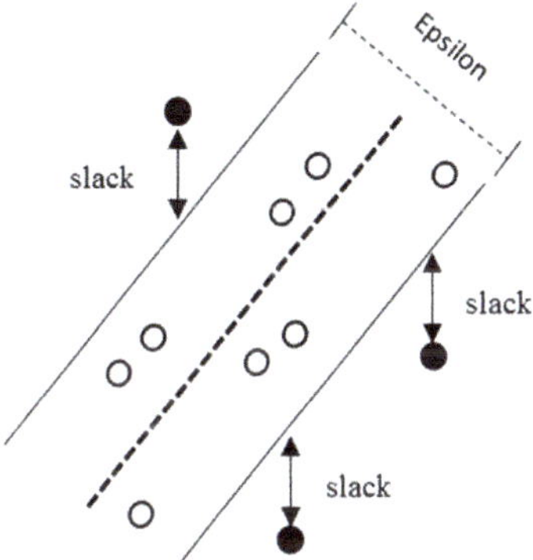

Figure 4. Procedure of support vector regression from an input sample.

3.2. ANN

An ANN is an information processing system based on the structure and function of biological neural networks. ANNs utilize connected artificial neurons and learn input parameters through these neurons to explain their unique behaviors, thereby performing nonlinear modeling. Multilayer perceptron (MLP) is the most commonly used ANN architecture; it consists of input, hidden, and output layers as shown in Figure 5 [35]. As shown in Equation (6), all the neuron connections in every layer of an ANN include weights and biases. The neuron values of previous layers are adjusted by varying weights and compensated through bias.

$$y_i = f(net) = f\left(\sum_{i=1}^{n} \omega_i x_i + b_j\right) \tag{6}$$

where x is the input value ($i = 1, \ldots, n$), w is the weight vector between neurons, b is the bias of the neuron, f is the activation function, and y is the output value. MLP is generally trained using an inverse propagation algorithm wherein interconnected weights of the network are repeatedly modified to minimize errors defined as the root mean square errors (RMSEs) [36]. Previous studies have reported that the Levenberg–Marquardt algorithm is suitable because it yields consistent results for the majority of ANNs [37].

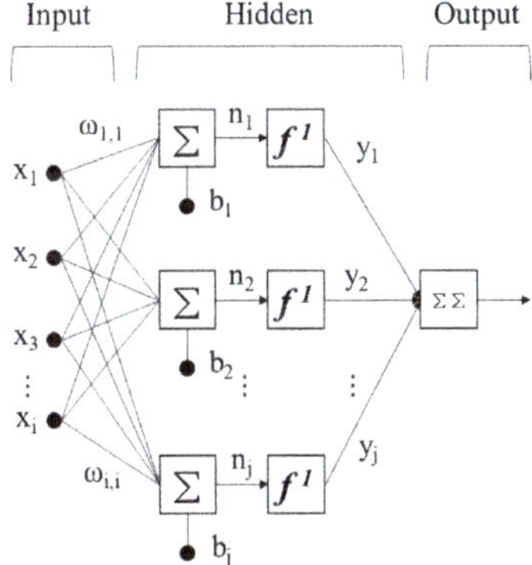

Figure 5. Procedure of an artificial neural network from an input sample.

3.3. Ensemble

In the ensemble method, numerous weak learners are combined to create a learner featuring strong predictive power. In this study, the regression tree method was used as the ensemble learner, and the boosting method was used for combining the models used. The boosting method performs slightly better than the bagging method and exhibits a higher predictive power. In addition, the boosting method can be used to compensate for shortcomings in learning the next regression tree by assigning weights to reduce errors between the prediction and output values of the first regression tree. The least-squares method was used for calculating the error. Least-squares boosting is a sequential ensemble method that sequentially builds a decision tree because it compensates for errors in the previous tree as shown in Figure 6; it is defined as Equation (7) [38,39]:

$$F_T(x) = \sum_{t=1}^{T} f_t(x) \tag{7}$$

where x is the input variable, f_t is the weak learner, and $F_T(x)$ is the strong learner.

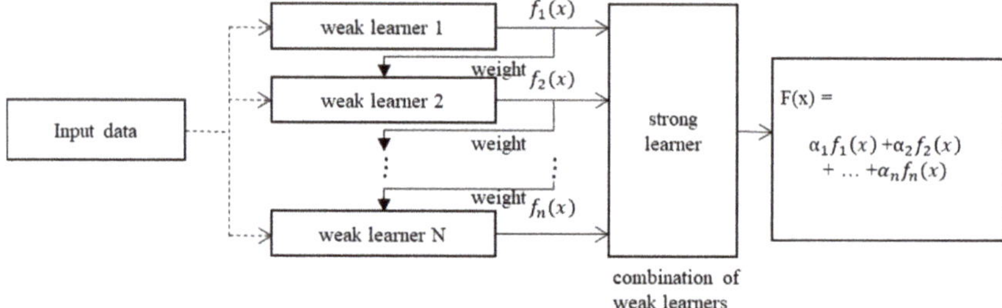

Figure 6. Procedure of an ensemble from an input sample.

Each weak learner produces a hypothesis $h(x_i)$ as the output for each training set sample. At each iteration step t, one weak learner is selected, and a coefficient α_t is assigned to minimize the sum of the training errors of the final t-step acceleration classifier, which is defined as follows:

$$E_t = \sum_i E[F_{t-1}(x_i) + \alpha_t h(x_i)] \tag{8}$$

where $F_{t-1}(x)$ is the accelerated learner generated up to the last training stage, $E(F)$ is the error function, and $f_t(x) = \alpha_t h(x)$ is the weak learner currently considered for addition to the final classifier. At each iteration step in the training process, a weight with the same value as the current error $E(F_{t-1}(x_i))$ for the training data set is assigned to the next data set, in order to compensate for the errors.

3.4. LR

LR of ML is a method of predicting values based on a linear function similar to the LR method [40]. However, because various optimization algorithms and cross-validation are employed, LR can achieve a higher predictive power than normal LR. In LR, the number of input data is indicated as n; each input datum is expressed as $x^{(i)} = \left[x_1^{(i)}, x_2^{(i)}\right]$ using index i, and the label (actual value) is expressed as $y^{(i)}$. During model training, the error is typically determined using a loss function, such as the expression defined in Equation (9):

$$loss\ function = l^{(i)}(\omega, b) = \frac{1}{2}\left(\hat{y}^{(i)} - y^{(i)}\right)^2 \tag{9}$$

where 1/2 (a constant value) is added to enhance the ease of computation by yielding a value of 1 when the 2D function is differentiated. The smaller the error value, the higher the similarity between the predicted value and the actual value; if the two values are equal, the error will be zero. LR is used for training a model by determining the model parameter $\omega^*, b^* = argminL(\omega, b)$, which minimizes the average loss of input data.

4. Results and Discussion

4.1. Experimental Variability of Static and Dynamic Tests

In this study, the coefficient of variation (COV), which is equal to the standard deviation (i.e., σ divided by the average values of the specimen set μ), was used to evaluate the experimental variability of the static and dynamic measurements of the concrete specimens. Outliers were confirmed using the modified Z-score method and excluded from the analyses [41]. The statistical parameters—COV and μ—of the obtained static and dynamic experimental results are presented in Table 2.

Table 2. Summary of statistical analysis.

Curing Age / Variable		fc (MPa) Mix 1	fc (MPa) Mix 2	Ec (MPa) Mix 1	Ec (MPa) Mix 2	Ed.Vp (MPa) Mix 1	Ed.Vp (MPa) Mix 2	Ed.Vs (MPa) Mix 1	Ed.Vs (MPa) Mix 2	Ed.LT (MPa) Mix 1	Ed.LT (MPa) Mix 2	Ed.TR (MPa) Mix 1	Ed.TR (MPa) Mix 2
Day 4	N	16	54	16	54	16	54	16	54	16	54	16	54
	μ	8.05	24.17	10,030	16,791	19,766	30,953	16,273	19,184	15,455	24,310	14,939	23,878
	COV	3.86	5.04	5.2	7.63	5.99	4.1	3.96	13.59	4.37	3.69	4.69	3.89
Day 7	N	9	55	9	55	9	55	9	55	9	55	9	55
	μ	10.05	29.49	11,526	18,190	23,367	32,912	15,664	22,036	17,262	26,129	16,876	25,252
	COV	2.41	3.63	2.99	6.57	2.17	3.73	13.17	8.33	2.56	2.31	2.51	2.62
Day 14	N	16	50	16	50	16	50	16	50	16	50	16	50
	μ	14.08	38.09	14,666	20,874	27,706	35,191	17,710	23,913	20,259	28,542	19,636	27,516
	COV	3.69	5.34	8.50	7.02	4.71	3.35	12.38	2.77	3.49	2.47	3.38	2.97
Day 28	N	32	50	32	50	32	50	32	50	32	50	32	50
	μ	19.19	43.94	16,699	23,085	30,830	36,870	21,019	25,329	23,705	30,375	22,769	29,670
	COV	4.35	4.78	7.54	7.12	3.73	2.52	12.58	1.87	3.65	2.13	3.89	2.68

The COV of *Ec* ranged from 2.99% to 8.50%, and the COV of *fc* ranged from 2.41% to 5.34%. When determining the COV of a slightly higher *Ec*, the opposite side of the test specimen was slightly skewed such that the deformation during compression was nonuniform. The error due to the test appeared to have a more significant effect on the *Ec* results than the *fc* results. In Figure 7 and Table 3, the theoretical (ACI 209R) and experimental values for *Ec* and fc are compared. As presented in Figure 7 and Table 3, concrete with FA and GBFS has characteristics such that the development of the initial elastic modulus and strength develop later than those of normal concrete, but the long-term elastic modulus and strength are increased. In addition, it has the advantage of being suitable for mass concrete because it generates less heat of hydration.

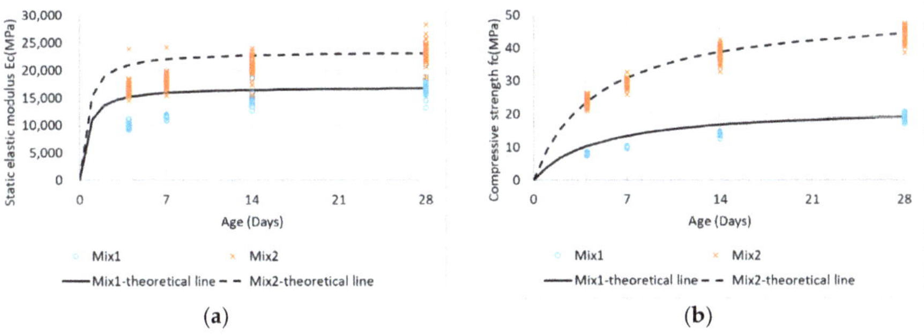

Figure 7. Comparison of experimental and theoretical values for Ec and fc. (**a**) Static elastic modulus; (**b**) Compressive strength.

Table 3. Comparison of experimental (mean) and theoretical values for Ec and fc.

Type	Static Elastic Modulus				Compressive Strength			
	Day 4	Day 7	Day 14	Day 28	Day 4	Day 7	Day 14	Day 28
Mix 1 (Theoretical)	15,182	15,913	16,429	16,699	10.27	13.37	16.73	19.14
Mix 1 (Experimental)	10,030	11,526	14,666	16,699	8.05	10.05	14.08	19.19
Mix 2 (Theoretical)	20,987	21,998	22,711	23,085	23.75	30.91	38.69	44.25
Mix 2 (Experimental)	16,791	18,190	20,874	23,085	24.17	29.49	38.09	43.94

The COV values of $Ed.Vp$ determined through the P-waves ranged from 2.17% to 5.99%, and the COV values of $Ed.LT$ and $Ed.TR$ obtained through the longitudinal and transverse resonance frequency tests were 2.13–4.37% and 2.51–4.69%. The COV values of $Ed.Vp$, $Ed.LT$, and $Ed.TR$ were reasonably consistent; however, the COV values of $Ed.Vs$ determined through the S-waves ranged from 1.87% to 13.59%, which were higher than those obtained using the other methods. This was because it was often difficult to accurately detect the initial arrival time of the S-waves in early-age concrete, owing to the interference between P-waves and S-waves. Furthermore, even when the S-wave transducer was used, low-amplitude P-waves always appeared with the S-waves in the time domain. This phenomenon was significant owing to the effects of moisture and voids as well as the relatively low compressive strength in the early-age concrete specimens. Therefore, these data sets were not included in the analyses. A thorough effort is required to determine the Vs of concrete with relatively low compressive strength.

4.2. Relationship between Static and Dynamic Elastic Moduli

Various equations have been proposed for predicting Ec using Ed. Lydon and Balendran proposed the following empirical Equation (10) [11]:

$$E_c = 0.83E_d(\mathrm{MPa}) \tag{10}$$

Another empirical Equation (11) was specified in British Standards BS 8110 Part 2 [12]:

$$E_c = 1.25E_d - 19,000(\mathrm{MPa}) \tag{11}$$

It should be noted that Equation (11) is not applicable to concrete containing a lightweight aggregate or more than 500 kg per m^3 of cement.

Popovics proposed a more general equation for lightweight concrete and normal-weight concrete considering the effect of density [13]:

$$E_c = \frac{446.09E_d^{1.4}}{\omega_c}(\mathrm{MPa}) \tag{12}$$

where ω_c is the density of hardened concrete (kg/m^3).

Figure 8 presents the relationship between Ed and Ec determined using the four methods; based on the figure, it is evident that Ed and Ec have a nonlinear relationship. In addition, the modulus of elasticity of the concrete, including FA and GBFS, is low in the early-age ranges and at low strength, so it may exhibit nonlinearity. Subsequently, Equations (10)–(12) and the general regression equations were substituted to compare the experimental results with the values obtained using the proposed equations. The equation recommended in BS 8110 Part 2 was located at the top of the data and fits the $Ed.Vp$ data; moreover, the Lydon–Balendran equation was located at the bottom of the data and fits the $Ed.Vs$ data. The Popovics equation and the general regression equation were centrally located and fit the $Ed.LT$ and $Ed.TR$ data. As shown in Figure 8, it can be confirmed that the four equations were suitable for predicting *Ed* values; however, as the existing equations were not limited to a specific method, the calculated and measured *Ec* values of the entire set of data were compared for each equation.

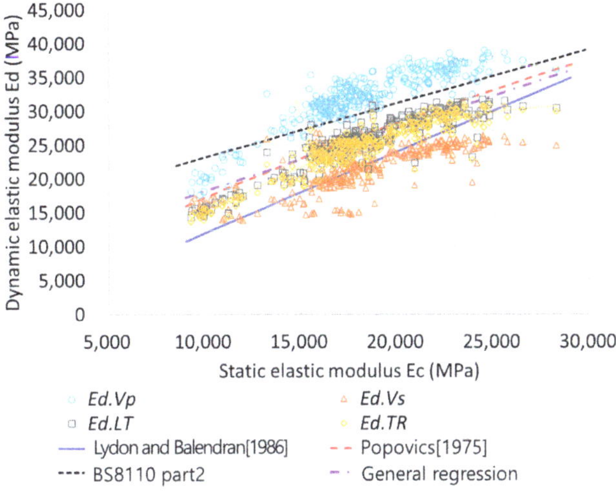

Figure 8. Relationship between *Ec* and *Ed*.

The MSE and MAPE were used to verify the errors between the predicted and measured values. The MSE and MAPE are defined by Equations (13) and (14), respectively:

$$MSE = \frac{1}{n} \sum_{i=1}^{n} (A_i - P_i)^2 \tag{13}$$

$$MAPE = \frac{1}{n} \sum_{i=1}^{n} \left| \frac{A_i - P_i}{A_i} \right| \tag{14}$$

where n is the number of experimental data, *Ai* is the measured value, and *Pi* is the predicted value. The MSE, MAPE, and R values for the calculated and measured *Ec* values of the entire data set are listed in Table 4. Among the four equations, the Popovics equation yielded the lowest MAPE value, and the error in its *Ec* prediction was small. For the general regression equation, the MSE value was lower than that for the Popovics equation; however, its MAPE was higher because *Ed* and *Ec* have a nonlinear relationship. Therefore, the results of the general regression equation at a low elastic modulus were higher than those predicted using the Popovics equation. Because of the nonlinear relationship between *Ed* and *Ec*, the general regression equation was considered to generate a larger error than the Popovics equation within the low elastic modulus range. In addition, as the MAPE and R values for *Ed.LT* and *Ed.TR* in the Popovics and general regression equations exhibited a small error range, the prediction error of *Ec* could be minimized when the Popovics equation was applied only to the resonance frequency test.

Table 4. Correlation analysis for predicted and measured *Ec* values obtained using the proposed equations and the general regression equation.

Equation Converting *Ed* into *Ec*	MSE	MAPE	R
General regression	2.09×10^7	20.38%	0.6306
Popovics [9]	2.20×10^7	19.22%	0.6019
Lydon and Balendran [7]	2.43×10^7	22.94%	0.6306
BS 8110 Part 2 [8]	5.22×10^7	37.46%	0.6306

As the existing equations were not applied to a specific *Ed*, the *Ec* value may have a large error value if the equations are incorrectly applied. General regression using individual *Ed* values was analyzed because the *Ed* measured using the four NDE methods possessed different characteristics. Figure 9 presents the overall regression results for each *Ed* value. The MSE, MAPE, and R values between the calculated *Ec* and measured *Ec* are listed in Table 5. The data for *Ed.Vp* and *Ed.Vs* exhibited a significant variance, resulting in MAPE values of 8.59% and 12.95%, respectively. However, the data for *Ed.LT* and *Ed.TR* exhibited a small variance; hence, their MAPE values were reduced to approximately 7%.

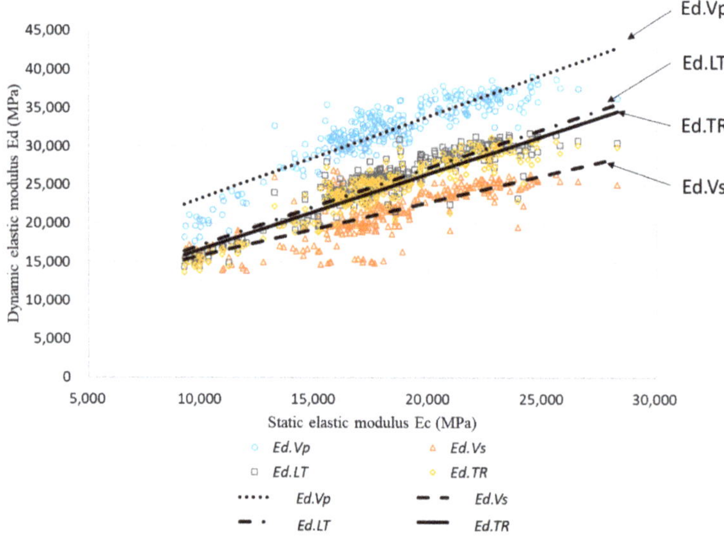

Figure 9. Comparison of individual *Ed* and *Ec* values.

Table 5. Analysis of correlation between predicted *Ec* and measured *Ec* using general regression for each variable.

Variable	MSE	MAPE	R
Ed.Vp	3.37×10^6	8.59%	0.8965
Ed.Vs	1.02×10^7	12.95%	0.7585
Ed.LT	2.51×10^6	7.06%	0.9198
Ed.TR	2.58×10^6	7.07%	0.9178

4.3. Comparison between Static and Prediction Elastic Moduli

Recently, studies have been conducted to improve prediction accuracy and data reliability, using ML methods. In this study, four ML methods were applied by combining the four predicted *Ed* value sets to develop a prediction model that is superior to existing models and the general regression model. SVM, ANN, ensemble, and LR were used as the ML methods, and *Ec* prediction and analyses were performed for 15 different sets of combinations. Table 6 lists the MSE and MAPE values of the predicted and measured *Ec* values for the 15 combinations using ML, where A denotes *Ed.Vp*, B denotes *Ed.Vs*, C denotes *Ed.LT*, and D denotes *Ed.TR*.

Table 6. *Ec* results for the 15 combinations predicted using four machine learning (ML) methods.

No.	Combination	SVM		ANN		Ensemble		Linear regression	
		MSE	MAPE	MSE	MAPE	MSE	MAPE	MSE	MAPE
1	A	2.75×10^6	7.16%	2.28×10^6	5.65%	1.90×10^6	5.47%	2.74×10^6	7.11%
2	B	6.25×10^6	10.76%	4.46×10^6	8.71%	3.46×10^6	8.06%	5.94×10^6	10.74%
3	C	2.14×10^6	5.55%	2.10×10^6	5.17%	1.42×10^6	4.50%	2.16×10^6	5.66%
4	D	2.20×10^6	5.63%	2.20×10^6	5.26%	1.50×10^6	4.52%	2.20×10^6	5.70%
5	A,B	2.68×10^6	6.83%	2.27×10^6	5.16%	1.70×10^6	5.32%	2.65×10^6	6.84%
6	A,C	2.19×10^6	5.52%	2.00×10^6	4.76%	1.31×10^6	4.26%	2.20×10^6	5.73%
7	A,D	2.22×10^6	5.67%	2.12×10^6	5.51%	1.41×10^6	4.35%	2.21×10^6	5.81%
8	B,C	2.10×10^6	5.46%	1.78×10^6	4.64%	1.26×10^6	4.18%	2.11×10^6	5.52%
9	B,D	2.13×10^6	5.50%	1.95×10^6	4.98%	1.30×10^6	4.26%	2.12×10^6	5.56%
10	C,D	2.14×10^6	5.54%	1.84×10^6	4.92%	1.32×10^6	4.31%	2.12×10^6	5.56%
11	A,B,C	2.14×10^6	5.46%	1.64×10^6	4.48%	1.22×10^6	4.12%	2.15×10^6	5.55%
12	A,B,D	2.15×10^6	5.53%	1.64×10^6	4.61%	1.26×10^6	4.11%	2.15×10^6	5.62%
13	A,C,D	2.19×10^6	5.50%	1.75×10^6	4.89%	1.27×10^6	4.25%	2.19×10^6	5.68%
14	B,C,D	2.08×10^6	5.45%	1.69×10^6	4.37%	1.24×10^6	4.20%	2.09×10^6	5.50%
15	A,B,C,D	2.13×10^6	5.45%	1.45×10^6	4.33%	1.19×10^6	4.03%	2.14×10^6	5.52%

For the results of the *Ec* prediction using the single *Ed*, C had the lowest MAPE value, and the prediction error increased in the order of D, A, and B. Depending on the applied ML method, the MAPE values differed by approximately 1–2%; however, the order remained the same. The MAPE values of individual variables predicted using the general regression of Table 4 ranged from 7 to 13%; however, the MAPE values of individual variables predicted using ML ranged from 4.5 to 8%. When the ML techniques were used, the reduction in the prediction errors was 2.5–5% greater than that when using general regression; hence, using ML in *Ec* prediction could result in a more accurate prediction. Based on the analysis of the combinations of two variables, the overall MAPE values were lower than those of single variables, and the BC combination was optimal for predicting *Ec*. The prediction accuracy improved because the longitudinal wave (C) and the transverse wave (B) complemented each other. For the combinations of three *Ed* variables, the ABC combination and the BCD combination exhibited the lowest MSE and MAPE values, respectively, based on the ML methods. The optimal combination varied depending on the applied ML method; however, the BC combination significantly contributed toward *Ec* prediction because BC was included in both optimal combinations. For the 15 combination sets, the accuracy of the ANN and ensemble methods improved as the number of variables increased, and the SVM and linear regression methods had similar prediction accuracies in two or more combination sets. Overall, the ensemble method yielded the highest accuracy for all combination sets, among the four methods.

Although an analysis of Table 5 confirmed that the BC combination contributed significantly toward the predicted *Ec* values, it was difficult to determine in detail the individual variables that contributed toward this prediction. Therefore, the relative importance (RI) using the weight of the ANN was calculated to determine the detailed contribution of each variable involved in the combination [42]:

$$RI_x = \sum_{y=1}^{m} \omega_{xy}\omega_{yz} \tag{15}$$

In Equation (15), RI is the relative importance of the variable, ω_{xy} is the weight between the input neuron and the hidden neuron, and ω_{yz} is the weight between the hidden neuron and output neuron.

Table 7 lists the RI values for 11 combination sets using the ANN method. The order of RI in the combination was C, D, A, and B, which was similar to the order of the MAPE for individual variables using ML and the order of general regression. Based on the RI analysis, the RI value of B had a minimal ratio; however, when B is combined with C, it should be used as a combination rather than a single variable because it improves the accuracy by 0.6%.

Table 7. Comparison of relative importance (RI) values in the ANN.

A/B	A/C	A/D	B/C	B/D	C/D
68%/32%	27%/73%	47%/53%	26%/74%	29%/71%	75%/25%

A/B/C	A/B/D	A/C/D	B/C/D	A/B/C/D	
17%/12%/71%	35%/9%/56%	29%/38%/33%	21%/41%/38%	11%/4/%/47%/38%	

Figure 10 displays the relationship between the measured Ec and the ratio of the predicted Ec to the measured Ec obtained using the four ML methods. The combinations were selected based on the ensemble method, which achieved the highest accuracy. Among all the ML methods, the ratios of the predicted values with two or more variables exhibited smaller variations than those of the single variables. The results generated via the ensemble method were closer to a ratio of 1, compared with those obtained using other methods, and the data distribution was concentrated. Additionally, in the existing linear prediction, it is difficult to consider the error according to age, but the error can be minimized with training by the ML method ANN and the ensemble method. Figure 11 compares the predicted values from the ensemble method with the experimental values for the concrete age. As confirmed from the figure, there is a good correlation, and some data at an early age are out of the line of equality because ultrasonic pulse velocity and resonance frequency methods are suitable for hardened concrete, so it is necessary to develop ML in consideration of concrete age in the future.

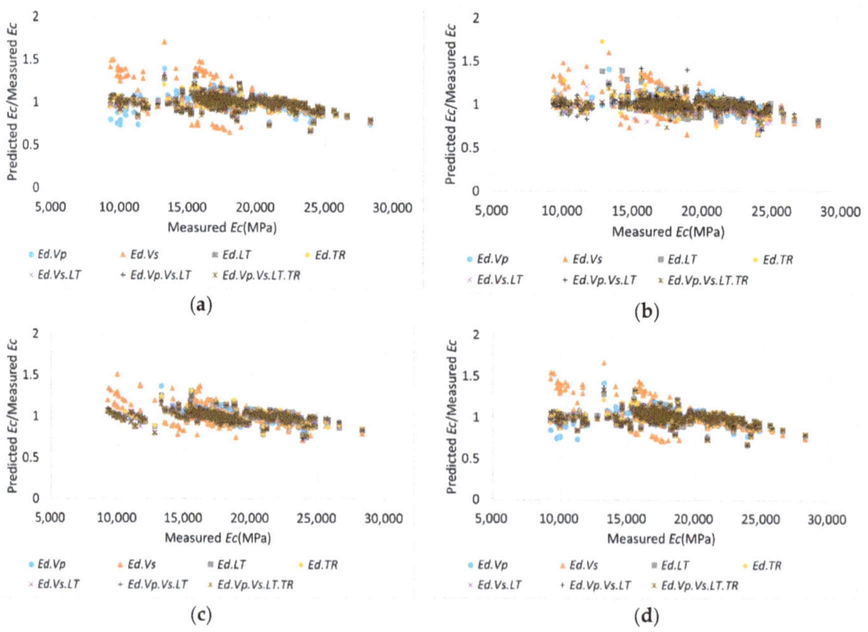

Figure 10. Relationship between measured Ec and the ratio of predicted Ec to measured Ec using ML methods: (**a**) SVM; (**b**) ANN; (**c**) Ensemble; (**d**) LR.

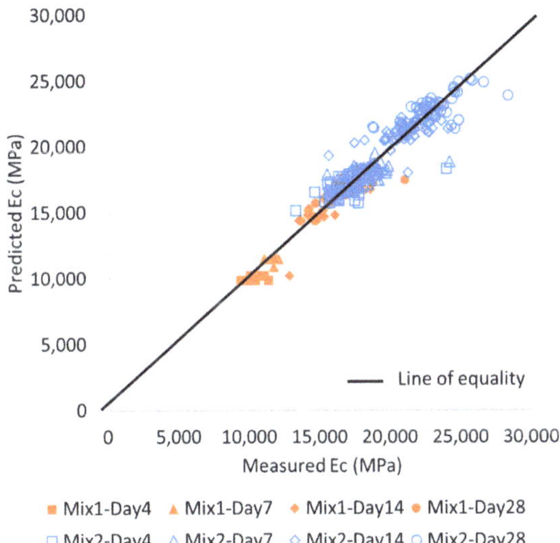

Figure 11. Relationship between measured *Ec* (experimental values) and predicted *Ec* (theoretical values) by the ensemble.

Table 8 lists the prediction results for each section of the seven ensemble combinations, where A denotes *Ed.Vp*, B denotes *Ed.Vs*, C denotes *Ed.LT*, and D denotes *Ed.TR*; the section was categorized into three ranges. The differences in error for each section were not large; however, the error was generally high at a low elastic modulus. In particular, the error for *Ec* predicted using B was high at a low modulus of elasticity, because the low elastic modulus data contained numerous early-age data. By contrast, based on the results presented in Table 6, the contribution of *Vs* was large because the accuracy improved when B was included in the combination. Additionally, it was useful for 0.4% when it was 15,000 MPa or more, except for the early-age elastic modulus. Therefore, it could improve the accuracy when performing resonance frequency tests and determining *Vs*, which are the most reliable and widely used methods.

Table 8. MAPE for each section of *Ec* predicted using the ensemble method.

Combination	0–15,000 (MAPE)	15,000–20,000 (MAPE)	20,000–30,000 (MAPE)	0–30,000 (MAPE)
A	5.43%	5.56%	5.33%	5.47%
B	17.20%	7.06%	5.81%	8.06%
C	4.78%	4.43%	4.51%	4.50%
D	4.73%	4.39%	4.64%	4.52%
B,C	5.25%	4.02%	4.00%	4.18%
A,B,C	5.31%	4.00%	3.79%	4.12%
A,B,C,D	4.67%	3.95%	3.89%	4.03%

4.4. Relationship between Static Elastic Modulus and Compressive Strength

The *Ec* of concrete is determined through static uniaxial compressive tests performed according to ASTM C469/C469M-14 requirements. However, *Ec* is typically determined using compression strength values based on the design codes. Therefore, various equations have been proposed to express the relationship between Ec and fc. Equation (16), which is proposed in the CEB-FIP and Eurocode 2 codes, is typically applied to normal and high-strength concretes [14]:

$$E_c = 22,000(f_c/10)^{1/3}(\text{MPa}) \tag{16}$$

Equation (17), which is recommended by the ACI 318 Committee for fc values less than 38 MPa, is expressed as follows [15]:

$$E_c = 0.43 f_c^{0.5} \omega_c^{1.5}(\text{MPa}) \tag{17}$$

where ω_c is the density ($1440 \le \omega_c \le 2560 \text{ kg/m}^3$).

Equation (18), suggested by the ACI 363 Committee, is used when the value of fc lies between 21 MPa and 83 MPa [16]:

$$E_c = \left(3320 f_c^{0.5} + 6900\right)(\omega_c/2300)^{1.5}(\text{MPa}) \tag{18}$$

Equation (19) was proposed by Noguchi et al.; correction factors are applied for ordinary and high-strength concretes based on extensive experimental data [43]:

$$E_c = k_1 k_2 33,500(f_c/60)^{1/3}(\omega_c/2400)^2(\text{MPa}) \tag{19}$$

where k_1 and k_2 are the correction factors for the aggregate and SCM, respectively.

Figure 12 depicts the relationship between Ec and fc in accordance with the ASTM C469 and ASTM C39/C39M-14a standards. Equations (16)–(19) were substituted for comparing the predicted values with the measured values. For Equation (19), $k_1 = k_2 = 0.95$ was employed because materials similar to crushed cobblestone and ground-granulated blast-furnace slag were used.

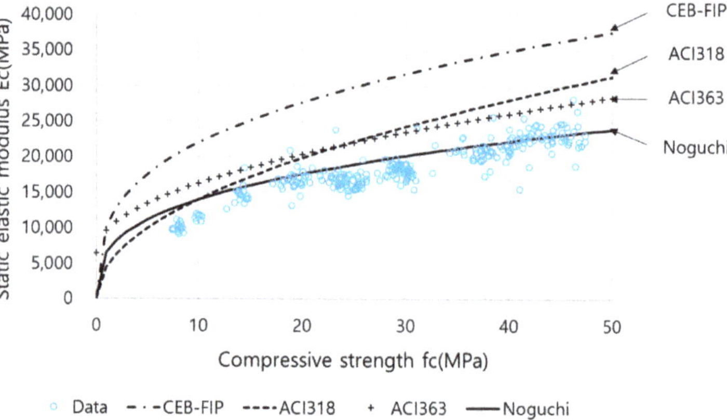

Figure 12. Comparison of the relationships between Ec and fc.

The MAPEs between the Ec calculated using Equations (16)–(18) and the measured Ec were 70%, 27%, and 35%, respectively. However, Equation (19) yielded an MAPE of 10%; hence, it was the most suitable equation for predicting Ec values. The equation proposed by Noguchi et al. was more suitable than the equations used for ordinary Portland cement, such as CEB-FIP, ACI 318, and ACI 369; this is because the aggregate and SCMs were similar and appropriate correction factors were applied. Therefore, Ec and fc exhibited a strong correlation, and the values were within a reasonable range.

4.5. Relationship between Dynamic Elastic Modulus and Compressive Strength

Figure 13 presents the relationship between fc and Ed. The relationship between Ed and fc was linear when the compressive strength exceeded 20 MPa; however, this relationship was nonlinear at lower strength values, especially due to the concrete mix with FA and GBFS. The fc value was calculated using the general regression equations for individual Ed values, and the results of the comparison with the measured fc values are listed in Table 9. As the fc values calculated using all the Ed values resulted in large MSE and MAPE values, it was unsuitable to use Ed values to compute fc. Based on the results of the analysis of individual Ed values, the data for A and B exhibited a large variance and MAPE values of 23.54% and 23.64%, respectively. However, the data for C and D exhibited a small variance and MAPE values of approximately 15% each. The MSE difference between A and B was more than twice the A value; however, their MAPE values were almost equal (Table 9). The reason for this difference was that A exhibited a significant variance for all ranges, whereas B exhibited a small variance, except for the early-age ranges.

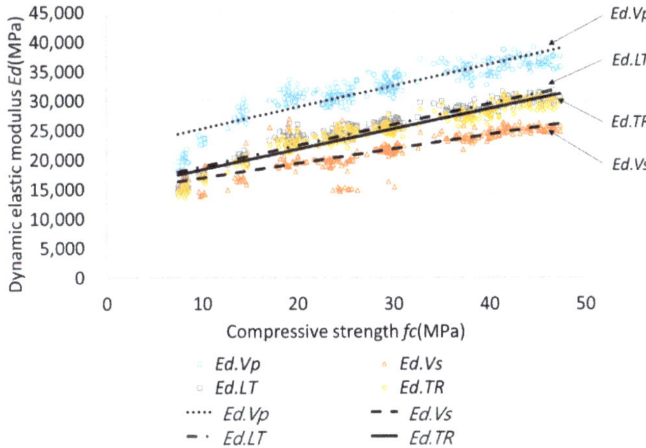

Figure 13. Analysis of general regression for individual Ed and fc.

Table 9. Comparison of MSE, MAPE, and R values for predicted fc and measured fc.

Variable	MSE	MAPE	R
All data	167.65	43.26%	0.6469
$Ed.Vp$	29.86	23.54%	0.8953
$Ed.Vs$	69.94	23.64%	0.7956
$Ed.LT$	12.98	15.24%	0.9501
$Ed.TR$	13.83	15.34%	0.9471

4.6. Comparison between Predicted and Measured Values of Compressive Strength

The MSE and MAPE values for the predicted and measured fc results for 15 combinations were obtained using four ML methods (Table 10 and Figure 14). For the fc prediction results using individual Ed values, the MAPE value of C was the lowest, and the prediction error increased in the order of D, A, and B. The MAPE values of the individual Ed predicted using the general regression equation presented in Table 9 ranged from 15.3% to 23.5%. However, the MAPE values of the individual Ed predicted using ML varied from 6.2% to 22.5%. The SVM and LR methods reduced the MAPE by 2%, as compared to the general regression. However, when the ANN and ensemble methods were used, the MAPE decreased by 8%. Therefore, the ANN and ensemble methods can be used to predict fc more accurately. The SVM and LR methods exhibited a significant variation in MAPE for two or more combinations, and the ANN and ensemble methods exhibited decreases in the MAPE as the number of variables increased. For the ANN and ensemble methods, the optimal combinations using two and three variables were BC, ABC, and BCD. Hence, the BC combination contributed significantly to the accurate prediction of *fc*, because all the optimal combinations included BC. This observation was similar to that for the predicted results of *Ec*.

Table 10. Predicted *fc* results for 15 combination sets using the four ML methods.

No.	Combination	SVM		ANN		Ensemble		Linear Regression	
		MSE	MAPE	MSE	MAPE	MSE	MAPE	MSE	MAPE
1	A	24.57	19.29%	17.92	12.22%	11.54	9.95%	24.23	18.48%
2	B	49.87	21.32%	29.50	15.47%	21.78	14.31%	44.77	22.59%
3	C	11.94	13.71%	10.75	7.65%	5.06	6.28%	11.84	13.12%
4	D	12.75	13.67%	10.99	8.23%	5.75	6.80%	12.48	13.18%
5	A,B	22.21	16.93%	13.50	8.86%	9.34	9.05%	21.86	16.76%
6	A,C	10.34	11.57%	6.26	6.88%	4.57	6.19%	10.26	11.57%
7	A,D	12.32	12.67%	7.61	7.32%	5.14	6.41%	11.94	12.32%
8	B,C	11.16	12.75%	5.23	5.57%	4.50	5.90%	11.13	12.52%
9	B,D	11.38	12.37%	5.18	6.61%	4.48	6.03%	11.34	12.53%
10	C,D	11.93	13.61%	5.47	5.94%	4.51	5.95%	11.81	13.10%
11	A,B,C	9.00	10.10%	4.94	6.20%	4.04	5.66%	8.92	10.48%
12	A,B,D	10.41	10.85%	5.10	6.23%	4.15	5.71%	10.07	10.88%
13	A,C,D	10.28	11.36%	5.06	5.91%	4.06	5.71%	10.26	11.54%
14	B,C,D	11.14	12.79%	4.77	5.84%	4.00	5.59%	11.02	12.47%
15	A,B,C,D	9.05	10.01%	4.07	5.19%	3.56	5.19%	8.80	10.36%

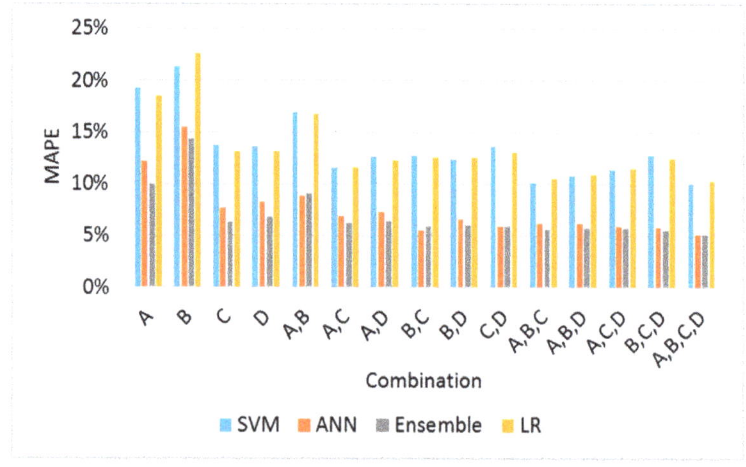

Figure 14. Comparison of MAPE for combinations using four ML methods.

Table 11 and Figure 15 list the RI values for two or more combinations using Equation (15) derived from the ANN method. The order of RI in the combination sets was C, D, A, and B, which was similar to the trend of the MAPE order for individual variables using ML, the order for the general regression, and the RI order of the *Ec* predicted values.

Table 11. Comparison of RI values using the ANN method.

A/B	A/C	A/D	B/C	B/D	C/D
56%/44%	16%/84%	21%/79%	15%/85%	8%/92%	87%/13%

A/B/C	A/B/D	A/C/D	B/C/D	A/B/C/D	
34%/15%/50%	40%/12%/48%	24%/47%/29%	6%/73%/20%	15%/9%/%/44%/32%	

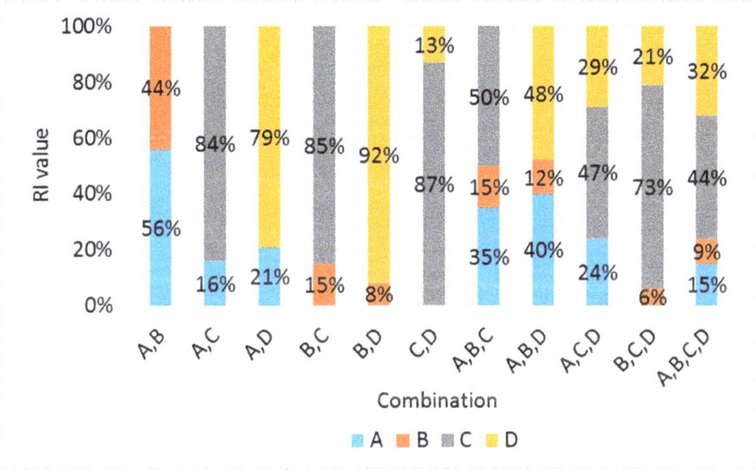

Figure 15. Comparison of RI values using the ANN method.

Figure 16 depicts the relationship between the measured *fc* and the ratio of the predicted *fc* to the measured *fc* using the four ML methods. The combination sets were selected based on the ensemble method, which achieved the highest accuracy. As B was affected by moisture and voids during the early ages, the MAPE of B was high for the early-age data range; however, at higher strength values (35–50 MPa), B yielded a lower MAPE than A. The results for the ANN and ensemble methods were close to a ratio of 1 for all the ranges, and the variance in the data was small and dense; this resulted in an excellent prediction accuracy. However, the results obtained using the SVM and LR methods had a large dispersion at strengths of 20 MPa and lower. This was because the nonlinearity for the range below 20 MPa was high, as shown in Figures 12 and 13. Furthermore, as the basic functions of SVM and LR are linear, the function might be generalized for the high-strength range, which exhibits linearity and contains numerous data. Therefore, errors in the prediction for low-strength ranges may be high. As a result, the existing linear predictions are difficult for considering the slow development of strength at a low age, but this can be overcome through the use of ML such as ANNs and ensembles and the combination of various *Eds*. On the other hand, the values predicted by ensemble are presented with the experimental results in Figure 17 for the comparison analysis. Similarly, to in Figure 12, it can be seen that the error in early-aged concrete is relatively large; thus, it can be recognized that this approach is more effective for hardened concrete after 14 days of age.

The MAPE values for various *fc* ranges of the seven combination sets predicted using the ensemble method are presented in Table 12 and Figure 18. The MAPE values for the ranges of 0–15 and 35–50 MPa were similar; however, those for the range of 15–35 MPa increased. For the 15–35 MPa strength range, because the later-age data of Mix 1 and the early-age data of Mix 2 were mixed, the prediction error may be higher, owing to the differences in the mix proportions and curing ages of the data set.

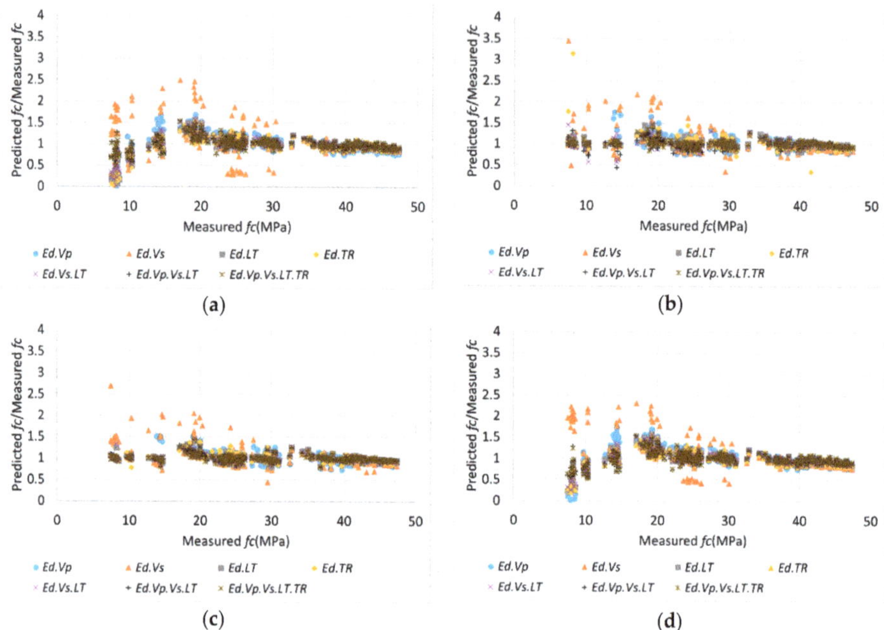

Figure 16. Relationship between the measured *Ec* and the ratio of predicted *fc* to measured *fc* using the four ML methods. (**a**) SVM; (**b**) ANN; (**c**) Ensemble; (**d**) LR.

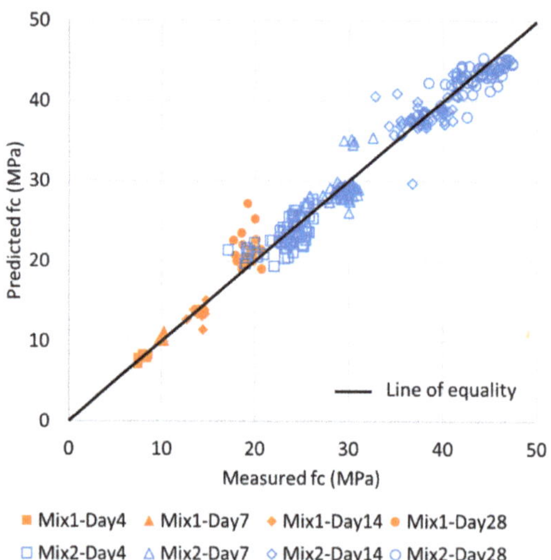

Figure 17. Relationship between measured *fc* (experimental values) and predicted *fc* (theoretical values) by the ensemble method.

Table 12. MAPE values for each section of *fc* predicted using the ensemble method.

Combination	MAPE (0–15 MPa)	MAPE (15–35 MPa)	MAPE (35–50 MPa)	MAPE (0–50 MPa)
A	9.80%	11.95%	7.04%	9.95%
B	36.62%	12.95%	6.89%	14.31%
C	3.44%	8.59%	4.06%	6.28%
D	4.98%	8.78%	4.64%	6.80%
B,C	4.24%	7.66%	3.99%	5.90%
A,B,C	3.98%	7.50%	3.65%	5.66%
A,B,C,D	3.96%	6.62%	3.57%	5.19%

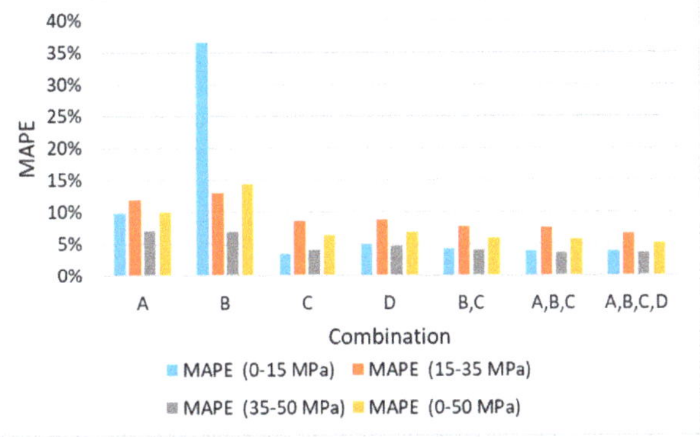

Figure 18. MAPE values for each section of *fc* predicted using the ensemble method.

In this study, the correlations between *Ed* and *Ec*, and *Ed* and *fc*, were analyzed using four types of *Ed*. Additionally, four ML methods—SVM, ANN, ensemble, and LR—were applied to analyze the accuracy and RI of each variable combination. The relationship between *Ed-Ec* and *Ed-fc* was nonlinear in the low-elastic-modulus and low-strength ranges. The *Ec* and *fc* values predicted using the ML methods were more accurate than those obtained using general regression. The ML methods were advantageous over general regression because they allow correction for nonlinearities in the ranges of the early-age strength and elastic modulus. The ensemble method achieved the best performance in predicting *Ec* and *fc*, among the four ML methods. C produced the highest accuracy when used for a single variable; however, it was difficult to predict *Ed* and *fc* as a single variable for B, because the variance in the data was significant, owing to the presence of moisture and voids at the early age. However, B could yield the best accuracy in predicting *Ec* and *fc* when combined with C.

5. Conclusions

Currently, it can be difficult or impossible to obtain sufficient specimens by destructive tests according to the ASTM C469 and C39 for the measurement of the Ec and fc required for the design, construction, and maintenance of a structure. Thus, it is necessary to predict them by the non-destructive test (NDT), and one of the most reliable NDT methods is to estimate Ec and fc from the dynamic elastic modulus by ultrasonic pulse velocity or resonance frequency tests. The Ec and fc have usually been predicted from Ed by classical equations or the regression analysis, but these cannot not give an accurate estimation due to the intrinsic effects of each test method and nonlinearity between Ec or fc and Ed. In this regard, this study presented a possibility to improve the estimation of Ec and fc from Ed by a machine learning (ML) algorism based on the results of four types of test on the 282 specimens, comparing with the classical equations such as the Lydon and Balendran equation, Popovics equation, and BS 8110, as well as the regression method. The optimum relationship between Ec or fc and Ed was derived by using the characteristics of Ed from ultrasonic velocity and resonance frequency test methods associated with four ML algorisms. The accuracy of predicting Ec and fc was compared with the general regression and ML based on four reliable Ed measurements, for the complete range of experimental data. The following conclusions have been drawn:

- Among the four Eds, A is the largest, C and D are located in the middle, and B has the smallest value. The correlation with Ec and fc was identified as C > D > A > B.
- The Lydon and Balendran equation yields similar values to the B, and the Popovics equation provides results that are similar to the C and D. In addition, BS 8110 Part 2 yields similar values to the A.
- For the Ec and fc prediction results, the application of ML improved the accuracy by 2.5–5% and 7–9%, respectively, as compared with general regression.
- When Ec was predicted using the four Ed values, the order of accuracy of a single variable was C > D > A > B. The combination sets (B, C), (A, B, C), and (A, B, C, D) yielded the highest accuracy. Moreover, only the combination of B and C attained an accuracy level of up to 4.18%. Although B contained large-variance data for the early-age concrete, it complemented the most widely used and reliable C method.
- When fc was predicted using the four Ed values, the order of accuracy of a single variable was C > D > A > B, which is similar to the order for the Ec values. The combinations of (B, C) and (B, D) were optimal, yielding accuracies of up to 5.90%. Although the variance in the B data was large for the prediction of fc, B was considered to be useful in supplementing the C and D prediction methods and predicting the resonance frequency test results.
- For predicting the Ec and fc values, the RI values of B/C were 26%/74% and 15%/85%, respectively. Although B did not contribute significantly, it influenced the improvement in accuracy by 0.6%. When compared with respect to the sections, it was useful for 0.4% when it was 15,000 MPa or more, except for early-age strength.
- With the existing linear prediction, it is difficult to overcome the differences in the characteristics of the eigen data of A, B, C, and D and use them as representative values. However, if ML is used, the representative value of Ed can be calculated through the experimental values of the ultrasonic pulse velocity and resonant frequency test, and it can be very advantageous for predicting Ec and fc. Throughout this study, C was the best of the four dynamic elastic moduli commonly used, and gave the highest contribution in various combinations. B contributed to predictive accuracy through a combination. The possibility of the use of two or more variables has been verified by ML; if various data on age, temperature, etc. are secured, it will be possible to use a correction for these and to generate more accurate mechanisms and predictive values.
- As shown in Figures 11 and 17, since this study is based on the test methods for the dynamic elastic modulus for hardened concrete, the estimation accuracy tends to be slightly lower for concrete with 4- or 7-day ages. Thus, for a further study, more consideration of other factors such as the type of binder, binder/cement ratio, age, curing temperature, etc. will be beneficial, as including more variables in the ML will help to improve the estimation of Ec and fc.

Author Contributions: J.Y.P. and S.-H.S. conceived and designed the experiments; Y.G.Y. and T.K.O. performed the experiments and analyzed the data; J.Y.P. and S.-H.S. contributed device/analysis tools; Y.G.Y. and T.K.O. wrote the paper. All authors have read and agreed to the published version of the manuscript.

Funding: This work was supported by Incheon National University Research Grant in 2019.

Conflicts of Interest: The authors declare no conflict of interest.

References

1. Mehta, P.K.; Monteiro, P.J.M. *Concrete-Microstructure, Properties, and Materials*, 4th ed.; McGraw-Hill Education: New York, NY, USA, 2013.
2. ASTMC666/C666M-15. *Standard Test Method for Resistance of Concrete to Rapid Freezing and Thawing*; ASTM International: West Conshohoken, PA, USA, 2015.
3. Popovics, J.S. A study of static and dynamic modulus of elasticity of concrete. In *ACI-CRC Final Report*; American Concrete Institute: Farmington Hills, MI, USA, 2008.
4. ASTM C597M-16. *Standard Test Method for Pulse Velocity through Concrete*; ASTM International: West Conshohoken, PA, USA, 2016.
5. ASTM C215-14. *Standard Test Method for Fundamental Transverse, Longitudinal, and Torsional Resonant Frequencies of Concrete Specimens*; ASTM International: West Conshohoken, PA, USA, 2016.
6. Subramaniam, K.V.; Popovics, J.S.; Shah, S.P. Determining elastic properties of concrete using vibrational resonance frequencies of standard test cylinders. *Cem. Concr. Aggreg. (ASTM)* **2000**, *22*, 81–89.
7. ASTM C469/C469M-14. *Standard Test Method for Static Modulus of Elasticity and Poisson's Ratio of Concrete in Compression*; ASTM International: West Conshohoken, PA, USA, 2014.
8. Lee, B.J.; Kee, S.H.; Oh, T.; Kim, Y.Y. Effect of Cylinder Size on the Modulus of Elasticity and Compressive Strength of Concrete from Static and Dynamic Tests. *Adv. Mater. Sci. Eng.* **2015**, *2015*, 580638. [CrossRef]
9. Jurowski, K.; Grzeszczyk, S. Influence of Selected Factors on the Relationship between the Dynamic Elastic Modulus and Compressive Strength of Concrete. *Materials* **2018**, *11*, 477. [CrossRef]
10. Lee, B.J.; Kee, S.H.; Oh, T.K.; Kim, Y.Y. Evaluating the Dynamic Elastic Modulus of Concrete Using Shear-Wave Velocity Measurements. *Adv. Mater. Sci. Eng.* **2017**, *2017*, 1651753. [CrossRef]
11. Lydon, F.D.; Balendran, R.V. Some observations on elastic properties of plain concrete. *Cem. Concr. Res.* **1986**, *16*, 314–324. [CrossRef]
12. British Standard Institute. *Structural Use of Concrete—Part 2: Code of Practice for Special Circumstance*; BS 8110-2:1995; BSI: London, UK, 1985.
13. Popovics, S. Verification of relationships between mechanical properties of concrete-like materials. *Mater. Struct.* **1975**, *8*, 183–191. [CrossRef]
14. EN 1992-1-2. *Eurocode 2: Design of Concrete Structures—Part 1-2: General Rules-Structure Fire Design*; European Committee for Standardization: Brussels, Belgium, 2004.
15. ACI Committee 318. *Building Code Requirements for Structural Concrete (ACI 318-11) and Commentary*; American Concrete Institute: Farmington Hills, MI, USA, 2014; p. 503.
16. ACI committee 363. State-of-the-art report on high-strength concrete. *ACI J. Proc.* **1984**, *81*, 364–411.
17. Chavhan, P.P.; Vyawahare, M.R. Correlation of Compressive strength and Dynamic modulus of Elasticity for high strength SCC Mixes. *IJETR* **2015**, *3*, 42–46.
18. Malhotra, V.M.; Berwanger, C. *Correlations of Age and Strength with Values Obtained by Dynamic Tests on Concrete, Mines Branch Investigation Rep. IR 70-40*; Department of Energy, Mines and Resources: Ottawa, ON, Canada, 1970.
19. Shkolnik, I.E. Effect of nonlinear response of concrete on its elastic modulus and strength. *Cem. Concr. Compos.* **2005**, *27*, 747–757. [CrossRef]
20. Chopra, P.; Sharma, R.K.; Kumar, M. Prediction of compressive strength of concrete using artificial neural network and genetic programming. *Adv. Mater. Sci.* **2016**, *2016*, 7648467. [CrossRef]
21. Han, I.J.; Yuan, T.F.; Lee, J.Y.; Yoon, Y.S.; Kim, J.H. Learned Prediction of Compressive Strength of GGBFS Concrete Using Hybrid Artificial Neural Network Models. *Materials* **2019**, *12*, 3708. [CrossRef] [PubMed]
22. Shih, Y.F.; Wang, Y.R.; Lin, K.L.; Chen, C.W. Improving non-destructive concrete strength tests using support vector machines. *Materials* **2015**, *8*, 7169–7178. [CrossRef] [PubMed]
23. Park, J.Y.; Yoon, Y.G.; Oh, T.K. Prediction of Concrete Strength with P-, S-, R-Wave Velocities by Support Vector Machine (SVM) and Artificial Neural Network (ANN). *Appl. Sci.* **2020**, *10*, 84. [CrossRef]
24. Chithra, S.; Kumar, S.R.R.S.; Chinnaraju, K.; Ashmita, F.A. A comparative study on the compressive strength prediction models for high performance concrete containing nano silica and copper slag using regression analysis and artificial neural networks. *Constr. Build. Mater.* **2016**, *114*, 528–535. [CrossRef]

25. Erdal, H.I.; Karakurt, O.; Namli, E. High performance concrete compressive strength forecasting using ensemble models based on discrete wavelet transform. *Eng. Appl. Artif. Intell.* **2013**, *26*, 1246–1254. [CrossRef]

26. Yuvaraj, P.; Ramachandra Murthy, A.; Iyer, N.R.; Sekar, S.K.; Samui, P. Support vector regression based models to predict fracture characteristics of high strength and ultra-high strength concrete beams. *Eng. Fract. Mech.* **2013**, *98*, 29–43. [CrossRef]

27. Yan, K.; Shi, C. Prediction of elastic modulus of normal and high strength concrete by support vector machine. *Constr. Build. Mater.* **2010**, *24*, 1479–1485. [CrossRef]

28. Cihan, M.T. Prediction of Concrete Compressive Strength and Slump by Machine Learning Methods. *Adv. Civil Eng.* **2019**, *2019*, 1–11. [CrossRef]

29. Young, B.A.; Hall, A.; Pilon, L.; Gupta, P.; Sant, G. Can the compressive strength of concrete be estimated from knowledge of the mixture proportions?: New insights from statistical analysis and machine learning methods. *Cem. Concr. Res.* **2019**, *115*, 379–388. [CrossRef]

30. ASTM C31/C31M-12. *Standard Practice for Making and Curing Concrete Test Specimen in the Field*; ASTM International: West Conshohoken, PA, USA, 2012.

31. ASTM C39/C39M-14a. *Standard Test Method for Compressive Strength of Cylindrical Concrete Specimens*; ASTM International: West Conshohoken, PA, USA, 2014.

32. Popovics, J.S.; Song, W.; Achenbach, J.D.; Lee, J.H.; Andre, R.F. One-sided stress wave velocity measurement in concrete. *J. Eng. Mech.* **1998**, *124*, 1346–1353. [CrossRef]

33. Sansalone, M.; Lin, J.M.; Streett, W.B. A procedure for determining P-wave speed in concrete for use in impact-echo testing using a P-wave speed measurement technique. *Mater. J.* **1997**, *94*, 531–539.

34. Vapnik, V.; Cortes, C. *Support-Vector Networks*; Kluwer Academic Publisher: Boston, MA, USA, 1995; Volume 20, pp. 273–297.

35. Rosenblatt, F. *Principles of Neuro Dynamics: Perceptrons and the Theory of Brain Mechanisms*; Spartan Books: Washington, DC, USA, 1962; pp. 29–51.

36. Rumerlhar, D.E. Learning representation by back-propagating errors. *Nature* **1986**, *323*, 533–536.

37. Daliakopoulos, I.N.; Coulibaly, P.; Tsanis, I.K. Groundwater level forecasting using artificial neural networks. *J. Hydrol.* **2005**, *309*, 229–240. [CrossRef]

38. Benkaddour, M.K.; Bounoua, A. Feature extraction and classification using deep convolutional neural networks, PCA and SVC for face recognition. *Traitement du Signal* **2017**, *34*, 77–91. [CrossRef]

39. Hansen, L.K.; Salamon, P. Neural network ensembles. *IEEE TPAMI* **1990**, *12*, 993–1001. [CrossRef]

40. Yan, X. Linear Regression Analysis: Theory and Computing. *Int. Stat. Rev.* **2010**, *78*, 134–159.

41. Barnett, V.; Lewis, T. *Outliers in Statistical Data*, 3rd ed.; JohnWiley & Sons: New York, NY, USA, 1994.

42. Ibrahim, O.M. A comparison of methods for assessing the relative importance of input variables in artificial neural networks. *Res. J. Appl. Sci.* **2013**, *9*, 5692–5700.

43. Noguchi, T.; Tomosawa, F.; Nemati, K.M.; Chiaia, B.M.; Fantilli, A.P. A practical equation for elastic modulus of concrete. *ACI Mater. J.* **2009**, *106*, 690–696.

Article

Non-Deterministic Assessment of Surface Roughness as Bond Strength Parameters between Concrete Layers Cast at Different Ages

Janusz Kozubal [1], Roman Wróblewski [1,*], Zbigniew Muszyński [2], Marek Wyjadłowski [1] and Joanna Stróżyk [1]

[1] Faculty of Civil Engineering, Wroclaw University of Science and Technology, 50-370 Wrocław, Poland; janusz.kozubal@pwr.edu.pl (J.K.); marek.wyjadlowski@pwr.edu.pl (M.W.); joanna.strozyk@pwr.edu.pl (J.S.)

[2] Faculty of Geoengineering, Mining and Geology, Wroclaw University of Science and Technology, 50-370 Wrocław, Poland; zbigniew.muszynski@pwr.edu.pl

[*] Correspondence: roman.wroblewski@pwr.edu.pl

Received: 15 May 2020; Accepted: 30 May 2020; Published: 3 June 2020

Abstract: The importance of surface roughness and its non-destructive examination has often been emphasised in structural rehabilitation. The presented innovative procedure enables the estimation of concrete-to-concrete strength based on a combination of low-cost, area-limited tests and geostatistical methods. The new method removes the shortcomings of the existing one, i.e., it is neither qualitative nor subjective. The interface strength factors, cohesion and friction, can be estimated accurately based on the collected data on a surface texture. The data acquisition needed to create digital models of the concrete surface can be performed by terrestrial close-range photogrammetry or other methods. In the presented procedure, limitations to the availability of concrete surfaces are overcome by the generation of subsequential Gaussian random fields (via height profiles) based on the semivariograms fitted to the digital surface models. In this way, the randomness of the surface texture is reproduced. The selected roughness parameters, such as mean valley depth and, most importantly, the geostatistical semivariogram parameter sill, were transformed into contact bond strength parameters based on the available strength tests. The proposed procedure estimates the interface bond strength based on the geostatistical methods applied to the numerical surface model and can be used in practical and theoretical applications.

Keywords: concrete; roughness; texture; close-range photogrammetry; bond strength; random field generation; semivariograms

---

1. Introduction

Concrete layers of considerably different ages are frequently used in new and existing (strengthening) structures. The predominant shear action appears in structural elements made from a combination of precast and cast in place concrete like, for example, floor systems, slab renovation or bridge decks. The same action and an even greater age difference appear in reinforced or repaired structures by increasing their thickness. The joints made during concreting are another example but neither the age of concrete layers at the interface is different nor is the shear dominant. Instead, tension is usually a major action.

There exist several reasons why research on the adhesive action of concrete is important:

- It requires standardisation due to its importance in practice.
- Aggressive contaminants may penetrate the delaminated interface and cause unfavourable conditions for durability.

- Element stiffness may be reduced and larger deformations appear, reducing serviceability.
- Load transfer between layers changes with the bond failure.
- Shrinkage is a factor that influences pure cohesive action and element deformations.

Concrete structures consisting of layers cast at various ages require a different approach than other concrete structures in design and construction. This approach includes contact surface preparation, construction stages, curing, load resistance, etc., and the most important factor is the bond strength of two concrete layers because it is critical for the durability and load-bearing capacity. Besides its material strength, the way the surface of the existing concrete is prepared has a significant influence on its element capacity and overall durability due to the random distribution of material and interface parameters over the contact surface.

Since the bond is a relatively low action, it is also sensitive to several factors, like roughness and tortuosity in the contact area, concrete composition, density, strength, porosity and micro-damages of the interfacial zone. Among those factors, surface roughness and its condition determine concrete-to-concrete interface behaviour. Being essential, substrate surface condition assessment is the subject of extensive research focused on friction and cohesive actions. Frictional action strongly depends on the interlock of aggregate particles [1] and thus is mostly activated after delamination. At the same time, existing reinforcement is activated and dowel action with compression that is normal to the interface appears. The cohesive action, originated by chemical and other connections, bonds the surfaces before their delamination and enables element continuity. With this in mind, the methods used to obtain internal forces in ultimate and serviceability limit states should be different. Linear elastic models are better suited to serviceability limit states with continuous stress distribution at the interface. However, this assumption is no longer valid in the ultimate limit state when delamination occurs.

According to the shear friction theory proposed in [2] and successfully used in design codes, the transfer mechanism at the concrete-to-concrete interface is ensured by friction and is a subject of simultaneous shear and compression forces as in [3], where the load transfer mechanism of shear forces consists of:

- Cohesion, due to mechanical interlocking between particles;
- Friction, due to the existence of compression stresses at the interface and due to the relative displacement between concrete parts;
- Dowel action, due to the deformation of the reinforcement bars crossing the interface.

Thus, the ultimate limit state shear stress at the interface v_{Rdi} is represented by the following equation:

$$v_{Rdi} = c \cdot f_{ctd} + \mu \cdot \sigma_n + \rho \cdot f_{yd} \cdot (\mu \cdot \sin \alpha + \cos \alpha) \tag{1}$$

where f_{ctd} is the design tensile strength of concrete (and is not specified if in precast or cast in place), σ_n is the external normal stress acting on the interface, ρ is the reinforcement ratio, f_{yd} is the yield strength of the reinforcement and α is the angle between the shear reinforcement and the shear plane. The other symbols, c—cohesion factor and μ—friction factor, are the factors that depend on the roughness of the interface and are classified according to three categories based on the surface finishing. The surface categories are, to some extent, arbitrary and can lead to an inaccurate assessment of cohesion and friction (usually too conservative). Moreover, it is not clear if the cohesion and friction are average or design values, so they are not characterized statistically. Taken together, the evaluation of the factors is only qualitative.

Model Code 2010 [4] improves surface qualification based on the average roughness R_a where a smooth surface is defined with $R_a < 1.5$ mm, rough with $R_a \geq 1.5$ mm and very rough/indented with $R_a \geq 3$ mm. A very smooth surface is not defined. However, as noted in [5], the average roughness R_a is not sensitive enough. It does not provide information on the local variability and different profiles can present the same R_a.

One of the first quantitative approaches is presented in [6,7]. In [6], the contact stress depends on the matrix proportions of maximum particle size and volumetric percentage of aggregate. Moreover, the contact areas are obtained from the probability density functions of the particles' (aggregate) occurrence in the surface plane, thus the method has a probabilistic approach. The model presented in [7] is based on concrete strength, crack width and maximum aggregate size. Therefore, both methods require the assessment of surface parameters.

Since shear friction theory is well established, recent research [8,9] has been focusing on the evaluation of the design coefficients to overcome qualitative assessment. These limitations have been overcome by the calculation of the c and μ factors based on surface roughness parameters such as the mean valley depth R_{vm}. Moreover, the design values of both the c and μ factors have been provided in [10]. The design parameters have also been derived for a trilinear model which is proposed in [11]. The model is based on a parametric analysis of the available tests. The design coefficients for monotonic and cyclic loading have been calibrated to obtain the target values of the reliability index. The proposed model is also consistent with [12].

Further improvement in surface evaluation methods would include joint micro- (roughness from 1 µm to 0.5 mm) and macro-texture (waviness from 0.5 mm to 50 mm) analysis. Waviness and roughness can be related to different frequencies and wavelengths. The accuracy of the close-range photogrammetry used in this paper is on the border of these intervals, with a predominance of waviness, but other methods are also available [13,14]. Contact, X-ray and optical methods like profilometry, photogrammetry and scanning can be used to measure the surface parameters through point, linear or surface measurements. In this paper, the 3D surface parameters are used to evaluate the shear strength parameters at the interface of the concrete layers.

Surface scanning in laboratory conditions, as presented in [8], can be a preliminary step before other tests; for example, pull-off, as presented in [15–17]. Destructive tests, such as tension, shear and a combination of shear and compression, are also possible [10]. Part of the extensive research described in [5] was performed with a profilometer and roughness parameters were obtained from the 2D analysis. The samples were then subjected to destructive tests to determine the shear strength. Maximum valley depth R_v was found to be the most adequate surface parameter and presented an almost linear correlation with the concrete bond strength in shear. However, in order to avoid problems with possible strong surface irregularities, mean valley depth R_{vm} was preferred.

One of the advantages of non-destructive methods is the reliable determination of waviness and roughness parameters from relatively small sample sizes. These measurement techniques are accurate and time-saving, but they may also be time-consuming when they are numerically post-processed. However, the possible evaluation of strength parameters based on the concrete surface examination is very promising. Moreover, it is possible to carry out the tests in situ or in laboratory conditions, and the in situ examination can be used as a quality check. This problem is also essential in structural rehabilitation and strengthening with composite materials [18].

The proposed roughness assessment was sought independently of the chosen measurement method. Attempts were made to eliminate subjective elements from the measurement and compose a flexible set of procedures. The whole method is compatible with surface monitoring tools used in civil engineering, such as 3D laser scanners and confocal microscopes, and resistant to possible human error.

Although there are extensive test data available, theoretical results do not follow the same degree of accuracy. Therefore, we established a method combining geostatistical image analysis and aggregate composition with close-range photogrammetry (CRP) and leading to the non-destructive method of bond strength assessment. Therefore, this paper aims to improve existing non-destructive methods of concrete surface examination with:

- More detailed investigation—both roughness and waviness are considered, so profile height and inclination variability are identified.
- Application of random surfaces, i.e., storing information on surface texture in a random image of a real surface.
- Introduction of statistical objective parameters instead of traditional roughness parameters in order to reproduce the randomness of any surface texture.

In the presented method, the experimental research and sampling may be limited due to the application of geostatistics which predicts data values at unsampled locations. Furthermore, the geostatistical image of a real surface is used to obtain a required texture/roughness parameter. Then, interface strength parameters are derived based on the available test data and the texture/roughness parameter.

The geostatistical methods [19–21] involved in predicting surface images are gaining importance in existing structures when examined surfaces cannot be easily accessed or used as a quality check. Moreover, apart from practical applications, it is possible to use the generated images in reliability problems, so we present the developed source code in the R language [22] to allow application of the presented algorithms.

2. Materials and Methods

2.1. Concrete Mix Composition

Concrete samples with the composition and density presented in Table 1 were used in the research. The cement used for the concrete mix was the CEM I 42.5R (Górażdże Cement S.A., Górażdże, Poland) with a specific surface area of 375.2 m^2/kg and the following compressive strength: 2-days 26.6 MPa and 28-days 56.0 MPa. Moreover, the cement consisted of 2.83% of sulfur trioxide, 0.044% of chloride ion, 0.61% of total alkali and 2.11% of ignition loss. The aggregate was composed of fine and coarse particles. The particle size content curve obtained from the sieve analysis is presented in Figure 1. While the fine aggregate consisted mostly of sand with a grain size of 1/2 mm, the sand constituted 40.2% of the aggregate by weight. Furthermore, the fine aggregate particles were natural sand and were usually rounded and well-rounded in the cubic or occasionally elongated form [23], and the surface texture of these particles was smooth. Sand grains were mainly individual minerals such as quartz and feldspar.

Table 1. Concrete mix composition.

Cement (C)	Plasticizer	Aggregate		Water (W)	W/C	Bulk Density	Density after 90 d
CEM I 42.5 R	Sikaplast 2545	fine d < 2 mm	coarse 2 mm < d < 16 mm	-	-	-	-
kg/m^3	kg/m^3	kg/m^3	kg/m^3	kg/m^3	-	kg/m^3	kg/m^3
458.13	5.55	666.37	999.56	187.42	0.41	2317.04	2153.02

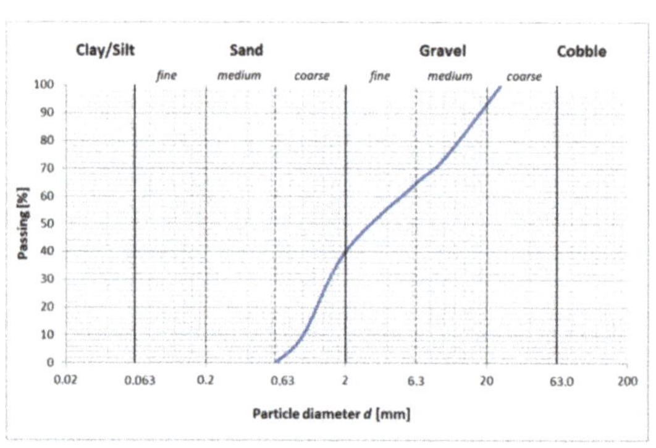

Figure 1. Particle size distribution of aggregates according to sieve analysis.

The remaining 59.8% was the coarse aggregate of crushed granite in two fractions: fine gravel (24.6%) and medium gravel (30.2%), as presented in Figure 2. These particles were angular and very angular in elongated and slightly flat forms with rough surface textures. Their shape index SI according to [24] was low, e.g., for medium gravel, SI = 16.

Figure 2. Granite fragments of coarse aggregates: medium and fine gravel.

2.2. Samples

Cube concrete samples of $150 \times 150 \times 150$ mm^3 were prepared. They matured in water for 28 days and after another 60 days, smaller samples were water cut (Figure 3) to a size of $120 \times 120 \times 30$ mm^3. The cut sample surfaces of 120×120 mm^2 were further processed. Sandblasting was chosen from many available surface treatment methods. This method is a popular process for concrete processing in technical applications and it transforms the surface randomly in an isotropic way. Sandblasting was performed using a standardized nozzle, pressure and aggregate grain size in order to repeat the experiment in the future. During the sandblasting with sand grains with d = 0.10–2.00 mm and pressure 0.6 N/mm^2, an attempt was made to maintain a uniform regime, i.e., constant distance of the nozzle from the surface and constant speed of its movement to obtain even surfaces, as presented in Figure 4.

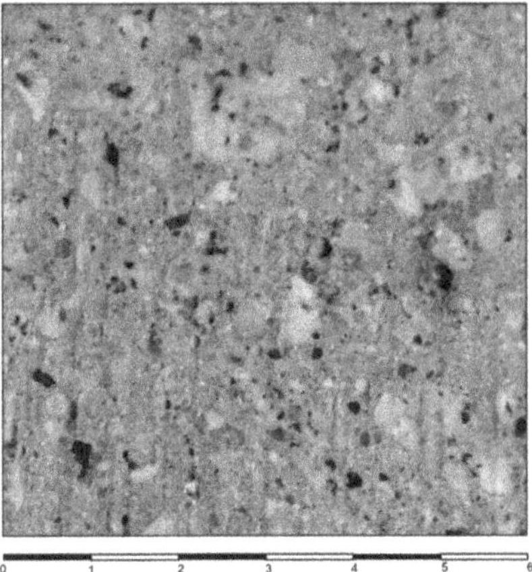

Figure 3. A 60×60 mm^2 fragment of the water cut concrete sample surface.

Figure 4. Stereoscopic image of the sandblasted surface (together, the left and right images create a 3D effect while keeping eyes as "looking into the distance").

The surface area ratios of the concrete components of the aggregate and cement matrix were estimated on the sand processed surfaces with 2D image analysis. The aggregate and the cement matrix ratios were assessed, as presented in Figure 5 and Table 2, in areas A, B, C and D. The ratio of the aggregate surface varied from 41.81% to 49.48%.

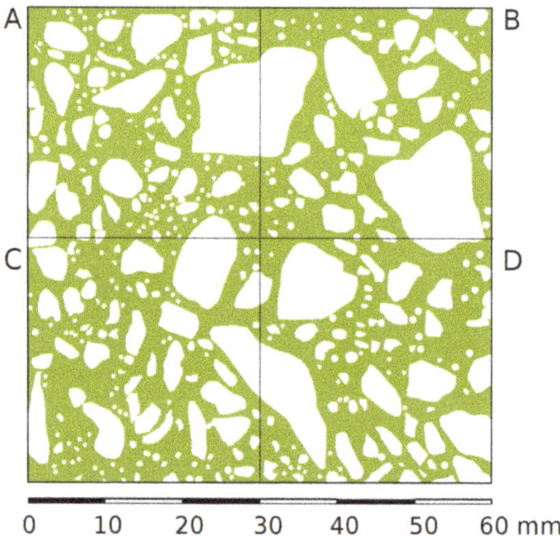

Figure 5. The 2D model of the concrete surface presented in Figure 3 (green—acement matrix, white—aggregates). (**A–D**): four segments of the sample.

Table 2. Sample presented in Figure 5—aggregate and cement matrix areas and ratios.

Area	Unit	Concrete Area	Aggregates Area	Cement Matrix Area
A	[mm^2]	900	433	467
	[%]	100	48.12	51.88
B	[mm^2]	900	445	455
	[%]	100	49.48	50.52
C	[mm^2]	900	376	524
	[%]	100	41.81	58.19
D	[mm^2]	900	416	484
	[%]	100	46.22	53.78
A + B + C + D	[mm^2]	3600	1671	1929
	[%]	100	46.41	53.59

Afterward, five surfaces of the five samples were used in the research. One of the surfaces, P0, was not processed (mold touching left) and the others (P20, P21, P22, P23) were sandblasted after water cutting.

2.3. Background of the Proposed Method

Based on samples of limited dimensions, the presented method detects surface parameters and extends the results to a larger concrete area. The developed approach includes a combination of the following methods:

- point cloud acquisition (with known coordinates (x,y,z) representing sample surface) with an adequate accuracy;
- surface assessment with use of geostatistical methods, i.e., fitting of theoretical semivariogram to empirical data;
- generation of Gaussian random field (via height profiles) based on the fitted semivariogram;
- computation of the roughness/texture parameters based on the generated profiles or surfaces;
- correlation analysis of the semivariogram parameters with shear strength factors throughout roughness/texture parameters.

In this paper, determination of the surface parameters, i.e., point cloud coordinates and profile height (x,y,z), was obtained using close-range photogrammetry. The entire procedure of the assessment of the concrete shear strength parameters is presented in Figure 6.

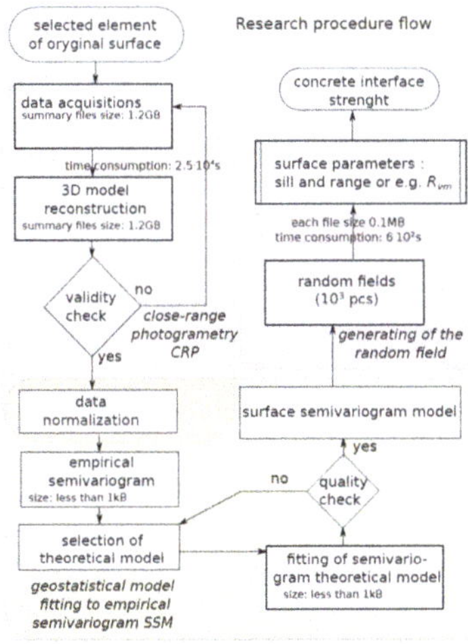

Figure 6. Block diagram of the shear strength parameter assessment.

2.4. Data Acquisition Employing Photogrammetry

Terrestrial close-range photogrammetry (CRP) is widely used in civil engineering. Professional photogrammetric cameras equipped with high-quality lenses take metric photos with known orientations in the adopted reference system. These photos include elements of internal orientation, which allows researchers to read the coordinates of the photographed objects. Based on these coordinates, it is possible to control the shape of structural elements and their location in space. Nowadays, another approach is emerging in photogrammetry which involves the use of the so-called non-metric cameras with cheaper lenses that are characterized by greater distortions (e.g., radial, tangential). Geometric distortions are minimized in the camera calibration process based on the analysis of many photos presenting special calibration charts (e.g., black and white chessboard), as presented in Figure 7a. The lack of elements of internal and mutual orientation is replaced by a large number of photos taken, which overlap to a large extent (covering almost the same area as the object). Automatic detection of the tie points on adjacent photos allows researchers to estimate the camera position and orientation in space. It allows them to build a 3D model of the measured object. To obtain the correct scale of the model, as well as to optimize the estimated camera position and provide appropriate georeferencing, the control points are established on the object. The known coordinates of the control points (usually determined using an electronic total station) also provide the possibility of checking the accuracy of the created 3D model of the object. The obtained model's accuracy depends on many of the following factors: size of the object, shooting distance, camera matrix resolution and lens parameters, number of control points, their arrangement and coordinate accuracy, as well as the lighting, colour and texture of the photographed object.

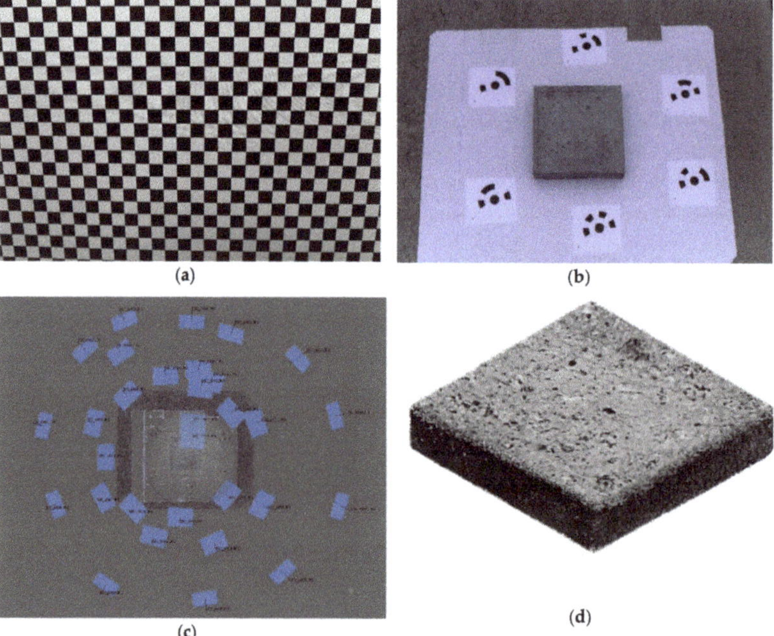

Figure 7. Stages of close-range photogrammetry (CRP) model preparation: (**a**) black and white chessboard for camera calibration; (**b**) a sample on the stand with markers (six coded control points); (**c**) estimated locations of the camera (blue rectangles); (**d**) dense point cloud obtained for the concrete sample.

In the literature, there are many examples of photogrammetry applications in construction and civil engineering. The usefulness of CRP in crack detection and the accurate 3D modelling of medieval bridge elements as well as for exterior material characterization were analysed in [25]. An example of a CRP application for the deformation analysis of a concrete beam during a loading experiment is presented in [26]. Precise estimation of the vertical deflections and horizontal displacements of the concrete beam during the load test in laboratory conditions showed that the photogrammetry technique was capable of monitoring both static and dynamic deformations [27]. Nowadays, photogrammetry is increasingly used for measurement of the vibrating structures. The most current trends in this technique are point tracking, digital image correlation and targetless approaches, as well as the integration of photogrammetry with other measurement techniques used in structural dynamics (e.g., interferometry) [28]. Both the point tracking technique and the development of a 3D model of the object based on a set of photos require an effective algorithm for recognizing the control point markers on the pictures. The aforementioned desire to integrate various measurement techniques usually involves combining photogrammetry with laser scanning. A coupled 3D laser scanning and digital image correlation system for the geometry acquisition of a railway tunnel was proposed and tested in [29]. When integrating measurement methods, attention should be paid to the accuracy achieved by the individual techniques and, in particular, this accuracy should be checked in real conditions prevailing at the construction site [30].

In the presented research, the photos of the concrete surface samples (120×120 mm^2) were taken with a Nikon D800 non-metric camera (Nikon CEE GmbH, Warsaw, Poland) with a 50 mm single-focal-length lens with a resolution of 7360×4912 pixels. To calculate the pre-calibration parameters (in Agisoft Lens software, Version 0.4.2 beta 64 bit (build 2399), Agisoft LLC, St. Petersburg, Russia [31]), 10 photos of a black and white chessboard were taken (Figure 7a). Next, for each concrete sample, several dozen photos were taken using a special flat stand and six markers (control points) (Figure 7b). The coordinates of the control points in the local coordinate system were determined using the least square method (LSM) through the adjustment of 15 linear measurements. The horizontal position accuracy of the control points after adjustment was no greater than 1 mm. For each concrete sample, the tie points on the photos were found with the highest accuracy requirement and all photos were aligned in the Agisoft PhotoScan Professional software (Version 1.2.4 64 bit (build 2399), Agisoft LLC, St. Petersburg, Russia) [32]. Examples of the estimated positions of the camera when pictures of the concrete sample were taken are presented in Figure 7c as blue rectangles. The obtained tie points were filtered using the values of reprojection error, reconstruction uncertainty and projection accuracy. Based on control points, the optimization of camera alignment, as well as the scaling and georeferencing of the samples, were performed. The last step was to generate a dense point cloud for each concrete sample, an example of which is presented in Figure 7d.

The CSV format files were generated from the results of the photogrammetric measurement of the concrete samples. Each file contained a collection of points representing the sample surface (3D model) supplemented with colour information components in the form of records (x,y,z,R,G,B). The collected z coordinates of the points were normalized in reference to an average value of z or in reference to a surface trend if necessary.

2.5. Geostatistical Models of the Sample Surfaces

To fit the theoretical semivariograms to empirical data from 3D surface models, geostatistical methods were used. The semivariograms measured the spatial autocorrelation of the acquired sample points. The measured height, i.e., the "z" values, were used to build an empirical semivariogram with the LSM and the Gauss–Newton algorithm as a nonlinear fitting method. The general form of the semivariograms according to Equation (2) was used [33], where for each h, half of the mean value of the squared difference $(z(d) - z(d + h))$ is defined as semivariance with a squared length unit.

$$\gamma(h) = \frac{1}{2N} \sum_{i=1}^{N} (z(d) - z(d + h))^2 \quad \{d, d + h\} \in \boldsymbol{P} \tag{2}$$

where $z(d)$ are normalized "z" coordinates at location d, h is the lag distance, i.e., distance that separates the two analysed locations and N—number of tested pairs $\{d, d + h\}$ from space $\boldsymbol{P}$.

For any two locations in a small distance on a surface, a small value of a measure $(z(d) - z(d + h))^2$ is expected. With the increasing lag distance h between points, the similarity of the measure decreased. This natural behaviour could be described by various theoretical relationships, as presented in Table 3 and Figure 8. The nugget model is used for discontinuous characteristics or areas of limited local extend. Spherical, exponential, Gaussian and linear-plateau models are important among the commonly used semivariograms in engineering applications.

Table 3. Basic semivariogram models.

Model Name	Semivariogram	Equation No.
nugget	$\gamma(h) = \begin{cases} 0 \Rightarrow h = 0 \\ s \Rightarrow h > 0 \end{cases}$	(3)
linear with sill	$\gamma(h) = \begin{cases} \frac{sh}{r} \Rightarrow h \leq r \\ s \Rightarrow h > r \end{cases}$	(4)
spherical	$\gamma(h) = \begin{cases} s\left[1.5\frac{h}{r} - \kappa\left(\frac{h}{r}\right)^3\right] \Rightarrow h \leq r \\ s \Rightarrow h > r \end{cases}$	(5)
exponential	$\gamma(h) = s\left(1 - e^{\frac{-h}{r}}\right)$	(6)
logarithmic	$\gamma(h) = \begin{cases} 0 \Rightarrow h = 0 \\ s \, \log(h + r) \Rightarrow h > 0 \end{cases}$	(7)

r—range [L], typically constant value limiting the zone of mutual correlation of points in the model, s—sill is the model constant [L^2], h—lag distance, variable in the model function [L], κ—model constant typical as 0.5 [·].

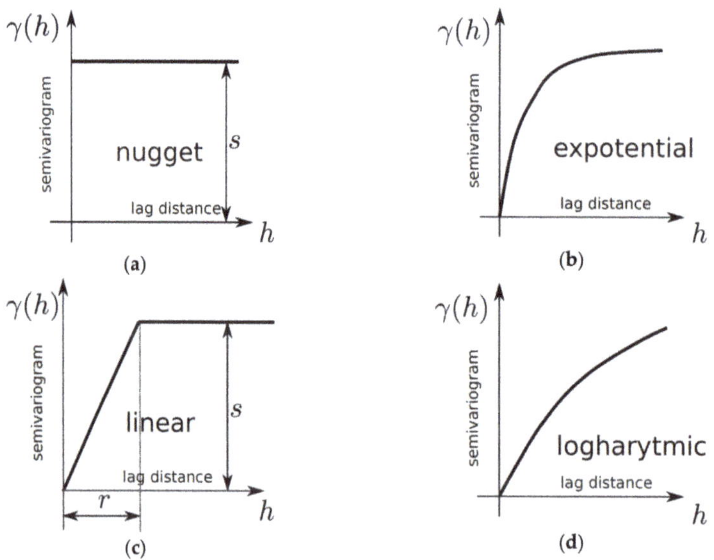

Figure 8. *Cont.*

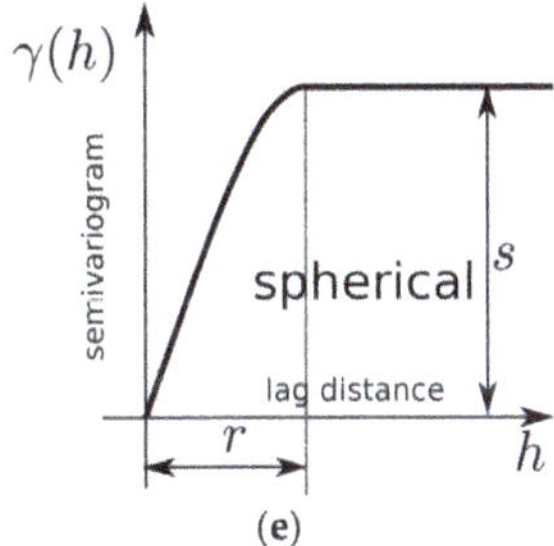

Figure 8. Theoretical semivariogram models (graphs): (**a**) nugget; (**b**) expotential; (**c**) linear; (**d**) logarithmic; (**e**) spherical.

Notation and semivariance fittings are presented in Figure 9, where the empirical points and fitted model are presented. When a pair of points appears in an interval, the lag distance between them increases linearly. This concept introduces a statistical measure of "z" variability based on a distance between points which is different from subjective deterministic measures. Moreover, some specific features of semivariograms are commonly used to describe natural observations:

- The distance where the model first flattens is known as the range r [L].
- Point or sample locations separated by a distance smaller than the range r are spatially autocorrelated, whereas locations farther apart than the range r are not.
- The value that the semivariogram model attains at the range r is called the sill s [L^2]. The partial sill is the sill minus the nugget [L^2].
- Theoretically, at zero separation distance (lag = 0 [L]), the semivariogram value is 0 [L^2]. However, at an infinitesimally small separation distance, the semivariogram often exhibits a nugget [L^2] effect, which is some value greater than 0. The nugget effect can be attributed to measurement error or spatial sources of variation at distances smaller than the sampling interval, or both. Natural phenomena can vary spatially over a range of scales. Variation at microscales smaller than the sampling distance will appear as part of the nugget effect.

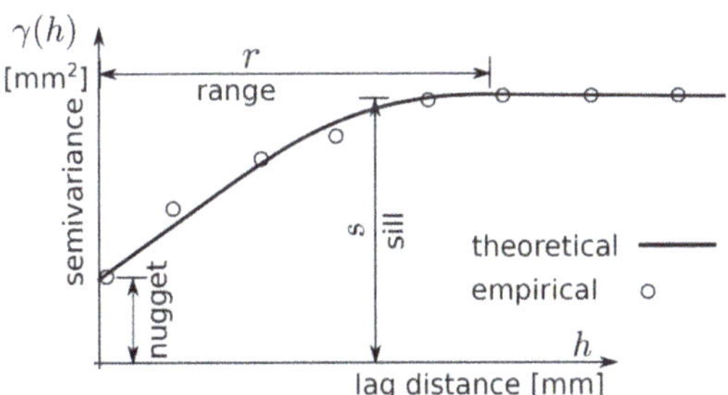

Figure 9. Semivariogam fitting and notation.

2.6. Geostatistical Model Fitting to Empirical Semivariogram—Surface Semivariogram Model (SSM)

Semivariogram modelling governs a step between spatial description and spatial prediction. Geostatistical analysis provides many semivariogram functions to model the empirical semivariogram, as presented in Table 3 and Figures 8 and 9.

The selection of a theoretical model and its fitting procedure is crucial to get a satisfactory prediction of unsampled points. The maximum likelihood or least square regression (including the weighted version) can be used to fit the experimental semivariance. A small and comparable sum of the deviations indicates comparable performance of the models. Therefore, the selection of the semivariogram model is a prerequisite for better performance. It is possible to have several models to choose from. The selection of a satisfactory model requires the balancing of goodness of fit and model complexity. Taken together, a likelihood ratio approach leading to a chi-square test was used in this study.

Semivariograms were prepared with two methods for the CRP normalized surface models:

- for each pair of points from the surface,
- for a combination of points along parallel lines in (0°, 45° or 90°).

At this stage, theoretical semivariograms were fitted to the empirical semivariograms with the use of LSM with uniform weights, as presented in Figure 10.

```
# preparing environment - loading libraries spatial data,
# geostatistical and  read of csv files
 library(sp)
 library(gstat)
 library(readr)
#read and collection data in structure and normalise
 dP20<-read_csv("P20.csv",col_names=TRUE,comment = "")
 mili<-10^3 # conversion meters to milimeters
 for(a in c(1:3)){  dP20[,a]<-(dP20[,a]-min(dP20[,a]))*mili}
 dP20$Z<-dP20$Z-mean(dP20$Z)
#spatial class of data
 coordinates(dP20)<-c("X","Y")
 class(dP20)
#presentation of raw height z
 spplot(dP20,zcol="Z")
#semivariogram of Z in space (X,Y)
 empvar <- variogram(Z~X+Y, dP20)
#fitting theoretical spherical model on empvar
 fit.variogram(empvar, vgm(, "Sph"))
#directional semivariograms
 dirempvariogram = variogram(Z~X+Y, dP20, alpha = c(0,45,90))
```

Figure 10. Algorithm No. 1 in R language. Data collection and semivariogram fitting to a sample surface (# starts description line).

Algorithm 1 in the R language [22] contains the main part of the fitting code for the spherical model, based on the *Gstat* package [34,35] and SP spatial pack library [36]. The fitted model according to Equation (5) is fully described with two parameters: r (range) and s (sill), for the zero value of the nugget effect. Both parameters were subsequently used in the process of the generation of a random field described in the next section.

2.7. Generation of the Random Field

As mentioned before, semivariograms were used to perform simulations of both the profile and the surface by using the random field theory for the Gaussian process, correlated with the semivariogram. For the field generation, the sequential simulation algorithm was used and coded in the R language [22], as presented in Figure 11. It was assumed that the modelled phenomenon could be described employing the Gaussian random field. The use of the Gaussian random field in Euclidean space assumes that a random process is stationary and isotropic. The symmetrical non-negative correlation function and zero mean value were also assumed. Moreover, the points of the generated samples were evenly distributed on the surfaces and on the profiles.

The random field generator presented in Figure 11 was also based on the *Gstat* package [34,35]. The sequential algorithm operates well on fields of large dimensions (i.e., more than 10^7 points). To approximate the conditional distribution at a location, this effective method uses only data and simulated values from a local neighborhood. The process is described by parameter n_{max}. The larger the n_{max}, the better approximation is with the time consumption increasing exponentially. The selection of the nearest n_{max} data or previously simulated points is performed in *Gstat* with a bucket quadrate neighbourhood search algorithm [37].

```
# sample without sand blasting
# parameters and field dimensions
 xdim<-10^3
 ydim<-10^3
 xy <- expand.grid(1:xdim, 1:ydim)
 names(xy) <- c("x","y")
       sill<-0.01
       range<-5
       numberofsim<-4
# method and model
       g.method<- gstat(formula=z~1, locations=~x+y,
                        model=vgm(sill,model="Sph",range), nmax=20)
# generator
       yyG <- predict(g.method , newdata=xy, nsim=numberofsim)
       coordinates(yyG)<-c("x","y")
       class(yyG)
#names of random process realisations
       names(yyG)<-c("pa","pb","pc","pd")
       spplot(yyG)
```

Figure 11. Algorithm No. 2 in R language. Random field generator. (# starts description line).

In the sequential Gaussian simulation, a random path through the locations was assumed because with a regular path, the local approximation of conditional distribution (using only the n_{max} nearest) may have caused spurious correlations. *Gstat* reuses expensive results (neighborhood selection and solution to the kriging equations) for each of the subsequent simulations when multiple realisations of the same field are requested. There is a considerable speed gain in the simulating of multi-fields in a single call compared to several hundred calls, each for simulating a single field. The random number generator used was the native random number generator of the R environment. Besides this, the reproduction of the sampling results (i.e., fixing randomness) was assured by using the random number seed with the *set.seed()* function.

As a result of the random field generation, a cloud of points with the coordinates (x,y,z) was obtained. Examples of the various generated random profiles based on the spherical model, that fits well with relatively even surfaces, are presented in Figure 12.

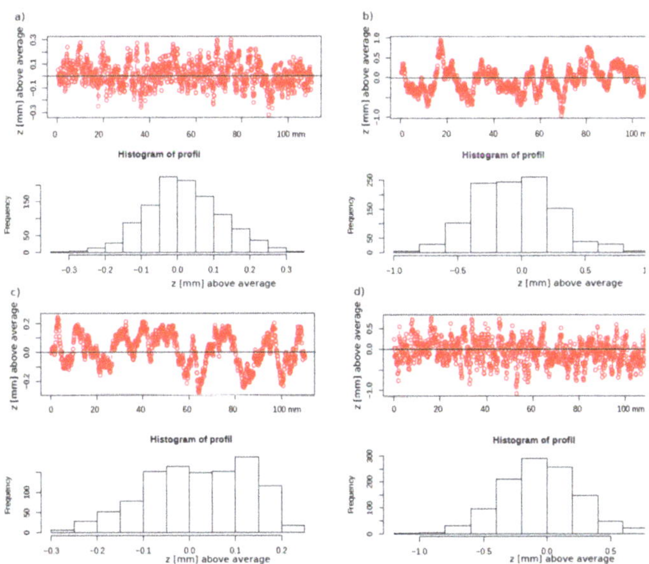

Figure 12. Examples of virtual profiles based on the spherical model with parameters: (**a**) s = 0.01 mm^2, r = 1 mm, (**b**) s = 0.1 mm^2, r = 5 mm, (**c**) s = 0.01 mm^2, r = 5 mm, (**d**) s = 0.1 mm^2, r = 1 mm. All profiles are supplied with histograms of z. (Horizontal axis - index is a vector of distance in 100 μm).

2.8. Surface Roughness Parameters

The random field containing the "z" coordinates obtained from the generated virtual surface could be used to calculate any traditional roughness parameters such as average roughness (R_a), mean peak height (R_{pm}), mean valley depth (R_{vm}) and many others. However, to explore the available experimental results, [5,10] the mean valley depth (R_{vm}) was used. R_{vm} is defined as the average of the maximum valley depth from each sampling length and is given by:

$$R_{vm} = \frac{1}{n} \sum_{i=1}^{n} v_i , \tag{8}$$

where v_i is the maximum valley depth at each sampling length i and n is the number of equal sampling lengths.

Then, a relationship between the roughness parameter (e.g., R_{vm}) and semivariogram parameter sill s (Figure 9) was obtained with the regression analysis algorithm presented in Figure 13. Evaluation of the roughness parameters is not a necessary step when full test data (i.e. (x,y,z) coordinates) are available because semivariogram parameters can be directly correlated with the strength test results.

```
# collection data from multiple calculations
 rez<-c()    # clear list of results
 sill<-c(1:20)/100 # sill={0.01,0.02,...,0.20} sqmm
 rang<-c(1:5)*5    # rang={5,10,15,20,25} mm
 for (v1 in sill) {
   for (v2 in rang ) {
     # multiParametric prepares  ntimes random field
     # generations and
   the average value is result
     rez<-rbind(rez,multiParametric(ntimes,v1,v2))}}
 # transform rezults to data frame
 rozw<-data.frame(rez)
 colnames(rozw)<-c("cfactor")
 #
 # fitting for expotential equation with three parameters {A,B,C}
 # the initial values are { A=0.5, B=2, C=0.27 }
 # and data frame is located
 # in rozw, defoult algorithm is Gauss-Newton, weights are
 # equal each other
 CohesionFIT<-nls(cfactor~B*sill^A+C,data=rozw,
 start=list(A=0.5,B=2,C=0.27))
```

Figure 13. Algorithm No. 3 in R language. Nonlinear least squares (NLS) estimates of the s parameters. NLS use as a default the Gauss–Newton algorithm.

2.9. Concrete Interface Strength

The final step in the presented method was the transformation of the roughness parameters and, more importantly, the semivariogram parameters into contact bond strength. Factors c and μ are required (Equation (1)) if [3] is used. According to the results presented in [5,10], relationships between the mean valley depth (R_{vm}) and the factors are expressed by the following:

$$c = 1.062 \, R_{vm}^{0.145}, \tag{9}$$

$$\mu = 1.366 \, R_{vm}^{0.041}, \tag{10}$$

However, relationships between the semivariogram parameter s (sill) and the c, μ factors can be achieved with the method presented in Figure 13.

3. Results and Discussion

The five concrete surface samples (P0, P20, P21, P22, P23) described in Section 2.2 were examined with the use of CRP. Based on the control points (Figure 7b), the optimization of the camera alignment, as well as the scaling and georeferencing of the models, were performed. The final number of tie points, as well as the obtained model accuracy, are summarized in Table 4. The total error calculated based on six control points did not exceed 0.32 mm and the total error of model scaling calculated based on six scale bars did not exceed 0.16 mm for all five concrete samples. The last step of CRP was to generate a CSV file with the dense point cloud for each concrete sample, as presented in Figure 14. The files contained sets of coordinates and color records (x,y,z,R,G,B). Discrete projection of the surface with a resolution of 100 points per 1 mm (2540 dpi) in the sample plane results in a single file size of approximately 40 MB.

Table 4. The number of tie points and assessment of obtained accuracy for the CRP models of the concrete surface samples.

Parameter	Sample Number				
	P0	P20	P21	P22	P23
Number of photos	31	32	26	23	22
Number of tie points	191,205	185,801	184,875	126,506	146,182
Total error of markers in mm	0.26	0.30	0.32	0.27	0.26
Total error of scale bars in mm	0.16	0.16	0.14	0.16	0.15
Original density of point model of concrete sample surface obtained from CRP after cutting to the size of 110 mm × 110 mm (points/mm^2)/(points/inch^2)	225/145,161	282/181,935	269/173,548	263/169,677	250/161,290

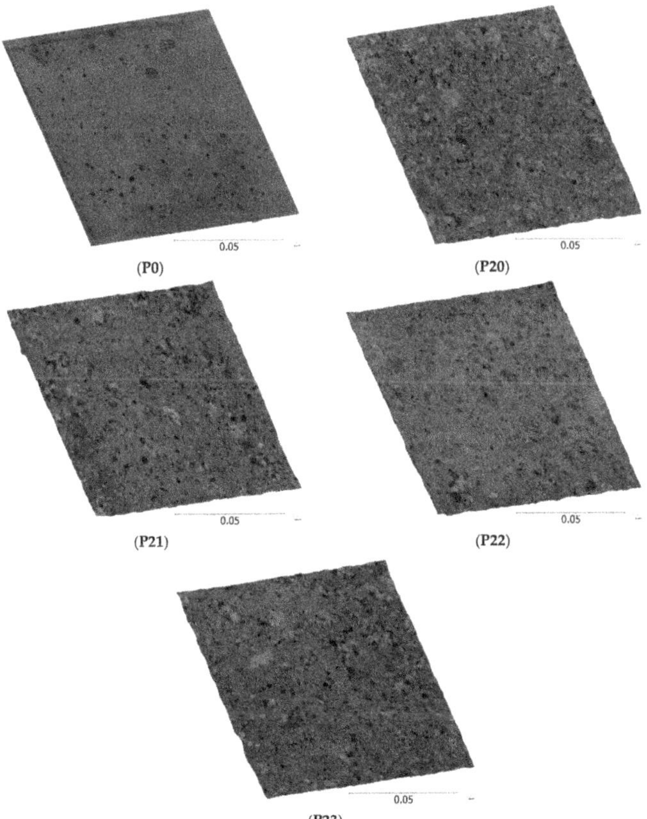

Figure 14. The perspective view of the segments of point clouds obtained from CRP as models of concrete surfaces for samples P0, P20, P21, P22 and P23 (the scale bars are given in meters).

The resulting three-dimensional surface models of the concrete samples were slightly jagged at the edges. For this reason, the concrete sample models were cut symmetrically on each side to a final size of 110 mm × 110 mm and further calculations were performed on the cut samples with the densities presented in Table 4.

For each sample, theoretical semivariograms were fitted to the empirical data with the use of LSM with uniform weights (Algorithm 1 in Figure 10). The "z" coordinates were normalized in reference to the average value and the final semivariogram type was selected based on the shape of the empirical semivariogram, as presented in Figure 15 and Table 5. This selection included the theoretical semivariogram type and method of point selection (each pair from surface or combination along parallel lines).

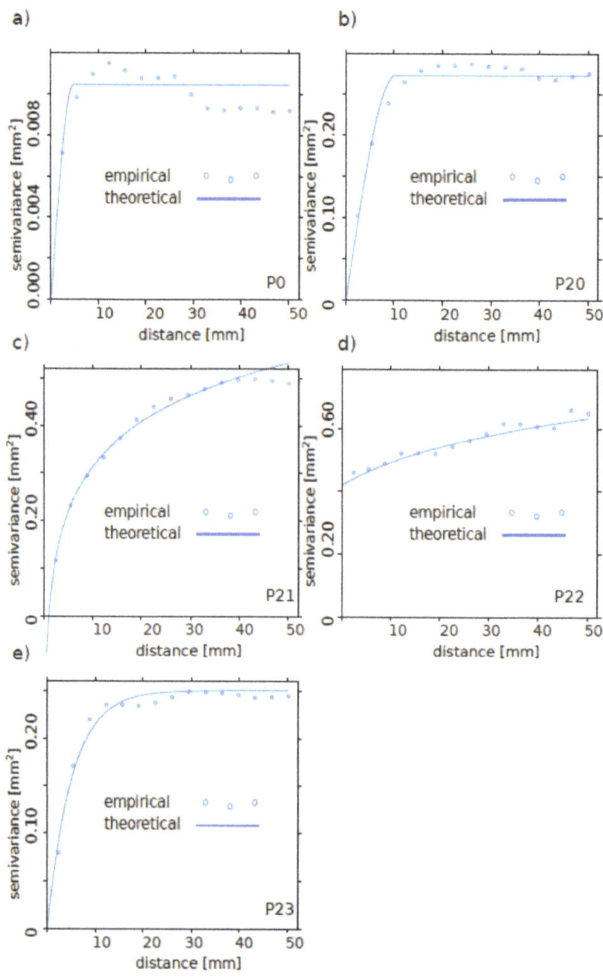

Figure 15. Empirical and theoretical semivariograms for samples: (**a**) P0, (**b**) P20, (**c**) P21, (**d**) P22, (**e**) P23.

Table 5. Semivariogam fitting results.

Sample Number	Semivariogram Type	Sill mm^2	Range mm	Method	Surface Type
P0	Spherical	0.0102	4.4744	each pair from surface	not processed, mold touching
P0	Spherical	0.0980	4.513	cop-x [1]	not processed, mold touching
P0	Spherical	0.0112	4.941	cop-x45 [2]	not processed, mold touching
P0	Spherical	0.0110	5.012	cop-y [3]	not processed, mold touching
P20	Spherical	0.2536	10.675	each pair from surface	sandblasting
P20	Spherical	0.2551	11.710	cop-x [1]	sandblasting
P20	Spherical	0.2510	9.913	cop-x45 [2]	sandblasting
P20	Spherical	0.2548	12.017	cop-y [3]	sandblasting
P21	Expotential	0.4832	8.6369	each pair from surface	sandblasting
P21	Expotential	0.4838	8.6367	cop-x [1]	sandblasting
P21	Expotential	0.4831	8.6353	cop-x45 [2]	sandblasting
P21	Expotential	0.4827	8.6367	cop-y [3]	Sandblasting
P22	Logarithmic	0.0791	0.5139	each pair from surface	Sandblasting
P23	Exponential	0.2366	10.4068	each pair from surface	Sandblasting

[1] cop-x—combination of points along parallel lines, along the x-direction; [2] cop-x45—combination of points along parallel lines, along 45 deg to the x-direction; [3] cop-y—combination of points along parallel lines, along the y-direction.

It is noteworthy that the surface processing type did not change the semivariogram model because the models for samples P0 and P20 were the same and the semivariogram model changed among sandblasted surfaces. This suggests that factors other than the observed parameters can contribute to the semivariogram type.

The theoretical semivariograms presented in Table 5 were subsequently used to generate virtual surfaces, as presented in Figures 16 and 17. Furthermore, from these results, the regression models of the mean valley depth R_{vm}, cohesion factor c and friction factor μ were approximated as a power function of sill s:

$$R_{vm} = B_1 \cdot s^{A_1} + C_1, c = B_2 \cdot s^{A_2} + C_2, \mu = B_3 \cdot s^{A_3} + C_3, \tag{11}$$

where (A, B, C) is a set of fitting vectors.

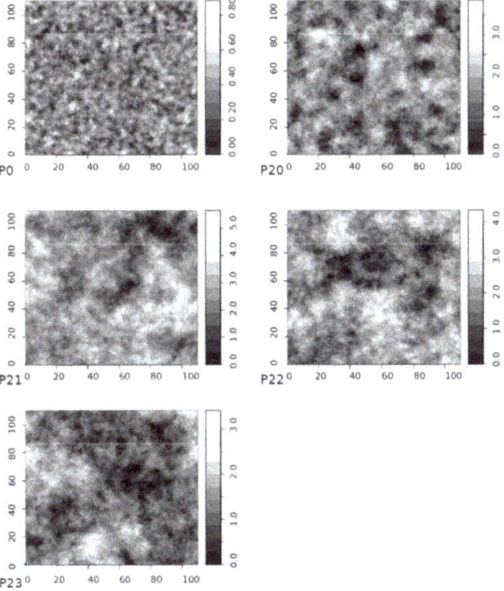

Figure 16. Single random field realisations of the original surfaces P0, P20, P21, P22, P23 (scale values are z coordinates in mm).

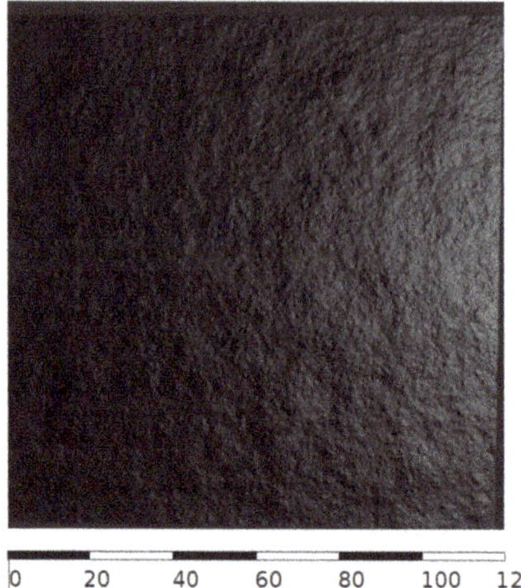

Figure 17. A 3D image of the random field realisation of the original surface P20.

In Table 6, the regression parameters can be found, and in Figure 18, the fitting results of factors c and μ can be found.

Table 6. Parameters of regression models R_{vm}, c and μ.

Factor/ Parameter	B	A	C	Residual Sum-of-Squares	Number of Iterations to Convergence	Achieved Convergence Tolerance
R_{vm}	0.752192	0.472575	−0.008818	0.006090	8	5.463×10^{-8}
c	0.58668	0.04188	−0.216210	0.000370	10	2.483×10^{-6}
μ	1.76614	−0.01068	2.78001	0.0003032	6	1.009×10^{-6}

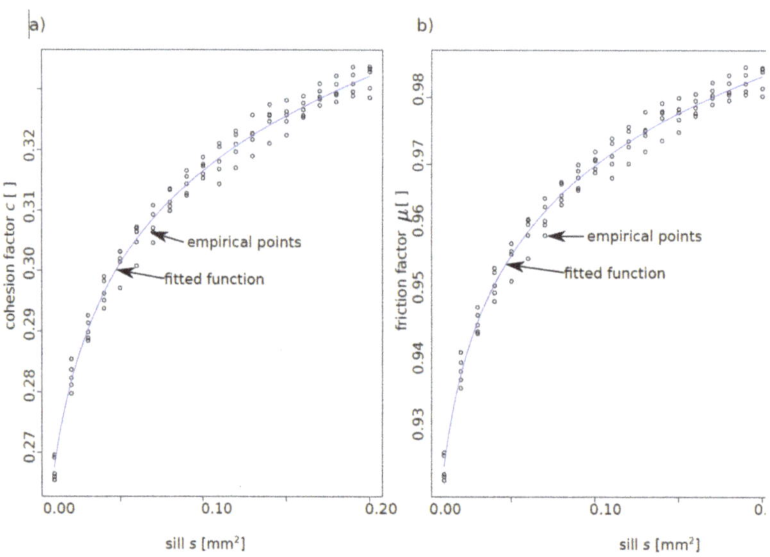

Figure 18. Results of fitting cohesion—c (**a**) and friction—μ (**b**) factors to the sill—s parameter.

The estimation of factors c and μ as functions of s changes Equation (1) to:

$$v_{Rdi} = c(s){\cdot}f_{ctd} + \mu(s){\cdot}\sigma_n + \rho{\cdot}f_{yd}{\cdot}(\mu(s){\cdot}\sin\alpha + \cos\alpha) < 0.5\,v\,f_{cd}, \tag{12}$$

where v_{Rdi} is the shear strength at the concrete-to-concrete interface; $c(s)$ and $\mu(s)$ are the factors that depend on the interface geostatistical parameter s model; ρ is the reinforcement ratio; σ_n is the external normal stress acting on the interface; v is a strength reduction factor; and f_{cd} is the concrete compressive strength.

If the reinforcement is absent, Equation (12) leads to:

$$v_{Rdi} = c(s){\cdot}f_{ctd} + \mu(s){\cdot}\sigma_n \tag{13}$$

or

$$\frac{v_{Rdi}}{f_{ctd}} = c(s) + \mu(s){\cdot}\frac{\sigma_n}{f_{ctd}}. \tag{14}$$

Equation (14)'s plots are presented in Figure 19 for the range of normal stress $\sigma_n \in [0;0.5]\,f_{ctd}$. This type of plot could be included in any formal design document.

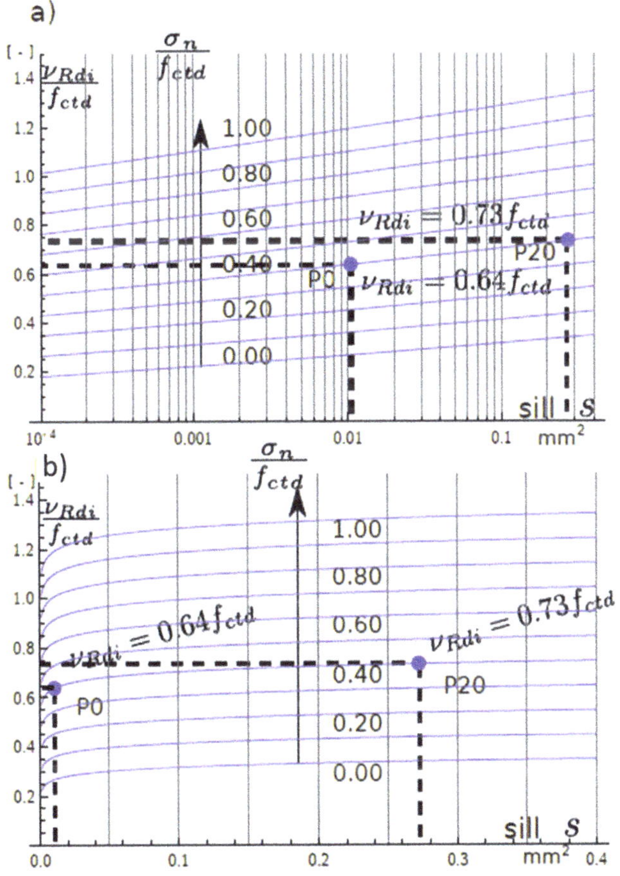

Figure 19. Plots of the relative interface shear strength as a function of the geostatistical sill—s parameter. (a) half-logaritmic scale of sill—s; (b) linear scale of sill—s.

In the absence of normal stress σ_n, the value of v_{Rdi}/f_{ctd} gives the cohesion factor c value which, according to [3], varies from 0.025 to 0.4 (and even 0.5 for indented surface). Since the surface of the sample P0 was not processed, it could be regarded as smooth or even very smooth according to [3] $c = 0.025 - 0.20$. The minimum value of c in Figure 19 is greater than 0.2 which indicates that the values of c in [3] might be too conservative. However, where the factor μ is concerned, it does not change very much with surface texture, according to [3] (0.5–0.7 and 0.9 for indented surface). Considering that $\sigma_n/f_{ctd} = 1$, the plots in Figure 19 give $c + \mu$ values, which for the smooth or very smooth surface is 0.525–0.8. However, in Figure 19 the minimum value of $c + \mu = 1.1$ which again might be too conservative.

4. Conclusions

Instead of traditional roughness parameters, objective statistical parameters were introduced. Thus, a unique random representation of the tested sample was achieved and the randomness of the surface texture was reproduced. Moreover, the sample can be reproduced with random field generation and used if access to the test surface is limited. This feature responds to the need for the non-destructive assessment of strength parameters solely based on concrete surface topography.

Although the developed procedure was tested on a limited number of samples, it exhibited versatile character. The method can be adjusted to the available equipment and field or laboratory conditions. Data collection to create a digital surface model can use any technique (with satisfactory accuracy) and the parameters can be adjusted to the available test data. Although, presented samples did not require much numerical effort, it should be noted that the point cloud size affects the time of numerical calculations.

There exist some limitations of the method applied in this paper, including the need for shadow-free illumination of the sampled element for correct CRP data collection and the use of licensed CRP software. Moreover, despite the simple in situ measurement procedure, the processing time for the model to be obtained can be too long and can affect the time of the construction work. An important feature of this method is a resistance to local texture changes, gaps, aggregate sharp edges, etc., but at the same time these imperfections are relativized and local heterogeneity diminishes because deterministic data are lost. In future research, the influence of surface processing on the semivariogram type should be examined. A full quantitative approach is recommended in the revision of design codes.

Author Contributions: Conceptualization, J.K., R.W. and M.W.; methodology, J.K., Z.M., M.W. and J.S.; software, J.K.; validation, R.W., Z.M., M.W. and J.S.; formal analysis, J.K., Z.M., M.W. and J.S.; investigation, J.K., R.W., Z.M., M.W. and J.S.; resources, R.W., Z.M. and J.S.; data curation, J.K., Z.M. and J.S.; writing—original draft preparation, J.K., R.W., Z.M., M.W. and J.S.; writing—review and editing, J.K., R.W., Z.M.; visualization, J.K., Z.M. and J.S.; supervision, J.K., R.W.; project administration, J.K., R.W. and M.W.; funding acquisition, J.K., R.W. All authors have read and agreed to the published version of the manuscript.

Funding: This research received no external funding.

Acknowledgments: The authors would like to thank Anna Pawlik and Anna Szalc for their help in making and curing the samples.

Conflicts of Interest: The authors declare no conflict of interest.

References

1. Cranston, W.B.; Kamiński, M.; Wróblewski, R. Aggregate interlock in cracked concrete with excessive crack width. *Arch. Civ. Eng.* **1996**, *42*, 177–193.
2. Birkeland, P.W.; Birkeland, H.W. Connections in Precast Concrete Construction. *ACI J. Proc.* **1966**, *63*, 345–368.
3. *EN 1992-1-1 Eurocode 2: Design of Concrete Structures -Part 1-1: General Rules and Rules for Buildings*; Comité Européen de Normalisation: Brussels, Belgium, 2004.
4. *Model Code 2010: Final Draft*; International Federation for Structural Concrete (fib): Laussane, Switzerland, 2012.

5. Santos, P.M.D. Assessment of the Shear Strength between Concrete Layers. In Proceedings of the 8th Fib International PhD Symposium in Civil Engineering, Copenhagem, Denmark, 20–23 June 2010.

6. Walraven, J.C.; Reinhardt, H.W. Theory and Experiments on the Mechanical Behaviour of Cracks in Plain and Reinforced Concrete Subjected to Shear Loading. *HERON* **1981**, *26*, 1–68.

7. Vecchio, F.J.; Collins, M.P. Modified Compression-Field Theory for Reinforced Concrete Elements Subjected To Shear. *J. Am. Concr. Inst.* **1986**, *83*, 219–231.

8. Santos, P.M.D.; Júlio, E.N.B.S. A state-of-the-art review on roughness quantification methods for concrete surfaces. *Constr. Build. Mater.* **2013**, *38*, 912–923. [CrossRef]

9. Mohamad, M.E.; Ibrahim, I.S.; Abdullah, R.; Abd Rahman, A.B.; Kueh, A.B.H.; Usman, J. Friction and cohesion coefficients of composite concrete-to-concrete bond. *Cem. Concr. Compos.* **2015**, *56*, 1–14. [CrossRef]

10. Santos, P.M.D.; Júlio, E.N.B.S. Interface shear transfer on composite concrete members. *ACI Struct. J.* **2014**, *111*, 113–121.

11. Figueira, D.; Sousa, C.; Calçada, R.; Serra Neves, A. Design recommendations for reinforced concrete interfaces based on statistical and probabilistic methods. *Struct. Concr.* **2016**, *17*, 811–823. [CrossRef]

12. *EN 1990. Eurocode–Basis of Structural Design*; Comité Européen de Normalisation: Brussels, Belgium, 2005.

13. Schabowicz, K.; Ranachowski, Z.; Jóźwiak-Niedźwiedzka, D.; Radzik, Ł.; Kudela, S.; Dvorak, T. Application of X-ray microtomography to quality assessment of fibre cement boards. *Constr. Build. Mater.* **2016**, *110*, 182–188. [CrossRef]

14. Schabowicz, K. Non-destructive testing of materials in civil engineering. *Materials* **2019**, *12*, 3237. [CrossRef]

15. Hoła, J.; Sadowski, Ł.; Reiner, J.; Stach, S. Usefulness of 3D surface roughness parameters for nondestructive evaluation of pull-off adhesion of concrete layers. *Constr. Build. Mater.* **2015**, *84*, 111–120. [CrossRef]

16. Courard, L.; Piotrowski, T.; Garbacz, A. Near-to-surface properties affecting bond strength in concrete repair. *Cem. Concr. Compos.* **2014**, *46*, 73–80. [CrossRef]

17. Sadowski, Ł.; Hoła, J. New nondestructive way of identifying the values of pull-off adhesion between concrete layers in floors. *J. Civ. Eng. Manag.* **2014**, *20*, 561–569. [CrossRef]

18. Trapko, T.; Musiał, M. PBO mesh mobilization via different ways of anchoring PBO-FRCM reinforcements. *Compos. Part B Eng.* **2017**, *118*, 67–74. [CrossRef]

19. Gaetan, C.; Guyon, X. *Springer Series in Statistics Spatial Statistics and Modeling*; Springer: Berlin/Heidelberg, Germany, 2010.

20. Wackernagel, H. *MultivariateGeostatistics. An Itroduction and Applications*; Springer: Berlin/Heidelberg, Germany, 2003.

21. Diggle, P.J.; Tawn, J.A. Model-based geostatistics. *Appl. Stat.* **1998**, *47*, 299–350. [CrossRef]

22. The R Project for Statistical Computing. Available online: https://www.R-project.org/ (accessed on 20 May 2020).

23. *EN ISO 14688-2. Geotechnical Investigation and Testing–Identification and Classification of Soil–Part 2: Principles for a Classification*; Comité Européen de Normalisation: Brussels, Belgium, 2017.

24. *EN 933-4. Tests for Geometrical Properties of Aggregates Part 4: Determination of Particle Shapes-Shape Index*; Comité Européen de Normalisation: Brussels, Belgium, 2008.

25. Arias, P.; Armesto, J.; Di-Capua, D.; González-Drigo, R.; Lorenzo, H.; Pérez-Gracia, V. Digital photogrammetry, GPR and computational analysis of structural damages in a mediaeval bridge. *Eng. Fail. Anal.* **2007**, *14*, 1444–1457. [CrossRef]

26. Psaltis, C.; Ioannidis, C. An automatic technique for accurate non-contact structural deformation measurements. In Proceedings of the International Archives of the Photogrammetry, Remote Sensing and Spatial Information Sciences -ISPRS Archives, Dresden, Germany, 25–27 September 2006.

27. Kwak, E.; Detchev, I.; Habib, A.; El-Badry, M.; Hughes, C. Precise photogrammetric reconstruction using model-based image fitting for 3D beam deformation monitoring. *J. Surv. Eng.* **2013**, *139*, 143–155. [CrossRef]

28. Baqersad, J.; Poozesh, P.; Niezrecki, C.; Avitabile, P. Photogrammetry and optical methods in structural dynamics–A review. *Mech. Syst. Signal Process.* **2017**, *86*, 17–34. [CrossRef]

29. Farahani, B.V.; Barros, F.; Sousa, P.J.; Cacciari, P.P.; Tavares, P.J.; Futai, M.M.; Moreira, P. A coupled 3D laser scanning and digital image correlation system for geometry acquisition and deformation monitoring of a railway tunnel. *Tunn. Undergr. Sp. Technol.* **2019**, *91*, 102995. [CrossRef]

30. Muszyński, Z.; Rybak, J.; Kaczor, P. Accuracy Assessment of Semi-Automatic Measuring Techniques Applied to Displacement Control in Self-Balanced Pile Capacity Testing Appliance. *Sensors* **2018**, *18*, 4067. [CrossRef]

31. Agisoft Lens. Version 0.4.2 beta 64bit (build 2399). Lens Calibration Software. Agisoft LLC. 2016. Available online: http://www.agisoft.com (accessed on 16 April 2018).
32. Agisoft PhotoScan Professional Edition. Version 1.2.4 64 bit (build 2399). Multi-View 3D Reconstruction. Agisoft LLC. 2016. Available online: http://www.agisoft.com (accessed on 16 April 2018).
33. Isaaks, E.H.; Srivastava, R.M. *An Introduction to Applied Geostatistics*; Oxford University Press: New York, NY, USA, 1989.
34. Pebesma, E.J. Multivariable geostatistics in S: The gstat package. *Comput. Geosci.* **2004**, *30*, 683–691. [CrossRef]
35. Pebesma, E.J.; Wesseling, C.G. Gstat: A program for geostatistical modelling, prediction and simulation. *Comput. Geosci.* **1998**, *24*, 17–31. [CrossRef]
36. Bivand, R.S.; Pebesma, E.; Gómez-Rubio, V. *Applied Spatial Data Analysis with R*; Springer: Berlin/Heidelberg, Germany, 2013; ISBN 9781461476177.
37. Clarkson, K.L. Fast algorithms for the all nearest neighbours problem. In Proceedings of the 24th Annual Symposium on Foundations of Computer Science (sfcs 1983), Tucson, AZ, USA, 7–9 November 1983; pp. 226–232.

Article

Mechanical Behavior and Failure Mode of Steel–Concrete Connection Joints in a Hybrid Truss Bridge: Experimental Investigation

Yingliang Tan [1], Bing Zhu [1,*], Le Qi [1], Tingyi Yan [1], Tong Wan [1] and Wenwei Yang [2]

[1] Department of Bridge Engineering, School of Civil Engineering, Southwest Jiaotong University, Chengdu 610031, China; yingliang_tan@my.swjtu.edu.cn (Y.T.); qileswjtu@126.com (L.Q.); tingyi_yan@yeah.net (T.Y.); wantongswjtu@163.com (T.W.)
[2] Department of Civil Engineering, School of Civil and Hydraulic Engineering, Ningxia University, Yinchuan 750021, China; nxyangww@126.com
* Correspondence: bing_zhu126@126.com; Tel.: +86-199-5028-4704

Received: 12 May 2020; Accepted: 2 June 2020; Published: 3 June 2020

Abstract: The core part of a hybrid truss bridge is the connection joint which combines the concrete chord and steel truss-web members. To study the mechanical behavior and failure mode of steel–concrete connection joints in a hybrid truss bridge, static model tests were carried out on two connection joints with the scale of 1:3 under the horizontal load which was provided by a loading jack mounted on the vertical reaction wall. The specimen design, experimental setup and testing procedure were introduced. In the experiment, the displacement, strain level, concrete crack and experimental phenomena were factually recorded. Compared with the previous study results, the experimental results in this study demonstrated that the connection joints had the excellent bearing capacity and deformability. The minimum ultimate load and displacement of the two connection joints were 5200 kN and 59.01 mm, respectively. Moreover, the connection joints exhibited multiple failure modes, including the fracture of gusset plates, the slippage of high-strength bolts, the local buckling of compressive splice plates, the fracture of tensile splice plates and concrete cracking. Additionally, the strain distribution of the steel–concrete connection joints followed certain rules. It is expected that the findings from this paper may provide a reference for the design and construction of steel–concrete connection joints in hybrid truss bridges.

Keywords: hybrid truss bridge; steel–concrete connection joint; mechanical behavior; failure mode; strain; static test

1. Introduction

Steel–concrete composite bridges are widely used because they combine the advantages of prestressed concrete box girder bridges and steel truss bridges. A composite bridge with steel truss members instead of traditional concrete webs attracted the attention of researchers, which was also known as a hybrid truss bridge (HTB). HTBs have a lighter weight and smaller beam height than prestressed concrete box girder bridges, enhancing the bridge span.

The Arbois Bridge [1], which was built in France in 1985, was the first attempt of this type of bridge in the world. Additionally, more representative HTBs have been built in France, such as Boulonnais Viaducts [2] and Bras de la Plaine Bridge [3]. Many hybrid truss bridges have been designed and built in other countries in Europe, including Lully Viaduct [4], Dreirosen Bridge [5], Europe Bridge [6], and Nantenbach Railroad Bridge [7,8]. In recent years, some countries in Asia have also begun to study this kind of bridge and put it into practice [9–15]. However, the study of HTBs in China started late, and the first HTB, Jiangshan Bridge, was built in 2012 [16]. Although the same type of structure was applied to several bridges (e.g., Minpu Bridge [17], Houhecun Bridge [18], and Deshenglu Bridge [19]), the research on HTB was not comprehensive.

The steel–concrete connection joint is considered the most important part of hybrid truss bridges, transferring the load between the concrete chord and steel truss web [11,17,20–24]. Therefore, some experimental and analytical studies on such joint structures have been reported. At the end of the 20th century and the beginning of the 21st century, Japanese scholars conducted various types of research to investigate the mechanical properties of connection joints [12,15,25–28]. Additionally, the stress transfer mechanism and mechanical characteristic of two types of steel–concrete connection joints were compared by Sato et al. [29]. The results showed that the connection joint with the perfobond rib (PBL) shear connectors had a greater bearing capacity than that of the joint with welded headed studs. In the early studies, the welding process was widely recommended and applied in the joint structures. However, the mechanical behavior was affected by the form of welding [29]. To reduce welding during construction, Jung et al. [1,11,13,20,22] proposed a new connection joint composed of connection plates and a connection bolt. Furthermore, experimental and numerical investigations were carried out on this new connection system and HTB girders to clarify the structural capacity, fatigue capacity and torsional behavior. Additionally, Zhou et al [18], Yin et al. [23,30] and Tan et al. [24] introduced another joint with high-strength bolts to decrease welding and conducted static model tests to investigate the connection performance of such joints. Their results clarified that the connection joint with high-strength bolts had excellent bearing capacities and safety reserves. However, the failure mode of the steel–concrete connection joint remains controversial. Zhou et al [18] reported that such joints failed because of the local buckling and fracture of gusset plates, while Yin et al. [23,30] found that the local buckling of steel truss-web members was one of the main failure modes. Moreover, what is less clear is the mechanical behavior of such joints at each loading step. In particular, the strain distribution rule of the main components is unclear. Hence, it is necessary to carry out the model test to investigate the mechanical behavior and failure mode of connection joints in detail.

In this paper, we sought to investigate the mechanical behavior and failure mode of steel–concrete connection joints. More specifically, this study aimed to determine the following specific research directions: (1) the ultimate bearing capacity and corresponding displacement of the proposed joints, (2) the typical failure mode of such joints, (3) the strain distribution rule of the main components, and (4) the key component of the steel–concrete connection joint to carry the external load. Therefore, we conducted static loading tests on two joint specimens with the scale of 1:3. Such experimental investigations of the proposed joint form may enrich the experimental data and provide an experimental reference for the design and construction of such joints in hybrid truss bridges.

2. Experimental Program

2.1. Specimens

As shown in Figure 1, two specimens with the scale of 1:3 were designed on the basis of the typical steel–concrete connection joint E9, which was selected from the preliminary design of the first hybrid continuous truss bridge in Chinese railway bridges.

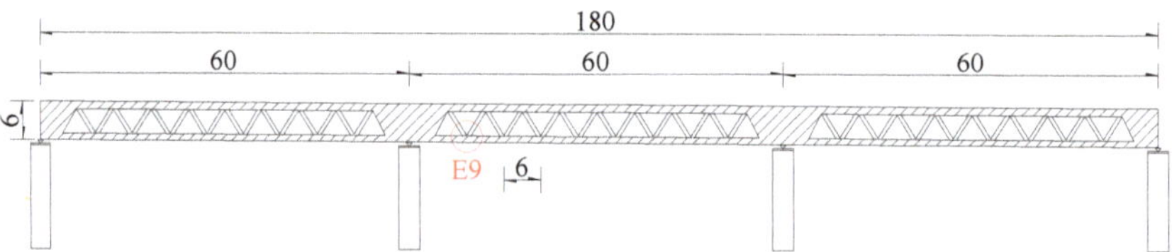

Figure 1. The most unfavorable joint (unit: m). E9: the ninth joint of the lower chord from left to right.

Figure 2 shows the schematic diagram of the joint specimens. Furthermore, Specimen 1 (Southwest Jiaotong University, Chengdu, China) and Specimen 2 (Southwest Jiaotong University, Chengdu, China) shared the same model parameters. The steel–concrete joint consists of the concrete chord (1764 mm × 334 mm × 367 mm), gusset plates ((CRTB) Tycoon Industrial Development Co., Ltd., Baoji, China) (764 mm × 629 mm × 16 mm) perforated 18 holes (40 mm in diameter), PBL shear connectors (Southwest Jiaotong University, Chengdu, China) (12 mm in diameter), steel truss-web members (268 mm × 184 mm × 20 mm), steel reinforcements (12 mm in diameter), and rectangular stirrups (8 mm in diameter). The material properties of C50 concrete (Sichuan Southwest Cement Co., Ltd., Chengdu, China), Q370qE steel (Baoshan Iron & Steel Co., Ltd., Shanghai, China), and HRB400 steel (Baoshan Iron & Steel Co., Ltd., Shanghai, China) are listed in Table 1.

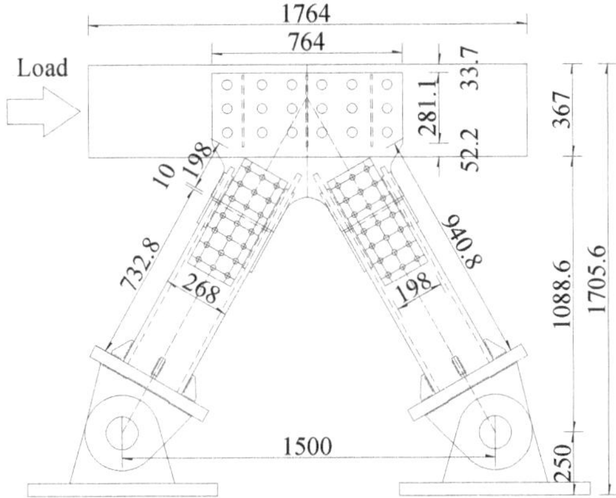

Figure 2. Schematic diagram of the specimen (unit: mm).

Table 1. Material properties of the test specimen.

Member	Material	f_y (MPa)	f_u or f_{cu} (MPa)	E_s or E_c (GPa)
Concrete	C50	N/A	61.3	34.6
Gusset plate	Q370qE	452	583	206
Steel web	Q370qE	439	557	206
Reinforcement	HRB400	458	640	203

f_y: yield strength of steel; f_u: tensile strength of steel; f_{cu}: compressive strength of concrete; E_s: elastic modulus of steel; E_c: elastic modulus of concrete. N/A: this item does not exist.

2.2. Experimental Setup and Testing Procedure

The static tests of the steel–concrete connection joints were carried out in the National Engineering Laboratory for Technology of Geological Disaster Prevention in Land Transportation, and the experimental setup is shown in Figure 3. To provide enough reaction force, two steel pedestals and a reaction device were installed on the ground. The steel pedestals with the hinge supported the connection joints. The horizontal load was offered by the loading jack with a capacity of 6300 kN, acting on one end of the concrete chord, whose direction was shown by the solid arrow in the Figure 3.

Before the formal multi-step loading, pre-loading was performed to avoid assembly clearance. The loading step of 400 kN was adopted within 0–2000 kN. Then, when the horizontal load was in the range of 2000–3000 kN, the loading step was reduced to 200 kN. Finally, the loading step of 100 kN was applied until the end of the loading procedure. The laser displacement sensor (D1) was used for measuring the horizontal displacement in the loading direction. Furthermore, the strain gauges were installed to monitor the strain of the main components, including the concrete chord, gusset plates, PBL shear connectors, and the steel truss-web members, as shown in Figure 4.

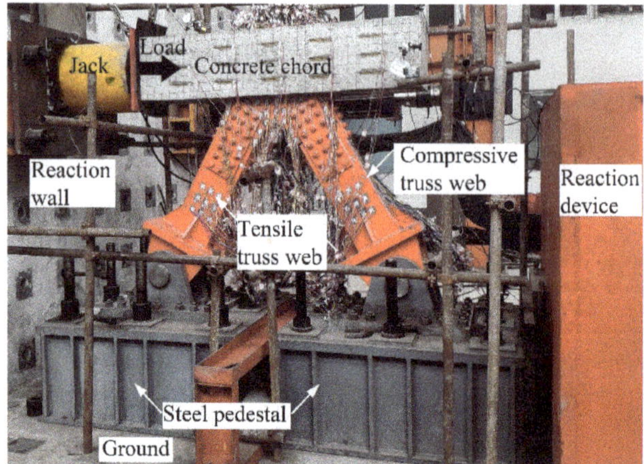

Figure 3. Experimental setup.

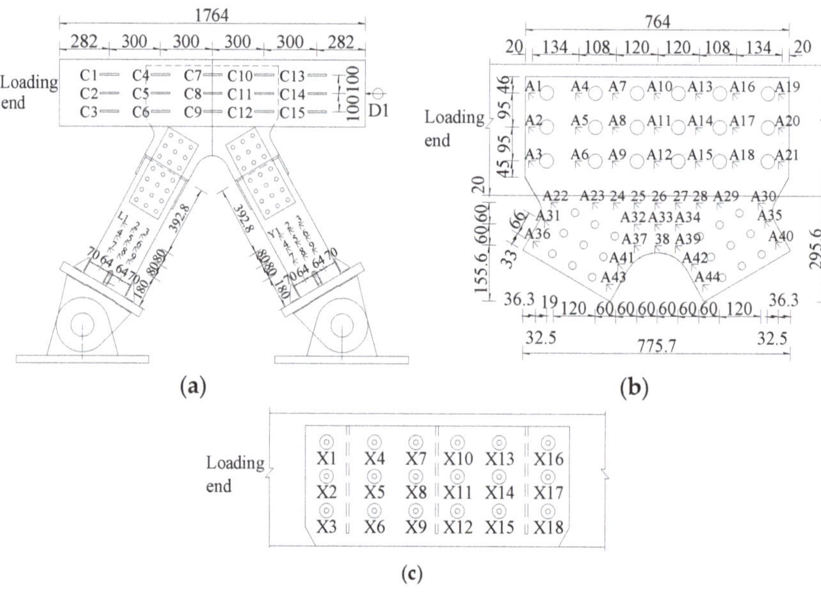

(a)   (b)

(c)

Figure 4. Arrangement of the displacement meters and strain gauges (unit: mm). (**a**) The concrete chord and steel truss-web members; (**b**) the gusset plates; and (**c**) the PBL shear connectors. C1: No. 1 strain gauge of concrete chord; A1: No. 1 strain gauge of gusset plates; X1: No. 1 strain gauge of PBL shear connectors.

3. Experimental Results and Discussions

3.1. Experimental Phenomena and Failure Modes

To precisely investigate the failure modes of the proposed connection joints, the experimental phenomena and data were recorded in detail. Figures 5 and 6 present the failure modes of Specimen 1 and Specimen 2, respectively. For Specimen 1, there was no macroscopic damage until the applied load reached 3800 kN. Then, the relative location between the gusset plates and the splice plates was changed on the tension side. Additionally, the initial crack with the length of 170 mm appeared on the concrete chord and which did not develop any more in the subsequent loading procedure. At the load of 4400 kN, local buckling of the splice plate on the compressive side was observed, as shown in Figure 5c. Then, in Figure 5b, the visible slips of high-strength bolts were noticed at the load of 4800 kN, which indicated that the high-strength bolts were out of action. When the applied load reached 5200 kN, a lot of cracks appeared on the concrete chord, as shown in Figure 5e. Finally, two noises were heard within seven seconds due to the fracture of the gusset plates and splice plates with hand holes (in Figure 5a,d), meanwhile, Specimen 1 lost its bearing capacity to resist the external load.

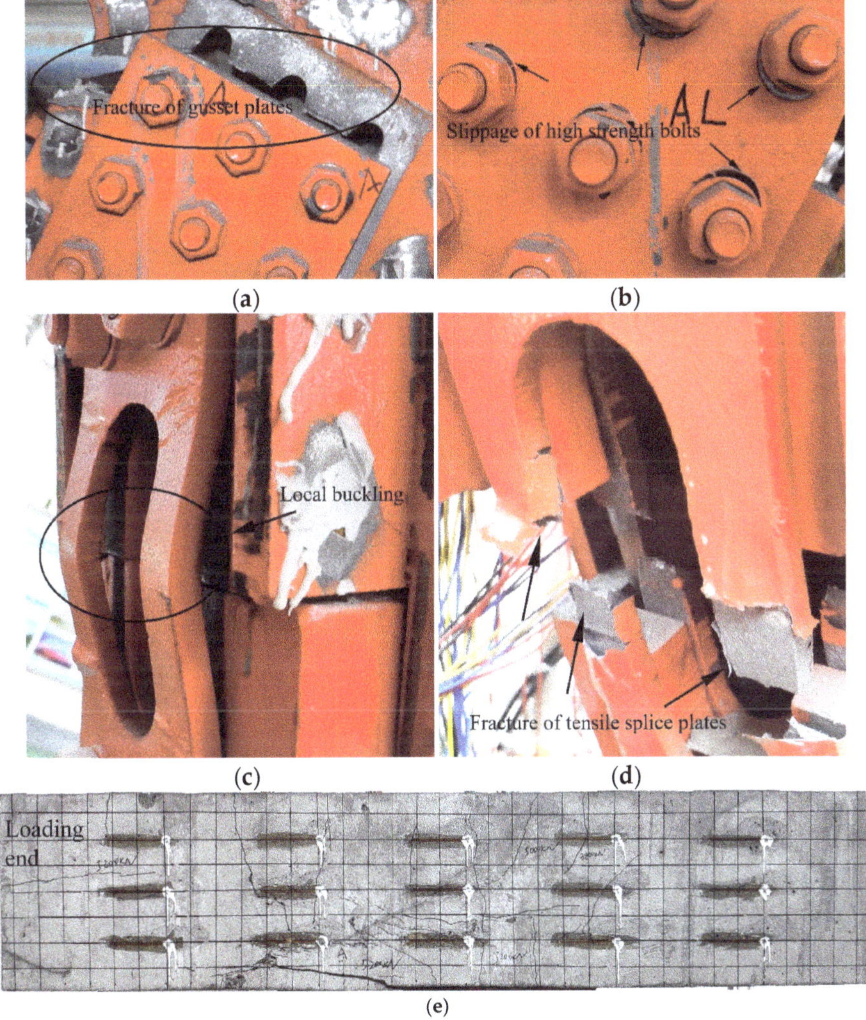

Figure 5. Failure modes of Specimen 1: (**a**) the fracture of gusset plates; (**b**) the slippage of high-strength bolts; (**c**) the local buckling of the compressive splice plates; (**d**) the fracture of the tensile splice plates; and (**e**) the concrete cracking.

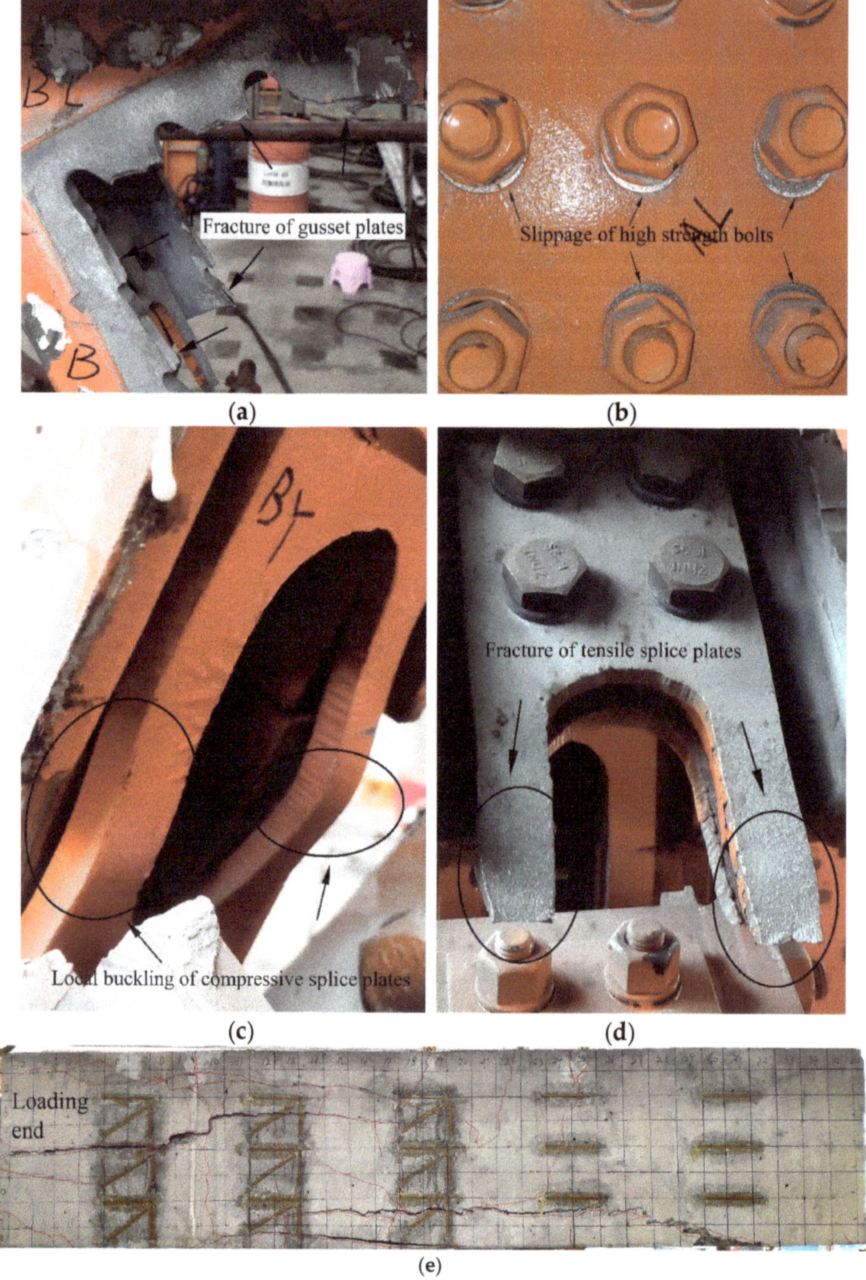

Figure 6. Failure modes of Specimen 2: (**a**) the fracture of the gusset plates; (**b**) the slippage of the high-strength bolts; (**c**) the local buckling of the compressive splice plates; (**d**) the fracture of the tensile splice plates; (**e**) the concrete cracking.

For Specimen 2, the first crack of the concrete chord emerged near the loading end at the load of 3400 kN, but this crack did not spread in the latter loading stages. Furthermore, the relative location malposition between the gusset plates and the splice plates with hand holes was also observed. Then, the local buckling of the compressive splice plates and the slippage of high-strength bolts were observed, as shown in Figure 6c,b. At the peak load of 5400 kN, it can be seen from Figure 6e that some cracks severely developed. Finally, the gusset plates suddenly broke completely causing a sound to be heard and Specimen 2 lost its bearing capacity, as shown in Figure 6a. Meanwhile, looking at Figure 6d, the splice plates with hand holes were also pulled off.

Based on the experimental observation of two specimens, the order of the destruction course of the proposed connection joints was the local buckling of the compressive splice plates, the slippage of high-strength bolts, the cracking of the concrete chord, the fracture of the gusset plates and the tensile splice plates. Clearly, two specimens showed similar test phenomena and failure modes. Nonetheless, the damage degree of Specimen 2 was greater than that of Specimen 1, for example, the gusset plates of Specimen 2 were pulled to the point of total fracture. The difference in the experimental results between the two specimens is attributable to the fact that the first fracture of Specimen 1 dissipated part of the energy, making Specimen 1 unable to continue to bear a greater load, and also causing the damage degree to be less than that of Specimen 2.

Compared with previous tests [11,18,23,30–32], one interesting finding from Table 2 was that the fracture of gusset plates and the slippage of high-strength bolts seemed to be two particular failure modes of such connection joints with high-strength bolts. A possible explanation for this might be that the cross-sectional area of the gusset plate was reduced due to the bolt holes, resulting in stress concentration. In addition, the two specimens in this study showed other failure modes, including the local buckling of compressive splice plates, the fracture of tensile splice plates, and concrete cracking. In particular, in contrast to previous studies, there were wide and long cracks on the concrete chord, which indicated that the performance of the concrete chord was also fully exerted in this study. To make the assembly process easy, it was necessary to drill hand holes on some splice plates, which caused these splice plates to buckle or fracture. This is difficult to avoid in the type of scale model unless these splice plates are not set here. Moreover, as Table 2 shows, for the steel–concrete connection joints with the high-strength bolts, the failure modes of such joints are related to the relative thickness of the gusset plates and truss-web members. For example, for specimens RGP (Rectangular gusset plate is used in the specimen), the increase in the thickness of steel truss-web members caused the failure mode to change from the failure of steel truss web to the failure of gusset plates and high-strength bolts. On the other hand, for the embedded type joints (joint type B), if the steel web members are not damaged, the specimens will fail due to the cracking of the concrete chord.

Table 2. Comparison of the failure modes.

Joint Type	Specimen	Concrete Chord (mm)	Gusset Plate (mm)	Truss Web (mm)	Failure Modes
	RGP-1 [23,30]	517 × 833	12	217 × 183 × 12	5,7
	RGP-2,3 [23,30]	517 × 833	12	217 × 183 × 22	1,2
A	SJ-1 [18]	367 × 400	8	217 × 183 × 12	8
	SJ-2 [18]	367 × 400	8	217 × 183 × 12	1,2
	Specimen 1	334 × 367	16	268 × 184 × 20	1,2,3,4,6
	Specimen 2	334 × 367	16	268 × 184 × 20	1,2,3,4,6
	JSGP-1 [23]	517 × 833	N/A	217 × 183 × 12	5
	JSGP-2 [23]	517 × 833	N/A	217 × 183 × 22	6
	PSGP-1 [23]	517 × 833	N/A	217 × 183 × 12	5
B	PSGP-2,3 [23]	517 × 833	N/A	217 × 183 × 22	6
	ZHJD-1,2 [31]	367 × 400	N/A	217 × 183 × 12	5,6
	EHT [11]	2200 × 250	N/A	Φ318 × 15	6
	T1 [32]	225 × 325	N/A	Φ180 × 8	6

Joint type A: the gusset plates and the truss-web members are connected by high-strength bolts. Joint type B: the truss-web members are partially embedded in the concrete chord, and there is no gusset plate. Please refer to the corresponding reference for the detailed definition of RGP, SJ, JSGP, PSGP, ZHJD, EHT, and T1. SJ-2: Strengthened the gusset plates. 1: Fracture of gusset plates. 2: Slippage of high-strength bolts. 3: Local buckling of compressive splice plates. 4: Fracture of tensile splice plates. 5: Local bucking of compressive web member. 6: Concrete cracking. 7: Yield of tensile web member. 8: Local buckling of gusset plates.

3.2. Load–Displacement Curves

Figure 7 presents the relation between the applied load and the horizontal displacement. As shown in Figure 7, two load–displacement curves were almost coincident. The corresponding displacement linearly increased with the increase in the applied load in the initial stage. Specimen 1 began to yield, and the curve started to flatten at the load of 3200 kN. With the increase in the load, the exposed gusset plates were pulled to fracture and the composite joint failed to carry the applied load. Hence, the peak load of 5200 kN was regarded as the bearing capacity of Specimen 1. In contrast to Specimen 1, the ultimate capacity of Specimen 2 reached 5400 kN.

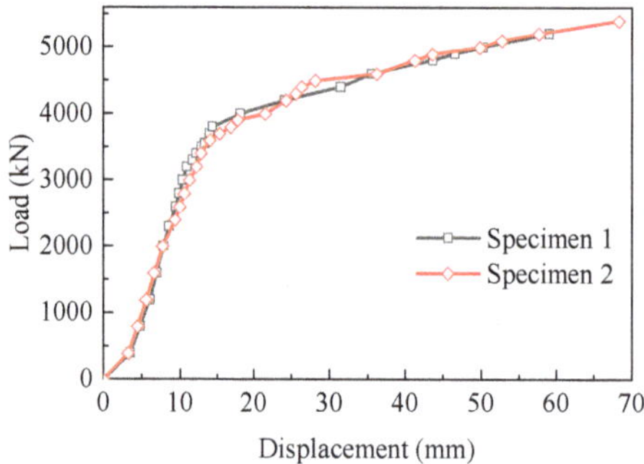

Figure 7. Load–displacement curves of the two specimens.

Table 3 provides a comparison of the characteristic loads and displacements between this study and previous studies [18,23]. RGP specimens and SJ specimens had a similar design to the specimens in this test, and they were all scale modes with a scale ration of 1:3. The minimum yield load and ultimate load of the specimens in this study were 3200 kN and 5200 kN, respectively. No significant difference in the yield load was found between the RGP specimens and the specimens in this study. However, the ultimate bearing capacity of the composite joints in this test was significantly greater than that of the RGP specimens and SJ specimens. Additionally, the minimum value of the ultimate displacement was 59.01 mm. The comparison of the results in Table 3 indicated that Specimen 1 and Specimen 2 had the greatest bearing capacity and deformability. For example, without considering the steel grade, the ultimate bearing capacity and displacement of Specimen 1 increased by 16% and 36%, respectively, compared with those of Specimen RGP-3, due to the increase in the thickness of the gusset plate from 12 to 16 mm.

Table 3. Experimental characteristics of the steel–concrete composite joint specimens.

Specimen	N_y	N_u	N_u/N_y	D_y	D_u	D_u/D_y	EI
RGP-1 [23]	3200	3500	1.09	27.03	-	-	11.84
RGP-2 [23]	3400	4200	1.24	23.50	42.79	1.82	14.47
RGP-3 [23]	3400	4500	1.32	23.43	43.46	1.85	14.51
SJ-1 [18]	2156	2548	1.18	16.13	34.49	2.14	13.37
SJ-2 [18]	2156	2940	1.36	15.92	40.12	2.52	13.54
Specimen 1	3200	5200	1.63	10.90	59.01	5.41	29.36
Specimen 2	3400	5400	1.59	12.89	68.31	5.30	26.38

N_y and D_y: yield load (kN) and the corresponding displacement (mm); N_u and D_u: ultimate load and the corresponding displacement; EI (the joint stiffness): N_y/D_y (10^4 kN /m).

3.3. Load–Strain Curves of the Concrete Chord

Specimen 1 was selected as the narrative object to avoid redundancy in the following sections. The strain results of the concrete chord are set out in Figure 8. To minimize the influence of the strain gauge failure, the average strain values of each section in the elastic state (2000 kN), elastic-plastic state (3500 kN), and closely ultimate state (5000 kN) are shown in Figure 8a. There is a clear trend of reducing in the compressive strain values. Evidently, concrete near Section A suffered from the biggish axial force. At the load of 2000 kN, the compressive strains of Section B, C and D were 50%, 15% and 1% of the strain of Section A (−663.14 με). The average strain values decreased quickly along the loading direction, because PBL shear connectors transferred force efficaciously from the concrete chord to the gusset plate. Moreover, the area of the concrete chord close to Section E was almost not subjected to the load. Figure 8b,c presents the strain values of the first row and the first column of gauges, respectively. There was a rapid decrease in the compressive strain values along the loading direction on the concrete chord. For example, the strain values of C1, C4, C7, C10 and C13 were −498.32 με, −291.49 με, −49.06 με, 29.82 με, and 6.73 με, respectively, at a load level of 2000 kN. Conversely, in the vertical direction, the strain values of C2 and C3 were 1.34 times and 1.66 times the strain of C1 (−498.32 με). There was a significant increase in the strains from the top to the bottom of the concrete chord.

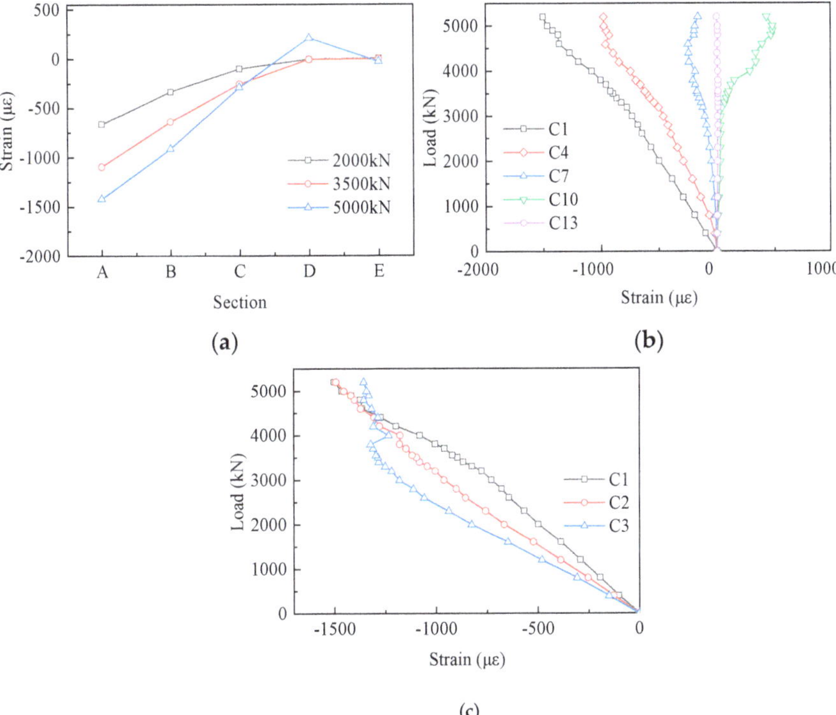

Figure 8. Strain results of the concrete chord: (**a**) the average strain values of each section; (**b**) the first row of the strain gauges; (**c**) the first column of the strain gauges.

3.4. Load–Strain Curves of the Gusset Plates

Figure 9 shows the load–strain curves of the gusset plates which were wrapped by the concrete chord. In the second row, the strain values of A2, A5, A8, A14, and A17 were 6%, 30%, 47%, 25%, and 6% of that of A11 at approximately 5000 kN, respectively. What stands out in Figure 9a,b is the rapid increase in strains from A2 to A11 and the significant decrease in that from A11 to A20. In other words, the strain results of the embedded gusset plates presented a trend of increasing first and then decreasing along the loading direction. Furthermore, almost all the points did not reach the yield strain (2194 με), except for A11. In the vertical direction, in Figure 9c there was an evident tendency of increasing from A4 to A6 (taking the second column strain gauges as an example). Moreover, the same phenomena occurred in other columns. Exceptionally, in Figure 9d compressive strains occurred at the first column of measuring points, indicating that the front ends of the gusset plates were biased towards compression at the early loading stage.

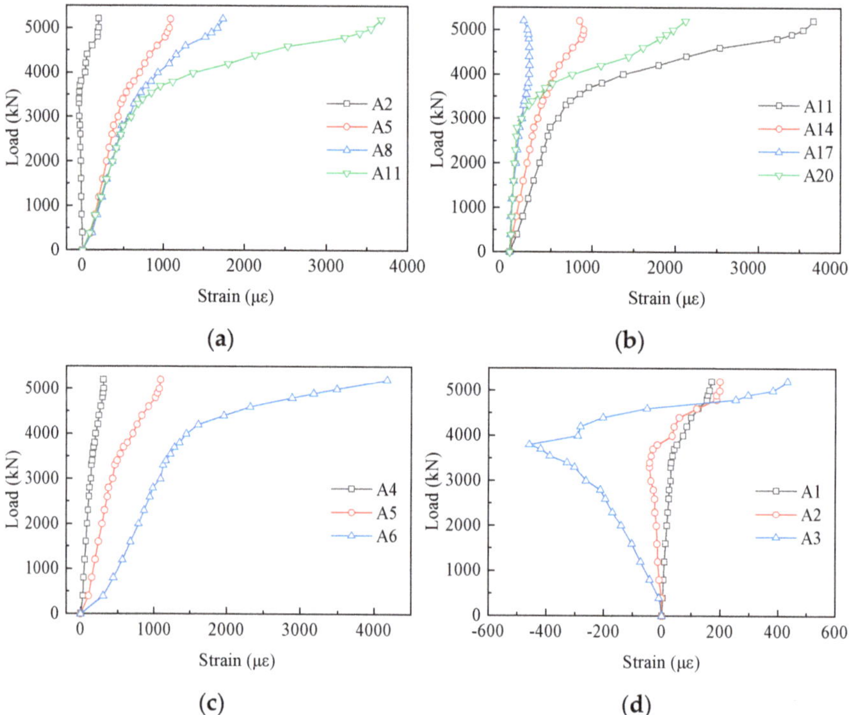

Figure 9. Load–strain curves of the embedded gusset plates: (**a**) A2 to A11 in the second row; (**b**) A11 to A20 in the second row; (**c**) the second column of strain gauges; and (**d**) the first column of the strain gauges.

Figure 10 presents the strain results of the gusset plates that were exposed to the air. Different from the strain results of the embedded gusset plates, the exposed gusset plates had a higher strain level. From the results in Figure 10a,b, the strains increased sharply after 3800 kN. Moreover, almost all the strain results of the fourth-row gauges exceeded the yield strain at the load of 4000 kN. Interestingly, the strains of the embedded gusset plates declined gradually from the middle to the edge in the loading direction. In the vertical direction, Figure 10c shows the load–strain curves of the middle row, where no significant variation tendency from A26 to A38 was evident. However, it should be noted that three measuring points were at a high-strain level. The strain of A26, A33, and A38 were bigger than the yield strain, when the load was about 3500 kN. Furthermore, the strain of those three points reached 8489 με, 8249 με and 10085 με, respectively, at the load of 4200 kN. In addition, the gauges could not function well soon afterward. These results suggest that the exposed part of the gusset plates was a key area for the steel–concrete composite truss joint to exert bearing capacity.

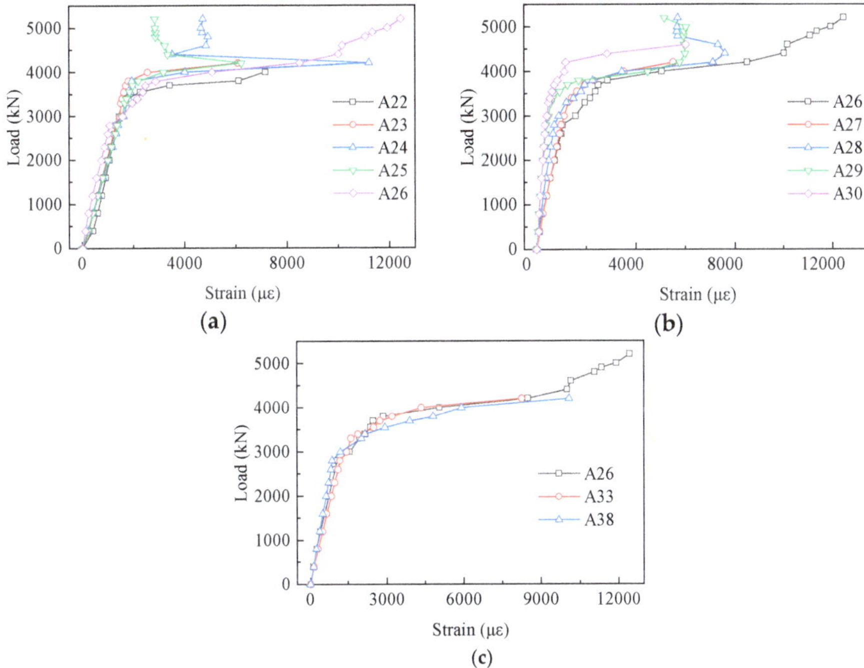

Figure 10. Load–strain curves of the exposed gusset plates: (**a**) A22 to A26 in the fourth row; (**b**) A26 to A30 in the fourth row; and (**c**) A26 to A38 in the middle column.

3.5. Load–Strain Curves of the Truss-Web Members

Figure 11 shows the measured strains of the tensile truss-web members. Almost all the strains of the tensile truss-web members were smaller than the yield strain during the whole loading process. From Figure 11a,b, it can be seen that the strain results at each loading stage and the trend of increasing were different. The slope of the load–strain curves decreased gradually, which showed that the strains of the tensile truss web increased along the loading direction. By contrast with L1, the tensile strain of L2 and L3 increased by 16% and 24% at the load of 3500 kN, respectively. In the vertical direction, Figure 11b shows that the strains increased from L1 to L7 and the growth trend was more obvious. Evidently, what stands out in the Figure is the growth of the strains in both directions.

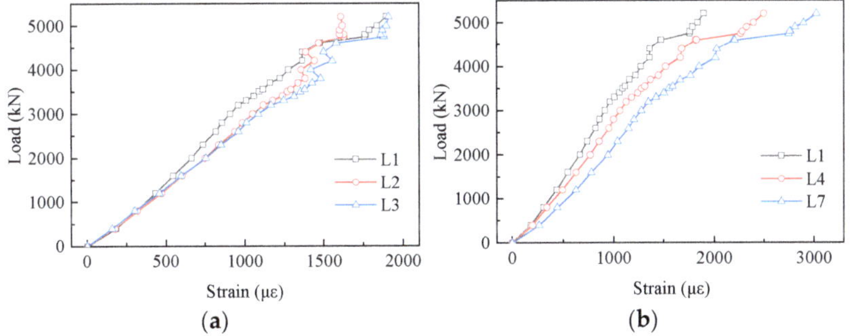

Figure 11. Load–strain curves of the tensile truss-web member: (**a**) L1, L2, L3; and (**b**) L1, L4, L7; L1: No. 1 strain gauge of tensile truss web.

The strain results of the compressive truss web are displayed in Figure 12. As shown in Figure 12a, the strain value decreased from Y1 to Y3 along the loading direction, which was contrary to the strain distribution of the tensile truss web. The striking observation to emerge from the strain comparison between the tensile truss web and the compressive truss web was that the area of truss web near the inside of the angle had greater strains. Moreover, the regularity of the strain distribution in the vertical direction was weaker in comparison with the tensile truss web.

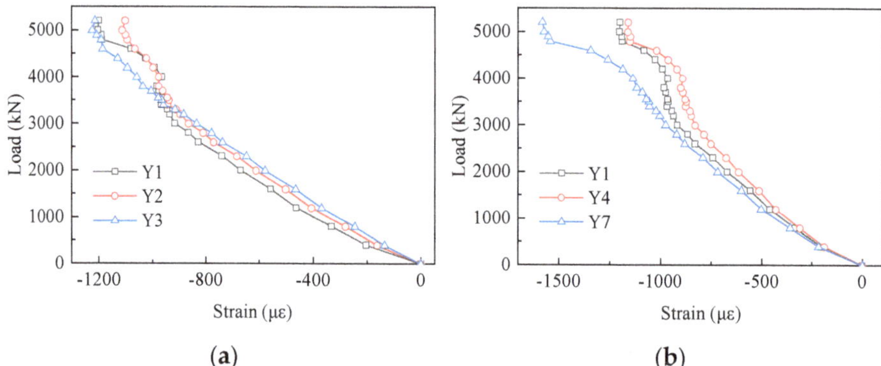

Figure 12. Load–strain curves of the compressive truss-web member: (**a**) Y1, Y2, Y3; and (**b**) Y1, Y4, Y7; Y1: No. 1 strain gauge of compressive truss web.

3.6. Load–Strain Curves of PBL Shear Connectors

Figure 13 displays the strain results of the PBL shear connectors. From this chart, we can see that no measuring points in the first row exceeded the yield strain during the loading process. Furthermore, there was a visible increase in the strains from X1 to X7 in the first row. Compared with X1, the strain of X7 increased by 71% at the load of 3500 kN. However, on the contrary, the strains decreased from X7 to X16. Compared with X7, the strains of X10, X13 and X16 reduced by 35%, 67% and 96% at the load of 3500 kN, respectively. The strains of the measuring points in the first row increased firstly and then decreased along the loading direction. In the vertical direction, in Figure 13b there was a clear trend of increase in the strain from X1 to X3. The strains of X2 and X3 were about 2.11 and 4.36 times that of X1 at the load of 3500 kN, respectively.

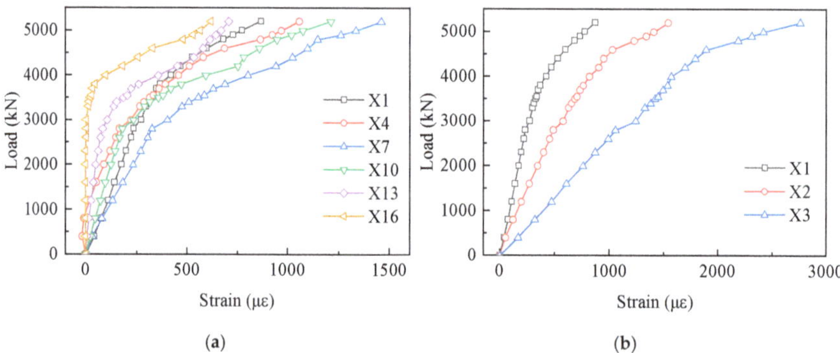

Figure 13. Load–strain curves of the perfobond rib (PBL) shear connectors: (**a**) X1 to X16 in the first row; and (**b**) X1 to X3 in the first column.

Figure 14 shows the load proportion of PBL shear connectors in the same row and the same column with different loads. In the first row, the load proportions from X1 to X16 were 26%, 14%, 35%, 18%, 7% and 0%, respectively, at the load of 2000 kN. The PBL shear connectors in the first row took on 14%, 18%, 25%, 20%, 12%, and 11% of the strains in the first row, when the load reached 5000 kN. Hence, in the same row, it can be seen that the first three PBL shear connectors in the first row bore the vast majority of the load in the initial stage. Moreover, the load proportion of the first three PBL shear connectors decreased gradually with the increase in the load. As shown in Figure 14b, at the load of 2000 kN, the load proportions of X1, X2, and X3 were 14%, 27%, and 59%, respectively. What stands out in Figure 14b is the rapid increase in the load proportion in the vertical direction. Moreover, the load proportion of X3 exceeded 50%. Figure 14c presents the load proportion of the third row of measure points in its column. The load transmitted by the third of points accounted for at least 60% of the total load transmitted by the same row of PBL shear connectors excluding X9. It was suggested that the third row of PBL shear connectors played the vital role in transferring the load between the concrete chord and the gusset plates.

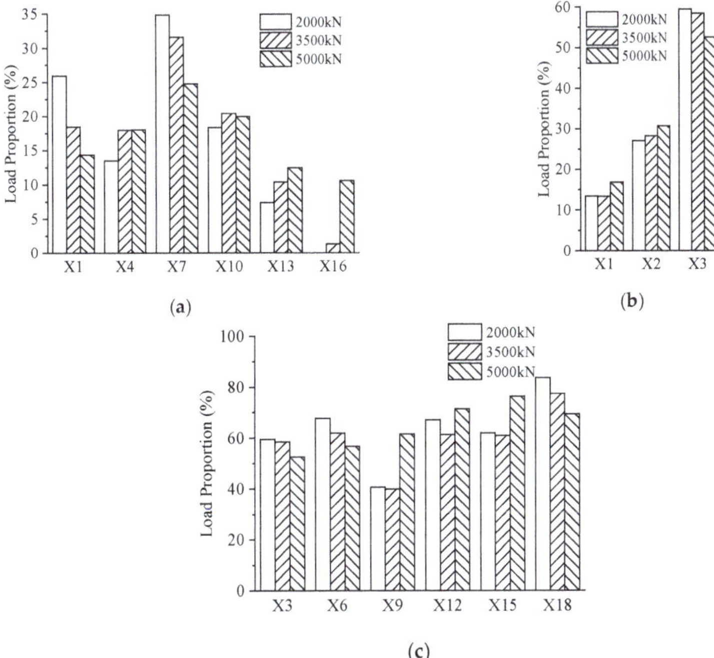

Figure 14. Load proportion of the PBL shear connectors: (**a**) X1 to X16 in the first row; (**b**) the first column; and (**c**) the load proportion of the strain gauges in the third row to their respective columns.

4. Conclusions

In this investigation, the purpose was to fully study the mechanical behavior and failure mode of steel–concrete connection joints by static model tests, especially to determine the strain distribution rule of the main components and the typical failure mode of such joints. The following conclusions can be drawn based on the present study.

1. Compared with the previous experimental results in Table 3, the present results indicated that the proposed joints showed excellent bearing capacity and deformability. The yield load of Specimen 1 and Specimen 2 was 3200 kN and 3400 kN, respectively. The ultimate load of Specimen

1 and Specimen 2 was 5200 kN and 5400 kN, respectively. Furthermore, the ultimate displacement of Specimen 1 was 59.01 mm and the ultimate displacement of Specimen 2 reached 68.31 mm.

2. In the experiments, Specimen 1 and Specimen 2 shared similar failure modes, such as the fracture of gusset plates, the slippage of high-strength bolts, the local buckling of compressive splice plates, the fracture of tensile splice plates and concrete cracking. In particular, the fracture of gusset plates, the slippage of high-strength bolts and concrete cracking were the most prominent failure modes. Moreover, it can be seen from the comparison of the failure modes between the proposed joints and the previous test specimens that the fracture of the gusset plates and the slippage of the high-strength bolts seemed to be two typical failure modes of such connection joints with high-strength bolts.

3. The strain distribution of the steel–concrete connection joint followed certain rules. For the concrete chord, the strains decreased quickly in the loading direction. On the contrary, the strains increased in the vertical direction. For the gusset plates, in the loading direction, the strains reduced gradually from the middle to the edge. In the vertical direction, strains of the embedded gusset plate enhanced from the top to the bottom, but the strain variation tendency of the exposed gusset plate was not obvious. For the steel truss-web members, the areas near the inside of the angle between the tensile truss web and the compressive truss web had greater strain values than in other areas. For PBL shear connectors, the strain distribution patterns were in keeping with the patterns of the embedded gusset plates. The third-row PBL shear connectors resisted most of the load rather than the other two-row connectors.

4. The exposed gusset plate was the key component of the steel–concrete connection joint to resist the load. The exposed gusset plate was at a higher strain level than the other members at the same loading step. Under the extreme condition, the exposed gusset plates of both specimens experienced a severe tensile fracture and were therefore unable to continue to bear the load. Furthermore, in the ultimate state, the fracture of the gusset plates prevented the specimen from continuing to bear the external load.

In this study, we focused on the experimental investigation of steel–concrete connection joints. However, the numerical analysis is also important for a full understanding of mechanical behavior and failure mode of such joints. We will perform the numerical analysis in another article in detail.

Author Contributions: Writing—original draft manuscript, Y.T.; conceptualization, Y.T., W.Y. and B.Z.; experiment operation, Y.T., L.Q., T.Y. and T.W.; data collection, Y.T., L.Q., T.Y. and T.W.; data analysis, Y.T., L.Q. and T.Y.; figures, Y.T. and L.Q.; writing—review and editing, Y.T., B.Z. and W.Y. All authors have read and agreed to the published version of the manuscript.

Funding: The research described in this paper was supported by the Scientific Research Service for the Lingang Yangtze River Bridge on the Zigong-Yibin Line of the newly built South Sichuan Intercity Railway (No. SRIG2019FW0001) and the Scientific Research and Development Project of the China Railway First Survey and Design Institute Group Co. Ltd. (No. 16–18).

Conflicts of Interest: The authors declare no conflict of interest.

References

1. Jung, K.H.; Lee, S.H.; Yi, J.W.; Kim, J.H.J.; Ha, J.H. Fatigue capacity of a new connection system for a prestressed concrete hybrid truss web girder. *Mag. Concr. Res.* **2012**, *64*, 665–672. [CrossRef]

2. Pham, X.T. A new generation of composite bridges, The Boulonnais Viaducts in France. *Proc. FIP Symp. Lisbon Port.* **1997**, *2*, 909–916.

3. Tanis, J.-M. Bras de la Plaine Bridge, Reunion Island, France. *Struct. Eng. Int.* **2003**, *13*, 259–262. [CrossRef]

4. Dauner, H.; Oribasi, A.; Wery, D. The Lully Viaduct, a composite bridge with steel tube truss. *J. Constr. Steel Res.* **1998**, *46*, 67–68. [CrossRef]

5. Bacchetta, A.; Dauner, H.G. The Dreirosenbridge over the Rhine at Basel. In Proceedings of the IABSE Congress Report, Lucerne, Switzerland, 1 January 2000; International Association for Bridge and Structural Engineering. pp. 440–447.

6. Reis, A.J.; Pedro, J.J.O. The Europe Bridge in Portugal: Concept and structural design. *J. Constr. Steel Res.* **2004**, *60*, 363–372. [CrossRef]

7. Liu, Y.; Xiong, Z.; Luo, Y.; Cheng, G.; Liu, G.; Yang, J. Double-composite rectangular truss bridge and its joint analysis. *J. Traffic Transp. Eng.-Engl. Ed.* **2015**, *2*, 249–257. [CrossRef]

8. Hanswille, G. Composite bridges in Germany designed according to Eurocode 4-2. In *Composite Construction in Steel and Concrete VI*; American Society of Civil Engineers: Reston, VA, USA, 2001; pp. 391–405.

9. Aoki, K.; Takatoku, Y.; Notoya, H.; Uehira, Y.; Kato, T.; Yamaguchi, T. Design and construction of Sarutagawa bridge and Tomoegawa bridge. *Bridge Found.* **2005**, *39*, 2–11.

10. Minami, H.; Yamamura, M.; Taira, Y.; Furuichi, K. Design of the Kinokawa viaduct composite truss bridge. In Proceedings of the 1st JSCE Congress, Composite Structures, Osaka, Japan, 13–19 October 2002; pp. 371–380.

11. Jung, K.H.; Kim, J.H.J.; Yi, J.W.; Lee, S.H. Development and Evaluation of New Connection Systems for Hybrid Truss Bridges. *J. Adv. Concr. Technol.* **2013**, *11*, 61–79. [CrossRef]

12. Furuichi, K.; Yamamura, M.; Nagumo, H.; Yoshida, K.; Eligehausen, R. Experimental study on a new joint for prestressed concrete composite bridge with steel truss web. In Proceedings of the International Symposium on Connections between Steel and Concrete, RILEM Publications SARL, Bagneux, France, 10–12 September 2001; pp. 1250–1259.

13. Jung, K.H.; Yi, J.W.; Kim, J.H.J. Structural safety and serviceability evaluations of prestressed concrete hybrid bridge girders with corrugated or steel truss web members. *Eng. Struct.* **2010**, *32*, 3866–3878. [CrossRef]

14. Yoshioka, T. The latest technologies of prestressed concrete bridges in Japan. In Proceedings of the JSCE-VIFCEA Joint Seminar on Concrete Engineering in Vietnam, Ho Chi Minh City, Vietnam, 8–9 December 2005; pp. 72–86.

15. Furuichi, K.; Hishiki, Y.; Yoshida, K.; Honda, T.; Yamamura, M.; Minami, H. The proposal of a design method of the joint structure for the steel/concrete truss bridge. *Proc. Jpn. Soc. Civ. Eng.* **2006**, *62*, 349–366. [CrossRef]

16. Weiguo, Y. Study on New Joint Structure in Steel Truss-web Bridge. *J. Highw. Transp. Res. Dev.* **2013**, *30*, 61–66.

17. Xue, D.; Liu, Y.; He, J.; Ma, B. Experimental study and numerical analysis of a composite truss joint. *J. Constr. Steel Res.* **2011**, *67*, 957–964. [CrossRef]

18. Zhou, L.; He, G. Model test for the end joint of long-span steel-concrete composite truss railway bridges. *China Civ. Eng. J.* **2012**, *45*, 92–99.

19. He, J.; Xin, H.; Liu, Y.; Ma, B.; Han, B. Mechanical Performance of Composite Truss Bridge with Double Decks. *IABSE Congr. Rep.* **2012**, *18*, 1025–1032. [CrossRef]

20. Choi, J.-H.; Jung, K.-H.; Kim, T.-K.; Kim, J.-H.J. Analytical and Experimental Studies on Torsional Behavior of Hybrid Truss Bridge Girders with Various Connection Joints. *J. Adv. Concr. Technol.* **2014**, *12*, 478–495. [CrossRef]

21. Liu, Y.; Xin, H.; He, J.; Xue, D.; Ma, B. Experimental and analytical study on fatigue behavior of composite truss joints. *J. Constr. Steel Res.* **2013**, *83*, 21–36. [CrossRef]

22. Jung, K.-H.; Lee, S.-H.; Yi, J.-W.; Choi, J.-H.; Kim, J.-H.J. Torsional Behavior of Hybrid Truss Bridge according to Connection Systems. *J. Korea Concr. Inst.* **2013**, *25*, 63–72. [CrossRef]

23. Yin, G.; Ding, F.; Wang, H.; Bai, Y.; Liu, X. Connection Performance in Steel-Concrete Composite Truss Bridge Structures. *J. Bridge Eng.* **2016**, *22*, 04016126. [CrossRef]

24. Tan, Y.; Zhu, B.; Yan, T.; Huang, B.; Wang, X.; Yang, W.; Huang, B. Experimental Study of the Mechanical Behavior of the Steel-Concrete Joints in a Composite Truss Bridge. *Appl. Sci.* **2019**, *9*, 854. [CrossRef]

25. Miwa, H. Experimental Study on the Mechanical Behaviour of Panel Joints in PC Hybrid Truss Bridges. *J. Struct. Eng.* **1998**, *44*, 1475–1483.

26. Sakai, Y.; Hosaka, T.; Isoe, A.; Ichikawa, A.; Mitsuki, K. Experiments on concrete filled and reinforced tubular K-joints of truss girder. *J. Constr. Steel Res.* **2004**, *60*, 683–699. [CrossRef]

27. Shoji, A.; Kondo, T.; Osugi, T.; Sonoda, K. A Study on Panel Point of a PC Bridge with Steel Truss Webs. *Proc. Jpn. Soc. Civ. Eng.* **2004**, *2004*, 75–89.

28. Toshio, N.; Satoru, O.; Toshiaki, K. A study on structural performance of joint in PC hybrid truss bridge. *Rep. Obayashi Corp. Tech. Res. Inst.* **2004**, *68*, 1–6.

29. Sato, Y.; Hino, S.I.; Sonoda, Y.; Yamaguchi, K.; Cheon, S.B. Study on load transfer mechanism of the joint in hybrid truss bridge. *J. Struct. Eng. A* **2008**, *54*, 778–785. [CrossRef]

30. Yin, G.; Wang, H. Study on ultimate bearing capacity of steel-concrete composite truss bridge bottom-chord joints. *China Civ. Eng. J.* **2016**, *49*, 88–95.
31. Zhou, L.; He, G. Experimental research on end joint of steel-concrete composite truss. *Balt. J. Road Bridge E* **2012**, *7*, 305–313. [CrossRef]
32. Maojun, D.; Zhao, L.; Jiandong, Z.; Duo, L. Experimental Study on New PBL-Steel Tube Joint for Steel Truss-Webbed Concrete Slab Composite Bridges. *J. Southeast Univ. (Nat. Sci. Ed.)* **2016**, *46*, 572–577.

Article

Laboratory Methods for Assessing the Influence of Improper Asphalt Mix Compaction on Its Performance

Michał Wróbel, Agnieszka Woszuk and Wojciech Franus *

Department of Geotechnical Engineering, Faculty of Civil Engineering and Architecture,
Lublin University of Technology, Nadbystrzycka 40, 20-618 Lublin, Poland; m.wrobel@pollub.pl (M.W.);
a.woszuk@pollub.pl (A.W.)
* Correspondence: w.franus@pollub.pl; Tel.: +48-815-384-416

Received: 6 May 2020; Accepted: 26 May 2020; Published: 29 May 2020

Abstract: Compaction index is one of the most important technological parameters during asphalt pavement construction which may be negatively affected by wrong asphalt paving machine set, weather conditions, or the mix temperature. Presented laboratory study analyzes the asphalt mix properties in case of inappropriate compaction. The reference mix was designed for AC 11 S wearing layer (asphalt concrete for wearing layer with maximum grading of 11 mm). Asphalt mix samples used in the tests were prepared using Marshall device with the compaction energy of 2×20, 2×35, 2×50, and 2×75 blows as well as in a roller compactor where the slabs were compacted to various heights: 69.3 mm (+10% of nominal height), 66.2 mm (+5%), 63 mm (nominal), and 59.9 mm (−5%) which resulted in different compaction indexes. Afterwards the samples were cored from the slabs. Both Marshall samples and cores were tested for air void content, stiffness modulus in three temperatures, indirect tensile strength, and resistance to water and frost indicated by ITSR value. It was found that either insufficient or excessive level of compaction can cause negative effect on the road surface performance.

Keywords: asphalt mix; compaction index; volumetric parameters; stiffness modulus; moisture resistance

---

1. Introduction

The quality of an asphalt road surface depends on many factors, including: The type of materials used (aggregates, asphalt, and others), climatic conditions, traffic load, the drainage system, and the method of construction [1,2]. Important factors directly determining the service life of the road surface are the features of the designed asphalt mix, such as: fatigue limit, resistance to rutting, or resistance to water and frost [3–5]. Other parameters, such as the density of the mineral-asphalt mixture or the air void content (porosity), are indirect. It is generally assumed that in case of dense-graded asphalt mix the content of air voids should be high enough to avoid rutting and low enough at the same time to limit water and dirt penetration into the surface structure [6]. In addition, the surface texture is related to the content of air voids in the wearing layer, which has a direct impact on the safety of road users [7]. Hence, the need for appropriate compaction of particular pavement layers has been known since the beginning of asphalt road construction technology. The main purpose of asphalt mix compaction is to achieve appropriate density to meet the requirements for physical and mechanical properties and to ensure a tight surface which will be exploited for the maximum possible life cycle [8]. It is a difficult process involving the selection of appropriate compaction parameters in order to achieve the assumed compaction index (CI) of the layer.

Under laboratory conditions, samples of asphalt mix can be compacted in various ways, depending on the purpose of the sample (type of tests), the applicable regulations, as well as the type of technology used [9]. The most commonly used devices are gyratory compactor, Marshall compactor, asphalt roller compactor, and a vibrating compactor [10]. These methods differ in many parameters, such as the pressure force, the way the force is transmitted, compaction time, and also the final shape of the obtained sample and the particles orientation inside [11]. In many European countries the basic method is Marshall compaction method which involves dropping the compacting hammer on a sample of asphalt mix in cylindrical form a certain number of times, which depends on the further use of the sample. Another way is to use a gyratory compactor, where the sample is compacted to the required density or a certain number of rotations, while being simultaneously compacted and sheared [12]. The method is essential in the case of projecting asphalt concrete mixes for very thin layers (BBTM). Moreover, A. Woszuk stated that in order to determine the operating temperatures of Warm Mix asphalt the gyratory compaction method is recommended [10]. According to the European standards [13], the sample placed in a cylindrical form with a diameter of 100 mm or 150 mm is subjected to a constant compressive stress of 600 kPa during the rotational movement, which is transmitted through the piston to parallel sample bases. During the compaction, the sample is rotated at a speed of 30 rpm and the angle of inclination is 1.25° (typical parameters in the Superpave system [14]). Such process better reflects the compaction in real conditions than Marshall method as well it is possible to determine significant coefficients concerning susceptibility to compaction of the mix: compaction coefficient K, mixture stability index MSI, and mixture resistance index MRI [15,16]. Georgiou et al. found that in the case of compacting in the gyratory device, the samples were segregated, whereas in the case of cores they had a low level of segregation [17]. Additionally, in order to obtain the internal structure of samples comparable to roller compaction, the compaction angle parameter should be appropriately selected, slightly higher than that specified by technical regulations. Taking into account the resistance to rutting, Tapkin and Keskin demonstrated that the samples made in the gyratory press were less susceptible to permanent deformations than those compacted using the Marshall method [18]. Lee et al. found that in case of crumb rubber asphalt mixes air void content was significantly dependent on the number of gyratory rotations and directly related to this rutting increased along with air voids [19]. In reference to the interlayer bonding, Jaskuła showed that the higher the compaction index, the higher was maximum shear strength between the layers [20]. The shearing and mechanical behavior is one of the major issues in many other fields like food industry, paints and coating industry, and also energy industry. In these areas hybrid gels featuring interpenetrating covalent can be applied, which exhibit both high toughness and recoverability of strain-induced network damage [21].

In the rutting test, however, samples made in a roller compactor are used [22]. Airey and Collop indicate that slab-compacted specimens are more closely correlated with cores sampled from the road surface than gyratory and vibratory-compacted specimens [11]. There are also self-compacting mixes, e.g., mastic asphalt, often used in bridge structures [23] or SCC cement concrete [24].

Under real construction conditions, the target compaction is carried out by road rollers of different sizes and types of drums, however the first compaction is carried out by the paver. A vibrating and heated board (table) levels and pre-compacts the mix and shapes longitudinal and transverse slopes. This is a very important process as it allows to achieve about 90% of the required compaction of the pavement layer. In order to effectively develop the road surface design, the compaction of the mixture in the laboratory must properly simulate the compaction in real conditions [25].

According to the European standard, the required level of density expressed as percentage refers to the volumetric density obtained for samples made in the Marshall compactor compared to the results obtained for samples cored from the pavement. The degree of compaction of the layer on site, in addition to the testing of the cores, can also be determined by nuclear and non-nuclear density gauges [26]. In relation to these types of density tests, Micaelo et al. noted that the nuclear method has greater fluctuations in results [27]. Another method may be the density measurement based on an analyzer using neural networks which has the ability to predict the density in real time during the compaction process [28,29].

During road construction (also in laboratory) there is a risk of improper layer compaction, which results from several factors: inappropriate choice of rollers, weather conditions, properties, and temperature of the asphalt mix. Praticò et al. also stated that volumetric properties are also influenced by the test method and even the diameter of the samples [30].

As shown in the presented literature analysis, relevant compaction significantly affects the quality of the asphalt road surface. Moreover, literature data clearly indicate that there are strong correlations between the compaction method and the mechanical properties of asphalt mix. The main aim of this article is to assess the influence of improper compaction (including under or over-compaction) on the quality and durability parameters of the road surface such as density, air void content, stiffness modulus, indirect tensile strength, and resistance to water and frost.

2. Experimental Materials

Asphalt Mix Components

The bitumen used in the tests was 50/70 asphalt from ORLEN Asfalt Sp. z o.o. (Plock, Poland). The properties of asphalt binder are presented in Table 1.

Table 1. Properties of the base bitumen.

Test	Specification	Result	Specification Limit
Penetration (25 °C, 0.1 mm)	EN 1426:2009	59	50–70
Viscosity at 135 °C (mPa·s)	ASTM D 4402	428	-
Softening point(°C)	EN 1427:2009	53	46–54
Penetration index	EN 12591:2010	−0.1	-

The aggregates used in laboratory tests were sand, dolomite, and limestone as a mineral filler. The dolomite used in the research comes from the Piskrzyn deposit (Kopalnie Dolomitu S. A.) while limestone from the Bukowa deposit (Lhoist Bukowa Sp. z o.o.). Both mines are located in the Świętokrzyskie Province (Poland). The sand used in the study comes from the Baranówka deposit. The microstructure of the initial carbonate raw materials (dolomites and limestone) used in the study is presented in Figure 1. The chemical composition of the minerals determined by the X-ray fluorescence spectrometry (XRF) method is presented in Table 2. For carbonate raw materials, the dominant chemical component is CaO, which is 80.94% for limestone filler and 39.62% for dolomite. In the case of the latter, MgO is also an important component, with a content of 14.76%. For quartz sand the main component is SiO_2 (92.99%). Silica is also present in dolomite aggregate and its content is 10.05%. The remaining chemical components in the tested raw materials are present in minor quantities. Mineral composition of used aggregates and filler identified using the X-ray diffraction (XRD) method is presented in Figure 2. The content of individual mineral phases was determined by the Rietveld refinement. For dolomite aggregate, the content of dolomite equals 87.6%, 1.2% of calcite and 11.2% of quartz. In case of limestone filler the calcite content is 98.3%. The mineral composition of the filler is supplemented with quartz in the amount of 1.7%. In the sand used for the study, the main component is quartz, which is 93.6%, followed by marginal amounts of potassium feldspar identified as albite (6.4%). SEM images of the materials used are shown in Figure 3. The shape of dolomite grains is shown in Figure 3a and can be described as angular, while the character of their surface is presented in Figure 3b and can be clearly marked by a three-way cleavage which results in rough character. Limestone filler and crushed sand are presented in Figure 3d respectively. The limestone grains reach a size of about 1μm and represent a fairly regular shape. Quartz grains present in the sand most often reach irregular shapes emphasized by a clear fracture.

Figure 1. Microphotographs of dolomite (**a**) and limestone (**b**).

Table 2. Chemical composition of the studied materials.

	Dolomite (% wt)	Limestone Filler (% wt)	Sand (% wt)
MgO	14.76	0.21	-
Al_2O_3	3.53	0.14	2.80
SiO_2	10.05	0.25	92.99
K_2O	1.42	-	0.82
CaO	39.62	80.94	0.25
TiO_2	0.24	-	0.09
MnO	0.23	-	-
Fe_2O_3	2.15	0.07	1.29
LOI	28.00	18.39	1.78

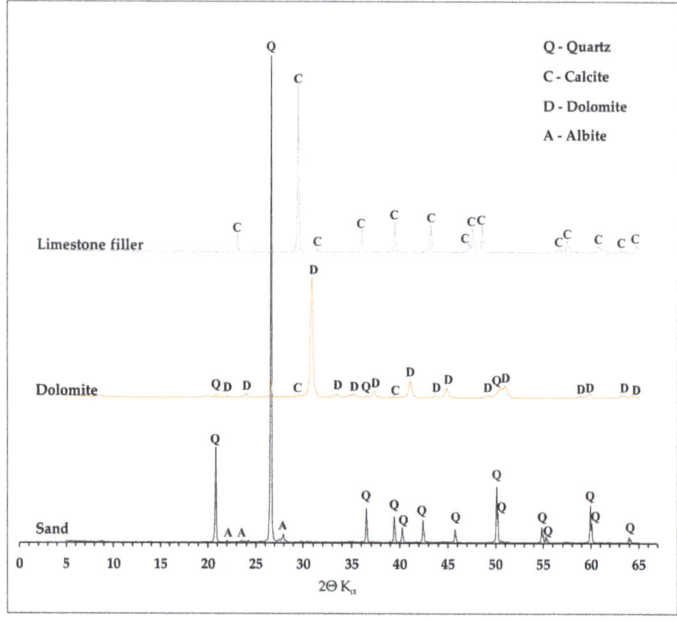

Figure 2. Mineral composition of used aggregates and filler.

Figure 3. Scanning electron microscopy (SEM) images of dolomite (**a**,**b**), limestone filler (**c**) and sand (**d**).

Aggregate grading was determined by dry sieving method [31]. In case of the filler air jet sieving test was applied [32]. The composition of the mineral mix was designed using the grading envelope method applicable in Poland [22].

The designed reference asphalt mix is intended for construction of asphalt concrete wearing layer (AC 11 S) on roads with traffic loads of KR category 1–2 in accordance with Polish technical standards [22]. The content of dosed asphalt was 5.7% in relation to the weight of the asphalt mix. Figure 4 shows the grading of the mineral mix. The composition of both mineral and asphalt is presented in Table 3.

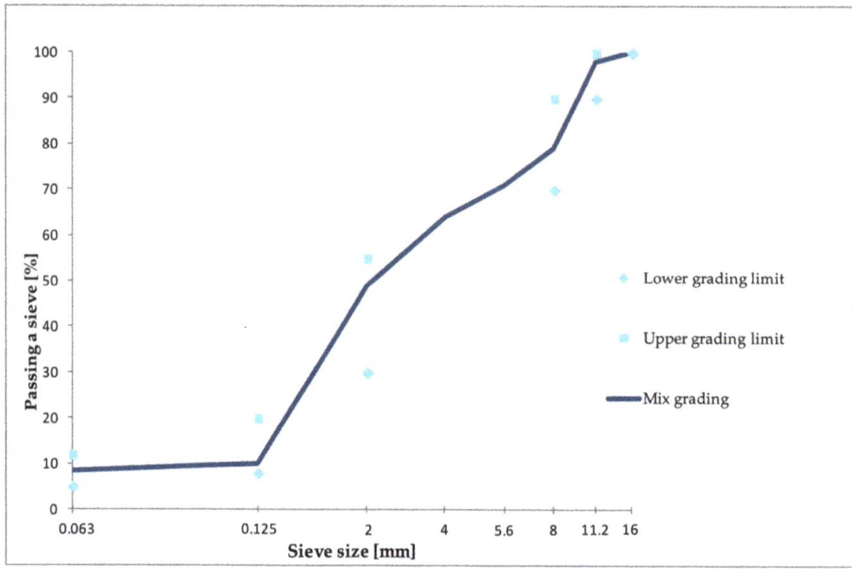

Figure 4. Grading of the mineral mix.

Table 3. Composition of mineral mix and asphalt mix.

Material Type	Percentage (%)	
	Mineral Mix	**Asphalt Mix**
Mineral filler	6.0	5.5
0/2 Quartz, fine aggregate	20.0	18.8
0/2 Dolomite, fine aggregate	26.0	24.6
2/8 Dolomite, coarse aggregate	26.0	24.6
8/11 Dolomite, coarse aggregate	22.0	20.8
50/70 Asphalt	-	5.7

3. Research Method

3.1. Samples Preparation

Samples made of reference asphalt mix were compacted in a Marshall device using 50 blows per side and named M3. The compaction temperature was 135 °C according to Polish requirements [19]. Samples were also compacted using different energies: 20, 35, and 75 blows per side, named M1, M2, M4 respectively. In order to reflect the compaction method in the real conditions prevailing during the placement of asphalt mix into the road surface and to assess the influence of the method on the properties samples were also made in a slab roller compactor that uses tilting pivoted circular sector with closed loop control system. To achieve various compaction indexes, samples were compacted to variable target heights: 63 mm, 66.2 mm, 69.3 mm, and 59.9 mm. The weight of the batch used for each slab was the same and was calculated taking into account the density of the asphalt mix, assuming that a 63 mm high slab sample should have a compaction index of 98% which is the minimum value required for compacting the asphalt mix under real conditions.

The attempt to produce a sample with a height of 56.7 mm (reduced by 10% compared to 63 mm) failed. The compaction device stopped at a slab height of 58.0 mm and the obtained sample was deformed, thus excluded from further testing.

Cylindrical cores were cut out of each slab for further testing and marked with symbols:

- Cores sampled from slab no. 1, 69.3 mm high (63 mm + 10%)—R1

- Cores sampled from slab no. 2, 66.2 mm high (63 mm + 5%)—R2
- Cores sampled from slab no. 3, 63 mm high—R3
- Cores sampled from slab no. 4, 59.9 mm high (63 mm − 5%)—R4

3.2. Test Methods

Volumetric parameters:

Compaction index was determined on core samples according to the following formula:

$$CI = \frac{\rho_{core}}{\rho_{Marshall}} \times 100 \ (\%)$$

where:

CI—compaction index [%]

ϱ_{core}—bulk density tested on core sample

$\varrho_{Marshall}$—bulk density determined on Marshall samples

Bulk density was determined using the B method—saturated surface dry (SSD) [33] and calculated for each sample according to the formula:

$$\rho_{bssd} = \frac{m_1}{m_3 - m_2} \times \rho_w$$

where:

ρ_{bssd}—bulk density (SSD) (kg/m^3)

m_1—mass of the dry specimen (g)

m_2—mass of the specimen in water (g)

m_3—mass of the saturated surface-dried specimen (g)

ρ_w—density of water at test temperature (kg/m^3)

Maximum density was determined experimentally according to the EN 12697:5:2018 standard [34] and calculated using the following formula:

$$\rho_m = \frac{m_2 - m_1}{1000 \times \left(V_{pp} - \frac{m_3 - m_2}{\rho_w}\right)}$$

where:

ρ_m—density of the asphalt mix (kg/m^3)

m_1—mass of the pycnometer (g)

m_2—mass of the pycnometer with the specimen (g)

m_3—mass of the pycnometer with the specimen and water (g)

ρ_w—density of water at test temperature (kg/m^3)

V_{pp}—Volume of the pycnometer filled to the measuring line (m^3)

Air void content was calculated according to the EN 12697-8:2018 [35] using the following formula:

$$V_m = \frac{\rho_m - \rho_b}{\rho_m} \times 100\%$$

where:

V_m—air void content in the asphalt mix specimen (%)

ρ_m—density of the asphalt mix (kg/m^3)

ρ_b—bulk density of the asphalt mix (kg/m^3)

ITSR:

Water and frost resistance tests of the asphalt concrete were performed based on the EN12697-12:2018 standard [36]. This test evaluates the effect of one freeze-thaw cycle of saturated mix asphalt samples on indirect resistance to stretching. Both the samples made in the Marshall compactor and the samples cut from the slabs were divided into two sets. The control series samples were conditioned in a laboratory on flat surface in room temperature of 20 ± 5 °C. The samples from the second set were conditioned in water at 40 °C for 72 h, then frozen at −18 °C for 16 h and re-conditioned in water at 25 °C for 24 h. After conditioning, the indirect tensile strength of all test pieces was tested according to EN 12697-23 at 25 °C. Based on the obtained results, an indicator of resistance ITSR was calculated:

$$ITSR = \frac{ITS_w}{ITS_d}$$

where:

ITSR—indirect tensile strength ratio (%)
ITSw—the average strength for the wet set of samples (kPa)
ITSd—the average strength for the dry set of samples (kPa)

Stiffness modulus:

The tests of the stiffness modulus were performed according to the EN 12697-26:2018 standard [37]. Controlled force was applied to each sample. The samples were subject to five dynamic loads applied to the sample vertically, along the diameter of the base. Force increase time, measured from zero to a maximum value, was 0.124 s. Maximum force generated a horizontal dislocation of sample equal to 5 μm. The sample, after performing the test, was rotated by 90° around the horizontal axis and tested again. A reliable stiffness modulus for each sample was a mean average out of two measurements. The final result was the arithmetic mean out of three tested samples. The tests were performed in three temperatures: 23 °C, 10 °C, −2 °C. The following Poisson's ratios were applied for the temperatures, respectively: 23 °C–0.4; 10 °C–0.3; −2 °C–0.25.

4. Results and Discussion

4.1. Bulk Density and Compaction Index

The results of bulk density of the tested samples are presented in Figure 5.

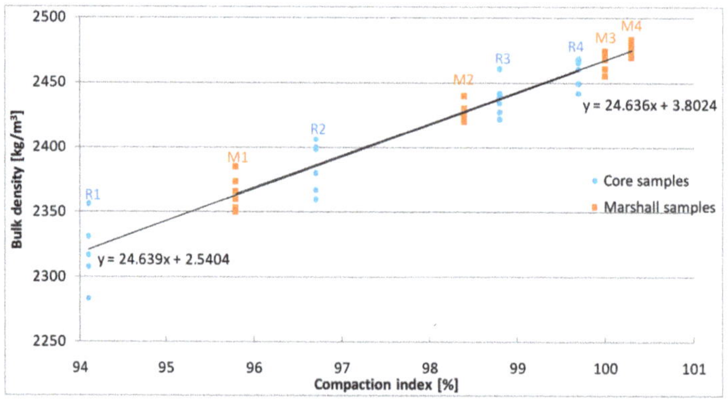

Figure 5. The results of bulk density.

Bulk density of the reference samples was 2467 kg/m^3. The results for Marshall samples varied from 2364 kg/m^3 to 2475 kg/m^3. The core samples cut from the slabs are characterized by bulk density ranging from 2321 kg/m^3 to 2459 kg/m^3. In Figure 5, a linear decrease in bulk density can be observed as the slab height increases. A similar trend can be observed for Marshall samples under decreasing compaction energy. The average results of bulk density were the basis for determining the compaction indexes listed in Table 4.

Table 4. Specimen determination, bulk density results, air void content results, and compaction index values.

Specimen Type	Height Change (%) Compaction Energy (Blows)	Bulk Density (kg/m^3)	Standard Deviation	Air void Content (%)	Standard Deviation	Compaction Index (%)
R1	+10 *	2321	25	8.3	1.0	94.1
R2	+5 *	2385	19	5.7	0.8	96.7
R3	0 *	2437	14	3.7	0.5	98.8
R4	−5 *	2459	11	2.8	0.4	99.7
M1	2 × 20 **	2364	5	6.5	0.5	95.8
M2	2 × 35 **	2427	8	4.1	0.3	98.4
M3	2 × 50 **	2467	7	2.5	0.3	100.0
M4	2 × 75 **	2475	13	2.2	0.2	100.3

* Height change of the slab samples (%); ** Compaction energy of Marshall samples (Blows)

According to the Polish technical regulations, the compaction index of asphalt concrete during the paving process is at least 98% [38]. On the other hand, slabs prepared in the laboratory for testing resistance to permanent deformations should have a compaction index ranging from 98% to 100% [22]. According to the assumptions, a change in the target height of slab compaction influenced the compaction index value. Samples cut from slabs with a height reduced by 5% (59.9 cm) had a compaction index of 99.7%. This value is within the requirements for both laboratory samples for further testing and for mixtures compacted with rollers during road surface preparation. Slabs of increased thickness were characterized by compaction index at the level of 97.3% and 94.3%, which indicates under-compaction of the mixture. Insufficiently compacted asphalt layers are more susceptible to deeper water penetration and more intensive asphalt oxidation and, as a consequence, to faster surface degradation. On the contrary, excessively compacted asphalt pavements are more susceptible to permanent deformations and low-temperature cracking [39].

The Marshall samples compacted using 50 blows per side are reference samples with a compaction rate of 100% in accordance with Polish requirements. As expected, the compaction index was dependent on the compaction energy. However, as in the case of the slab compactor, this parameter was found to be not very susceptible to compaction. Although similar correlations of density and compaction index were obtained. Proper compaction of the tested mixture is significantly influenced by the applied filler with regular grain shapes. As indicated by Zulkati et al. if the filler has a large enough diameter and a regular shape, it acts as a friction-lubricate agent. This facilitates a faster and smoother reorientation movement of larger aggregates, thereby resulting in high compaction susceptibility [40]. Conversely, Melloti et al. noticed that irregular shaped filler has a negative effect on the workability of asphalt mix [41].

As shown by previous studies, particle orientation and general aggregate structure is significantly different in samples compacted by different methods. Slab samples are characterized by an even particle size distribution across, while samples produced using other methods of laboratory compaction are susceptible to circumferential particle orientation. Contact and interlocking of aggregates, depending on the shape of the aggregate, also has a significant impact on the compaction. Use of angular aggregate particles means achieving more contact points and more uniform distribution of internal forces, with a better interconnection between elements and improvement to fatigue performance as well as permanent deformation resistance [42]. In the tests, crushed aggregates were used, which in combination with the asphalt binder guarantee the durability of the created bonds. Moreover, applied aggregates have an angular shape, which was confirmed by SEM analysis. Finally, the orientation and segregation of aggregates affects both the air void content and the mechanical properties of the asphalt mix [11].

4.2. Density And Air Void Content of the Asphalt Mix

The results of the air void content calculations are presented in Figure 6.

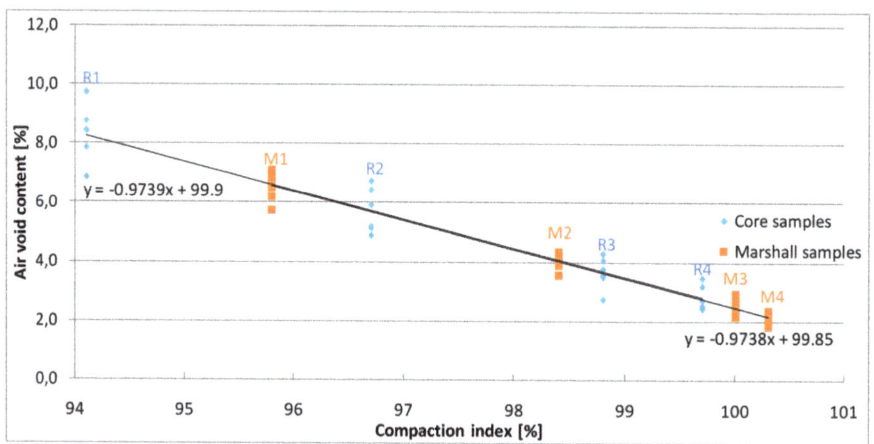

Figure 6. The results of air void content calculations.

Air void content was calculated according to the maximum density of the asphalt mix that was 2530 kg/m^3 and the bulk density of individual samples. The values of the parameter determined on Marshall samples varied from 2.2% to 6.5% and was adequate to the number of blows (Table 4). The air void content of AC 11 S reference samples made was 2.5%. According to technical regulations [19], asphalt concrete can be used for the wearing course if the samples' porosity is between 1 and 3%. The analysis of the obtained results shows that increasing the compaction energy above the standard requirements results in a slight change in the content of air voids; with an increase in compaction by 50% (from 50 × 2 to 75 × 2), a decrease in voids was from 2.5% to 2.2% and the compaction index was 100.3%. On the other hand, the reduction in the number of blows during the compaction of samples results in a considerable increase in free spaces. As indicated by Lucas Júnior et al. in samples with high air void content, the mutual blocking of aggregates is not fully possible because of their shape and texture. As a consequence, the functional properties of the produced mix deteriorate [43].

The content of air voids determined for samples cut out of slabs was from 2.8% to 8.3%. The value increased with the decrease of the compaction index, which is the expected relation. The core samples with a 99.7% compaction index were characterized by a free space content of 2.8%, which was closest to the results obtained for reference samples. It was also noted that in samples with the 98.8% compaction index, the free air void content was above the upper limit according to Polish requirements and was 3.7%, however the requirement of 1 to 3% porosity refers to samples made using the Marshall method. In case of surfaces made of AC 11 S mix compacted with rollers in real conditions, the technical recommendations [38] indicate that this layer should be compacted at a rate above 98% and contain between 1 and 4.5% air voids. It can be concluded that the samples made in a roller compactor with a compaction rate of at least 98% meet the requirement for the porosity of the asphalt layer. Nevertheless, the results of these tests are based on the assumed simulation of compaction by rolling in the laboratory, and the differences in the results are the effect of different slabs thicknesses. Under real conditions, the correct compaction of each layer is expected. In addition, one of the important elements influencing the proper effect of this process is the temperature of the asphalt mix—the lower the larger the voids in the layer, which is confirmed by both volumetric studies [16,44] and x-ray computed tomography analyses [45]. Wang L et al. noticed that not only the percentage content of air voids affects the performance of the surface, but also the spatial distribution and pore diameter [46]. Jiang W et al. proved that the characteristics of air voids are significantly influenced by aggregate gradation [47]. Moreover, the structure of aggregate, number of contact points, and orientation of each aggregate are dependent on the compaction methods and conditions [48].

4.3. Water and Frost Resistance

The results of the indirect tensile strength test are shown in Figure 7; Figure 8 as well as Table 5. The strength obtained for the samples compacted using Marshall method was in the range of 504 kPa to 648 kPa in the dry set and 421 kPa to 598 kPa in the wet set. It was observed that the results were dependent on the compaction energy, whereas in the case of 2×75 blows the strength increased only by 3% in the dry set and 4% after freezing cycle comparing to the reference samples.

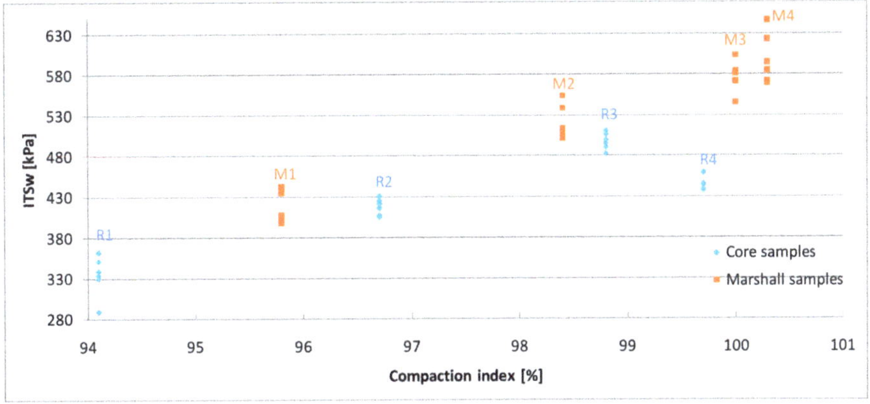

Figure 7. The results of indirect tensile strength test for the wet set.

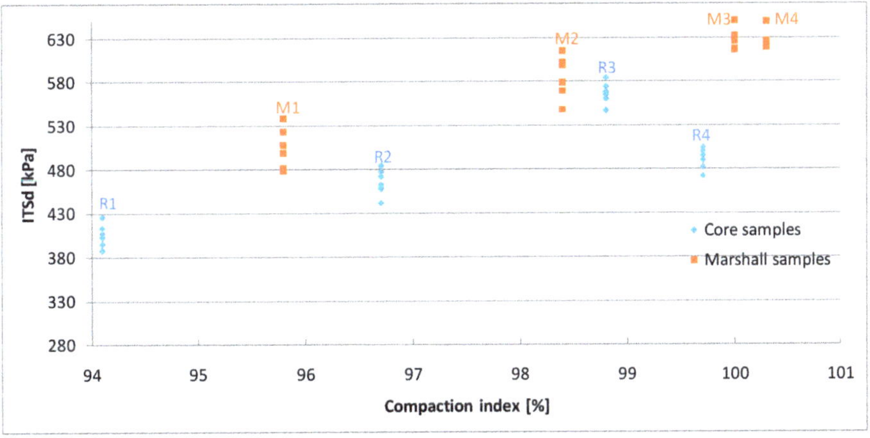

Figure 8. The results of indirect tensile strength test for the dry set.

Table 5. Average results of indirect tensile strength for the wet set and standard deviations.

	Compaction Index (%)	ITS_w (kPa)	Standard Deviation	ITS_d (kPa)	Standard Deviation	ITSR (%)
Cored samples	94.1	334	25	406	13	82.4
	96.7	418	10	466	15	86.6
	98.8	497	10	567	13	87.8
	99.7	448	10	491	12	91.2
Marshall samples	95.8	421	19	504	23	83.5
	98.4	520	22	585	24	88.9
	100	577	19	629	12	91.7
	100.3	598	30	648	33	92.3

Considering the core samples in general, it was noted that the strength increased with the growth of the compaction index, however the strength of the samples with the highest compaction index (99.7%) turned out to be lower than expected. This is probably due to the over-compaction of the asphalt mix. A further consequence of excessively "dense" mixtures is a decrease in resistance to permanent deformations and low-temperature cracking [39]. It was also observed that asphalt mix has reduced strength characteristics if the air void content is above the required limits. This is particularly the case for R1 samples where the porosity was the highest (8.3%) and the strength characteristics were lowest, 406 kPa for dry set and 334 kPa after the freezing cycle respectively. The acceptable limit of voids content was also exceeded in R4 samples, which is reflected in their strength.

Correlations between the compaction methods and the obtained strength results were observed. At a similar value of compaction index Marshall samples were more resistant than cores, in a similar range from 54 kPa to 91 kPa on average in dry set and from 33 kPa to 87 kPa after freezing cycle. The difference may be caused by change in aggregate interlock pattern in the particular method [49]. Thus, the strength of the samples is directly related to the compaction process, which significantly affects all properties of the asphalt mix.

The results of water and frost resistance tests are presented in Table 4. The average ITSR parameter for reference Marshall specimens was 91.7%. In case of the other Marshall samples the value varied from 83.5% for the least compacted to 92.3% for the most compacted ones. R4 core samples, which on the basis of the test results were classified as over-compacted had high resistance to water and frost of 91.2%. The same was observed for most dense Marshall samples. This is probably the effect of very low air void content limiting water absorption in a natural way, which resulted in a slight decrease in strength of samples subjected to one freeze-thaw cycle. As predicted, the lowest ITSR value was obtained on R1 samples that were much under-compacted and had the highest air void content. Thus, moisture damage appeared, which is associated with both loss of cohesion and loss of adhesion [50]. In reference to the nominal value of Marshall compaction energy, R3 samples was less frost resistant than expected, even though they were properly compacted and had the highest strength among the samples cut out of slabs. However, in tests of asphalt mix properties there is not always a positive correlation between strength and frost resistance measured by ITSR [51,52].

As is well-known, moisture damage resistance of asphalt mix is strongly related to the aggregate-binder adhesion. Among the aggregates used, dolomites with CaO content of 39.62% prevailed. Alkaline aggregates are characterized by excellent adhesion to asphalt, which results not only in high water resistance but also in good strength parameters. Lucas Júnior et al. indicates that the indirect tensile strength also depends on aggregate sphericity and texture [43].

The mixture also contains filler with a very high lime content. It is naturally hydrophilic material with a tendency to form strong bonds with hydrophobic organic compounds such as bitumen [53].

4.4. Stiffness Modulus

The results of stiffness modulus tests are presented in Figures 9–11 and Table 6. The values of the parameter determined on the Marshall samples varied from 1384 MPa to 2197 MPa in 23 °C, 4404 MPa to 6201 MPa in 10 °C, 11,075 MPa to 14,446 MPa in −2 °C. Compaction index was found to have a linear correlation to the increasing stiffness modulus.

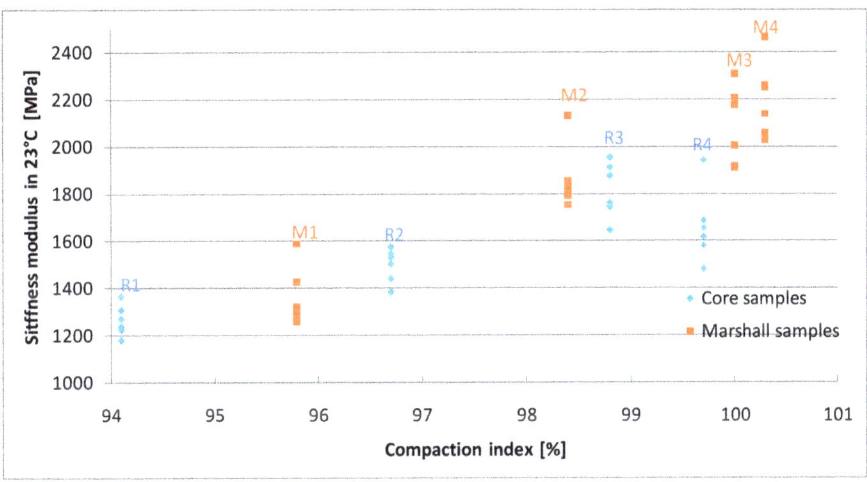

Figure 9. The results of stiffness modulus tests in 23 °C.

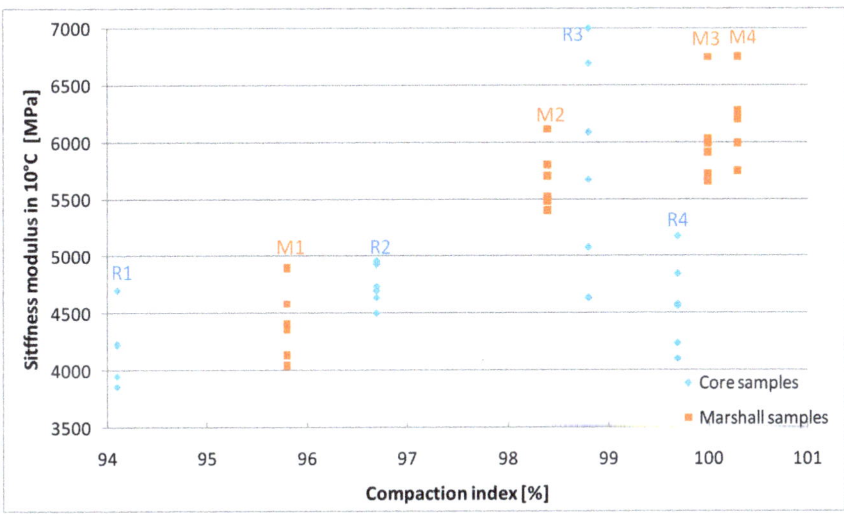

Figure 10. The results of stiffness modulus tests in 10 °C.

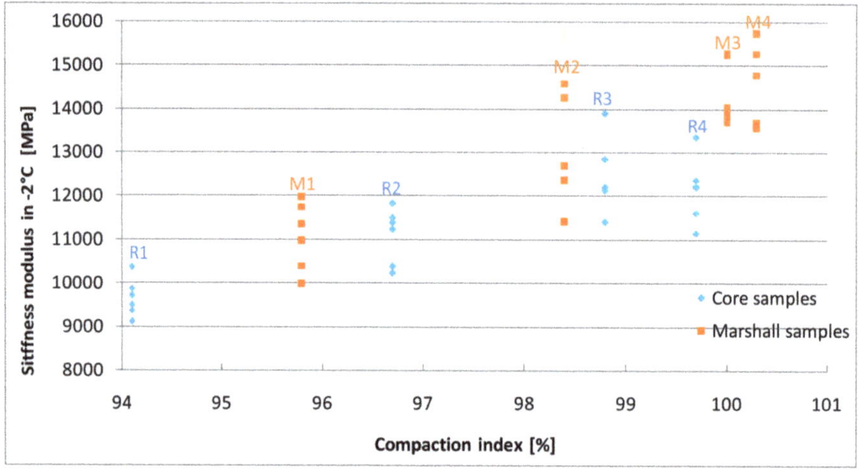

Figure 11. The results of stiffness modulus tests in −2 °C.

Table 6. Average results of stiffness modulus in and standard deviations.

	Compaction Index (%)	Stiffness in 23 °C (MPa)	Standard Deviation	Stiffness in 10 °C (MPa)	Standard Deviation	Stiffness in −2 °C (MPa)	Standard Deviation
Cored Samples	94.1	1262	65	4283	363	9661	441
	96.7	1496	72	4750	173	11,103	636
	98.8	1814	117	5864	915	12,742	1013
	99.7	1660	155	4590	393	12,162	746
Marshall Samples	95.8	1384	122	4404	312	11,075	771
	98.4	1857	137	5669	266	12,951	1226
	100	2086	166	6012	390	14,331	733
	100.3	2197	159	6201	334	14,446	958

The stiffness of the samples cut out of the slabs increased as the density index increased, but only to a certain level. In case of 99.7% compaction index core samples, results were 1660 MPa in 23 °C, 4590 MPa in 10 °C, and 12162 MPa in −2 °C and were lower than the values obtained for samples with 98.8% CI value. These results confirm previous assumptions that R4 samples with the highest compaction index (99.7%) were over-compacted, whereas in case the Marshall samples compacted with increased energy a decrease in stiffness did not occur. Moreover, for similar compaction indexes, samples made using the Marshall method had higher stiffness than samples cored from slabs, which is consistent with the research of Hartman et al. [54]. This may be due to different compaction energies and the size and type of mold in the individual methods [11].

High stiffness modulus asphalt mixes are also characterized by higher resistance to rutting at positive temperatures [55]. Based on the analysis of the test results obtained, it can be concluded that over-compacted mixtures not only have lower strength but are also less resistant to high temperatures, including the development of permanent deformations. At negative temperatures, on the other hand, higher values of the asphalt mix stiffness modulus are expected.

Asphalt changes its properties as a result of oxidation processes and, as time passes, asphalt surfaces become even stiffer and consequently more susceptible to low temperature cracks and premature degradation [39,56]. Therefore, controlling the stiffness of the designed asphalt mix is a factor that can prevent early defects of the surface. It is particularly important in the case of uncertain compaction process and adaptation of additives softening the asphalt binder [57–59].

In the conducted research the core samples that occurred to have the lowest stiffness in −2 °C (9661 MPa) are characterized by the lowest compaction index (94.3%), the highest porosity (8.3%), and the lowest frost resistance (ITSR = 82.4%).

The results clearly indicate that the stiffness modules tend to decrease as the voids increase, which is the expected relation [9,56].

5. Conclusions

The research presented in the article concerned the assessment of the influence of improper compaction of mineral-asphalt mixtures determined by the compaction index on the properties of the produced mix, and in particular on the resistance to water and frost. The studies were carried out on samples compacted using Marshall method with different compaction energies as well as on samples cut from slabs which were compacted to different target heights. The mineral-asphalt mix was an asphalt concrete designed for the wearing course of an asphalt pavement—AC 11 S.

The analysis of the obtained results indicates that inappropriate compaction of asphalt mixes is reflected in the results of physico-mechanical properties. The core samples with the required compaction index of at least 98% also had an air void content in the range corresponding to the technical recommendations for asphalt pavement layers, i.e., from 1 to 4.5%. However, the free space content increased as the compaction index decreased, which is in line with expectations.

As a general rule, compared to core samples, Marshall samples were characterized by higher indirect tensile strength and stiffness modulus at similar compaction index values. Although the Marshall method is very popular, it does not reflect compaction on the construction site well due to the vertical-only impact movement, in contrast to the roller compactor. Despite the 50% increase in compaction energy, the Marshall samples also do not exhibit features typical of excessively compacted samples such as reduced strength and stiffness.

The results of the obtained tests indicate that under-compacted asphalt mixes are characterized by the lowest indirect tensile strength both in dry state and after soaking as well as the lowest stiffness modulus regardless of the test temperature. Also the ITSR value was the lowest for these samples, which confirms that insufficiently compacted asphalt layers are more vulnerable to water and frost. The increased air voids content allows water molecules to penetrate deeper, which combined with the freeze-thaw cycles, causes premature degradation of the pavement.

When compacting asphalt mixes, over-compaction is particularly dangerous. As indicated by the analysis of the results, the compacted mix may have a compaction index and air void content within the required limits. However, there may be a loosening of the aggregate and damage to the contact points, which results in reduced resistance to weather conditions leading to the destruction of the road surface. Also the analysis of the test results indicates a decrease in strength and stiffness for over-compacted core samples.

In the compaction analysis, the properties of the aggregates used should also be taken into account, such as grain shape, surface character, or mineralogical composition. Crushed aggregates with a rough surface and regular grain shapes ensure proper mutual interlocking and a durable connection on the contact with the asphalt binder.

In order to identify the primary cause of a change in the physico-mechanical properties of abnormally compacted mineral-asphalt mixtures, it is necessary to analyze the damage at the contact between aggregate and asphalt as well as conduct computer simulations including finite element analysis.

Author Contributions: Conceptualization: A.W., M.W. and W.F.; methodology: A.W. and W.F.; formal analysis: A.W. and M.W.; investigation: M.W. and W.F.; writing: A.W., M.W. and W.F.; writing—review and editing A.W., M.W. and W.F.; visualization M.W.; supervision: A.W.; funding acquisition A.W. All authors have read and agreed to the published version of the manuscript.

Funding: This research was funded by the National Centre for Research and Development of Poland grant number LIDER/5/0013/L-9/17/NCBR/2018.

Conflicts of Interest: The authors declare no conflict of interest.

References

1. Cui, S.; Blackman, B.; Kinloch, A.J.; Taylor, A.C. Durability of asphalt mixtures: Effect of aggregate type and adhesion promoters. *Int. J. Adhes. Adhes.* **2014**, *54*, 100–111. [CrossRef]
2. Roseen, R.M.; Ballestero, T.; Houle, J.; Briggs, J.F.; Houle, K.M. Water Quality and Hydrologic Performance of a Porous Asphalt Pavement as a Storm-Water Treatment Strategy in a Cold Climate. *J. Environ. Eng.* **2012**, *138*, 81–89. [CrossRef]
3. Sol-Sánchez, M.; Fiume, A.; Moreno-Navarro, F.; Rubio-Gámez, M. Analysis of fatigue cracking of warm mix asphalt. Influence of the manufacturing technology. *Int. J. Fatigue* **2018**, *110*, 197–203. [CrossRef]
4. Du, Y.; Chen, J.; Han, Z.; Liu, W. A review on solutions for improving rutting resistance of asphalt pavement and test methods. *Constr. Build. Mater.* **2018**, *168*, 893–905. [CrossRef]
5. Jiang, Y.; Lin, H.; Han, Z.; Deng, C. Fatigue Properties of Cold-Recycled Emulsified Asphalt Mixtures Fabricated by Different Compaction Methods. *Sustainability* **2019**, *11*, 3483. [CrossRef]
6. Brown, E.R. Density of Asphalt Concrete: How Much is Needed? *Transp. Res. Rec. J. Transp. Res. Board* **1990**, *1282*, 27–32.
7. Praticò, F.; Vaiana, R. A study on volumetric versus surface properties of wearing courses. *Constr. Build. Mater.* **2013**, *38*, 766–775. [CrossRef]
8. Finn, F.N.; Epps, J.A. *Compaction of Hot Mix Asphalt Concrete*; Research Report 214-21; Texas Transportation Inst.: College Station, TX, USA, 1980.
9. Woszuk, A.; Franus, W. Properties of the Warm Mix Asphalt involving clinoptilolite and Na-P1 zeolite additives. *Constr. Build. Mater.* **2016**, *114*, 556–563. [CrossRef]
10. Woszuk, A.; Franus, W. A Review of the Application of Zeolite Materials in Warm Mix Asphalt Technologies. *Appl. Sci.* **2017**, *7*, 293. [CrossRef]
11. Airey, G.; Collop, A. Mechanical and structural assessment of laboratory and field-compacted asphalt mixtures. *Int. J. Pavement Eng.* **2014**, *17*, 50–63. [CrossRef]
12. Mallick, R.B. Use of Superpave Gyratory Compactor To Characterize Hot-Mix Asphalt. *Transp. Res. Rec. J. Transp. Res. Board* **1999**, *1681*, 86–96. [CrossRef]
13. *EN 12697-31; Bituminous Mixtures - Test Methods for Hot Mix Asphalt. Specimen Preparation by Gyratory Compactor*; Polish Committee for Standardization: Warsaw, Poland, 2007.
14. *Superpave Mixture Design Guide*; U.S. Dep. Transportation Fed. Highw. Adm.: Washington, DC, USA, 2001.
15. Tao, M.; Mallick, R.B. Effects of Warm-Mix Asphalt Additives on Workability and Mechanical Properties of Reclaimed Asphalt Pavement Material. *Transp. Res. Rec. J. Transp. Res. Board* **2009**, *2126*, 151–160. [CrossRef]
16. Chomicz-Kowalska, A.; Gardziejczyk, W.; Iwański, M.M. Moisture resistance and compactibility of asphalt concrete produced in half-warm mix asphalt technology with foamed bitumen. *Constr. Build. Mater.* **2016**, *126*, 108–118. [CrossRef]
17. Georgiou, P.; Sideris, L.; Loizos, A. Evaluation of the effects of gyratory and field compaction on asphalt mix internal structure. *Mater. Struct.* **2015**, *49*, 665–676. [CrossRef]
18. Tapkın, S.; Keskin, M. Rutting analysis of 100 mm diameter polypropylene modified asphalt specimens using gyratory and Marshall compactors. *Mater. Res.* **2013**, *16*, 546–564. [CrossRef]
19. Lee, S.-J.; Amirkhanian, S.; Putman, B.J.; Kim, K.W. Laboratory Study of the Effects of Compaction on the Volumetric and Rutting Properties of CRM Asphalt Mixtures. *J. Mater. Civ. Eng.* **2007**, *19*, 1079–1089. [CrossRef]
20. Jaskula, P. Influence of compaction effectiveness on interlayer bonding of asphalt layers. In Proceedings of the 9th International Conference Environmental Engineering, Vilnius, Lithuania, 22–23 May 2014. [CrossRef]
21. Xin, H.; Brown, H.R.; Naficy, S.; Spinks, G.M. Time-dependent mechanical properties of tough ionic-covalent hybrid hydrogels. *Polymer* **2015**, *65*, 253–261. [CrossRef]
22. *Asphalt surfaces on national Roads, Technical requirements WT-2 part I*; General Directorate for National Roads and Motorways: Warsaw, Poland, 2014.
23. Luo, S.; Qian, Z.; Yang, X.; Wang, H. Design of gussasphalt mixtures based on performance of gussasphalt binders, mastics and mixtures. *Constr. Build. Mater.* **2017**, *156*, 131–141. [CrossRef]
24. Esquinas, Á.R.; Ramos, C.; Jiménez, J.R.; Fernández, J.; De Brito, J.M.C.L. Mechanical behaviour of self-compacting concrete made with recovery filler from hot-mix asphalt plants. *Constr. Build. Mater.* **2017**, *131*, 114–128. [CrossRef]

25. Peterson, R.L.; Mahboub, K.; Anderson, R.M.; Masad, E.; Tashman, L. Comparing Superpave Gyratory Compactor Data to Field Cores. *J. Mater. Civ. Eng.* **2004**, *16*, 78–83. [CrossRef]

26. Smith, B.C.; Diefenderfer, B.K. Comparison of Nuclear and Nonnuclear Pavement Density Testing Devices. *Transp. Res. Rec. J. Transp. Res. Board* **2008**, *2081*, 121–129. [CrossRef]

27. Micaelo, R.; Azevedo, M.C.; Ribeiro, J. Hot-mix asphalt compaction evaluation with field tests. *Balt. J. Road Bridg. Eng.* **2014**, *9*, 306–316. [CrossRef]

28. Commuri, S.; Zaman, M. A novel neural network-based asphalt compaction analyzer. *Int. J. Pavement Eng.* **2008**, *9*, 177–188. [CrossRef]

29. Commuri, S.; Mai, A.T.; Zaman, M. Neural Network–Based Intelligent Compaction Analyzer for Estimating Compaction Quality of Hot Asphalt Mixes. *J. Constr. Eng. Manag.* **2011**, *137*, 634–644. [CrossRef]

30. Praticò, F.; Vaiana, R.; Moro, A. Dependence of Volumetric Parameters of Hot-Mix Asphalts on Testing Methods. *J. Mater. Civ. Eng.* **2014**, *26*, 45–53. [CrossRef]

31. *EN 933-1 Tests for geometrical properties of aggregates. Determination of particle size distribution. Sieving method*; Polish Committee for Standardization: Warsaw, Poland, 2012.

32. *EN 933-10 Tests for Geometrical Properties of Aggregates. Assessment of Fines. Grading of Filler Aggregates (Air Jet Sieving)*; Polish Committee for Standardization: Warsaw, Poland, 2009.

33. *EN 12697-6 Bituminous mixtures - Test Methods for Hot Mix Asphalt. Determination of bulk density of bituminous specimens*; Polish Committee for Standardization: Warsaw, Poland, 2012.

34. *EN 12697-5 Bituminous mixtures-Test methods. Determination of the maximum density*; Polish Committee for Standardization: Warsaw, Poland, 2018.

35. *EN 12697-8 Bituminous mixtures-Test methods. Determination of void characteristics of bituminous specimens*; Polish Committee for Standardization: Warsaw, Poland, 2018.

36. *EN 12697-12 Bituminous mixtures-Test methods. Determination of the water sensitivity of bituminous specimens*; Polish Committee for Standardization: Warsaw, Poland, 2018.

37. *EN 12697-26 Bituminous mixtures-Test methods. Stiffness.*; Polish Committee for Standardization: Warsaw, Poland, 2018.

38. *Asphalt surfaces on national Roads, Technical requirements WT-2 part II*; General Directorate for National Roads and Motorways: Warsaw, Poland, 2016.

39. Vacková, P.; Valentin, J.; Kotoušová, A. Impact of lowered laboratory compaction rate on strength properties of asphalt mixtures. *Innov. Infrastruct. Solut.* **2017**, *3*, 1–8. [CrossRef]

40. Zulkati, A.; Diew, W.Y.; Delai, D.S. Effects of Fillers on Properties of Asphalt-Concrete Mixture. *J. Transp. Eng.* **2012**, *138*, 902–910. [CrossRef]

41. Melotti, R.; Santagata, E.; Bassani, M.; Salvo, M.; Rizzo, S. A preliminary investigation into the physical and chemical properties of biomass ashes used as aggregate fillers for bituminous mixtures. *Waste Manag.* **2013**, *33*, 1906–1917. [CrossRef]

42. Cheung, L.W.; Dawson, A.R. Effects of Particle and Mix Characteristics on Performance of Some Granular Materials. *Transp. Res. Rec. J. Transp. Res. Board* **2002**, *1787*, 90–98. [CrossRef]

43. Júnior, J.L.O.L.; Babadopulos, L.F.A.L.; Soares, J.B. Effect of aggregate shape properties and binder's adhesiveness to aggregate on results of compression and tension/compression tests on hot mix asphalt. *Mater. Struct.* **2020**, *53*, 1–15. [CrossRef]

44. Bernier, A.; Zofka, A.; Josen, R.; Mahoney, J. Warm-Mix Asphalt Pilot Project in Connecticut. *Transp. Res. Rec. J. Transp. Res. Board* **2012**, *2294*, 106–114. [CrossRef]

45. Ren, J.; Xing, C.; Tan, Y.; Liu, N.; Liu, J.; Yang, L. Void distribution in zeolitewarm mix asphalt mixture based on X-ray computed tomography. *Materials* **2019**, *12*, 1888. [CrossRef] [PubMed]

46. Wang, L.B.; Frost, J.D.; Shashidhar, N. Microstructure Study of WesTrack Mixes from X-Ray Tomography Images. *Transp. Res. Rec. J. Transp. Res. Board* **2001**, *1767*, 85–94. [CrossRef]

47. Jiang, W.; Sha, A.; Xiao, J. Experimental Study on Relationships among Composition, Microscopic Void Features, and Performance of Porous Asphalt Concrete. *J. Mater. Civ. Eng.* **2015**, *27*, 04015028. [CrossRef]

48. Xing, C.; Xu, H.; Tan, Y.; Liu, X.; Ye, Q. Mesostructured property of aggregate disruption in asphalt mixture based on digital image processing method. *Constr. Build. Mater.* **2019**, *200*, 781–789. [CrossRef]

49. Sarsam, S.I.; Al-Obaidi, M.K. Assessing the Impact of Various Modes of Compaction on Tensile Property and Temperature Susceptibility of Asphalt Concrete. *Int. J. Sci. Res. Knowl.* **2014**, *2*, 297–305. [CrossRef]

50. Hamedi, G.H.; Nejad, F.M. Evaluating the Effect of Mix Design and Thermodynamic Parameters on Moisture Sensitivity of Hot Mix Asphalt. *J. Mater. Civ. Eng.* **2017**, *29*, 04016207. [CrossRef]

51. Sanij, H.K.; Meybodi, P.A.; Hormozaky, M.A.; Hosseini, S.; Olazar, M. Evaluation of performance and moisture sensitivity of glass-containing warm mix asphalt modified with zycothermTM as an anti-stripping additive. *Constr. Build. Mater.* **2019**, *197*, 185–194. [CrossRef]

52. Woszuk, A.; Bandura, L.; Franus, W. Fly ash as low cost and environmentally friendly filler and its effect on the properties of mix asphalt. *J. Clean. Prod.* **2019**, *235*, 493–502. [CrossRef]

53. Nciri, N.; Shin, T.; Lee, H.; Cho, N. Potential of Waste Oyster Shells as a Novel Biofiller for Hot-Mix Asphalt. *Appl. Sci.* **2018**, *8*, 415. [CrossRef]

54. Hartman, A.M.; Gilchrist, M.D.; Walsh, G. Effect of Mixture Compaction on Indirect Tensile Stiffness and Fatigue. *J. Transp. Eng.* **2001**, *127*, 370–378. [CrossRef]

55. Ahmed, K.; Irfan, M.; Ahmed, S.; Ahmed, A.; Khattak, A. Experimental Investigation of Strength and Stiffness Characteristics of Hot Mix Asphalt (HMA). *Procedia Eng.* **2014**, *77*, 155–160. [CrossRef]

56. Halle, M.; Rukavina, T.; Domitrovic, J. Influence of temperature on asphalt stiffness modulus. In Proceedings of the 5th Eurasphalt Eurobitume Congress, Istanbul, Turkey, 13–15 June 2012; pp. 13–15.

57. Woszuk, A.; Wróbel, M.; Franus, W. Application of Zeolite Tuffs as Mineral Filler in Warm Mix Asphalt. *Materials* **2019**, *13*, 19. [CrossRef]

58. Abdullah, M.E.; Kamaruddin, N.H.M.; Daniel, B.D.; Hassan, N.A.; Hainin, M.R.; Tajudin, S.A.A.; Madun, A.; Mapanggi, R. Stiffness modulus properties of hot mix asphalt containing waste engine oil. *ARPN J. Eng. Appl. Sci.* **2016**, *11*, 14089–14091.

59. Woszuk, A.; Wróbel, M.; Franus, W. Influence of Waste Engine Oil Addition on the Properties of Zeolite-Foamed Asphalt. *Materials* **2019**, *12*, 2265. [CrossRef]

 *materials*

Article

Assessment of the Mechanical Properties of ESD Pseudoplastic Resins for Joints in Working Elements of Concrete Structures

Dominik Logoń *, Krzysztof Schabowicz and Krzysztof Wróblewski

Faculty of Civil Engineering, Wrocław University of Science and Technology, Wybrzeże Wyspiańskiego 27, 50-370 Wrocław, Poland; krzysztof.schabowicz@pwr.edu.pl (K.S.); k.wroblewski@wp.pl (K.W.)
* Correspondence: dominik.logon@pwr.edu.pl

Received: 9 May 2020; Accepted: 24 May 2020; Published: 26 May 2020

Abstract: Concrete structure joints are filled in mainly in the course of sealing works ensuring protection against the influence of water. This paper presents the methodology of testing the mechanical properties of ESD pseudoplastic resins (E-elastic deformation, S-strengthening control, D-deflection control) recommended for concrete structure joint fillers. The existing standards and papers concerning quasi-brittle cement composites do not provide an adequate point of reference for the tested resins. The lack of a standardised testing method hampers the development of materials universally used in expansion joint fillers in reinforced concrete structures as well as the assessment of their properties and durability. An assessment of the obtained results by reference to the reference sample has been suggested in the article. A test stand and a method of assessing the mechanical properties results (including adhesion to concrete surface) of pseudoplastic resins in the axial tensile test have been presented.

Keywords: ESD resin; expansion joint; quasi-plastic material; energy absorption

---

1. Introduction

Expansion joints in building structures move in various directions. The most frequent direction of displacement is the direction perpendicular to the edge of the expansion joint. Depending on temperature changes, we observe either the opening, expansion or the narrowing, and closing down of the expansion joint gap [1–3]. The widening of the expansion joint gap directly causes the elongation of the material filling in the joint, and if the strength limits of the filler material are exceeded, it results in the irreversible damage causing the loss of sealing reliability. The increasing width of the expansion gap may also result in the filler material breaking off from the concrete surface, which also causes the loss of water tightness. The material can be fed by gravity pouring or by means of pressure injection pumps. The reacting resin hardens and sets, becoming permanently flexible mass, which during the cyclic work of the expansion joint should expand or shrink depending on the changes in expansion joint width, as shown in Figure 1.

Numerous papers have emphasised the dependence of physical and mechanical properties of tested resins on ambient temperature [4–6]. The discrepancy of the obtained results has also been attributed to the impact of resins' adhesion to materials [7]. It has been found that the substrate's humidity has a significant influence on test results, which makes it necessary to ensure strict control of the testing conditions [8,9]. A review of relevant literature shows that resin tests are carried out on various test stands [10–12], which indicates the need to modify the standards to correspond to the scope of conducted tests. The lack of standardised procedures taking into account the possibility of assessing properties makes it impossible to compare the results obtained at various research centres [13–15].

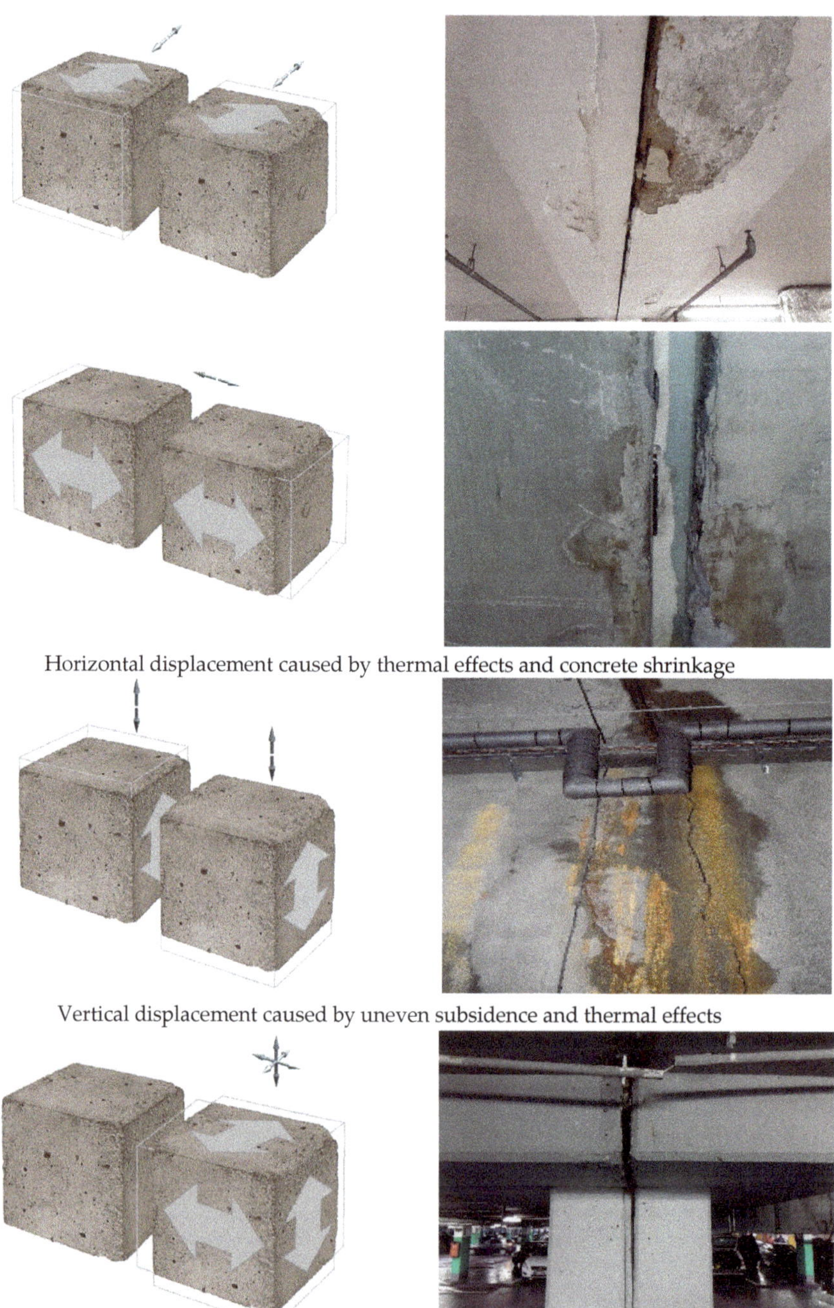

Figure 1. Models showing possible displacements of structural members at expansion joint.

The materials showing elastic properties are subjected to tests that provide data regarding the maximum elongation of the material in question or the maximum breaking force. A lot of standards enable the basic test to be carried out, i.e., the axial tensile test [16–26]. It is worth noting that none of the current standards is dedicated to expansion joint quasi-plastic filler materials, and the existing methods of assessment of the adhesion to the substrate do not refer to pseudoplastic behaviour under load.

In the course of research [27,28], an analysis has been conducted with respect to curves obtained in a static tensile test taking into account multi-axial stress and strain of elastic materials. The paper [29] described a methodology of tests of elastomeric bearings in complex states of strain according to the requirements of the standards PN-EN 1337-3 [30] and PN-ISO 37 [16]. The authors presented a design for a modern stand for biaxial tensile test for elastomers exceeding the scope of application of the aforementioned standards. The conducted tests cannot be adapted for the purposes of assessment of mechanical properties of pseudoplastic resins intended to be used as expansion joint fillers, including their adhesion to the concrete substrate.

The method of conducting tests and assessment of the deformation capacity dedicated to quasi-brittle cement composites that is proposed in numerous standards and papers [31–36] is not sufficient for the assessment of the mechanical properties of pseudoplastic resins functioning as expansion joint fillers. The existing hyperelastic models for the description of the behaviour of non-elastic materials (including hyperelastic resins) do not, however, take into account the impact of a large number of variables determining the usefulness of ESD materials (E-elastic deformation, S-strengthening control, D-deflection control) as expansion joint fillers. There is no information on the adhesion of materials to concrete substrates, and in particular on the strengthening control and deflection control areas [37,38].

The method proposed in [31–33] for the assessment of ESD quasi-brittle cement composites has been used in our own work. Those works present a possibility of assessing the mechanical properties of materials based on the stress–strain correlation in any case of the loads recording the values (force, deformation and absorbed energy), but do not include information regarding changes in the width of gaps in expansion joints during an axial tensile test.

In this work, the aforementioned method has been modified for the purpose of assessing the mechanical properties of ESD quasi-plastic resins in the tensile strength tests—taking into account the assessment of their adhesion to the concrete surface.

2. Testing

2.1. Materials Used for Tests

In this article, the following specimens has been prepared for testing the mechanical properties of ESD pseudoplastic resins recommended as concrete structure joint fillers:

- specimen 0C—standard, reference cement mortar with the following composition and parameters: cement CEM I 32.5, compressive strength f_c = 4.48 MPa, mortar class M4, three-point bending tensile strength f_{tb} = 0.14 MPa (distance between supports 150 mm).
- resins used as a filler in the expansion joint model, as presented in Table 1:

  1A—resin based on thixotropic acrylic for injections,
  2A—resin based on acrylic mixed with water,
  3A—resin based on polyurethane mixed with water,
  4A—polyurethane-based resin,
  5A—resin based on elastic epoxy,
  6A—bitumen-based resin.

Table 1. Resin specimens—tensile strength f_t elongation at break ε, density, viscosity and pH.

Specimen	Tensile Strength f_t (MPa)	Elongation at Break ε (%)	Density (kg/dm^3)	Viscosity (mPas)	pH
1A	>0.4	>240	0.95–2.09	<50	5–11
2A	>0.6	~50	0.99–2.09	<50	5–11
3A	>0.6	>80	0.99–1.16	<200	ca. 7
4A	ca. 0.7	ca. 50	0.99–1.10	80–155	-
5A	>4.0	ca. 70	0.97–1.11	500–1700	-
6A	ca. 0.2	ca. 500	0.99–1.10	1750–3500	-

As shown in Table 1, six types of resins have been selected for tests. Those resins are universally used and available on the construction market and, more importantly, are used as expansion joint fillers. Each resin is characterised by different properties and technical parameters. The resins were fed into the gaps of the expansion joint model by means of the so-called gravity pouring. Subsequently, the expansion joint models were put aside until the resins hardened, i.e., for a period of 24 h. The entire testing process was divided into several series of tests. Each series of tests was conducted for a different resin. In total, tests for six different filler resins were carried out.

The axial tensile test for resins has been conducted in accordance with PN-EN ISO 37 standard [16] using the Instron 33R strength testing machine.

The testing procedure included the proper preparation of specimens. Resins in Table 2 were placed on a flat surface 400×300 mm, circa 5 mm thick. Afterwards, with the use of a template, shaped elements were cut out—the so-called "oars"—which were then placed, one by one, in the strength testing machine, as shown in Figure 2.

The testing was carried out until the moment of breaking of the resin specimens for which the results of maximum elongation at break and maximum breaking force were obtained.

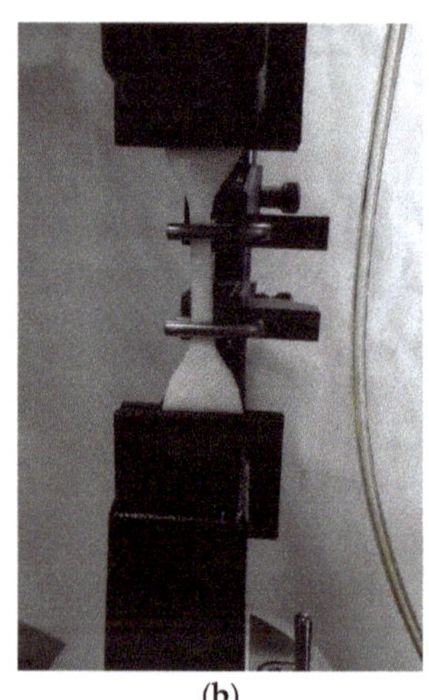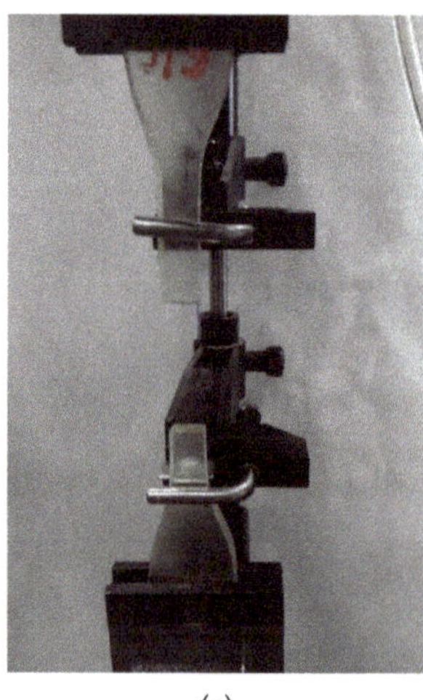

(a)   (b)   (c)

Figure 2. Resin testing according to [16]—axial tensile test: (**a**) view of the strength testing machine—test stand, (**b**) view of the shaped resin elements, the so-called "oars", placed in the clamps and (**c**) view of the completed test at the moment of breaking of the resin specimen.

Examples of diagrams of stress as a function of deformation in axial tensile test—5 samples of resin 1 (Table 2) are presented in Figure 3.

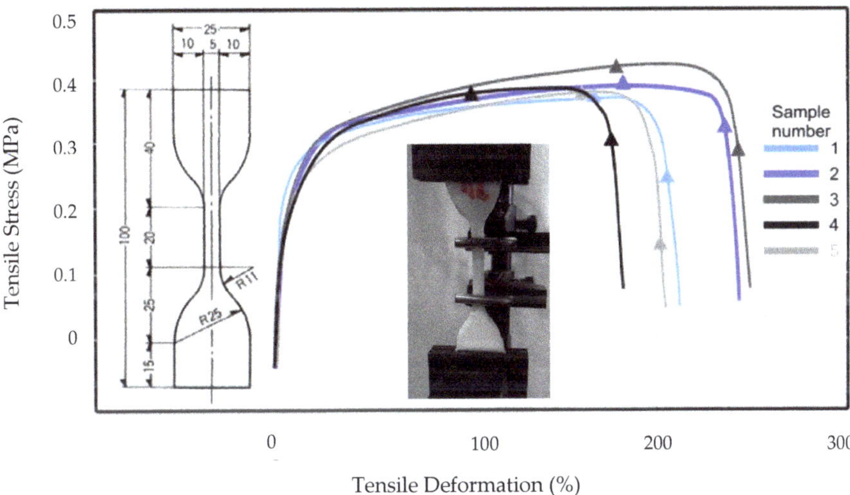

Figure 3. The tensile stress-deformation curves as a function of deformation in axial tensile test of resin 1, Table 2 (5 samples - based on thixotropic acrylic) according to PN-EN ISO 37 [16].

The obtained results of the axial tensile test for the analysed resins according to PN-EN ISO 37 [16] are presented in Table 2.

Table 2. Results from the axial tensile test for the analysed resins according to PN-EN ISO 37 [16].

Tested Resin	Specimen No.	Load at Maximum (N)	Tensile Stress at Maximum (MPa)	Load at Break (N)	Tensile Stress at Break (MPa)	Tensile Strain for F_{max} (%)	Tensile Strain at Break (%)
Resin 1	1	10.56	0.406	8.066	0.310	199.7	241.8
	2	10.90	0.425	9.626	0.376	201.2	257.7
	3	10.86	0.430	7.969	0.316	179.5	246.0
	4	10.41	0.421	8.237	0.33	157.4	236.3
	5	11.08	0.420	5.437	0.206	193.7	236.5
Average values		10.76	0.421	7.867	0.308	186.3	243.7
Resin 2	1	6.77	0.529	2.106	0.165	6.5	37.8
	2	7.88	0.737	2.409	0.225	1.1	17.3
	3	6.13	0.605	2.289	0.203	3.4	23.8
Average values		6.93	0.624	2.268	0.198	3.67	26.3
Resin 3	1	28.88	0.716	23.457	0.582	174.0	175.3
	2	28.62	0.664	8.435	0.196	162.3	160.3
	3	28.22	0.681	23.259	0.561	167.7	186.2
	4	27.22	0.623	24.968	0.571	186.3	197.3
	5	28.28	0.743	25.995	0.683	170.4	178.1
Average values		28.24	0.685	21.223	0.519	172.1	179.4
Resin 5	1	481.29	6.398	481.286	6.398	115.2	115.2
	2	466.59	5.938	466.588	5.938	102.2	102.2
	3	446.46	5.539	215.259	2.670	89.7	89.1
	4	468.19	5.792	467.858	5.788	96.7	96.9
	5	450.42	5.980	450.136	5.976	103.6	103.8
Average values		462.59	5.929	416.226	5.354	101.5	101.4

2.2. Preparation of Specimens for Tests

The expansion joint model prepared for the tests was a model in which class C37 concrete specimens with $100 \times 100 \times 100$ mm dimensions were used. The tests were carried out in laboratory conditions, with a temperature of 20 °C and stable air humidity. The preparation of the expansion joint model consisted in arranging two concrete specimens in parallel with each other, Figure 4.

For all expansion joint model specimens, a 10-mm-wide gap was prepared, and each of the resins and the reference cement mortar was then poured into that gap. In the case of all the fillers, there was the same method of preparation - mechanical cleaning of the concrete specimens surface (1A,2A,3A,4A,5A,6A–Table 1) and pouring the resin into the expansion joint.

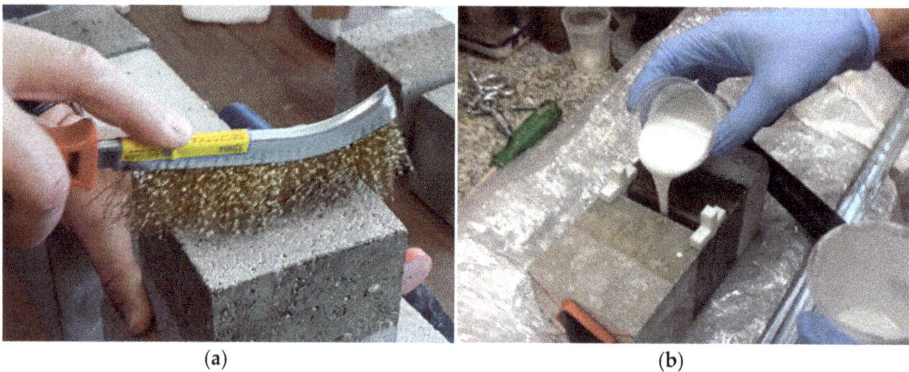

(a) (b)

Figure 4. Preparation of a specimen for tests: (**a**) cleaning the surface and (**b**) pouring the resin into the expansion joint.

2.3. Description of the Test Stand

The expansion joint model specimens were tested using a Hounsfield 10K-S strength testing machine (Tinius Olsen TMC-United States) and the Horizon numerical processing software (ver.1). The adopted crosshead speed was 5 mm/min. A view of the test stand is presented in Figure 5.

Owing to the use of clamps on a harness (jointed system)—the expansion joint model displacement in one direction, perpendicular to the side surfaces of the gap, was simulated. The expansion joint model tension reflects the actual behaviour of filler materials in repaired reinforced concrete structures. The displacements in perpendicular direction are smaller and are not the main cause of destruction of pseudoplastic fillers in expansion joints.

The aim of the tests of expansion joint models filled with sealing resins was to determine the force–deformation correlation serving as the basis for the assessment of the usefulness of resins as expansion joint fillers. The test was based on the measurement of the displacement of crosshead. The stress–strain tests took into account the weight of the lower concrete cube. After the test, the resin condition after the break and the percentage degree of the resin adhesion to the concrete substrate were assessed visually.

Figure 6 shows an exemplary load–deformation curve for an expansion joint model specimen marked as 1AR5 (resin 1, humidity or contamination of the substrate A: dry, substrate cleaned manually R, specimen 5) with photographs depicting the testing process. Manual cleaning concrete surface results in less resin adhesion (than mechanical cleaning).

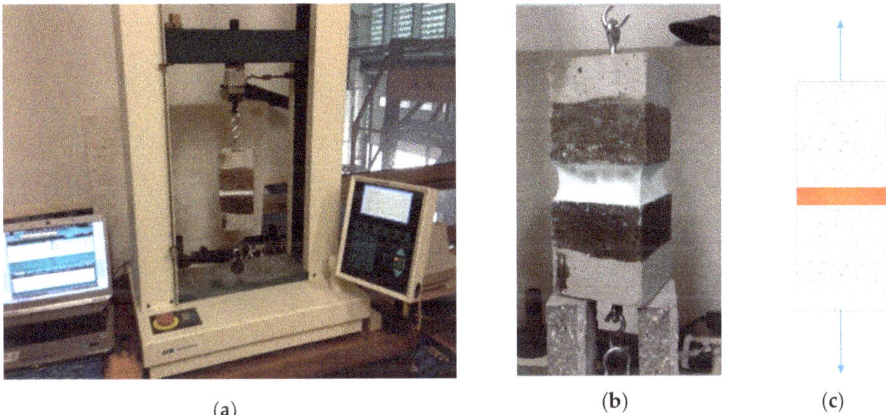

Figure 5. A view of the test stand: (**a**) testing machine, (**b**) close-up of a specimen during testing and (**c**) the adopted tension method.

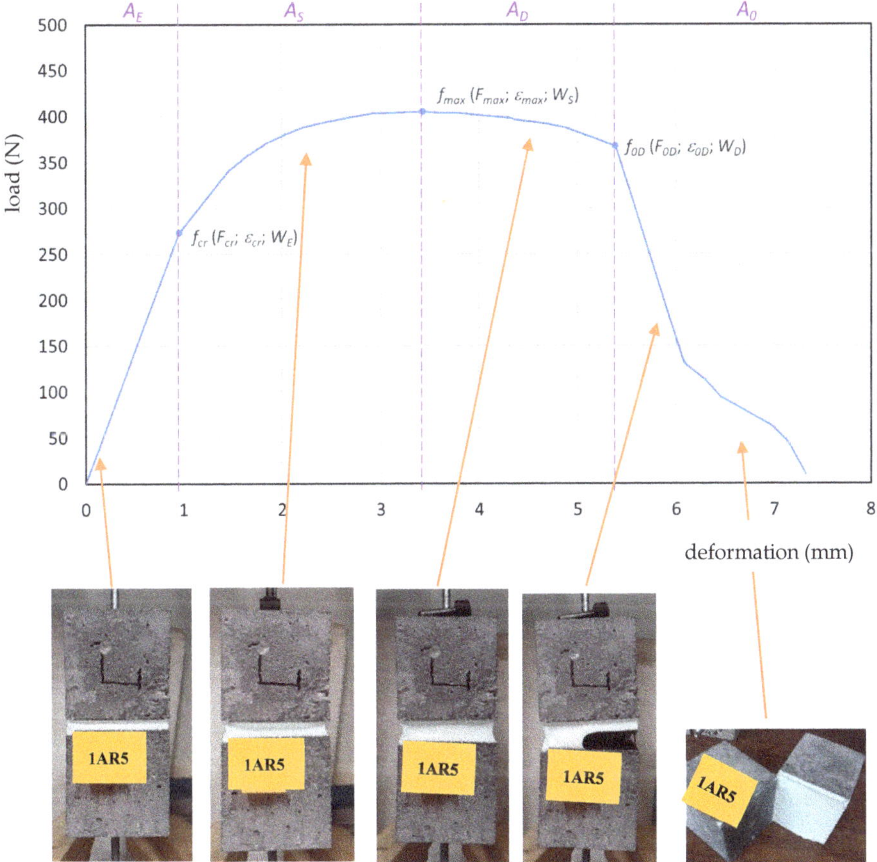

Figure 6. An exemplary diagram of the load–deformation curve for an expansion joint model specimen AR5 with photographs depicting the testing process—manual cleaning surface.

2.4. Diagram of Properties Assessment for ESD Pseudoplastic Materials

Figure 7 shows a diagram for the assessment of mechanical properties based on the load–deformation curve of ESD pseudoplastic materials (*E*-elastic deformation, *S*-strengthening control, *D*-deflection control) in an axial tensile test according to [31–33].

The drawing outlines the adhesion loss area A_{0X} in which the resin has been pull off from the concrete surface.

Characteristic areas have been determined: A_E-elastic deformation (f_{cr}, the proportionality range-Hooke's law), A_S-strengthening control (the area between f_{cr}, and the occurrence of maximum stress f_{max}), A_D-deflection control (the area between f_{max} and f_d) and A_P-propagation area.

Point f_d corresponds to the ability to carry stress f_{cr} and is a determinant of the optimum ESD pseudoplastic filler.

Any point $f_X(F_X, \varepsilon_X, W_X)$ on the load–deformation curve has been defined with the use of a corresponding force F_X, deformation ε_X and absorbed energy W_X (area under the curve).

The characteristic points f_X ending each of the areas A_X marked respectively: f_{cr}, f_{max}, f_d.

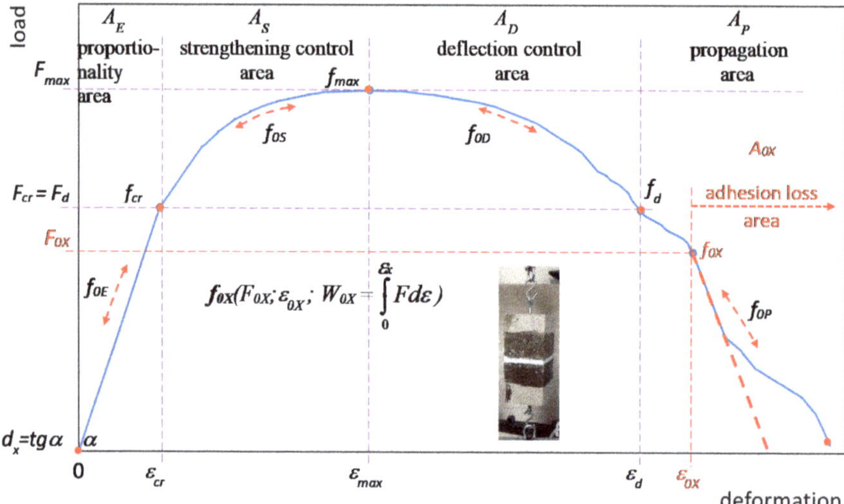

Figure 7. A diagram for the assessment of mechanical properties based on the load–deformation curve of ESD pseudoplastic materials in an axial tensile test.

What has been identified is the pulling off or breaking off of the pseudoplastic filler at any point $f_{OX}(F_{OX}, \varepsilon_{OX}, W_{OX})$ and the correlating area A_{OX} (point f_{OE} in the proportionality area A_{OE}, point f_{OS} in the strengthening area A_{OS} and point f_{OD} in the deflection control area A_{OD}—additionally point f_{OP} in the propagation area A_{OP}).

Point f_{OX}, where the pull off/destruction of the material was recorded, was considered to be the end point of the deformation capacity assessment. If there was no pulling off the resin, point f_d ending the deflection control range served as the determinant of the end of the test.

The assessment of the deformation capacity of the materials used in the Hooke's law range has been determined as $d_x = tg\alpha$.

3. Test Results

The presentation of the results of 0C1, 0C2, 0C3—a model with M4 cement mortar filler-and the averaged result serving as the reference 0C for the tested resins, is shown in Figure 8.

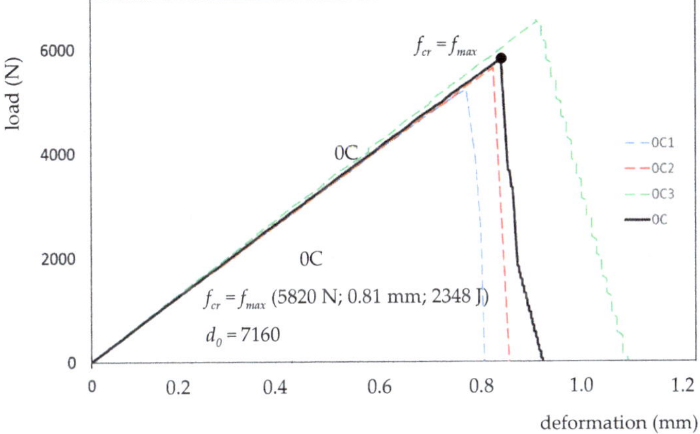

Figure 8. Presentation of load–deformation curves for expansion joint model specimens 0C1, 0C2, 0C3 and the reference specimen 0C in the axial tensile test.

For the averaged reference specimen *0C*, the following have been determined: f_{cr}(force-F_{max}; deformation-ε_{max}; absorbed energy-W_{max}), and deformation capacity d_0. Also $f_{cr} = f_{max}$ (5820 N; 0.81 mm; 2348 J) and $d_0 = 7160$ (tgα force-deformation correlation from the Hooke's law range) have been determined. The results are presented in Table 3.

Figure 9 shows a diagram for the reference specimen and the "strongest" resin 5A (based on elastic epoxide). Specimen 5A shows slightly higher deformation capacity $d_0 = 6509$ (lower tgα than the reference specimen $d_0 = 7160$), a larger proportionality area f_{cr} and, additionally, a significant strengthening control area to point f_{0S}, at which there was a catastrophic, rapid break of the resin.

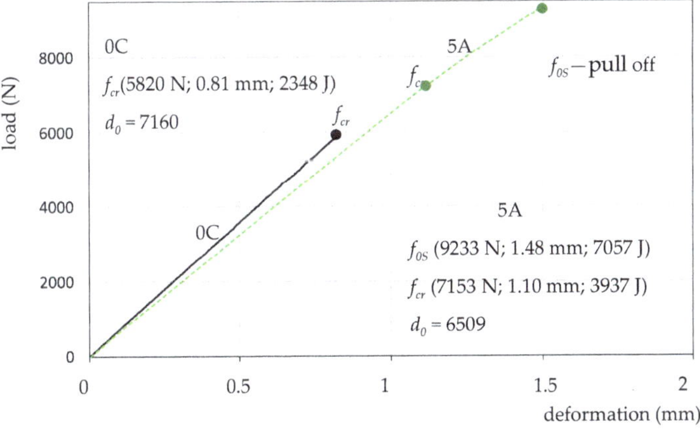

Figure 9. The load–deformation curve for specimen *5A* and specimen *0C*.

Figure 10 presents collective curves for resins 1A, 2A, 3A, 4A, 5A, 6A and the reference specimen 0C. Each resin was tested on at least three specimens, out of which the most representative one was selected for comparison purposes. It was not necessary to average the diagrams due to the deformation of the moment of resin destruction.

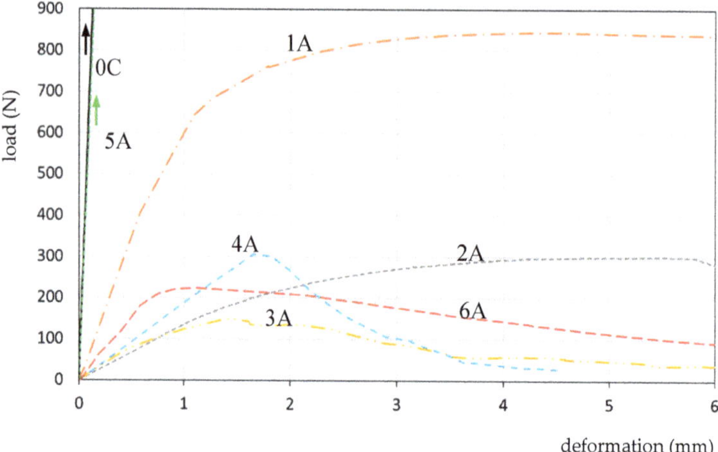

Figure 10. The load–deformation curves for specimens 1A, 2A, 3A 4A 5A, 6A and the reference specimen 0C.

For each of the presented resins, characteristic points f_X and f_{0X} (moment of destruction/pulling off the resin) were determined, which is shown in Table 3.

Table 3. Presentation of the results (characteristic points on the force–deformation curve) for the reference specimen 0C and tested resins 1A, 2A, 3A, 4A, 5A, 6A.

Specimen	Point	F (N)	ε (mm)	W (J)	d_x (N/mm)
0C	f_{cr}	5820	0.81	2348	7160
1A	f_{cr}	424	0.59	124	715
	f_{max}	868	4.27	3005	-
	f_{0D}	699	16.12	12370	-
2A	f_{cr}	163	1.21	98	135
	f_{max}	324	5.60	1323	-
	f_{0D}	323	5.96	1456	-
3A	f_{cr}	93	0.57	26	165
	f_{0S}	173	1.57	164	-
4A	f_{0E}	330	1.77	291	186
5A	f_{cr}	7153	1.10	3937	6509
	f_{0S}	9233	1.48	7057	-
6A	f_{cr}	203	0.64	63	311
	f_{max}	247	1.17	189	-
	f_d	203	2.95	603	-

The obtained data show more precisely the behaviour of expansion joint filler materials in a tensile test, enabling the assessment and comparison of various resins. The presented data enable the assessment of the behaviour of resins in each of the proportionality, strengthening control and deflection control areas. A tabular presentation of data enables the characteristic points f_X from a number of tests to be averaged for the purpose of the assessment of ESD pseudoplastic materials.

Figure 11 is a presentation of load–deformation curves for the reference specimen 0C and resins 1A, 2A, comparing the obtained values of mechanical properties with the defined reference specimen.

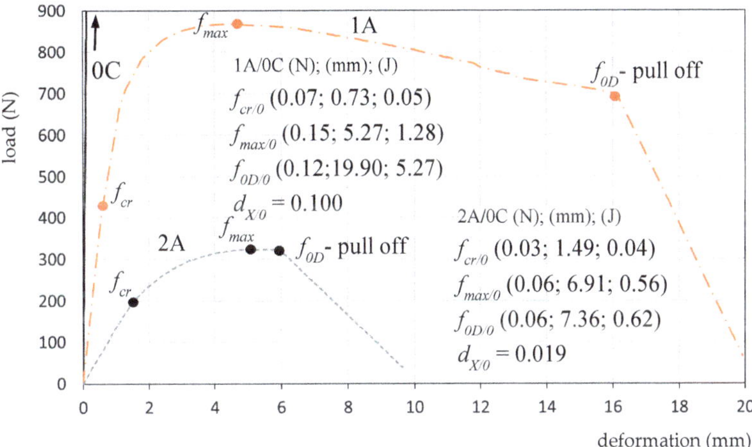

Figure 11. The load–deformation curves for specimens 1A, 2A and specimen 0C.

Table 4 contains a comparison of the obtained values of the mechanical properties of the tested resins 1A, 2A, 3A, 4A, 5A and 6A in relation to the reference specimen 0C.

Table 5 presents the results of resins 1A, 2A, 3A, 4A, 5A, and 6A compared to the linear correlation of specimen 5A, with the largest Hooke's law area.

The proposed method of describing points on the force–deformation curve enables the comparison of any selected points f_{x1} (or areas A_{X1}) and their comparison with any selected points f_{x2} (areas A_{X2}) chosen for the analysis of the obtained effects: $f_{x1}/f_{x2}(F_{X1}/F_{X2}, \varepsilon_X\ 1/\varepsilon_{X2}, W_{X1}/W_{X2})$.

The results presented in Tables 4 and 5 indicate the possibility of juxtaposing freely the data selected for analysis, enabling the comparison of multiples of the achieved effects on various test stands. However, it is recommended to quickly introduce a standardised testing procedure that will enable an accurate comparison of the results obtained at various research centres.

Examples of deformations of resins 1A, 2A and 3A filling the expansion joints before and after the loss of adhesion to the substrate (point f_{0X}) in an axial tensile test is presented in Figure 12.

A linear decrease in the load–deformation correlation in an axial tensile test of the presented model indicates the pulling off the fillers from the substrate. That process may be more or less dynamic. In the case of resin 5A, there was a catastrophic break.

Table 4. Resins 1A, 2A, 3A, 4A, 5A and 6A compared with the reference specimen *0C*.

Specimen	$f_{x/0}$	$F_{x/0}$	$\varepsilon_{x/0}$	$W_{x/0}$	$d_{x/0}$
1A/0C	$f_{cr/0}$	0.07	0.73	0.05	0.100
	$f_{max/0}$	0.15	5.27	1.28	-
	$f_{0D/0}$	0.12	19.90	5.27	-
2A/0C	$f_{cr/0}$	0.03	1.49	0.04	0.019
	$f_{max/0}$	0.06	6.91	0.56	-
	$f_{0D/0}$	0.06	7.36	0.62	-
3A/0C	$f_{cr/0}$	0.02	0.70	0.01	0.023
	$f_{0S/0}$	0.03	1.94	0.07	-
4A/0C	$f_{0E/0}$	0.06	2.19	0.12	0.026
5A/0C	$f_{cr/0}$	1.23	1.36	1.68	0.909
	$f_{0S/0}$	1.59	1.83	3.01	-
6A/0C	$f_{cr/0}$	0.03	0.79	0.03	0.043
	$f_{max/0}$	0.04	1.44	0.08	-
	$f_{d/0}$	0.03	3.64	0.26	-

Table 5. Resins 1A, 2A, 3A, 4A, 5A and 6A compared to the linear correlation of specimen 5A.

Specimen	$f_{x/cr}$	$F_{x/cr}$	$\varepsilon_{x/cr}$	$W_{x/cr}$	$d_{x/cr}$
1A/5A	$f_{cr1/cr}$	0.059	0.536	0.031	0.110
	$f_{max1/cr}$	0.121	3.882	0.763	-
	$f_{0D1/cr}$	0.098	14.655	3.142	-
2A/5A	$f_{cr1/cr}$	0.023	1.100	0.025	0.021
	$f_{max1/cr}$	0.045	5.091	0.336	-
	$f_{0D1/cr}$	0.045	5.418	0.370	-
3A/5A	$f_{cr1/cr}$	0.013	0.518	0.007	0.025
	$f_{0S1/cr}$	0.024	1.427	0.042	-
4A/5A	$f_{0E/cr}$	0.046	1.609	0.074	0.029
5A/5A	$f_{cr1/cr}$	1.000	1.000	1.000	1.000
	$f_{0S1/cr}$	1.291	1.345	1.792	-
6A/5A	$f_{cr/cr}$	0.028	0.582	0.016	0.048
	$f_{max/cr}$	0.035	1.064	0.048	-
	$f_{d/cr}$	0.028	2.682	0.153	-

Resins 2A and 1A were characterised by a less rapid loosening process—as seen in the diagram. Resins 4A, 3A and 6A, after a partial loosening, were characterised by the greatest capacity for deformation and energy absorption in the destruction propagation area. In that area, the loss of tightness of the joints occurs, which is why that range is not taken into consideration when interpreting the obtained results.

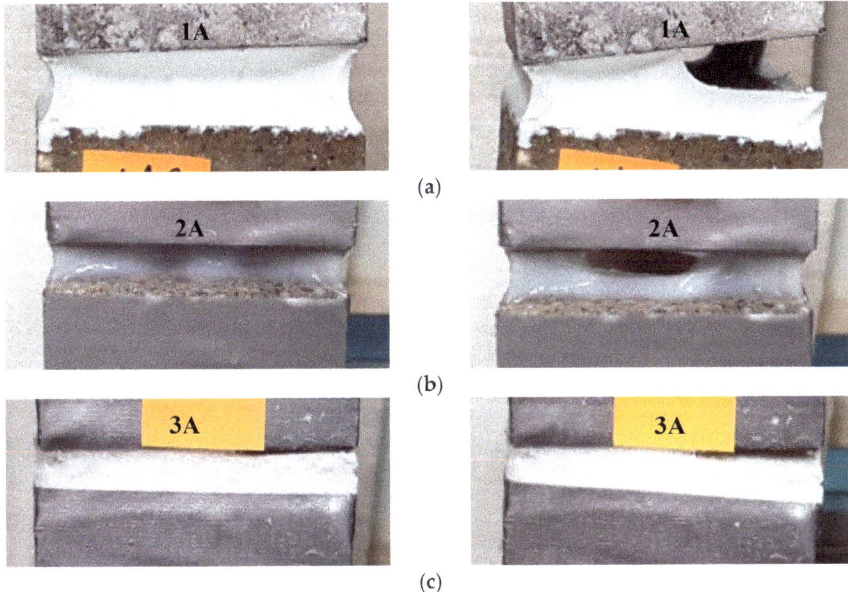

Figure 12. Examples of specimens of resins filling the expansion joints before and after the loss of adhesion to the surface (point f_{0X}) in an axial tensile test: (**a**) resin 1A, (**b**) resin 2A and (**c**) resin 3A.

4. Discussion of the Results

The assessment of tensile tests of a model of an expansion joint filled with ESD resin mass is made possible by the recording of data: force/stress, corresponding deformation and work-as the quantity of absorbed energy (surface area under the stress–strain curve), Figure 7. The characteristics of each point f_x on the load–deformation curve are presented in the form of f_x (force/load; deformation; absorbed energy), Figures 8–11.

The proportionality area A_E determines the quality of the filler, whereas the areas of strengthening control A_S and deflection control A_D (characterised by a greater ability to carry stress than the Hooke's law proportionality area) contribute to the improvement of durability of sealants in the expansion joints of concrete structures, and constitute an additional safety range.

The characteristics of ESD pseudoelastic resins serving as sealants in expansion joint gaps indicate an increased deformation capacity in areas A_S and A_D—resulting in a considerable quantity of absorbed energy compared to the proportionality area A_E. The strengthening control area A_S is more significant in that type of fillers (it does not generate a destruction of the filler structure).

The deflection control area A_D, with loads greater than those occurring in the proportionality area and comparable to those in the strengthening control area, additionally enables an effective implementation of the applicability range of the ESD resin material in expansion joints. That effect is also controlled by the possible appearance of the first forms of destruction of the structure on the side surface of the pulled material (it usually cannot be seen with a naked eye).

The area of propagation, weakening A_P, with the decreasing stress and increasing deformation, is characterised by the moment of the resin specimen being pulled off from the concrete surface together with the absorption of another portion of energy in the resin structure destruction process. The area of propagation (of carrying stress that is lower than the critical one) is not recommended for ESD pseudoplastic sealing resins (it may also be assessed).

The area of the loss of adhesion A_0 (Figure 7), with the decreasing stress and increasing deformation, is characterised by the moment of the resin specimen being pulled off from the concrete surface together with the destruction of the resin structure. The deformations breaking and destroying the structure of the material result in a significant linear decrease in stress (linear decrease in the force–deformation correlation). What determines the maintenance of reliability of the whole system, i.e., the expansion joint gap filled with sealing resin, is the adhesion control area A_0. The loosening of the resin from the concrete surface is considered an emergency situation. The loss of adhesion of the sealing resin in the expansion joint is, from the perspective of the maintenance of the sealant's reliability, the moment when water tightness is lost. At that moment, the water influencing the whole system is able to permeate through the filled expansion joint.

From the perspective of the sealant's reliability, it is assumed that the loss of adhesion should occur as late as possible. The breaking off, loosening of ESD pseudoplastic resins should take place in the propagation area A_P, which is not significant as regards the values of carried stress.

What is an important component of the analysis of the obtained results is the correct determination of the Hooke's law range (f_{cr}). The linear correlation should be determined on large-scale diagrams. It should be noted that the correlation should be determined from a section between 50% f_{cr} and f_{cr}. The force-deformation correlation in the initial range is not linear due to the specimen's arrangement on the slings and, therefore, should not be taken into account (when interpreting the test results).

A need to standardise the testing procedure has been emphasised. Because the tests are conducted on differing test stands (e.g., with different crosshead speed, ambient temperature, etc.), the obtained data cannot be compared with one another. Until relevant standard provisions are adopted, it is proposed to compare the results as a multiple of the established reference item. It may be a specimen with defined parameters or e.g., the Hooke's law range of the best of the tested materials.

The obtained results of the tests of resins presented in Table 3 enable a preliminary assessment of the possibility of using them as expansion joint fillers. Depending on the obtained deformations, an appropriate filler can be selected for a specific case of an expansion joint on a structural element. Thus:

a.   If the expansion joint gaps expand by 5% in relation to the original width, then practically all the resins can be used as fillers for such a joint. All resins 1A, 2A, 3A, 4A, 5A and 6A meet that condition in area A_E.

b.   If the expansion joint gaps expand by 10% in relation to the original width, then practically all the resins can be used as fillers for such a joint. Resins 2A, 4A and 5A meet that condition in area A_E, and resins 1A, 3A and 6A—in area A_S.

c.   If the expansion joint gaps expand by 15% in relation to the original width, resin 5A cannot be used to fill such a joint. Resins 1A, 2A. 3A and 4A meet that condition in area A_S, and resin 6A—in area A_D.

d.   If the expansion joint gaps expand by 50% in relation to the original width, resins 3A, 4A, 5A and 6A cannot be used to fill such a joint. Resin 2A meets that condition in area A_S, and resin 1A—in area A_D.

e.   If the expansion joint gaps expand by 100% in relation to the original width, then only resin 1A can be used as a filler for such a joint. Resin 1A meets that condition in area A_D.

A comparison of the tested resins with the reference specimen 0C (Table 4) confirms the conclusions presented above. An analysis of the results in relation to a predefined reference item, based on a multiple of changes, enables a comparison of the results obtained at various research centres—until an applicable standard (norm) is established. In summary, it should be noted that only specimens 1A and 2A could be classified as ESD pseudoplastic resins. Specimen 1A shows lower deformation capacity ($d_x = 715$) in comparison with specimen 2A ($d_x = 135$), with a much greater ability to carry stress and higher quantity of absorbed energy in the areas: elasticity, strengthening control and deflection control.

What is worth noting is the comparison of the results of the proposed testing method with the traditional method of assessing the mechanical properties of resins presented in Figure 3 and Table 2. The presented data show that resin 1 (Figure 3) has different ESD mechanical parameters compered to resin 1A (Figure 11), which leads to various conclusions regarding the assessment of the usefulness of pseudoplastic materials proposed as fillers in working elements of concrete structures—taking into account the assessment of their adhesion to the concrete surface.

The comparison of Figures 6 and 11 shows that mechanical cleaning of the concrete surface 1A results in better resin adhesion than manual preparation of the concrete surface 1AR5.

As mentioned above, the existing hyperelastic models used for a description of the correlation of non-elastic materials (including hyperelastic resins) do not take into account the impact of a large number of variables determining the usefulness of ESD materials as expansion joint fillers. In particular, they do not take into consideration the materials' adhesion to concrete surfaces and the strengthening control and deflection control areas.

5. Conclusions

An assessment has been proposed of the mechanical properties of ESD pseudoplastic materials for working joints of concrete structures, by means of a description f_X (force, deformation and absorbed energy) on a force–deformation curve, in an axial tensile test.

It has been suggested that there is a need for a standardised test stand together with a reference specimen to enable a comparison of the obtained values of mechanical properties with a multiple of the established reference specimen. Such an assessment enables a comparison of the results of tests carried out at various research centres.

Quasi-plastic ESD fillers are recommended for joints in working concrete structures. The proportionality area (the range of stress, deformation and absorbed energy) determines the quality of the filler, whereas the strengthening control and deflection control areas contribute to the improvement of durability of concrete fillers and serve as an additional protection for sealants.

For ESD materials, it is not recommended to assess the destruction propagation area as that range generates too large deformation of pseudoplastic materials—with abilities to carry stress lower than f_{cr} (Hooke's law range).

Based on the proposed tests and analyses, it has been found that resin 1A—based on thixotropic acrylic—is the best ESD pseudoplastic specimen.

Author Contributions: K.S. conceived and designed the experimental work and completed the experiments; D.L. analysed the test results and performed editing; and K.W. prepared the specimens and analysed the test results. All authors have read and agreed to the published version of the manuscript.

Funding: This research received no external funding.

Conflicts of Interest: The authors declare no conflict of interest.

References

1. Hajduk, P. Design of expansion joints in industrial floors and the most frequent causes of their damage-part I. *Build. Eng.* **2016**, *1*, 1–4. (In Polish). Available online: http://www.inzynierbudownictwa.pl/technika,materialy_i_technologie,artykul,projektowanie_dylatacji_podlog_przemyslowych_oraz_najczestsze_przyczyny_ich_uszkodzen___cz_i,8806 (accessed on 7 May 2020).
2. Lohmeyer, G.; Eberling, K. Betonböden fur Produktions- und Lagerhallen: Planung, Bemessung, Ausfuhrung. *Verl. Bud Tech.* **2012**, 608. Available online: https://shop.verlagbt.de/bauplanung-ausfuehrung/betonboeden-fuer-produktions-und-lagerhallen.html (accessed on 7 May 2020).
3. ACI 360R-10. Guide to Design of Slabson Ground, Reported by ACI Committee 360. Available online: https://www.concrete.org/Portals/0/Files/PDF/Previews/360R-10web.pdf (accessed on 7 May 2020).
4. Ferrier, E.; Rabinovitch, O.; Michel, L. Mechanical behaviour of concrete resin/adhesive–FRP structural assemblies under low and high temperatures. *Constr. Build. Mat.* **2016**, *127*, 1017–1028. [CrossRef]
5. Firmo, J.P.; Correia, J.R.; Pitta, D.; Tiago, C.; Arruda, M.R.T. Experimental characterization of the bond between externally bonded reinforcement (EBR) CFRP strips and concrete at elevated temperatures. *Cem. Concr. Compos.* **2015**, *60*, 44–54. [CrossRef]

6.   Patterson, B.A.; Busch, C.E.; Bratcher, M.; Cline, J.; Harris, D.E.; Masser, K.A.; Fleetwood, A.L.; Knorr, D.B., Jr. Influence of temperature dependent matrix properties on the high-rate impact performance of thin glass fiber reinforced composites. *Compos. Part B* **2020**, *192*. [CrossRef]

7.   Savaş Erdem, S.; Dawson, A.R.; Thom, N.H. Impact load-induced micro-structural damage and micro-structure associated mechanical response of concrete made with different surface roughness and porosity aggregates. *Cem. Concr. Res.* **2012**, *42*, 291–305. [CrossRef]

8.   Ozturk, H. Fracture mechanics interpretation of thin spray-on liner adhesion tests. *Int. J. Adhes. Adhes.* **2012**, *34*, 17–23. [CrossRef]

9.   Lee, A.T.; Michel, M.; Ferrier, E.; Benmokrane, B. Influence of curing conditions on mechanical behaviour of glued joints of carbon fibre-reinforced polymer composite/concrete. *Constr. Build. Mater.* **2019**, *227*, 116385. [CrossRef]

10.   Hassan, S.A.; Gholami, M.; Ismail, Y.S.; Sam, A.R.M. Characteristics of concrete/CFRP bonding system under natural tropical Climate. *Constr. Build. Mat.* **2015**, *77*, 297–306. [CrossRef]

11.   Du, J.; Bu, Y.; Shen, Z. Interfacial properties and nanostructural characteristics of epoxy resin in cement matrix. *Constr. Build. Mat.* **2018**, *164*, 103–112. [CrossRef]

12.   Li, Y.; Peng, S.; Miao, J.-T.; Zheng, L.; Zhong, J.; Wua, L.; Weng, Z. Isotropic stereolithography resin toughened by core-shell particles. *Chem. Eng. J.* **2020**, *394*, 124873. [CrossRef]

13.   Opelt, C.V.; Coelho, L.A.F. On the pseudo-ductility of nanostructured epoxy resins. *Polym. Test.* **2019**, *78*, 105961. [CrossRef]

14.   Gao, Y.; Romero, P.; Zhang, H.; Huang, M.; Lai, F. Unsaturated polyester resin concrete: A review. *Constr. Build. Mat.* **2019**, *228*, 116709. [CrossRef]

15.   Bellini, C.; Polini, W.; Sorrentino, L.; Turchetta, S. Mechanical performances increasing of natural stones by GFRP sandwich structures. *Proced. Struct. Integr.* **2018**, *9*, 179–185. [CrossRef]

16.   Norma ISO 37:2017. Rubber, Vulcanized or Thermoplastic—Determination of Tensile Stress-Strain Properties, Szwajcaria. Available online: https://www.iso.org/standard/68116.html (accessed on 7 May 2020).

17.   Norma AS 1145.1-2001, Determination of Tensile Properties of Plastics Materials General Principles, Australia. Available online: https://www.standards.org.au/standards-catalogue/sa-snz/manufacturing/pl-010/as--1145-dot-1-2001 (accessed on 7 May 2020).

18.   Norma, JIS K 6251. Rubber, Vulcanized or Thermoplastic-Determination of Tensile Stress-Strain Properties, Japonia. 2017. Available online: https://global.ihs.com/doc_detail.cfm?document_name=JIS%20K%206251&item_s_key=00256583 (accessed on 7 May 2020).

19.   Norma AS/NZS 1660.2.3:1998. Test Methods for Electric Cables, Cords and Conductors Insulation, Extruded Semi-Conductive Screens and Non-Metallic Sheaths—Methods Specific to PVC and Halogen Free Thermoplastic Materials, Australia. Available online: https://www.saiglobal.com/PDFTemp/Previews/OSH/as/as1000/1600/16603.pdf (accessed on 7 May 2020).

20.   Norma DIN 53504. Testing of Rubber—Determination of Tensile Strength at Break, Tensile Stress at Yield, Elongation at Break and Stress Values in a Tensile Test, Niemcy. 2017. Available online: https://global.ihs.com/doc_detail.cfm?document_name=DIN%2053504&item_s_key=00028732 (accessed on 7 May 2020).

21.   Norma ASTM D412-16. Standard Test Methods for Vulcanized Rubber and Thermoplastic Elastomers—Tension, USA. 2016. Available online: https://global.ihs.com/doc_detail.cfm?rid=BSD&document_name=ASTM%20D412 (accessed on 7 May 2020).

22.   Norma ISO 6259-3:2015. Thermoplastics Pipes—Determination of Tensile Properties—Part 3: Polyolefin Pipes, Szwajcaria. 2015. Available online: https://www.iso.org/standard/62421.html (accessed on 7 May 2020).

23.   Norma BS 2782-3:Methods 320A to 320F:1976. Methods of Testing Plastics. Mechanical Properties. Tensile Strength, Elongation and Elastic Modulus, Wielka Brytania. 1976. Available online: https://shop.bsigroup.com/ProductDetail?pid=000000000000037491 (accessed on 7 May 2020).

24.   Norma ASTM D882-18. Standard Test Method for Tensile Properties of Thin Plastic Sheeting, USA. 2018. Available online: https://www.astm.org/Standards/D882 (accessed on 7 May 2020).

25.   Norma ASTM D412-16. Standard Test Methods for Vulcanized Rubber and Thermoplastic Elastomers-Tension, USA. 2016. Available online: https://www.astm.org/Standards/D412 (accessed on 7 May 2020).

26. Norma ISO 1798:2008. Flexible Cellular Polymeric Materials—Determination of Tensile Strength and Elongation at Break, Szwajcaria. 2008. Available online: https://www.iso.org/standard/41059.html (accessed on 7 May 2020).

27. Sułowski, M.; Ciesielka, M.; Jurczak-Kaczor, P. Tension curve analysis in technical education. In Proceedings of the III Conference e-Technologies in Engineering Education eTEE'2016, Stanisław Staszic AGH University of Science and Technology, Krakow, Poland, 11 April 2016. (In Polish). Available online: http://yadda.icm.edu.pl/yadda/element/bwmeta1.element.baztech-704015f0-eed1-4406-9703-bade596497b2/c/ZN_WEiA_PG_48-13.pdf (accessed on 7 May 2020).

28. Ziobro, J. Multiaxial stress and strain analysis for rubber based on natural rubber NR. *Science Journals of Rzeszów University of Technology* 288. *Mechanics* **2013**, *85*, 197–206. (In Polish). Available online: http://oficyna.prz.edu.pl/download/hGKmo6Cj9JUSAlejUodH9BIX4-NEpbd299dEd_LWQoIk99NiUALQ,rVIXoyCjMyG2EefjUfODQ6dih5MgJHLSw1LhMsZTY6MQdoIG1AcTJaPA/mech-85-02-13-pw-1.pdf (accessed on 7 May 2020).

29. Lechman, M.; Mazurczuk, R.; Fedorczyk, Z. Tests of elastomers in complex states of strain. *ITB Q.* **2010**, *152*. (In Polish). Available online: http://yadda.icm.edu.pl/yadda/element/bwmeta1.element.baztech-article-BTB2-0060-0060/c/Lechman.pdf (accessed on 7 May 2020).

30. Norma PN-EN 1337-3:2005. Structural Bearings–Elastomeric Bearings. (In Polish). Available online: https://sklep.pkn.pl/pn-en-1337-3-2005d.html (accessed on 7 May 2020).

31. Logoń, D. FSD cement composites as a substitute for continuous reinforcement. In *Eleventh International Symposium on Brittle Matrix Composites BMC-11*; Andrzej, M., Ed.; Brandt Warsaw: Institute of Fundamental Technological Research: Warsaw, Poland, 28–30 September 2015; pp. 251–260. Available online: http://www.ippt.pan.pl/en/home/72-aktualnosci/ippt-info/1436-brittle-matrix-composites-bmc-12-at-ippt-pan.html (accessed on 7 May 2020).

32. Logoń, D. Identification of the Destruction Process in Quasi Brittle Concrete with Dispersed Fibers Based on Acoustic Emission and Sound Spectrum. *Materials* **2019**, *12*, 2266. [CrossRef] [PubMed]

33. Norma ASTM 1018. Standard Test Method for Flexural Toughness and First Crack Strength of Fiber–Reinforced Concrete. Philadelphia, USA Vol.04.02. 1992. Available online: https://www.astm.org/Standards/C1018.htm (accessed on 7 May 2020).

34. Norma EN 14651. Test Method for Metallic Fibre Concrete. Measuring the Flexural Tensile Strength (Limit of Proportionality (LOP), Residual). Available online: https://shop.bsigroup.com/ProductDetail/?pid=000000000030160504 (accessed on 11 July 2005).

35. JCI. Japan Concrete Institute Standard. Method of Test for Bending Moment–Curvature Curve of Fiber-Reinforced Cementitious Composites, S-003-2007. (2003) Method of Test for Load-Displacement Curve of Fiber Reinforced Concrete by Use of Notched Beam. (2003) Method of Test for Fracture Energy of Concrete by Use of Notched Beam. Available online: https://www.jci-net.or.jp/j/jci/study/jci_standard/JCI-S-003-2007-e.pdf (accessed on 7 May 2020).

36. Logoń, D. Hybrid reinforcement in SRCC concrete. In Proceedings of the Third International Conference on Sustainable Construction Materials and Technologies SCMT3, Kyoto, Japan, 18–21 August 2013; paper ID: e154. Available online: http://www.claisse.info/2013%20papers/data/e154.pdf (accessed on 7 May 2020).

37. Bergstrom, J. *Mechanics of Solid Polymers Theory and Computational Modeling*; Elsevier: Amsterdam, The Netherlands, 2015; p. 520. Available online: https://www.elsevier.com/books/mechanics-of-solid-polymers/bergstrom/978-0-323-31150-2 (accessed on 7 May 2020).

38. Zhao, Z.; Mu, X.; Du, F. Modeling and Verification of a New Hyperelastic Model for Rubber-Like Materials. *Math. Probl. Eng.* **2019**. [CrossRef]

 *materials* 

Article

Mechanical and Non-Destructive Testing of Plasterboards Subjected to a Hydration Process

Zbigniew Ranachowski [1], Przemysław Ranachowski [1], Tomasz Dębowski [1], Adam Brodecki [1], Mateusz Kopec [1], Maciej Roskosz [2], Krzysztof Fryczowski [3], Mateusz Szymków [4], Ewa Krawczyk [4] and Krzysztof Schabowicz [4,*]

[1] Experimental Mechanics Division, Institute of Fundamental Technological Research, Polish Academy of Sciences, Pawińskiego 5B, 02-106 Warszawa, Poland; zranach@ippt.pan.pl (Z.R.); pranach@ippt.pan.pl (P.R.); tdebow@ippt.pan.pl (T.D.); abrodec@ippt.pan.pl (A.B.); mkopec@ippt.pan.pl (M.K.)

[2] Faculty of Mechanical Engineering and Robotics, AGH University of Science and Technology, Aleja Mickiewicza 30, 30-059 Kraków, Poland; mroskosz@agh.edu.pl

[3] Faculty of Energy and Environmental Engineering, Silesian University of Technology, ul. Akademicka 2A, 44-100 Gliwice, Poland; krzysztof.fryczowski@polsl.pl

[4] Faculty of Civil Engineering, Wrocław University of Science and Technology, Wybrzeże Wyspiańskiego 27, 50-370 Wrocław, Poland; Mateusz.Szymkow@pwr.edu.pl (M.S.); krawczyk.em@gmail.com (E.K.)

* Correspondence: krzysztof.schabowicz@pwr.edu.pl

Received: 29 April 2020; Accepted: 20 May 2020; Published: 23 May 2020

Abstract: The aim of this study was to investigate the effect of plasterboards' humidity absorption on their performance. Specimens' hydration procedure consisted of consecutive immersing in water and subsequent drying at room temperature. Such a procedure was performed to increase the content of moisture within the material volume. The microstructural observations of five different plasterboard types were performed through optical and scanning electron microscopy. The deterioration of their properties was evaluated by using a three-point bending test and a subsequent ultrasonic (ultrasound testing (UT)) longitudinal wave velocity measurement. Depending on the material porosity, a loss of UT wave velocity from 6% to 35% and a considerable decrease in material strength from 70% to 80% were observed. Four types of approximated formulae were proposed to describe the dependence of UT wave velocity on board moisture content. It was found that the proposed UT method could be successfully used for the on-site monitoring of plasterboards' hydration processes.

Keywords: plasterboards; moisture content; hydration processes; mechanical properties; ultrasound measurements

---

1. Introduction

The hydration of calcium sulphate hemihydrates (stucco reaction) nowadays has great industrial and economic importance. Based on the data published by Eurogypsum [1], ca. 5 million tonnes of these products are used every year in the European building industry. The plasterboards are used in renovations, as well as for building constructions. They could be used for both, humidity- and fire-resistant applications. Humidity-resistant boards are usually reinforced with glass fiber mesh. In the novel method presented in [2], nano silver protecting coating was described to enhance the plaster performance. On the other hand, fireproofing panels used for building fire protection systems, usually have low water content and silica fillers in their structure. The article [3] presents an example of silica filler application in shrinkage controlling and, consequently, reducing cracks. Novel, synthetic gypsum boards were characterized by their enhanced performance. This by-product was described and tested in [4]. Despite years of research, a fully convincing description of the stucco hydration reaction was not presented. The conversion of hemihydrate ($CaSO_4 \cdot 0.5H_2O$) to gypsum

($CaSO_4 \cdot 2H_2O$) was studied mainly through quenching reactions or indirect methods [5,6]. Additionally, the $CaSO_4$-H_2O system could have up to five phases [7]. Hemihydrates, prepared from the dehydration of gypsum, exist in two forms—α and β. The most popular β-form is generated via dry calcining (120–180 °C). The α-form is prepared under hydrothermal conditions, involving high pressure—up to 8 bar. Both forms were described as structurally undistinguishable, but with different crystal habits [8]. α-stucco is much more expensive and difficult to generate. It is used wherever mechanical strength is of prime importance. The much less expensive β-form is produced and applied on a considerably larger scale. β-stucco is widely used in the production of plasterboards. Gypsum product is similar for both stucco forms from which is it produced; however, the reaction process is different. α-stucco induction time is shorter and conversion percentage is higher. It should be emphasized that the reaction never occurs with 100% efficiency. In order to increase the reaction rate and improve the hardness of the product, gypsum seed crystals are usually added to the stucco and water mixture. This does not affect the amount of water needed. In the case of low concentrations of seeds, the rate of crystal growth is greater than the rate of nucleation. When concentrations of added seeds are higher (above 0.5% *w/w*), the situation is reversed and nucleation is faster [9]. The stages of the hydration process are well recognised—a comparatively rapid dissolution of hemihydrate and a subsequent, slower precipitation of $CaSO_4 \cdot 2H_2O$ [10]. Nevertheless, description of the kinetics and mechanism of the whole hydration process was not fully investigated yet. It is known that the plasterboards are made of partially dehydrated gypsum of high porosity (5% ÷ 40% by vol.) so its density is low but the mechanical strength remains at the acceptable level. The noteworthy physical effect caused by the further unwanted hydration of the plasterboard volume is the loss of their elasticity and strength. Plasterboards products are exploited in the environment with increased humidity, which reduces their durability significantly. Additionally, the occurrence of organic mould growth in hydrated plasterboards might create severe health issues.

The following papers were devoted to testing procedures of plasterboards. During the inspection and diagnostics of surfaces made of gypsum boards, the method presented in reference [11] was found to be useful. This study presents a data system supporting the inspection and diagnostics of partition walls and wall cladding assembled using the drywall construction method. Furthermore, the classification of anomalies and their probable causes was described. In reference [12], the system supporting the inspection and diagnostics of partition walls was described and additional research on the correlation matrix was supplemented. In [13], the physical and mechanical properties of innovative panels were determined. These energy-saving panels were successfully used in India. In addition to traditional panels, composites are increasingly appearing in building constructions. The experimental testing of the compressive strength of gypsum wall panels was described in [14] and the physical and mechanical properties of panels were determined. The above article, as well as [15], describes composite panels consisting of gypsum, plaster and fiberglass. Currently, the world's awareness of society is growing more and more frequently, apart from the properties described in the above articles, the materials must be ecological. In [16], a recycling process for gypsum plasterboard was presented. It comprised of grinding and calcining the residue of drywall sheets and further resulted in the obtaining of the product of acceptable quality. It was concluded that, after rehydration, it was possible to use only the gypsum waste to mould solid specimens. This experiment showed that product merely discarded in the environment could be also recycled. The influence of humidity on the properties of concrete was investigated by using contact and non-contact methods and was reported in [17], but, so far, no similar research in relation to plasterboards was performed.

2. Materials and Methods

Five sets of specimens, six pieces of each, made of five different plasterboards types (labelled A, B, G, I and K) were prepared for the examination. All the specimens were 12.5 mm thick. The following board types were investigated:

- A—Standard board for indoor installations;
- B—Low-quality, standard-type board rejected after quality control by the producer;
- G—Fire-resistant type;
- I—Acoustic-insulating type;
- K—Humidity-resistant type.

Figure 1 shows the general view of five types of square 200 × 200 mm plasterboard specimens, prepared for examination. The black arrows on the board surfaces denote the fabrication direction.

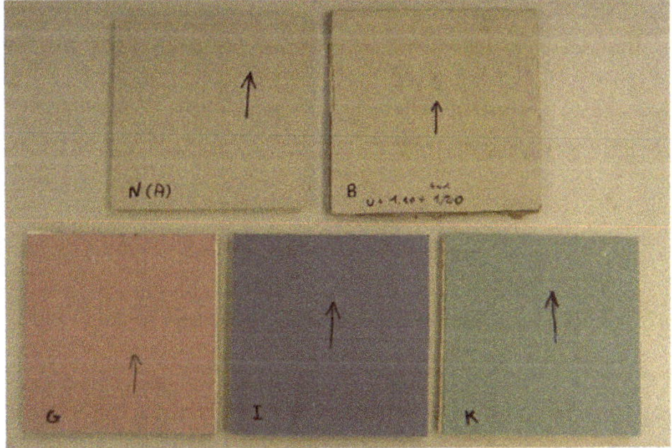

Figure 1. View of five types of square 200 × 200 mm plasterboard specimens prepared for examination.

The microstructural observations of all types of investigated boards were analysed by using optical light microscopy (AM4113ZTL 1.3 Megapixel Dino-Lite Digital Microscope with integral LED lighting, (Keyence, Mechelen, Belgium) and a low vacuum scanning electron microscope JEOL JSM6460 LV (Jeol, Tokyo, Japan) with an Energy Dispersive Spectroscopy (EDS) analyser, fabricated by JEOL.

The following experimental procedure was performed to assess the deterioration of mechanical properties due to the hydration process. The specimens made of five types of plasterboard were initially subjected to mechanical testing in the as-received state. The other set of specimens was immersed in water at room temperature for 10 min, dried at the same temperature for 60 min and finally directed to mechanical testing. The third set of specimens was subjected to the hydrating procedure twice and then the flexural test was performed. Applying the procedure described above, the three levels of hydration were examined: (1) dried in standard room conditions, (2) hydrated with lower intensity and (3) hydrated with higher intensity. The attempts to repeatably introduce the different states of hydration were not successful. The controlling of the specimens' water intake was performed by consecutive weighting.

The method of moisture content measurement was performed as follows. The plasterboard specimens in the as-received state were dried for 1 hour, applying a hot air fan, and stored at room temperature in dry conditions for 24 hours. Then, the specimens were weighted and their mass in dry state m_0 was determined. The mass in hydrated state m_{h1} and m_{h2} was determined 60 min after 10 min of hydration (or, after two cycles of hydration). The moisture content m_{c1} after one cycle of hydration and the moisture content m_{c2} after two cycles of hydration were given by the following formulas:

$$m_{c1} = (m_{h1} - m_0) \cdot 100/m_0; \ m_{c2} = (m_{h2} - m_0) \cdot 100/m_0 \qquad (1)$$

The A-type boards have been designed for usage in the compartments where the relative humidity did not exceed 70%. The usage of the K-type boards in the increased humidity environment (up to 85%) was allowed within a period not exceeding ten hours. Gypsum plasterboards were composed of a plaster core encased in and firmly bonded to outer paper liners. Due to its fabrication method, the lining and also the board as a whole showed significant anisotropic properties [17–20]. Therefore, the mechanical properties of plasterboard sheets were determined in parallel and perpendicular to the sheet production direction, applying standard EN 520 [21]. Conventionally, a three-point bending test of plasterboards is the established procedure for their mechanical properties assessment. Square-shaped specimens were tested in this research. The procedure consisted of the loading of two specimens sets, two pieces each, in two directions mentioned above. The flexural strength of plasterboard was defined at maximal force registered by the loading system during the three-point bending. The bending test of investigated specimens was performed by using a hydraulic MTS-858 testing machine (MTS, Eden Prairie, Minnesota, USA) of 250 kN capacity, as was shown in Figure 2. The maximal force registered by the loading system was recalculated to the stress applied to the specimens, and, therefore, at 100 N of load, corresponded with 750 kPa of stress. The investigated material was brittle and had a porosity of 20%–40%. Therefore, the measurements of the modulus of elasticity were highly dispersed and not reliable.

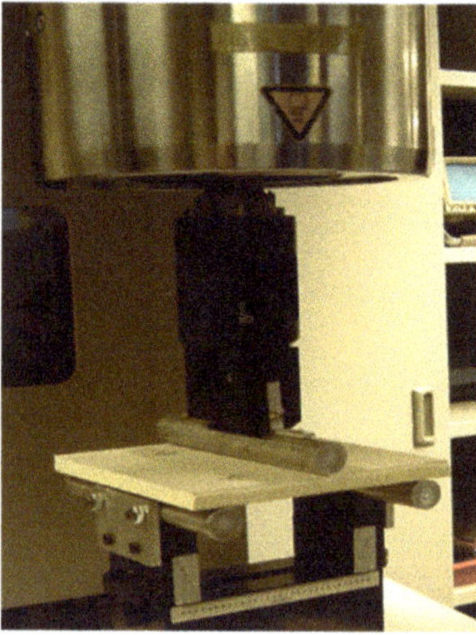

Figure 2. Experimental setup for the 3-point bending test of plasterboards.

Ultrasonic longitudinal wave velocity c_L was determined in all specimens to find the dependence of wave velocity on the degree of hydration. The specifications of the investigated specimens are shown in Tables 1 and 2.

Table 1. Mechanical properties of tested plasterboard compositions.

Board Symbol	Remarks	Board Thickness [mm]
A	standard board for indoor installations	12.5
B	low quality board of standard type	12.5
G	fire resistant type	12.5
I	acoustic insulating type	12.5
K	humidity resistant type	12.5

Table 2. Mechanical properties of tested plasterboard compositions (cont.).

Board Symbol	Apparent Density [kg/m^3]	Flexural Strength F$_{max}$ [N], Declared by the Manufacturer— Parallel and Perpendicular to the Sheet Length
A	648	>550/>210
B	648	–
G	880	>550/>210
I	1024	>550/>210
K	672	>550/>210

In this study, a new, non-destructive method of the plasterboard hydration assessment was proposed. The selection of such a method was based on research presented in [22,23], where the deterioration of material elasticity in different building material, including the brittle matrix with air pores, were determined by using the ultrasound testing (UT) method. The dilatation of UT wavelength measured in the investigated material was, in any case, less than 2 mm, which is at least six times smaller than the material thickness. This was the main reason for the propagation of the longitudinal wave model adoption and neglection of the possible Lamb wave. Therefore, the change of UT longitudinal wave velocity c_L was proposed to be a measure of plasterboard mechanical properties deterioration due to occurring hydration. In large-scale objects of small thicknesses, such as the plasterboards, the following dependence of UT longitudinal wave velocity c_L and the scalar modulus of elasticity E could be used:

$$c_L = \sqrt{\frac{E}{\rho(1-v^2)}} \tag{2}$$

where ρ—bulk material density; v—Poisson ratio.

The dependence of c_L on concrete humidity was investigated by using contact and non-contact methods [24]. The UT waves' velocity determined by the authors remained in the range of 3500–4500 m/s. It was found that, in concrete, c_L increases alongside increases in humidity storage. No remarkable loss of the concrete elasticity modulus due to humid conditions of its storage was observed. However, in highly porous and moisture-sensitive gypsum, a decrease in elasticity modulus and mechanical strength occurred with increased humidity content. Subsequently, a decreasing c_L was found. This tendency was further examined in presented paper. Due to the high attenuation of UT waves in gypsum (2 to 5 dB per mm of thickness) and its lower specific density in relation to concrete, the authors have applied a dedicated instrumentation set to c_L measure in this material. c_L was determined with the accuracy higher than 0.5%, measuring the time of propagation T of the elastic wave front across the board of known thickness d with application of the formula $c_L = d/T$. The investigation was performed by using the computerized ultrasonic material tester UTC110, produced by Eurosonic (Vitrolles, France) [25]. It was reported in [24] that low-frequency ultrasound (200–600 kHz) is reasonable for the testing of building materials, where the specimens' thicknesses do not exceed several centimeters. However, such low frequency range ultrasound cannot be used for the testing of millimeter-thick boards due to excessively long wavelength and further errors in c_L measurement. The series of preliminary tests performed by the authors have revealed that the required sensitivity of ultrasound parameters to investigate the structural properties of gypsum boards was achieved when the ultrasonic wavelength is comparable to the dimensions of local voids and pores. This wavelength λ reminded in the following relation to the frequency f of emitting source and propagation velocity c_L of travelling ultrasonic waveform:

$$\lambda = \frac{c_L}{f} \tag{3}$$

Thus, taking into consideration the propagation velocity of 900 ÷ 1800 m/s registered in plasterboards, the authors recommend the application of frequency of 1 MHz to achieve the propagation of waveforms in the range of 0.9–1.8 mm. The instrumentation used for measuring included transmitting and receiving transducers of type Videoscan, fabricated by Olympus [26]. These transducers allow one to emit the ultrasonic beam of 19 mm in diameter at 1 MHz frequency. That modern transducer type was designed for coupling with low-density (i.e., 600–2000 kg/m^3) materials and thus exhibiting a low acoustic impedance of 10 MegaRayl. The contact between the rough surface of the board and the transducers' face was achieved by using 0.6-mm-thick polymer jelly interfacing foil (Olympus PM-4-12, US Division of Olympus, Waltham, Massachusetts, US. The investigation was not aimed to recover particular defects in the specimens. Therefore, a relatively wide penetrating UT beam was used. The use of higher intensity of the beam was achieved to suppress the remarkable attenuation of the material. The custom-designed holder with articulated joints and compressing spring was prepared for the correct coupling of the transducers and investigated board surface. The detailed view of the holder is presented in Figure 3.

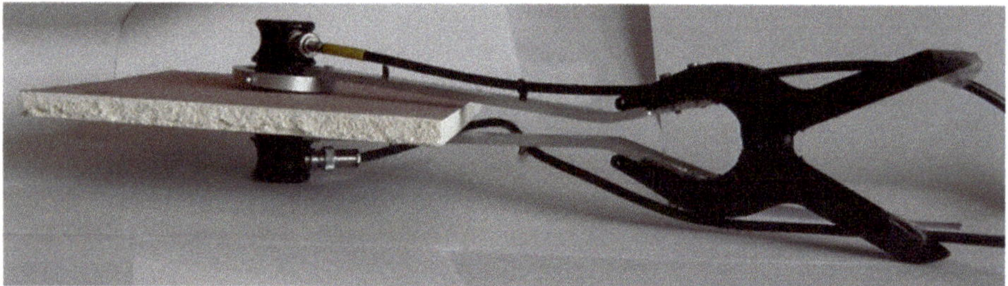

Figure 3. Detailed view of the custom designed holder for correct coupling between the ultrasonic transducers and both sides of the plasterboard specimen.

3. Results and Discussion

3.1. Microstructural Characterization of Plasterboards in the As-received Condition

The essential stage of plasterboard fabrication is a foaming process in which the porosity of gypsum increases up to a required level. This process is driven by the addition of the surface-active agents [9]. Such addition result in a formation of a complex structure with large, mostly spheroidal macropores, smaller pores of irregular shape, being interlaced with blade and needle shaped crystals of calcium sulphate. Most macropores were measured with ca. 100 μm of diameter; however, smaller and larger structures were also visible. Calcium sulphate crystals were only observed by using SEM. The distribution and size of the macropores were assessed by using optical microscopy. The microstructures of investigated specimen types are presented in Figure 4.

The dominant macropores with a diameter ranging from 20 to 200 μm were uniformly distributed in a significant volume of specimen derived from the standard plasterboard designed for indoor installations (Figure 4). This microstructure resulted in a low product density (648 kg/m^3).

The B-type specimen was characterized by small pores (10–20 μm) of irregular shape that appear in considerable quantity. These pores were also present in SEM micrographs. In both cases, they appeared as 'black speckles' because these regions do not reflect visual light nor electrons. Small pores situated at the walls of the foamed gypsum form the openings among the cavities. Such material behaviors modify the overall system of closed porosity into the open porosity system. This resulted in increased moisture permeability as well as in reduced UT longitudinal wave velocity. The microstructure of G-type, fire-resistant plasterboard was characterized by less regular shapes of macropores. The distribution of macropores was more sparse within the compact matrix, resulting in an increase in specific density (880 kg/m^3). Specimens of type I and K were characterized by a considerably higher presence of smaller pores with diameters between 10 and 20 μm, immersed in the compact matrix with increased contents of very small gypsum crystals. This resulted in the remarkable growth of a specific density and a much higher compactness of these structures.

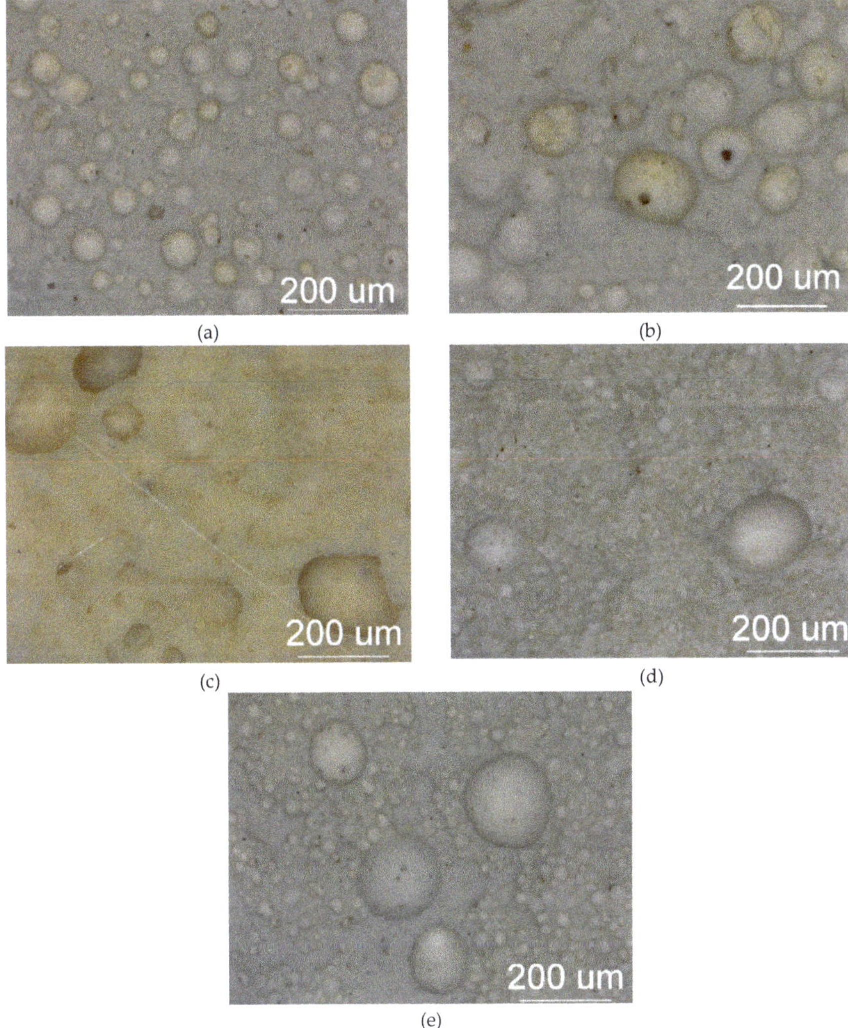

Figure 4. Macrostructure images of investigated specimen types: (**a**) type A, (**b**) type B, (**c**) type G, (**d**) type I, (**e**) type K.

Figure 5 shows a detailed view of a regular- and irregular-shaped pore system of a standard plasterboard core. The micrograph of the foamed gypsum matrix was made by using low vacuum SEM. The variety of pore sizes was presented.

Figure 6 presents the microstructure of investigated plasterboard types. Figure 6a presents a micrograph of a standard A-type plasterboard. The structure of this material consists of large macropores within the matrix of blade- and needle-shaped crystals. The volume of small, irregular-shaped pores is relatively low. The microstructure of the B-type specimen (Figure 6b) consists of the high number of irregular-shaped micropores within the open porosity system. The microstructure of the fire-resistant G-type plasterboard (Figure 6c) is generally similar to the standard one; however, it is more compact and it is reinforced with glass fibers. A reduced number of macropores occurred in acoustic insulating plasterboard I type (Figure 6d). Increased humidity penetration resistance was achieved through the large regions of high compact gypsum containing very small and densely spaced crystals, as presented in Figure 6f. The data presented in Table 3 include the weight percentage of elements for boards of type A and G. The mass percentage volume values, i.e., 55% wt. for oxygen, 19% wt. for sulphur and 23% wt. for calcium, presented a good agreement with theoretical stoichiometric values specified for $CaSO_4 \cdot 2H_2O$. In the case of a gypsum molecule, these theoretical stoichiometric percent values for oxygen, sulphur and calcium were 55.8, 18.6 and 23.3, respectively. The enlisted data confirmed that the plasterboard cores were produced by using pure, technical gypsum raw material. Increased silicon content observed in the board of type G resulted from the presence of glass-fibre reinforcement.

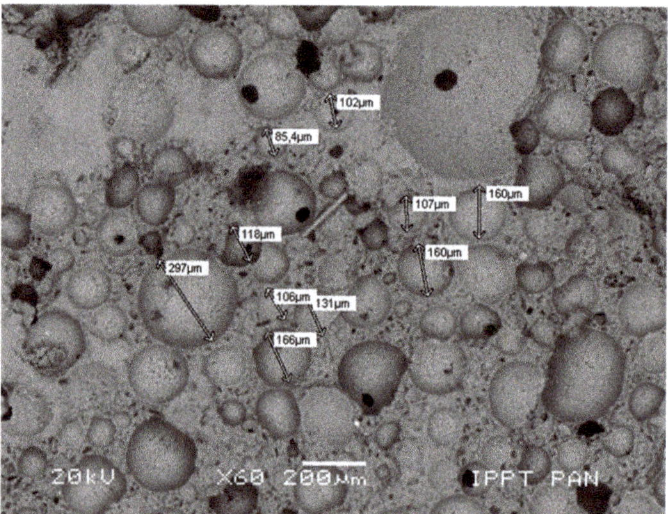

Figure 5. Microstructure of foamed gypsum matrix, with various sizes of pores.

Table 3. Chemical composition of the plasterboards made of materials A and G.

wt. (%)	O	Si	S	Ca
A material	55.32	0.74	19.15	23.49
G material	56.54	1.20	19.52	22.74

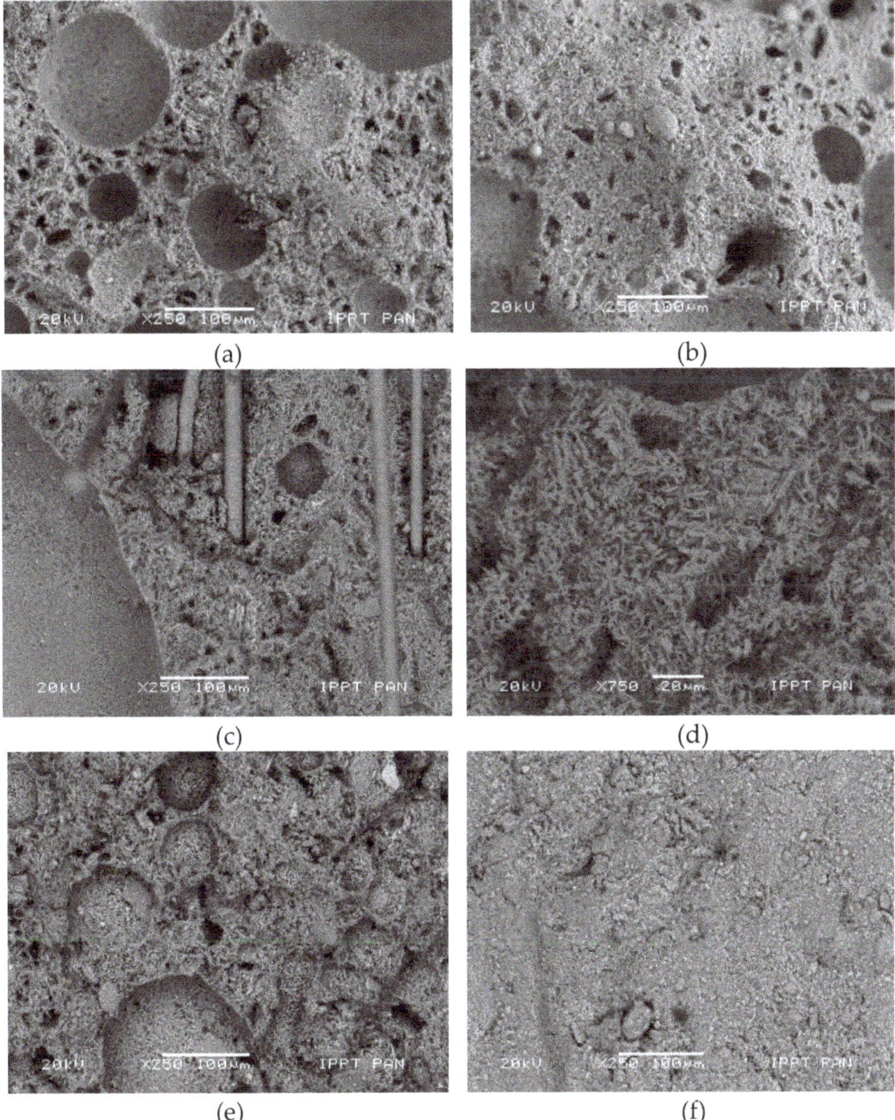

Figure 6. Microstructure of investigated specimens: (**a**) type A, (**b**) type B, (**c**) and (**d**) type G—low and high magnifications of the glass-fiber-reinforcement region—(**e**) type I, (**f**) type K.

3.2. The Effect of Hydration on the Mechanical Properties of Plasterboards

The results of all three-point bending tests are presented in Figures 7–11. In these figures, the blue curves represent the behavior of the specimens in the as-delivered state, the green ones—after first hydration cycle—and the brown curves—after two hydration cycles. The results were recalculated to determine the max. stress S [MPa] occurring in the central specimens' cross sections, using the standard formula:

$$S = (3Fl)/(2bh^2) \tag{4}$$

F stands for the loading force, l—for the length of the support span, b—specimen width, h—specimen thickness there. The averaged results of all mechanical tests were listed in Table 4.

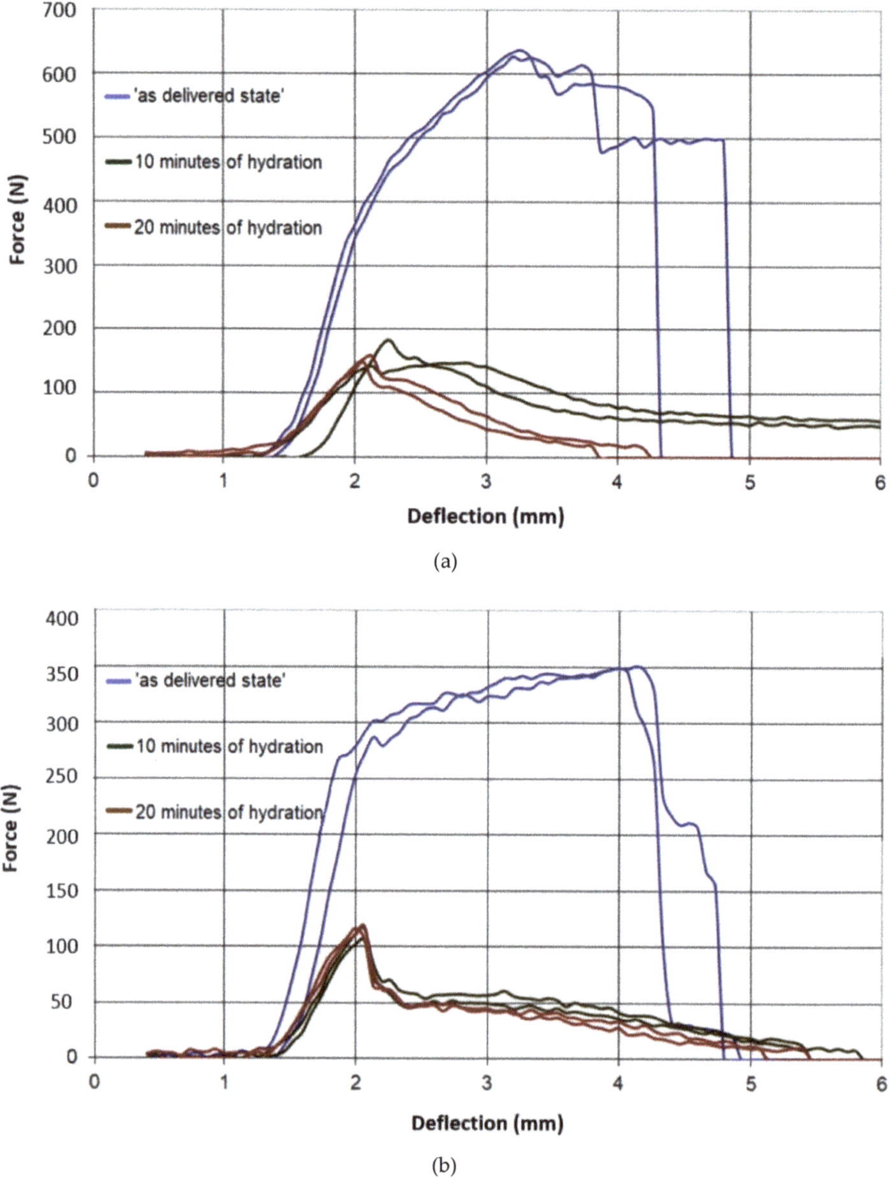

Figure 7. Force–deflection curves of the tested specimens of type A, (**a**) parallel to the sheet production direction, (**b**) perpendicular to sheet production direction.

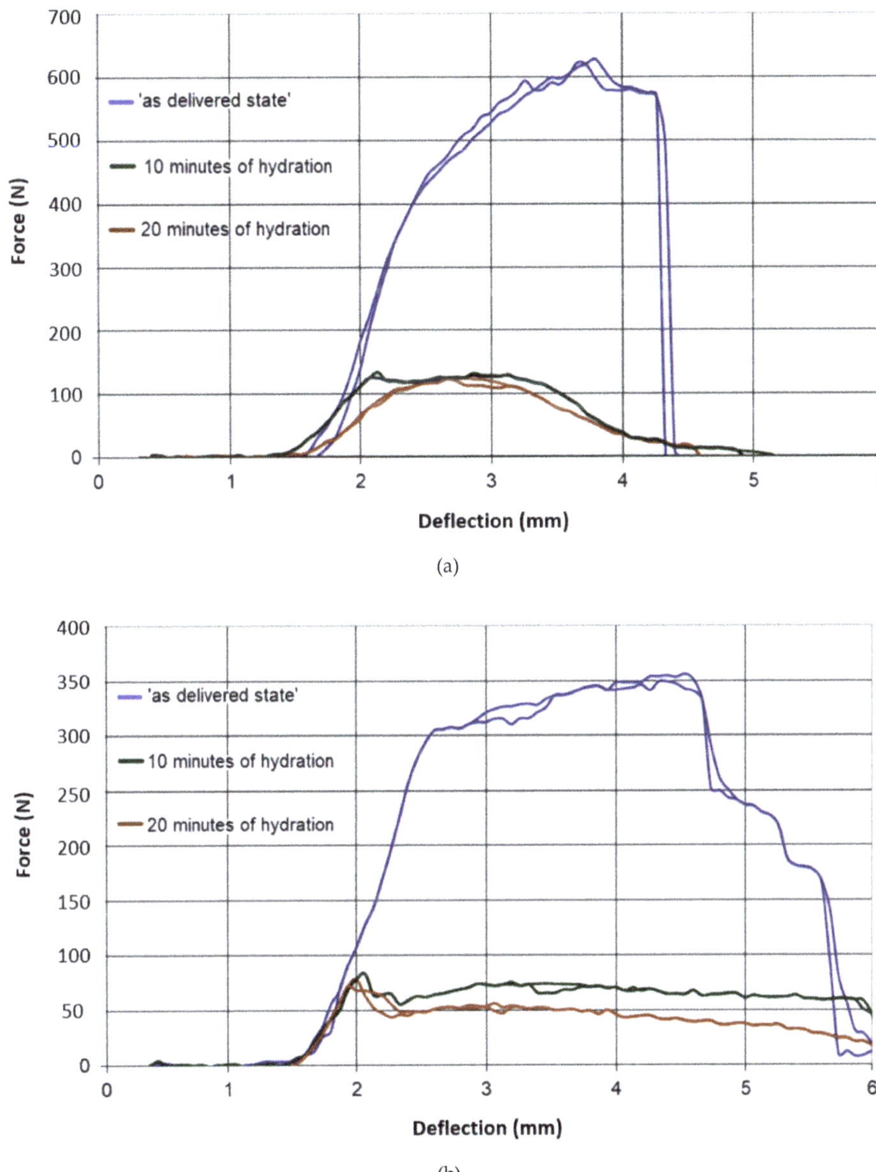

(a)

(b)

Figure 8. Force–deflection curves of the tested specimens of type B, (**a**) parallel to the sheet production direction, (**b**) perpendicular to sheet production direction.

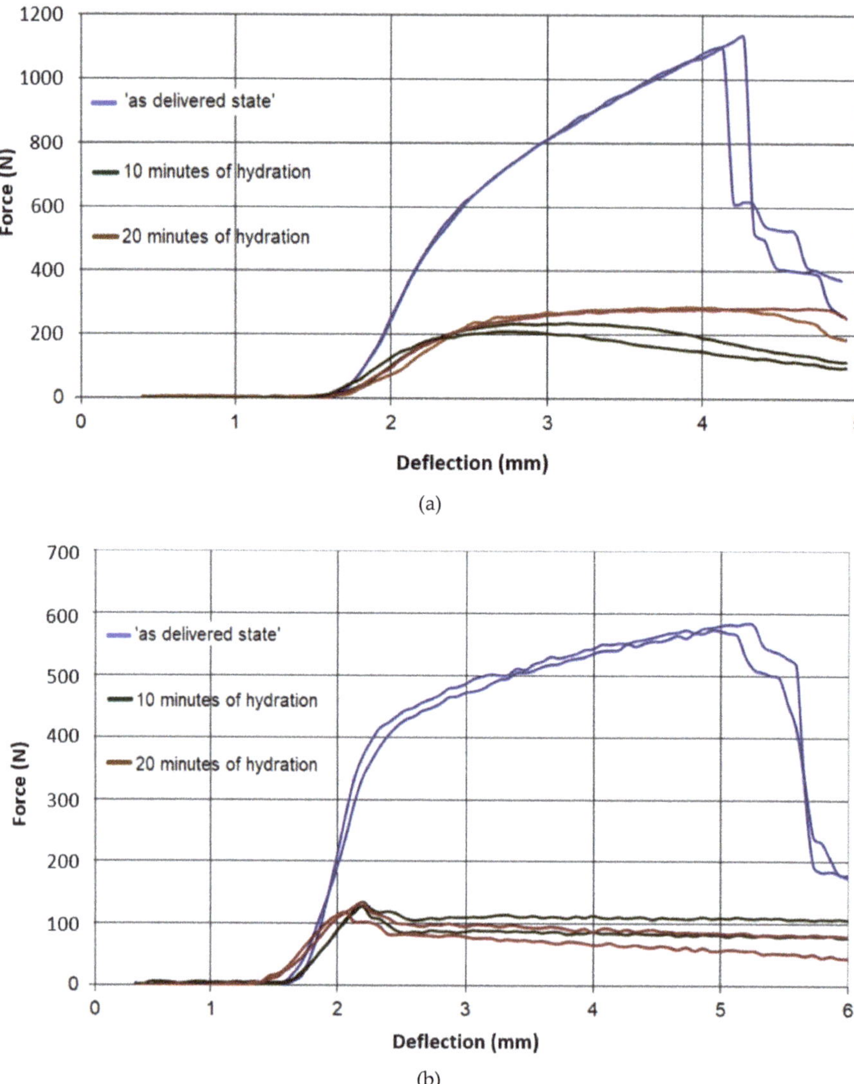

Figure 9. Force–deflection curves of the tested specimens of type G, (**a**) parallel to the sheet production direction, (**b**) perpendicular to sheet production direction.

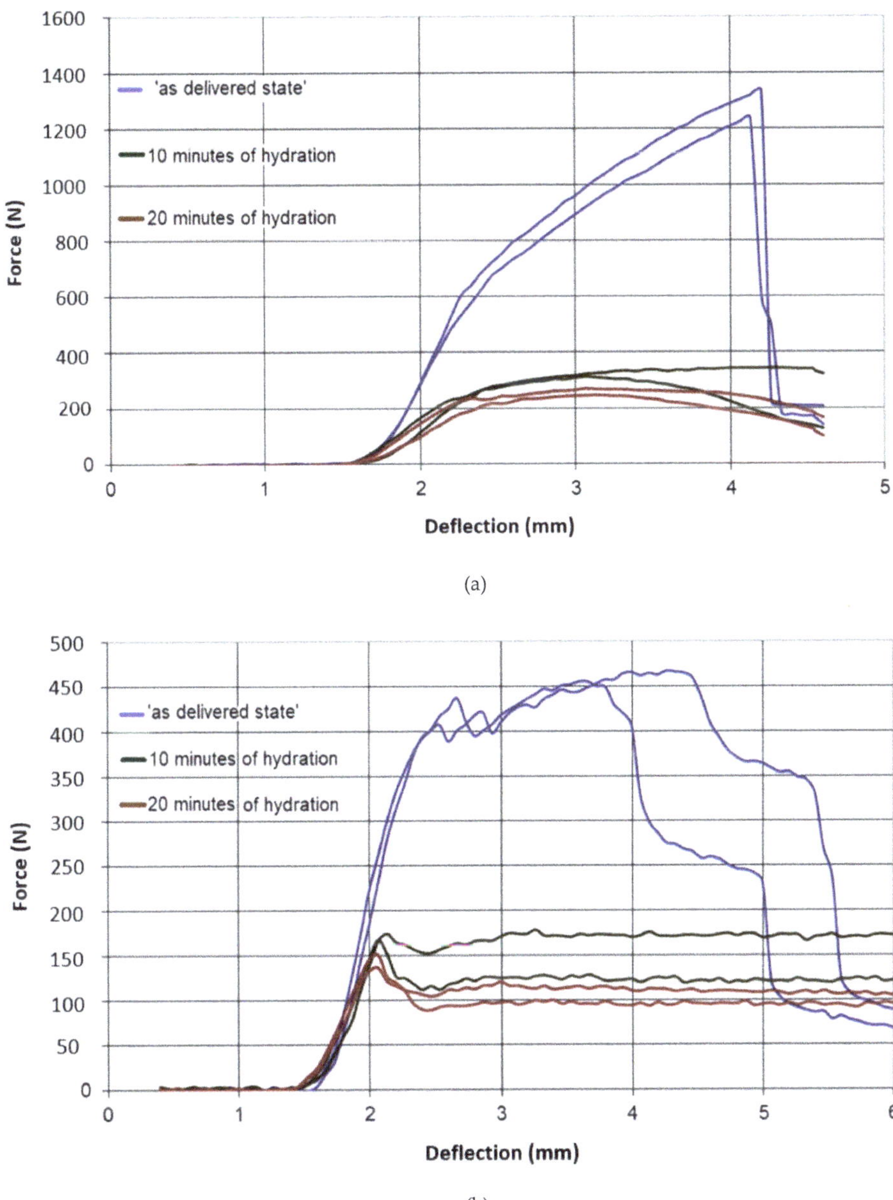

(a)

(b)

Figure 10. Force–deflection curves of the tested specimens of type I, (**a**) parallel to the sheet production direction, (**b**) perpendicular to sheet production direction.

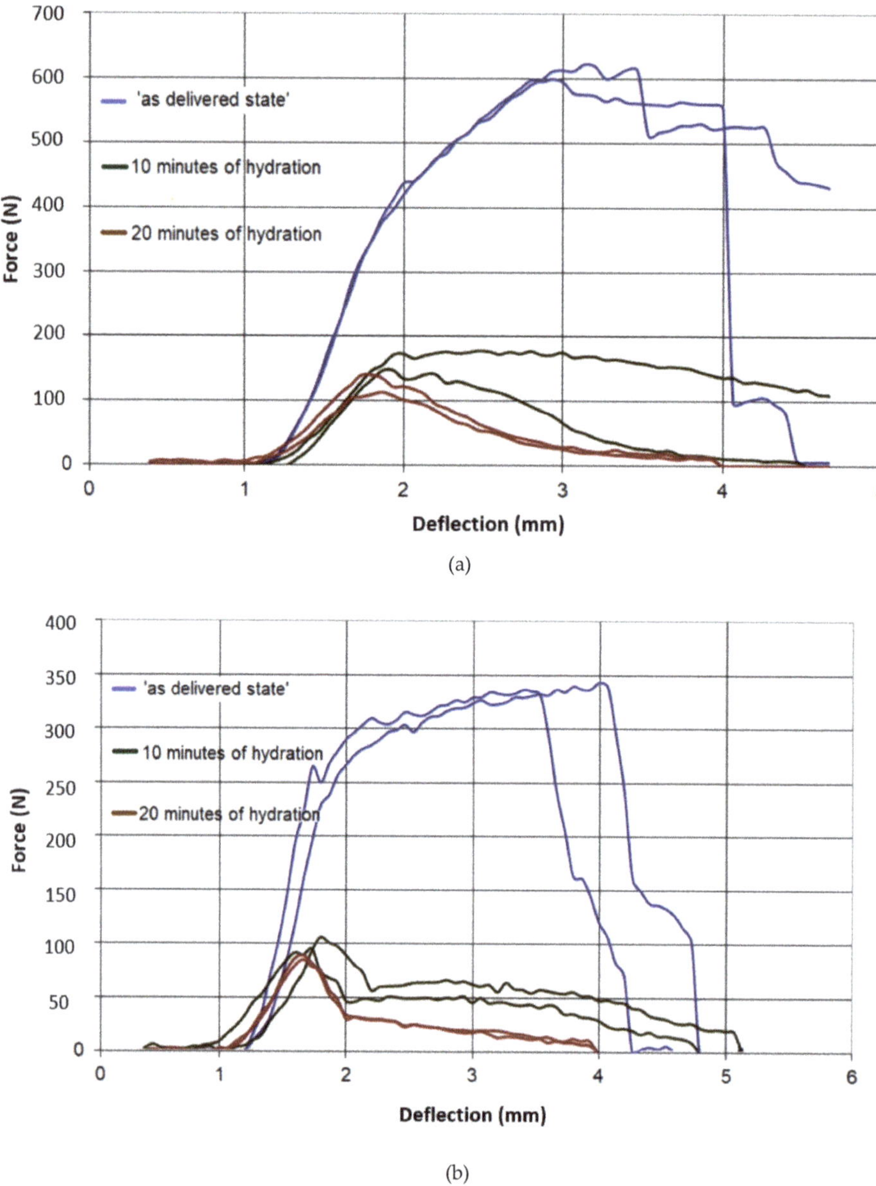

Figure 11. Force–deflection curves of the tested specimens of type K, (**a**) parallel to the sheet production direction, (**b**) perpendicular to sheet production direction.

The difference between the results of the mechanical tests obtained after the first and second hydration cycles was relatively small. However, the loss of mechanical strength after the first hydration was significant. Such loss may indicate that, after two hydration cycles, nearly all possibilities to introduce the moisture to the boards pore system were exhausted.

Table 4. Results of the mechanical tests performed on investigated plasterboard types.

Board Type	Max. Stress in the Initial State, ‖ to Sheet Production Direction [MPa]	Max. stress in the Initial State, ⊥ to Sheet Production Direction [MPa]	Max. stress after First Hydration, ‖ to Sheet Production Direction [MPa]	Max. Stress after First hydration, ⊥ to Sheet Production Direction [MPa]	Max. Stress after Second Hydration ‖ to Sheet Production Direction [MPa]	Max. Stress after Second Hydration ⊥ to Sheet Production Direction [MPa]
A	4.74	2.62	1.24	1.29	1.15	0.84
B	4.69	2.64	1.01	0.62	0.95	0.58
G	8.34	4.34	2.14	0.98	1.68	0.94
I	9.65	3.46	2.48	1.27	1.93	1.08
K	4.59	2.50	1.22	0.75	0.95	0.65

* The standard deviation of all obtained mechanical test results did not exceeded 5%.

It was observed in Figures 7 and 8 that both type A and B in the as-delivered condition have similar mechanical strengths. However, the B-type boards, characterized by higher open porosity system, were capable to absorb more moisture. Such absorption caused a higher loss of mechanical strength in both of the loading directions. The glass-fiber-reinforced boards of G and I type had a higher initial strength, as compared to different types of boards. However, the hydration process led to the deterioration of bond connection between the fibers and matrix and the subsequent loss of specimens' strength (Figures 9 and 10). The decrease in the mechanical properties for about 70%–80% was observed, regardless of the level of hydration. It was found that I and K types absorbed ca. 5% water by wt. and others (A,B) absorbed more than 30%.

3.3. Assessment of Ultrasonic Wave Velocity c_L Loss after Hydration Tests—Discussion

The results of UT tests are presented in Figure 12. For each plasterboard type, eight measurements at different board locations of accidental choice were performed and then the average wave velocity was determined. The standard deviation of a single series of measurements was included in the range of 2% ÷ 3% from the average value of the readings. The procedure was performed on the boards (1) in the as-delivered condition, (2) after immersing in water at room temperature for 10 min and 60 min of drying, and (3) after twice immersing for 10 min and 60 min of drying. These procedures were performed in order to obtain two states of the hydration, i.e., 'the moderate hydration' and 'the strong hydtation' in a sufficiently repetitive manner. Any intermediate state was not possible to achieve. It could be observed from the Figure 12, different values of c_L for different board types when the board humidity was zero. The results obtained during tests (2) and (3) take into account the loss of longitudinal wave velocity UT c_L in all types of analysis. The amount of this loss depended on the plate's microstructure and varied from 6% to 35%. The accuracy of UT measurements was estimated as +/− 2% and could be further improved by performing more repetitions for each measurement. It is worth mentioning that the time of one series of UT measurements was estimated to 5 min, which makes a good recommendation for that method for the in-situ application.

Based on the results of the UT measurements, it was possible to propose four approximate formulas (Formulas (5)–(8)) to describe the UT wave velocity (km/s) dependence on the moisture content (%). The correlation coefficient calculated for these approximations was greater than 0.9. The testing of standard board (specimens of type A) showed that hydration procedures increase the material moisture content to 31%. Within that range of M_c , the following formula is valid:

$$c_L = -0.0002 \left(2M_c\right)^2 - 0.0026 \, M_c + 1.8008 \tag{5}$$

The fire-resistant board exhibited a lower water permeability during the tests and thus UT wave velocity measurements were higher in this material. Therefore, within the range of M_c up to 19%, the following formula is valid for the specimens of type G:

$$c_L = -0.0012 \ (2M_c)^2 - 0.0122 \ M_c + 1.8971 \tag{6}$$

Boards with a compact matrix of K and I types were least permeable for the humidity and they finally achieved about 5% of the moisture content after the tests. In this case, the following formula with increased linear term is proposed:

$$c_L = \ 0.0096 \ (2M_c)^2 - 0.073 \ M_c + 1.735 \tag{7}$$

The low-quality board (B) characterized by a greater contribution of open porosity exhibited the lowest value of UT wave velocity. The specific type of microstructure caused the highest increase in the material moisture content up to 47%. The theoretical model of influence of open and closed porosity on c_L is more widely discussed in [25]. In that last case, quadratic polynomial function was unsuccessful due to the necessary function increase for small moisture content, which was not observed. Therefore, it is proposed to neglect the square term of the equation and apply the following formula:

$$c_L = \ -0.0093 \ M_c + 1.4092 \tag{8}$$

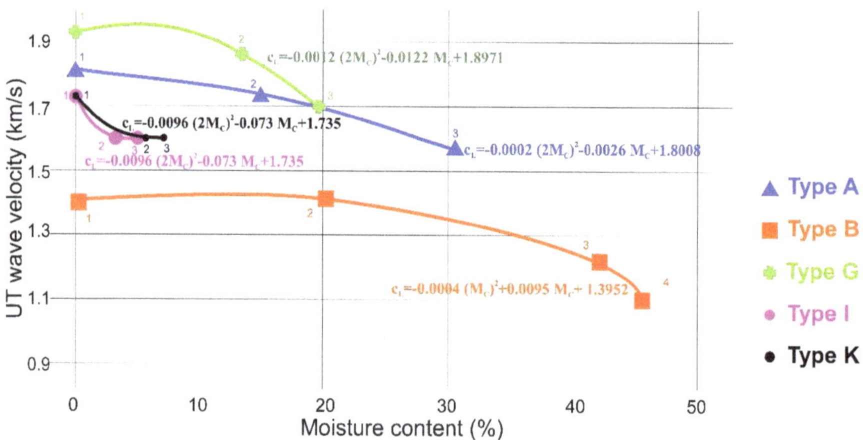

Figure 12. Results of the ultrasound testing (UT) tests. The UT wave velocity was measured in all specimens types before and after the hydration tests.

4. Conclusions

The effect of the hydration processes on the mechanical strength of five types of the porous gypsum plasterboards was investigated throughout this research. Three-point bending tests have revealed the strength loss of 70%–80% for all tested specimens due to hydration. Subsequent Ultrasonic (UT) measurements revealed that longitudinal wave velocity c_L decreased after hydration. Based on the plasterboard porosity volume [27], four types of approximated formulae were proposed in which the UT wave velocity (km/s) dependence on the moisture content (%) was described.

The dedicated UT transducers with low acoustic impedance and a solid-state polymer jelly interface were capable of achieving the required propagation parameters of UT waves and further determined their velocity in investigated plasterboard. The registered changes of c_L values were remarkably greater than the standard deviation of single series of measurements, included in the range of 2%÷3% from the average value of the readings. It was found that the ultrasonic method could be successfully used for plasterboards hydration processes monitoring.

In authors' opinions, more research work is required to determine the relationship between ultrasonic wave velocity and mechanical properties. That relationship strongly depends on the plasterboard type. For the standard indoor installation board type, it can be concluded that approx. a 70% loss of mechanical strength denotes a 10% loss of UT wave velocity. For I and K types, an 80% loss of mechanical strength denotes an 8% loss of UT wave velocity.

Author Contributions: Z.R. prepared the UT instrumentation, analyzed UT test results and performed paper editing; P.R. performed the optical microscopic observations; T.D. prepared the specimens; A.B. performed the SEM observations; M.K. performed the specimen hydration procedure and proofreading; M.R. analyzed UT test results, K.F. prepared the specimens; M.S. completed the experiments; E.K. completed the experiments; K.S. conceived and designed the experimental work. All authors have read and agreed to the published version of the manuscript.

Funding: This research received no external funding.

Conflicts of Interest: The authors declare no conflict of interest.

References

1. Calcium Sulphate Consortium. Available online: http://www.eurogypsum.org/the-gypsum-industry/use-of-gypsum-in-buildings/ (accessed on 2 April 2020).
2. Shirakawa, M.A.; Gaylarde, C.C.; Sahao, H.D.; Lima, J.R.B. Inhibition of Cladosporium growth on gypsum panels treated with nanosilver particles. *Int. Biodeterior. Biodegrad.* **2013**, *85*, 57–61. [CrossRef]
3. Baux, C.; Melinge, Y.; Lanos, C.; Jauberthie, R. Enhanced gypsum panels for fire protection. *J. Mater. Civ. Eng.* **2008**, *20*, 71–77. [CrossRef]
4. Kania, T.; Nowak, H. Gypsum partitions in medical rooms. *MATEC Web Conf.* **2018**, *174*. [CrossRef]
5. Abuasi, H.A.; McCabe, J.F.; Carrick, T.E.; Wassel, R.W. Displacement rheometer: A method of measuring working time and setting time of elastic impression materials. *J. Dent.* **1993**, *21*, 360–366. [CrossRef]
6. Winkler, M.M.; Monaghan, P.; Gilbert, J.L.; Lautenschlager, E.P. Comparison of four techniques for monitoring the setting kinetics of gypsum. *J. Prosthet. Dent.* **1998**, *79*, 532–536. [CrossRef]
7. Van Driessche, A.E.S.; Stawski, T.M.; Benning, L.G.; Kellermeier, M. Calcium sulfate precipitation throughout its phase diagram. In *New Perspectives on Mineral Nucleation and Growth*; Van Driessche, A., Kellermeier, M., Benning, L., Gebauer, D., Eds.; Springer: Cham, Switzerland, 2017; Available online: https://www.springerprofessional.de/en/calcium-sulfate-precipitation-throughout-its-phase-diagram/11950112 (accessed on 12 May 2020).
8. Clifton, J.R. Thermal analysis of calcium sulfate dihydrate and supposed and forms of calcium sulfate hemihydrate from 25 to 500 °C. *J. Res. Natl. Bur. Stand. A Phys. Chem.* **1972**, *76A*, 41–49. Available online: https://nvlpubs.nist.gov/nistpubs/Legacy/SP/nbsspecialpublication305supp4.pdf (accessed on 12 May 2020). [CrossRef]
9. Gurgul, S.J.; Seng, G.; Wiliams, G.R. A kinetic and mechanistics study into the transformation of calcium sulfate hemihydrate to dehydrate. *J. Synchrotron Radiat.* **2019**, *26*, 774–784. [CrossRef]
10. Isern, E.R.; Messing, G.L. Direct foaming and seeding of highly porous, lightweight gypsum. *J. Mater. Res.* **2016**, *31*, 2244–2251. [CrossRef]
11. Gaiao, C.; de Brito, J.; Silvestre, J.D. Gypsum plasterboard walls: Inspection, pathological characterization and statistical survey using an expert system. *Mater. Constr.* **2012**, *62*, 285–297. [CrossRef]
12. Gaiao, C.; de Brito, J.; Silvestre, J. Inspection and diagnosis of gypsum plasterboard walls. *J. Perform. Constr. Facil.* **2011**, *25*, 172–180. [CrossRef]
13. Kaftanowicz, M.; Krzemiński, M. Multiple-criteria analysis of plasterboard systems. *Procedia Eng.* **2015**, *111*, 364–370. [CrossRef]
14. Alagusankareswari, K.; Jenitha, G.; Sastha Arumuga Pandi, S. Experimental study on empty glass fibre reinforced gypsum panel. *Indian J. Sci. Res.* **2018**, *17*, 217–220. Available online: https://www.researchgate.net/profile/Jenitha_Ganapathy/publication/325711122_EXPERIMENTAL_STUDY_ON_EMPTY_GLASS_FIBRE_REINFORCED_GYPSUM_PANEL/links/5b1f77d3aca272277fa76954/EXPERIMENTAL-STUDY-ON-EMPTY-GLASS-FIBRE-REINFORCED-GYPSUM-PANEL.pdf (accessed on 12 May 2020).

15. Wu, Y.-F. The structural behavior and design methodology for a new building system consisting of glass fiber reinforced gypsum panels. *Constr. Build. Mater.* **2009**, *23*, 2905–2913. [CrossRef]

16. Erbs, A.; Nagalli, A.; Mymrine, V.; Carvalho, K.Q. Determination of physical and mechanical properties of recycled gypsum from the plasterboard sheets. *Cerâmica* **2015**, *61*, 360. [CrossRef]

17. Petrone, C.; Magliulo, G.; Manfredi, G. Mechanical properties of plasterboards: Experimental tests and statistical analysis. *J. Mater. Civ. Eng.* **2016**, *28*, 1–12. [CrossRef]

18. Petrone, C.; Magliulo, G.; Lopez, P.; Manfredi, G. Out-of-plane seismic performance of plasterboard partitions via quasi-static tests. *Bull. N. Z. Soc. Earthq. Eng.* **2016**, *49*, 125–137. Available online: https://discovery.ucl.ac.uk/id/eprint/1496935/1/Petrone%20et%20al%20-%20BNZSEE%20-%20accepted%20version.pdf (accessed on 12 May 2020). [CrossRef]

19. Gencel, O.; del Coz Diaz, J.J.; Sutcu, M.; Koksal, F.; Rabanal, A.; Marinez Barrera, G.; Brostow, W. Properties of gypsum composites containing vermiculite and polypropylene fibers: Numerical and experimental results. *Energy Build.* **2014**, *70*, 135–144. [CrossRef]

20. Oliver-Ramirez, A.; Garcia-Santos, A.; Neila-Gonzalez, F.J. Physical and mechanical characterization of gypsum boards containing phase change materials for latent heat storage. *Mater. Constr.* **2011**, *61*, 465–484. [CrossRef]

21. BSEN 520:2004 + A1 2009. Gypsum Plasterboards—Definitions, Requirements and test Methods. Available online: https://Infostore.saiglobal.com/en-us/Standards/EN-520-2004-A1-2009-346166_SAIG_CEN_CEN_791645/ (accessed on 2 April 2020).

22. Ranachowski, Z.; Ranachowski, P.; Dębowski, T.; Gorzelańczyk, T.; Schabowicz, K. Investigation of structural degradation of fiber cement boards due to thermal impact. *Materials* **2019**, *12*, 944. [CrossRef]

23. Schabowicz, K.; Gorzelańczyk, T.; Szymków, M. Identification of the degree of fibre-cement boards degradation under the influence of high temperature. *Autom. Constr.* **2019**, *101*, 190–198. [CrossRef]

24. Berriman, J.; Purnell, P.; Hutchins, D.A.; Neild, A. Humidity and aggregate content correction factors for air-coupled ultrasonic evaluation of concrete. *Ultrasonics* **2005**, *43*, 211–217. [CrossRef] [PubMed]

25. MISTRAS Products & Systems. Available online: http://www.eurosonic.com/en/products/ut-solutions/utc-110.html (accessed on 19 November 2018).

26. Olympus Ultrasonic Transducers. Available online: http://www.olympus-ims.com/en/ultrasonic-transducers/contact-transducers/ (accessed on 2 April 2020).

27. Zielinski, T.G. Microstructure-based calculations and experimental results for sound absorbing porous layers of randomly packed rigid spherical beads. *J. Appl. Phys.* **2014**, *116*, 034905. [CrossRef]

 *materials*

Article

Identification of the Destruction Model of Ventilated Facade under the Influence of Fire

Krzysztof Schabowicz [1], Paweł Sulik [2] and Łukasz Zawiślak [1,*]

[1] Faculty of Civil Engineering, Wrocław University of Science and Technology, Wybrzeże Wyspiańskiego 27, 50-370 Wrocław, Poland; krzysztof.schabowicz@pwr.edu.pl

[2] Instytut Techniki Budowlanej, Filtrowa 1, 00-611 Warszawa, Poland; p.sulik@itb.pl

* Correspondence: lukasz.zawislak@pwr.edu.pl; Tel.: +48-534-205-6614

Received: 8 April 2020; Accepted: 18 May 2020; Published: 22 May 2020

Abstract: Ventilated facades are becoming an increasingly popular solution for external part of walls in the buildings. They may differ in many elements, among others things: claddings (fiber cement boards, HPL plates, large-slab ceramic tiles, ACM panels, stone cladding), types of substructures, console supports, etc. The main part that characterizes ventilated facades is the use of an air cavity between the cladding and thermal insulation. Unfortunately, in some aspects they are not yet well-standardized and tested. Above all, the requirements for the falling-off of elements from ventilated facades during a fire are not precisely defined by, among other things, the lack of clearly specified requirements and testing. This is undoubtedly a major problem, as it significantly affects the safety of evacuation during a fire emergency. For the purposes of this article, experimental tests were carried out on a large-scale facade model, with two types of external-facade cladding. The materials used as external cladding were fiber cement boards and large-slab ceramic tiles. The model of large-scale test was 3.95 m × 3.95 m, the burning gas released from the burner was used as the source of fire. The test lasted one hour. The facade model was equipped with thermocouples. The cladding materials showed different behavior during the test. Large-slab ceramic tiles seemed to be a safer form of external cladding for ventilated facades. Unfortunately, they were destroyed much faster, for about 6 min. Large-slab ceramic tiles were destroyed within the first dozen or so minutes, then their destruction did not proceed or was minimal. In the case of fiber cement boards, the destruction started from the eleventh minute and increased until the end of the test. The authors referred the results of large-scale test to testing on samples carried out by other authors. The results presented the convergence of large-scale test with samples. External claddings was equipped with additional mechanical protection. The use of additional mechanical protection to maintain external cladding elements increases their safety but does not completely eliminate the problem of the falling-off of parts of the facade. As research on fiber cement boards and large-slab ceramic tiles presented, these claddings were a major hazard due to fall-off from facade.

Keywords: ventilated facades; large-scale facade model; fire safety; fiber cement board; large-slab ceramic tile

---

1. Introduction

Ventilated facades are a type of external part of multilayer wall, which has a construction part and is usually a masonry or concrete wall, but it can also be a wooden or steel structure. The wall is then fitted with insulation, with consoles holding the elements of the substructure and the external cladding. This external cladding protects the aforementioned layers against environmental influences and gives the final shape and appearance of the facade. There is an air gap between the external cladding and the insulation, also known as a ventilation gap. The width of the air gap in ventilated facades ranges from 20 to 50 mm [1,2], some sources also provide higher values, e.g., from 40 to 100 mm [3]. Ventilated facades allow for shaping external claddings from various materials, structures, textures or colors. Due to their high aesthetics, ventilated facades are increasingly often used as external parts of newly built buildings, but they are also perfect for buildings undergoing renovations. External cladding elements can have very large individual elements. The standard dimension for fiber cement boards is 1250 mm × 3100 mm [1] and for HPL 1850 mm × 4100 mm [1].

Regulations control a number of requirements for external walls, such as ensuring appropriate thermal insulation, the requirements for durability and protection of the building according to [4,5] and safety of use in environmental and emergency situations. One of the most important requirements a building must meet in emergency situations, is the impact of fire, is to ensure the possibility of evacuation of users and work of rescue teams [5]. External walls with external facade cladding must ensure, among other things, sufficient durability in emergencies, i.e., prevent facade elements from falling off. This problem is widely known throughout Europe.

The European Commission in [5], presented analysis of today's requirements of the falling-off of parts of facades in countries. In [5] standardization proposals were also presented, among others of falling-off cladding elements during a fire. Propose two methods for assessing falling elements of the façade, the first one was dependent on fire class:

- No falling-off of parts larger than 1 kg and 0.1 m^2 (class F1);
- No falling-off of parts larger than 5 kg and 0.4 m^2 (class F2);
- No burning particles at all (class D0);
- Limited duration burning debris <20 s (class D1).

There was presented also alternative test method: falling parts are limited to a maximum of 1 kg and an area of 0.1 m^2 for each individual piece. Until the standards are harmonized in Europe, manufacturers of entire ventilated façade systems, designers and contractors [6–8] will have a problem delivering these products within the European Union.

There are many standards in the world for the large-scale facade test [9–12]. They are based on the spread of fire from a recess/hole, simulating the window openings of a room in a real building. The fire source is located there, defined by the normal temperature action curve. The flames escape from the recess affecting the external facade cladding and other wall elements. Individual standards differ in details, i.e., the type of fire source: wood crib [9–11] or propane-butane gas [12], dimensions of the recess/hole, test time, shape of the facade model in large scale and its dimensions. The comparison of individual standards for the test of large—scale facade models in the scope of fire safety is presented by Smolka in [13]. Due to the growing awareness of the phenomenon of fire spreading on the external part of the elevation and a number of threats caused by this phenomenon, the European Commission started an attempt to harmonize the testing standards [11].

Sulik and Kinowski [14,15] analyzed the large-scale facade models at the impact of fire. Ventilated facades with different external claddings have been assumed as initial conditions. The external claddings include fiber cement boards and high-pressure laminate (HPL) panels, ceramic tiles, natural stone and synthetic stone marmoglass (glass conglomerate), layered steel ACM panels (aluminum composite material). During the research, a number of dependencies were noted, among others: that the way the external facade cladding is installed has an impact on its safety. Mechanical assembly is safer than adhesive assembly. The requirements for the test duration are usually 60–120 min. After 30 min the destruction of the cladding is minimized or even stopped in some cases. The claddings that have been positively tested in this type of test are fiber cement boards and ACM panels, in which case the falling-off of parts were up to 2 kg. In the case of ceramic tiles, the falling-off of parts were also of acceptable weight, but they were sharp and posed a risk to evacuating people. In the case of stone cladding and marmoglass, the weight of the falling elements was several or even several dozen kilograms. Sulik and Kinowski [6] also presents the verification of fire safety of glass facades. The results indicate a problem with the falling-off of elements of facades.

It should also be emphasized that fiber cement boards are not well identified when it comes to the conditions of fire and high temperatures. Szymków's research presented in the study [16] showed that fibers in fiber cement boards are destroyed at 230 °C only after 3 h. Destruction of such boards during the bending test takes place through brittle high-energy cracking. Other fiber-reinforced materials show similar behavior. The study [17] shows that the decrease of compression strength of concrete and fiber-reinforced concrete at temperatures up to 300 °C is about 10%. On the other hand, in the case of fiber-reinforced cement composites, the modulus of rupture increases with the temperature increase up to about 300 °C [18]. Temperatures up to about 300 °C for fiber-reinforced materials are relatively safe in a short period of time. Their destruction takes place only after a longer period of time, usually after several hours. Szymków in his work [16] also carried out tests on samples of fiber cement boards at 400 °C. At this temperature, the samples showed much lower stability and were destroyed much faster. The results had large discrepancies because, depending on the manufacturer, components, manufacturing technology, they "withstood" for up to several minutes. Some were destroyed even in a shorter period of time. Unfortunately, in case of fire, the temperature impact on external facade claddings may reach a value locally even up to 900 °C (the external curve provides a value of 660 °C), and such experimental tests are not available for fiber cement boards. Looking at the above analogies of other materials with the use of fibers, interesting conclusions are contained in the study [17], where the tests for concrete and fiber-reinforced concrete were performed. It was found that the temperature of 800 °C reduces the compressive strength of concrete and fiber-reinforced concrete class C30/37 by over 90%. In the case of high-quality concrete and fiber-reinforced concrete of class C60/75, the compression strength decreases by more than 90%, only after the samples are heated at 1000 °C. At 500 and 600 °C, the samples without the addition of fibers were destroyed during their annealing, whereas those with the addition of polypropylene fibers retained their residual bending strength [18]. The study [19] also showed a positive effect of using fibers to increase the bending strength of beams subjected to the normative fire temperature curve. The fibers have a positive effect on increasing the load capacity of the elements under the influence of high temperatures, e.g., in an emergency situation, such as a fire.

The authors of this study and [20–26], due to the lack of scientific literature on the problem of destruction of fiber cement boards used as external facade claddings in ventilated facades, have attempted to analyze this issue. This analysis was based on the large-scale model of facade. This issue is important because the popularity and demand for ventilated facades is increasing, and unfortunately the problem quoted by the authors concerns the safety of these facades in an exceptional situation, i.e., the impact of fire.

To sum up, it can be argued that today practically no type of material used for external facade claddings (except for steel sheets) ensures that the condition specified in the Regulation [5] is met. It is therefore necessary to apply some kind of compromises and, above all, to harmonize the testing standards and analysis of these results.

2. Materials and Model of Ventilated Facade

In order to solve the scientific task, a model was prepared to reproduce the facade of the building, made in the so-called large-scale model. The model was constructed with reference to the existing wooden frame building systems.

The analyzed wall was a part of the wooden panel and modular skeleton construction system. The construction of the elements consisted of a wooden skeleton with a glass wool insulation material filling.

The subject of the research verification was the layout of the ventilated facade, which included two variants of external claddings: fiber cement boards and large-slab ceramic tiles. The external cladding was attached to steel consoles. The supporting structure was a wooden. Between the posts there was a layer of thermal insulation made of glass wool. On the inside and outside of the wall there was a layer of 12.5 mm plasterboard. The substructure fastening the external cladding was made of an aluminum grate with a section of L60 × 40 mm × 2 mm, screwed to steel consoles. The consoles were fixed to the system skeleton wall (model's supporting element) by means of 8 × 60 steel disc screws.

The layout of the aluminum substructure and the consoles transferring loads from the aluminum substructure to the system skeleton wall is shown in Figure 1.

The external cladding of the ventilated facade consisted of fiber cement boards with a thickness of 8 mm and density of 1700 kg/m^3 and large-slab ceramic tiles with a thickness of 5.6 mm and density of 2855 kg/m^3. The external facade claddings attached to the aluminum grate were made with the use of the system adhesive technology and additionally with the use of steel mechanical connectors, i.e., perforated steel tapes attached to the substructure. The dilatation between individual boards was 8 mm. Figure 1 shows the division scheme of the external cladding and indicates the material of which they were made, namely, the fiber cement boards in the left part and the large-slab ceramic tiles in the right part.

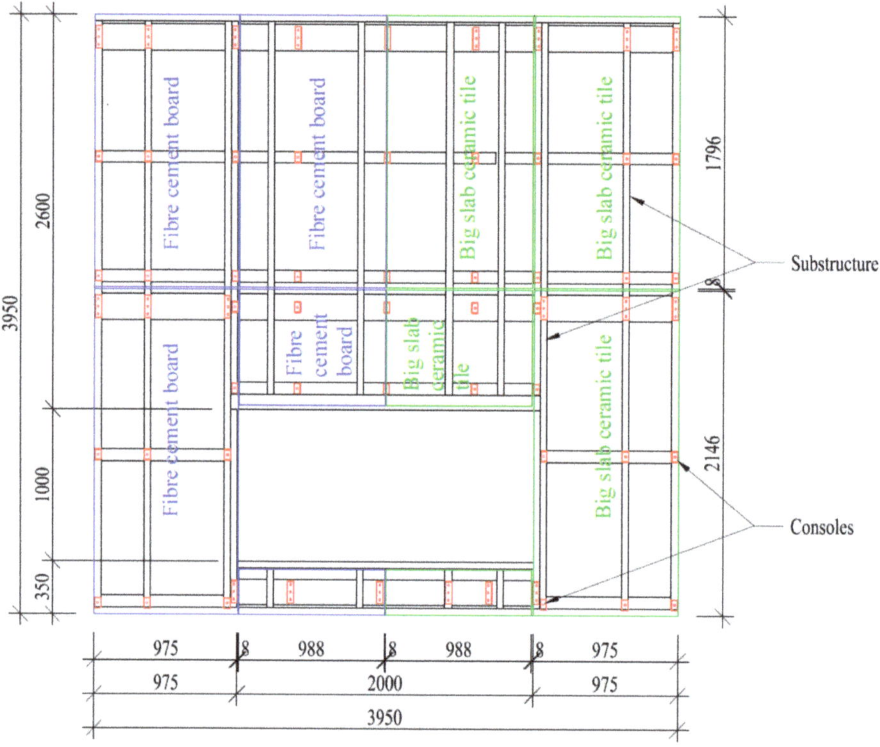

Figure 1. Scheme of the aluminum substructure, console arrangement, external cladding panels arrangement and material indication.

Selected details of the tested ventilated facade are shown in Figure 2. Total dimensions of the tested model of ventilated facade were 3950 mm × 3950 mm, air gap width 38 mm, dimensions of fire hole 2000 mm × 1000 mm.

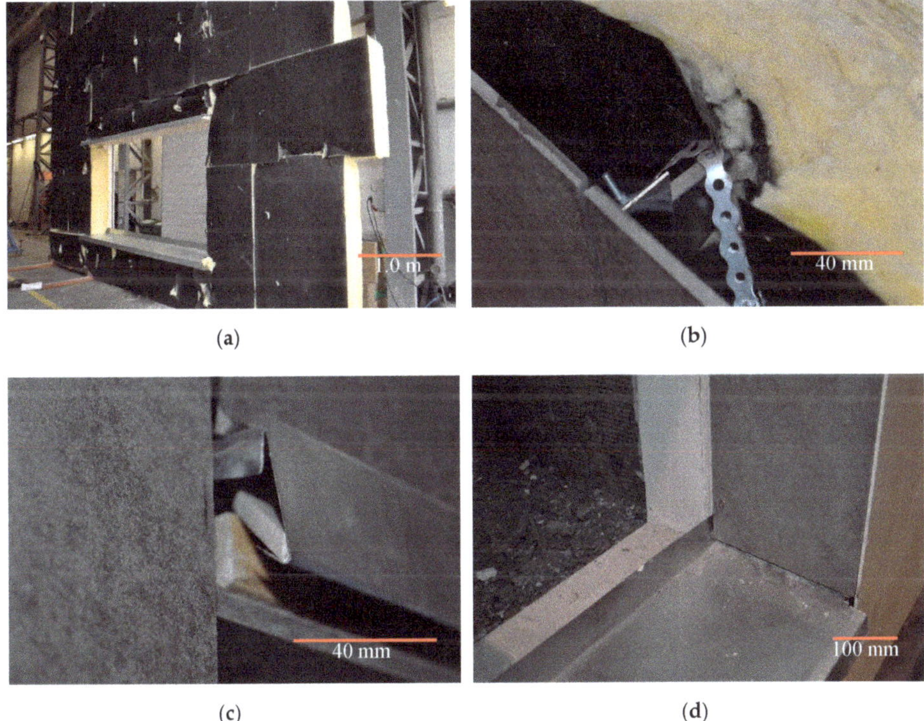

Figure 2. Selected details of the ventilated facade model under test: (**a**) thermal insulation system; (**b**) the detail of perforated tape connection with the console; (**c**) the detail of closing sheet above the window opening; (**d**) the detail of sill system.

The scenario of the ventilated facade fire assumed that the flames would escape through a window opening from a room located directly behind the facade, inside the building. In order to map the room from which the flame was emitted, a recess was made in the model of the facade where the source of fire was placed. The burner parameters were selected in such a way as to reproduce the standard fire in the room, defined in the fire resistance test standard [20]. The fire was mapped with a gas burner releasing propane-butane gas at the speed of 3.8 m/s. The air from the recess in which the so-called sand burner was placed, supplied by the furnace installation had the possibility of inflow from the side of the opening, through a technical solution ensuring laminar air inflow, the so-called vent. In order to verify and identify damage to the external cladding, four thermocouples were installed. Thermocouples send information regarding temperatures as a function in time. Influence of high temperature was one of the negative external environmental effects on buildings elements. Thermocouples were placed in the gaps between the panels: two in the part of fiber cement boards and two in the part of large-slab ceramic tiles. The location of the thermocouples is shown in Figure 3. A ventilated facade with open joints was made, where additional gaps (gaps between individual external cladding panels) can provide air circulation.

Figure 3. Ventilated facade model with location of thermocouples.

3. The Test and Results

The test was carried out in a closed hall, with an ambient temperature of 18.9 °C and relative humidity of 60.7%. The test started with setting the burner and calibrating the gas release. The first 5 min of the fire revealed smoke/charring of the external cladding, no falling-off of the external cladding elements (Figure 4a).

The first falling-off of elements were observed in the 6th minute of the test, i.e., large-slab ceramic tiles elements started to fall off (Figure 4b). The subsequent minutes of high temperature impact caused greater destruction of the external cladding, particularly visible in the part where large-slab ceramic tiles were located. On 11th minute, the degradation of fiber cement boards started significantly. Figure 4c shows the initial stages of expansion of the fiber cement boards. The destruction of the large-slab ceramic tiles slowed down around 20th minute. The places where the temperature impact was highest were destroyed, and in the remaining places, large-slab ceramic tiles "tolerated" high temperatures quite well. Large-slab ceramic tiles had the greatest degradation in the first several minutes of fire. No further degradation of ceramic sinters was noticed.

These materials destruct themselves also in another way. In the case of large-slab ceramic tiles, which have a uniform structure, the destruction takes place by cracking and falling-off of pieces of elements. In the case of fiber cement boards and their nonuniform structure, the boards, despite significant warping (deplanation) caused by high temperature (e.g., Figure 4e), do not fall off among others due to the good tensile properties of the fibers. Only complete destruction of the fibers causes the elements to fall off.

The test revealed cracking and falling-off of fragments of both fiber cement boards and large-slab ceramic tiles. This mainly concerned the external cladding located above the opening, i.e., above the fire source. Some of the claddings have detached from the grate but have not fallen off and hang on perforated steel tapes mechanically attached to the cladding. The maximum mass of a single component which fell off and dropped during the test was 1.15 kg. This was due to a safety system using a steel perforated tape and mechanical fasteners. Degradation of the large-slab ceramic tiles took place during the first few minutes, then this part of the facade was stable. Fiber cement boards behaved differently. The first minutes showed the stability of the fiber cement boards. From 11th minute, the boards started to show significant degradation progressing practically until the end of the test. If perforated steel tapes were not used, the cladding elements falling off the facade would probably be of considerable dimensions.

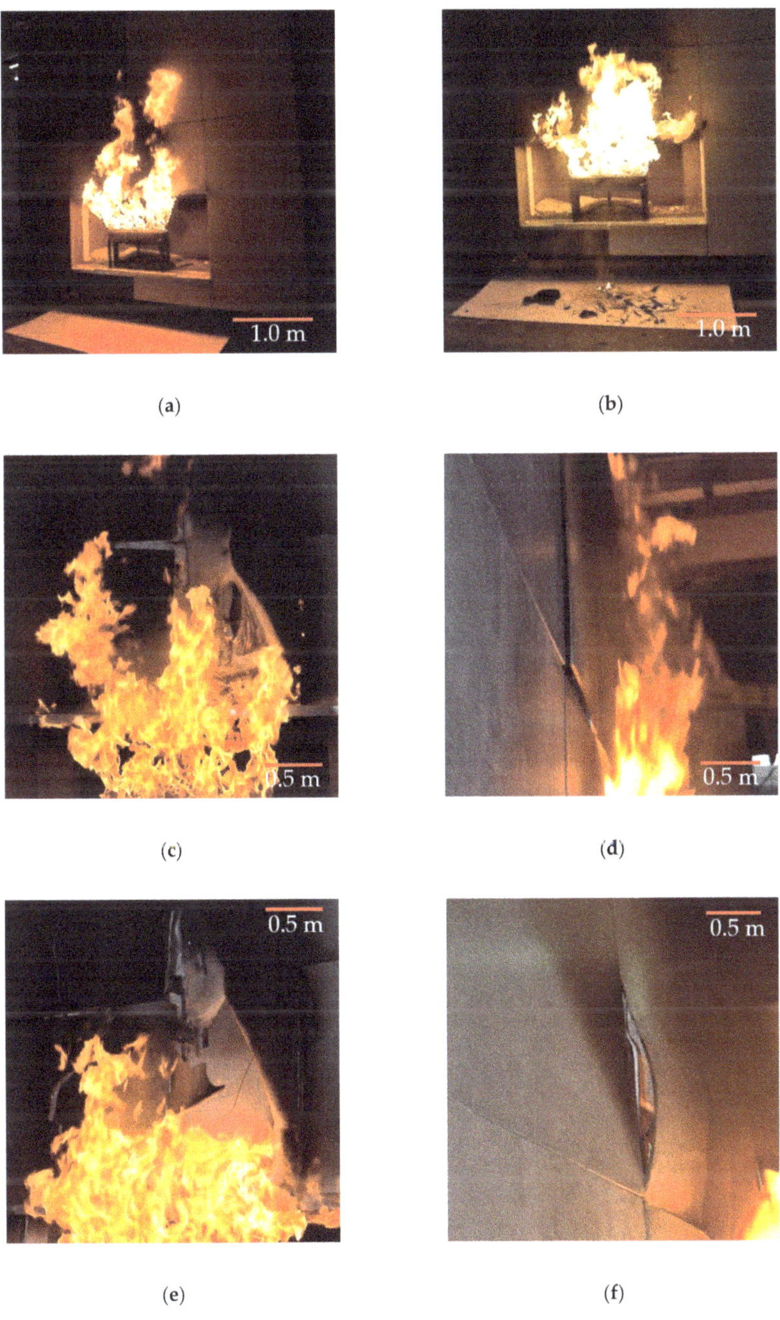

Figure 4. *Cont.*

(g)   (h)

Figure 4. Ventilated facade model view in subsequent test minutes.: (**a**) 1st minute of the test; (**b**) 6th minute of the test; (**c**) 11th minute of the test; (**d**) 21st minute of the test; (**e**) 31st minute of the test; (**f**) 41st minute of the test; (**g**) 56th minute of the test; (**h**) 61st minute of the test.

The horizontal upper splay, made of 0.5 mm thick steel sheet, deformed, but its location has not changed. The aluminum grate directly above the recess with the fire source, except in places directly sheltered by the steel top splay was burnt at a height of up to 1660 mm. Glass mineral wool was melted directly above the recess at a height of up to 1400 mm. The steel fasteners and consoles remained in place, as did the lower splay. Figure 5 shows the condition of the facade after the fire source was extinguished.

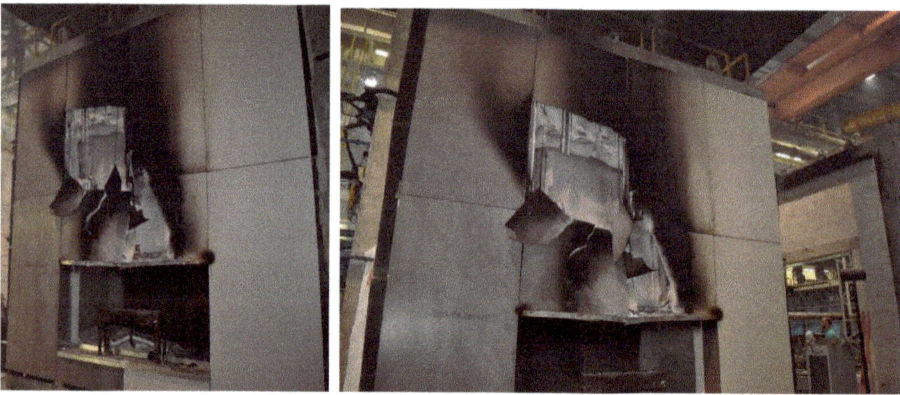

Figure 5. Ventilated facade model view after the fire source was extinguished and the test was completed.

The results of temperature measurement from the thermocouples located in the part of the fiber cement boards, as shown in Figure 3, are presented as a function of time in Figure 6 for thermocouples TE1 and TE2. On the other hand, the results for TE3 and TE4 thermocouples located in the part of large-slab ceramic tiles are shown in Figure 7. The results presented for TE1 and TE2 thermocouples show high instability and temperature variations. This was due to the high development of degradation over time of fiber cement boards, especially after full development of the fire. The locations of the greatest temperature fluctuations in Figure 6 can be associated with cracking or detachment of external cladding elements.

Thermocouples TE3 and TE4, as opposed to thermocouples located in a part of fiber cement boards, initially showed higher temperature increase. The power of the fire was constant, the difference in temperature was caused by the heating of the elements. In 6–7 min, TE3 thermocouples showed high stability—large-slab ceramic tiles were quickly destroyed in the central part above the fire recess. TE4 thermocouple initially showed high instability and temperature fluctuations. This was caused by the influence of lower temperature on large-slab ceramic tiles in this part of the facade. TE4 thermocouple became stable in about 15th minute when the ceramic sinters were no longer damaged by high temperature. The damage is shown in Figure 4c. This state or a slightly altered state, was maintained practically until the end of the test.

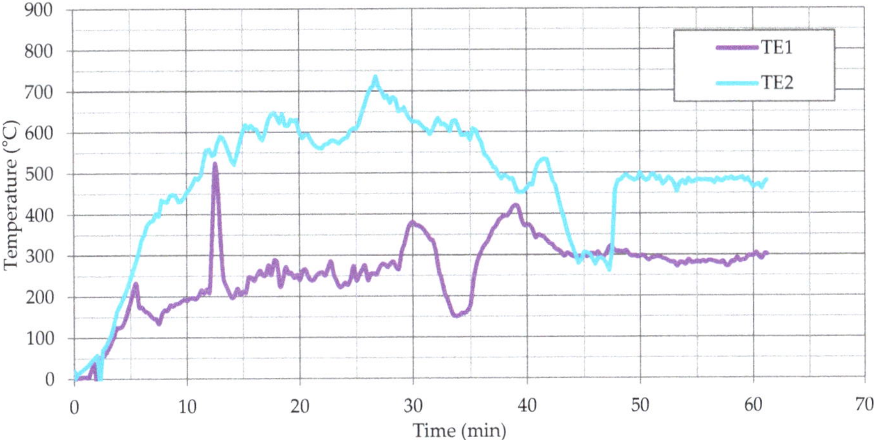

Figure 6. Temperature measurement results for TE1 and TE2 thermocouples.

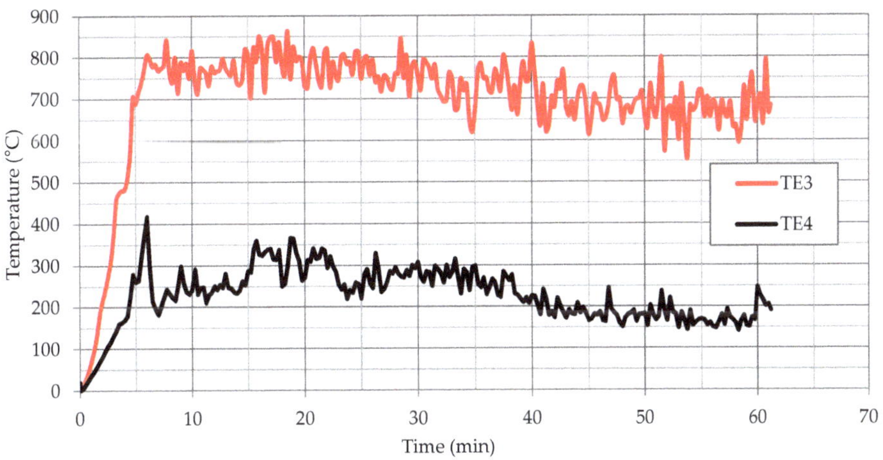

Figure 7. Temperature measurement results for TE3 and TE4 thermocouples.

4. Discussion

The test of a model of a ventilated facade made of two different external claddings showed significant material differences. Large-slab ceramic tiles very quickly, in the first few minutes or so, were destroyed by high temperatures, then their degradation does not deepen. The course of events in the part made of fiber cement boards looks different. In the initial stage of the fire, the panels show high resistance—the first few minutes or so. About 11 min, the first signs of degradation of fiber cement boards are visible (Figure 4c). Then the state of the degradation was deepened until the final destruction. Fiber cement boards have exhausted their load-bearing capacity during high temperature exposure. They were held on the structure only by perforated steel tapes. The authors wanting to refer the results from large-scale test to testing on samples by Szymków [1], had to specify the method of reference of the results. The temperature in the large-scale test was much higher than in Szymków's research [1], the time function will be not representative. The authors determined that as the most reflecting value will be function of integral.

The integral corresponding to the temperature function from time was determined for all thermocouples. The following results were obtained:

- the integral for TE1 (°C × min) was 15,745.0 (°C × min);
- the integral for TE2 (°C × min) was 29,475.1 (°C × min);
- the integral for TE3 (°C × min) was 42,686.2 (°C × min);
- the integral for TE4 (°C × min) was 13,912.0 (°C × min).

These results are shown in Figure 8 together with temperature diagrams for all thermocouples: TE1, TE2, TE3, TE4. In this graph we can observe a linear growth pattern of the integral for TE3 and TE4 thermocouples (thermocouples placed in a part of the cladding made of large-slab ceramic tiles). In addition, in the part with ceramic sinters the fire force was much higher. The maximum temperature for TE3 was 862.7 °C and for TE2 thermocouple measuring the temperature in the area of fiber cement boards it was 735.5 °C. The much higher temperatures in the large-slab ceramic tiles area were caused by the much faster destruction of this material, the "release" of access to the thermocouples. The falling-off of large-slab ceramic tiles made easier way to burn mineral wool, this allows maintain higher temperature. In addition, it was likely that ceramic sinters insulate thermocouples much worse than fiber cement boards. The temperature course for TE1 and TE2 thermocouples contains much more disturbances and significant faults. In addition, the integral function has a visible refraction in 34th and 39th minutes for TE1 thermocouple and 42nd and 48th minutes for TE2 thermocouple.

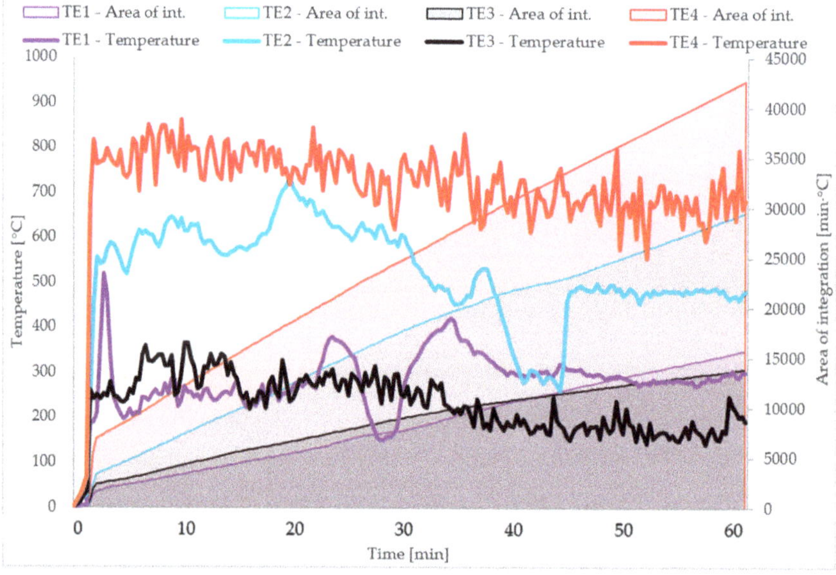

Figure 8. Temperature measurement results for thermocouples and the increasing integral for the temperature-time function.

Szymków in [16] carried out tests to identify the degree of destruction of fiber cement boards under high temperatures. He analyzed the influence of temperature 400 °C in a given unit of time. Different samples of 20 mm × 100 mm fiber cement boards were subjected to high temperature influence in 1 to 15 min. The samples differed in technical parameters, composition, production technology and area of application and were tested at different times. Characteristic parameters of individual sample series given by the manufacturer are presented in Table 1.

Table 1. Characteristic parameters of individual sample series.

Boards Series Designation	Board Thickness (mm)	Board Color	Application	Bulk Density (g/cm^3)	Pressing During Production	Modulus of Rupture *MOR* (MPa)
SERIE A	8.0	Natural color	External	1.60	Yes	25
SERIE B	341.97	Dyed in the mass	External	1.60	Yes	30
SERIE C	139.37	Dyed in the mass	External	1.65	Yes	30
SERIE D	133.65	Natural color	External	1.70	Yes	20
SERIE E	281.84	Natural color	external	1.20	Partially	12

For the purposes of the articles, selected test results from [16] were used. Fracture energy W_f and modulus of rupture *MOR* were analyzed. The results are presented in Table 2.

Table 2. Aggregate summary of averaged values of fracture energy W_f and modulus of rupture MOR for panels, under the influence of high temperature 400 °C from 1 to 15 min [1].

Boards Series Designation	Fracture Energy W_f (J/m^2)	Modulus of Rupture *MOR* (MPa)
SERIE A2—1 min of influence	372.82	20.89
SERIE A3—2.5 min of influence	341.97	23.05
SERIE A4—5 min of influence	139.37	17.68
SERIE A5—7.5 min of influence	133.65	15.50
SERIE B2—1 min of influence	281.84	23.30
SERIE B3—2.5 min of influence	302.99	23.80
SERIE B4—5 min of influence	301.84	25.56
SERIE B5—7.5 min of influence	309.52	30.29
SERIE B6—10 min of influence	229.02	23.22
SERIE B7—12.5 min of influence	134.53	14.13
SERIE B8—15 min of influence	86.94	12.53
SERIE C2—1 min of influence	1375.14	35.75
SERIE C3—2.5 min of influence	1396.19	44.61
SERIE C4—5 min of influence	320.86	22.44
SERIE C5—7.5 min of influence	276.12	17.63
SERIE C6—10 min of influence	30.67	7.98
SERIE D2—1 min of influence	297.11	23.27
SERIE D3—2.5 min of influence	298.18	26.36
SERIE D4—5 min of influence	233.90	23.90
SERIE D5—7.5 min of influence	204.31	21.31
SERIE D6—10 min of influence	152.94	15.31
SERIE E2—1 min of influence	736.44	14.41
SERIE E3—2.5 min of influence	555.62	11.89
SERIE E4—5 min of influence	250.92	10.11

As the results shown in the study [16] show, for fiber cement boards at high temperature of 400 °C, the fracture energy increased in the initial phase. In case of B, C, D series boards, the fracture energy increased from 2% to 9%. Then a decrease in the fracture energy was visible for all types of fiber cement boards. The same was true for modulus of rupture *MOR*, which increased from 3% to 20% in the initial period and then decreases. Table 3 shows the aggregate summary of the temperature function integral from the maximum time that a sample could be tested for the A–E series of boards.

Table 3. Aggregate summary of the temperature function integral from the maximum time [1].

Boards Series Designation	Fracture Energy W_f (J/m^2)	Modulus of Rupture *MOR* (MPa)	Function Integral (°C × min)
SERIES A5—7.5 min of influence	133.65	15.50	3000
SERIES B8—15 min of influence	86.94	12.53	6000
SERIES C6—10 min of influence	30.67	7.98	4000
SERIES D6—10 min of influence	152.94	15.31	4000
SERIES E4—5 min of influence	250.92	10.11	2000

For the purpose of further analyses, the average value of the function integral for all series was assumed to be 3800 (°C × min). The average value of the function integral for all series corresponds to the increasing integral for TE1 thermocouple in about 20.5 min and the increasing integral for TE2 thermocouple in about 12.75 min. This is reflected in Figure 6; Figure 8. The time of 12.75 min approximately coincides with the point of significant fault in the temperature graph for TE1 thermocouple. This corresponds to the beginning of the destruction in 11th minute of the tested model (Figure 5). Fiber cement boards detach into parts, probably obscuring the TE1 thermocouple. Then the fiber cement panels were continuously destroyed until the end of the test. This corresponds to integral 42,686.2 (°C × min). The tests shown in [16] end with the integral value of 6000 (°C × min). The results obtained during the test in question indicate identical tendencies to the behavior of the samples presented in [16].

5. Conclusions

The model of large-scale ventilated facade is a huge source of knowledge about its behavior during a fire. The problem of destruction of the external cladding in the case of fiber cement boards and large-slab ceramic tiles has not been sufficiently recognized so far and such studies as presented in the article indicate trends in the "behavior" of the facade and these claddings.

Fiber cement boards pose a great threat to the safety of use in case of flames escaping from window openings to facades during a fire. Large-slab ceramic tiles appear to be a safer form of external cladding for ventilated facades. Unfortunately, they were destroyed much quicker, i.e., starting from the 6th minute. The danger of falling elements passes after a dozen or so minutes of fire. In the case of fiber cement boards, the visible destruction starts from about 11th minute and runs throughout the whole period of high temperature impact. Falling-off of elements in the case of fiber cement boards were large sizes, even despite the use of additional protections. In the case of standard mechanical or adhesive fastening, fiber cement boards would pose an even greater threat.

In the next part of the article compared behavior of fiber cement boards on samples test with the large-scale facade test. The temperature integral was taken as the comparative value of the samples with the large-scale façade test. The results of both tests show convergent results. The samples were a good alternative initial verification of the facade cladding behavior in fire conditions in global analysis.

The authors are planning next research in this field, developing the model and using more thermocouples. The tests will be carried out on a greater number of different types of claddings.

The next research steps, helping to better solve the problems of ventilated facades and to increase their safety, should, according to the authors, concern the samples of fiber cement panels tested at a temperature closer to the actual fire, i.e., about 550–650 °C (Figure 6). Such tests should also be carried out for ceramic sinters, which have not yet been described in the literature.

Author Contributions: K.S. conceived and designed the experimental work; P.S. prepared the specimens, completed the experiments; Ł.Z. analyzed the test results and performed study editing. All authors discussed the results, commented on the manuscript, wrote the paper, and did the review editing. All authors have read and agreed to the published version of the manuscript.

Funding: This research received no external funding.

Conflicts of Interest: The authors declare no conflict of interest.

References

1. Schabowicz, K. *Elewacje wentylowane. Technologia produkcji i metody badania płyt włóknisto-cementowych*; Oficyna Wydawnicza Politechniki Wrocławskiej: Wrocław, Poland, 2018; p. 214.
2. *Guideline for European Tehcnical Approval of Kits for External Wall Claddings—Part II: Cladding Components, Associated Fixings, Subframe and Possible Insulation Layer*; European Technical Approval Guidelines; European Organisation for Technical Assessment: Brussels, Belgium, 2012.
3. Ibañez-Puy, M.; Vidaurre-Arbizu, M.; Sacristán-Fernández, J.A.; Martín-Gómez, C. Opaque Ventilated Façades: Thermal and energy performance review. *Renew. Sustain. Energy Rev.* **2017**, *79*, 180–191. [CrossRef]
4. *Guideline for European Tehcnical Approval of Kits for External Wall Claddings—Part I Ventilated Cladding Kits Comprising Cladding Components and Associated Fixings*; European Technical Approval Guidelines; European Organisation for Technical Assessment: Brussels, Belgium, 2012.
5. European Commission. *Development of a European Approach to Assess the Fire Performance of Facades*; European Commission: Brussels, Belgium, 2018.
6. Sulik, P.; Kinowski, J. *Bezpieczeństwo Użytkowania Elewacji*; Materiały Budowlane: Ustryki Dolne, Poland, 2014; pp. 38–39.
7. Kosiorek, M. *Analiza Wybranych Wymagań Dotyczących Bezpieczeństwa Pożarowego*; Materiały Budowlane: Ustryki Dolne, Poland, 2014; Volume 7, pp. 2–3.
8. Sędłak, B.; Kinowski, J.; Sulik, P.; Kimbar, G. The risks associated with falling parts of glazed façades. *Open Eng.* **2018**, *8*, 147–155. [CrossRef]
9. *BS 8414-1:2015+A1:2017 Fire Performance of External Cladding Systems. Test Method for Non-Loadbearing External Cladding Systems Applied to the Masonry Face of a Building, Building Research Establishment*; BSI: London, UK, 2017.
10. Polish Committee for Standardization. *Bezpieczeństwo Pożarowe ścian Zewnętrznych Budynków*; Polish Committee for Standardization: Warsaw, Poland, 2001.
11. European Commission. *Procedure European Organisation for Technical Assessment No 761/PP/GRO/IMA/19/1133/11140*; European Organisation for Technical Assessment: Brussels, Belgium, 2019.
12. *Reaction-To-Fire Tests for Façades—Part 2: Large-Scale Test*; International Organization for Standardization: Geneva, Switzerland, 2019; ISO 13785-2:2002.
13. Smolka, M.; Anselmi, E.; Crimi, T.; Le Madec, B.; Móder, I.F.; Park, K.W.; Rupp, R.; Yoo, Y.-H.; Yoshioka, H. Semi-natural test methods to evaluate fire safety of wall claddings: Update. In *Proceedings of the MATEC Web of Conferences*; EDP Sciences: Les Ulis, France, 2016; Volume 46, p. 1003.
14. Sulik, P.; Sędłak, B.; Kinowski, J. Study on critical places for maximum temperature rise on unexposed surface of curtain wall test specimens. In *Proceedings of the MATEC Web of Conferences*; EDP Sciences: Les Ulis, France, 2016; Volume 46, p. 2006.
15. Kinowski, J.; Sędłak, B.; Roszkowski, P.; Sulik, P. Impact of the method of fixing facade cladding on their behavior in fire conditions. *Mater. Bud.* **2017**, *8*, 204–205.
16. Szymków, M. Identification of the degree of destruction of fibre cement boards under the influence of high temperatur. Ph.D. Thesis, Wydział Budownictwa Lądowego i Wodnego, Politechnika Wrocławska, Wrocławska, Poland, 2018.
17. Bednarek, Z.; Drzymała, T. Wpływ temperatur występujących podczas pożaru na wytrzymałość na ściskanie fibrobetonu. *Zesz. Nauk. SGSP* **2008**, *36*, 61–84.
18. Drzymała, T.; Ogrodnik, P.; Zegardło, B. Wpływ oddziaływania wysokiej temperatury na zmianę wytrzymałości na zginanie kompozytów cementowych z dodatkiem włókien polipropylenowych. *Tech. Transp. Szyn.* **2016**, *12*, 82–86.
19. Al-Attar, A.A.; Abdulrahman, M.B.; Hamada, H.M.; Tayeh, B.A. Investigating the behaviour of hybrid fibre-reinforced reactive powder concrete beams after exposure to elevated temperatures. *J. Mater. Res. Technol.* **2020**, *9*, 1966–1977. [CrossRef]
20. Weghorst, R.; Hauze, B.; Guillaume, E. Determination of fire performance of ventilated facade systems on combustible insulation using LEPIR2. In Proceedings of the 14th International Fire and Engineering Conference Interflam, Windsor, UK, 4–6 July 2016; pp. 1–12.

21. Adamczak-Bugno, A.; Gorzelańczyk, T.; Krampikowska, A.; Szymków, M. Stosowanie metod nieniszczących przy remontach, modernizacjach i wzmocnieniach obiektów budowlanych. *Bad. Nieniszcz. Diagn.* **2017**, *3*, 20–23.

22. Bełzowski, A. *Degradacja Mechaniczna Kompozytów Polimerowych: Oficyna Wydawnicza*; Oficyna Wydawnicza Politechniki Wrocławskiej: Wrocław, Poland, 2002; p. 176.

23. Bezerra, E.M.; Joaquim, A.P.; Savastano, H. Some properties of fibre-cement composites with selected fibers. In Proceedings of the Conferencia Brasileira de Materiais e Tecnologias Não Convencionais: Habitações e Infra-Estrutura de Interesse Social Brasil-NOCMAT 2004, Pirassununga, SP, Brasil, 29 October–3 November 2004; pp. 34–43.

24. Kolaitis, D.I.; Asimakopoulou, E.K.; Founti, M.A. A Full-scale fore test to investigate the fire behaviour of the "ventilated facade" system. In Proceedings of the Interflam 2016, Windsor, UK, 4–6 July 2016; pp. 1–12.

25. Bentchikou, M.; Guidoum, A.; Scrivener, K.; Silhadi, K.; Hanini, S. Effect of recycled cellulose fibres on the properties of lightweight cement composite matrix. *Constr. Build. Mater.* **2012**, *34*, 451–456. [CrossRef]

26. Polish Committee for Standardization. *Fire Resistance Tests. Part 1. General Requirements*; Polish Committee for Standardization: Warsaw, Poland, 2012.

Article

Tests of Concrete Strength across the Thickness of Industrial Floor Using the Ultrasonic Method with Exponential Spot Heads

Bohdan Stawiski [1] and Tomasz Kania [2],*

[1]   Faculty of Environmental Engineering and Geodesy, Wrocław University of Environmental and Life Sciences, pl. Grunwaldzki 24, 50-363 Wrocław, Poland; bohdan.stawiski@upwr.edu.pl

[2]   Faculty of Civil Engineering, Wrocław University of Science and Technology, Wybrzeże Wyspiańskiego 27, 50-370 Wrocław, Poland

*   Correspondence: tomasz.kania@pwr.edu.pl

Received: 31 March 2020; Accepted: 28 April 2020; Published: 2 May 2020

Abstract: The accepted methods for testing concrete are not favorable for determining its heterogeneity. The interpretation of the compressive strength result as a product of destructive force and cross-section area is burdened with significant understatements. It is assumed erroneously that this is the lowest value of strength at the height of the tested sample. The top layer of concrete floors often crumble, and the strength tested using sclerometric methods does not confirm the concrete class determined using control samples. That is why it is important to test the distribution of compressive strength in a cross-section of concrete industrial floors with special attention to surface top layers. In this study, we present strength tests of borehole material taken from industrial floors using the ultrasonic method with exponential spot heads with a contact surface area of 0.8 mm^2 and a frequency of 40 kHz. The presented research project anticipated the determination of strength for samples in various cross-sections at the height of elements and destructive strength in the strength testing machine. It was confirmed that for standard and big borehole samples, it is not possible to test the strength of concrete in the top layer of the floor by destructive methods. This can be done using the ultrasonic method. After the analysis, certain types of distributions of strength across concrete floor thickness were chosen from the completed research program. The gradient and anti-gradient of strength were proposed as the new parameters for the evaluation of floor concrete quality.

Keywords: concrete floors; compressive strength; strength distribution; industrial floors; ultrasound tests

---

1. Introduction

Composite materials belong to an exceptional group because it is difficult to control grain distribution in the material being formed. The most characteristic example of such a composite is concrete, the most broadly used among them in the construction industry. Tests of this material concern both spatial structure (porosity, tightness, pore, and grain distribution) and its basic parameter, that is, compressive strength. Modifications of this composite aim at the improvement of its homogeneity [1–4].

The industrial floor belongs to those building structures in which concrete should be of the highest quality in the top layer because it is exposed to considerable abrasion, local pressure, impacts, etc. Often, the top layer of the concrete floor quickly crumbles (Figure 1) and strength testing using sclerometric methods does not confirm the concrete class determined on control samples [5].

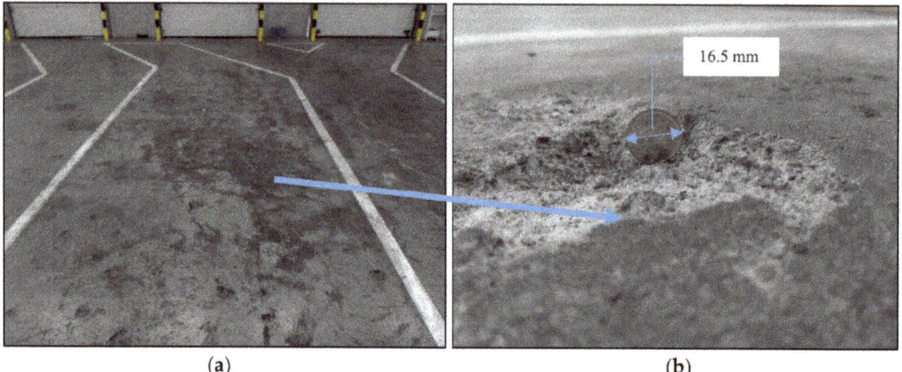

(a)   (b)

Figure 1. Example of crumbling on the top layer of concrete industrial floor: (**a**) General view; (**b**) close-up of cracked surface.

Strength tests on borehole material of small diameter and height, cut out from various depths, indicate often considerable weakening of concrete in the top layer compared to strength determined on samples cut out from a deeper layer. This problem was noticed in the European standard [6], highlighting that in the top layer, compressive strength can be lower than in the bottom layer by as much as 25%. However, the given value is based on tests of samples from a height not lower than 50 mm. Excessive reduction of samples is not acceptable due to the thickness of the maximum aggregate fraction. Practically, it can be assumed that minimum samples are 50 mm high and before cutting off of the top layer approximately 60 mm high. Compressive strength determined on such samples using the destructive method refers to concrete at a depth of 30 mm from the floor top. Many things indicate that most of the unfavorable phenomena accumulate in this 30 mm small layer. Tests on this type of structure should focus first on this layer. Since in the available scientific literature, there are few test results concerning distributions of compressive strength across the thickness of industrial floors in thin cross-sections taking into account the top layers, the authors of this study undertook this task. Recent articles [7–9] shows that acoustic non-destructive methods are suitable for testing concrete strength and heterogeneity. The concrete strength tests presented in this article were performed using the ultrasound method using spot heads. The ultrasonic wave velocities were correlated with the results of destructive tests. The obtained results from tests concerning strength distributions of floors were arranged by types. Additional characteristics of concrete were proposed based on the strength change rate using mathematical bases of quality evaluation.

2. Literature Review

In calculation theories, concrete is usually treated as a homogeneous entity, specified often as quasihomogeneous, at least in the macroscopic sense. The question of the actual heterogeneity of concrete remains open for several reasons. The most important reason is that the accepted test methods do not favor recognition of its heterogeneity. According to the binding standard [10], they are conducted on big material samples: cubes with dimensions 150 mm × 150 mm × 150 mm or cylinders with a diameter of 150 and a height of 300 mm, which blur porosity changes at the height of sample elements.

The heterogeneity of strength for elements with a large height dimension, such as pillars and vertical partitions formed in the built-in position, was the subject matter of a publication. Research showed a weakening of concrete in their horizontal cross-sections along with an increase in the measurement height. Tests conducted during the period from 1962 until 1978 in Japan [11] on concrete elements 4 m high formed in the built-in position indicated a 10% to 20% strength increase in the bottom zone and a 10% to 30% strength decrease in the top zone versus middle cross-sections. Giaccio and Alberto Giovambattista [12] showed a 30% reduction of concrete compressive strength in a water dam formed in a vertical position along with an increase in the measurement height. The water draining capacity measured experimentally in concrete used to form these elements was more than 10%.

Khatib and Mangat showed that the cement pastes formed horizontally can have the pore volume twice as large near the top layer than near the bottom layer. Pores near the top layer are also considerably larger (from 46% to 98% of pores above 0.1 µm) than pores near the bottom layer (from 38% to 80% respectively) [13]. The weakening of concrete structures in top cross-sections along the direction of forming and strengthening in lower cross-sections is effected mainly by: the designed aggregate composition, type and quality of cement, quantity of fly ash, consistency of the concrete mix, use of chemical additions binding water, and the vibration method applied to the formed concrete element. The intensity of water drainage from the concrete mix and the rate of desegregation of its components depends on these factors. This phenomenon is called 'bleeding' [14–17]. It is a spontaneous process that is the result of the difference in densities between the binding agent, aggregate composition, and water. In the technical literature [18,19], depending on the type and method of making the elements, two types of bleeding are distinguished:

- internal, which is characteristic for stocky cross-sections formed in the vertical position,
- superficial, which is observed in the case of laying floors and elements of small thickness in the horizontal position.

Topics regarding strength distribution at the height of stocky cross-sections formed in a vertical position are researched, described, and characterized in the technical and scientific literature. Superficial bleeding, the homogeneity characteristics of near-surface structures, and strength distribution for concrete floors in thin layers are not topics broadly described in the scientific literature. Information about mathematical criteria used to describe strength variability in industrial floors is also missing. Cases of superficial damage to the floors which are the subject of this study are frequent, therefore the authors undertook such research.

3. Materials and Test Methods

The research project assumed that the analysis would be carried out on samples from borehole materials taken from various regions of the country to avoid any potential material-related errors connected with, for example, the aggregate source. The floors were constructed of concrete class C25/30 and C30/37. Borehole materials with diameters 80 and 100 mm were provided, 6 pieces from each tested floor. They came from floors that were not superficially hardened, from floors with surface hardened using mineral agents and also with resin flooring (Figure 2).

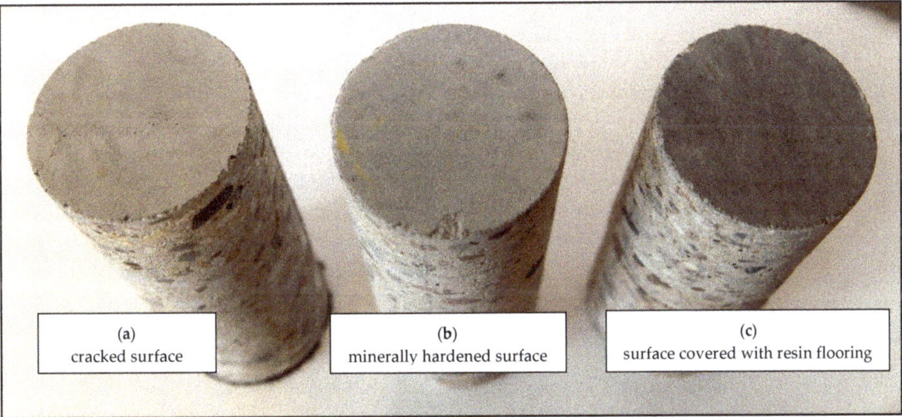

Figure 2. Borehole materials from the tested floors: (a) cracked surface; (b) hardened surface without damage; (c) resin flooring on concrete.

The hardened layer in certain areas of the halls became loosened which is regarded as an emergency condition (Figure 3).

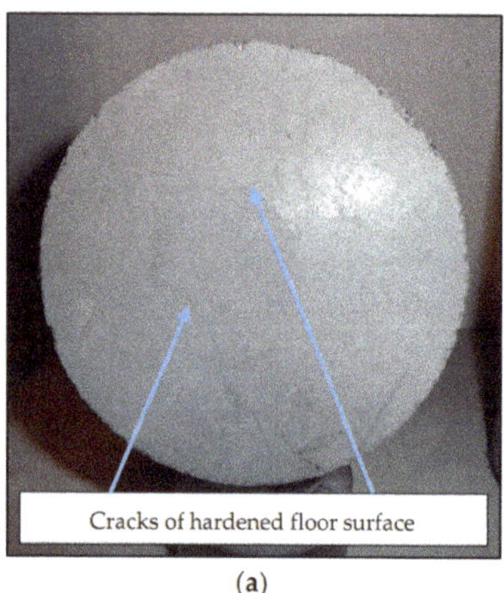

(**a**)

(**b**)

Figure 3. Tested borehole material from floors with a hardened top layer with damage: (**a**) cracked surface; (**b**) loosened surface.

The presented damage indicates substrate hardening of low strength (weak concrete).

The compressive strength of concrete samples in their various cross-sections was determined with the use of the non-destructive method calibrated with the destructive tests of the samples. This type of examination is being widely used for the range of building materials, such as concrete, wood, steel, ceramics, and for the diagnostics of building structures [20–25]. Tests were carried out with the use of the ultrasound method based on longitudinal wave velocities [26]. Dependencies between ultrasonic pulse velocity (*UPV*), elastic modulus (*E*), and Poisson's ratio (*v*) were researched and described [27,28]. Passing wave velocity (C_L) is proportional to the square root of the dynamic modulus of elasticity (E_d), and inversely proportional to the square root of its density (ρ) in Equation (1):

$$C_L = (E_d/\rho \cdot (1 - v_d)/((1 + v_d) \cdot (1 + 2v_d)))^{1/2} \ (\text{km/s}) \tag{1}$$

In Equation (1), v_d is the dynamic Poisson's ratio. The dependency in Equation (1) applies to homogeneous and isotropic materials. Concrete is a heterogeneous material. High attenuation in concrete limits the *UPV* method to low frequencies (up to 120 kHz). Under this condition, ultrasounds do not interact with most concrete inhomogeneities [29] and it can be regarded as a homogeneous material [30]. Scientific literature shows a strict relationship between concrete compressive strength (f_c) and *UPV* [27,28,30–36]. This dependency is also described in standards [7,37,38]. Komlos and others [36] stated that UPV non-destructive method of strength testing requires the calibration with the results of destructive tests. Authors research conclusions are the same [36,39]. In the presented research, UPV tests were performed using Unipan 543 tester (company, Warsaw, Poland) and spot heads with frequency 40 kHz. The test stand used to measure ultrasonic wave velocity on the tested concrete borehole material is presented in Figure 4. The results of spot heads testing are presented in the studies [35,39,40].

At the beginning of the tests, the borehole materials were first cleaned and dried to air-dry moisture according to the standard [6]. Testing cores were stored in laboratory conditions for 14 days before the tests. Then, in the first stage of the tests, ultrasound test planes were marked along the borehole materials (by height) at a distance of 10 mm from each other. In the marked planes, the velocity of ultrasonic wave transmission was tested in two directions, I and II, perpendicular to each other (Figure 5).

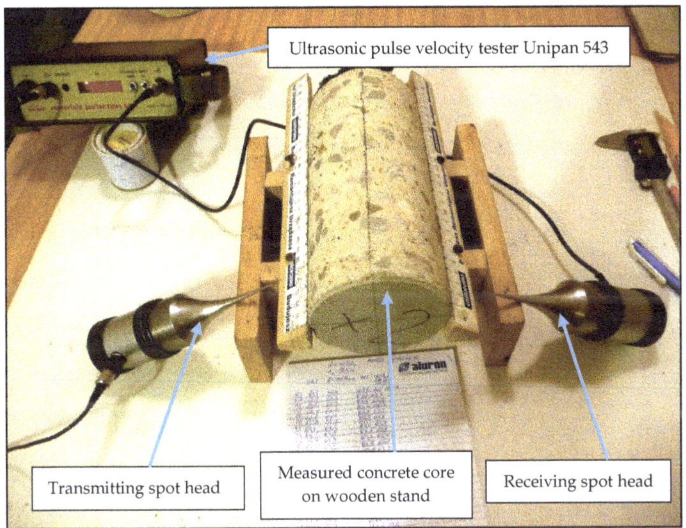

Figure 4. Test stand for measuring ultrasound velocity.

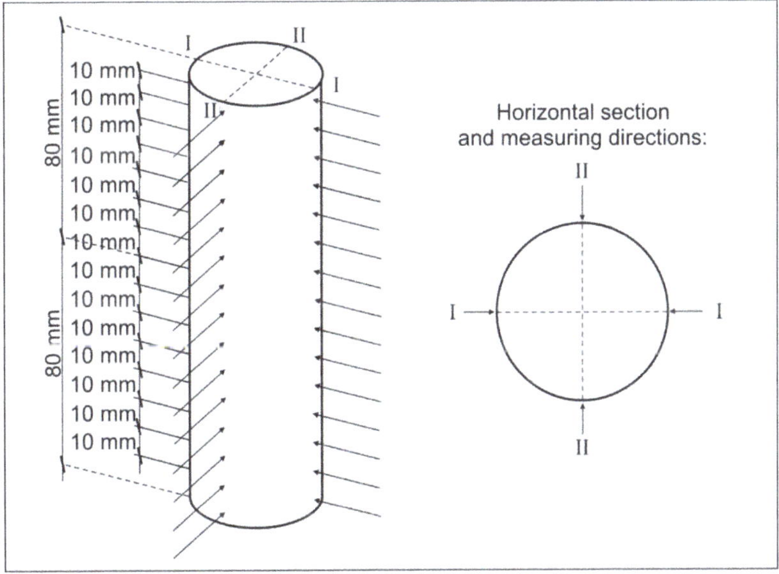

Figure 5. Location of the measurement planes at borehole heights—1st stage of the tests.

In subsequent batches, the top zone of the floor at the thickness of 3–4 cm was examined in more detail. The planes in which measurements were conducted were compacted, at first every 5 mm, and in certain series every 2 mm (Figure 6).

Such densely located planes for measurement of the ultrasonic impulse were possible because the diameter of the ends of the spot heads used was only 1 mm (Figure 7).

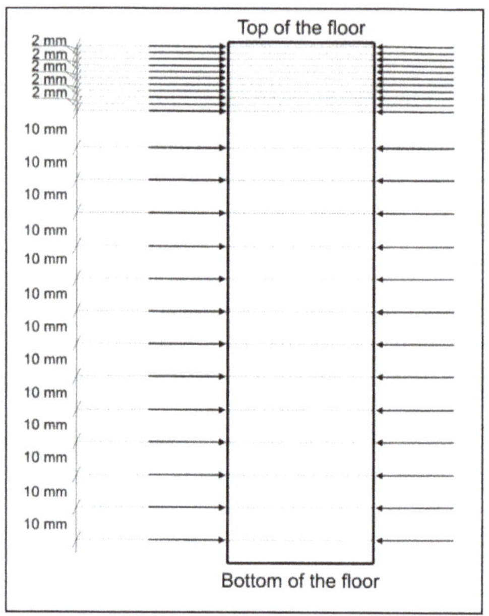

Figure 6. Location of the measurement planes at borehole height—2nd stage of the tests.

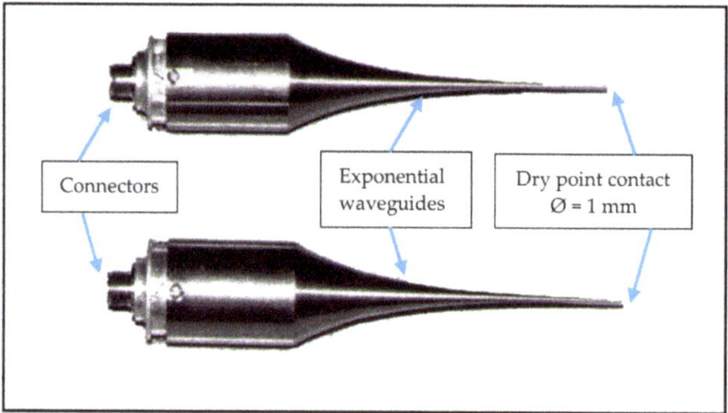

Figure 7. Spot heads with exponential waveguide where the surface area of contact with concrete is 0.8 mm^2.

Boreholes were cut thus obtaining samples with $\varphi = h$ (length equal to their diameter). To the pulse velocity determined in the middle of each sample height, compressive strength determined on the strength machine as a relation of destructive force P (N) to the surface area of cross-section A (mm^2) of Equation (2) was assigned:

$$f_c = P/A \text{ (MPa)} \tag{2}$$

Destructive strength f_c (MPa) and mean ultrasonic wave velocities C_L (km/s) from two testing directions (Figure 5, cross-sections, directions I and II), specified in the middle of the height of the cylindrical sample were the basis for determination of the dependency between ultrasonic pulse velocity and concrete strength for the given series of samples. Scaling curves established hypothetically were used to convert the rate of ultrasound wave in the given cross-section at the borehole height into concrete compression strength in this cross-section. The selection of the dependency f_c-C_L was made independently for each of the tested concretes taking into account different aggregate, different cement, different additions, and also different conditions of execution not known in detail.

4. Results

4.1. Calibration of Ultrasound Pulse Velocity-Compression Strength Curves Based on the Strength Machine Tests

After the measurements of ultrasonic pulse velocities of the tested samples, they were cut and tested in uniaxial loading on the strength machine. On that basis, hypothetical scaling curves were chosen.

Scaling Equation (3) was approximated for the measurements of the first concrete type shown in Section 4.2:

$$f_c = 0.1983 \cdot C_L^{\,4.3081} \ (\text{MPa}) \tag{3}$$

The approximation of the scaling curve in Equation (3) was performed based on the results of destructive tests conducted on cylindrical samples with dimensions 10 cm × 10 cm. The results of destructive tests, measured pulse velocities, and strength values calculated using Equation (3) are presented in Table 1.

Table 1. Results of uniaxial destructive tests, measured pulse velocities, and strength values calculated with the use of the chosen hypothetical scaling curve.

Ordinal Number	Ultrasound Longitudinal Wave Velocity C_L (km/s)	Compression Strength (MPa)		Difference f_c from Equation (3)—$f_{c,\,\varnothing10/10}$ (MPa)
		Destructive Test $f_{c,\,\varnothing10/10}$	f_c from Equation (3) $(0.1983 \cdot C_L^{\,4.3081})$	
1	3.460	40.24	41.66	1.42
2	3.459	43.22	41.61	−1.61
3	3.404	39.60	38.83	−0.77
4	3.372	38.11	37.28	−0.83
5	3.362	37.55	36.81	−0.74
6	3.288	32.88	33.44	0.56
7	3.282	34.02	33.18	−0.84
8	3.191	30.99	29.40	−1.59
9	3.190	31.77	29.36	−2.41
10	3.172	28.17	28.65	0.48
11	3.168	28.42	28.49	0.07
12	3.167	28.64	28.46	−0.18
13	3.108	27.12	26.24	−0.88
14	3.092	25.16	25.66	0.50
15	3.078	25.21	25.17	−0.04
16	3.07	25.22	24.89	−0.33
17	3.02	23.38	23.19	−0.19
18	2.838	17.22	17.74	0.52
19	2.742	14.62	15.30	0.68
Mean value	3.182	30.081	29.755	−0.33

Scaling curves for the remaining tested boreholes of the floors were approximated in the same way. Scaling curves established hypothetically were used to convert the rate of ultrasound wave in the given cross-section at the borehole height into concrete compression strength in this cross-section.

4.2. Tested Types of Strength Distributions in Cross-Section of Concrete Borehole Materials

The tested borehole materials provided an answer to the question of the distribution of concrete strength across the thickness of industrial floors in the tested facilities. Ignoring minor fluctuations of strength at different levels resulting from random positioning of aggregate or small defects in the concrete structure, or errors in measurement of the distance (the head ends got inside surface pores), all results can be grouped according to the summary below.

4.2.1. Concrete "Homogeneous" across Its Thickness with Superficial Weakening

Passing times t (µs), wave velocities C_L (km/s), and compressive strengths f_c (MPa) in planes parallel to the surface of the boreholes are presented in Table 2. Calculation of compressive strength from ultrasound wave velocities was done with the use of Equation (3). The borehole materials were taken from an industrial floor 22 cm thick. Results are presented starting with the ordinal number 1 (5 mm from the top of the floor) in the direction of the bottom of the floor.

Table 2. Results of concrete ultrasound velocity test in borehole No. 1.

Ordinal Number	Height from the Bottom of the Floor (cm)	Ultrasound Netto Passing Time in Direction I-I $t_{n\,I\text{-}I}$ (µs)	Ultrasound Netto Passing Time in Direction II-II $t_{n\,II\text{-}II}$ (µs)	Mean Ultrasound Netto Passing Time t_n (µs)	Ultrasound Wave Velocity C_L (km/s)	Concrete Compression Strength f_c (MPa)
1	21.5	35.3	36.2	35.75	2.615	12.47
2	21.0	29.8	32.0	30.90	3.026	23.39
3	20.5	29.4	29.3	29.35	3.186	29.20
4	20.0	30.2	30.2	30.20	3.096	25.81
5	19.5	30.5	30.9	30.70	3.046	24.06
6	19.0	30.1	30.6	30.35	3.081	25.27
7	18.5	28.6	29.8	29.20	3.202	29.84
8	18.0	29.5	29.7	29.60	3.159	28.15
9	17.0	29.4	29.2	29.30	3.191	29.40
10	16.0	28.5	28.4	28.45	3.286	33.36
11	15.0	28.4	29.9	29.15	3.208	30.08
12	14.0	28.8	29.8	29.30	3.191	29.40
13	13.0	29.1	28.7	28.90	3.235	31.18
14	12.0	29.4	29.4	29.40	3.180	28.96
15	11.0	29.2	29.9	29.55	3.164	28.34
16	10.0	28.5	29.7	29.10	3.213	30.28
17	9.0	28.5	28.8	28.65	3.264	32.40
18	8.0	28.3	29.4	28.85	3.241	31.43
19	7.0	29.2	29.5	29.35	3.186	29.20
20	6.0	28.3	30.2	29.25	3.197	29.63
21	5.0	28.8	29.3	29.05	3.219	30.52
22	4.0	29.0	29.2	29.10	3.213	30.28
23	3.0	29.7	29.8	29.75	3.143	27.54
24	2.0	29.2	28.6	28.90	3.235	31.18
25	1.0	29.5	29.2	29.35	3.186	29.20

Distribution of compressive strength in the tested concrete at the sample height is presented in Figure 8.

Fluctuations across the average strength values (marked with the thin blue line in Figure 8) were relatively small. Its strong decrease begins at 19 cm from the bottom. Concrete is homogeneous across the floor thickness, only the top layer, approximately 10–20 mm, was weakened. The compressive strength of concrete in the top layer of the floor drops from the value of 25–30 MPa down to the value of 12 MPa at the measurement height of 10 mm.

4.2.2. "Homogeneous" Concrete with Weakening across the Top Layer 30–50 mm Thick

Calculation of compressive strength from ultrasound wave velocities was done with the use of Equation (3). The borehole materials were taken from an industrial floor 21 cm thick. Results are presented in Table 3 starting with the ordinal number 1 (5 mm from the top of the floor) in the direction of the bottom of the floor.

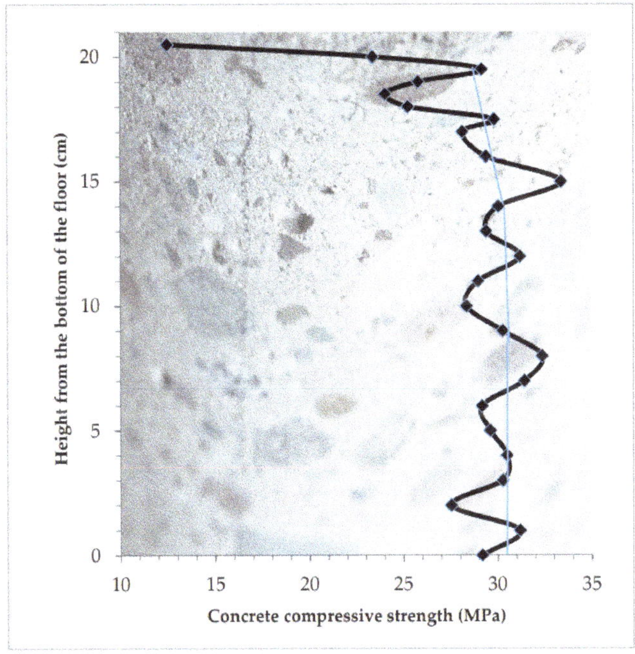

Figure 8. Concrete "homogeneous" across the floor thickness, only the thin top layer is considerably weakened.

Table 3. Results of concrete ultrasound velocity test in borehole No. 2.

Ordinal Number	Height from the Bottom of the Floor (cm)	Ultrasound Netto Passing Time in Direction I-I $t_{n\ I\text{-}I}$ (µs)	Ultrasound Netto Passing Time in Direction II-II $t_{n\ II\text{-}II}$ (µs)	Mean Ultrasound Netto Passing Time t_n (µs)	Ultrasound Wave Velocity C_L (km/s)	Concrete Compression Strength f_c (MPa)
1	20.5	31.3	30.5	30.90	3.040	24.24
2	20.0	31.1	30.1	30.60	3.070	25.28
3	19.5	30.4	30.3	30.35	3.096	26.19
4	19.0	29.6	30.7	30.15	3.116	26.95
5	18.5	29.7	30.1	29.90	3.142	27.94
6	18.0	28.4	28.8	28.60	3.285	33.83
7	17.0	28.4	28.9	28.65	3.279	33.58
8	16.0	28.4	28.3	28.35	3.314	35.14
9	15.0	27.6	28.4	28.00	3.355	37.07
10	14.0	27.6	28.1	27.85	3.373	37.94
11	13.0	28.0	28.3	28.15	3.337	36.22
12	12.0	27.8	28.0	27.90	3.367	37.64
13	11.0	28.0	28.0	28.00	3.355	37.07
14	10.0	28.6	28.5	28.55	3.291	34.09
15	9.0	27.6	28.5	28.05	3.349	36.78
16	8.0	27.4	28.3	27.85	3.373	37.94
17	7.0	27.6	27.6	27.60	3.404	39.44
18	6.0	27.7	27.9	27.80	3.379	38.23
19	5.0	27.4	27.8	27.60	3.404	39.44
20	4.0	27.7	27.9	27.80	3.379	38.23
21	3.0	27.6	27.5	27.55	3.410	39.75
22	2.0	28.1	28.6	28.35	3.314	35.14
23	1.0	27.4	26.6	27.00	3.480	43.35
24	0.2	27.4	26.0	26.70	3.519	45.49

The distribution of compressive strength for the tested concrete at the height of the sample is presented in Figure 9.

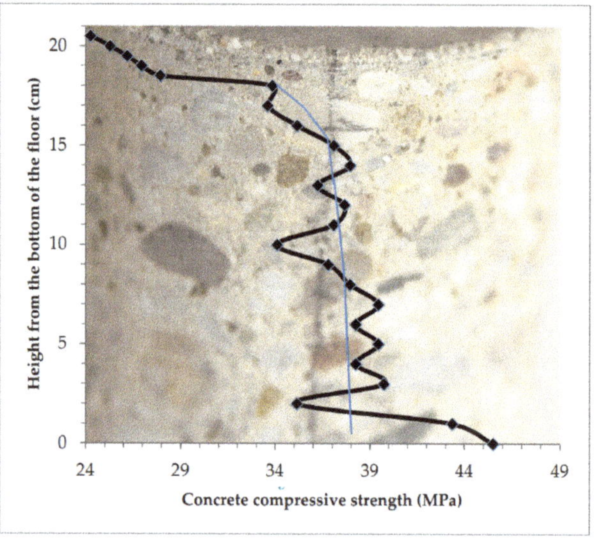

Figure 9. Concrete across the thickness of the floor relatively homogeneous, weakened thick top layer.

The concrete was relatively homogeneous across the floor thickness. The weakened top layer was 30–50 mm thick. Compressive strength in this layer of concrete drops from the value of 34–40 MPa down to the value of 24 MPa in the near-surface cross-section.

The bottom layer of the sample was significantly strengthened. Measurements made at a height of 2 mm from the bottom of the sample showed an increase in strength to 46 MPa from 35–40 MPa in close layers above.

4.2.3. Concrete Strength Is Changing across Entire Section with a Quick Weakening of the Top Layer

The borehole materials were taken from an industrial floor 15 cm thick. Calculation of compressive strength from ultrasound wave velocities was done with the use of Equation (4).

$$f_c = 0.123 \cdot C_L{}^{4.3081} \; (\text{MPa}) \tag{4}$$

Results are presented in Table 4, starting with the ordinal number 1 (5 mm from the top of the floor) in the direction of the bottom of the floor.

Table 4. Results of concrete ultrasound velocity test in borehole No. 3.

Ordinal Number	Height from the Bottom of the Floor (cm)	Ultrasound Netto Passing Time in Direction I-I $t_{n\,I\text{-}I}$ (µs)	Ultrasound Netto Passing Time in Direction II-II $t_{n\,II\text{-}II}$ (µs)	Mean Ultrasound Netto Passing Time t_n (µs)	Ultrasound Wave Velocity C_L (km/s)	Concrete Compression Strength f_c (MPa)
1	14.5	34.2	34.6	34.40	2.733	9.35
2	14.0	34.4	34.4	34.40	2.733	9.35
3	13.0	34.0	35.8	34.90	2.693	8.78
4	12.0	29.7	30.7	30.20	3.113	16.38
5	11.0	28.2	28.2	28.20	3.333	22.00
6	10.0	26.9	28.5	27.70	3.394	23.77
7	9.0	29.2	27.0	28.10	3.345	22.34
8	8.0	28.6	26.7	27.65	3.400	23.95

Table 4. *Cont.*

Ordinal Number	Height from the Bottom of the Floor (cm)	Ultrasound Netto Passing Time in Direction I-I $t_{n\ I\text{-}I}$ (μs)	Ultrasound Netto Passing Time in Direction II-II $t_{n\ II\text{-}II}$ (μs)	Mean Ultrasound Netto Passing Time t_n (μs)	Ultrasound Wave Velocity C_L (km/s)	Concrete Compression Strength f_c (MPa)
9	7.0	27.2	26.6	26.90	3.494	26.97
10	6.0	26.9	27.2	27.05	3.475	26.33
11	5.0	27.5	27.2	27.35	3.437	25.11
12	4.0	27.8	26.1	26.95	3.488	26.75
13	3.0	25.9	26.5	26.20	3.588	30.21
14	2.0	26.4	26.9	26.65	3.527	28.07
15	1.0	25.8	26.3	26.05	3.608	30.97
16	0.5	25.7	26.6	26.15	3.595	30.46

The concrete across the floor thickness changed its strength along with the change in the measurement plane. The top layer with a high thickness (up to 50 mm) became weakened very quickly in the direction of the floor top. The distribution of compressive strength for the tested concrete at the height of the sample is presented in Figure 10.

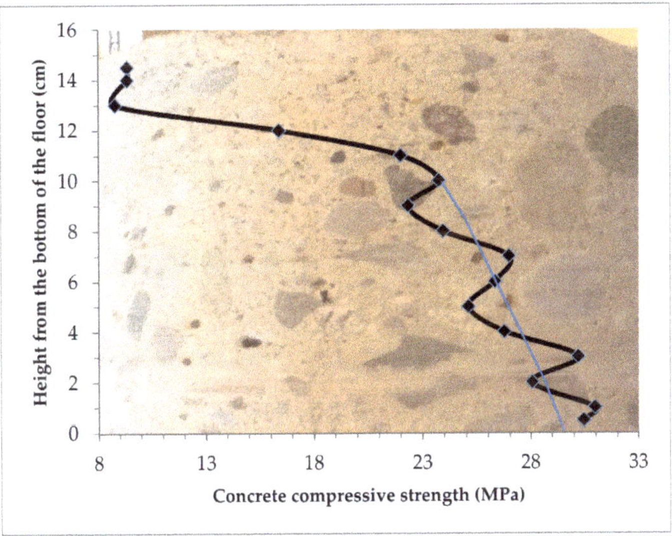

Figure 10. The change of concrete strength across the floor thickness was continuous. The weakened top layer was very thick.

4.2.4. Concrete Reinforced Superficially to the Expected Value Using the Mineral Powder

To improve the hardness of the top layer of the floor, powders made of hardening materials based on a cement binding agent and strong filler were used. If the strength of the hardened concrete had characteristics in the vertical cross-section similar to the one described in Section 4.2.1 (only a thin 10–15 mm top layer of the floor was weakened), a correctly made hardening layer may balance the shortage of strength. The floor will achieve the designed quality on its top layer. In the presented example top layer of concrete floor has been reinforced using mineral powder to the average level in the vertical cross-section, in an expected way. After tests of the borehole samples of the examined floor on the strength testing machine, the scaling curve described using Equation (3) was adjusted.

The result of measurements regarding ultrasonic wave velocity and compressive strength in the sample's cross-section is presented in Table 5.

Table 5. Results of concrete ultrasound velocity test in borehole No. 4.

Ordinal Number	Height from the Bottom of the Floor (cm)	Ultrasound Netto Passing Time in Direction I-I $t_{n\,I\text{-}I}$ (µs)	Ultrasound Netto Passing Time in Direction II-II $t_{n\,II\text{-}II}$ (µs)	Mean Ultrasound Netto Passing Time t_n (µs)	Ultrasound Wave Velocity C_L (km/s)	Concrete Compression Strength f_c (MPa)
1	19.8	28.5	28.8	28.65	3.274	32.83
2	19.5	28.7	29.7	29.20	3.213	30.28
3	19.0	29.6	31.0	30.30	3.097	25.84
4	18.5	29.8	32.4	31.10	3.021	23.22
5	18.0	29.4	32.5	30.95	3.038	23.79
6	17.5	28.0	29.6	28.80	3.259	32.19
7	17.0	28.2	29.2	28.70	3.269	32.62
8	16.5	27.5	29.0	28.25	3.323	35.00
9	16.0	27.3	28.9	28.10	3.341	35.83
10	15.0	27.8	28.2	28.00	3.350	36.25
11	14.0	27.4	28.8	28.10	3.340	35.78
12	13.0	28.5	29.2	28.85	3.252	31.89
13	12.0	27.7	29.3	28.50	3.294	33.71
14	11.0	27.7	29.4	28.55	3.288	33.44
15	10.0	27.9	28.9	28.40	3.304	34.15
16	9.0	28.0	28.6	28.30	3.315	34.64
17	8.0	28.9	28.8	28.85	3.251	31.85
18	7.0	28.8	27.6	28.20	3.328	35.23
19	6.0	28.4	27.2	27.80	3.376	37.47
20	5.0	29.0	28.3	28.65	3.274	32.83
21	4.0	27.8	27.8	27.80	3.374	37.38
22	3.0	28.0	28.2	28.10	3.338	35.69
23	2.0	28.0	28.3	28.15	3.332	35.42
24	1.0	27.7	27.8	27.75	3.380	37.67
25	0.2	26.6	27.8	27.20	3.450	41.14

Distribution of strength in cross-section of the sample taken from the floor with the hardening layer made as expected (Table 5) is presented in Figure 11.

Figure 11. The strength of the floor reinforced superficially to the expected value using the mineral powder.

The superficially weakened floor, with the compressive strength of 25–30 MPa, strengthened in the near-surface zone up to the designed value of 30 MPa. Repair efficiency depends on the weakest cross-section under the strengthened layer and its ability to transfer shear stress arising because of the shrinkage of the strengthened layer. In the lower zone of the sample (2 mm from the bottom of the floor) the concrete was strengthened to a value of over 40 MPa. The concrete strengthening zone was about 20 mm thick.

4.2.5. Concrete Floors under Emergency Conditions with Top Layer Hardened using Mineral Powder

As far as concrete floors strengthened (and hardened) using the mineral powder under emergency conditions are concerned, three types of strength distribution were analyzed:

- the case of strengthening the top layer with too small a quantity of mineral powder,
- the case of strengthening the floor with too weak a top layer of concrete,
- the case of excessive strengthening of the top layer of the floor.

Table 6 presents the results of measurements concerning ultrasonic wave velocity in cross-section of the floor strengthened with too small a quantity of mineral powder. The borehole materials were taken from a floor 16.5 cm thick. Calculation of compressive strength from ultrasound wave velocities was done with the use of Equation (5):

$$f_c = 6.147 \cdot C_L{}^2 - 18.172 \cdot C_L + 10.786 \ (\text{MPa}) \tag{5}$$

Table 6. Results of concrete ultrasound velocity test in borehole No. 5.

Ordinal Number	Height from the Bottom of the Floor (cm)	Ultrasound Netto Passing Time in Direction I-I $t_{n\,I\text{-}I}$ (µs)	Ultrasound Netto Passing Time in Direction II-II $t_{n\,II\text{-}II}$ (µs)	Mean Ultrasound Netto Passing Time t_n (µs)	Ultrasound Wave Velocity C_L (km/s)	Concrete Compression Strength f_c (MPa)
1	16.2	30.3	28.3	29.30	3.3560	19.03
2	16.0	30.0	28.8	29.40	3.3450	18.78
3	15.8	30.4	29.2	29.80	3.3000	17.76
4	15.6	31.0	29.0	30.00	3.2780	17.27
5	15.4	30.9	29.2	30.05	3.2730	17.16
6	15.2	30.9	29.6	30.25	3.2510	16.68
7	15.0	31.1	29.7	30.40	3.2350	16.33
8	14.8	30.6	29.8	30.20	3.2560	16.79
9	14.6	30.8	29.4	30.10	3.2670	17.03
10	14.4	30.2	29.8	30.00	3.2780	17.27
11	14.2	29.0	28.3	28.65	3.4320	20.82
12	14.0	28.6	28.0	28.30	3.4750	21.87
13	13.0	27.2	27.5	27.35	3.5960	24.93
14	12.0	27.0	27.0	27.00	3.6420	26.14
15	11.0	26.6	26.8	26.70	3.6830	27.24
16	10.0	26.9	26.6	26.75	3.6760	27.05
17	9.0	26.7	27.2	26.95	3.6490	26.32
18	8.0	26.4	26.6	26.50	3.7110	28.00
19	7.0	26.6	25.8	26.20	3.7530	29.17
20	6.0	26.1	26.6	26.35	3.7320	28.58
21	5.0	26.6	26.5	26.55	3.7040	27.81
22	4.0	25.8	25.3	25.55	3.8490	31.91
23	3.0	26.6	26.4	26.50	3.7110	28.00
24	2.0	26.6	26.9	26.75	3.6760	27.05
25	1.0	26.3	26.6	26.45	3.7180	28.20

Distribution of strength in cross-section of a sample taken from the floor in which too small a quantity of hardening powder was used is presented in Figure 12.

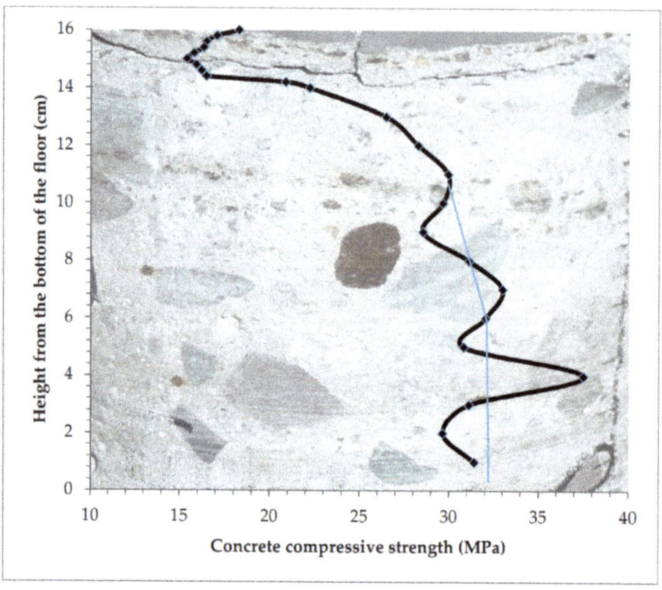

Figure 12. Too weak ineffective strengthening of the industrial floor.

In this case, the compressive strength drops from a value close to 30 MPa at a depth of 5 cm from the top of the floor to 16 MPa in the zone close to the surface of destruction. The destruction occurred in the weakest layer of concrete, where strength reaches a value of not more than 16.3 MPa. In the reinforced surface plane, the strength increases to 19 MPa at a depth of 3 mm from the top of the floor. The mineral strengthening of the floor, in this case, should reach a depth of 25–30 mm, at which the concrete strength was at a minimum level of 20–25 N/mm^2.

The strengthening will be ineffective also in the case in which concrete below the zone affected by the mineral powder is too weak and the entire strengthened layer is loosened. Measurements of ultrasonic wave velocity in cross-section of a sample taken from such a floor are presented in Table 7. The dependency between wave velocity C_L and strength f_c described using Equation (6) was used for calculations of the distribution of strength in the sample's cross-section.

$$f_c = 112.880 \cdot C_L{}^2 - 379.850 \cdot C_L + 324.7 \text{ (MPa)} \tag{6}$$

Table 7. Results of concrete ultrasound velocity test in borehole No. 6.

Ordinal Number	Height from the Bottom of the Floor (cm)	Ultrasound Netto Passing Time in Direction I-I $t_{n\,I\text{-}I}$ (µs)	Ultrasound Netto Passing Time in Direction II-II $t_{n\,II\text{-}II}$ (µs)	Mean Ultrasound Netto Passing Time t_n (µs)	Ultrasound Wave Velocity C_L (km/s)	Concrete Compression Strength f_c (MPa)
1	21.0	47.5	49.5	48.54	2.122	26.90
2	20.0	49.4	51.0	50.20	2.051	20.50
3	19.0	60.4	62.0	61.21	1.682	5.14
4	18.0	46.7	47.0	46.84	2.199	35.20
5	17.0	46.7	47.6	47.12	2.185	33.70
6	16.0	47.4	48.5	47.96	2.147	29.52
7	15.0	46.7	47.7	47.16	2.184	33.49
8	14.0	47.6	47.1	47.35	2.175	32.50
9	13.0	48.1	47.4	47.77	2.156	30.42
10	12.0	46.7	46.7	46.71	2.205	35.92
11	11.0	46.8	46.3	46.55	2.212	36.82

Table 7. *Cont.*

Ordinal Number	Height from the Bottom of the Floor (cm)	Ultrasound Netto Passing Time in Direction I-I $t_{n\,I\text{-}I}$ (µs)	Ultrasound Netto Passing Time in Direction II-II $t_{n\,II\text{-}II}$ (µs)	Mean Ultrasound Netto Passing Time t_n (µs)	Ultrasound Wave Velocity C_L (km/s)	Concrete Compression Strength f_c (MPa)
12	10.0	47.9	46.5	47.20	2.182	33.28
13	9.0	48.9	47.6	48.25	2.134	28.18
14	8.0	49.6	47.7	48.66	2.116	26.38
15	7.0	48.4	47.7	48.07	2.142	29.00
16	6.0	46.2	47.6	46.91	2.195	34.82
17	5.0	48.8	45.2	46.99	2.192	34.39
18	4.0	50.4	46.2	48.34	2.130	27.78
19	3.0	47.4	47.2	47.27	2.179	32.92
20	2.0	45.7	46.6	46.16	2.231	39.09
21	1.0	45.7	45.7	45.66	2.255	42.18

The distribution of strength in cross-section of the tested sample is presented in Figure 13.

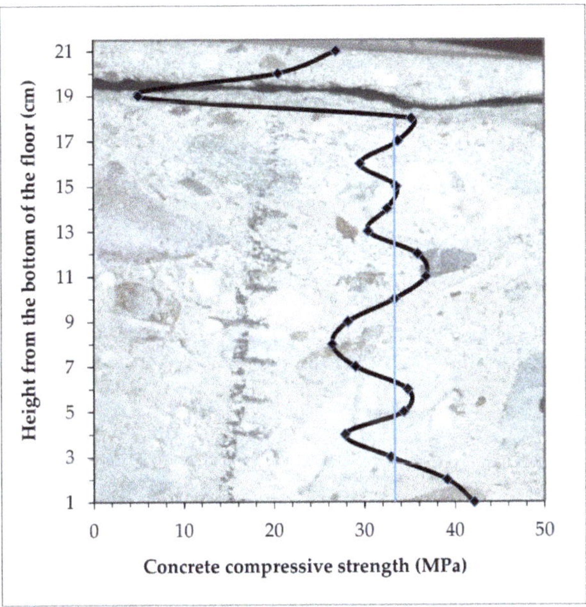

Figure 13. Strengthening that was ineffective because the concrete under the hardening layer had a very low strength of approximately 5 MPa.

In this situation, the strengthening was ineffective because directly under the mineralized layer, with a strength of 20 MPa, the concrete was very weak, its strength decreases down to the value of approximately 5 MPa near the surface of destruction, and then increases up to the designed value of 30 MPa.

The floor reaches the highest strength value at the last measuring point, 1 cm deep from the bottom edge. The reinforcement was even at a thickness of 4 cm from the bottom, where the measured strength was 28 MPa and reaches 40 MPa in the plane at the bottom of the sample. The effect of concrete strengthening in the lower planes of industrial floors (Figures 9, 11 and 13) usually results from aggregate segregation, which occurs in the process of vibrating the concrete mix in gravitational field forces.

A similar effect of damage to the hardened floor occurs if excessive powder is used. Measurements regarding ultrasonic wave velocity for a borehole material from such a floor are presented in Table 8. The calibrated dependency of the approximation described using Equation (7) was used for calculation of compressive strength in the cross-section of the sample.

$$f_c = 24.780 \cdot C_L - 33.600 \text{ (MPa)} \tag{7}$$

Table 8. Results of concrete ultrasound velocity test in borehole No. 7.

Ordinal Number	Height from the Bottom of the Floor (cm)	Ultrasound Netto Passing Time in Direction I-I $t_{n\,I-I}$ (µs)	Ultrasound Netto Passing Time in Direction II-II $t_{n\,II-II}$ (µs)	Mean Ultrasound Netto Passing Time t_n (µs)	Ultrasound Wave Velocity C_L (km/s)	Concrete Compression Strength f_c (MPa)
1	21.0	47.0	55.0	51.00	2.950	39.50
2	20.0	55.5	54.9	55.20	2.726	33.94
3	19.0	79.0	79.2	79.10	1.902	13.53
4	18.0	59.0	59.4	59.20	2.541	29.38
5	17.0	59.9	59.2	59.55	2.526	29.01
6	16.0	60.6	59.4	60.00	2.508	28.54
7	15.0	58.5	58.6	58.55	2.570	30.07
8	14.0	59.2	58.8	59.00	2.550	29.59
9	13.0	60.0	58.1	59.05	2.548	29.54
10	12.0	57.9	58.0	57.95	2.596	30.73
11	11.0	58.6	59.4	59.00	2.550	29.59
12	10.0	59.9	58.6	59.25	2.539	29.32
13	9.0	58.8	59.0	58.90	2.554	29.70
14	8.0	61.3	58.3	59.80	2.516	28.74
15	7.0	61.0	59.0	60.00	2.508	28.54
16	6.0	58.0	58.8	58.40	2.576	30.24
17	5.0	56.6	58.3	57.45	2.619	31.29
18	4.0	75.8	64.0	69.90	2.152	19.74
19	3.0	60.9	44.0	52.45	2.868	37.48
20	2.0	52.3	52.3	52.30	2.877	37.68
21	1.0	52.0	51.4	51.70	2.910	38.51

The distribution of strength in the cross-section of the tested sample is presented in Figure 14.

Figure 14. Homogeneous concrete across the thickness of the floor was weakened in the top zone and was strengthened excessively (strength growth from 13 to 40 MPa).

In this case, the strength of the top layer exceeded the designed value. The measured strength of the borehole material in the middle of the sample's height has a value of approximately 30 MPa. Considering the possible weakening in the upper zone, the concrete was excessively strengthened with mineral powder. After strengthening, in the near-surface zone, the strength value increased up to 40 MPa, and below the strengthened zone it drops down to the value of 13.5 MPa. In the destruction zone, in which it was impossible to make a measurement, the value of strength was lower than 13.5 MPa—a value measured from its nearest surface (2 cm from the top of the floor). Considerable shrinkage in the place of contact between the layer with increased strength and the weakened layer lead to loosening (Figures 14 and 15).

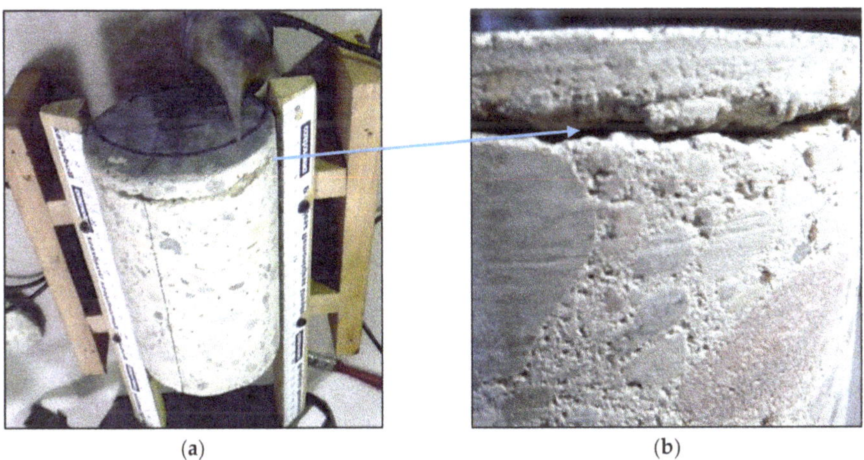

(a)   (b)

Figure 15. Loosening of the floor from the weak concrete; (**a**) view of the entire sample; (**b**) close-up to the loosened zone.

The various situations involving floor concrete weakening in the top layer shown above, as well as successful and unsuccessful strengthening using hardening powder, explain the causes of the defects and the effects of the 'repairs'. The examination of concrete strength distribution across the thickness of the floor was possible thanks to the applied ultrasound method with spot heads.

The interpretation of the results obtained from tests of concrete strength distributions across the thickness of the floor presented above is satisfactory, however, no reference documents or admissible values regarding strength variability in a cross-section of concrete slab elements are available concerning this. The subsequent section presents a proposal of additional criteria for assessing concrete in floors and an attempt to indicate mathematical criteria ensuring the basis for evaluation of strength changes across the thickness of the tested concrete layer based on the strength values and the rate of change of strength values obtained from measurements of ultrasonic wave velocities.

4.3. Proposal for Evaluating Concrete Quality in the Floor Based on Its Strength Gradient and Anti-Gradient

The gradient of the scalar strength field indicates the direction of the quickest growth of strengths in individual points. The modulus (that is 'length') of each vector in such a field is equal to the rate of strength field growth in the given direction. The vector opposite to the gradient is sometimes called an anti-gradient. The strength gradient is a vector value and its unit in the SI system is pascal per meter (Pa/m). Due to the range of the measured strength gradient values for industrial floors, the unit MPa/cm is used for the description of them.

In the considered case of the strength field test, you can refer to a vertical strength gradient that corresponds to a strength change in line with the distance from the bottom of the slab (with the height of the measurement plane). The strength gradient between two measurement planes 1 and 2 can be expressed using Equation (8):

$$\nabla f_c = (f_{c,2} - f_{c,1})/(Z_2 - Z_1) \ (MPa/cm) \tag{8}$$

where $f_{c,2}$ is concrete strength in plane 2 (closer to the top surface of the slab); $f_{c,1}$ is concrete strength in plane 1 (closer to the bottom surface of the slab); Z_2—distance of the second measurement plane from the bottom of the slab; Z_1—distance of the first measurement plane from the bottom of the slab.

Figure 16 shows the calculated gradients between various measurement planes for the borehole material for which measurement results are presented in Table 4.

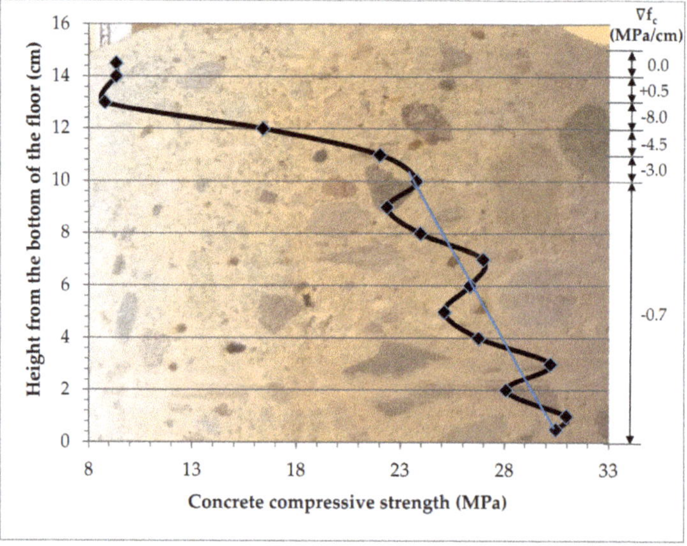

Figure 16. Concrete strength gradients at various floor depths.

At a depth of 0 to 10 cm counting from the bottom of the slab, ∇f_c is 0.7 MPa/cm apart from local fluctuations, and in layers located closer to the top surface it changes quickly (−3.0; −4.5; −8.0 MPa/cm). In the zone of powder-based hardening, the strength begins to increase (∇f_c = +0.5 MPa/cm). Changes in the strength of concrete in the areas of fluctuation around the average value (marked with a blue line in Figure 16) are related to the arrangement of grains on the path of ultrasound waves in concrete and are not the subject of consideration. A zone in which there is a rapid decrease or increase in strength is analyzed, e.g., after strengthening and applied to a thin top layer, 20 to 50 mm, rarely thicker. When the anti-gradient is large, the decrease in concrete strength in the surface layer is very large and the floor will need repair. It is proposed to test the gradient in the following variants:

- $\nabla f_{c,min}$—from the level of the beginning of a rapid decrease in strength (as above) upwards, to the zone of strengthening (increase in strength value) in the case of hardened floors or to the floor surface when there is no strengthening zone;
- $\nabla f_{c\,min,10\,mm}$—on a thickness of 10 mm up from the place of onset of a rapid decrease in strength in the near-surface area (20–50 mm from the top of the floor);
- $\nabla f_{c,max}$—gradient of strength increase in the hardening zone, relevant to the floors treated with a surface hardening agent.

A summary showing described strength gradient parameters together with measured extreme and mean values of compressive strength of tested samples is presented in Table 9.

Table 9. Characteristics of strength gradients and measured strength values of the tested samples.

Ordinal Number	Characteristics of Floor Strength Distribution	Gradient of Strength Decrease in Near Surface Zone (MPa/cm)		Gradient of Strength Increase in Hardening Zone (MPa/cm)	Compressive Strength in the Entire Cross-Section (MPa)		
		$\nabla f_{c,min}$	$\nabla f_{c\ min,10\ mm}$	$\nabla f_{c,max}$	Max $f_{c,max}$	Min $f_{c,min}$	Average $f_{c,av}$
1	Homogeneous with weakening of 10 mm of the top layer, (Table 2, Figure 8)	−16.7	−16.7	- *	33.38	12.48	28.42
2	Homogeneous with weakening of 35 mm of the top layer (Table 3, Figure 9)	−3.8	−6.9	- *	45.49	24.24	35.29
3	Variable across the entire cross-section with quick weakening of 50 mm of the top layer (Table 4, Figures 10 and 15)	−2.8	−7.6	- *	30.97	8.78	22.55
4	Surface strengthened to the designed value of 30 MPa from 22 MPa (Table 5, Figure 11)	−9.0	−9.0	7.1	41.14	23.22	33.45
5	Poorly hardened surface with low strength (Table 6, Figure 12)	−4.3	−5.1	2.4	31.91	16.33	23.09
6	Hardened surface with too low strength (Table 7, Figure 13)	−30.1	−30.1	10.9	42.18	5.14	30.86
7	Excessively hardened surface with low strength (Table 8, Figure 14)	−15.8	−15.8	13.0	39.5	13.53	30.22

*—floor without surface hardening.

Apart from the minimum, mean, and maximum values of compressive strength in cross-section of industrial floors, the strength variability dynamics can be, in this case, described using the value of strength gradient and anti-gradient, which is especially important for the distribution of strength for near-surface concrete layers. The floor should be made using concrete of a specified class and with strength variability gradient at as low a level as possible, guaranteeing that its performance features are maintained during use. The quality of concrete in an industrial floor can be in this case specified using not one but two parameters. The presented examples of strength distribution in the floors indicate that crumpling and cracking in industrial floors used in a standard way usually occur in the following cases:

- if their minimum compressive strength measured in cross-section is considerably lower than the designed value;
- if their compressive strength in the surface zone is lower than its service load;
- if the strength gradient for the top, hardened layer exceeds the value of shear strength of concrete. Shrinkage in the place of contact between the level of concrete with increased and weakened strength leads to loosening.

Presented measurements of concrete floors require further research, also with the use of different measurement tools. A discussion of a wide pool of strength distribution results is needed, based on which it may be possible to determine the limit values of the measured concrete strength gradients that can be used to standardize the quality assessment of industrial concrete floors. In the future, it is also planned to carry out measurements of ultrasonic wave velocity and strength distribution in the different concrete constructions formed horizontally to compare them with the results presented in this article. The obtained values of reduced compressive strength in the upper zone of concrete elements are also important in the case of bent reinforced concrete slabs, in which it is advisable that the top zone is not weakened.

5. Conclusions

Despite meeting the requirements for industrial floors tested by the standard method, local fluctuations in compressive strength values in surface cross-sections are a frequent reason for their failure conditions.

In industrial floors made of concrete without surface hardening, the weakest layer is the top layer 10–50 mm thick. The strength of such thin layers can be tested using the ultrasound method with spot head on borehole materials taken from the structure.

Determination of the distribution and dynamics of concrete compressive strength changes throughout the entire cross-section allows, for example, for the establishing of the thickness of the layer which must be removed during repair of the floor. The reconstructed layer must be laid on concrete having the necessary strength (typically $f_c \geq 20$ MPa).

Concrete floors analyzed in the article can be divided into three types based on the curve showing the distribution of strength across their thickness. In this case, the one-parameter evaluation of concrete in the floor, based on the standard destructive compressive strength tests, is not sufficient. The addition of a parameter, which is a strength gradient, was proposed. Floor concretes can be controlled based on the tests of concrete compressive strength class and its gradients across the floor thickness.

Author Contributions: Conceptualization, B.S. and T.K.; methodology, B.S.; validation, B.S. and T.K.; investigation, T.K. and B.S.; resources, T.K. and B.S.; writing—original draft preparation, B.S. and T.K.; writing—review and editing, T.K. and B.S.; visualization, T.K.; supervision, B.S. All authors have read and agreed to the published version of the manuscript.

Funding: This research received no external funding.

Conflicts of Interest: The authors declare no conflict of interest.

Materials **2020**, *13*, 118

References

1. Guiming, W.; Yun, K.; Tao, S.; Zhonghe, S. Effect of water–binder ratio and fly ash on the homogeneity of concrete. *Constr. Build. Mat.* **2013**, *38*, 1129–1134.
2. Caijun, S.; Zemei, W.; Jianfan, X.; Dehui, W.; Zhengyu, H.; Zhi, F. A review on ultra high performance concrete: Part I. Raw materials and mixture design. *Constr. Build. Mat.* **2015**, *30*, 741–751.
3. Rhazi, J.; Hassaim, M.; Ballivy, G.; Hunaidi, O. Effects of concrete non-homogeneity on Rayleigh waves dispersion. *Mag. Concr. Res.* **2002**, *54*, 193–201. [CrossRef]
4. Maj, M.; Ubysz, A.; Hammadeh, H.; Askifi, F. Non-Destructive Testing of Technical Conditions of RC Industrial Tall Chimneys Subjected to High Temperature. *Materials* **2019**, *12*, 2027. [CrossRef] [PubMed]
5. Stawiski, B.; Kania, T. Effect of manufacturing technology on quality of slabs made of regular concrete. In Proceedings of the Conference People, Buildings and Environment, Lednice, Czech Republic, 7–9 November 2012; Hanák, T., Ed.; Institute of Structural Economics and Management, Faculty of Civil Engineering: Brno, Slovakia, 2012; Volume 2, pp. 391–400. Available online: http://www.fce.vutbr.cz/ekr/PBE/PBE2012_Proceedings.pdf (accessed on 1 March 2020).
6. EN 13791. *Evaluation of Concrete Compression Strength in Structures and Prefabricated Concrete Products*; European Committee for Standardization: Brussels, Belgium, 2008.
7. Schabowicz, K. Modern acoustic techniques for testing concrete structures accessible from one side only. *Arch. Civ. Mech. Eng.* **2015**, *15*, 1149–1159. [CrossRef]
8. Schabowicz, K. Non-Destructive Testing of Materials in Civil Engineering. *Materials* **2019**, *12*, 3237. [CrossRef]
9. Szymanowski, J.; Sadowski, Ł. The influence of the addition of tetragonal crystalline titanium oxide nanoparticles on the adhesive and functional properties of layered cementitious composites. *Compos. Struct.* **2019**, *233*, 111636. [CrossRef]
10. EN 206+A1:2016-12. *Concrete—Requirements, Properties, Production and Conformity*; European Committee for Standardization: Brussels, Belgium, 2016.
11. Tanigawa, Y.; Baba, K.; Mori, H. Estimation of concrete strength by combined non-destructive testing method. *Mater. Sci.* **1984**, *82*, 57–76.
12. Giaccio, G.; Giovambattista, A. Bleeding: Evaluation of its effects on concrete behavior. *Mater. Constr.* **1986**, *19*, 265–271. [CrossRef]
13. Khatib, J.M.; Mangat, P.S. Porosity of cement paste cured at 45 °C as a function of location relative to casting position. *Cem. Concr. Comp.* **2003**, *25*, 97–108. [CrossRef]
14. Loh, C.K.; Tan, T.S.; Yong, T.S.; Wee, T.H. An experimental study on bleeding and channelling of cement paste and mortar. *Adv. Cem. Res.* **1998**, *10*, 1–16. [CrossRef]
15. Larish, M.D. Fundamental Mechanisms of Concrete Bleeding in Bored Piles. In Proceedings of the Concrete 2019 Conference, Sydney, Australia, 8–11 September 2019.
16. Kog, Y.C. Integrity Problem of Large-Diameter Bored Piles. *J. Geotech. Geoenviron. Eng.* **2009**, *135*, 237–245. [CrossRef]
17. Vanhove, Y.; Khayat, K.H. Forced Bleeding Test to Assess Stability of Flowable Concrete. *ACI Mater. J.* **2016**, *113*, 753–758. [CrossRef]
18. Powers, T.C. *The Properties of Fresh Concrete*; John Wiley & Sons, Inc.: New York, NY, USA, 1968.
19. *ACI 302.1R-80 Guide for Concrete Floor and Slab Construction*; American Concrete Institute: Detroit, MI, USA, 2015.
20. Jasinski, R.; Drobiec, Ł.; Mazur, W. Validation of Selected Non-Destructive Methods for Determining the Compressive Strength of Masonry Units Made of Autoclaved Aerated Concrete. *Materials* **2019**, *12*, 389. [CrossRef] [PubMed]
21. Nowak, R.; Orlowicz, R.; Rutkowski, R. Use of TLS (LiDAR) for Building Diagnostics with the Example of a Historic Building in Karlino. *Buildings* **2020**, *10*, 24. [CrossRef]
22. Nowak, T.; Karolak, A.; Sobótka, M.; Wyjadłowski, M. Assessment of the Condition of Wharf Timber Sheet Wall Material by Means of Selected Non-Destructive Methods. *Materials* **2019**, *12*, 1532. [CrossRef]
23. Schabowicz, K.; Gorzelańczyk, T.; Szymków, M. Identification of the Degree of Degradation of Fibre-Cement Boards Exposed to Fire by Means of the Acoustic Emission Method and Artificial Neural Networks. *Materials* **2019**, *12*, 656. [CrossRef]

24. Stawiski, B.; Kania, T. Determination of the influence of cylindrical samples dimensions on the evaluation of concrete and wall mortar strength using ultrasound method. *Procedia Eng.* **2013**, *57*, 1078.

25. Stawiski, B.; Kania, T. Examining the distribution of strength across the thickness of reinforced concrete elements subject to sulphate corrosion using the ultrasonic method. *Materials* **2019**, *12*, 2519. [CrossRef]

26. Bogas, J.A.; Gomes, M.G.; Gomes, A. Compressive strength evaluation of structural lightweight concrete by non-destructive ultrasonic pulse velocity method. *Ultrasonics* **2013**, *53*, 962–972. [CrossRef]

27. Brunarski, L. Estimation of concrete strength in construction. *Build. Res. Inst. Quat.* **1998**, *2–3*, 28–45.

28. Anugonda, P.; Wiehn, J.S.; Turner, J.A. Diffusion of ultrasound in concrete. *Ultrasonics* **2001**, *39*, 429–435. [CrossRef]

29. Sansalone, M.; Streett, W.B. *Impact-Echo Nondestructive Evaluation of Concrete and Masonry*; Bullbrier Press: Ithaca, NY, USA, 1997.

30. Breysse, D. Nondestructive evaluation of concrete strength: An historical review and a new perspective by combining NDT methods. *Constr. Build. Mater.* **2012**, *33*, 139–163. [CrossRef]

31. Facaoaru, I. Contribution à i'étude de la relation entre la résistance du béton à la compression et de la vitesse de propagation longitudinale des ultrasons. *RILEM* **1961**, *22*, 125–154.

32. Leshchinsky, A. Non-destructive methods instead of specimens and cores, quality control of concrete structures. In Proceedings of the International Symposium, RILEM, Ghent, Belgium, 12–14 June 1991; E&FN SPON: London, UK, 1991; pp. 377–386.

33. Bungey, J.H. The validity of ultrasonic pulse velocity testing of in-place concrete for strength. *NDT Int.* **1980**, *13*, 296–300. [CrossRef]

34. Szpetulski, J. Testing of compressive strength of concrete in construction. *Constr. Rev.* **2016**, *3*, 21–24. (In Polish)

35. Gudra, T.; Stawiski, B. Non-destructive strength characterization of concrete using surface waves. *NDT E Int.* **2000**, *33*, 1–6. [CrossRef]

36. Komlos, K.; Popovics, S.; Nurnbergerova, T.; Babal, B.; Popovics, J.S. Ultrasonic Pulse Velocity Test of Concrete Properties as Specified in Various Standards. *Cem. Concr. Compos.* **1996**, *18*, 357–364. [CrossRef]

37. PN-B-06261. *Non-Destructive Testing of Structures; Ultrasound Method of Testing Compressive Strength of Concrete*; Polish Committee for Standardization: Warsaw, Poland, 1974.

38. EN 12504-4. *Testing Concrete-Part 4: Determination of Ultrasonic Pulse Velocity*; European Committee for Standardization: Brussels, Belgium, 2004.

39. Stawiski, B.; Stawiski, M. Testing characteristics of directional ultrasound probes with geometrically defined waveguides. *NDT Bull.* **2000**, *10*, 17–19.

40. Stawiski, B. *Ultrasonic Testing of Concrete and Mortar Using Point Probes*; Monograph; Wrocław University of Technology Printing House: Wrocław, Poland, 2009; ISSN 0324-9875.

Article

Material Analysis of Steel Fibre Reinforced High-Strength Concrete in Terms of Flexural Behaviour. Experimental and Numerical Investigation

Czesław Bywalski *, Maciej Kaźmierowski, Mieczysław Kamiński and Michał Drzazga

Faculty of Civil Engineering, Wroclaw University of Science and Technology, Wybrzeże Wyspiańskiego 27, 50-370 Wrocław, Poland; maciej.kazmierowski@pwr.edu.pl (M.K.); mieczyslaw.kaminski@pwr.edu.pl (M.K.); michal.drzazga@pwr.edu.pl (M.D.)
* Correspondence: czeslaw.bywalski@pwr.edu.pl; Tel.: +48-71-320-3397

Received: 13 February 2020; Accepted: 30 March 2020; Published: 1 April 2020

Abstract: The paper presents the results of tests for flexural tensile strength ($f_{ct,fl}$) and fracture energy (G_f) in a three-point bending test of prismatic beams with notches, which were made of steel fibre reinforced high-strength concrete (SFRHSC). The registration of the conventional force–displacement (F–δ) relationship and unconventional force-crack tip opening displacement (CTOD) relationship was made. On the basis of the obtained test results, estimations of parameters $f_{ct,fl}$ and G_f in the function of fibre reinforcement ratio were carried out. The obtained results were applied to building and validating a numerical model with the use of the finite element method (FEM). A non-linear concrete damaged plasticity model CDP was used for the description of the concrete. The obtained FEM results were compared with the experimental ones that were based on the assumed criteria. The usefulness of the flexural tensile strength and fracture energy parameters for defining the linear form of weakening of the SFRHSC material under tension, was confirmed. Own equations for estimating the flexural tensile strength and fracture energy of SFRHSC, as well as for approximating deflections (δ) of SFRHSC beams as the function of crack tip opening displacement (CTOD) instead of crack mouth opening displacement (CMOD), were proposed.

Keywords: high-strength concrete; steel fibres; flexural tensile strength; fracture energy; numerical analysis

---

1. Introduction

Steel fibre reinforced high-strength concrete (SFRHSC), in comparison with plain high-strength concrete, is marked by a more quasi-plastic character and increased resistance to cracking on bending [1,2]. The following should be considered to be the most important parameters of dispersed reinforcement structure, which can inhibit cracking in the cement matrix: volume of fibres in the composite (V_f) [3], slenderness of fibres (λ) [4,5], fibre material characteristics, spatial distribution of fibres in concrete [6], as well as adhesion of fibres to cement matrix, resulting from mechanical anchorage, adhesion, and friction [7]. The addition of steel fibres to concrete might improve some of its mechanical parameters, including compressive strength (f_c) [8,9] and the modulus of elasticity (E_c) [3,8,10]. The use of steel fibres in concrete also significantly increases its flexural tensile strength ($f_{ct,fl}$) [3,4,9,11]. This improvement is insignificant in the range of proportional strains and clear after they are exceeded, both before reaching the breaking load value and after exceeding it. Knowing the mechanical response of the structure under loading conditions is crucial for the SFRHSC beams in bending.

The behaviour of steel fibre reinforced concrete (SFRC) or SFRHSC beam in three-point bending test can be presented while using Tlemat's proposal [7], which assumes a four-phase scope of work.

Figure 1 shows the cross section of SFRC or SFRHSC beam before failure. Particular phases, depending on the load level, differ in the idealized distribution of stress blocks that result from the development of cracks in the fracture process zone (FPZ). The individual steps can be described, as follows: Phase 1—linear-elastic relationship between force (F) and deflection, no cracking in the zone of tensile stresses (zone Z1), slight influence of the slenderness of fibres (λ) on the value of the transferred load (F), maximum value of tensile stress being dependent on the concrete and volume of fibres in the composite (V_f) [12]; Phase 2—the initiation of the micro-cracking process (0.1–0.2 mm), forces that result from the load are taken over by dispersed reinforcement in the zone Z2, the beginning of stress reduction in the FPZ, slight development of the fibre-pulling out process after crossing the limit load, change of the linear-elastic force–deflection relationship into the elastic-plastic relationship; Phase 3—further propagation of the cracking, development of fibre-pulling out process (zone Z3), continuation of force transmission by the fibres (zone Z2), raising the position of the neutral axis (NA); and, Phase 4—negligible transmission of forces by fibres (most of the fibres pull out from the cement matrix, zone Z4), no tensile stresses in the zone Z4, further possibility of load transfer provided by fibres in the zones Z2 and Z3 until the cracking reaches the neutral axis (NA).

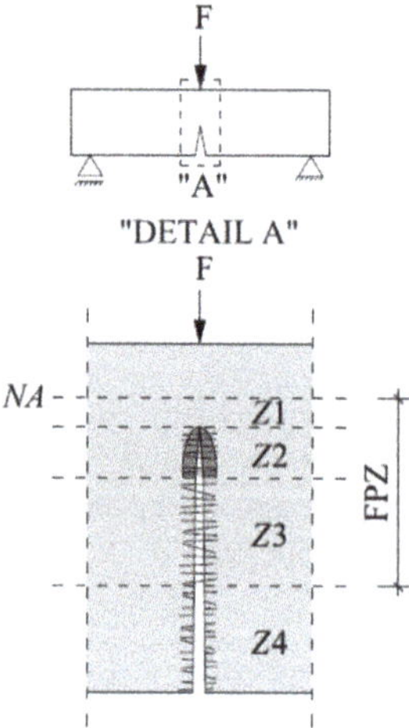

Figure 1. Cross-section of the three-point bending steel fibre reinforced concrete (SFRC) or steel fibre reinforced high-strength concrete (SFRHSC) beam before failure.

Despite numerous studies, the method of testing and assessment of SFRHSC properties after cracking is still under discussion [11,13]. Technical committees propose various methods for determining the fracture energy and residual stresses. The lack of full standardization hinders the design and full use of this type of material in the construction industry. Therefore, there is a well-founded need for research in this direction. Differentiation applies to: loading method (three- or four-point bending test), shape and geometry of the researched element (beams, circular plates, cubes, etc.), or measured values (deflections, CMOD, CTOD).

The fracture resistance of SFRHSC is determined on the basis of standardized experimental tests [14]. According to [15], force–deflection (F–δ) or force-crack mouth opening displacement (F–CMOD) relationship in a three-point bending test are recorded. F–CMOD relationship analysis is widely used. It is performed by means of a clip gauge extensometer or a linear variable differential transformer (LVDT) [15]. The Italian National Research Council [16] recommends assessing the properties after cracking of steel fibre reinforced concrete based on the F–CTOD relationship. The analysis of the F–CTOD relationship is more experimentally difficult to perform as compared to F–CMOD. However, registering the force-width relationship of a crack opening is valuable from the point of view of conducting numerical analyses and cracking energy. Namely, it enables the direct comparison of experimental and numerical results and definition of a numerical material model (e.g., determination of the allowable crack width with tension in the concrete damaged plasticity model). In turn, Bencardino [14] showed that an assessment of fracture resistance could be carried out on the basis of force-crack tip opening displacement relationship (F–CTOD). Bencardino recorded F–CTOD and F–CMOD relationships for HSC and SFRHSC beams at the same time. Almost identical shape and size of F–CTOD and for F–CMOD curves could be observed. However, research of Bencardino must be continued to generalize the conclusions. Taking the above into account, the results of tests of F–CTOD relationship are also presented in the paper. In this way, the database of literature related to F–CTOD relationships has been extended. It also allowed for proposing equations for the approximation of SFRHSC beams deflections in CTOD function instead of CMOD [15].

The purpose of the article is to assess the flexural tensile strength ($f_{ct,fl}$) and fracture energy (G_f) of high-performance concrete with the addition of steel fibres (SFRHSC). The above parameters are essential when designing SFRHSC structural elements. The fracture energy of the SFRHSC composites is determined on the basis of standardized experimental tests [15]. The F–δ or F–CMOD relationships for three-point bending beams are recorded. Experimental tests were carried out in accordance with standard [15], and extended by registering the F–CTOD relationships, in order to achieve the above objectives. Additionally, the estimation of $f_{ct,fl}$ and G_f quantities as a function of fibre reinforcement degree was made, and equations were developed, which contribute to the extension of standard records [15]; the standard currently allow an approximation of deflections of steel fibre reinforced concrete beams only as a function of CMOD. The obtained experimental results of the beams were compared with the numerical results (finite element method (FEM)) in quantitative and qualitative terms in order to assess the usefulness of the parameters of flexural tensile strength ($f_{ct,fl}$) and fracture energy (G_f), while also defining the linear form of weakening of the SFRHSC material under tension. The knowledge of the mechanical response of the structure under load is essential from the designer's point of view concerning bending beams.

2. Experimental Research

2.1. Materials and Methods

The scope of research included making three series of composites (A, B, and C), which were differentiated by the content of steel fibres. Series A (control) did not contain any fibres; Series B contained 1.0% steel fibres (78 kg/m³); and, Series C included 1.5% steel fibres (118 kg/m³). The scope of tests and sample specifications are presented in Table 1. The samples used for strength tests were cured and protected according to [17].

Table 1. Scope of tests and specification of samples.

Measured Parameter	Sample Type	Sample Dimensions (mm)	Series			Total
			A	**B**	**C**	
$f_{ct,fl}$, $f_{ct,fl,L}$	beam	150 × 150 × 600	3	3	3	9
f_c	cube	150 × 150 × 150	3	3	3	9
E_c	cylinder	150 × 300	5	5	5	15

The concrete mix was designed for the strength class C55/67 [18]. The following components were used in order to produce the concrete mixture: portland cement CEM-I 52.5N, natural washed aggregate with a fraction of 0–2 mm, broken granite aggregate with fractions 2–8 mm and 8–16 mm, polymer superplasticizer (Sika ViscoCrete 5-600 for high-performance concrete; properties: strong liquefaction (class S4/S5), no chlorides, increased early, and final strength of concrete mix, density 1.07 kg/dm³, PH 4.4 ± 1,0), and fly ash from hard coal (specific surface—3610 cm²/g; penetrability for the sieve with mesh 0.045 mm—38.50% [19]; mineralogical composition: SO_3—0.38%, CaO—2.44%,

SiO$_2$—51.72%, Al$_2$O$_3$—25.15%, and Fe$_2$O$_3$—5.21%). The water-cement ratio of the mixture was 0.3 (w/c = 0.3). Straight steel fibres with a length of 20 mm (l) and a diameter of 0.3 mm (d) were added to the concrete samples, series B and C. The tensile strength of the fibres was 2000 MPa, while the Young's modulus was 200 GPa. In order to obtain uniform consistency of concrete mix for all series, while maintaining a constant value of the w/c coefficient, test samples were prepared on the basis of which the optimal superplasticizer content was determined in individual composite series (A, B, and C). The consistency test was carried out using the drop cone method [20]. The indicator of the degree of mix liquidity ranged from 155 to 165 mm (consistency class S3/S4 [20]). Table 2 presents the composition of the concrete mix for individual series. All the components of the mixture were dosed by weight with an accuracy of 1% (without steel fibres and superplasticizer). The volume of each series of concrete mix was 453.2 dm^3 (in accordance with the assumed programme of research with a wider scope). The order of dispensing the ingredients was as follows: the aggregate was mixed first, and then cement and fly ash were added. After thorough mixing, water was added, followed by a superplasticizer. Steel fibres were the last ingredient added to the mix. The total mixing time was 490 s. The concrete preparation equipment included: mixers to make trial batches, electronic scales for dosing the components of the concrete mix with an accuracy of 5 g and 0.1 g, a plate mixer (concrete plant), containers for transporting concrete, and a vibration table for compacting the concrete mix. The steel fibres and superplasticizer were dosed by weight with an accuracy of 0.1 g. Other components of the mixture were dosed with an accuracy of 5 g.

Table 2. Concrete mixture composition.

Material		Series		
		A	**B**	**C**
Steel fibres	kg/m^3	0	78	118
	% (V$_f$) *	0	1.0	1.5
Superplasticizer	kg/m^3	8.25	9.65	10.80
Cement-CEM I 52.5R	kg/m^3		550	
Sand 0–2 mm	kg/m^3		600	
Granite aggregate 2–8 mm	kg/m^3		490	
Granite aggregate 8–16 mm	kg/m^3		590	
Fly ash	kg/m^3		30	
Coefficient w/c	-		0.30	

* $V_f = W_f/\rho_f$ (W_f—fibre content in mass units in 1 m^3; ρ_f—fibre material density in kg/m^3).

Experimental studies were carried out to determine the flexural tensile strength of HSC and SFRHSC, as well as their proportionality limit ($f_{ct,fl,L}$) and the behaviour of beams in conditions of the exceeded critical limit load. This is why parallel relationships: force–deflection (F–δ) and force-crack tip opening displacement (F–CTOD) were observed. Characteristics of the tested composites, compressive strength (f_c), and modulus of elasticity (E_c) tests were carried out, owing to the need to obtain more accurate material. The obtained results were used for numerical analyses of HSC and SFRHSC beams.

The testing of the flexural tensile strength was carried out in accordance with [15]. During the study, parallel relationships: force–deflection (F–δ) and force-crack tip opening displacement (F–CTOD) were recorded. The adopted research program aimed, among others, at comparing these relationships.

Flexural tensile strength ($f_{ct,fl}$) and the corresponding limit of proportionality ($f_{ct,fl,L}$) were determined according to dependence (1), taking, for calculations, the limit load (F_{max}) and the load corresponding to deflection ($F_{0.05}$) respectively [15]. Figure 2 shows the test setup project. Figure 3a shows the beam prepared for testing, and Figure 3b depicts the detailed arrangement of the sensors for the measurement of top notch opening (CTOD) and deflection (δ).

$$\sigma_{ct,fl} = \frac{3F_i l}{2bh_{sp}^2},$$
(1)

where:

l—beam span;
b—beam width; and,
h_{sp}—beam height less notch height.

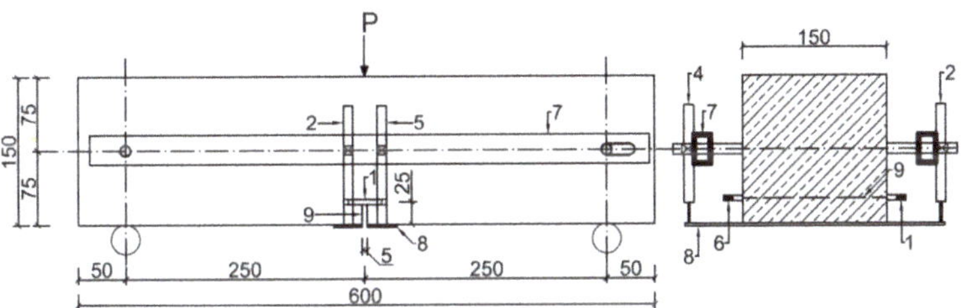

Figure 2. Test setup project: 1 and 6—sensors for measuring notch opening width (CTOD), 2–5—sensors for measuring deflection measurement (δ), 7—aluminum profile, 8—aluminum flat bar, and 9—notch.

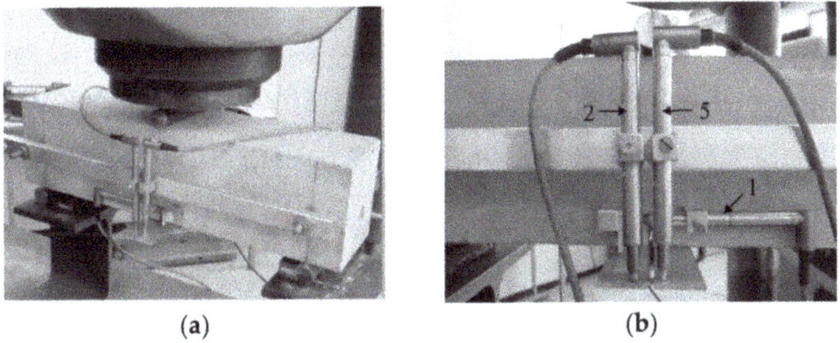

(a)   **(b)**

Figure 3. Flexural tensile strength test ($f_{ct,fl}$): (**a**) C2 beam; (**b**) location of sensors for measuring deflections (No. 2 and 5) and notch opening (No. 1) in C1 beam.

Measurement data CTOD and δ were recorded by means of extensometers with a range of 10 mm and accuracy of 0.001 mm, which were placed symmetrically along the longitudinal axis of the beam. The testing of beams was carried out at the ZD100 strength testing station (VEB Werkstoffprufmaschinen, Leipzig, Germany) with the computerized recording of synchronized results (time, F and δ). The frequency of records was 10 Hz.

Crack tip opening displacement (CTOD) means the width of a crack at the top of the technological notch. Figures 2 and 3 indicate the location of the crack opening measurement (sensor No. 1). The height of the technological notch is 25 ± 1 mm, while the width is 4.98 mm (<5 mm). These values were adopted in accordance with the standard for testing fibre reinforced concrete [15]. The cuts were made while using a power saw with a guide, after prior marking. The dimensions of the technological notch were controlled with an electronic caliper.

The compressive strength tests (f_c) were carried out in accordance with the standard [21] and the modulus of elasticity was measured according [22].

All of the tests were carried out after 28 days from preparing the composites.

2.2. Experimental Results and Discussion

2.2.1. Results of Tests for Mechanical Features

The results of flexural tensile strength tests ($f_{ct,fl}$), limit of proportionality ($f_{ct,fl,L}$), compressive strength (f_c) and the modulus of elasticity (E_c) with the arithmetic mean of results $\bar{x}$, and standard deviation for each series (s) are presented in Table 3.

Table 3. Test results.

Series	$f_{ct,fl}$ (MPa)			$f_{ct,fl,L}$ (MPa)			f_c (MPa)			E_c (GPa)		
	Result	$\bar{x}$	s	Result	$\bar{x}$	s	Result	$\bar{x}$	s	Result	$\bar{x}$	s
A	5.32 5.69 5.91	5.64	0.30	- - -	-	-	76.41 76.97 77.34	76.91	0.47	34.50 36.90 33.47 36.17 37.35	35.68	1.64
B	8.86 11.34 12.79	11.00	1.99	5.34 6.42 6.65	6.14	0.70	85.19 87.16 88.17	86.84	1.52	40.35 41.07 39.80 42.99 41.20	41.08	1.21
C	11.52 12.80 14.96	13.09	1.74	6.42 7.27 7.33	7.01	0.51	95.19 95.44 96.04	95.56	0.44	37.77 41.61 43.73 42.56 43.12	41.76	2.36

The average density of composites (ρ) in each series was: A—2389 kg/m^3, B—2493 kg/m^3, and C—2522 kg/m^3.

When analysing the results that are contained in Table 3, it can be seen that, with an increasing volume of steel fibres (V_f) in the composite, the flexural tensile strength ($f_{ct,fl}$) also significantly increases. For B series beams, the average increase in the value $f_{ct,fl}$ when compared to that obtained for series A beams (control series) was 95.0%, and for C series beams 132.1%. It should be noted that the results of flexural tensile strength ($f_{ct,fl}$) of SFRHSC beams can be burdened with a large spread of results (V = 20–30%) [23,24]. This might be due to, among other factors, the effect of the scale (different surfaces of the beams' breakthrough) or the efficiency of anchoring the fibres in the cement matrix. The average value of stresses defining the conventional limit of proportionality of bending beams was for series B—6.14 MPa and for series C—7.01 MPa, while the coefficient of variation was 11.4% and 7.3%, respectively.

Figure 4 presents the results of flexural tensile strength ($f_{ct,fl}$), depending on the fibre reinforcement ratio ($V_f l/d$) and a linear regression Equation (2) with the determination coefficient of R^2 = 86%, which is a measure of the quality of model fit. The root from the variance of regression equation estimators was 1.15 for the directional coefficient and 0.80 for the free expression, respectively. It should be noted that a large confidence interval of the directional regression coefficient was obtained (2.72), owing to the small size of the sample (nine results). On the basis of the statistical test carried out (Student's t-distribution), for the assumed level of significance α = 5%, the obtained regression coefficient of the model is statistically significant ($p < \alpha$, where p is the probability of accepting the null hypothesis of the parameter).

$$f_{ct,fl} = f'_{ct,fl} + 7.54\left(\frac{V_f l}{d}\right), \tag{2}$$

where: $f'_{ct,fl}$—the flexural tensile strength of HSC.

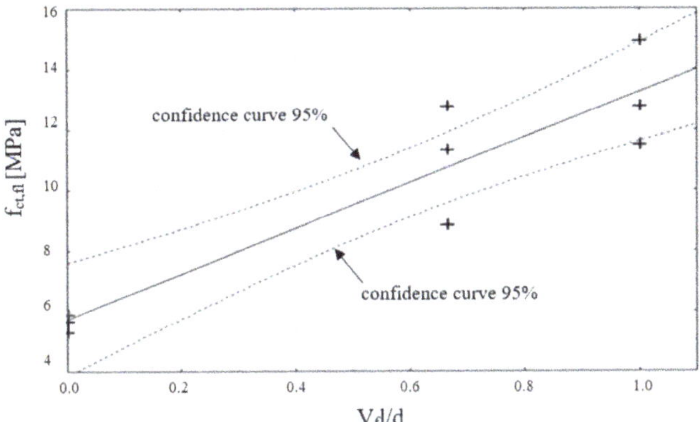

Figure 4. Flexural tensile strength results ($f_{ct,fl}$), depending on the fibre reinforcement ratio ($V_f l/d$), the regression curve, and 95% confidence curves.

As part of the analysis of the model's fit to the empirical data, the authors also considered various parametric models with two explanatory variables in the form of compressive strength (f_c) and fibre reinforcement ($V_f l/d$), which are available in the literature on the subject [3,4,25,26]. Consequently, a non-linear regression Equation (3) with a coefficient $R^2 = 86.4\%$ was obtained. The results of flexural tensile strength ($f_{ct,fl}$), depending on the compressive strength (f_c) and fibre reinforcement ratio ($V_f l/d$), as well as a graphical presentation of the Equation (3) are shown in Figure 5. It should be noted that the regression equation applies to the range of variables under consideration (compressive strength 76–96 MPa; slenderness of fibres 66.7; and, fibre content by volume 0; 1.0; 1.5%).

$$f_{ct,fl} = 0.66\sqrt{f_c} + \frac{6.88 V_f l}{d} \tag{3}$$

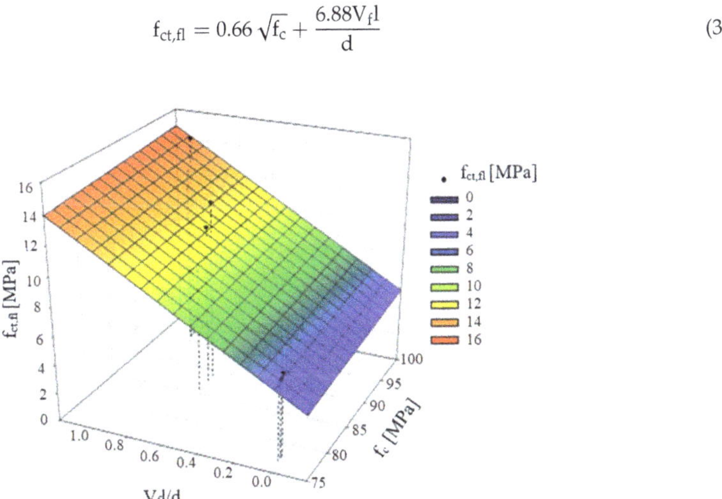

Figure 5. Flexural tensile strength results ($f_{ct,fl}$) depending on compression strength (f_c) and fibre reinforcement ratio ($V_f l/d$).

The roots of the compressive strength estimate and the fibre reinforcement ratio of model (3) were 0.09 and 1.21, respectively. Statistically significant structural parameters of the model ($p < \alpha$) were obtained on the basis of the Student's *t*-test, for the assumed level of significance $\alpha = 5\%$.

Volumetric addition of steel fibres (V_f) in series B (1.0%) and C (1.5%) resulted in an increase in the compression strength (f_c) in relation to series A (control) by 12.9% and 24.3%, respectively.

The obtained increase in compressive strength (f_c) of SFRHSC concretes is confirmed in the scientific literature on the subject [3,4,27], which reports, for the addition of fibres (V_f) in the range of 0–1.5%, an increase in compressive strength in the case of SFRHSC from 10% to 20%, as compared with HSC.

The presence of steel fibres in the composite also increased the value of the modulus of elasticity (E_c). For the B and C series, the increase in the modulus of elasticity in relation to the reference series (series A) was 15.0% and 17.0%, respectively. It was observed that the increase in fibre content from 1.0% to 1.5% (V_f) in the composite (C series results) resulted in only a slight increase in the modulus of elasticity (by 2.0%). A similar increase in the modulus of elasticity (E_c) of SFRHSC in relation to HSC can be found in [8,28,29]. The increase ranges from 7.0% to 27.0%, for the addition of steel fibres in the range of 0–1.5% in the composite. It should be noted that, along with the increase in the content of steel fibres in the composite (V_f), the compressive strength (f_c) [10,30] and modulus of elasticity (E_c) [31,32] are not always significantly increased.

2.2.2. F–δ and F–CTOD Relationship

Figure 6 shows the relationships of force–deflection (F–δ), in particular SFRHSC beams, and in Figure 7 the same relationship can be seen for HSC beams. The deflection values (δ) were the arithmetic mean of the sensor readings (Nos. 2–5, according to Figure 2). The force–deflection relationships that were obtained on the basis of the results of the arithmetic mean of beams in particular series (A, B and C) are shown in Figures 7 and 8 (Figure 7—due to the smaller range of HSC beams deformation in relation to SFRHSC beams). The second type of relationship, namely force-crack tip opening displacement (F–CTOD), which was measured during the HSC and SFRHSC beam tests, is shown in Figures 9 and 10, respectively. The values of individual beams CTOD placed on the graph are the mean of the sensors nos. 1 and 6 (see Figure 2). The F–CTOD relationships that were obtained on the basis of the arithmetic mean of the results in particular series are shown in Figure 11 (series A, B, and C) and Figure 10 (series A only).

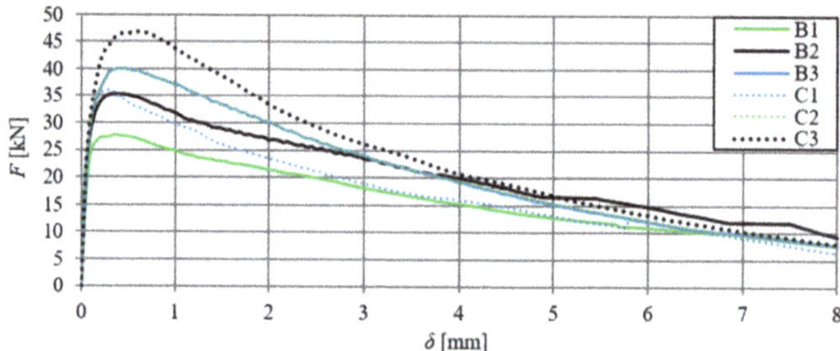

Figure 6. Force–deflection relationships for SFRHSC beams (B and C series, curve C2 coincides with curve B3).

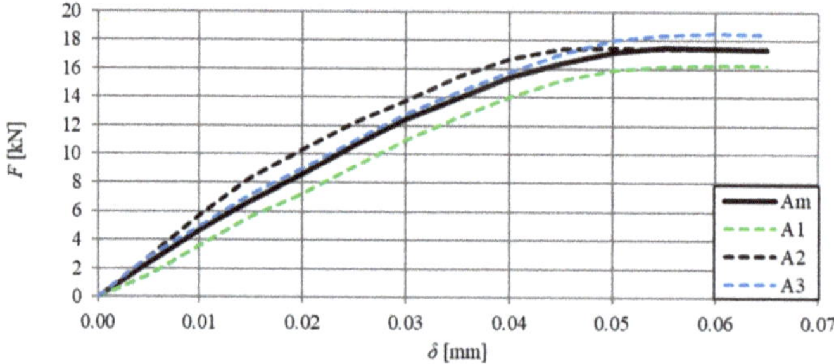

Figure 7. Force–deflection relationships for HSC beams (series A) and curve F–δ obtained on the basis of the arithmetic mean of results (continuous line).

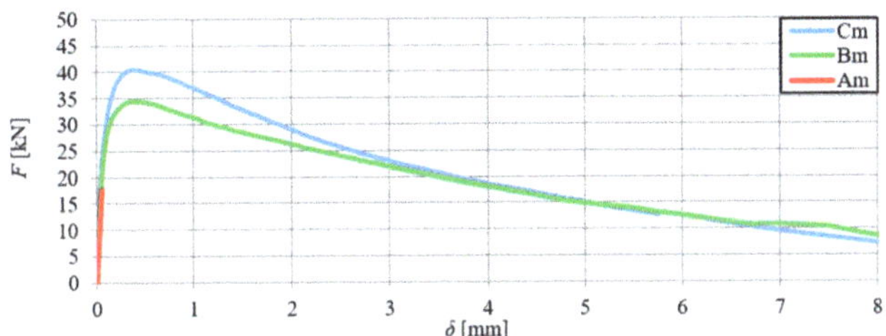

Figure 8. Force–deflection relationships obtained on the basis of the arithmetic mean of the results for beams from series A, B, and C.

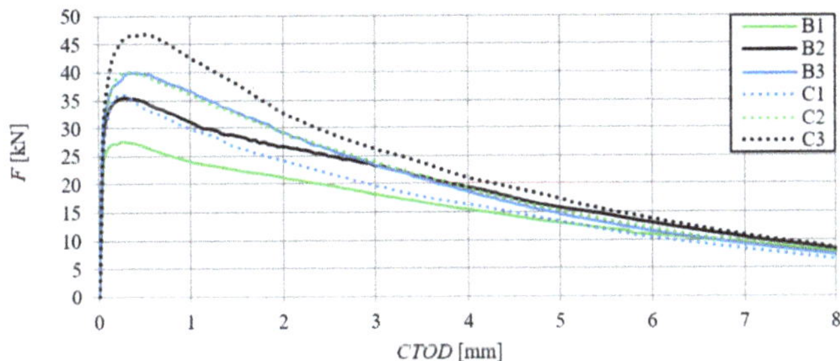

Figure 9. Force-crack tip opening displacement relationships of SFRHSC beams (B and C series).

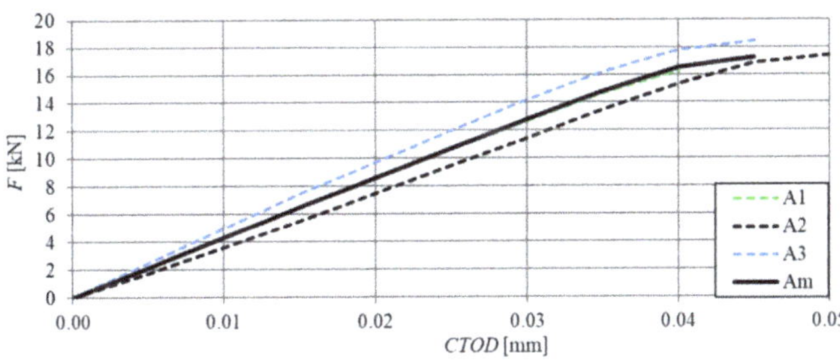

Figure 10. Force-crack tip opening displacement relationships of HSC beams (series A) and the F–CTOD curve obtained on the basis of the arithmetic mean of the results (continuous line, curve A1 coincides with curve Am).

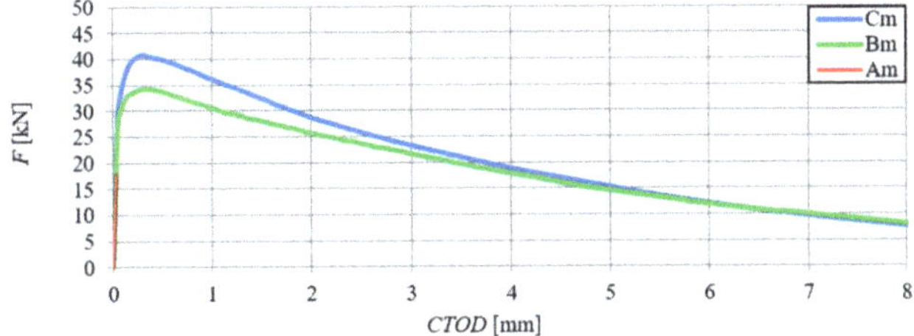

Figure 11. Force-crack tip opening displacement relationships obtained on the basis of the arithmetic mean of the results for beams in series A, B and C.

When analysing Figures 6 and 9, one can observe minimal differences in the shape and size of curves describing F–δ and F–CTOD relationships (curves overlap). Figure 12 presents deflections of the B series beams, depending on the crack tip opening displacement (CTOD) and the δ–CTOD relationship obtained on the basis of the arithmetic mean of the series results. Analogous results are presented in Figure 13 for C-series beams. The δ–CTOD relationships that were obtained on the basis of the arithmetic mean of series B and C results can be approximated by linear Equations (4) and (5) (R^2 = 99%), respectively. These equations could be a contribution to the extension of the standard records [15], which presently allow for an approximation of deflections (δ) of beams that were modified with steel fibres only as a function of crack mouth opening displacement (CMOD).

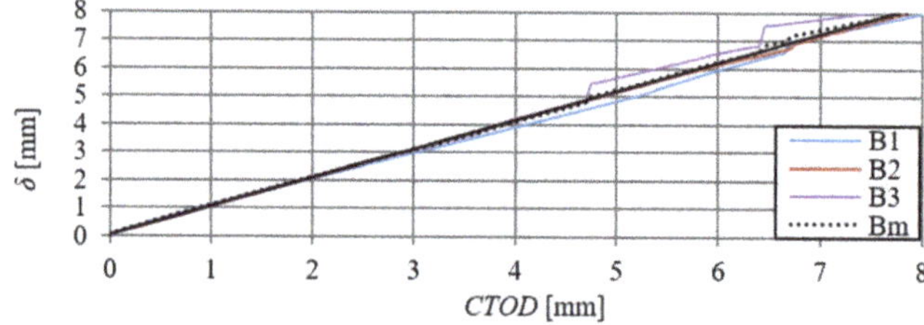

Figure 12. Deflections–crack tip opening displacement relationship of B-series beams and δ–CTOD curve obtained on the basis of the arithmetic mean of results (Bm).

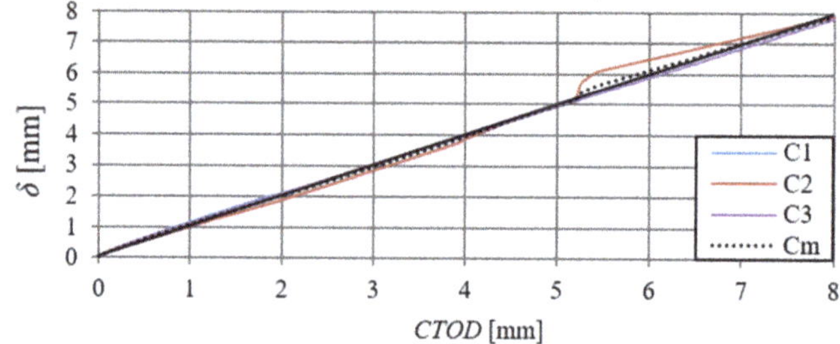

Figure 13. Deflection–crack tip opening displacement relationships of C-series beams and the relationship δ–CTOD obtained on the basis of the arithmetic mean of results (Cm).

$$\delta = 1.03\text{CTOD} + a, \tag{4}$$

where: a—constant term; a = 0.02, if CTOD > 0; a = 0, if CTOD = 0.

$$\delta = 0.99\text{CTOD} + b, \tag{5}$$

where: b—constant term; b = 0.05, if CTOD > 0; b = 0, if CTOD = 0.

It should be noted that the regression Equations (4) and (5) were determined for specific laboratory and material conditions of the composite. In addition, they relate to the considered range of variables, and they should be verified in the future on a larger number of research results.

Figure 14 shows the damaged HSC and SFRHSC beams (owing to a similar failure mode of beams, only C series is presented). The presence of fibres affects the destruction characteristics of SFRHSC beams, which do not suddenly divide in contrast to HSC ones. After they are "destroyed", SFRHSC beams are able to carry a set load (usually smaller than the breaking load) until the broken parts are "separated".

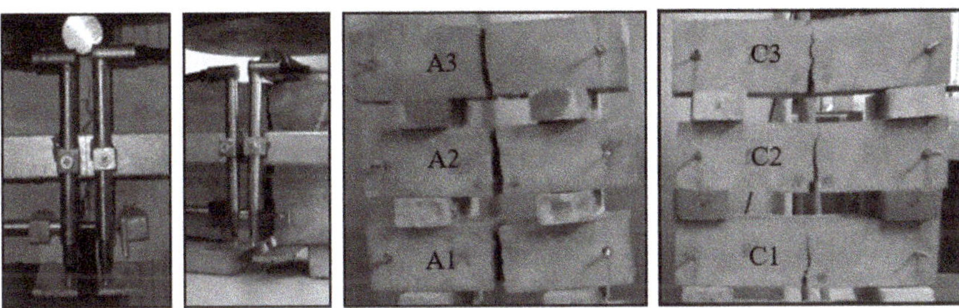

Figure 14. Damaged beams; from left: beam C2; beam A1 (no fibres); series A; series C.

2.2.3. Fracture energy of HSC and SFRHSC

The fracture energy, which is marked by the range of post-critical stress–elongation of concrete dependence (σ–w), can be treated according to Hilleborg's proposal and the RILEM technical committee [33,34], as the material constant of the composite (other methods for calculating fracture energy are also applicable [35]). It determines the amount of energy that is required to produce a crack with a unit area that can be calculated while using the Equation (6).

$$G_f = \frac{1}{bh_{sp}}\left[\int_0^{\delta_u} F(\delta)d\delta + m\left(1 - \alpha^2\right)g\delta_u\right],\tag{6}$$

where: m—beam mass; g—gravitational acceleration; δ_u—maximum deflection recorded during the test; $\alpha = 1 - L/2l$; L—beam length; b, h_{sp}, l—explanations identical to Equation (1).

Table 4 shows the results of the fracture energy for particular beams, along with the arithmetic mean ($\bar{x}$) and the standard deviation (s) for each series. Figure 15 shows the influence of fibre reinforcement ratio ($V_f l/d$) upon the value of beam fracture energy (G_f) and the polynomial regression Equation (7) with the coefficient $R^2 = 96\%$ ($p < \alpha$). The roots from the estimators' variance of the polynomial regression model (7) were, respectively: 3.15; 3.21; and, 0.57. On the basis of the Student's *t*-test, for the assumed level of significance $\alpha = 5\%$, the obtained regression coefficients for the model are statistically significant ($p < \alpha$).

$$G_f = -10.72\left(\frac{V_f l}{d}\right)^2 + 19.68\left(\frac{V_f l}{d}\right) + 0.04.\tag{7}$$

Table 4. HSC and SFRHSC beams fracture energy results along with the arithmetic mean (x) and standard deviation (s) for each series (kNm/m^2).

Beam	Result	$\bar{x}$	s
A1	0.036		
A2	0.043	0.0412	0.004
A3	0.044		
B1	7.051		
B2	9.033	8.394	1.16
B3	9.098		
C1	7.688		
C2	9.103	8.996	1.26
C3	10.195		

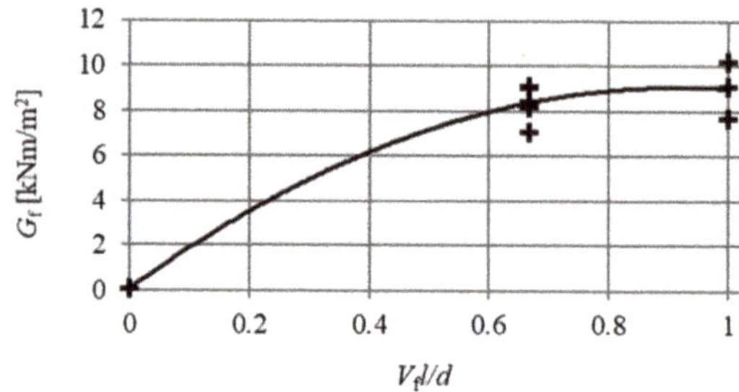

Figure 15. HSC and SFRHSC beams' fracture energy results (G_f) depending on fibre reinforcement ratio ($V_f l/d$) and polynomial regression Equation (7).

It should be noted that regression Equation (7) was determined for specific material conditions of the composite, and it applies to the considered range of variables (V_f, l, d and type of fibre). It should be remembered that the length, diameter, type, and volume of fibre strongly influence the fracture energy.

3. Numerical Analysis of Beams

3.1. Assumptions for the Numerical Model and Geometrical Model

Numerical analyses of SFRHSC beams were performed to assess the usefulness of the flexural tensile strength and fracture energy parameters for defining the linear form of weakening of the SFRHSC material under tension, using a concrete damaged plasticity (CDP) model. To this end, quantitative and qualitative analyses were performed. Numerical calculations of HSC and SFRHSC beams were made in the Abaqus/Standard [36] program while using the CDP model.

The following assumptions were made in the numerical calculations: the influence of steel fibres on the behaviour of the composite was taken into account on the basis of calculated composite fracture energy (G_f); attention was focused on physical nonlinearity describing HSC and SFRHSC; nonlinear effects were included using the Newton–Raphson increment-iterative method; and, the calculations were performed in the plane state of strain.

The discretization of the beam in the numerical model was performed with the use of 1539 finite elements. Plane stress elements were used, four-node with reduced integration (CPS4R), located above the notch (in the place of stress concentration–crack propagation), with a mesh size of 5 mm (dimension determined by the notch width), and three-node (CPS3), located on the remaining area of variable dimensions (5–25 mm). The mesh of finite elements and boundary conditions for the beam, which were used in numerical calculations, are shown in Figure 16. Load of the beam was carried out in a kinematic way through the task of displacement (Figure 16, point A), changeable in time. The displacement point in the model was located in accordance with the place of force effects during experimental beam testing. The displacement value was $\delta = 8$ mm for SFRHSC beams and $\delta = 0.1$ mm for HSC beams. The size of the load increment in the range and $0.001, 1E - 10$ the maximum number of load steps of 5000 were assumed.

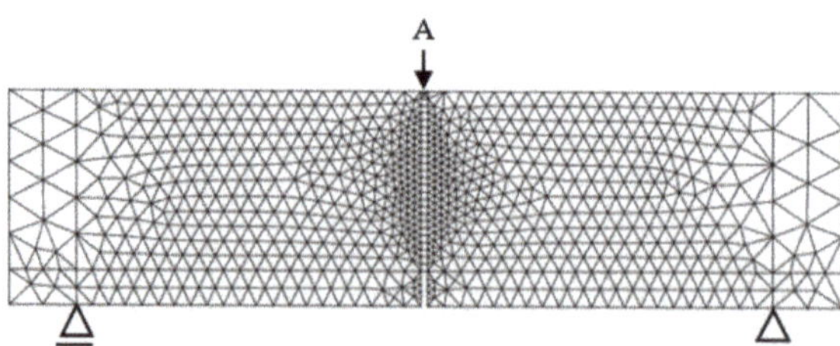

Figure 16. Finite-element mesh and boundary conditions for the beam used in numerical calculations.

The use in numerical simulations, in the non-linear σ_t–ε_t relationship (weakening of the material), leads to highly sensitive results for the discretization of finite elements [36,37] (ambiguous response at the level of structure depends on the deformation location zone); hence, alternatively, the form of material weakening can be described with the use of fracture energy (G_f). In this case, the brittle damage is described by the dependence $\overline{\sigma}_t - u_t^{-pl}$ ($\overline{\sigma}_t$—effective stress), instead of $\overline{\sigma}_t - \varepsilon_t^{-pl}$, which is initiated when the crack (associated with the deformation location) begins to propagate, after the material reaches stress that is equal to the axial tensile strength (σ_{t0}). With the opening of the crack (the displacement increases), the tensile stress decreases up to the zero value, which corresponds to the maximum displacement ($u_{t,max}^{-pl}$). The dissipation of energy in the creation of cracks with a unitary field is expressed by the formula:

$$G_f = \int_0^{u_{t,max}^{-pl}} \overline{\sigma}_t \, du_{t,max}^{-pl}. \tag{8}$$

The above dependence shows that the energy that is released during cracking (G_f) is equal to the value of the area under the curve $\overline{\sigma}_t - u_t^{-pl}$, and the curves themselves may assume different forms of weakening, e.g., linear or exponential.

The force-displacement at the construction level does not depend on the finite element discretization given the above relationships [37]. Due to its simplicity, this modelling method is often used in commercial FEM programs [38]. The shortcomings of the method involve limiting the location zone to one row of elements and reducing its width along with the compaction of the mesh, instead of remaining constant. It should be noted that the above limitations do not affect the mathematical model of non-linear deformations related to cracking or damage [36,37]. An improved continuous medium model can be used in order to eliminate irregularities in the mathematical model for weakened material (getting rid of the pathological dependence of the discrete solution on the mesh) [39,40].

3.2. Material Model

For the numerical analysis of HSC and SFRHSC beams, the default values of the model parameters were adopted, defining its operation in a complex stress state (β, ϵ, f, K_c, μ) [36], and they are shown in Table 5. It has been assumed that according to the standard [18], the Poisson's ratio of uncracked concrete equals 0.2.

Table 5. The default parameters of the concrete damaged plasticity (CDP) model in the complex stress state.

Name of Parameter	β	ϵ	f	K_c	μ
Value	36°	0.1	1.16	0.667	0

In Table 5: β—the pitch of the hyperbolic asymptote of the Drucker–Prager surface to the hydrostatic axis, as measured in the meridional plane; ϵ—eccentricity of the plastic potential, being a small positive value characterizing the speed at which the hyperbole of the plastic potential is approaching its asymptote; f—a number defining the quotient of the limit compressive stress in the biaxial state to the limit compressive stress in the uniaxial state; K_c—a parameter defining the shape of the surface of the plastic potential on the deviatorial plane, which depends on the third invariant of the stress state; and, μ—a viscous parameter, allowing to barely exceed the surface of plastic potential, in some sufficiently small steps of the task.

Density (ρ), the modulus of elasticity (E_{cm}), and compressive strength (f_{cm}) of individual composite series were adopted on the basis of experimental tests (see point 3.1). In identifying the parameters in the uniaxial state of compression (σ_c–ε) of A, B, and C beams, it was assumed that concrete behaves in a linear elastic manner to linear stress 0.6 f_{cm} [41], and it is then strengthened to the stresses equal to f_{cm}, which corresponds to strains that are equal to $\varepsilon_{c,in} = 1.5‰$.

The material degradation associated with crushing of concrete was ignored ($d_c = 0$), owing to the obtained failure mode of beams triggered off by cracking during tension. By tension, linear form of composite weakening ($\sigma_t^- - u_t^{-pl}$) was assumed, as in Figure 17. In this case, the constitutive parameter defining destruction during tension, destructive displacement (u_t^{-pl}), was determined according to the relationship [36]:

$$u_{t,max}^{-pl} = \frac{2G_f}{\sigma_{t0}}.$$ (9)

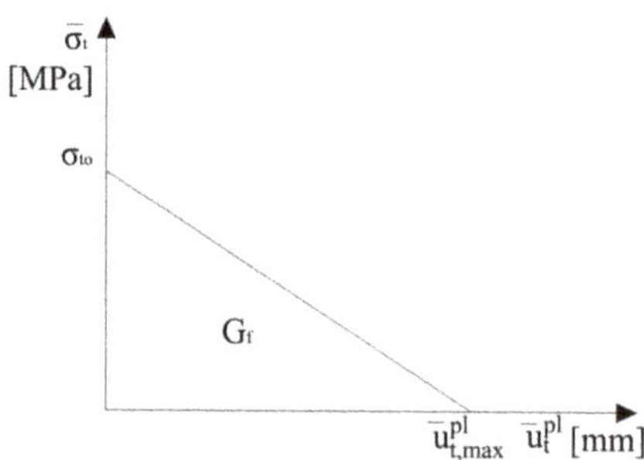

Figure 17. Linear weakening during tension.

For calculations $u_{t,max}^{-pl}$, the average values of fracture energy of individual series of beams (A, B, and C) were taken; they come from Table 4. Owing to the lack of own tests of axial tensile strength of SFRHSC samples, the parameter σ_{t0} was determined on the basis of the results of the proportionality limit ($f_{ct,fl,L}$) [15] and then, owing to the complexity of the problem of determining the relationship σ_{t0}–$f_{ct,fl,L}$ as well as the small number of experimental studies in this respect, a simplification was made that consisted of accepting the estimate based on the Raphael dependency [42], based against on Navier's hypothesis (dependence obtained on the basis of tests for concrete without fibres):

$$\sigma_{t0} = 0.75 f_{ct,fl,L}.$$ (10)

In the case of HSC beams, the parameter σ_{t0} was calculated in an analogous way, with flexural tensile strength ($f_{ct,fl}$). An arbitrary stress of $0.01\ \sigma_{t0}$ was accepted and material degradation of $d_t = 0.95$ was assumed in order to avoid loss of convergence due to zero stress and stiffness. Table 6 presents details of the material parameters implemented for each series of beams.

Table 6. Material constants adopted for the CDP model for different composite series.

Series	The Law of Strengthening in Compression		The Law of Weakening in Tension		
	σ_c (MPa)	$\varepsilon_c^{-in}\cdot 10^{-3}$	σ_t (MPa)	u_t^{-pl} (mm)	d_t
A	46.15	1.29	4.23	0	0
	76.91	1.50	0.042	0.04	0.95
B	52.10	1.27	4.60	0	0
	86.84	1.50	0.046	3.648	0.95
C	57.34	1.37	5.25	0	0
	95.56	1.50	0.053	3.424	0.95

3.3. Results of Numerical Analyses

A comparative analysis was carried out in order to verify the numerical model, which included the relationships F–δ and F–CTOD (quantitative analysis) and crack propagation (qualitative analysis). The analysis was performed on the basis of the arithmetic mean of beam results for each series. In Figures 18 and 19, the static equilibrium path (F–δ) of the FEM model was compared with the experimental results that refer to the HSC and SFRHSC beams, respectively. In turn, numerical and experimental F–CTOD curves were compared in Figures 20 and 21.

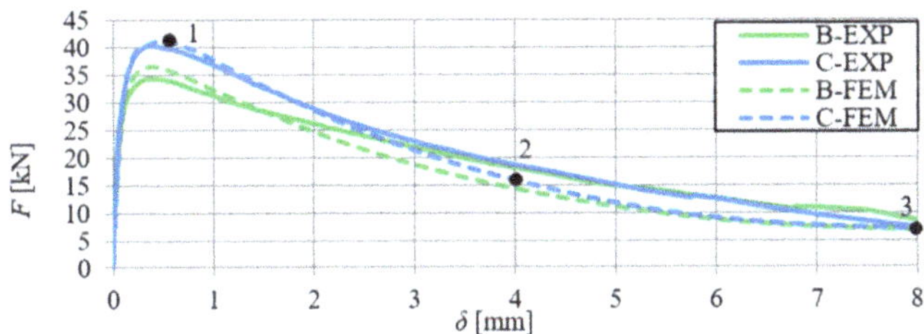

Figure 18. Comparison of numerical and experimental force–deflection relationship of SFRHSC beams in a three-point bending tests (B and C series).

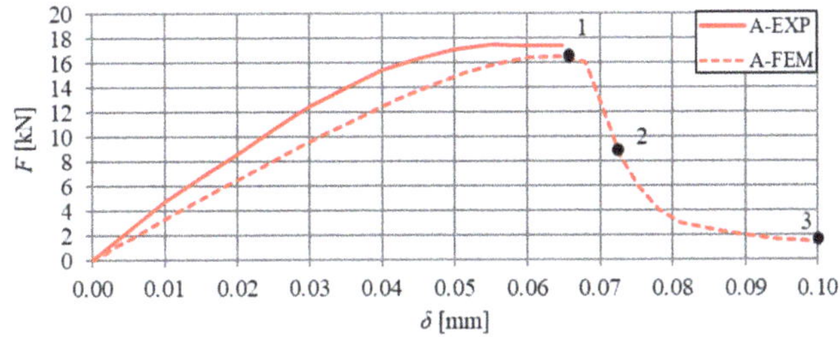

Figure 19. Comparison of numerical and experimental force–deflection relationship of HSC beams in three-point bending tests (series A).

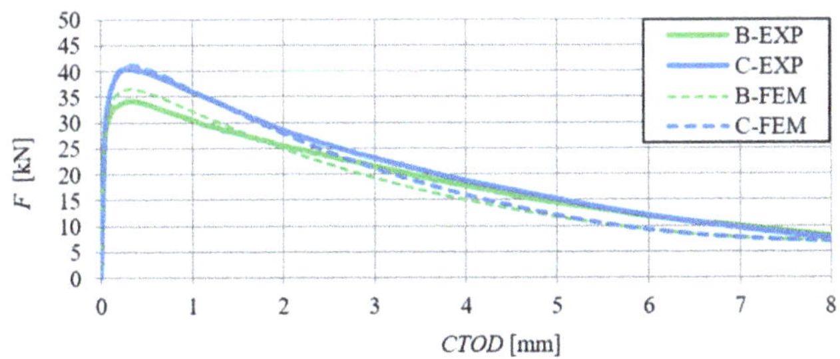

Figure 20. Comparison of numerical and experimental force-crack tip opening displacement relationship of SFRHSC beams in three-point bending tests (B and C series).

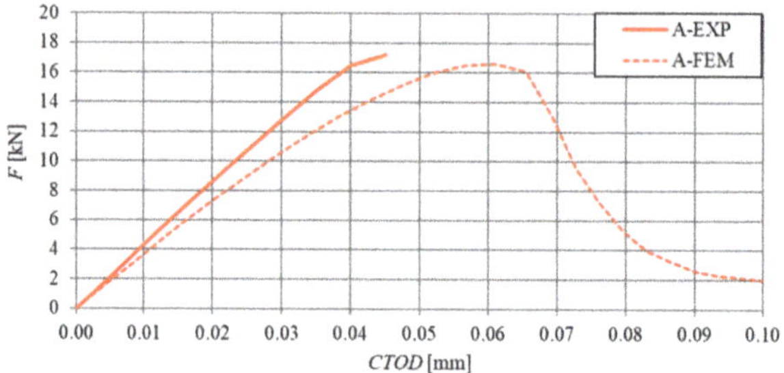

Figure 21. Comparison of numerical and experimental force-crack tip opening displacement relationship of HSC beams in three-point bending tests (series A).

The comparative analysis of the relationships F–δ and F–CTOD confirms reliable compatibility of the obtained results of experimental and numerical test on SFRHSC beams. Compatibility can be observed both for the limit load (load-bearing capacity) and the behaviour of beams in the beyond-elastic range. In the case of HSC beams, there are some differences in the elastic work of the structure. This could be due to the influence of the test methodology on the value of the modulus of elasticity (E_c). The obtained numerical results confirm the overall compatibility of the computational model with assumptions regarding the material strength hypothesis. The differences between the experimental stiffness of the beams and the stiffness of the beams in the numerical model are also illustrated in the graphs. It confirms the correct selection of the degradation variable in the concrete model d_t and the right assumption of simulating the fibres in the material based on the fracture energy.

The comparative analysis of the cracking in the beams, for the clarity of the argument, was limited to two cases: series A and C (the largest volume share of fibres in the composite and their absence). The analysis was made on the basis of damage map comparison, being defined by changes in the value of parameter DAMAGET (d_t degradation of stiffness illustrating the destruction of material). The damage maps were prepared for the characteristic points of the static balance paths of HSC and SFRHSC beams (points 1, 2, and 3 marked in Figures 18 and 19), which are shown in Figures 22 and 23, respectively. The FEM images of the damaged HSC and SFRHSC beams were compared to images that were obtained during the experiment (Figures 22 and 23).

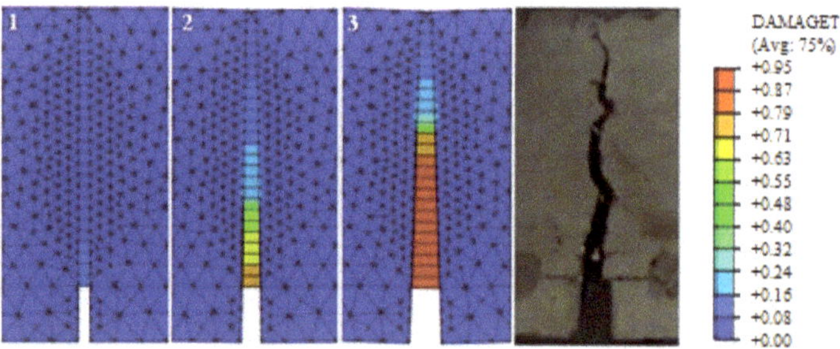

Figure 22. Image of finite element method (FEM) destruction corresponding to the characteristic points of the static equilibrium path (see Figure 18) of SFRHSC beams (series C); the final image of the destruction of the tested C2 beam and the legend for the material damage map.

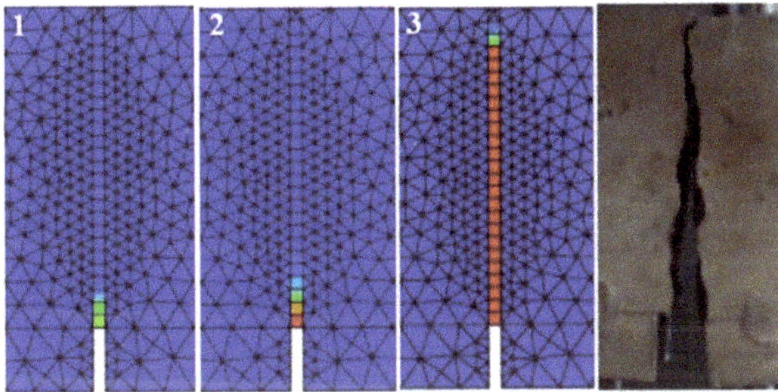

Figure 23. Image of FEM destruction corresponding to the characteristic points of the static equilibrium path (see Figure 19) of the HSC beams (series A) and the final picture of destruction of the tested A1 beam.

Since, in the CDP model, it is not possible to shape cracks in a discrete way, taking into account the chipping of the material, damage maps (DAMAGET) should be identified with the gradual exclusion of finite elements from cooperation. In this way, their specific 'gluing' and further participation in the transmission of deformations to adjacent elements takes place. This imperfection does not have a significant impact upon the behaviour of the entire research element (convergence of static equilibrium pathways) [43].

In numerical analyses, the obtained damage images qualitatively correspond to the crack propagation images, which were obtained during experimental studies (Figure 22—SFRHSC beam and Figure 23—HSC beam). Moreover, the analysis of damage maps that were defined by the parameter d_t made it possible to follow the process of crack formation and development with the increasing load of the model. For the SFRHSC beam model (C series), the development of the fracture process zone (crack range) was "smooth" (successive elimination of finite elements from cooperation), which should be identified as crack propagation due to gradual force transmission through the fibres and crack propagation was rapid and violent in the case of the HSC beam (A series).

4. Final Conclusions

On the basis of the presented research results, performed analyses, and literature review, the following conclusions were formulated:

- The tests that were carried out on beams have shown that together with the increase in the volumetric amount of fibres (V_f) in the beam, flexural tensile strength increases considerably ($f_{ct,fl}$).
- Laboratory tests carried out showed considerably higher (over two hundred times) fracture energy (G_f) in the case of SFRHSC beam as opposed to HSC beams.
- The presence of fibres affects the destruction characteristics of SFRHSC beams, which do not suddenly divide in contrast to HSC ones. After they are "destroyed", the SFRHSC beams are able to carry a set load (usually smaller than the breaking load) until the broken parts are "separated".
- Regression Equations (2), (3), and (7) with determination coefficients $R^2 > 85\%$ and statistically significant structural parameters were obtained as a result of statistical analysis of the test results ($f_{ct,fl}$, f_c and G_f). These equations can be used to estimate the flexural tensile strength ($f_{ct,fl}$) and fracture energy (G_f) for SFRHSC beams. In view of the sample size (≤ 9), the proposed regression equations should be verified in the future by performing experimental tests on a larger number of samples.
- For SFRHSC beams, the force–deflection (F–δ) and force-crack tip opening displacement (F–CTOD) relationships were almost identical (shape and area under the graph). The relationship between the deflection and crack tip opening displacement (δ–CTOD) obtained for the mean results of

B- and C-series beams is described by linear regression Equations (4) and (5). These equations could be a contribution to the extension of the standard [15], which now allow for approximating the deflections (δ) of steel-fibre-modified beams only as a function of crack mouth opening displacement (CMOD).

- Numerical analysis of HSC and SFRHSC beams with the use of the CDP model showed high compliance with the experimental results in terms of quantity (conformity of the F–δ and F–CTOD relationships) and quality (image of the cracking).
- Minor differences between the results of laboratory tests and numerical analyses of HSC and SFRHSC beams confirm the reliability of the parameters adopted, experimentally and theoretically determined, and describing the elastic-plastic material model used in numerical simulations.
- The basic material parameter used in numerical simulations (CDP model), which is employed to describe the behaviour of SFRHSC beams in the beyond-elastic range is the fracture energy (G_f), takes into account the presence of fibres in the composite. To determine the energy, it is necessary to know, for example, the relationship: F–δ, F–CMOD or F–CTOD.

Author Contributions: Conceptualization, C.B. and M.K. (Maciej Kaźmierowski); Investigation, M.K. (Maciej Kaźmierowski); Methodology, C.B.; Supervision, M.K. (Mieczysław Kamiński); Validation, C.B.; Visualization, M.K. (Maciej Kaźmierowski) and M.D.; Writing—original draft, M.K. (Maciej Kaźmierowski). All authors have read and agreed to the published version of this manuscript.

Funding: This research received no external funding.

Conflicts of Interest: The authors declare no conflict of interest.

References

1. Mudadu, A.; Tiberti, G.; Germano, F.; Plizzari, G.; Morbi, A. The effect of fiber orientation on the post-Cracking behavior of steel fiber reinforced concrete under bending and uniaxial tensile tests. *Cem. Concr. Compos.* **2018**, *93*, 274–288. [CrossRef]
2. Prebhakumari, K.; Jayakumar, P. Fracture Parameters of Steel Fibre Reinforced High Strength Concrete by Size Effect Method. *Int. J. Sci. Eng. Res.* **2013**, *4*, 254–259.
3. Fantilli, A.; Chiaia, B.; Gorino, A. Fiber volume fraction and ductility index of concrete beams. *Cem. Concr. Compos.* **2016**, *65*, 139–149. [CrossRef]
4. Lantsoght, E. How do steel fibers improve the shear capacity of reinforced concrete beams without stirrups? *Compos. Part B Eng.* **2019**, *175*, 107079. [CrossRef]
5. Berkowski, P.; Kosior-Kazberuk, M. Effect of Fiber on the Concrete Resistance to Surface Scaling Due to Cyclic Freezing and Thawing. *Procedia Eng.* **2015**, *111*, 121–127. [CrossRef]
6. Eik, M.; Puttonen, J.; Herrmann, H. An orthotropic material model for steel fibre reinforced concrete based on the orientation distribution of fibres. *Compos. Struct.* **2015**, *121*, 324–336. [CrossRef]
7. Tlemat, H.; Pilakoutas, K.; Neocleus, K. Modelling of SFRC using Inverse Finite Element Analysis. *Mater. Struct.* **2006**, *39*, 197–207. [CrossRef]
8. Ezeldin, A.; Balaguru, P. Normal and high strength fiber reinforced concrete under compression. *J. Mater. Civil Eng.* **1992**, *4*, 415–429. [CrossRef]
9. Abbass, W.; Khan, M.; Mourad, S. Evaluation of mechanical properties of steel fiber reinforced concrete with different strengths of concrete. *Constr. Build. Mater.* **2018**, *168*, 556–569. [CrossRef]
10. Sivaraja, M.; Kandasamy, S.; Thirumurugan, A. Mechanical strength of fibrous concrete with waste rural materials. *J. Sci. Ind. Res.* **2010**, *69*, 308–312.
11. Kazmierowski, M. Deflections of Double-Span Reinforced Concrete Beams Made of High-Strength Concrete Modified with Steel Fibers. Ph.D. Thesis, Wroclaw University of Technology, Wroclaw, Poland, July 2019.
12. Banthia, N. Fiber Reinforced Cements and Concretes. *Can. J. Civ. Eng.* **2001**, *28*, 879–880.
13. Hameed, R.; Turatsinze, A.; Duprat, F.; Sellier, A. Metallic fiber reinforced concrete: Effect of fiber aspect ratio on the flexural properties. *J. Eng. Appl. Sci.* **2009**, *4*, 67–72.
14. Bencardino, F. Mechanical Parameters and Post-Cracking Behaviour of HPFRC according to Three-Point and Four-Point Bending Test. *Adv. Civ. Eng.* **2013**. [CrossRef]

15. European Committee for Standardization. *Test Method for Metallic Fibered Concrete–Measuring the Flexural Tensile Strength (Limit of Proportionality (LOP), Residual)*; EN 14651:2005; European Committee for Standardization: Brussels, Belgium, 2005.

16. National Research Council. *Guide for the Design and Construction of Fiber-Reinforced Concrete Structures*; CNR-DT 204/2006; National Research Council: Rome, Italy, 2007.

17. European Committee for Standardization. *Testing Hardened Concrete, Part 2: Making and Curing Specimens for Strength Tests*; EN 12390-2:2000; European Committee for Standardization: Brussels, Belgium, 2000.

18. Polish Committee for Standardization. *Eurocode 2: Design of Concrete Structures-Part 1-1: General Rules and Rules for Buildings*; PN-EN 1992-1-1: 2008; Polish Committee for Standardization: Warsaw, Poland, 2010.

19. European Committee for Standardization. *Fly Ash for Concrete–Part 1: Definitions, Specifications and Conformity Criteria*; EN 450-1: 2012; European Committee for Standardization: Brussels, Belgium, 2012.

20. European Committee for Standardization. *Testing Fresh Concrete, Part 2: Slump-Test*; EN 12350-2:2009; European Committee for Standardization: Brussels, Belgium, 2009.

21. European Committee for Standardization. *Testing Hardened Concrete, Part 3: Compressive Strength of Test Specimens*; EN 12390-3:2009; European Committee for Standardization: Brussels, Belgium, 2009.

22. European Committee for Standardization. *Testing Hardened Concrete, Part 13: Determination of Secant Modulus of Elasticity in Compression*; EN 12390-13:2013; European Committee for Standardization: Brussels, Belgium, 2013.

23. Lambrechts, A. Performance classes for steel fibre reinforced concrete: Be critical. In Proceedings of the BEFIB 2008: 7th RILEM International Symposium on Fibre Reinforced Concrete, Chennai, India, 17–19 September 2008.

24. Ponikiewski, T.; Katzer, J. Mechanical Properties and Fibre Density of Steel Fibre Reinforced Self-Compacting Concrete Slabs by Dia and Xct Approaches. *J. Civ. Eng. Manag.* **2017**, *23*, 604–612. [CrossRef]

25. Wafa, F.; Ashour, S. Mechanical Properties of High-Strength Fiber Reinforced Concrete. *ACI Mater. J.* **1992**, *89*, 449–455.

26. Kaiss, S.; Al-Azzawi, K. Mechanical Properties of High-Strength Fiber Reinforced Concrete. *Eng. Technol. J.* **2010**, *28*, 2442–2453.

27. Sumathi, A.; Saravana, K. Strength Predictions of Admixed High Performance Steel Fiber Concrete. *ChemTech* **2014**, *6*, 4729–4736.

28. Eren, O.; Marrar, K.; Celik, T. Effect of Silica Fume and Steel Fibers on Some Properties of High-Strength Concrete. *Constr. Build. Mater.* **1997**, *11*, 373–382. [CrossRef]

29. Lin, W.; Wu, Y.; Cheng, A.; Chao, S.; Hsu, H. Engineering Properties and Correlation Analysis of Fiber Cementitious Materials. *Materials* **2014**, *7*, 7423–7435. [CrossRef]

30. Gao, J.; Sun, W.; Morino, K. Mechanical properties of steel fibre reinforced high strength light weight concrete. *Cem. Concr. Compos.* **1997**, *19*, 307–313. [CrossRef]

31. Bayramov, F.; Tasdemir, C.; Tasdemir, M. Optimisation of steel fibre reinforced concretes by means of statistical response surface method. *Cem. Concr. Compos.* **2004**, *26*, 665–675. [CrossRef]

32. Bae, B.; Choi, H.; Choi, C. Correlation Between Tensile Strength and Compressive Strength of Ultra High Strength Concrete Reinforced with Steel Fiber. *J. Korea Concr. Inst.* **2015**, *27*, 253–263. [CrossRef]

33. Hillerborg, A.; Modéer, M.; Petersson, P. Analysis of crack formation and crack growth in concrete by means of fracture mechanics and finite elements. *Cem. Concr. Res.* **1976**, *6*, 773–780. [CrossRef]

34. RILEM TC-FMC. Determination of fracture energy of mortar and concrete by means of three-Point bend test on notched beams. *Mater. Struct.* **1985**, *18*, 285–290. [CrossRef]

35. Spagnoli, A.; Carpinteri, A.; Ferretti, D.; Vantadori, S. An experimental investigation on the quasi-Brittle fracture of marble rocks. *Fatigue Fract. Eng. Mater. Struct.* **2016**, *8*, 956–968. [CrossRef]

36. *Abaqus User's Manual*; Version 6.14; Dassault Systèmes Simulia Corp: Johnston, RI, USA, 2014.

37. Pamin, J.; Winnicki, A. Computational models of materials: Damage, strain localization, applications. In *Modern Structural Mechanics with Applications to Civil Engineering*; Polish Academy of Sciences: Warsaw, Poland, 2015; pp. 259–278.

38. *Diana Finite Element Analysis User's Manual. Theory*; Release 9.6; TNO DIANA BV: Delft, The Netherlands, 2014.

39. Wosatko, A.; Genikomsou, A.; Pamin, J.; Polak, M.A.; Winnicki, A. Examination of two regularized damage-plasticity models for concrete with regard to crack closing. *Eng. Fract. Mech.* **2018**, *194*, 190–211. [CrossRef]

40. Wosatko, A.; Winnicki, A.; Polak, M.A.; Pamin, J. Role of dilatancy angle in plasticity-Based models of concrete. *Arch. Civ. Mech. Eng.* **2019**, *19*, 1268–1283. [CrossRef]
41. Jankowiak, T. *Failure Criteria for Concrete under Quasi-Static and Dynamic Loadings*, 1st ed.; Politechnika Poznanska: Poznan, Poland, 2011; pp. 47–96. ISBN 978-83-7143-981-0.
42. Raphael, J. Tensile strength of concrete. *Concr. Int.* **1984**, *81*, 158–165.
43. Sledziewski, K. Experimental and numerical studies of continuous composite beams taking into consideration slab cracking. *Maint. Reliab.* **2016**, *18*, 578–589. [CrossRef]

Article

Study of the Microstructure Characteristics of Three Different Fine-grained Tailings Sand Samples during Penetration

Yueqi Shi *, Changhong Li and Dayu Long

Key Laboratory of Ministry of Education for High-Efficient Mining and Safety of Metal Mines, University of Science and Technology Beijing (USTB), Beijing 100083, China; lch@ustb.edu.cn (C.L.); ldayu2018@163.com (D.L.)
* Correspondence: syq_shi@126.com

Received: 3 March 2020; Accepted: 27 March 2020; Published: 30 March 2020

Abstract: This paper explores the microstructural evolution characteristics of tailings sand samples of different types of infiltration failure during the infiltration failure process. The homemade small infiltration deformation instrument is used to test the infiltration failure characteristics of the tailings sand during the infiltration failure process. Evolutionary characteristics of the internal microstructure pores and particle distribution were also studied. Using CT (computerized tomography) technology to establish digital image information, the distribution of the microscopic characteristics of the particle distribution and pore structure after tailing sand infiltration were studied. Microscopic analysis was also performed to analyze the microscopic process of infiltration and destruction, as well as to see the microscopic structural characteristics of the infiltration and destruction of the total tailings. The test results show that there are obvious differences in the microstructure characterization of fluid soil and piping-type infiltration failures. Microstructure parameters have a certain functional relationship with macrofactors. Combining the relationship between macrophysical and mechanical parameters and microstructural parameters, new ideas for future research and the prevention of tailings sand infiltration and failure mechanisms is provided.

Keywords: X-ray microtomography; microstructure characteristics; infiltration damage

1. Introduction

Tailings sand is a solid waste material with a low content of useful components that remains after crushing and sorting the ore. This material is not suitable for further sorting under current economic and technological conditions. As a raw material for building tailings dams, it is a man-made three-phase dispersion medium with special structural characteristics as well as physical and mechanical properties [1,2]. The particle size of fine-grained tailings is very small, and thus, its specific surface area is large; its reactivity is also high [3,4]. Therefore, compared with ordinary tailings, the enrichment of fine-grained tailings can easily lead to lens bodies or a weak tailings dam interlayer. Furthermore, the strength of tailings sand, as sandy soil, is not high. When infiltration damage occurs, tailings sand will leak intensively in weak places, such as the dam foundation and abutment, forming voids inside the dam body, and even causing the local collapse of the dam body (tailings dam break). Scholars worldwide have carried out numerous experimental studies on the percolation and failure of natural soils and have made some progress in the study of the mechanical properties of tailings. However, most of them have discussed the macromechanical properties of tailings and the deformation and evolution of the fine content and microstructure. Studies on the characteristics and mechanical behavior of tailings are still scarce [5–13]. At the same time, high-resolution X-ray microtomography (micro-CT) has been widely used as a nondestructive technique that can perform three-dimensional (3D) imaging to analyze the internal characteristics of objects [14]. Although this technology has been widely used to investigate the microstructures of rocks and soils, it has not been applied to tailings sand.

This paper mainly studies the characteristics of the microstructural changes of different infiltration and destruction types in the tailings sand seepage process and provides new ideas for the study of tailings sand infiltration, failure mechanisms, and prevention measures. Taking the fine-grained tailings in Makeng, Fujian, as the research object, a homemade small osmotic deformation instrument was used to test the osmotic failure characteristics of tailings sand samples with fine contents of 30%, 50%, and 70%. The evolution of the internal microscopic pores and the particle distribution of fine-grained tailings sand during osmotic failure were characterized. During the process of osmotic failure, the head was stepwise loaded with four heads of water until the sample was damaged, and X-ray microtomography (micro-CT) was used to scan the tailings sand samples under the four head pressures. Using the VGStudio max 3.0 and Avizo 9.0.1 visualization software provided by Sanying Precision, Tianjin, China, the three-dimensional (3D) reconstruction of the scanned sample data was performed, the digital image information was established, and the microscopic characteristics of the particle distribution and pore structure of the tailings sand after infiltration and destruction were analyzed. During the infiltration and destruction process of granular tailings, the characteristics of the internal microstructure voids, and the changes in the particles are summarized. The characteristics of the infiltration and destruction of fine-grained tailings are summarized.

2. Materials and Methods

2.1. Test Tailings Sand Material

All of the tailings sand materials used in the tests were sampled from Fujian's Makeng iron tailings pond. The gradation curve is shown in Figure 1. Figure 1 shows that the fine grain content, with grain sizes less than 0.075 mm, is between 15% and 50%, and this material is referred to as silty sand (SM). At the same time, the coefficient of nonuniformity (Cu) of the tailings sand material is greater than 5; that is, the sizes of the coarse and fine particles in the tailings sand are very different, and the fine particles can easily fill in the pores formed by the coarse particles to form a better skeleton structure.

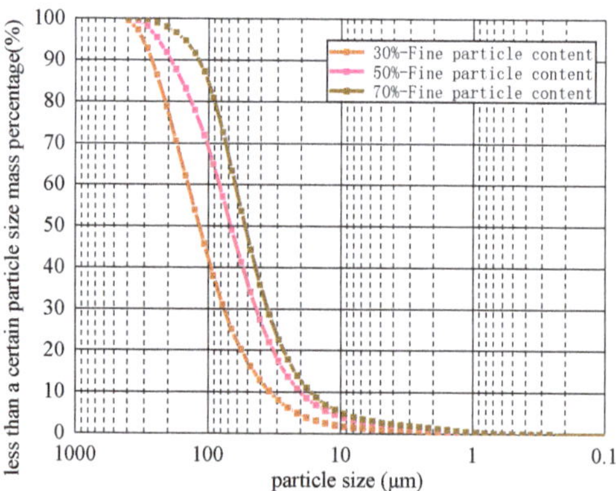

Figure 1. Gradation curve of tailings sand.

2.2. Test Device and Scheme

2.2.1. Small Osmotic Deformation Instrument

The test device used is a set of homemade small osmotic deformation instruments improved by an ordinary osmometer, as shown in Figure 2. The instrument consists of a test container, a pressurization device, a water supply device, and other parts. The test container is a transparent acrylic round tube with a height of 50 mm, an inner diameter of 10 mm, and an outer diameter of 14 mm. Furthermore, the device is a testing and filling device for tailings sand so that test phenomena can be conveniently observed. Since the particle size of the tailings sand used in this test is relatively small, and its permeability coefficient is small, the pressure device is a millimeter-scaled variable head pipe. The water supply device is a transparent water supply bottle.

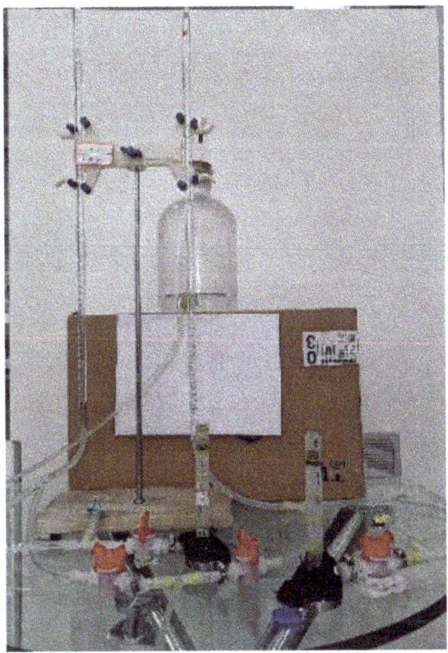

Figure 2. Small osmotic deformation instrument.

2.2.2. X-ray Micro-CT

The instrument used for tailings sand sample testing is a nano Voxel-3000 micro-CT provided by Sanying Precision, Tianjin, China, as shown in Figure 3. The instrument has advanced nondestructive 3D imaging and image analysis capabilities, with secondary optical magnification and a spatial resolution of up to 0.5 μm. Using this equipment, the 3D visualization of the internal microstructure of materials can be nondestructively investigated.

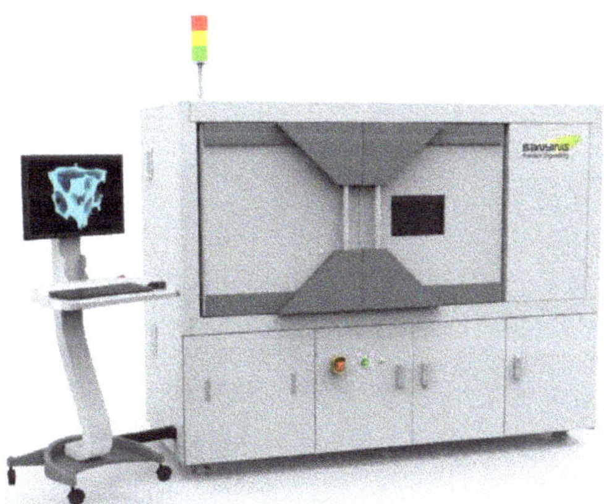

Figure 3. nanoVoxel-3000 micro-CT.

In this test, the test voltage of the micro-CT equipment is 140 kV, the test current is 70 μA, the exposure time is 0.42 s, the standoff distance (SoD) from the source to the sample is 13.6 mm, and the distance from the source to the flat panel detector (SDD) is 279.6 mm. The resolution is 6.16 μm, which can distinguish the characteristics of pores with pore sizes greater than 6.16 μm and tailings sand particles. This experiment adopts the continuous scanning method of a cone-beam with a scanning rate of 0.25°/frame and a total of 1440 projections. Numerous layers (1536) of the sample are cut longitudinally, and each layer thickness is 6.16 μm. Finally, 1536 two-dimensional (2D) slice images of 1800 × 1800 pixels are obtained. The parameters of the CT scan of the tailings sample are shown in Table 1.

Table 1. CT scan parameters.

Sample Name	Test Voltage/kV	Test Current/μA	Exposure Time/s	Resolution/μm	Pixel	X-ray Source to Sample Distance (sod)/mm	Distance from X-ray Source to Flat Panel Detector (sdd)/mm
Tailings sand sample	140	70	0.42	6.16	1800 × 1800 × 1536	13.6	279.6

2.2.3. Test Method and Process

To obtain the process of tailings sand infiltration and failure, the test was simulated by gradually increasing the head difference.

According to the critical head penetration failure heads of three different fine-grained tailings sand samples, the heads under test loading are divided into four grades. The top three grades are as follows: level 2 is the head value without seepage damage, level 3 is the critical head damage head value of the sample, and level 4 is the head value after seepage failure. After each stage is loaded stably, a CT scan is performed on the sample pressure relief, as shown in Figure 4. The meso-infiltration characterization inside the sample is obtained. After scanning, the head pressure of the seepage is increased to the next head value. According to the recorded experimental phenomena, the distribution of the head value for each stage of the sample is shown in Table 2.

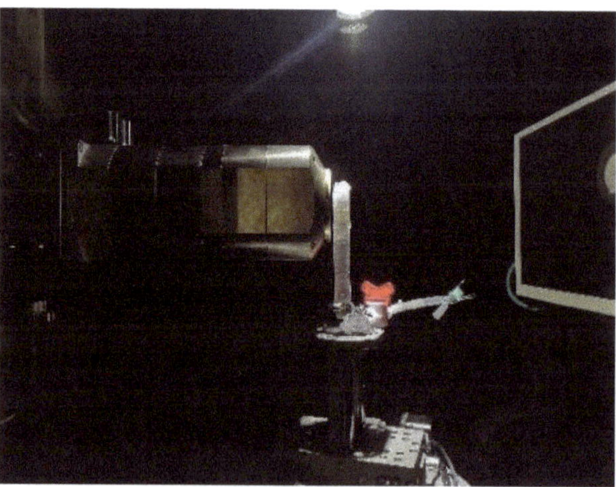

Figure 4. CT scan of the tailings sand sample.

Table 2. Head loading values at various levels.

Head Level	Loading Head Level/cm		
	30% Fine Content	50% Fine Content	70% Fine Content
1	29.0	32.0	48.0
2	37.0	43.0	52.0
3	39.0	46.0	56.0
4	42.0	51.0	58.0

During the test, attention is paid to whether the test surface has water turbidity, bubbling, or fine particle bounce, whether there is a bulge or a rise on the sample surface, and whether there are faults in the sample [15]. When the average velocity of seepage suddenly increases and is unstable or when concentrated seepage occurs on the pipe wall, the test is stopped.

During the test, the values of the seepage velocity and permeability coefficient are recorded as the head difference gradually increases (see Table 3).

Table 3. Seepage velocity and permeability coefficient during the test.

Fine Content	Head Level	Flow Rate cm^3/s	Permeability Coefficient cm/s
30%	1	8.34×10^{-3}	9.95×10^{-4}
	2	1.23×10^{-2}	1.14×10^{-3}
	3	2.78×10^{-2}	2.45×10^{-3}
	4	8.34×10^{-2}	6.52×10^{-3}
50%	1	7.59×10^{-3}	8.18×10^{-4}
	2	1.30×10^{-2}	1.04×10^{-3}
	3	2.98×10^{-2}	2.23×10^{-3}
	4	5.96×10^{-2}	4.01×10^{-3}
70%	1	6.95×10^{-3}	6.30×10^{-4}
	2	1.30×10^{-2}	8.60×10^{-4}
	3	1.67×10^{-2}	1.02×10^{-3}
	4	6.95×10^{-2}	3.84×10^{-3}

When the 30% fine-grained sample is loaded with a third-level head, a slight floating soil phenomenon appears on the top of the sample. When a fourth-level head is loaded, seepage failure occurs. The seepage at the top of the sample is very turbid, and the soil is completely jacked up. This phenomenon is a piping-type infiltration failure. When the 50% fine-grained sample is loaded with a third-level head, a slight floating soil phenomenon appears on the top of the sample, and microcracks appear in the middle and lower parts of the sample. When a fourth-level head is loaded, complete infiltration failure occurs. The seepage flow at the top of the sample is very turbid, the flow velocity increases sharply, and the lower cracks further expand. Fine particle bounces appear at the top of the sample, which suggests that fluid-type infiltration failure occurs. When the 70% fine particle sample is loaded with a second-level head, the flow rate increases sharply. The surface of the tailings sample exhibits a slight support phenomenon, and there are visible cracks in the lower part of the sample. When the third-level head is loaded, complete percolation failure occurs. The seepage at the top of the sample becomes very turbid, the flow velocity increases sharply, and the lower cracks further expand. Fine-grained beating appeared on the top of the sample, and it was judged that a fluid-type infiltration failure occurred.

3. Test Results and Analysis

3.1. Three-Dimensional Reconstruction of the CT Scan of Fine-Grained Tailings Sand

Cone shadows are generated in the upper and lower slices of the data generated by the CT scan, which affects the subsequent 3D reconstruction and quantitative characterization of the pore structure. Therefore, the CT layers need to be cropped in the VGStudio max 3.0 visualization software provided by Sanying Precision, Tianjin, China to obtain 1000 layers of 1800 × 1800 pixels. The 2D slice map was imported into the Avizo software for the 3D reconstruction for further analysis. The 2D slice image of the tailings sand sample obtained by cone-beam scanning is affected by various types of system noise and artifacts. The gray image obtained by the local mean filter needs to be filtered and denoised [16]. Then, the artifacts are effectively removed by the Avizo software provided by Sanying Precision, Tianjin, China, and the scanned data are reconstructed by the software's iterative reconstruction algorithm to obtain a 3D map of the tailings sand sample under the head pressures of 1–4, as shown in Figure 5. Afterward, the image segmentation processing is performed according to the commonly used Otsu algorithm [17], and the grayscale image is binarized. Finally, the 3D pore structure and tailings sand particle structure in the data volume are extracted by the Avizo software, and a 3D pore and tailings sand particle network model of the pores and tailings sand particle spatial arrangement is generated from the micro-CT image.

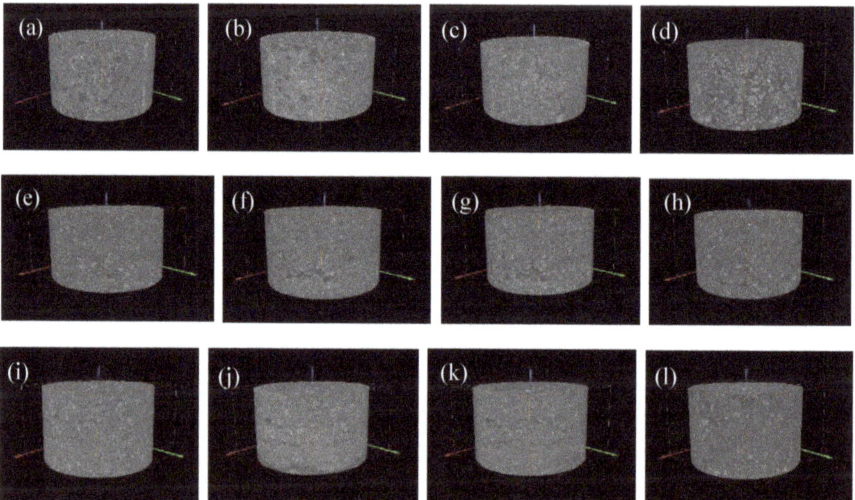

Figure 5. Three-dimensional reconstruction of the three fine-grained tailings samples. (**a**) First-level head of 30% fine content; (**b**) Second-level head of 30% fine content; (**c**) Third-level head of 30% fine content; (**d**) Fourth-level head of 30% fine content; (**e**) First-level head of 50% fine content; (**f**) Second-level head of 50% fine content; (**g**) Third-level head of 50% fine content; (**h**) Fourth-level head of 50% fine content; (**i**) First-level head of 70% fine content; (**j**) Second-level head of 70% fine content; (**k**) Third-level head of 70% fine content; (**l**) Fourth-level head of 70% fine content.

3.2. Characterization of the Microstructure of the Pores during Infiltration Failure of Tailings Sand

Based on the previous digital image processing, to obtain the pore structure model of the tailings sand samples, the grayscale image is binarized using image threshold segmentation, as shown in Figure 6a–x. The volume rendering module of the Avizo 3D visualization software is used to directly perform the 3D reconstruction of the pore structure of the sample, and the overall 3D pore microstructure of the three fine-grained tailings sand samples is obtained, as shown in Figure 7.

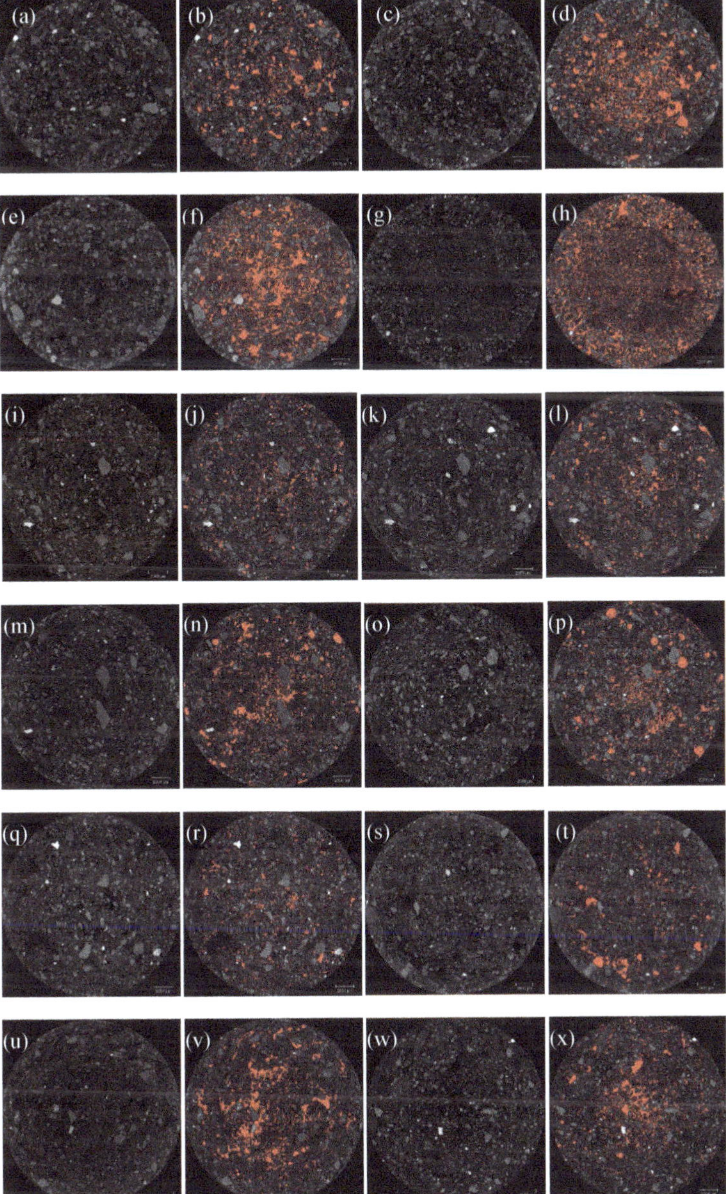

Figure 6. Pore threshold segmentation of the three fine-grained tailings sands. (**a,b**) First-level head of 30% fine content; (**c,d**) Second-level head of 30% fine content; (**e,f**) Third-level head of 30% fine content; (**g,h**) Fourth-level head of 30% fine content; (**i,j**) First-level head of 50% fine content; (**k,l**) Second-level head of 50% fine content; (**m,n**) Third-level head of 50% fine content; (**o,p**) Fourth-level head of 50% fine content; (**q,r**) First-level head of 70% fine content; (**s,t**) Second-level head of 70% fine content; (**u,v**) Third-level head of 70% fine content; (**w,x**) Fourth-level head of 70% fine content.

It can be seen from Figure 7 that the overall pore structures of the three fine-grained tailings sand samples are fully developed, the pores are unevenly distributed throughout the sample, and the pore shape changes are diverse at the surface of the pore structure. Small pores gradually penetrate into large pores, and the pore structure under the head pressure of level 4 is the most obvious.

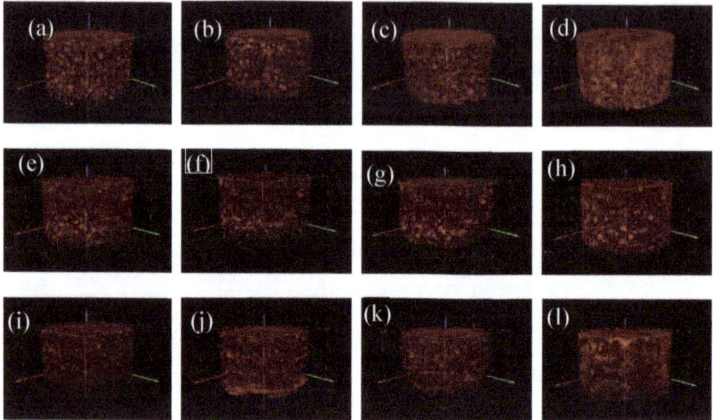

Figure 7. The overall pore structure of the three fine-grained tailings sand samples. (**a**) First-level head of 30% fine content; (**b**) Second-level head of 30% fine content; (**c**) Third-level head of 30% fine content; (**d**) Fourth-level head of 30% fine content; (**e**) First-level head of 50% fine content; (**f**) Second-level head of 50% fine content; (**g**) Third-level head of 50% fine content; (**h**) Fourth-level head of 50% fine content; (**i**) First-level head of 70% fine content; (**j**) Second-level head of 70% fine content; (**k**) Third-level head of 70% fine content; (**l**) Fourth-level head of 70% fine content.

3.2.1. Characterization of the Pore Number and Volume

Based on the 3D model of the tailings sand pore data obtained by Avizo 3D visualization software, the label analysis command in Avizo software is used to statistically analyze the sizes of the pores in the tailings sand sample, and the target pore size and number are filtered by the analysis filter command.

Based on the actual resolution of the tailings sand sample scanning, the pore size of the tailings sand sample is divided into five categories for statistical analysis: $D \geq 80\ \mu m$, $40\ \mu m \leq D < 80\ \mu m$, $20\ \mu m \leq D < 40\ \mu m$, $10\ \mu m \leq D < 20\ \mu m$, and $0\ \mu m \leq D < 10\ \mu m$. The statistical results are shown in Tables 4–6 and Figures 8–10.

Table 4. Statistics of the number of pores in the seepage process with the 30% fine content.

Head Level	Pore Size D/μm					Total Number of Pores	Total Connected Pore Volume (μm³)
	0–10	10–20	20–40	40–80	>80		
1	210,450	129,578	35,057	9092	4557	388,734	3.9×10^{10}
2	166,833	138,197	40,293	9347	3983	358,653	6.05×10^{10}
3	133,557	156,602	41,543	10,553	4702	346,957	6.5×10^{10}
4	81,410	140,923	44,661	24,151	6170	297,315	8.34×10^{10}

Table 5. Statistics of the number of pores in the seepage process with the 50% fine content.

Head Level	Pore Size D/μm					Total Number of Pores	Total Connected Pore Volume (μm³)
	0–10	10–20	20–40	40–80	>80		
1	126,404	87,362	44,256	23,020	9373	290,415	1.04×10^{10}
2	151,873	103,059	31,162	22,938	8224	317,256	2.32×10^{10}
3	142,467	97,344	41,445	19,447	6450	307,153	2.75×10^{10}
4	283,942	184,156	55,797	18,723	5370	547,988	3.47×10^{10}

Table 6. Statistics of the number of pores in the seepage process with the 70% fine content.

Head Level	Pore Size D/μm					Total Number of Pores	Total Connected Pore Volume (μm³)
	0–10	10–20	20–40	40–80	>80		
1	144,629	92,459	51,268	27,397	8370	324,123	7.41×10^9
2	121,473	79,449	28,819	12,602	3745	246,088	1.43×10^{10}
3	239,642	169,238	62,833	20,485	2605	494,803	4.8×10^{10}
4	555,965	321,095	83,504	19,136	2809	982,509	5.74×10^{10}

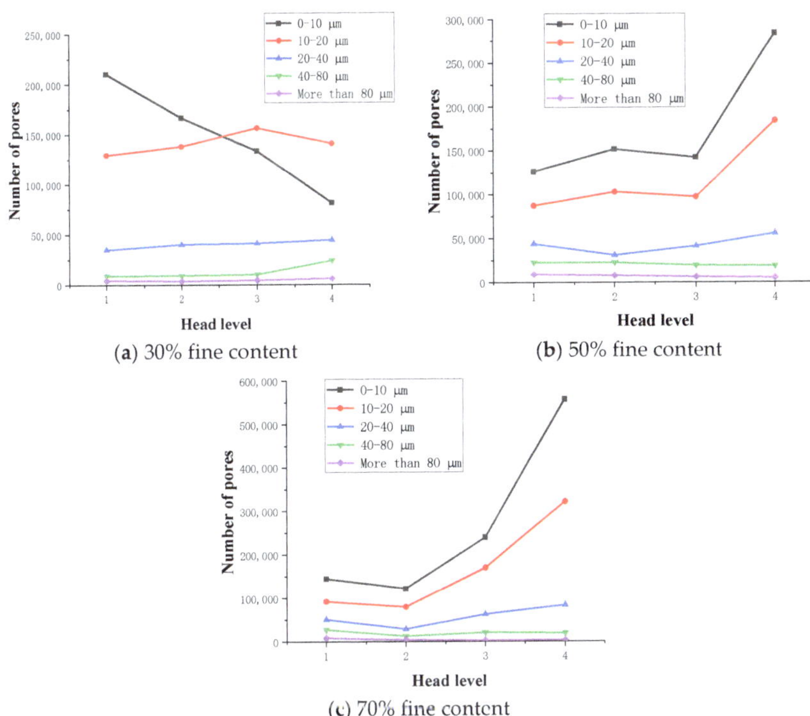

Figure 8. Variation in the pore number. (**a**) 30% fine content; (**b**) 50% fine content; (**c**) 70% fine content.

As seen from the above chart:

(1) For samples with a 30% fine content, pores ranging from 0–10 μm decrease significantly during the seepage process, while pores larger than 10 μm show a relatively increasing trend, indicating that as fine particles flow and migrate, small disconnected pores are formed prior to the seepage. When larger pores are formed, the connected pores are developed, the permeability of the sample is enhanced, and the particle migration capacity is improved.

(2) For 50% fine particles, pores ranging from 0–20 μm increase during the seepage process, and pores larger than 20 μm increase during the seepage process, indicating that during this process, the fine particles migrate into the large pores due to seepage forces. In this case, the large pores are divided into several small pores, which reduce the development speed of the connected pores, thus leading to a decrease in the percolation channels inside the sample. As the percolation pressure increases, fine particles clearly flow and migrate. At this time, the number of 0–20 μm pores shows a downward trend, whereas the 20–40 μm medium pores increase. As the total number of pores increases, the number of large pores greater than 40 μm is significantly reduced, resulting in greater interaction between soil particles and fluid damage to the sample as a whole.

(3) During the seepage process, the sample with a 70% fine content has more fine content, and the connected pores are not developed. All pores are reduced during the early stage of seepage, and the force between soil particles is significantly increased. Fluid soil damage rapidly occurs, and the number of pores rapidly increases after the damage.

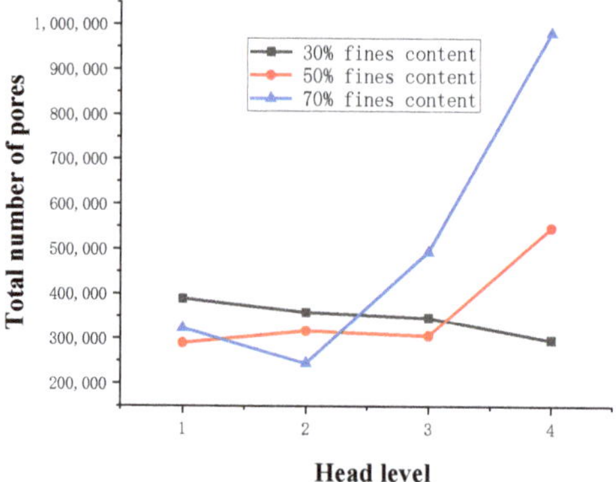

Figure 9. Change in the total pore quantity.

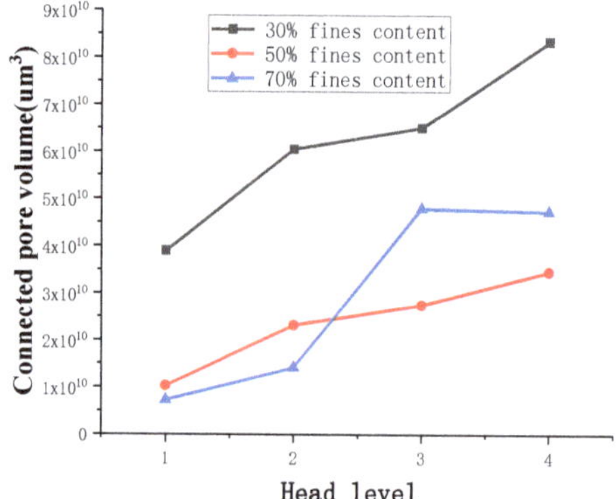

Figure 10. Total connected pore volume change.

Through a comparative analysis, we have determined the following:

(1) The total number of pores in the 30% fine-grained sample shows a decreasing trend during the seepage process. The small pores are connected to become large pores, which enhances the sample's permeability, and connected seepage channels are formed inside the sample. The connected pores are fully developed.

(2) In the 50% fine particle sample, the total number of pores first increases and then decreases during the seepage process. After the damage occurs, the number of pores rapidly increases. At the beginning of the seepage, the fine particle migration divided the large and medium pores into smaller pores, resulting in numerous pores. During the later stage of percolation, with increasing permeation force, the connected pores develop to a certain extent inside the sample, forming a certain percolation channel, and the particles begin to migrate with the permeation force, resulting in a rapid decrease in the number of micropores with sizes ranging from 0–20 μm. The number of medium pores of 20–40 μm increases, the number of large pores larger than 40 μm decreases, the total number of pores decreases, and the connected pores develop slowly.

(3)   During the seepage process, the total number of pores in a 70% fine-grained sample is relatively small, the soil particles and the seepage force increase rapidly, and the sample quickly undergoes fluid soil failure. After breaking the ring, the number of pores increases rapidly. The connected pore volume also increases rapidly.

3.2.2. Evolution Characteristics of the Connected Pores

On the basis of obtaining the overall pore structure of three fine-grained tailings sand samples, the connected pore structure of the samples with 1–4 head pressures along the seepage direction is extracted by the Avizo 3D visualization software (see Figure 11). Then, the volume of the connected pores is counted.

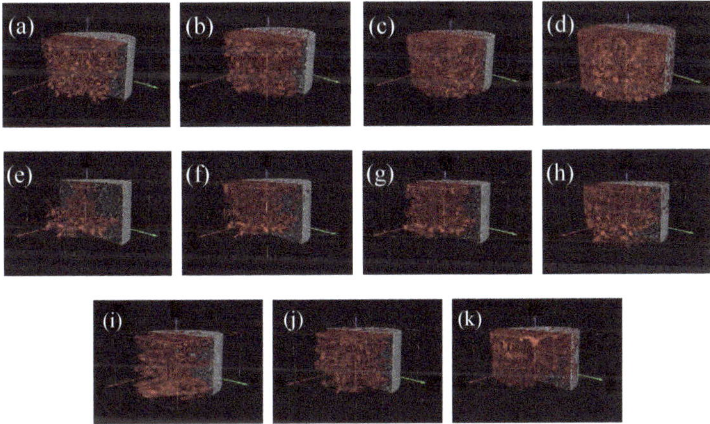

Figure 11. Connected pore structure under three levels of the water content at the three levels of fine particle content. (**a**) First-level head of 30% fine content; (**b**) Second-level head of 30% fine content; (**c**) Third-level head of 30% fine content; (**d**) Fourth-level head of 30% fine content; (**e**) First-level head of 50% fine content; (**f**) Second-level head of 50% fine content; (**g**) Third-level head of 50% fine content; (**h**) Fourth-level head of 50% fine content; (**i**) Second-level head of 70% fine content; (**j**) Third-level head of 70% fine content; (**k**) Fourth-level head of 70% fine content.

With increasing head pressure, the connected pore volume of the 30% fine particle sample gradually increases. Fine particles migrate and flow under the action of the seepage, small pores easily penetrate into large pores, and the increase is obvious. The connected pore structure of the 50% fine-grained tailings sand sample runs through the entire tailings sand sample along the seepage direction because this sand sample is finer than the 30% fine-grained tailings sand sample. Increasing the particle content reduces the pores between the particles, and thus, the connected pores passing through the entire sample are relatively insignificant at the first head pressure. As the head pressure gradually increases, the connected pores increasingly develop around them. This behavior indicates that as the seepage force increases, the fine particles migrate and flow to expand the small pores into larger pores, increasing the connectivity of the sample. As the connectivity increases, the damage to the sample is more serious, and the development of the connected pores is more obvious.

The 70% fine-grained tailings sand sample did not pass through the entire sample through the pores under the first-level head. As the head pressure is increased, the fine particles migrate and flow. At this time, the connected pores penetrate the entire sample. The 70% fine-grained tailings sand sample has a smaller proportion of pores than the 30% and 50% fine-grained tailings sand samples due to the large proportion of fine content, and the connectivity of the pores is poor. From level 2 to level 4 heads, with increasing seepage force, the connected pores develop from the sample center to the surrounding area. As the head pressure increases, the connected pore volume tends to increase first and then decrease. The 70% fine content tailings sand sample has a large proportion of fine content, and the pores between the particles are small. When the sample begins to infiltrate and break, the volume of the connected pores does not increase significantly. From the 2nd to the 3rd head, due to the continuous increase in the seepage force, a large number of fine particles migrate and flow, which makes the volume of the connected pores increase sharply to 4.80×10^{10} μm^3. From grade 3 to grade 4 heads, the volume of the connected pores slowly decreases, indicating that the tailings sand sample under the grade 4 head pressure is severely damaged. When a CT scan is carried out when the head pressure is unloaded, a large number of fine particles are uniform due to settlement.

3.2.3. Change Characteristics of the Porosity Layer by Layer

Based on the Avizo 3D visualization software, the pores are extracted after the threshold segmentation of three kinds of fine-grained tailings sand samples, and statistical analyses of the layer-by-layer porosity of the 1000-layer 2D slice map of three kinds of fine-grained tailings sand samples at various head pressures are performed separately. The layer-by-layer porosity statistics of the three fine-grained samples are shown in Figure 12.

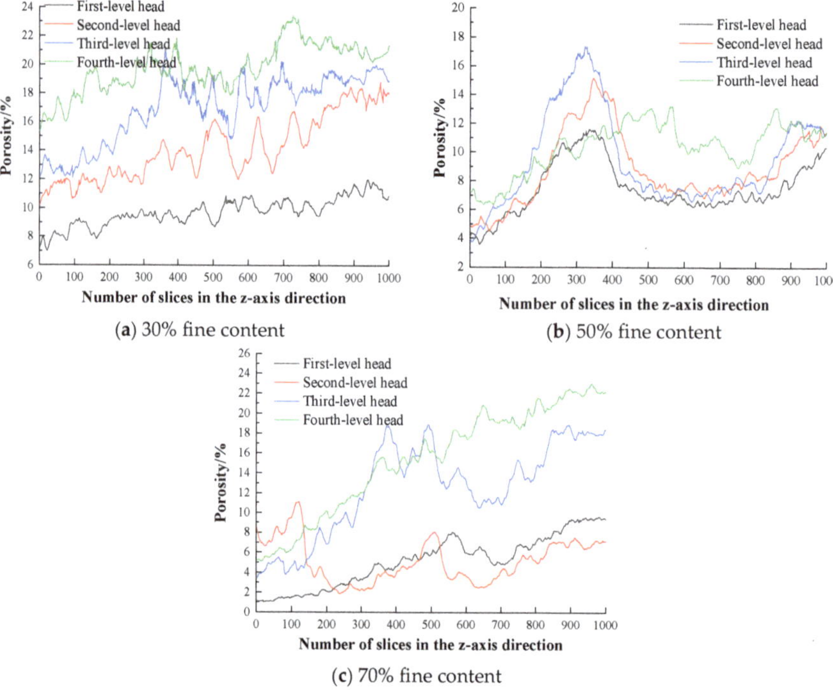

Figure 12. Layer-by-layer porosity statistics of the three fine-grained tailings sand samples.

The following conclusions can be drawn from Figure 12:

(1)   With an increasing head pressure of the 30% fine-grained samples, the porosity of the samples in the same area shows an increasing trend from layer to layer. An increase in the seepage force affects the porosity of the tailings sample at each layer. The effect of the rate is obvious. With an increase in the number of slice layers in the z-axis direction, that is, an increase in the number of 2D slice layers along the seepage direction, the tailings sand samples at all the levels of the water head show an increasing trend in the porosity of each layer.

(2) The 50% fine content tailings sand sample gradually increases with head pressure, and the average porosity gradually increases. Compared with the 30% fine-grained tailings sand sample, the porosity of each layer is smaller, indicating that the increase in the fine particle content has a significant effect on the porosity of each layer. It can be seen from the figure that the porosity of the tailings sand sample gradually increases from 200 to 400 layers, while the porosity of the 400 to 600 layers does not change much. This finding indicates that when the water level is between 1 and 2, the sample connected air does not develop, the fine particles do not migrate significantly in the 400–600 layer, and the head porosity of the 200–400 layer of the sample decreases sharply when the head pressure increases to a level 3 head, while the 400–600 layer decreases layer by layer. The porosity increases sharply, indicating that the sample suffered fluid damage.

(3) For the 70% fine content tailings sand sample, the head pressure and average porosity gradually increase. It can be seen from the layer-by-layer porosity curve under the level 2 head pressure that the porosity suddenly increases and then decreases in the areas at approximately 100 and 500 layers, indicating that the migration and accumulation of fine particles in this area are most obvious. When the head pressure is increased to level 3, the layer-by-layer porosity of the tailings sand sample has a sharp increase compared to the layer-by-layer porosity at the level 2 head pressure. Regarding the fluid soil penetration damage, the peak-to-layer porosity appeared in the 300–600-layer area, indicating that the tailings sand sample is the most damaged in this area.

3.3. Microstructure Distribution Characteristics of the Particle Size and the Particle Size during Percolation Failure of the Tailings Sand

Based on the Avizo 3D visualization software, the data obtained from CT scans of the three fine-grained tailings sand samples were binarized after threshold segmentation, and tailings sand particles with the highest gray values were extracted, as shown in Figure 13.

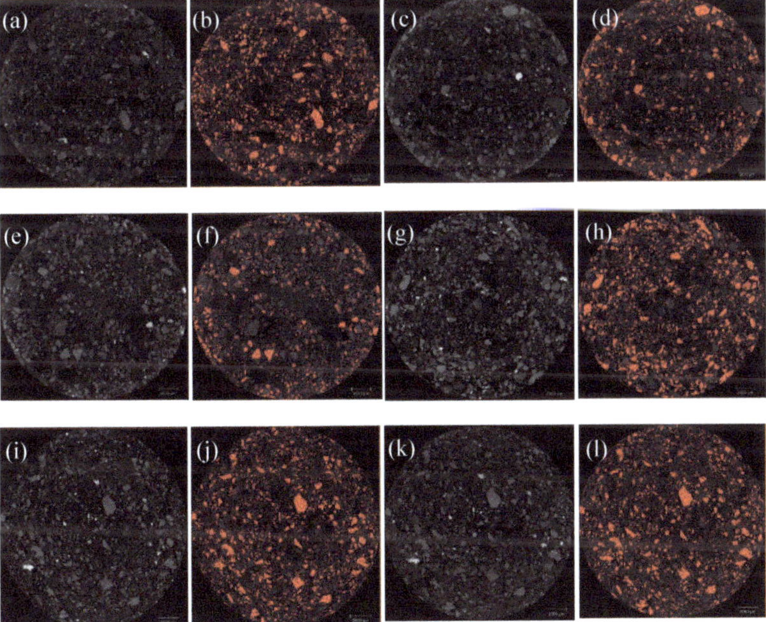

Figure 13. *Cont.*

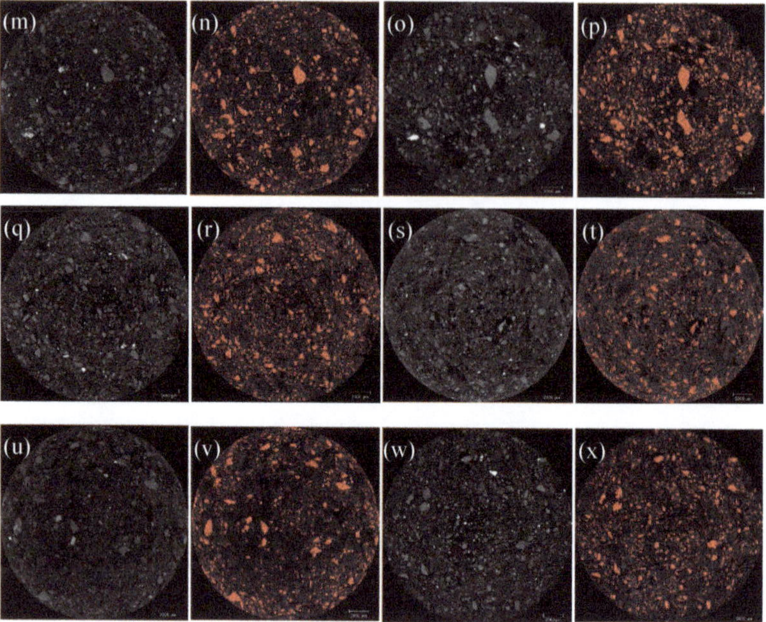

Figure 13. Threshold segmentation of the tailings sand particles. (**a**,**b**) First-level head of the 30% fine content; (**c**,**d**) Second-level head of the 30% fine content; (**e**,**f**) Third-level head of the 30% fine content; (**g**,**h**) Fourth-level head of the 30% fine content; (**i**,**j**) First-level head of the 50% fine content; (**k**,**l**) Second-level head of the 50% fine content; (**m**,**n**) Third-level head of the 50% fine content; (**o**,**p**) Fourth-level head of the 50% fine content; (**q**,**r**) First-level head of the 70% fine content; (**s**,**t**) Second-level head of the 70% fine content; (**u**,**v**) Third-level head of the 70% fine content; (**w**,**x**) Fourth-level head of the 70% fine content.

Based on the data of tailings sand particles obtained after threshold segmentation, 200 layers of 2D slices were used as a region, and a total of 1000 layers were divided into five regions through the Extract Subvolume command in the 3D visualization software of Avizo. Particle size distributions of 10–20 μm, 20–40 μm, and 40–75 μm were counted.

3.3.1. Thirty Percent Fine Content Sample

The statistical results are shown in Tables 7–10 and Figure 14.

Table 7. Variation in the number of fine particles percolated with the 30% fine content (0–10 μm).

Head Level	Layers					Total Number of Particles
	0–200 Layers	200–400 Layers	400–600 Layers	600–800 Layers	800–1000 Layers	0–10 μm
1	26,653	20,791	17,301	17,441	18,944	101,130
2	19,751	14,573	14,862	15,978	19,242	84,406
3	16,894	14,363	13,917	13,750	16,062	74,986
4	18,444	15,048	14,558	14,469	12,000	74,519

Table 8. Variation in the number of fine particles percolated with the 30% fine content (10–20 μm).

Head Level	Layers					Total Number of Particles
	0–200 Layers	200–400 Layers	400–600 Layers	600–800 Layers	800–1000 Layers	10–20 μm
1	19,241	16,522	14,149	13,905	14,608	78,425
2	15,282	12,091	12,460	13,044	16,546	69,423
3	12,337	10,511	11,703	10,406	14,108	59,065
4	14,183	10,871	9400	10,753	11,156	56,363

Table 9. Variation in the number of fine particles percolated with the 30% fine content (20–40 μm).

Head Level	Layers					Total Number of Particles
	0–200 Layers	200–400 Layers	400–600 Layers	600–800 Layers	800–1000 Layers	20–40 μm
1	10,450	10,572	10,384	10,157	10,312	51,875
2	10,248	9187	9322	9704	10,299	48,760
3	7881	7220	7186	7160	9973	39,420
4	7423	7080	7315	7237	8605	37,660

Table 10. Variation in the number of fine particles percolated with the 30% fine content (40–75 μm).

Head Level	Layers					Total Number of Particles
	0–200 Layers	200–400 Layers	400–600 Layers	600–800 Layers	800–1000 Layers	40–75 μm
1	7721	7394	7518	7758	8038	38,429
2	7472	7144	7095	7247	8036	36,994
3	6086	5661	5388	5582	6844	29,561
4	5591	5814	6387	6161	6644	30,597

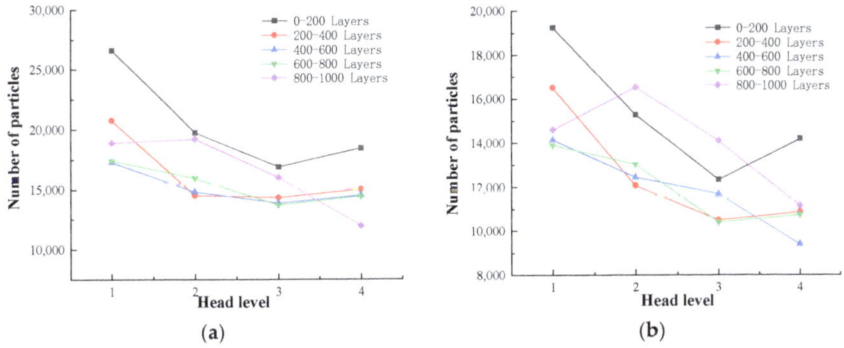

Figure 14. *Cont.*

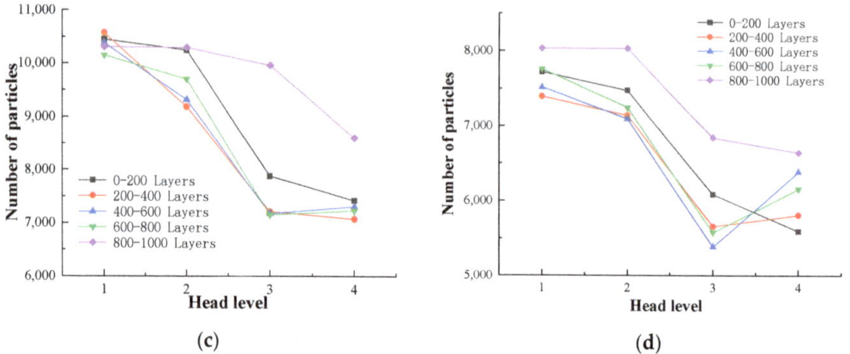

Figure 14. Variation in the number of fine particles percolated with the 30% fine content. (**a**) Particle size: 0–10 μm; (**b**) Particle size: 10–20 μm; (**c**) Particle size: 20–40 μm; (**d**) Particle size: 40–75 μm.

From the above chart, the following can be observed:

(1) As the water pressure level increases, the tailings particles with particle sizes of 0 to 75 μm in the same area generally show a gradual decrease in the direction of seepage. The ground moves in the direction of seepage. In the first three levels of the head difference, as the head difference increases, the particle migration becomes more obvious, and the particle size of the fine-grained tailings continues to increase. There is a proportional relationship between the two.

(2) In the first water pressure level to the second water pressure level, as the infiltration damage starts to occur, the tailings particles with particle sizes of 0–20 μm in the same area decrease significantly, and the tailings particles with particle sizes of 20–75 μm decline very slowly. The tailings particles with particle sizes of 0 to 20 μm are most sensitive to the seepage forces during the initial stage of infiltration and destruction. The seepage channels formed during the initial stage of infiltration and destruction are relatively small but can form good flow migration in the seepage channel. However, the tailings with particle sizes of 20 to 75 μm are larger, and thus, the flow is hindered in the pore channel. Additionally, in the 0–400 layers at the bottom of the sample, the particles migrate. This phenomenon is significant, and the downward trend of particles above 400 layers is obviously weakened. The analysis shows that some fine particles at the bottom of the sample migrate to fill the pore space above the 400 layers, and because the head difference is small, the particles stay in the space, resulting in more than 400 layers of the sample. Furthermore, the content of fine particles in the space of 0–20 μm increases, which is an upward trend compared with the 200–400 layers.

(3) From the second water pressure level to the third water pressure level, the 0 to 10 μm particle size tailings particles in the 0–800-layer sample have a slower downward trend, and the 20 to 75 μm size tailings particles begin to rapidly decline. It can be seen that during the infiltration failure process, the seepage channels and pores continue to develop and expand to make the tailings particles with particle sizes of 20 to 75 μm fully flow. In the 800–1000-layer sample, the fine particle content is from 20–75 μm. The upper-level head has already migrated, and thus, the number of particles does not decrease significantly. In contrast, the migration of fine-grained tailings with particle sizes of 40–75 μm is more obvious than that of the 20–40 μm particles, which confirms that the fine-grained tailings sand has migrated. The particle size migration is proportional to the head difference.

(4) From the third water pressure level to the fourth water pressure level, the tailings particles with particle sizes from 0 to 75 μm generally show an increasing trend. The seepage channels and pores of the tailings sample are completely destroyed. After pressure relief, the CT scans show

the obvious tailings particle sedimentation in the seepage channel, indicating that a penetrating infiltration failure channel is formed inside.

3.3.2. Fifty Percent Fine Content Sample

The statistical results of the 50% fine content are shown in Tables 11–14 and Figure 15.

Table 11. Variation in the number of fine particles percolated with the 50% fine particle content (0–10 µm).

Head Level	Layers					Total Number of Particles
	0–200 Layers	200–400 Layers	400–600 Layers	600–800 Layers	800–1000 Layers	0–10 µm
1	23,766	18,299	17,785	16,910	16,568	93,328
2	26,602	19,188	18,753	19,365	16,355	100,263
3	18,741	14,446	16,426	16,607	15,548	81,768
4	42,318	28,355	19,772	24,064	23,177	137,686

Table 12. Variation in the number of fine particles percolated with the 50% fine particle content (10–20 µm).

Head Level	Layers					Total Number of Particles
	0–200 Layers	200–400 Layers	400–600 Layers	600–800 Layers	800–1000 Layers	10–20 µm
1	33,729	18,744	19,050	18,225	17,541	107,289
2	24,766	19,084	19,974	19,998	16,777	100,599
3	18,999	15,798	17,927	17,105	15,843	85,672
4	35,974	27,324	22,419	25,302	24,038	135,057

Table 13. Variation in the number of fine particles percolated with the 50% fine particle content (20–40 µm).

Head Level	Layers					Total Number of Particles
	0–200 Layers	200–400 Layers	400–600 Layers	600–800 Layers	800–1000 Layers	20–40 µm
1	18,156	15,488	16,652	15,824	15,339	81,459
2	17,645	15,627	16,806	16,350	15,042	81,470
3	15,878	13,677	15,650	14,216	13,299	72,720
4	19,534	17,989	18,778	19,091	18,482	93,874

Table 14. Variation in the number of fine particles percolated with the 50% fine particle content (40–75 µm).

Head Level	Layers					Total Number of Particles
	0–200 Layers	200–400 Layers	400–600 Layers	600–800 Layers	800–1000 Layers	40–75 µm
1	9920	9387	10,290	10,190	10,485	50,272
2	9221	8956	9944	10,209	10,179	48,509
3	9459	8609	9782	9287	8721	45,858
4	6976	7723	9674	10,053	9544	43,970

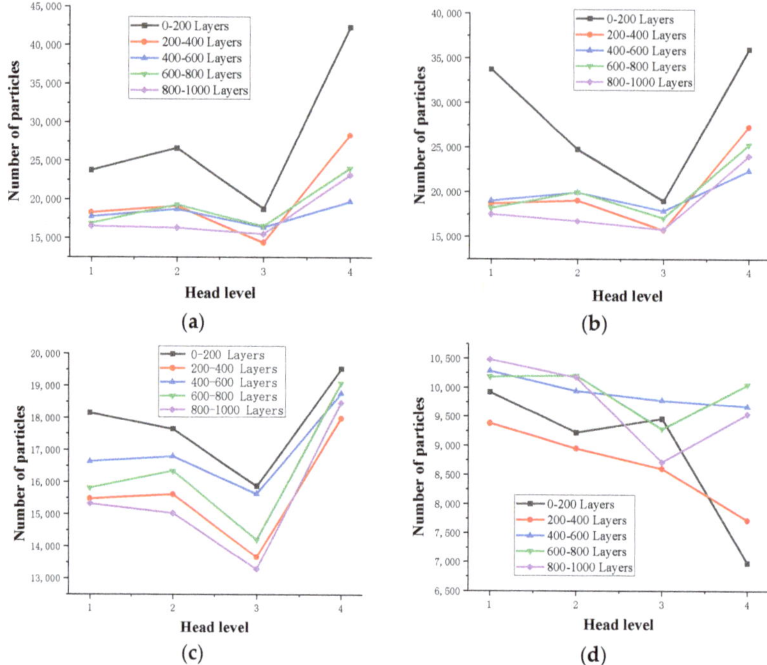

Figure 15. Variation in the number of fine particles percolated with the 50% fine content. (**a**) Particle size: 0–10 μm; (**b**) Particle size: 10–20 μm; (**c**) Particle size: 20–40 μm; (**d**) Particle size: 40–75 μm.

From the above chart, the following can be observed:

(1) In the first water pressure level to the second water pressure level, with increasing seepage pressure, the tailings particles with particle sizes from 0 to 10 μm in the entire sample begin to increase, indicating that there is 0 in the sample. The connected pore structure of 0–10 μm demonstrates a significant particle migration phenomenon. The fine particle content of 10–20 μm rapidly decreases at the bottom of the scanned sample, and the number increases in areas that are more than 200 layers larger than the scanned area, indicating that the internal pore size of the sample is greater than 10 μm. The pores are not developed, and the particles remain in the pores after migration within this size range, resulting in an increase in the fine particle content of pores of 10–20 μm. Fine particles larger than 20 μm show a certain downward trend as a whole. However, the downward trend is not obvious and almost remains the same, indicating that the connected pores larger than 20 μm in diameter do not develop inside the sample, and there are fewer internal seepage channels.

(2) From the second water pressure level to the third water pressure level, the permeability increases, the internally connected pores of the sample begin to develop, the fine particles of each particle size level begin to migrate and flow, and the overall content decreases.

(3) At the third water pressure level to the fourth water pressure level, the sample is damaged, and the internal pores are fully developed. As the pressure is released, the fine particles at each particle size level in the sample rapidly increase.

3.3.3. Seventy Percent Fine Content Sample

The statistical results of the 70% fine content are shown in Tables 15–18 and Figure 16.

Table 15. Variation in the number of fine particles percolated with the 70% fine particle content (0–10 μm).

Head Level	Layers					Total Number of Particles
	0–200 Layers	200–400 Layers	400–600 Layers	600–800 Layers	800–1000 Layers	0–10 μm
1	87,828	391	766	40,270	32,733	161,988
2	19,479	22,673	19,724	20,699	15,542	98,117
3	34,604	26,022	24,874	24,237	23,683	133,420
4	45,777	27,944	22,124	21,434	22,105	139,384

Table 16. Variation in the number of fine particles percolated with the 70% fine particle content (10–20 μm).

Head Level	Layers					Total Number of Particles
	0–200 Layers	200–400 Layers	400–600 Layers	600–800 Layers	800–1000 Layers	10–20 μm
1	55,980	841	1118	34,591	29,263	121,793
2	16,891	20,360	17,716	18,240	13,863	87,070
3	32,021	26,572	25,585	26,746	23,308	134,232
4	39,505	28,175	23,089	22,083	21,968	134,820

Table 17. Variation in the number of fine particles percolated with the 70% fine particle content (20-40 μm).

Head Level	Layers					Total Number of Particles
	0–200 Layers	200–400 Layers	400–600 Layers	600–800 Layers	800–1000 Layers	20–40 μm
1	23,315	5911	3267	23,898	21,525	77,916
2	12,441	15,808	13,941	14,219	11,125	67,534
3	23,148	21,702	20,354	22,096	18,821	106,121
4	26,506	23,883	20,851	19,905	18,711	109,856

Table 18. Variation in the number of fine particles percolated with the 70% fine particle content (40–75 μm).

Head Level	Layers					Total Number of Particles
	0–200 Layers	200–400 Layers	400–600 Layers	600–800 Layers	800–1000 Layers	40–75 μm
1	7115	8975	6980	11,679	11,576	46,325
2	7160	9582	8743	8640	6796	40,921
3	11,566	11,140	10,648	11,728	10,212	55,294
4	12,209	12,674	11,978	11,141	10,387	58,389

(1) In the first water pressure level to the second water pressure level, with increasing seepage pressure, the particle sizes of 0-10 μm, 10–20 μm, and 20–40 μm are exhibited within the range of 0–200 layers. The tailings particles decrease rapidly, while the fine content in the 200–600-layer increases, indicating that the fine content at the bottom migrated to the middle of the sample and that the internal pores of the sample are not well developed during the initial stage of percolation. No continuous percolation channel is formed.

(2) Under the second water pressure level to the third water pressure level, the seepage pressure increases, the migration force between particles increases, the overall content of the fine particles of each particle size increases, the internal pressure increases, and the connected pores begin to develop slowly. The sample will soon be damaged by liquid soil.

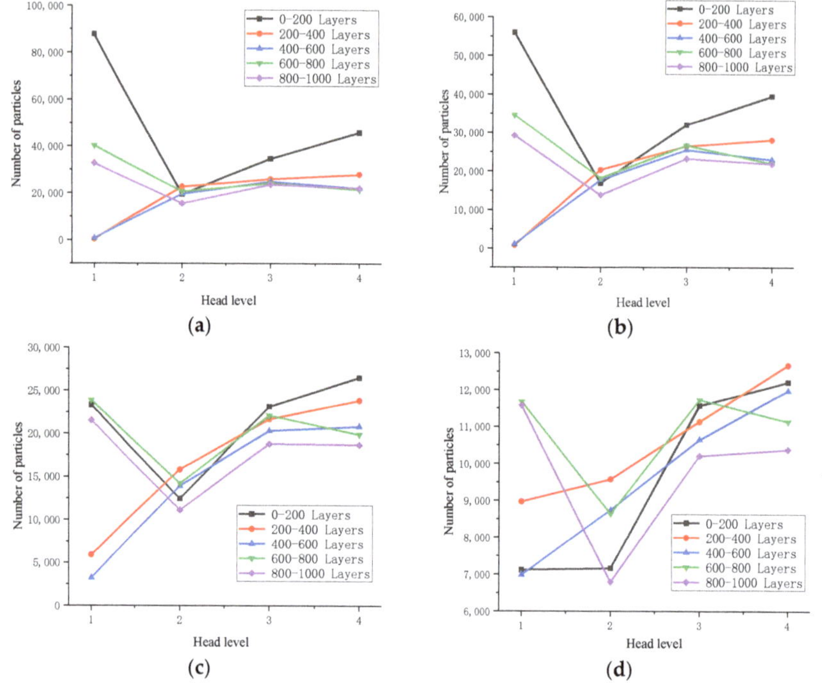

Figure 16. Variation in the number of the fine particles percolated with the 70% fine content. (**a**) Particle size: 0–10 μm; (**b**) Particle size: 10–20 μm; (**c**) Particle size: 20–40 μm; (**d**) Particle size: 40–75 μm.

3.3.4. Changes in the Total Value of the Fine Content of the Three Samples

The statistics regarding the changes in the fine contents in three kinds of fine-grained tailings sand samples during percolation are shown in Tables 19–22 and Figure 17.

Table 19. Fine particle content change during seepage (0–10 μm).

Head Level	30% Fine Content	50% Fine Content	70% Fine Content
1	101,130	93,328	161,988
2	84,406	100,263	98,117
3	74,986	81,768	133,420
4	74,519	137,686	139,384

Table 20. Fine particle content change during seepage (10–20 μm).

Head Level	30% Fine Content	50% Fine Content	70% Fine Content
1	78,425	107,289	161,988
2	69,423	100,599	98,117
3	59,065	85,672	133,420
4	56,363	135,057	139,384

Table 21. Fine particle content change during seepage (20–40 μm).

Head Level	30% Fine Content	50% Fine Content	70% Fine Content
1	51875	81459	77916
2	48760	81470	67534
3	39420	72720	106121
4	37,660	93,874	109,856

Table 22. Fine particle content change during seepage (40–75 μm).

Head Level	30% Fine Content	50% Fine Content	70% Fine Content
1	38,429	50,272	46,325
2	36,994	48,509	40,921
3	29,561	45,858	55,294
4	30,597	43,970	58,389

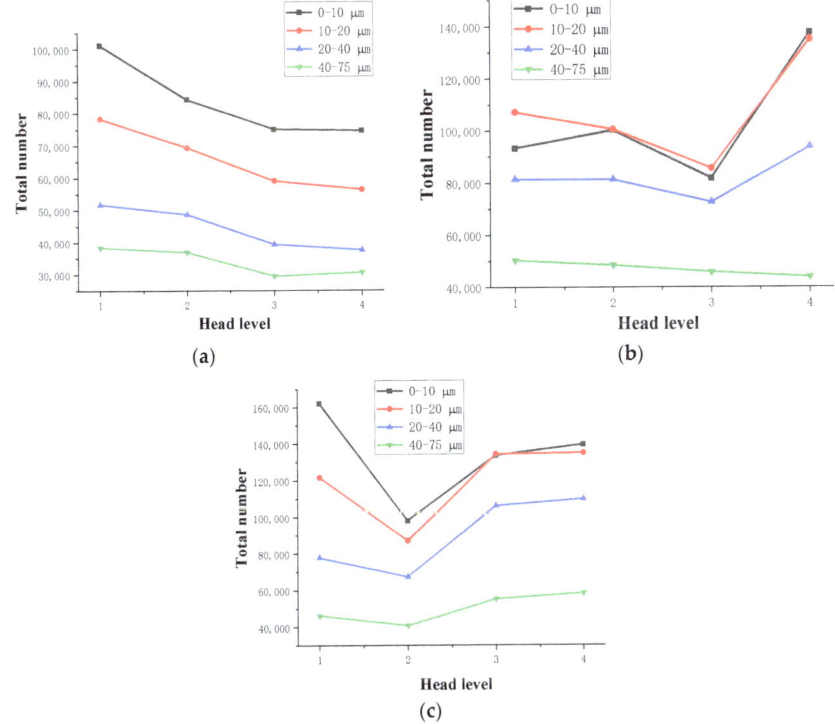

Figure 17. The number of fine particles with different particle sizes. (**a**) 30% fine content; (**b**) 50% fine content; (**c**) 70% fine content.

(1) It can be seen that the total content of all fine particles in the 30% fine particle sample decreases during percolation, and the fine particles flow out during percolation. Clearly, a piping-type infiltration failure has occurred.

(2) Regarding the 50% and 70% fine particle contents, all of the fine particle contents of the samples decreased first and then increased rapidly, and the 70% fine particle content sample changes significantly more than the 50% fine particle content sample. At the beginning of the percolation, the fine particles continuously flow out in the seepage direction. As the seepage pressure increases,

the fine particles do not have enough pores to flow, and the soil breaks down. After settlement, the fine particle content increases significantly.

(3) Through a comparative analysis, it can be observed that during the percolation of the sample, fine particles migrate along the percolation direction, and the smaller the particles are, the more obvious the migration. The difference is that the number of fine particles will affect the type of osmotic failure of the tailings sand sample. The tailings sand with a fine content near 30% will undergo piping-type osmotic failure, and its fine particle content will continue to decrease. At the 30% fine content, the tailings sand will undergo osmotic failure, its fine particle content will decrease, and then all of the particles will flow.

Comparing the fine particle content migration with the different fine particle contents and different particle diameters, it can be seen that the occurrence of fine particle tailings sand permeation failure is closely related to the fine particle content migration and the formation of the seepage channels. The relationship between the changes in the fine particle content of different particle sizes and the head pressure during the seepage process of tailing sand samples with different fine particle contents is shown in Figure 18.

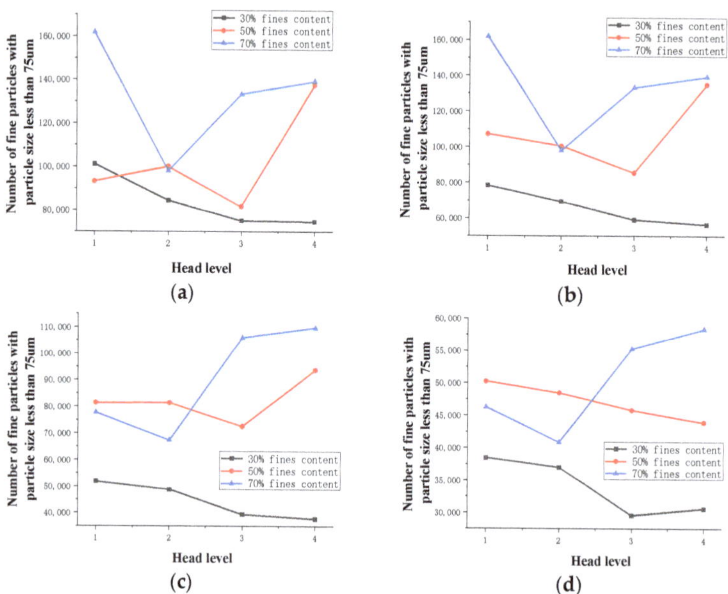

Figure 18. Variation in the particle number during infiltration with different fine contents. (**a**) Particle size: 0–10 μm; (**b**) Particle size: 10–20 μm; (**c**) Particle size: 20–40 μm; (**d**) Particle size: 40–75 μm.

From Figure 18, the following can be observed:

(1) The content of the 30% fine particles shows a downward trend, which indicates that the internal seepage channels and connected pore volumes are constantly expanding, and the tailings sand samples undergo piping-type infiltration failure.

(2) The content of 50% of fine particles shows a decreasing trend at the beginning of the seepage, but the overall change is not large. With increasing water pressure, the sample suffers soil damage, the internal seepage channels are not obvious, and the connected pore volume slowly increases.

(3) For the 70% fine particle content, during the early stage of seepage, due to a large number of fine particles, the 0–20 μm fine particle content rapidly decreases, but the internally connected

pore volume grows less. As the seepage pressure increases, the flow-type soil infiltration is quickly damaged.

During the seepage process, the changes in the number of fine-grained particles under the heads at all levels are shown in Table 23 and Figure 19.

Table 23. Fine particle content change during seepage (0–75 μm).

Head Level	30% Fine Content	50% Fine Content	70% Fine Content
1	269,859	332,347	408,023
2	239,583	330,842	293,643
3	203,031	286,018	429,067
4	199,139	410,587	442,449

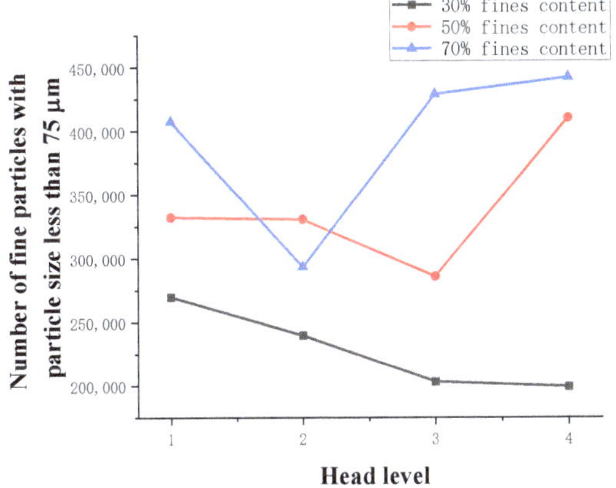

Figure 19. Fine particle content as a function of the number of infiltration processes.

From the above chart, the following can be observed:

(1) For the 30% fine particle content, the fine particle content of the sample decreases continuously at all levels of the seepage pressure until it stabilizes, and piping damage occurs

(2) For the 50% fine particle content, the total number of fine particles of the sample slowly decreases at the beginning of the seepage. As the seepage pressure increases, the rate of fine particles decreases gradually. After the infiltration failure of fluid soil occurs, the fine particle content rapidly increases due to settlement.

(3) For the 70% fine particle content of the sample at the beginning of the seepage, due to a large number of fine particles, the total number of fine particles decreases rapidly. After infiltration and destruction of the raw fluid soil, the fine particle content increases rapidly due to sedimentation.

(4) Through comparative analysis, it can be found that for samples with a small number of fine particles, since the connected pores are relatively developed, there are continuous fine particle seepage channels, and the fine particle content is also small. During the seepage process, fine particles will continue to migrate. The type of osmotic failure caused by effluent flow is a piping failure. For samples with a large number of fine particles, the internal communication pores are poorly developed, there is no continuous percolation channel, and the sample has a large number of fine particles. With increasing water pressure, cracks will appear in the sample, the whole flow will flow in the direction of seepage, and the soil will be damaged.

Both the relationship between different proportions of fine-grained tailings sand samples and the initial number of fine particles and the relationship between different proportions of fine-grained tailings sand samples and the average number of fine-grained contents per unit volume are analyzed, as shown in Figures 20 and 21.

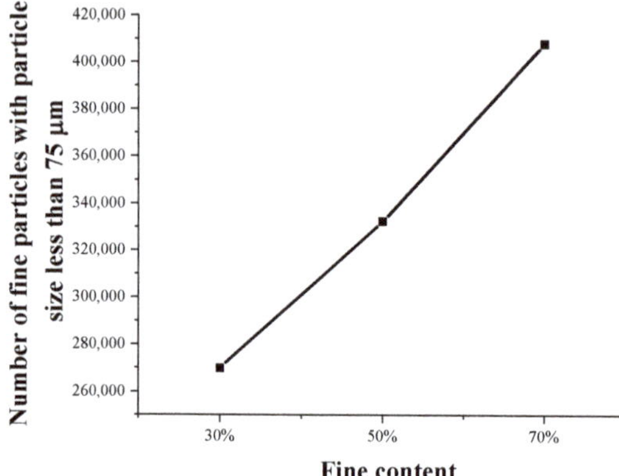

Figure 20. Relationship between the proportion of the fine content and the initial content of fine particles.

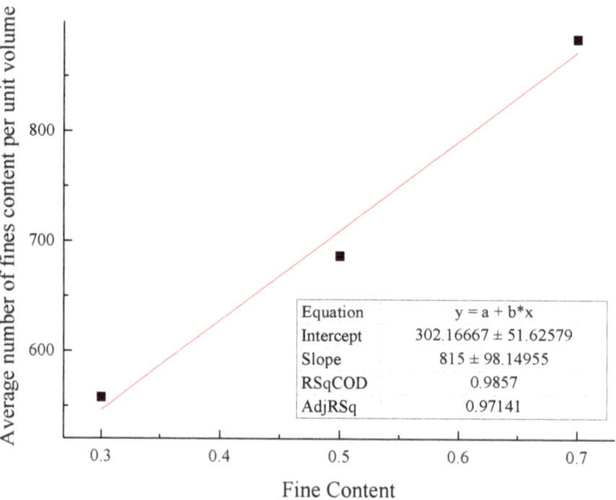

Figure 21. Relationship between the proportion of the fine content and the average content of fine particles per unit volume.

It can be seen that there is a clear linear relationship between the content of fine particles and the total number of fine particles. At the same time, the analysis of the fine particle content in the unit volume is the ratio of the fine particle content of the sample to the sample volume. Through data fitting, it can be found that the percentage of fine particle content and the fine particle content in the unit volume exhibit a very obvious linear correlation. Therefore, the fine particle content per unit volume can be used as the mesoparameter of samples with different fine particle contents.

3.4. Macroscopic and Mesofactor Analysis of the Seepage Process

Based on the macrophysical and mechanical properties, a comparative analysis of the parameters of the microstructure characterization of the tailings sand sample is performed. The macro- and mesofactor comparison table is shown in Table 24. The relationship between the fitted macro- and mesofactors is shown in Figures 22–26.

Table 24. Macro- and mesofactor comparison table.

Fine Content	Fine Content Total Number of Particles	Average Number of Fine Content Per Unit Volume	Average Porosity %	Total Volume of Initial Connected Pores	Permeability Coefficient (cm·s⁻¹)	Internal Friction Angle $\phi/°$	Cohesion c/kPa
30%	269,859	558	9.74342101	3.90×10^{10}	1.02×10^{-4}	40.4	31.52
50%	332,347	687	7.35860292	1.04×10^{10}	2.90×10^{-5}	38.6	35.12
70%	408,023	884	4.81893178	7.41×10^{9}	1.49×10^{-5}	36.7	42.36

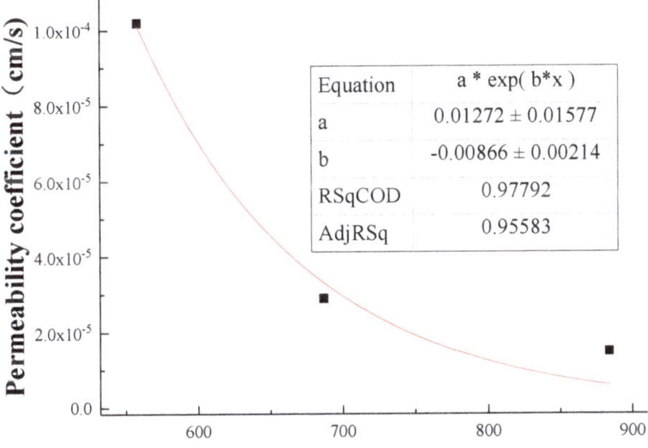

Figure 22. Fitting curve of the fine particle content and permeability coefficient per unit volume.

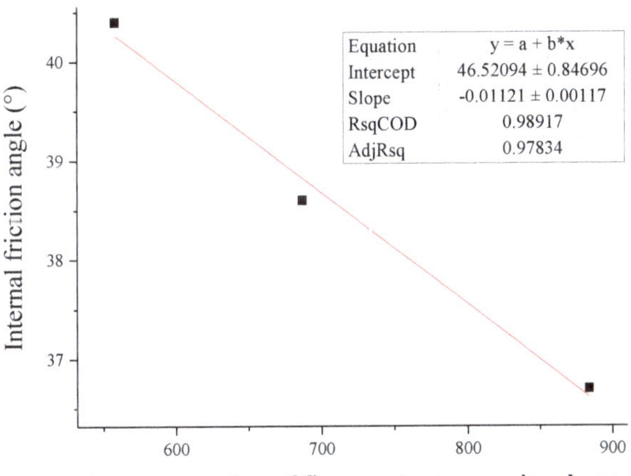

Figure 23. Fitting curve of the fine particle content and internal friction angle per unit volume.

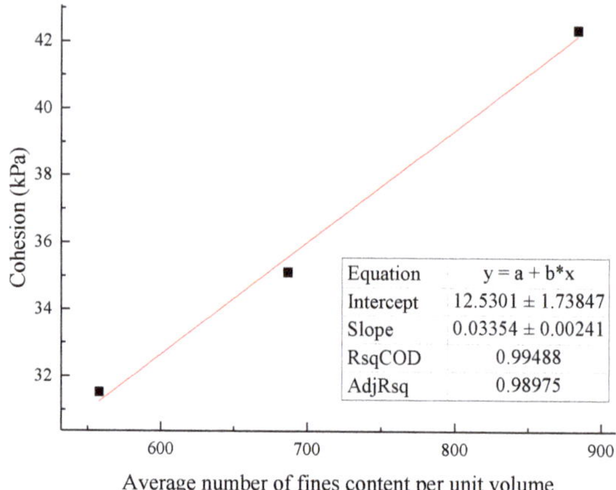

Figure 24. Fitting curve of the fine particle content and cohesion per unit volume.

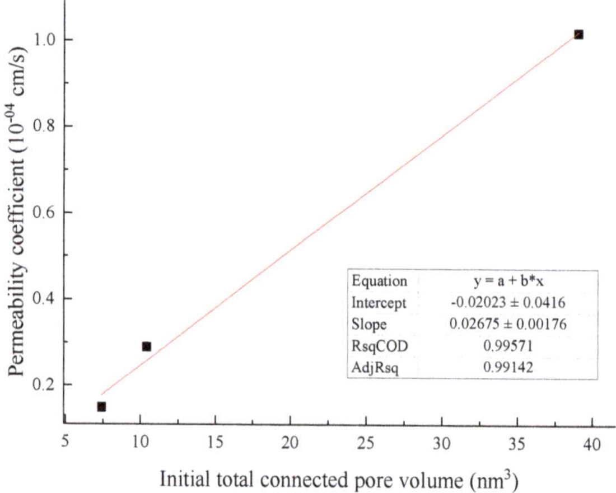

Figure 25. Fitting curve of the connected pore volume and permeability coefficient.

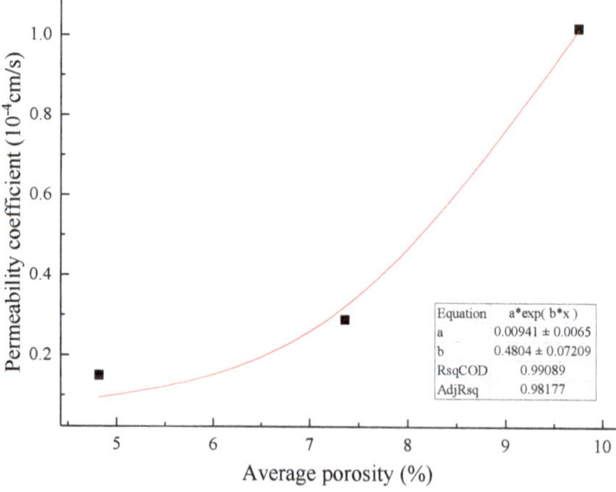

Figure 26. The fitting curve of the average porosity and permeability coefficient.

The following conclusions can be drawn from the above chart:

(1) Analyzing the data relationship between the content of fine particles per unit volume and the permeability coefficient, internal friction angle, and cohesion that affect the seepage and failure of the tailings sand sample macroscopically, the following data characteristics can be found:

With increasing fine particle content per unit volume, the permeability coefficient and the internal friction angle gradually decrease. The former shows a more obvious exponential function relationship, while the latter shows a more obvious positive correlation with a linear function. For the fitting formula, see Equations (1) and (2).

$$k = a_1 \times e^{b_1 n} \tag{1}$$

where k is the internal friction angle, cm·s^{-1}; n is the fine particle content per unit volume; a_1 is a material parameter, which can be obtained by fitting the experimental data; and b_1 is a material parameter, which can be obtained by fitting the experimental data.

$$\varphi = a_2 + b_2 \times n \tag{2}$$

where φ is the permeability coefficient, °; n is the connected pore volume, μm^3; a_2 is a material parameter, which can be obtained by fitting the experimental data; and b_2 is a material parameter, which can be obtained by fitting the experimental data.

As the content of fine particles in a unit volume increases, the cohesion gradually increases, and the two show a relationship in the form of an exponential function. The fitting formula is shown in Equation (3):

$$c = a \times e^{bn} \tag{3}$$

where c is the cohesion, kPa; n is the fine particle content per unit volume; a is a material parameter, which can be obtained by fitting the experimental data; and b is a material parameter, which can be obtained by fitting the experimental data.

(2) The connected pore volume and the average porosity of the sample can be used as the macro- and micropore structure characteristics of the sample. The data relationship analysis reveals the following data characteristics:

The initial volume of connected pores has a positive linear correlation with the permeability coefficient. The fitting formula is shown in Equation (4).

$$k = a + b \times V \tag{4}$$

where k is the permeability coefficient, cm·s^{-1}; V is the connected pore volume, μm^3; a is a material parameter, which can be obtained by fitting the experimental data; and b is a material parameter, which can be obtained by fitting the experimental data.

The fitting curve of the average porosity and permeability coefficient shows an exponential relationship, and it's fitting formula is shown in Equation (5).

$$k = a_1 \times n^{b_1} \tag{5}$$

where k is the permeability coefficient, cm·s^{-1}; n is the average porosity of the sample, %; a_1 is a material parameter, which can be obtained by fitting the experimental data; and b_1 is a material parameter, which can be obtained by fitting the experimental data.

Through the analysis of the above data, it can be found that the number of fine particles in a unit volume and the connected pore volume can be used as microstructure parameters to establish a relationship with the macroscopic physical and mechanical properties of the sample. The observation angle is combined with the macrophenomena for comparative analysis.

4. Discussion

4.1. Piping-Type Infiltration Failure

Piping is a very complicated failure process. Some scholars believe that the soil mainly consists of two parts; one is the skeleton of the soil, and the other is the movable fine particles contained in the soil skeleton. Therefore, under the action of water flow, through the pores in the soil skeleton, fine particles move with the action of water flow [18].

When the drag force of the water flow in the soil [19–22] breaks through the fine static balance between the soil particles, the effective stress between the skeletons decreases and the loss of fine particles occurs inside the soil, forming a large number of infiltration quicksand channels. The rearrangement and deposition of soil particles lead to changes in the microstructure and mechanical properties of the soil [23], and the stress redistribution between the framework particles further reduces the overall stability of the soil, causing piping failure. The development of piping is a process that gradually erodes from downstream particle loss to upstream particle loss [24]. Some scholars have also proposed a capillary model for piping in non-cohesive soils, suggesting that the permeability during piping is significantly affected by fine particle loss [25].

Summarizing the microscopic structural characteristics of the piping-type infiltration failure sample during the seepage process, it can be found that the total amount of fine particles in the sample shows a downward trend, which explains the phenomenon of the continuous loss of fine particles. Its layer-by-layer porosity also increases uniformly with the flow of seepage, and the volume of connected pores also increases, but the number of pores decreases, indicating that small micropores are connected to large pores due to the loss of fine particles. The skeleton structure of the sample was changed, which led to changes in the mechanical and permeability characteristics.

4.2. Soil-Type Infiltration Failure

The mechanism of soil-type infiltration failure can be analyzed with the dominant flow produced by the seepage in the sample [26]. Dominant flow refers to the concentrated and rapid flow of water in the pores in a few areas, while the velocity in other areas is much smaller than the velocity in the fast channel [27]. The formation and development of the dominant flow have spatiotemporal variability [28,29].

In this test, the spatial variability of the dominant flow in the sample created a cyclic process of "formation → development → disappearance → reformation" of the dominant flow. At the beginning of the seepage test, the seepage force is small, the dominant flow is not obvious, and fine particles cannot be dragged to migrate. As the head pressure is gradually increased, the seepage force begins to increase, the fine particles begin to redistribute under the drag of the seepage force, and the permeability coefficient of the sample begins to increase slowly. The randomness of the direction of the pore flow velocity is represented by the non-uniformity of permeation in the macrostructure. Under the non-uniform percolation, large local pores and weak links appear in the sample, and thus, the dominant flow is generated. The generation of the dominant flow provides a channel for the movement of fine particles, and the fine particles gradually move towards the upper part of the channel under the drag of the current. The fine particles in the soil around the dominant flow channel are lost, and the structure is destroyed. The coarse particles will collapse, thus blocking the dominant flow channel and forming an anti-filtration layer, and the dominant flow channel will disappear. The permeability coefficient of the reverse filtration layer is small, and the permeation flow rate is also low. There will be the continuous accumulation of particles, which makes the permeability coefficient of the reverse filtration layer smaller and smaller, and the osmotic pressure on both sides of the reverse filtration layer gradually increases. When the osmotic pressure reaches a critical value, a new dominant flow channel is formed. This explains the abrupt change in the layer-by-layer porosity of the mesostructure that seeps through; that is, the formation of the dominant flow causes the fine particles to migrate to the upper part of the channel and the layer-by-layer porosity increases. Coarse particles collapse, the dominant flow disappears, fine particles begin to accumulate, and the porosity decreases continuously from layer to layer. At the same time, the phenomenon that the coarse particle size at the bottom of the sample increases and the fine particle size decreases sharply when the soil failure occurs in the sample is also explained.

At the beginning of the formation of the dominant flow, the overall permeability coefficient of the soil does not change much, and the space for expansion of the dominant flow is limited. Therefore, the dominant flow can develop and transfer only in local areas. The formation and development of the dominant flow are accompanied by the seepage deformation and seepage failure of the soil body. The dominant flow will continue to find weaker areas in the soil body to form new dominant flows. In the process of slowly increasing the permeability coefficient, new and larger dominant streams continue to emerge. In the cycle of continuous formation → development → disappearance → reformation of the dominant flow, the soil seepage deformation gradually deepens, the soil structure is gradually destroyed, and the macroscopic infiltration velocity in the soil gradually increases. When the critical hydraulic gradient of the soil is reached, at the same time, the soil body was damaged by liquid soil. The structure of the sample has been severely disturbed after the failure of fluid soil occurred, and more microcracks have been formed in the sample, making it easier to move the particles in the soil. The content of fine particles increases sharply, and the number of pores also increases dramatically.

5. Conclusions

(1)　There are significant differences in the microscopic structural characterization between piping-type infiltration failure and fluid soil-type infiltration failure. The 30% fine-grained tailings sand sample has a type of piping failure, and the porosity gradually increases from layer to layer during the seepage process. Small pores of 0 to 10 μm are easily penetrated into pores larger than 10 μm. The total pores, both the number and the number of fine particles, showed a decreasing trend. The infiltration failure types for between 50% and 70% of the fine-grained tailings sand samples were fluid soil-type, and the porosity layer by layer in the middle of the sample significantly increased during the seepage process. The total pore number increased first and then decreased, while the fine particle content decreased first and then increased sharply.

(2)　The piping-type infiltration failure and connected pore development are significantly stronger than the fluid soil-type infiltration failure. The evolution characteristics of the connected pores of the piping-type osmotic failure and the fluid-type osmotic failure are basically similar; that is, the connected pore structure develops from the sample center to the surroundings.

(3)　It is found through experiments that there is a certain functional relationship between the microstructure index of tailings sand, micromechanics, and the permeability index. The fine particle content per unit volume demonstrates a relationship in the form of an exponential function with the permeability coefficient and cohesion. The fine particle content per unit volume has a linear negative relationship with the internal friction angle. The initial total connected pore volume has a positive linear correlation with the permeability coefficient. The average porosity has an exponential relationship with the permeability coefficient. Combining the relationship between macrophysical and mechanical parameters and microstructure parameters, a connection has been established for the comparative analysis of macrophenomena from the microstructure to the infiltration and destruction of tailings sand in the future.

Author Contributions: Conceptualization, Y.S. and C.L.; methodology, Y.S. and D.L.; software, Y.S. and D.L.; validation, Y.S.; formal analysis, Y.S.; investigation, Y.S.; resources, Y.S.; data curation, Y.S. and D.L.; writing—original draft preparation, Y.S.; writing—review and editing, C.L.; visualization, Y.S.; supervision, C.L.; project administration, C.L.; funding acquisition, C.L. All authors have read and agreed to the published version of the manuscript.

Funding: This research is supported by the National Key Research and Development Plan of China (No.2017YFC0804609&No.2018YFC0808402).

Conflicts of Interest: The authors declare that they have no conflicts of interest regarding the publication of this paper.

References

1. Wang, K.; Yang, P.; Lü, W.; Bu, L.; School of Civil and Resource Engineering, University of Science and Technology Beijing; Environment and Sustainability Institute, University of Exeter; Beijing Key Laboratory of Information Service Engineering, Beijing Union University. Status and development for the prevention and management of tailings dam failure accidents. *Chin. J. Eng.* **2018**, *40*, 526–539.
2. Li, C.; Bu, L.; Chen, L. Research situation of the disaster-causing mechanism of tailing dams and its developing trend. *Chin. J. Eng.* **2016**, *38*, 1039–1049.
3. Zhang, G.; Tong, L.; Li, Z. Discussion on treatment method and development and utilization of ultrafine tailings. *Met. Mine* **2017**, *38*, 171–177.
4. Yu, C.; Cheng, J. Statistical analysis of physical and mechanical parameters of fine tailings. *Subgrade Eng.* **2015**, *4*, 95–100.
5. Yin, G.; Wei, Z.; Wan, L.; Zhang, D. Experimental study on reinforcement model of fine tailings dam. *Chin. J. Rock Mech. Eng.* **2005**, *24*, 1030–1034.
6. Yin, G.; Jing, X.; Wei, Z.; Li, X. Model test and field test study on seepage characteristics of crude and fine tail sand dams. *Chin. J. Rock Mech. Eng.* **2010**, *s2*, 3710–3718.
7. Yin, G.; Zhang, Q.; Wei, Z.; Wang, W.; Geng, W. Experimental study on pore water migration characteristics and mechanism of fine structure of tailings. *J. Rock Mech. Eng.* **2012**, *31*, 71–79.
8. Zhang, Q.; Wang, Y.; Li, G.; Yin, G.; Wang, W.; Zhong, Y.; Yang, H. Numerical simulation of particle flow in meso-mechanical mechanism of tailings dam deformation. *J. Chongqing Univ.* **2015**, *38*, 71–79.
9. Qiao, L.; Qu, C.; Cui, M. Analysis of influence of fine particle content on engineering properties of tailings. *Rock Soil Mech.* **2015**, *36*, 923–927.
10. Zhang, C.; Ma, C.; Yang, C.; Chen, Q.; Pan, Z. Effect of particle size on shear strength and stability of tailings. *Chin. J. Geotech. Eng.* **2019**, *41*, 145–148.
11. Zhang, D.; Liu, Y.; Wu, S.; Wang, C. Failure mechanism analysis of tailings dam slope based on discrete-continuous coupling. *Geotech. Eng.* **2014**, *36*, 1473–1482.
12. Bu, L.; Zhou, H.; Li, C. Three-dimensional stability analysis of fine grained tailings dam with complex terrain by means of up-stream method electronic. *J. Geotech. Eng.* **2016**, *10*, 3905.
13. Xie, Y. Experimental Study on Seepage Failure Characteristics of Tailings Dam. Master's Thesis, Southwest University of Science and Technology, Mian Yang, China, 2015.
14. Zhang, P.H.; Lee, Y.I.; Zhang, J.L. A review of high-resolution X-ray computed tomography applied to petroleum geology and a case study. *Micron* **2019**, *124*, 102702. [CrossRef]
15. Han, F. Comparative Experimental Study on Seepage Failure Mechanism of Two Types of Dam Materials in Northern Shan'xi. Master's Thesis, Northwest A & F University, Yang Ling, China, 2017.
16. Saxena, N.; Hofmann, R.; Alpark, F.O.; Dietderich, J.E.; Hunter, S.; Day-Stirrat, R.J. Effect of image segmentation & voxel size on micro-CT computed effective transport & elastic properties. *Mar. Pet. Geol.* **2017**, *86*, 972–990.
17. Otsu, N. A threshold selection method from gray-level histogram. *IEEE Trans. Syst. Man. Cybern.* **1979**, *9*, 62–66. [CrossRef]
18. Zhou, X.; Jie, Y.; Li, G. Numerical simulation of piping based on coupling seepage and pipe flow. *Rock Soil Mech.* **2009**, *30*, 3154–3158.
19. Liu, J. *Lessons and Lessons from Seepage Control Theory of Earth-rock Dam*; China Water Conservancy and Hydropower Press: Beijing, China, 2006.
20. Chen, Q.; Liu, L.; He, C.; Zhu, F. Judgment method of piping type in missing coarse-grained soil. *Rock Soil Mech.* **2009**, *8*, 2249–2253.
21. Lu, L. Experimental Study on Coarse Granular Permeability and Permeability Law. Master's Thesis, Sichuan University, Chengdu, China, 2006.
22. Liu, Z.; Miao, T. Assessment for the noncohesive piping-typed soils. *Rock Soil Mech.* **2004**, *7*, 1072–1076.
23. Luo, Y.; Su, B.; Sheng, J.; Zhan, M. New understandings on piping mechanism. *Chin. J. Geotech. Eng.* **2011**, *33*, 1895–1902.

24. Ding, X.; Chen, L.; Hu, H.; Wen, L.; Li, Y. Experimental Study on the Piping Erosion Mechanism of Nonuniform Sand. *Henan Sci.* **2015**, *33*, 1967–1972.

25. Liu, Z.; Yue, J.; Miao, T. Capillary-Tube Model for Piping in Noncohesive Soils and its Application. *Chin. J. Rock Mech. Eng.* **2004**, *23*, 3871–3873.

26. Huang, D.; Chen, J.; Chen, L.; Wang, S. Laboratory Model Test Study on the Mechanism of Homogeneous Non-cohesive Soil Flow. *Chin. J. Rock Mech. Eng.* **2015**, *s1*, 3424–3431.

27. Liang, Y.; Chen, J.; Chen, L. Numerical simulation model for pore flows and distribution of their velocity. *Chin. J. Geotech. Eng.* **2011**, *33*, 1104–1109.

28. Quisenberry, V.L.; Phillips, R.E.; Zeleznik, J.M. Spatial distribution of water and chloride macropore flow in a well-structured soil. *Soil Sci. Soc. Am. J.* **1994**, *58*, 1294–1300. [CrossRef]

29. Wildenschild, D.; Jensen, K.H.; Villholth, K.; Illangasekare, T.H. A laboratory analysis of the effect of macropores on solute transport. *Ground Water* **1994**, *32*, 381–389. [CrossRef]

Article

Detection of Flaws in Concrete Using Ultrasonic Tomography and Convolutional Neural Networks

Marek Słoński [1,*], Krzysztof Schabowicz [2] and Ewa Krawczyk [2]

[1] Faculty of Civil Engineering, Cracow University of Technology, Warszawska 24, 31-155 Kraków, Poland
[2] Faculty of Civil Engineering, Wrocław University of Science and Technology, Wybrzeże Wyspiańskiego 27, 50-370 Wrocław, Poland; Krzysztof.Schabowicz@pwr.edu.pl (K.S.); krawczyk.em@gmail.com (E.K.)
* Correspondence: Marek.Slonski@pk.edu.pl; Tel.: +48-12-628-2562

Received: 27 February 2020; Accepted: 26 March 2020; Published: 27 March 2020

Abstract: Non-destructive testing of concrete for defects detection, using acoustic techniques, is currently performed mainly by human inspection of recorded images. The images consist of the inside of the examined elements obtained from testing devices such as the ultrasonic tomograph. However, such an automatic inspection is time-consuming, expensive, and prone to errors. To address some of these problems, this paper aims to evaluate a convolutional neural network (CNN) toward an automated detection of flaws in concrete elements using ultrasonic tomography. There are two main stages in the proposed methodology. In the first stage, an image of the inside of the examined structure is obtained and recorded by performing ultrasonic tomography-based testing. In the second stage, a convolutional neural network model is used for automatic detection of defects and flaws in the recorded image. In this work, a large and pre-trained CNN is used. It was fine-tuned on a small set of images collected during laboratory tests. Lastly, the prepared model was applied for detecting flaws. The obtained model has proven to be able to accurately detect defects in examined concrete elements. The presented approach for automatic detection of flaws is being developed with the potential to not only detect defects of one type but also to classify various types of defects in concrete elements.

Keywords: concrete; non-destructive testing; ultrasounds; ultrasonic tomography; acoustic methods; defects; diagnostic; detection; convolutional neural networks; transfer learning

---

1. Introduction

Concrete structural members need to be diagnosed at different times and for different reasons. For many years, attempts have been made to investigate geometrical and material imperfections in concrete members by exploiting the propagation of elastic waves. The current trend in the diagnosis of these elements is to apply a non-destructive testing equipment that allows one to obtain an image of the inside of the examined elements for early detection of flaws [1,2]. Acoustic techniques have seen greater attention because a clear shift has been toward acquiring more information about tested elements from acoustic signals [3–8]. Among the recently developed acoustic techniques, ultrasonic tomography (UT) stands out. Since it is still a relatively new approach, it has been used in a limited number of case studies. In Reference [9], it is recommended to identify defects in concrete members by means of the nondestructive ultrasonic tomography technique. The designers of the multi-probe measuring antenna, used in the ultrasonic tomograph, proposed to use ultrasonic tomography for the location of air voids in concrete members [10,11]. Samokrutov et al. [12–15] described the multiprobe antenna and its possible uses and presented an image-generating algorithm. On the basis of laboratory studies carried out using the tomograph, Schabowicz and Suvorov [16,17] introduced a change into the image-generating algorithm whereby it became possible to isolate the surface wave signals from the total picture of the wave and to remove noise for the benefit of the location of air voids in concrete members available from one side only. The other notable applications include practical use of ultrasonic tomography to test a unilaterally accessible concrete shell of heat pipe carrying tunnel [18] and to test non-destructive assessment of masonry pillars [19].

Currently, the recorded images of the inside of the examined concrete members, obtained from ultrasonic tomography, are inspected mainly manually, which can be expensive, time-consuming, and can be prone to errors. Moreover, there are no standards available, which could make it possible to objectively interpret the results obtained in this way. In that situation, a data-driven approach can be helpful because recent developments in computer hardware and software together with the corresponding advances in deep learning algorithms and availability of large datasets make fully automatic analysis of images possible [20–25]. Among deep learning algorithms, convolutional neural networks (CNNs) are currently the main research tools for automatic image analysis. CNNs go back to the 1980s, but, initially, they were mainly used for optical character recognition (OCR) [26,27]. At present, CNNs are commonly used for automatic extraction of information from data with a grid-like structure such as video and audio and for object detection, image classification, and image segmentation [28]. The use of CNN covers wide spectrum of applications such as in medical images analysis [29] and image-based civil infrastructure inspection and a condition assessment [30–32]. They are also applied in the context of ultrasound tomography for automatic analysis of ultrasound medical images [33,34]. On the other hand, the authors of this paper, are unaware of any applications of CNNs for automatic detection of flaws in concrete elements based on scans from ultrasonic tomography.

With the above in mind, this article contains a full description of the innovative methodology developed by the authors. The main objective of this work is, therefore, set to investigate the novel application of ultrasonic tomography and a convolutional neural network (CNN) for automating the assessment of defects and flaws in concrete elements. However, because it is the first attempt to tackle the flaws detection problem, this work involves few limitations. First, only one type of defect is considered. This assumes that the images with defects, obtained from ultrasonic tomography, belong to one category only. Second, in the experiments, only the images with easily visible flaws are considered. In future experiments, however, these two limitations will be considered to get a fully automatic system for detection and classification of defects in concrete elements.

The remainder of the paper is organized as follows. A brief description of ultrasonic tomography is presented. This is followed by a short introduction to artificial neural networks and a brief overview of convolutional neural networks. These provide the theoretical basis of the work undertaken in this paper. Next, we briefly discuss the proposed methodology for flaws detection. This is followed by a description of specific samples. The dataset is prepared, the neural network model is applied, numerical experiments are carried out, and the results of flaw detection are obtained. Lastly, a short discussion of the results and final conclusions are given.

2. Ultrasonic Tomography

Elastic wave stimulation in the tested element is called the ultrasonic tomography technique. Elastic wave stimulation in the tested element is called the ultrasonic tomography technique. The excitation source is a multi-head antenna. It contains several dozen integrated ultrasonic heads that generate 50 kHz ultrasonic pulses. This frequency is the most suitable for concrete and similar materials [11]. It is also used to receive and send ultrasonic signals with a maximum range not exceeding 2500 mm, which is the most reliable range for the ultrasonic tomographs currently in use [11]. The validity of using ultrasonic tomography for concrete elements results from the fact that concrete has a highly heterogenite inner structure, which causes a high level of structural noise and a fundamental ultrasonic signal damping during ultrasonic testing [11].

Figure 1 shows the new ultrasonic tomograph. The tomograph includes a multi-head ultrasonic antenna with a computer and a customized software suitable for graphic scan recording.

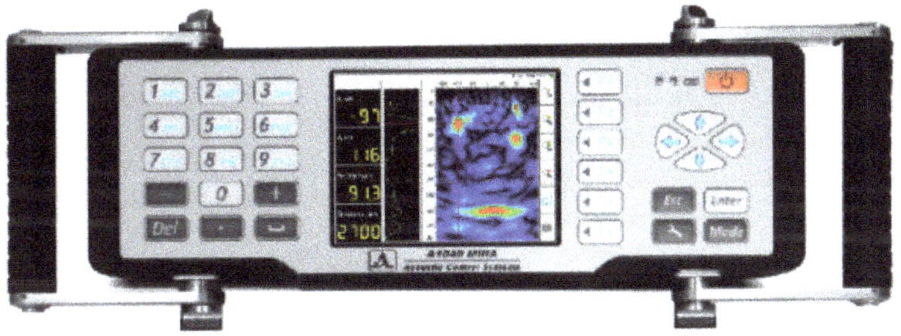

Figure 1. Ultrasonic tomograph: top and bottom view.

Each head is telescopically mounted separately in the antenna, and the head adapts to the testing surface. Thanks to dry contact, no coupling agents or special surface preparation for testing is required. The distance between the heads is 30 mm and 40 mm, respectively, both vertically and horizontally. Figure 2 shows the results of ultrasonic tomography tests in the form of C, D, and B scans and a 3D scan for a concrete element with a defect modeled by PVC (Poly Vinyl Chloride) pipes filled with air because ultrasonic tomography is more sensitive to air voids than to PVC pipes filled with air. The diameter of PVC pipes was 25 and 35 mm.

The results of testing using the ultrasonic tomography are then collected in a matrix table. This matrix is three-dimensional and is subsequently processed by the special software. In that way, it is possible to receive three scans: C, D, B, and 3D view in the mutually perpendicular directions, as shown in Figure 2d. Figure 2e shows names of three mutually perpendicular cross sections (scans) of the tested object and coordinate system used with tomograph antenna.

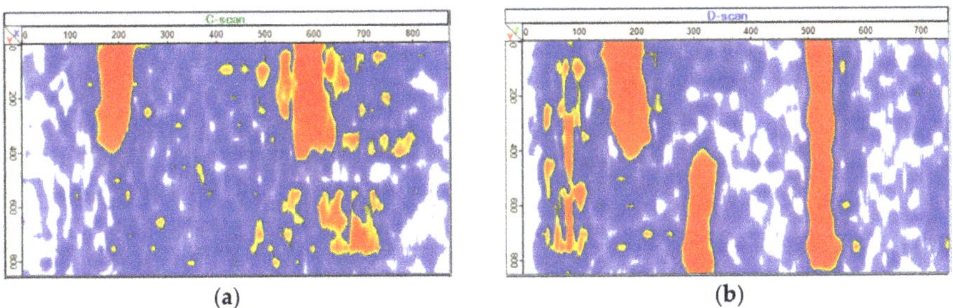

(a)   (b)

Figure 2. *Cont.*

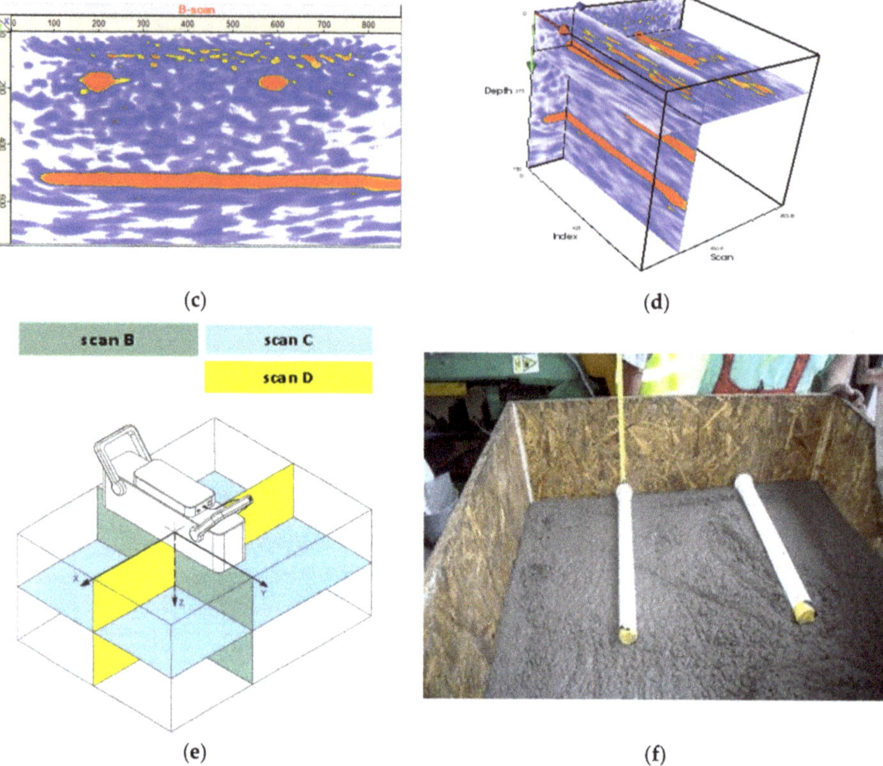

Figure 2. Example of ultrasonic tomography scans for a concrete member filled with air PVC pipes: (**a**) exemplary scan C, (**b**) exemplary scan D, (**c**) exemplary scan B, (**d**) 3D scan, (**e**) names of cross sections (scans) of tested object and coordinate system used with a tomograph antenna, and (**f**) the view of the embedded PVC.

3. Convolutional Neural Networks

The convolutional neural network (CNN) is a special type of a layered feed-forward artificial neural network (ANN), designed for processing signals in the form of multiple arrays like visual and audio signals [21]. The standard layered feed-forward neural network architecture is called a multi-layer perceptron (MLP) [35]. It has been used for many years but has several shortcomings. Up to the mid-2000s, the most popular neural network architecture had one or two hidden layers (i.e., shallow network). If the network has more than two hidden layers, the network is called a deep neural network (DNN).

The typical neural network like MLP or CNN has an input layer, at least one hidden layer with nonlinear units (neurons), and an output layer with linear units (for regression problems) or nonlinear units (for classification problems). A unit (neuron) computes a weighted sum of its inputs called activation of the unit and then the activation is sent to the activation function, which is, in general, an S-shaped function such as a sigmoid function or a rectified linear unit (ReLU) function.

The number of inputs in the input layer is equal to the total number of features in the input dataset. For example, for RGB images (three channels) with 150×150 pixels for each channel, the number of inputs is 67,500. The number of outputs depends on the problem at hand. For example, for the binary classification problem, there is only one output. For multi-class classification, the number of outputs corresponds to the number of classes.

A CNN has a slightly different architecture. It consists mainly of two types of hidden layers, which are a convolutional layer and a pooling layer. There is also a fully-connected (dense) layer that forms the network outputs. These layers are stacked to form a convolutional neural network structure. The structure of a typical CNN with 11 layers is presented in Figure 3.

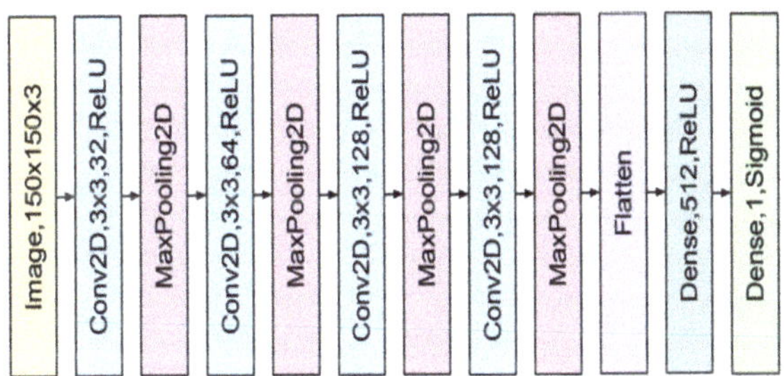

Figure 3. Example of the convolutional neural network (CNN) structure for processing the RGB image of 150×150 pixels in size. The CNN model consists of four convolutional layers with various number of the 3×3 kernel and the ReLU activation function (blue rectangles) with each followed by a pooling layer (red rectangles) and two fully-connected (dense) layers (green rectangles) after a flattened operation.

The aim of the convolutional layer is to detect the input image local features such as horizontal, vertical, or diagonal edges. It is done by using a convolution operation or by computing cross-correlation more strictly [21].

The pooling layer is used for dimensionality reduction of the processed data by combining values of small clusters (matrices) of input data. The most common pooling operation is defined by taking the maximal value. Additionally, the flattened operation is needed to change the input matrix to the output vector, which was then processed by the fully-connected (dense) layer. The fully-connected layer processes the input data by sending data from every unit (neuron) in the previous layer to every unit in another layer.

A layered neural network is qualified by using the backpropagation algorithm to efficiently compute the gradient of a loss function and the mini-batch stochastic gradient descent algorithm (SGD) for learning the weights of the neural network model. During training of convolutional neural networks with many parameters (several millions), especially on a small dataset, the main issue is overfitting the model to the training dataset. There are several methods and techniques to cope with overfitting [36].

One of the most effective techniques is image data augmentation [37]. This technique allows us to expand the training dataset via a number of random transformations such as rotation, shift, zoom, and flipping, which produce visually similar images. Another regularization technique is the dropout method, which is a technique to improve training process performance by switching off (setting to zero) some number (usually 50%) of randomly selected weights [38].

Another problem with learning the large convolutional neural network is the extra-long time needed for a successful training process. One of the possible solutions is to apply transfer learning. Transfer learning is a method used to adapt the pre-trained model to another dataset [39]. It is often done with fine-tuning, which is a technique to improve the performance of the neural network by adapting selected weights and by applying additional training to several convolutional layers of CNN while the rest of the layers are preserved.

There are many commonly used pre-trained CNN models (for example, AlexNet, VGG-16, U-Net), which can be used in other problems by applying transfer learning. The pre-trained network has parameters that were computed on a very large training dataset from a specific domain or task. For example, the benchmark dataset called ImageNet is used for testing novel CNN structures and training algorithms [40].

4. Experimental Study

4.1. Testing Methodology

The methodology for detecting flaws in concrete elements using ultrasonic tomography and convolutional neural networks is shown in the form of a flowchart in Figure 4 and is described in detail below.

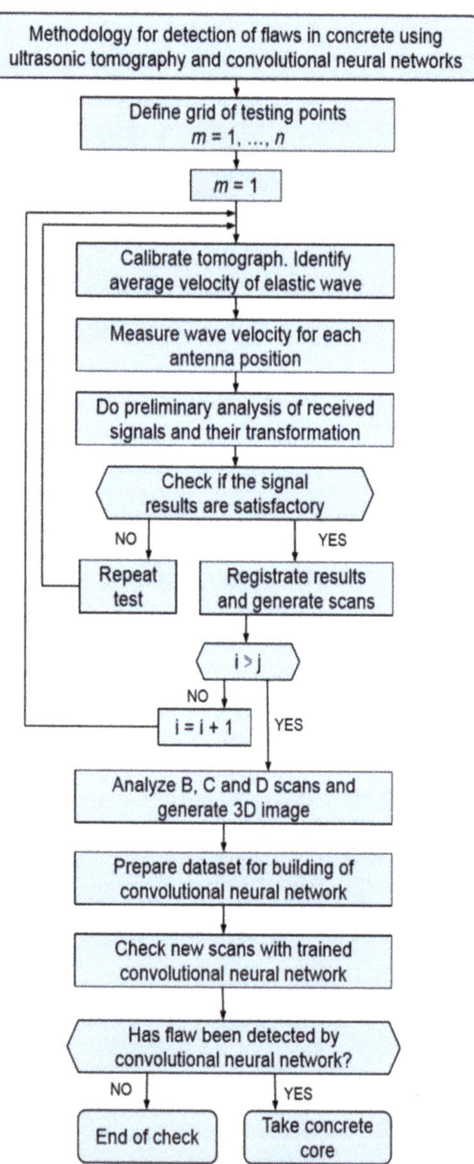

Figure 4. Flowchart presenting methodology for detecting flaws in concrete elements using ultrasonic tomography and convolutional neural networks.

The first step in the testing methodology consists of marking a grid of measuring points i = 1 … j evenly spaced at every 50 mm, with a minimum distance of 50 mm from the edge of the tested concrete member. The spacing can be increased to 100 mm if the surface of the investigated member is considerable. The tomograph is then calibrated by repeatedly measuring ultrasonic wave (signal) velocity and computing its mean value.

At subsequent steps, ultrasonic wave velocity is measured in each antenna position in each of the testing points. During measurements, a preliminary analysis of the ultrasonic signals is performed to find out if the thickness of the member can be identified or the defect can be detected on this basis. If this is not the case, the measured signals are transformed using the customized software. The transformation consists of compiling the registered data for a given measuring point.

If the results are acceptable, they are recorded. Lastly, flat scans B, C, and D are obtained in the three mutually perpendicular directions, showing the inside of the investigated concrete member. By using the dedicated software, it is also possible to build a three-dimensional scan. The next step is analyzing scans B, C, and D, and forming a set of data used for building (training and testing) the convolutional neural network (CNN). Lastly, the trained CNN is presented with a new scan and the network analyzes the new scan and decides if the scan contains a flaw or not.

The presented methodology of measuring and processing obtained results using ultrasonic tomography and convolutional neural networks can be useful when developing a prototype of a computer vision system to automatically detect flaws in concrete elements.

4.2. Dataset for Building Convolutional Neural Networks

For the purpose of this work, ten $1000 \times 1000 \times 1000$ mm^3 concrete cubic specimens were prepared. The specimens were made of C25/30 concrete based on aggregates with 8-mm maximum grading. Next, the specimens were tested in the laboratory using the technique of ultrasonic tomography and the images containing B-scans were recorded. The images were cropped to prepare a dataset, which was used to train a convolutional neural network. A CNN-based detection model was built using a dataset containing only 246 B-scans of concrete elements with flaws (52%) and without flaws (48%). The dataset was divided into three subsets: for training (56%), for validation (22%), and for testing (22%). Table 1 contains the number of training, validation, and testing samples, respectively.

Table 1. Number of training, validation, and testing samples.

Dataset	With Flaws	Without Flaws	Total
Training	71	66	137
Validation	29	26	55
Testing	28	26	54
Total	128	118	246

In Figure 5, images of B-scans of concrete elements without flaws from the training set are shown.

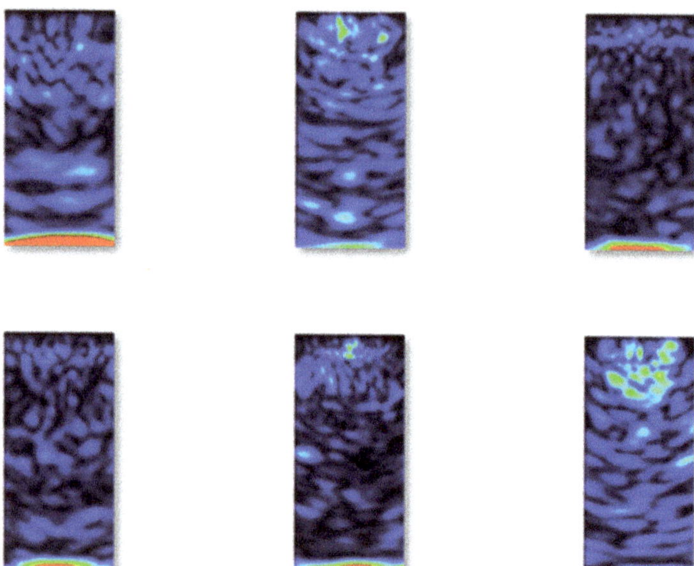

Figure 5. Selected examples of B-scans without flaws from the training set.

Similarly, in Figure 6, images of B-scans of concrete elements with flaws from the training set are shown. Comparing the images, it can be easily seen that the images taken from concrete elements with flaws contain red oval-shaped parts that correspond to the location of the flaws.

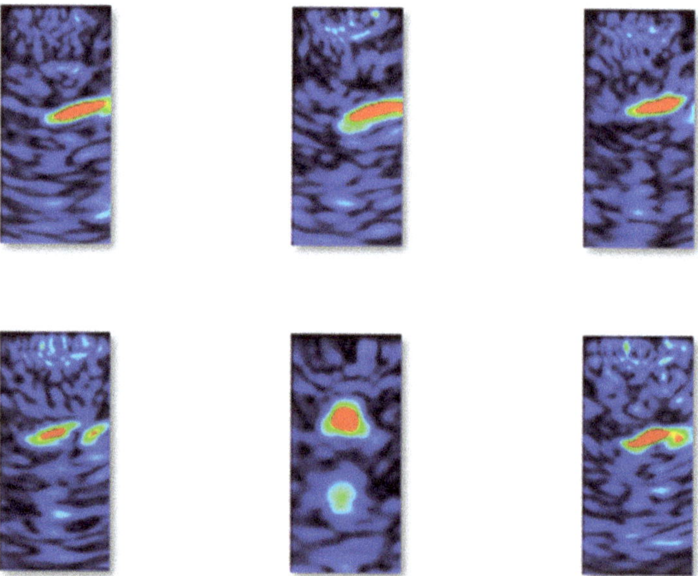

Figure 6. Selected examples of B-scans with flaws from the training set.

Selected examples of B-scans without flaws from the validation set are shown in Figure 7.

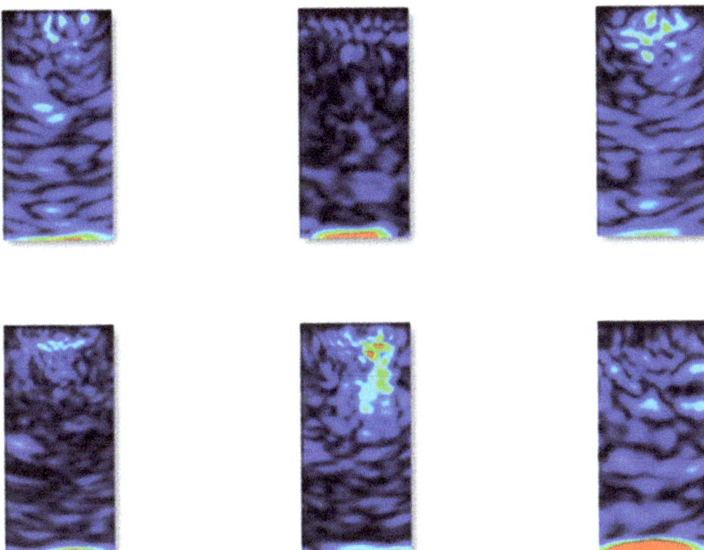

Figure 7. Examples of B-scans without flaws from the validation set.

Selected examples of B-scans with flaws from the validation set are shown in Figure 8.

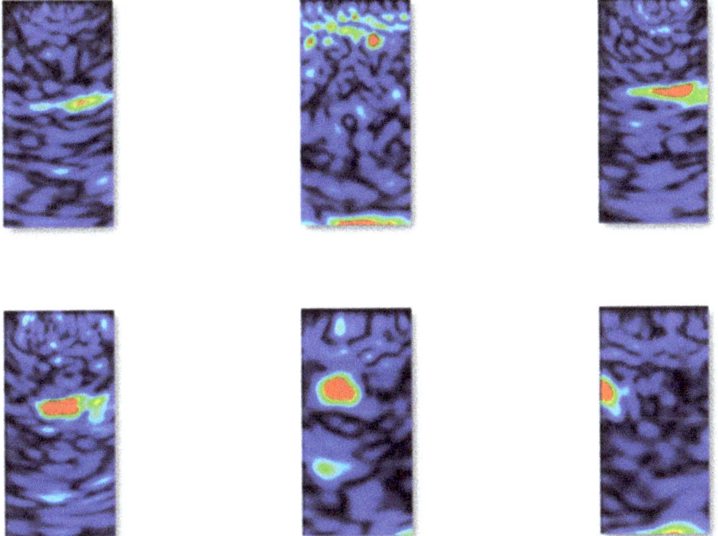

Figure 8. Examples of B-scans with flaws from the validation set.

Selected examples of B-scans without flaws from the testing set are shown in Figure 9.

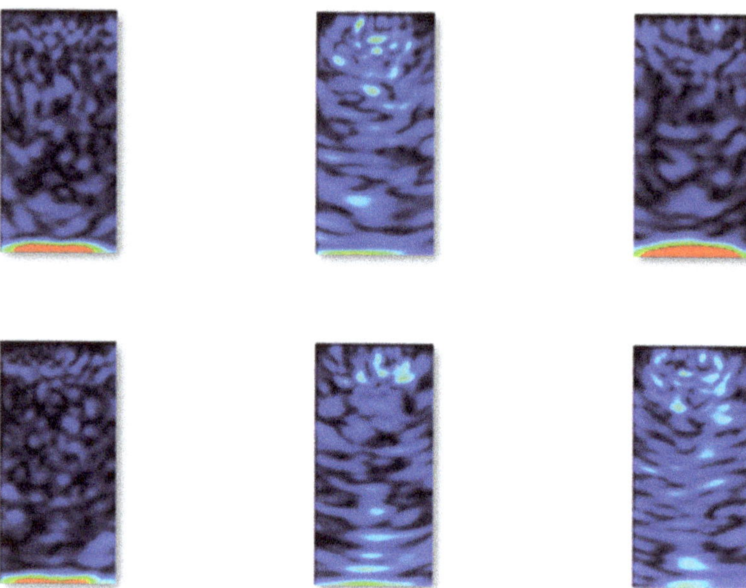

Figure 9. Examples of B-scans without flaws from the testing set.

Lastly, selected examples of B-scans with flaws from the testing set are shown in Figure 10.

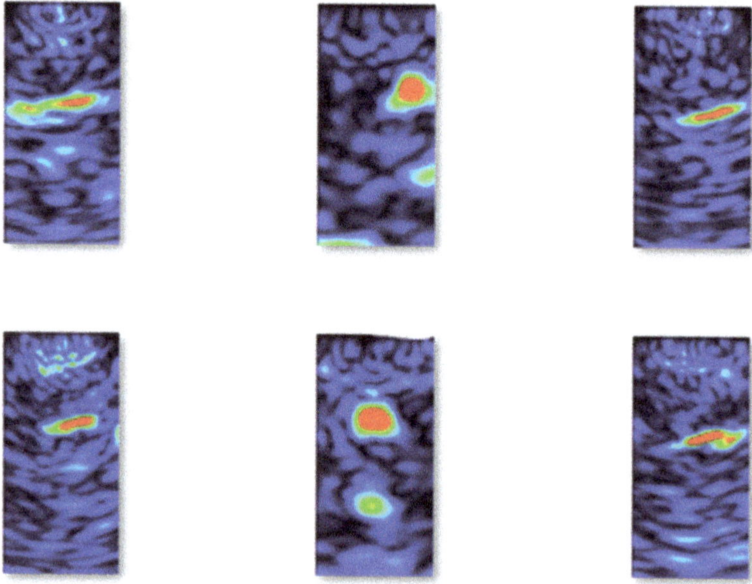

Figure 10. Examples of B-scans with flaws from the testing set.

4.3. Experimental Setup

In this work, to achieve better results in automatic flaw detection, we adopted transfer learning and used a pre-trained convolutional neural network called VGG-16. It was proposed in 2014 by K. Simonyan and A. Zisserman from the Visual Geometry Group at the University of Oxford [41].

VGG-16 was trained on more than a million images from the ImageNet database [40]. This database contains over 14 million images belonging to 1000 classes and was designed for the development of new deep learning algorithms for visual object recognition. VGG-16 was the winner of the ImageNet Challenge in 2014. The network consists of 21 layers including 13 convolutional layers with a filter size of 3 × 3. The VGG-16 architecture is shown schematically in Figure 11.

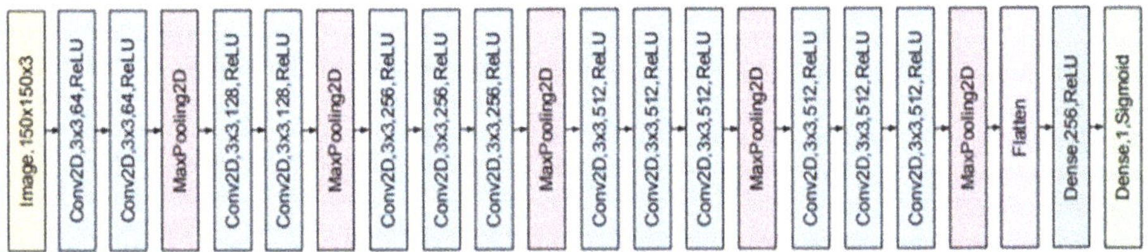

Figure 11. Structure of a pre-trained convolutional neural network with 21 layers (VGG-16).

We use the convolutional base of the pretrained network trained with an image augmentation binary classifier set on the top of the convolutional base. We also apply fine-tuning, introduced in Section 3, for training weights in the last three convolutional layers to improve the performance of the classifier. Table 2 contains the basic parameters such as the number of layers, the total number of parameters, and the size of the considered network.

Table 2. Basic parameters of the applied network.

CNN Model	#Convolutional Layers	#Layers	#Parameters
VGG-16	13	21	16,812,353

The input for the network was the normalized image of the CT scan. The input image was then processed by several convolutional and max pooling layers. The kernels in the convolutional layers have the same size of 3 × 3 and a different number of filters. All convolutional layers used the ReLU activation function. Max pooling layers used a stride of size 2. After all convolutional and max pooling layers, the flattening operation is applied and then the resulting vector of features is processed by a fully connected layer with a ReLU function. Lastly, the input is classified by the last layer, which is fully connected and has the sigmoid function. The convolutional neural network was trained by using the Adam optimizer to optimize the cross-entropy loss function.

The training process includes only fine-tuning of the last block of convolutional layers during 100 epochs of regularized training with image augmentation and dropout (50%). The CNN model is trained, validated, and tested by applying the corresponding datasets presented in Section 4.2.

The numerical experiments were prepared in the Python ecosystem using Keras library [42]. The computations were performed on a Dell Inspiron 15 laptop computer with 64-bit Windows 10, 32 GB RAM memory, Quad-Core Intel Core i7 processor, and NVIDIA GeForce GTX 1060 Ti (4 GB) graphics processing unit (GPU).

5. Results

In this section, the training process and the results of the experiments as well as the final performance of the convolutional neural network (CNN) for image-based detection of flaws in concrete elements are presented.

In Figure 12, the training and validation losses for the network applied in the experiments are presented. From the plots, it can be observed that the minimal training loss achieved for the network with image augmentation and fine-tuning are 0.04 and 0.06. In Table 3, the lowest training and validation losses and training time for one epoch are given.

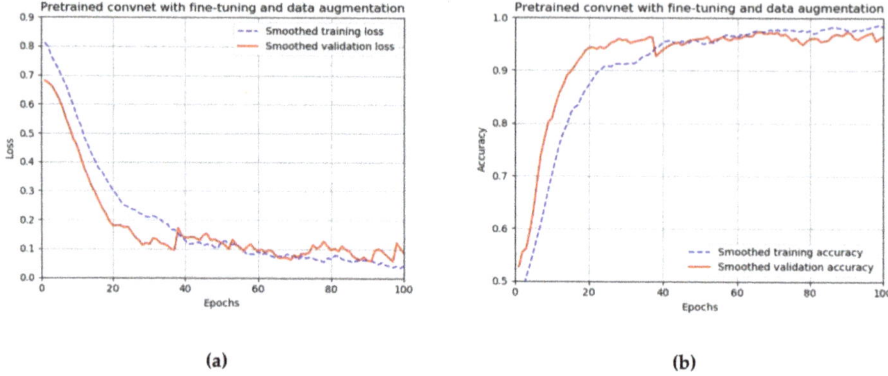

(a)   (b)

Figure 12. Performance of convolutional neural network (CNN): (**a**) training and validation losses, (**b**) training and validation accuracies.

Table 3. Training and validation losses and training time for one epoch.

Convolutional Neural Network (CNN) Model	Training Loss	Validation Loss	Time (s)
VGG-16	0.04	0.06	18

In Figure 12, the training and validation accuracies during training within 100 epochs are presented. It can be seen that the best results obtained in terms of accuracy are similar to the corresponding results in terms of loss. The training accuracy after 100 epochs is 98% while the maximal validation accuracy is 97%. Table 4 shows the highest training and validation accuracies. The table also contains the times needed to perform one epoch of training. The training time for the large pre-trained CNN with image augmentation and fine-tuning is 18 s per epoch.

Table 4. Training and validation accuracies and training time for one epoch.

CNN Model	Training Accuracy	Validation Accuracy	Time (s)
VGG-16	98%	97%	18

After the training process, it is possible to visualize the filters of the convolutional and pooling layers. It can be useful to better understand how the CNN model represents the visual information from the training dataset. In Figure 13, the image shows the visualization of 32 filters from the first convolutional layer, which is shown in the form of merged sub-images (2 rows and 16 columns). The size of each filter is 150 × 150 px.

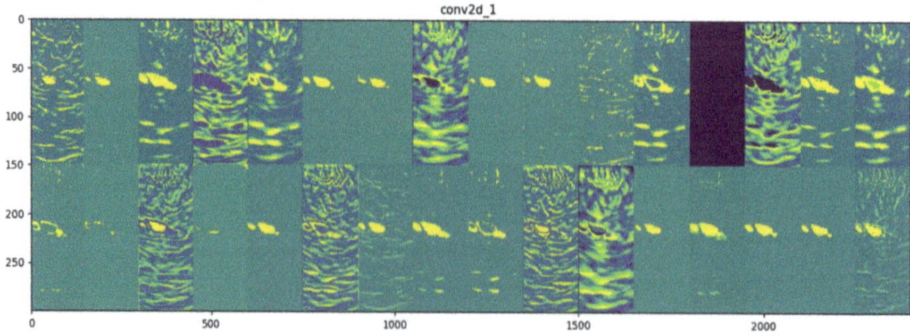

Figure 13. Visualization of 2 × 16 filters from the first convolutional layer.

In Figure 14, the image shows the visualization of the corresponding 32 filters after the first pooling layer. The size of each filter is 75×75 px.

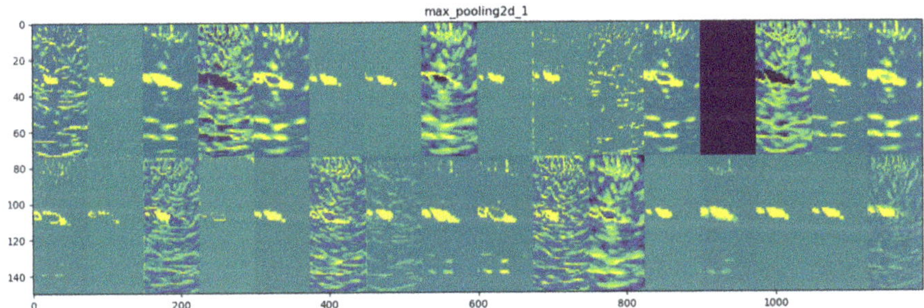

Figure 14. Visualization of 2×16 filters after the first max pooling layer.

Lastly, the trained convolutional neural network was checked against generalization capacity by using the testing dataset. After analysing the new B-scans by the network after fine-tuning, the generalization accuracy for the final network was close to 99%.

6. Discussion

The presented work is concerned with the development of novel methodology for fully automated analysis of images of the inside of the examined concrete elements obtained from an ultrasonic tomograph. The proposed methodology is based on the combination of the ultrasound tomography technique and the convolutional neural network (CNN). The CNN model is applied at the final stage of the developed methodology for classifying images into two categories: "element without a flaw" and "element with a flaw."

Because the main challenge was to collect enough data, instead of building the CNN model from scratch, we applied a pre-trained convolutional neural network called VGG-16 trained on the ImageNet dataset. The parameters of the pre-trained model were fine-tuned by using the collected images and the transfer learning technique. Moreover, two regularization techniques were applied to avoid problems with overfitting and obtain a satisfactory generalization property of the final CNN model. The first technique was image data augmentation by using basic transformations of an image-like rotation, horizontal flipping, and shifts. The second technique was network weights pruning called dropout. In the process of building the neural model, only 246 RGB images were used. After 100 epochs of training, the network achieved 98% accuracy on the training set with a loss of 0.05. The validation accuracy was 97% with a loss of 0.09. Lastly, the generalization accuracy was close to 99%, which confirms that the CNN model was built properly.

Analysis of the obtained results shows that it is very possible to detect defects in concrete samples with very high accuracy. In the provided example, the concrete cubic laboratory specimens were tested using ultrasonic tomography and the trained convolutional neural network. The experiments were carried out according to the proposed testing approach.

7. Conclusions

The innovative methodology of non-destructive detection of defects in concrete elements using ultrasound tomography and a convolutional neural network has been presented in detail. There are two main stages for detecting defects in concrete elements, according to the presented methodology. The first of these is to carry out non-destructive testing using ultrasound tomography techniques. The second stage is automatic defect detection using a convolutional neural network. Interesting results on building practices are given in the provided case study.

- A practical example of the application of the proposed methodology was presented,

- By using two regularization techniques and transfer learning, we were able to significantly reduce the risk of overfitting the model during training,
- In the end, we obtained the CNN-based classifier, which is able to generalize well on unseen images,
- The accuracy of detecting flaws using convolutional neural networks has been confirmed,
- The usefulness of the presented methodology for non-destructive identification of defects in concrete elements was shown,
- The proposed solution can serve as a prototype in the context of future practical applications.

The benefit of the proposed methodology is based on the fact that the combination of ultrasonic tomography and convolutional neural networks offer a method to build a computerized system for complete automatic detection of flaws in concrete elements when given a very small number of recorded image datasets. Based on this research experience, the authors have found that the developed methodology has practical applications in future automatic inspection systems.

In the future, because this was the first attempt to tackle the problem of automatic flaws detection, the limitation and challenges that were faced in this work will be addressed. First, we considered only one type of defect simulated by a PVC pipe embedded in the concrete element. As a result, the images with such a defect belong to one category only. Second, only the images with easily visible flaws were taken into account. In future work, the presented limitations will be considered by testing concrete specimens with other defects to be able to get closer to an automatic system for detection and classification of flaws in concrete elements.

Author Contributions: M.S. completed the experiments and analysed the test results. K.S. conceived and designed the experimental work. E.K. prepared the specimens and performed paper editing. All authors have read and agreed to the published version of the manuscript.

Funding: This research received no external funding.

Conflicts of Interest: The authors declare no conflict of interest.

References

1. Hoła, J.; Schabowicz, K. State-of-the-art non-destructive methods for diagnostic testing of building structures—Anticipated development trends. *Arch. Civ. Mech. Eng.* **2010**, *10*, 5–18. [CrossRef]
2. Schabowicz, K. Non-destructive testing of materials in civil engineering. *Materials* **2019**, *12*, 3237. [CrossRef] [PubMed]
3. Schabowicz, K. Modern acoustic techniques for testing concrete structures accessible from one side only. *Arch. Civ. Mech. Eng.* **2015**, *15*, 1149–1159. [CrossRef]
4. Rucka, M.; Wilde, K. Experimental study on ultrasonic monitoring of splitting failure in reinforced concrete. *J. Nondestruct. Eval.* **2013**, *32*, 372–383. [CrossRef]
5. Woźniak, T.Z.; Ranachowski, Z.; Ranachowski, P.; Ozgowicz, W.; Trafarski, A. The application of neural networks for studying phase transformation by the method of acoustic emission in bearing steel. *Arch. Metall. Mater.* **2014**, *59*, 1705–1712. [CrossRef]
6. Rucka, M.; Wilde, K. Ultrasound monitoring for evaluation of damage in reinforced concrete. *Bull. Pol. Acad. Sci. Tech. Sci.* **2015**, *63*, 65–75. [CrossRef]
7. Łazarska, M.; Woźniak, T.; Ranachowski, Z.; Trafarski, A.; Domek, G. Analysis of acoustic emission signals at austempering of steels using neural networks. *Met. Mater. Int.* **2017**, *23*, 426–433. [CrossRef]
8. Wojtczak, E.; Rucka, M. Wave Frequency Effects on Damage Imaging in Adhesive Joints Using Lamb Waves and RMS. *Materials* **2019**, *12*, 1842. [CrossRef]
9. Bishko, A. Improvement of imaging at small depths for acoustic tomography of reinforced concrete objects. In Proceedings of the 6th International Conference on Non-Destructive Testing and Technical Diagnostics in Industry, Moscow, Russia, 15–17 May 2007; Mashinostroenie; Moscow, Russia.
10. Bishko, A.; Samokrutov, A.A.; Shevaldykin, V.G. Ultrasonic echo-pulse tomography of concrete using shear waves low-frequency phased antenna arrays. In Proceedings of the 17th World Conference on Nondestructive Testing, Shanghai, China, 25–28 October 2008.

11. Kozlov, V.N.; Samokrutov, A.A.; Shevaldykin, V.G. Thickness Measurements and Flaw Detection in Concrete Using Ultrasonic Echo Method. *J. Nondestruct. Test. Eval.* **1997**, *13*, 73–84. [CrossRef]

12. Samokrutov, A.A.; Kozlov, V.N.; Shevaldykin, V.G.; Meleshko, I.A. Ultrasonic defectoscopy of concrete by means of pulse-echo technique. In Proceedings of the 8th European Conference for Non-Destructive Testing, Barcelona, Spain, 17–21 June 2002.

13. Samokrutov, A.A.; Kozlov, V.N.; Shevaldykin, V.G. Ultrasonic testing of concrete objects using dry acoustic contact. Methods, instruments and possibilities. In Proceedings of the 5th International Conference on Non-Destructive Testing and Technical Diagnostics in Industry, Moscow, Russia, 16–19 May 2006.

14. Samokrutov, A.A.; Shevaldykin, V.G. Ultrasonic Tomography of Metal Structure Using the Digital Focused Antenna Array Methods. *Russ. J. Nondestruct. Test.* **2011**, *47*, 16–29. [CrossRef]

15. Shevaldykin, V.G.; Samokrutov, A.A.; Kozlov, V.N. Ultrasonic Low-Frequency Short-Pulse Transducers with Dry Point Contact. Development and Application. In Proceedings of the International Symposium Non-Destructive Testing in Civil Engineering (NDT-CE), Berlin, Germany, 16–19 September 2003.

16. Schabowicz, K.; Suvorov, V. Nondestructive testing and constructing profiles of back walls by means of ultrasonic tomography. *Russ. J. Nondestruct. Test.* **2014**, *50*, 2. [CrossRef]

17. Schabowicz, K. Methodology for non-destructive identification of thickness of unilaterally accessible concrete elements by means of state-of-the-art acoustic techniques. *J. Civ. Eng. Manag.* **2013**, *19*, 325–334. [CrossRef]

18. Zielińska, M.; Rucka, M. Non-Destructive Assessment of Masonry Pillars using Ultrasonic Tomography. *Materials* **2018**, *11*, 2543. [CrossRef] [PubMed]

19. Schabowicz, K. Ultrasonic tomography—The latest nondestructive technique for testing concrete members—Description, test methodology, application example. *Arch. Civ. Mech. Eng.* **2014**, *14*, 295–303. [CrossRef]

20. Schmidhuber, J. Deep learning in neural networks: An overview. *Neural Netw.* **2015**, *61*, 85–117. [CrossRef]

21. Goodfellow, I.; Bengio, Y.; Courville, A. *Deep Learning*; MIT Press: Cambridge, MA, USA, 2016.

22. Cichocki, A.; Poggio, T.; Osowski, S.; Lempitsky, V. Deep learning: Theory and practice. *Bull. Pol. Acad. Sci. Tech. Sci.* **2018**, *66*. [CrossRef]

23. Higham, C.F.; Higham, D.J. Deep learning: An introduction for applied mathematicians. *Siam Rev.* **2018**, *61*, 860–891. [CrossRef]

24. Marcus, G. Deep learning: A critical appraisal. *arXiv* **2018**, arXiv:1801.00631.

25. Krizhevsky, A.; Sutskever, I.; Hinton, G.E. Imagenet classification with deep convolutional neural networks. *Adv. Neural Inf. Process. Syst.* **2012**, 1097–1105. [CrossRef]

26. Fukushima, K. Neocognitron: A self-organizing neural network model for a mechanism of pattern recognition unaffected by shift in position. *Biol. Cybern.* **1980**, *36*, 193–202. [CrossRef]

27. LeCun, Y.; Bengio, Y.; Hinton, G. Deep learning. *Nature* **2015**, *521*, 436–444. [CrossRef] [PubMed]

28. Khan, S.; Rahmani, H.; Shah, S.A.A.; Bennamoun, M. A guide to convolutional neural networks for computer vision. *Synth. Lect. Comput. Vis.* **2018**, *8*, 1–207. [CrossRef]

29. Litjens, G.; Kooi, T.; Bejnordi, B.E.; Setio, A.A.; Ciompi, F.; Ghafoorian, M.; Van Der Laak, J.A.; Van Ginneken, B.; Sánchez, C.I. A survey on deep learning in medical image analysis. *Med. Image Anal.* **2017**, *42*, 60–88. [CrossRef] [PubMed]

30. Koch, C.; Georgieva, K.; Kasireddy, V.; Akinci, B.; Fieguth, P. A review on computer vision based defect detection and condition assessment of concrete and asphalt civil infrastructure. *Adv. Eng. Inform.* **2015**, *29*, 196–210. [CrossRef]

31. Słoński, M. A comparison of deep convolutional neural networks for image-based detection of concrete surface cracks. *Comput. Assist. Methods Eng. Sci.* **2019**, *26*, 105–112.

32. Spencer, B.F.; Hoskere, V.; Narazaki, Y. Advances in Computer Vision-Based Civil Infrastructure Inspection and Monitoring. *Engineering* **2019**, *5*, 199–222. [CrossRef]

33. Jarosik, P.; Lewandowski, M. The feasibility of deep learning algorithms integration on a GPU-based ultrasound research scanner. In Proceedings of the 2017 IEEE International Ultrasonics Symposium, Washington, DC, USA, 6–9 September 2017; pp. 1–4.

34. Liu, S.; Wang, Y.; Yang, X.; Lei, B.; Liu, L.; Li, S.X.; Ni, D.; Wang, T. Deep Learning in Medical Ultrasound Analysis: A Review. *Engineering* **2019**, *5*, 261–275. [CrossRef]

35. Bishop, C.M. *Neural Networks for Pattern Recognition*; Oxford University Press: Oxford, UK, 1995.

36. Kukačka, J.; Golkov, V.; Cremers, D. Regularization for deep learning: A taxonomy. *arXiv* **2017**, arXiv:1710.10686.
37. Shorten, C.; Khoshgoftaar, T.M. A survey on image data augmentation for deep learning. *J. Big Data* **2019**, *6*, 60. [CrossRef]
38. Srivastava, N.; Hinton, G.; Krizhevsky, A.; Sutskever, I.; Salakhutdinov, R. Dropout: A simple way to prevent neural networks from overfitting. *J. Mach. Learn. Res.* **2014**, *15*, 1929–1958.
39. Yosinski, J.; Clune, J.; Bengio, Y.; Lipson, H. How transferable are features in deep neural networks? *Adv. Neural Inf. Process. Syst.* **2014**, *27*, 3320–3328.
40. Deng, J.; Dong, W.; Socher, R.; Li, L.J.; Li, K.; Fei-Fei, L. Imagenet: A large-scale hierarchical image database. In Proceedings of the IEEE Conference on Computer Vision and Pattern Recognition, Miami, FL, USA, 20–25 June 2009; pp. 248–255. [CrossRef]
41. Simonyan, K.; Zisserman, A. Two-Stream Convolutional Networks for Action Recognition in Videos. CoRR,abs/1406.2199. Published in Proc. NIPS. 2014. Available online: https://papers.nips.cc/paper/5353-two-stream-convolutional-networks-for-action-recognition-in-videos.pdf (accessed on 20 February 2020).
42. Chollet, F. *Deep Learning with Python*; Manning Publications, Co.: Shelter Island, NY, USA, 2018.

 *materials*

Article

Fiber Bragg Sensors on Strain Analysis of Power Transmission Lines

Janusz Juraszek

Faculty of Materials, Civil and Environmental Engineering, University of Bielsko-Biala, 43-309 Bielsko-Biala, Poland; jjuraszek@ath.bielsko.pl; Tel.: +48-338-279-191

Received: 5 March 2020; Accepted: 25 March 2020; Published: 27 March 2020

Abstract: The reliability and safety of power transmission depends first and foremost on the state of the power grid, and mainly on the state of the high-voltage power line towers. The steel structures of existing power line supports (towers) have been in use for many years. Their in-service time, the variability in structural, thermal and environmental loads, the state of foundations (displacement and degradation), the corrosion of supporting structures and lack of technical documentation are essential factors that have an impact on the operating safety of the towers. The tower state assessment used to date, consisting of finding the deviation in the supporting structure apex, is insufficient because it omits the other necessary condition, the stress criterion, which is not to exceed allowable stress values. Moreover, in difficult terrain conditions the measurement of the tower deviation is very troublesome, and for this reason it is often not performed. This paper presents a stress-and-strain analysis of the legs of 110 kV power line truss towers with a height of 32 m. They have been in use for over 70 years and are located in especially difficult geotechnical conditions—one of them is in a gravel mine on an island surrounded by water and the other stands on a steep, wet slope. Purpose-designed fiber Bragg grating (FBG) sensors were proposed for strain measurements. Real values of stresses arising in the tower legs were observed and determined over a period of one year. Validation was also carried out based on geodetic measurements of the tower apex deviation, and a residual magnetic field (RMF) analysis was performed to assess the occurrence of cracks and stress concentration zones.

Keywords: monitoring FBG; power transmission tower; civil engineering

---

1. Introduction

Power transmission lines extend over hundreds of kilometers. The collapse of one or more towers may have serious consequences, resulting in a local shutdown or even a blackout. Therefore, the systematic testing, maintenance and repair of tower elements are necessary. Unfortunately, power line operators often fail to do so. This especially applies to older structures which have been in use for over 30 years. The repair or replacement of a tower structure element is difficult because it is usually done manually or using a small winch. The tower's location on a steep mountainside or an island surrounded by water often creates an additional difficulty.

Periodic testing is performed by measuring deformation or the tower apex deviation. An interesting analysis of a high-voltage power line tower structure deformation is presented in [1]. The analysis concerned towers (pylons) with a height of 28–32.2 m supporting a double 110 kV power line. The structure was made of S235 steel. The measurements were performed using a new FOTON photogrammetric system [2–4]. The biggest measured deviation in the pylon apex totaled 480 mm. The downside of the solution is that the photographing device should be perpendicular to the tested object and far away from it. Moreover, fog and poor lighting conditions can render the measurement impossible. Such measurements can be made periodically but not continuously, and they are troublesome in practice. Usually, they are performed once every few years and continuous measurements are impossible. A new 3D photogrammetric technique of measuring the tower deviation was introduced by Xiao et al. [5]. Special markers were used and photographs were taken as towers were subjected to loads. However, this method requires a high computing power to determine the deformation of tested elements based on a series of photographs. Another interesting method for automatic supervision is the efficient extraction and classification of three-dimensional (3D) targets of electric power transmission facilities based on regularized grid characteristics computed from point cloud data acquired by unmanned aerial vehicles (UAVs) [6]. A special hashing matrix was constructed first to store the point cloud after noise removal by a statistical method which calculated the local distribution characteristics of the points within each sparse grid. Next, power lines were extracted by the neighboring grids' height similarity estimation and linear feature clustering. After that, by analyzing features of the grid in the horizontal and vertical directions, the transmission towers in candidate tower areas were identified. Unfortunately, this method does not enable a precise assessment of the tower deviation or strains in the tower legs. The system can be useful for monitoring long rows of power line towers and making a preliminary assessment of the tower apex deviation. An analysis of tower failures was also conducted by Davies [7]. It was concluded that 31% of tower failures were due to design errors, 29% were caused by ice, strong winds accounted for 19% and other causes were responsible for 11%. The towers analysed herein have been in use for more than 70 years, which confirms that they are not burdened with design errors—these would have revealed themselves by now. The American (ASCE) and British power transmission tower design standards were analyzed by Rao [8]. The author assessed them as very conservative with regard to experiments. The forces determined according to nonlinear theories turned out to be smaller than originally assumed, which may lead to the oversizing of structures.

Numerical as well as numerical-and-experimental simulations of power line towers were also performed, which were described, for example, in the works of Yin [9] and Lam and Yin [10] on experimentally verified numerical simulations of towers using the finite element method (FEM). One of the methods of monitoring the state of power line towers is vibration analysis [11]. This idea concerns measurements of the response frequency function of the tested tower. If damage occurs, the resonance peak is shifted. If the tower loses stiffness due to cracks, the frequency of vibrations decreases. Such changes can be considered indications of damage. However, the precise localization and identification of damage require not only accurate numerical models to simulate the tower vibrations, but also adequately high computing power. The significant researchers in this area are Al-Bermami and Kitiponrchai [12]. They simulated the occurrence of damage in two towers in Australia. A review of the design practices of the 1990s in this regard is presented in [13]. Considering current climate changes, the infrastructure of power transmission towers should be strengthened using appropriate stiffening devices [14]. At a later stage of the research, Al-Bermami applied a nonlinear methodology of anticipating the occurrence of structural damage in towers [15]. The above-mentioned impact of wind is one of the most typical loads. It may cause power lines to vibrate and damage to towers. Because weather conditions are now increasingly dramatic, the problem is becoming more and more serious. Interesting findings are presented by Frei et al. [16], who investigated the structural stability of power transmission lines. They based the stability assessment on the static analysis and the dynamic analysis. It follows that low modal frequencies become smaller before instability is reached (before a failure occurs), which may be an early warning sign.

An analysis of damage to a power transmission tower due to strong wind was conducted in [17,18]. Both static and dynamic simulations were performed. The analysis covered multi-member towers with a total height of $59 + 17.5 = 76.5$ m. The critical strength values were compared to the American (ASCE) and the British standard and to the DL/T 5154–2012 guidelines [19,20].

Depending on the standard, they are included in the range of 250–270 MPa for the tower legs and 100–140 MPa for the diagonal elements. The results of the analyses indicate that more emphasis should be put on the design of diagonal elements. The analyzed tower was designed using steel angle sections. A combination of an interesting analysis of a full-scale test and a numerical simulation of a power transmission tower with a height of 46.05 m and a 7.14 m × 7.14 m base under the impact of wind is presented in [21–23].

The full-scale test is the most reliable way to reflect the power transmission tower's mechanical properties. Obviously, the experimental cost is very high and the test is only possible for new tower types [24–26].

Ma et al. [27] proposed using an FBG sensor as a small aerometer to measure the wind speed. The essence of the sensor operation was the bending of the sensor's active element under a wind load. The sensor deformation was proportional to the wind speed. The lowest measurable wind speed totaled 3 m/s and the measurement error was smaller than 0.1 m/s. The device worked very well in difficult environmental conditions. FBG sensors are mainly used as strain–tension sensors. Most of the FBG sensors available on the market enable the connection of a few dozen sensors, each with a different wavelength (differing by 5 nm), to one telecommunication optical fiber. The use of optical multiplexes makes it possible to connect and build measurment networks with the capacity of several thousand FBG sensors. The important thing is that optical fibers are resistant to electromagnetic noise, which is crucial in measurements involving high voltage. The advantage of the FBG technology is the possibility of using high sampling frequencies of up to 2 kHz, which enables the analysis of dynamic as well as long-lasting quasi-static processes. The works present interesting investigations of deformations of complex metal structures using FBG sensors, as well as an innovative combination of deformation measurement techniques based on optical fibers and the Residual Magnetic Field method, which make it possible to identify stress concentration zones in a steel structure in advance [28,29]. The issue presented in this paper concerns the diagnostics of power transmission towers which have been in service for about 70 years and are located in especially difficult geotechnical conditions. The towers are truss structures with a height of up to 32 m. In the literature, there are no works on the diagnostics of this type of tower located in especially difficult terrain conditions, where deformation processes occur over a long period of time. The author proposes the use of appropriately designed FBG sensors enabling the analysis of strains in the tower legs with a parallel use of the residual magnetic field (RMF) method which makes it possible to detect the structure stress concentration zones.

2. Materials and Methods

A system monitoring the deformation of the structure of power transmission towers was built. The system was based on fiber Bragg grating (FBG Sylex, Bratislava, Slovakia) strain sensors. The essence of the solution was to introduce FBG sensors in certain points of the tower structure as a kind of "nervous system". In the case of supporting structures of power transmission lines, the appropriate places were those with the highest bending moment and the biggest compressive stresses. The places were pre-determined using numerical simulations and they were located at the tower base. The entire optical fiber system intended for the strain-and-stress state measurement was duly calibrated. Special grips had been designed earlier to make it possible to fix the strain sensor to the selected tower leg repeatedly. The FBG strain sensor is presented in Figure 1.

Figure 1. The fiber Bragg grating (FBG) strain sensor.

Based on preliminary testing, the measuring base L = 1000 mm was adopted. Using special grips, the sensor axis was placed in the center of gravity of the tower leg angle section.

The measuring part of each strain sensor was a Bragg grating with a specific wavelength embedded in an optical fiber. One sensor was used for each tower leg, i.e., there were four sensors per tower. Due to the optical interrogator traceability, a difference of at least 5 nm between the wavelengths of each

Bragg grating had to be kept. A change in strains in the tested structural element, i.e., in the tower leg, caused a change in the wavelength in the Bragg grating. This relation is described by the following formula (1):

$$\Delta\varepsilon = \left[\frac{\lambda_a - \lambda_0}{\lambda_0} - B\,(T_a - T_0) \right] A^{(-1)}$$ (1)

where:

$\Delta\varepsilon$ strain shift;

λ_a actual strain wavelength;

λ_0 initial strain wavelength;

T_a actual temperature;

T_0 initial temperature;

A and B calibration parameters. $A = 7.8648540 \times 10^{-7}$ [$\mu\varepsilon^{-1}$]; $B = 5.99813502 \times 10^{-6}$ [°C].

Another important factor was to select an appropriate initial tension of the sensor. The sensor's initial tension was set precisely using right-and-left nuts with a fine thread. Apart from the dedicated optical fiber sensors, the system included a 2 kHz FBG-800 optical interrogator (Fibre Optic Sensing & Sensing Systems–FOS&S, Geel, Belgium), a recorder, special software, a multiplier and telecommunication fibers. An option was also possible with a wireless transfer of measurement results from the interrogator. The constructed optical fiber system intended for the measurement of strains arising in the tower legs was first tested in laboratory conditions. The tests consisted of setting the known force and strain values to a fragment of a tower leg cut-out together with the FBG sensor fixed on it. The fixing manner ensured that the cut-out response was identical with that of the actual tower leg. Considering the required spacing between the traverses of the strength-testing machine with Accuracy Class 0.5, the tests were performed in the Research and Supervisory Center of Underground Mining. The diagram of the tension and compression test rig is presented in Figure 2. Special grips were designed to fix the angle section in the jaws of the strength-testing machine and to enable the angle section buckling in relation to the axis characterized by the lowest value of the second moment of inertia. In the first stage, tests calibrating the measuring system were carried out in the following sequence:

(a) testing the tower structural elements together with installed FBG SC-01 sensors with a measuring base of 1000 mm. The testing was carried out on the tower leg fragment in the form of 80 mm × 80 mm × 8 mm and 75 mm × 75 mm × 6 mm angle sections with special grips fixing the optical fiber sensors and enabling the appropriate initial tension;

(b) for the assumed constant 5 kN increment in force set by the strength-testing machine, strains corresponding thereto and established experimentally using the FBG SC-01 sensor were determined.

Figure 2. Laboratory stand intended for the calibration of the strain-state measuring system.

The testing results proved unequivocally that the optical fiber sensor and the method of fixing the sensor enabled a correct measurement of strains arising in the tested angle section of the tower leg.

The angle section with the attached grips and the installed FBG sensor was subjected to a tensile force and a compressive force included in the same range as in the tower leg real conditions. The results of the testing are illustrated by the load-strain characteristic plotted for the tested leg (cf. Figure 3).

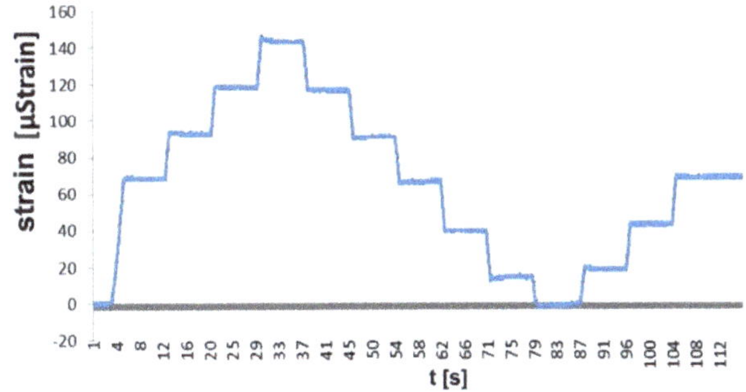

Figure 3. Calibration of the optical fiber transducer on the strength-testing machine.

For every 5 kN increment in force, there was an increment in strain totaling 23 µstrain. It can be observed that constant increments in load result in constant increments in strain. The testing results prove unequivocally that the optical fiber sensor and the method of fixing the sensor to the tower leg angle section enabled a correct measurement of strains in the case of both tensile and compressive forces.

Using the results of the laboratory testing, performed on the strength-testing machine (XB-OTS-600, 100 kN, Dongguan Xinbao Instrument Co., Ltd., Guangdong, China), of the angle sections used as the power line tower supporting elements and of the designed grips with the FBG sensors fixed to them, the angle sections were verified and re-designed to obtain the final form, as presented in Figure 4.

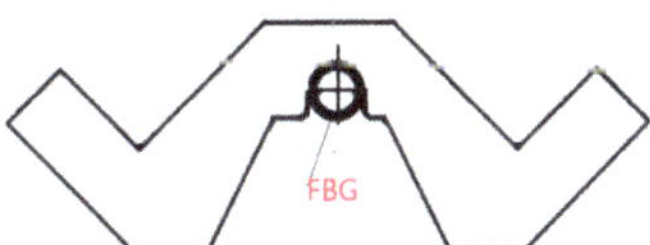

Figure 4. Designed grips for the fixing of FBG SC-01 and SC-02 sensors.

The change in design made the assembly of the grips easier and ensured an appropriate and constant initial tension of the sensors at a level of about 600 µstrain.

3. Results of Strain Testing of High-Voltage Power Line Towers

The strain testing results were composed of two parts. The first presented the results of measurements of strains arising in the tower legs under a simulated short-term lateral load of 500 N. The load was applied in the direction of the tower two axes (I, II) and along its diagonals (III). A diagram illustrating the loading is presented in Figure 5.

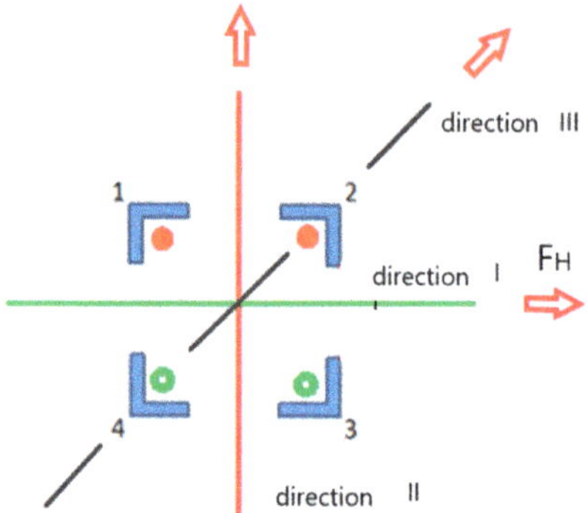

Figure 5. Direction of the load force in tower A.

The load applied to the tower in direction II resulted in compressive strains of −14 μstrain in legs 1 and 2. A tensile strain of 10 μstrain arose in the legs on the tower's opposite side. The strain measurement results are presented in Figure 6a,b.

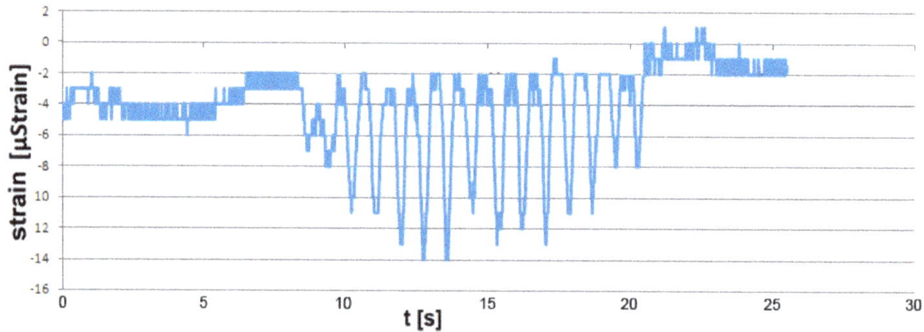

(**a**) Tower A in axis directions II

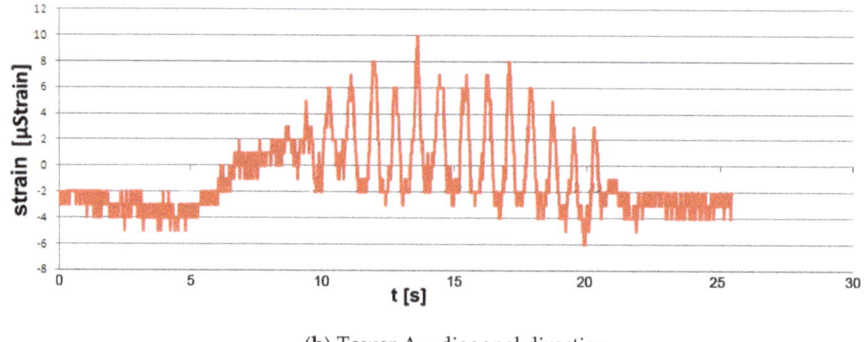

(**b**) Tower A—diagonal direction

Figure 6. Strains in the legs of tower A under a lateral load: (**a**) in the tower axis direction, (**b**) in the tower diagonal direction.

This meant the occurrence of tensile stresses of 2.1 MPa and compressive stresses of about −2.82 MPa. If a load was applied in direction III, the values of strain for leg XP1 total −19 μstrain, and for the leg located diagonally, 14 μstrain. The highest strain values were recorded if the tower was loaded along the axis in relation to which the tower apex was shifted. The recorded strain values totaled 21 μstrain (cf. Figure 7), which, based on physical relations, meant a tensile stress of about 4.5 MPa.

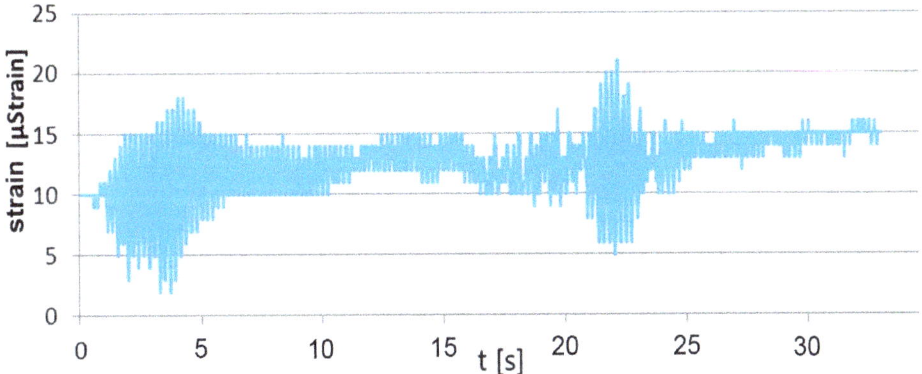

Figure 7. Highest strain values under a lateral load.

The second part of the testing concerned the observation of strains arising in selected legs of the tower over a period of one year. In the case of the legs of tower A, located on an island in a gravel mine area, strains rose in the winter period to 230–270 μstrain and then returned, in the case of leg 1, to the initial level of 150 μstrain. In the case of tower B, the course of the changes was similar, but the strain values did not return to the initial state and kept at the level of 230 μstrain. The tower leg was shortened, which meant that the stresses arising in the two legs were compressive. The strain values for the other two legs were similar but with the opposite sign, which meant that there was tension. By determining real strain values in the tower legs, it was possible to define the performance of the tower foundation. The tower rotation axis (direction II) at the tower base is marked in Figure 5. It passed between the legs subjected to compression and those subjected to tension. The hypothesis was also confirmed by the measurements of the tower apex deviation (cf. subsection on validation). Using physical relations (Hooke's law), the values of stresses arising in the tower's individual legs were determined. They are presented in Figure 8.

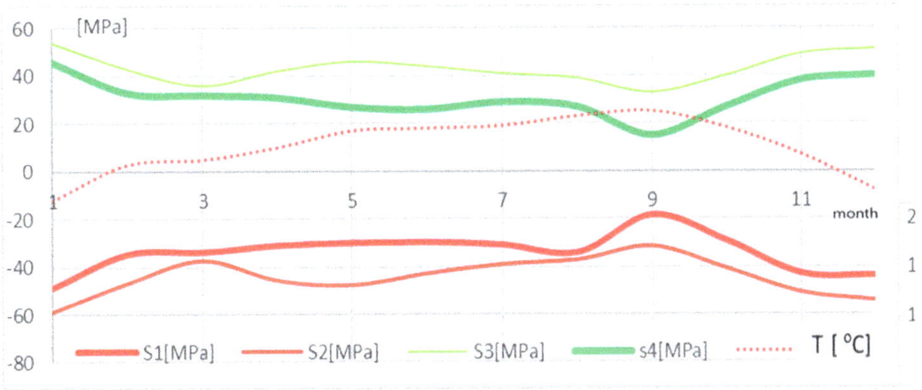

Figure 8. Stress and temperature distribution in the legs of tower A located on an island in a lake.

The changes in stresses were correlated with the changes in temperature (cf. Figure 8). Analyzing the temperature history during the observation, it can be noticed that stresses in the tower varied with changes in temperature. A drop in temperature involved a rise in the compressive stress value in the tower legs (to observation point 11). The rise in temperature that occurred in the post-winter period caused an increase in compressive stresses. In leg 1, stresses were increased by about 20 MPa, whereas in leg 2 the rise in compression was higher and totaled about 25 MPa. In summer, temperatures were higher and according to the observation results, stresses in leg 1 remained at the same level, whereas compressive stresses in leg 2 dropped to the value of about −50 MPa. The highest calculated value of compressive stress totaled 60 MPa. The stresses were compared to allowable stress values for this type of tower structure, which were defined precisely in [17]; they range from 150 to 250 MPa.

The tower located in the mountains did not display such regularity (cf. Figure 9).

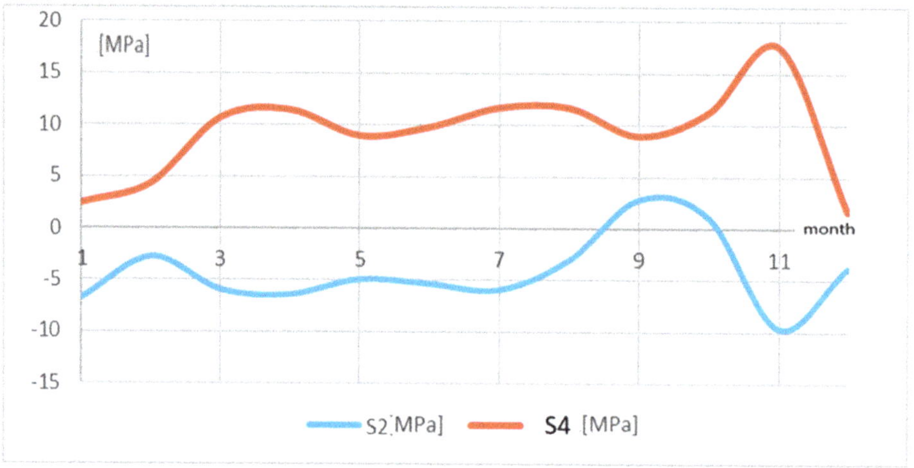

Figure 9. Stress and temperature distribution in the legs of tower B located in the mountains.

This is especially visible in the case of leg 2, for which the relations were the opposite, i.e., a rise in temperature in the initial period involved a drop in stresses by 15 MPa and a change in the nature of the stress, whereas a temperature drop changed the sign of stresses and increased them to the value of 5 MPa. It can be seen that as leg 4 was subjected to tension, leg 4 was compressed. It can be supposed that this is due to the landform features and# the fact that the terrain was unstable and wet. Most probably, the tower foundation rotated in relation to the axis along one of the tower base diagonals. Moreover, each of the tower legs had a separate foundation. The tower's hypothetical rotation at the tower base is presented in Figure 5 (direction III). In such a situation, one side of the tower structure was subjected to tension and the other to compression. This was also confirmed by geodetic measurements of the tower apex.

4. Validation

Validation by means of geodetic measurements was carried out on the tower apex deviation from the plumb line. The method of direct projection onto a leveling staff placed on the tower foundation was applied. A Leica TCR 407 tacheometer (LEICA, Wetzlar, Germany) and a measuring staff were used. The measurements revealed a tower axis deviation from the plumb line of 3 and 13.5 cm, respectively, in the direction parallel (component X) and perpendicular (component Y) to the course of the 110 kV high-voltage power transmission line (Figure 10).

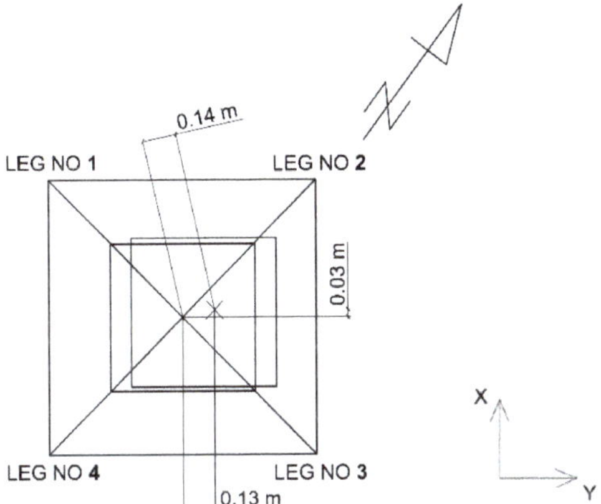

Figure 10. Tower A rotation and translation.

The results of the measurements of the tower deviation confirmed the results of the measurements of strains in the tower legs. They indicated that the tower structure inclined towards legs 2 and 3 and caused higher compressive stresses.

The tower legs were also tested using the residual magnetic field (RMF) method. The method is described in [28,29] and makes it possible to detect stress concentration zones in ferromagnetic elements, such as steel, used to build the tower legs. Stress concentration zones occur in areas where the magnetic indicator—the gradient of the normal and the tangential component along the measuring path—reaches a value exceeding 10 [(A/m)/mm.] No such areas were found in the case of the tower legs under consideration. Moreover, the method enables the detection of fractures and microcracks. In these particular areas, the residual magnetic field's normal component becomes zero. Despite the long in-service time, the RMF testing did not detect any cracks in the tower leg. The distribution of the gradient of the tangential component magnetic field along the tower A leg is presented in Figure 11.

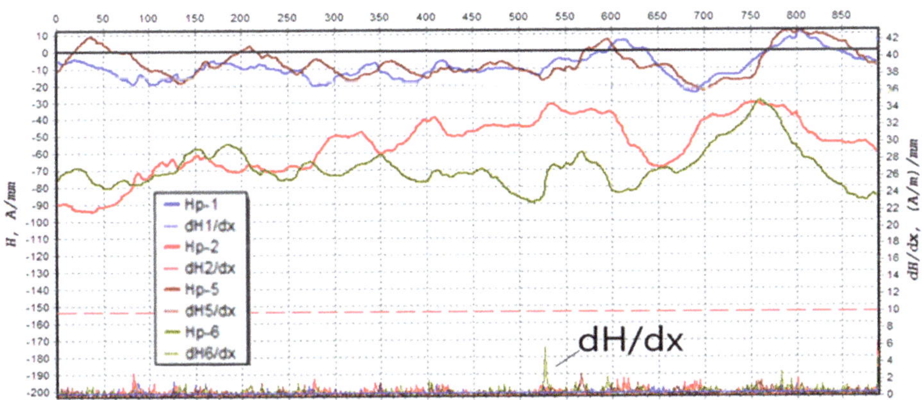

Figure 11. Distribution of the gradient dH/dx tangent component of magnetic field.

5. Conclusions

The conducted research proved an essential influence of difficult geotechnical and environmental conditions of the power line tower foundation on the strain state in the tower legs. The testing of supporting structures of high-voltage power transmission lines by means of optical FBG systems provided a lot of essential information about the real values of strains and stresses occurring in the legs of the analyzed towers of overhead power transmission lines. This is of particular importance in the case of lines which have been in use for many years. The obtained information may also contribute to an improvement in the operating safety of power line towers. The presented fiber-based system intended for the measurement of strains enabled dynamic measurements (horizontal load) as well as long-term measurements. The results of the strain-state measurements of the tower legs were confirmed by geodetic measurements of the tower apex deviation from the plumb line. Using various research methods, e.g., combining the FBG system to measure strains with the assessment of the tower's strain-and-stress state using the RMF method, made it possible to evaluate the performance of a tower structure in very difficult geotechnical conditions.

Funding: This research received no external funding.

Acknowledgments: The technical support from Tauron Company.

Conflicts of Interest: The author declares no conflict of interest.

References

1. Ličev, L.; Hendrych, J.; Tomeček, J.; Čajka, R.; Krejsa, M. Monitoring of excessive Deformation of Steel Structure Extra-High Voltage Pylons Periodica Polytechnica. *Civ. Eng.* **2018**, *62*, 323–329.
2. Ličev, L. Systém FOTOMNG, architektura, funkce a použití. *Ben Praha* **2015**, *230*, 114567.
3. Ptáček, J. *Filtrovaná Zpětná Projekce—Rozšířený Popis*; Univerzita Pa-lackého v Olomouci: Olomouc, Czech Republic, 2013.
4. Lichev, L.; Tomechek, J.; Hendrych, J.; Cajka, R.; Krejsa, M. New methods of evaluation of deformation structure extra-high voltage pylons. In Proceedings of the 15th International Multidisciplinary Scientific Geoconference and EXPO—SGEM 2015, Albena, Bulgaria, 16–25 June 2015; Volume 1, pp. 215–224.
5. Xiao, Z. Rapid three-dimension optical deformation measurement for transmission tower with different loads. *Opt. Lasers Eng.* **2010**, *48*, 869–876. [CrossRef]
6. Zhang, R. Automatic Extraction of high-Voltage Power Transmission Object from UAV Lidar Point Clouds. *Remote Sens.* **2019**, *11*, 2600. [CrossRef]
7. Davies, D. *North American Tower Failures: Causes and Cures*; Consolidated Engineering Inc.: Evansville, IN, USA, 2011.
8. Prasad Rao, N.; Samuel Knight, G.M.; Mohan, S.J.; Lakshmanan, N. Studies on failure of transmission line towers in testing. *Eng. Struct.* **2012**, *35*, 55–70. [CrossRef]
9. Yin, T. Dynamic reduction-based structural damage detection of transmission tower utilizing ambient vibration data. *J. Eng. Struct.* **2009**, *31*, 2009–2019. [CrossRef]
10. Lam, H. Dynamic reduction-based structural damage detection of transmission towers: Practical issues and experimental verification. *J. Eng. Struct.* **2001**, *33*, 1459–1478. [CrossRef]
11. Boller, C. *Encyclopedia of Structural Health Monitoring*; John Wiley and Sons Inc.: Hoboken, NJ, USA, 2009.
12. Al-Bermani, F.; Kitipornchai, S. Nonlinear analysis of transmission towers. *J. Eng. Struct.* **1992**, *14*, 139–151. [CrossRef]
13. Al-Bermani, F.; Kitipornchai, S. Nonlinear finite element analysis of latticed transmission towers. *J. Eng. Struct.* **1993**, *15*, 259–269. [CrossRef]
14. Al-Bermani, F. Upgrading of transmission towers using a diaphragm bracing system. *J. Eng. Struct.* **2004**, *26*, 735–744. [CrossRef]
15. Al-Bermani, F. Failure analysis of transmission towers. *J. Eng. Fail. Anal.* **2009**, *16*, 1922–1928. [CrossRef]
16. Fei, Q. Structural health monitoring oriented stability and dynamic analysis of a long-span transmission tower-line system. *J. Eng. Fail. Anal.* **2012**, *20*, 80–87. [CrossRef]
17. Zhang, J.; Xie, Q. Failure analysis of transmission tower subjected to strong wind load. *J. Constr. Steel Res.* **2019**, *160*, 271–279. [CrossRef]
18. Xie, Q.; Sun, L. Failure mechanism and retrofitting strategy of transmission tower structures under ice load. *J. Constr. Steel Res.* **2012**, *174*, 26–36. [CrossRef]
19. ASCE. *Design of Latticed Steel Transmission\Structures American*; Society of Civil Engineers: Reston, WV, USA, 1997.

20. Standard, B. *Lattice Towers and Masts—Part 3: Code of Practice for Strength Assessment of Members of Lattice Towers and Masts*; British Standards Institution: London, UK, 1999.

21. Prasad Rao, N.; Knight, G.M.S.; Lakshmanan, N.; Iyer, N.R. Investigation of transmission line tower failures. *Eng. Fail. Anal.* **2010**, *17*, 1127–1141. [CrossRef]

22. Szafran, J.; Rykaluk, K. A full-scale experiment of a lattice telecommunication tower under breaking load. *J. Constr. Steel Res.* **2016**, *120*, 160–175. [CrossRef]

23. Fu, X.; Wang, J.; Li, H.N.; Li, J.X.; Yang, L.D. Full-scale test and its numerical simulation of a transmission tower under extreme wind loads. *J. Wind Eng. Ind. Aerodyn.* **2019**, *190*, 119–133. [CrossRef]

24. Asgarian, B.; Eslamlou, S.D.; Zaghi, A.E.; Mehr, M. Progressive collapse analysis of power transmission towers. *J. Constr. Steel Res.* **2016**, *123*, 31–40. [CrossRef]

25. Shukla, V.K.; Selvaraj, M. Assessment of structural behavior of transmission line tower using strain gauging method. *Int. J. Steel Struct.* **2017**, *17*, 1529–1536. [CrossRef]

26. Xie, Q.; Cai, Y.; Xue, S. Wind-induced vibration of UHV transmission tower line system: Wind tunnel test on aero-elastic model. *J. Wind Eng. Ind. Aerod.* **2017**, *171*, 219–229. [CrossRef]

27. Ma, G. A passive optical fibre anemometr for wind Speed measurement on high-voltage overhead transmission lines. *IEEE Trans. Instrum. Meas.* **2012**, *61*, 539–544. [CrossRef]

28. Juraszek, J. Hoisting machine brake linkage strain. *Arch. Min. Sci.* **2018**, *63*, 583–597.

29. Juraszek, J. Residual magnetic field non-destructive testing of gantry cranes. *Materials* **2019**, *12*, 564. [CrossRef] [PubMed]

 *materials*

Article

Experimental Investigations of Timber Beams with Stop-Splayed Scarf Carpentry Joints

Anna Karolak *, Jerzy Jasieńko, Tomasz Nowak and Krzysztof Raszczuk

Faculty of Civil Engineering, Wroclaw University of Science and Technology, Wybrzeze Wyspianskiego 27, 50-370 Wrocław, Poland; jerzy.jasienko@pwr.edu.pl (J.J.); tomasz.nowak@pwr.edu.pl (T.N.); krzysztof.raszczuk@pwr.edu.pl (K.R.)
* Correspondence: anna.karolak@pwr.edu.pl; Tel.: +48-71-3203879

Received: 4 February 2020; Accepted: 17 March 2020; Published: 21 March 2020

Abstract: The paper presents the results of an experimental investigation of stop-splayed scarf joints, which was carried out as part of a research programme at the Wroclaw University of Science and Technology. A brief description and the characteristics of scarf and splice joints appearing in historical buildings are provided, with special reference to stop-splayed scarf joints (so-called 'Bolt of lightning') which were widely used, for example, in Italian renaissance architecture. Analyses and studies of scarf and splice joints in bent elements presented in the literature are reviewed, along with selected examples of analyses and research on tensile joints. It is worth noting that the authors in practically all the cited literature draw attention to the need for further research in this area. Next, the results of the authors' own research on beams with stop-splayed scarf joints, strengthened using various methods, e.g., by means of drawbolts (metal screws), steel clamps and steel clamps with wooden pegs, which were subjected to four-point bending tests are presented. Load-deflection plots were obtained for load-bearing to bending of each beam in relation to the load-bearing of a continuous reference beam. A comparative analysis of the results obtained for each beam series is presented, along with conclusions and directions for further research.

Keywords: carpentry joints; scarf and splice joints; stop-splayed scarf joints ('bolt of lightning'); static behaviour; experimental research

---

1. Introduction

Wood has been used for centuries in different buildings and structures, from residential buildings, through sacral architecture, to defensive structures for settlements and towns or complex engineered structures [1]. A large number of these structures, which have survived tens or hundreds of years, now require interventions to maintain or improve their technical condition. A key factor in their development is related to joints, which enabled building elements to be connected into a single whole and for loading to be shared between elements. Historical joints pay testimony to the highly developed techniques and craftsmanship of builders at that time [2]. It should also be noted that, to date, there are few reliable references that include detailed specifications related to forming carpentry joints [3], as their creators relied mainly on their own experience and tradition.

As a rule, the behaviour of joints has a large influence on the global outcome of the structure as a whole, especially with respect to internal forces. This is why carrying out a detailed analysis of the structure as a whole requires analyses of the behaviour of joints. A simple example serves to demonstrate the role played by correctly formed scarf and splice joints connecting construction elements. In the case of vertical loading, an element with a splice joint formed on an horizontal surface can transfer the bending moment up to a quarter of the moment transferred in a continuous beam, whereas the same joint formed on a vertical surface, with the same load, can transfer a moment of half the value of the moment transferred in the continuous beam. Analyses presented in the literature [4] suggest that in the case of a nibbed scarf joint, which is the most effective solution for transferring shearing stress [3], the moment that can be transferred is only one-third of the load in the continuous beam.

To a large degree, it is the technical condition of joints that shapes the ability of a structure to transfer loading, its static behaviour and deformations. Over time, carpentry joints in historic wooden structures are worn down or even completely destroyed by loading and other external factors, which can present serious danger. The damage or destruction of a joint can threaten the safety of the whole structure due to a considerable weakening of segments where elements are connected to one another. Understanding the static behaviour of joints in wooden structures allows for a detailed and comprehensive analysis of the whole system and helps in making decisions concerning corrective measures or reconstruction work, which is acceptable from the point of view of conservation doctrine [5,6].

Researchers distinguish many different types of carpentry joints in historic buildings, depending on their function and form (inter alia [7]). One group includes scarf joints and splice joints which enable a connection along the length of two elements. These were applied when the available material could not cover the whole beam length. Typically, whenever possible, the joints were applied in the least stressed sections, as the connecting elements at the point of contact could not bear a load greater than that in continuous sections [8]. Aside from extending foundation beams and capping beams in the building frames of historical structures, scarf and splice joints were used to extend roof frame elements, such as purlins or rafter beams. Until the introduction of glulam wood, this was a universal method for extending wood elements [9]. Scarf and splice carpentry joints are used today to restore historical joints or replenish material in historical elements of special value [10]. Examples of scarf and splice joints found in historical structures are presented in Figure 1.

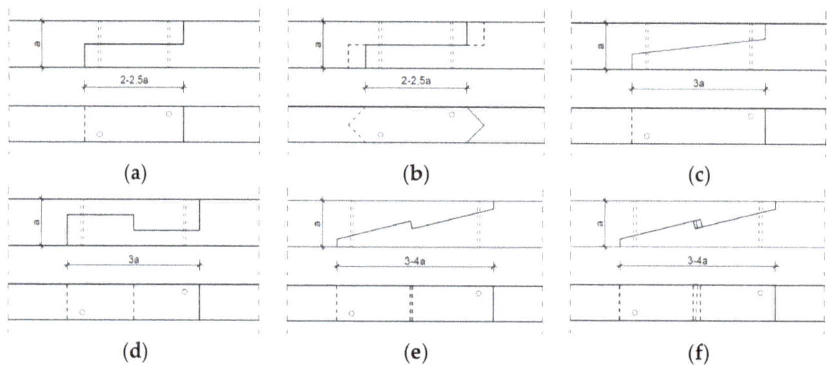

Figure 1. Various forms of scarf and splice joints: (**a**) splice joint, (**b**) nibbed splice joint, (**c**) nibbed scarf joint, (**d**) tabled splice joint, (**e**) stop-splayed scarf joint ('Bolt of lightning'), (**f**) stop-splayed and tabled scarf joint with key.

The type of joint applied in a given connection is related to the function to be performed or the type of loading to be transferred. For example, a nibbed scarf joint (Figure 1c) would be used where the connection is stressed vertically and located near a support. In such a situation, the joint is responsible for transferring shear stress. In situations where tensile elements are connected, such as rafter beams, a different type of joint needs to be applied. For tensile elements, stop-splayed and tabled scarf joints with a key (Figure 1f) were used.

2. Stop-Splayed Scarf Joints ('Bolt of Lightning')

Stop-splayed scarf joints (presented inter alia in [11–15] and referred to as 'Bolt of lightning', 'Trait-de-Jupiter') were widely used in historical buildings. They constituted a sophisticated type of connection along the length of elements in the form of a scarf joint. In historical structures, there are also elements connected along their whole length with a so-called stop-splayed scarf joint or a composite beam (built-up beams, composite beams with a teethed joint), described inter alia in [16–18].

Stop-splayed scarf joints ('Bolt of lightning') were used in ancient times, for example, in the Roman construction of bridges, and later, in roof beam constructions and wooden ceiling construction elements right up to the end of the 19th century. A special period of development for this method of joining beam elements was that of the Italian Renaissance, which was the time of such masters as Leon Battista Alberti (Figure 2a) [19] and Leonardo da Vinci (Figure 2b) [16]. This method of connecting elements along their length was applied especially in elements subjected to tensile and bending forces. Typically, the joints were wedged, which was supposed to help in transferring loading and to ensure a tight fitting joint. Today, they are used mainly to strengthen and repair historical structures. Examples of stop-splayed scarf joints ('Bolt of lightning') and so-called composite beams are presented below (Figures 2–4).

(a)   **(b)**

Figure 2. Sketches presenting scarf and splice joints in wooden beams (**a**) according to Leon Battista Alberti, (**b**) according to Leonardo da Vinci.

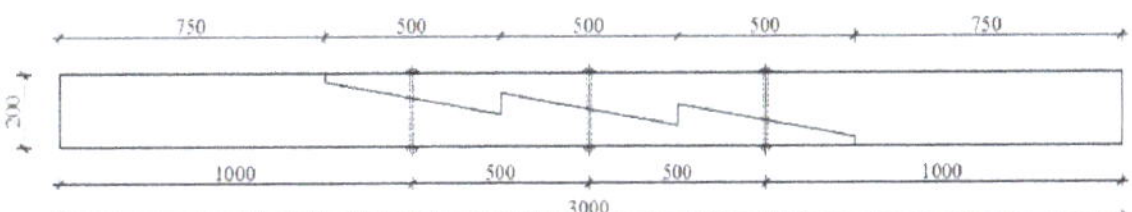

Figure 3. Sketch of a stop-splayed scarf joint ('Bolt of lightning') with dimensions of an actual element, drawing based on data from [11].

(a)   **(b)**

Figure 4. Examples of scarf and splice joining of wooden beams in existing buildings: (**a**) building in Italy, (**b**) building in Poland—13th century Czocha Castle in Sucha.

The few available descriptions of research concerning typical stop-splayed scarf joints ('Bolt of lightning') are presented in [16–18,20–23].

An analysis of the behaviour of a stop-splayed scarf joint subjected to tensile forces is presented in Figure 5 (drawing based on data from [20]). Different methods of strengthening the joints using timber pegs and steel pins were also analysed.

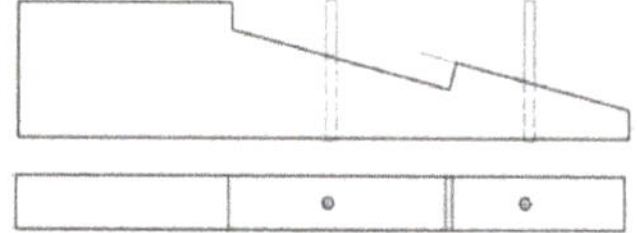

Figure 5. Geometry of a stop-splayed scarf joint ('Bolt of lightning').

An increase in estimated stiffness for the joints with strengthening was noted: 41% for the joint strengthened with wooden pegs and 52% for the joint strengthened with steel pins in relation to joint without strengthening was noted. Load-deflection plots were prepared and failure modes were described. Attention was paid to the difference in the static behaviour of the joints in relation to the material used for strengthening (wood, metal).

The results of research carried out at the University of Bath in the UK by a team consisting of Walker, Harris, Hirst and others on the static behaviour of scarf joints, inter alia stop-splayed scarf joints, which are universally found in historical buildings across England, are presented in [21]. The joints studied were under-squinted butt in halved scarf with two pegs, side-halved and bridled with two pegs, stop-splayed and tabled scarf with key and four pegs, and face-halved and bridled scarf with four pegs (Figure 6) (drawing based on data from [21]).

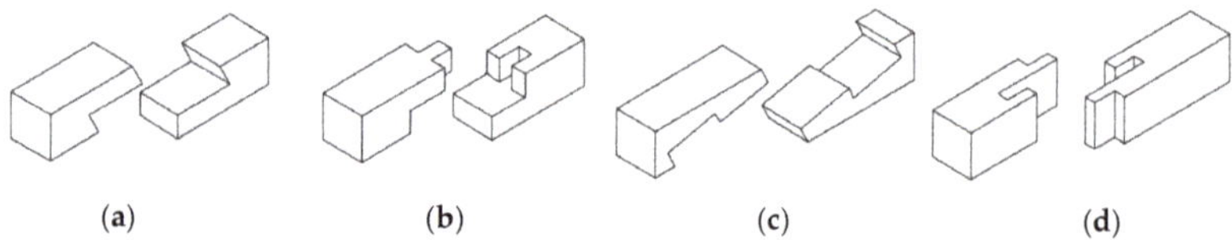

(**a**)   (**b**)   (**c**)   (**d**)

Figure 6. Scarf joints analysed in research: (**a**) under-squinted butt in halved scarf with two pegs, (**b**) side-halved and bridled with two pegs, (**c**) stop-splayed and tabled scarf with key and four pegs, (**d**) and face-halved and bridled scarf with four pegs.

Experimental testing [21] was carried out on model beams of 2.5 m in length joined by means of the joints listed above and 1.5 m continuous beams in order to compare results. The elements were subjected to a four-point vertical bending test and lateral bending test to provide for pure bending. Load-deflection plots were determined, and on the basis of a comparison of results for the different joint types and also the parameters for the continuous element, it was possible to determine the performance factor describing the relationship of loading and stiffness of the composite beam to that of the continuous beam. The stop-splayed scarf joint ('Bolt of lightning') displayed the greatest stiffness and load-bearing (28% in relation to the continuous beam) with bending in the vertical plane.

For the case of bending of the elements joined along their length with the stop-splayed scarf joint ('Bolt of lightning'), i.e., for the so called composite beams (or built-up beams), analyses were presented by Mirabella-Roberti and Bondanelli in [16]. Based on a numerical analysis, the authors identified the most probable locations of stress concentrations, especially in the vicinity of joining planes.

Rug et al. [17,18] present principles for developing and gauging the beams described above, which have been presented in the literature up to the 1970s. Today, it is difficult to find any principles which can provide a basis for constructing or restoring such building elements. For this reason, the University of Eberswalde in Germany carried out experimental research aimed at determining the load-bearing capacity of such elements, describing their static behaviour in terms of displacement resulting from applying loading. The research [17,18] was carried out on models constructed at a 1:1 technical scale (the dimensions adopted were the same as those in an existing rafter beam roof in a tower of a German church) and also on 1:2 scale models (Figure 7). Bending tests were carried out (in accordance with EN408), achieving an average loading at the level of approximately 57 kN; the deflection plot and flexibility modulus of the joint in the technical scale were determined in accordance with EN26891.

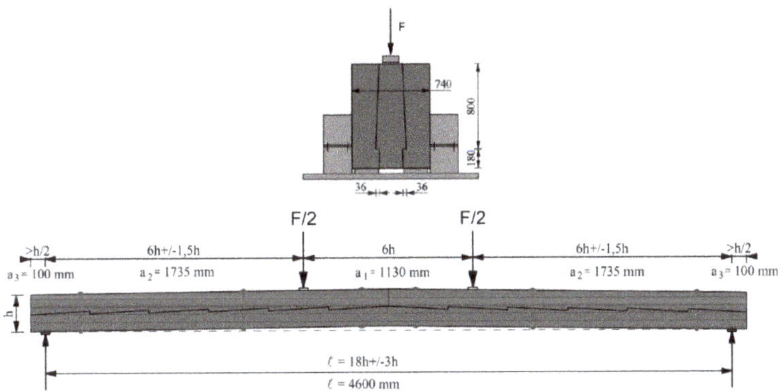

Figure 7. Model of a composite beam connected with a stop-splayed scarf joint and models used for testing of the joint itself and the element as a whole, image adapted from [17] with permission.

Sangree and Schafer [22,23] presented their research and numerical analysis carried out in Baltimore, USA on scarf joints with key found in traditional wooden constructions, e.g., in the Morgan Bridge, which was the subject of their analysis. They tested halved and tabled scarf joints and stop-splayed scarf joints with key (Figure 8 drawings based on data from [22,23]).

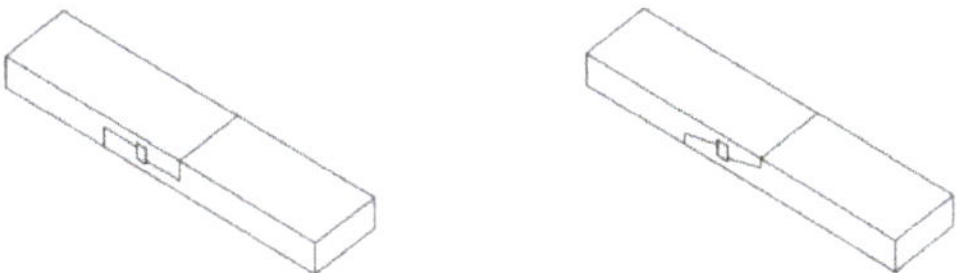

Figure 8. Models representing the geometry of scarf joints.

The joints were analysed as elements operating under complex loading conditions, i.e., tensile bending. In the case of the stop-splayed scarf joints with key [23], it was determined that the orientation of the key had the greatest influence on the static behaviour of the joint, as it causes a vertical pressure to the grain. In addition, special attention was paid to the presence of drawbolts as essential for sustaining the joint. In such cases, it was possible to obtain shear failure parallel to the grain, which made it possible to withstand higher stress levels.

The research presented to date in the literature concerning the static behaviour of carpentry joints has focused mainly on tenon and notched joints. There has been decidedly less research on joints typically subjected to bending (or bending and tensile stress or bending and compressive stress). It is worth noting that practically all researchers who have been concerned with this topic (in Germany inter alia Rug [17,18], in the UK Hirst et al. [21], and also in The Czech Republic Kunecky et al. [24–29] and Fajman et al. in [30–35]) underscore that there is a shortage of research for appropriately describing the static behaviour of such joints, and so proposing the most beneficial methods for repairing or strengthening them.

As a consequence (as part of a research project financed by the National Science Centre), experimental testing was carried out on the static behaviour of stop-splayed scarf joints subjected to bending. The goal was to determine the load-bearing capacity and stiffness of the joint subjected to testing and to determine the dependency of the type of joint and the method of sustaining it and its load-bearing capacity.

3. Experimental Tests

3.1. Static Schemes and Investigations Procedure

For the purposes of experimental testing, technical scale beam models were made from pine wood (Pinus sylvestris L.) of 360 cm in length and with a cross-section measuring 12 cm × 18 cm. Testing involved four series, each with three models. Series A included continuous beams as references, whereas series E, F, and G included beams with stop-splayed scarf joints in the horizontal plane. The series with joints differed from one another in terms of the methods uses to strengthen the joint, i.e.:

- series E—beams with stop-splayed scarf joints and two drawbolts (double-sided tooth plate connectors type C10 (Geka) + M12 screws);
- series F—beams with stop-splayed scarf joints and wooden keys made from oak wood and additional steel clamps;
- series G—beams with stop-splayed scarf joints and flat steel clamps and steel tie-rods.

With respect to the geometry of the joints, these were based on data obtained for real structures and on data from the literature. The schematics and views of the various models used for testing are presented in Figures 9–11.

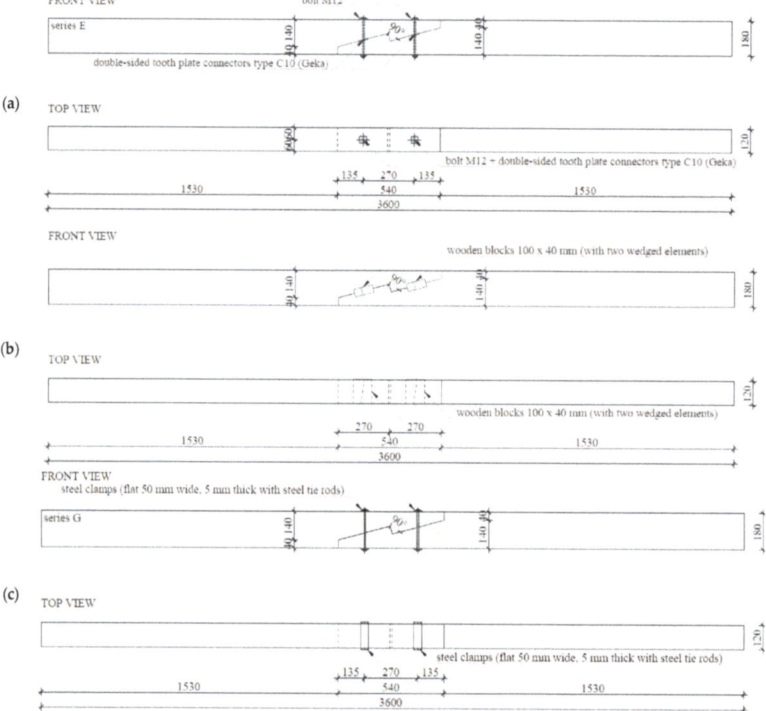

Figure 9. Schemes of the experimental testing models: (**a**) series E—with 2 drawbolts (double-sided tooth plate connectors type C10 + M10 screws), (**b**) series F—with wood inserts (and additional steel clamps), (**c**) series G—with steel clamps.

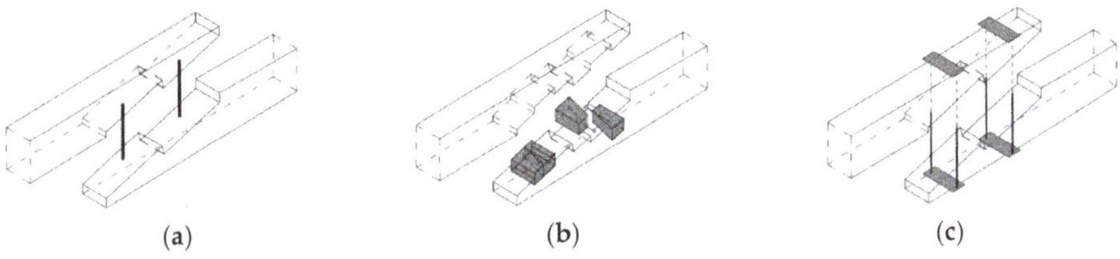

Figure 10. Axiometric schematics of joints used in the experimental models: (**a**) series E—with 2 drawbolts (double-sided tooth plate connectors type C10 + M10 screws), (**b**) series F—with wood inserts (and additional steel clamps), (**c**) series G—with steel clamps.

Figure 11. View of an example beam in each series: (**a**) series E—with 2 drawbolts (double-sided tooth plate connectors type C10 + M10 screws), (**b**) series F—with wood inserts (and additional steel clamps), (**c**) series G—with steel clamps.

In order to determine the load-bearing and the load-deflection plot, the beam was subjected to four-point bending tests, in accordance with the standard procedure described in [36]. A schematic of the experimental testing is provided in Figure 12 (drawing based on data from [36]).

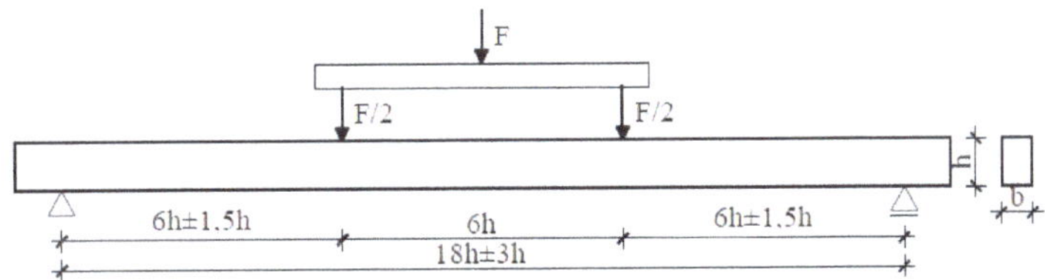

Figure 12. Scheme of the four-point bending test in accordance with standard procedure.

The experimental testing was carried out at the Building Construction Laboratory of the Faculty of Civil Engineering of the Wroclaw University of Science and Technology. An electronically-controlled linear hydraulic jack, the Instron 500 (Instron®, Norwood, MA, USA), was used. The results were registered using the MGC plus measurement system made by the Hottinger Baldwin Messtechnik (HBK GmbH, Darmstadt, Germany). The measurement equipment used in the experimental testing was calibrated to at least class 1 accuracy.

The beams were freely supported at both ends. The span between the axes of the supports was 3.24 m. The supports included a fork support, which ensured that there was no loss in flexural static (lateral buckling). The beams were loaded symmetrically with a loading force applied at two points, thanks to which pure bending was obtained in the central part of the element. The speed of application of the loading was 5 mm/min. A schematic and view of the testing stand is presented in Figures 13 and 14. Registration of the strains observed in the materials was carried out by means of a series of RL 300/50 strain gauge.

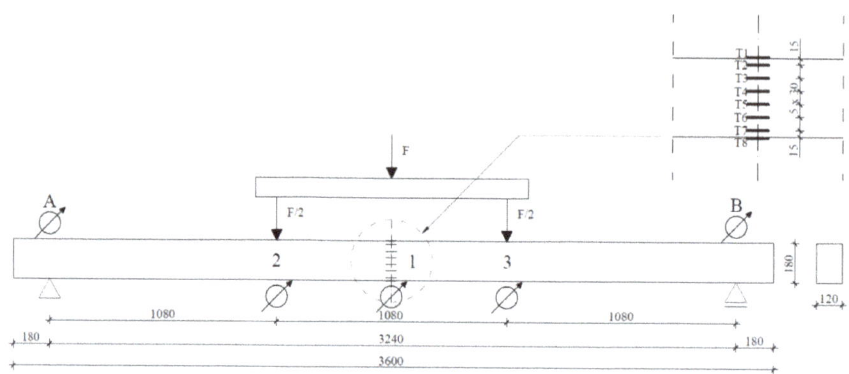

Figure 13. Scheme of the experimental testing site showing locations of the strain gauges.

Figure 14. View of experimental stand for testing with an example beam.

Additionally, during the course of the testing, the wood moisture was determined using a resistance hygrometer (FMW moisture meter) to take measurements in several locations on each tested beam. The moisture content of the elements was kept close to the required standard of 12% [36,37].

3.2. Results

All the beam models (altogether 12 elements) were loaded at a constant speed right up to their failure. The failure of beams in series A resulted from the shearing of fibres in the bottom part of the central area of the beam where earlier there were no visible cracks (except for cracks in the vicinity of knots in the beams). Sudden failure with no prior signs occurred. Generally, the failure of beams in series E, F, and G involved a loosening of the joint in the lower zone of the central part of the beam and the appearance of cracks and fractures on the edges of the joint and in points weakened as a result of earlier flaws (such as knots or primary cracks of the beam). The tested joint was unsymmetrical. In the right part of the joint, the beam elements were bent and pressed to each other. In the left part of the joint, the elements were also bent, but the bottom element was not pressed and the joint was loosened. Destruction occurred in the left part of the joint by breaking the upper element (visible cracks above the edge of the joint). The cross-section of the destruction was in the joint area. Failure resulted from delamination due to stretching across the fibres. The failure views of selected experimental models are presented in Figure 15.

Figure 15. Failure views for selected beams (a) A01, (b) E03, (c) F01, (d) G02.

Table 1 presents the value of the ultimate forces F_u obtained for each of the beam series. Ratio of mean destruction force for specified series to reference beam series is F_{uX}/F_{uA}, where X stands for E, F, or G (and expresses mean destructive force for this series).

Table 1. Ultimate forces for beams in each series.

		Beam Series			
		A	**E**	**F**	**G**
		44.95	13.21	14.52	9.09
Ultimate force Fu	[kN]	37.52	13.79	17.58	16.64
		55.74	11.00	7.03	17.06
Mean ultimate force for each beam series $\overline{F_u}$	[kN]	46.07	12.67	13.04	14.26
Standard deviation s	[kN]	9.16	1.47	5.43	4.49
Variation coefficient v	[%]	19.89	11.62	41.61	31.45
Ratio of mean destruction force for specified series to reference beam series		1.00	0.27	0.28	0.31

Load-deflection plots (in the central part of the beam span, i.e., in point 1) for beams in the various series are presented in Figure 16.

Deformations in the area of point 1 in the central part of the beams are presented in Figure 17. The deformation profiles are consistent with expected curves: for beam A01, the standard image of deformation in the cross-section for the continuous beam was obtained, whereas for beams E02, F02, and G02, it was noted that the typical curve shape for the composite cross-section was as presented inter alia in [18].

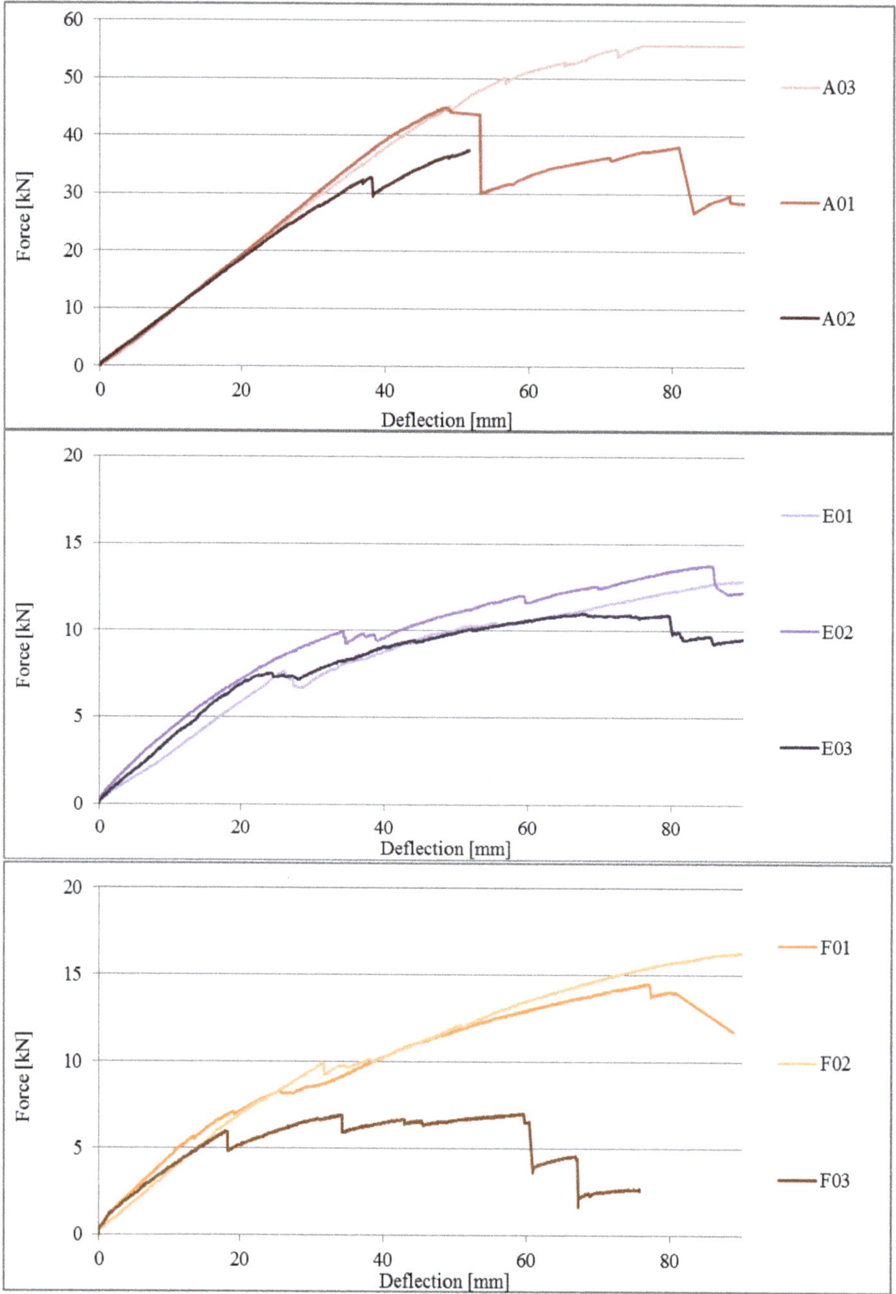

Figure 16. *Cont.*

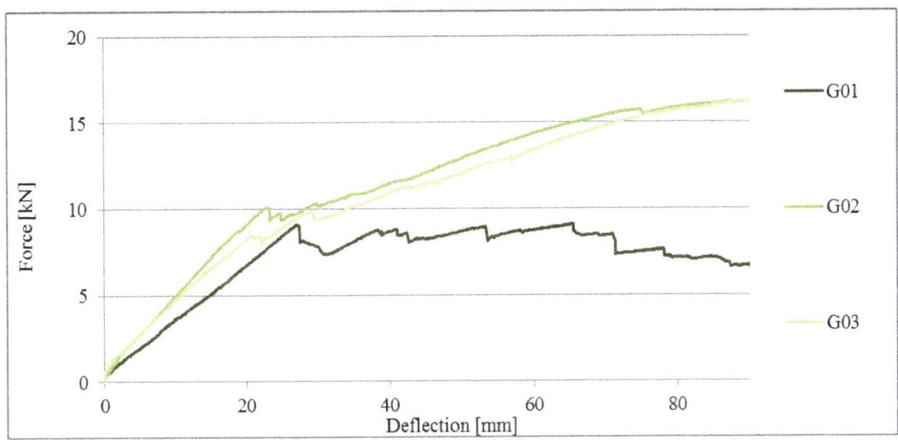

Figure 16. Load-deflection plots for beams in series A, E, F, and G.

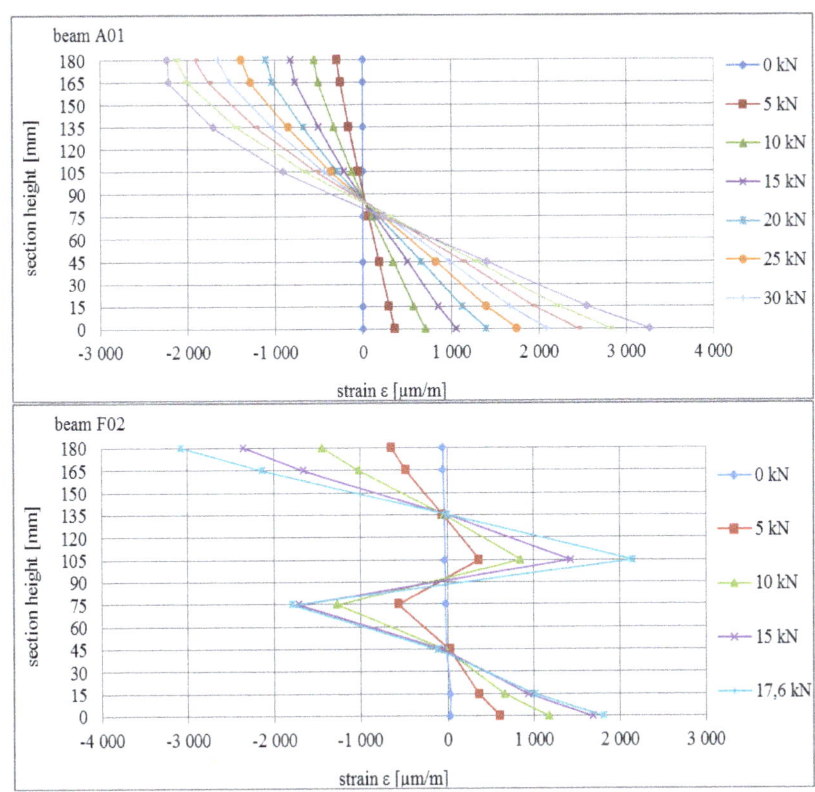

Figure 17. *Cont.*

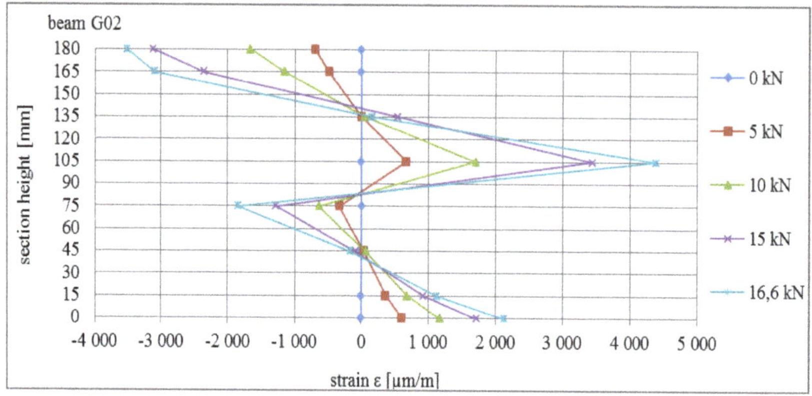

Figure 17. Strain profile in the cross-section for selected beams subjected to bending.

3.3. Analysis of Results

The mean ultimate force value for beams in series A (continuous beams) was 46.07 kN. The value of the mean load bearing for bending for this series was 24.88 kNm. Compared to the reference beams, beams joined with stop-splayed scarf joints, i.e., beams in series E, F and G, achieved lower load-bearing levels, which were comparable to one another. The highest values were obtained for series G beams with a mean ultimate force of 14.26 kN (load-bearing for bending of 7.70 kNm), which constituted 31.0% load-bearing in relation to reference beams. The largest variation in results was obtained for this series (variation coefficient exceeds 30%). Beams of series E and F obtained lower ultimate force values: series E—12.67 kN (load-bearing for bending 6.86 kNm), series F—13.04 kN (load bearing for bending 7.04 kNm) which constitutes respectively 27.5% and 28.3% load-bearing in relation to the reference beam. The results of series E were characterised by the smallest variation with an indicator of not much more than 10%. A comparison of the values of ultimate force values for beams in series A, E, F, and G is presented in Table 2 and in the curve in Figure 18.

Table 2. Comparison of load-bearing for beams in the various series.

		Beam Series			
		A	E	F	G
Mean ultimate force for each beam series $\overline{F_u}$	[kN]	46.07	12.67	13.04	14.26
Mean load-bearing for bending for each beams series $\overline{M_R}$	[kNm]	24.88	6.84	7.04	7.70
Ratio of load-bearing for specified series to reference beam series		1.00	0.27	0.28	0.31

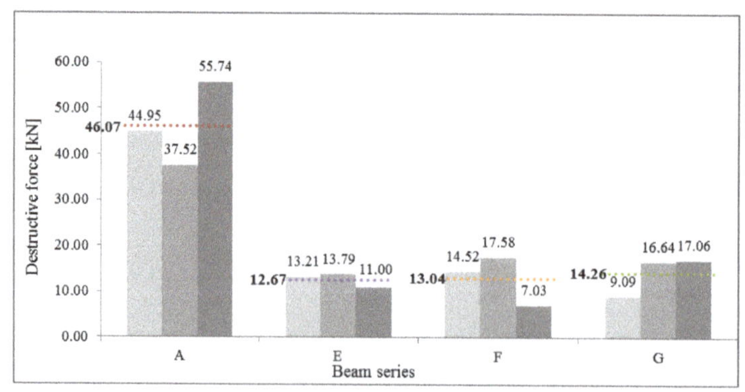

Figure 18. Comparison of values obtained for ultimate force values for beams of series A, E, F, and G.

As part of the comparative analysis, in Figure 19, a comparison of the load-deflection plots of the beams of series E, F, and G to the reference beams of series A is presented. It should be noted that the tested beam series attained similar values for final deflection, but with different levels of force, which are several times higher for series A, whereas the values for beams of series E, F, and G were similar to one another.

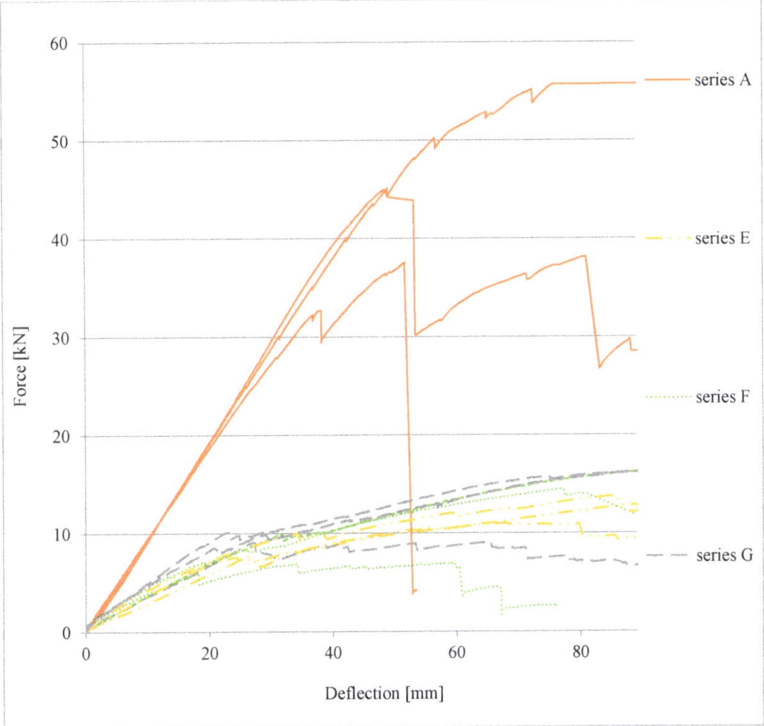

Figure 19. Comparison of load-deflection plots of beams of series E, F, G to the reference beam of series A.

An analysis of the graph (curves for series E, F, and G) shows a change in the nature of the static behaviour of the joint when strain η is equal to approximately 60% of the elastic to ductile state. Estimations on the basis of the load-deflection plots of the 'stiffness parameter' in the elastic state, calculated as the ratio of loading force value to deflection (1) for the beams representing series A, E, F, and G, are as follows: for A01—0.94 kN/mm, for E01—0.31 kN/mm, for F01—0.32 kN/mm, and for G02—0.45 kN/mm.

$$k = \mathrm{tg}\alpha = \frac{F_i}{u_i}\,[\mathrm{kN/mm}] \tag{1}$$

It should be noted that the parameters for beam E01 and F01 are close to one another and amount to approximately 30% of the stiffness parameter for the continuous beam A01, whereas in the case of beam G02, the parameter amounts to approximately 50% of the reference value. The most important test results obtained for the beams from the specified series in relation to the reference beam (continuous beam) are presented in Table 3.

Table 3. Presentation of results for beams from specified series.

Beam Type (Strengthening of the Joint)	Beam Series			
	A	E	F	G
	Reference-Continuous Beam	Drawbolts	Wooden Pegs (keys)	Steel Clamps
Mean ultimate force $\overline{F_u}$ [kN]	46.07	12.67	13.04	14.26
Mean load-bearing in bending $\overline{M_R}$ [kNm]	24.88	6.84	7.04	7.70
Ratio of load-bearing as compared to reference beam series	1.00	0.27	0.28	0.31
Mean deflection u1 in the middle of the span with a load of 10 kN [mm]	12.34	37.44	32.48	24.59
Ratio of deflection as compared to reference beam series	1.00	3.03	2.63	1.99
Mean stiffness parameter k [kN/mm]	0.94	0.29	0.32	0.39
Stiffness parameter as compared to the reference beam series	1.00	0.32	0.35	0.43

4. Conclusions

Stop-splayed scarf joints are common in historical structures where there are elements that are subject to bending, tensile stress, and bending with tensile stress, which are to be found primarily in roof framing elements, but also less commonly in wooden ceilings. When appropriately strengthened, e.g., with steel clamps or screws, etc., these joints can transfer bending loads. These types of strengthened joints have not been described to date in the literature.

The obtained test results can be helpful in the design and strengthening of this type of joint, especially in historical buildings.

Experimental testing carried out on technical-scale models presented the static behaviour of such connections subjected to bending load compared to bending of a continuous reference beam. The bending load-bearing levels obtained for beams with joints compared to the reference beam amounted to approximately 30% (27.5–31% depending on the type of strengthening method used). The stiffness levels obtained in laboratory experimental testing amounted to 30–40% in comparison to the reference beam. The lowest load-bearing capacity and stiffness were obtained for beams in series E, which were strengthened with drawbolts, and higher values for beams strengthened with steel clamps and wooden pegs.

Despite a small sample from a statistical point of view, the obtained test results confirm the results of other researchers regarding the level of load-bearing capacities observed in beams with joints of a similar type.

It is important to underscore that the laboratory testing involved three models in each series, which resulted in large variations in results in some series (as, for example, with the beams in series F, where the variation coefficient for ultimate force amounted to approximately 40%). It can be concluded that the high variation coefficient obtained resulted from primary flaws in the material, and not from the behaviour of the joints themselves. As a natural material, wood is characterised by large variations due to the presence of knots, cracks etc.

The experimental investigation that was carried out allowed us to obtain a description of the static behaviour of stop-splay scarf joints subjected to bending. For more precise analyses and conclusions, numerical analyses and further laboratory testing are recommended. Given the failure mechanisms observed and the deformations in the lower edges of the joint, it is recommended that in further experimental testing, consideration is given to testing asymmetrical joint strengthening, with greater strengthening in the zone, where the largest deformations were obtained.

As a next step of the research programme, numerical analyses and further experimental investigations in the laboratory of beams with different types of joints will be carried out on the described joints. The main goal of this research is to determine the solution of optimizing the design of carpentry joints in bent wooden elements.

Author Contributions: Conceptualization, J.J. and T.N.; methodology, J.J., T.N.; formal analysis, A.K., K.R., T.N.; investigation, A.K., T.N.; resources, A.K., K.R.; data curation, A.K.; writing—original draft preparation, A.K.; writing—review and editing, K.R., T.N.; supervision, J.J.; project administration, A.K., T.N.; funding acquisition, A.K. All authors have read and agreed to the published version of the manuscript.

Funding: This research was funded by the National Science Centre, grant number 2015/19/N/ST8/00787.

Acknowledgments: The authors would like to thank DWE Publisher for giving the licence and permission to use them by authors in the article.

Conflicts of Interest: The authors declare no conflict of interest. The funders had no role in the design of the study; in the collection, analyses, or interpretation of data; in the writing of the manuscript, or in the decision to publish the results.

References

1. Nowak, T.; Karolak, A.; Sobótka, M.; Wyjadłowski, M. Assessment of the Condition of Wharf Timber Sheet Wall Material by Means of Selected Non-Destructive Methods. *Materials* **2019**, *12*, 1532. [CrossRef] [PubMed]
2. Jasieńko, J.; Nowak, T.; Karolak, A. Historical carpentry joints. *Wiad. Konserw. J. Herit. Conserv.* **2014**, *40*, 58–82.
3. Branco, J.M.; Descamps, T. Analysis and strengthening of carpentry joints. *Constr. Build. Mater.* **2015**, *97*, 34–47. [CrossRef]
4. Yeomans, D. *The Repair of Historic Timber Structures*; Thomas Telford Ltd.: London, UK, 2003; Volume 3, p. 7.
5. Kłosowski, P.; Lubowiecka, I.; Pestka, A.; Szepietowska, K. Historical carpentry corner log joints—Numerical analysis within stochastic framework. *Eng. Struct.* **2018**, *176*, 64–73. [CrossRef]
6. Corradi, M.; Osofero, A.I.; Borri, A. Repair and Reinforcement of Historic Timber Structures with Stainless Steel—A Review. *Metals* **2019**, *9*, 106. [CrossRef]
7. Šobra, K.; Fonseca Ferreira, C.; Riggio, M.; D'Ayala, D.; Arriaga, F.; Aira, J.R. A new tool for the structural assessment of historic carpentry joints. In Proceedings of the 3rd International Conference on Structural Health Assessment of Timber Structures—SHATIS'15, Wrocław, Poland, 9–11 September 2015.
8. Ross, P. *Appraisal and Repair of Timber Structures*; Thomas Telford Ltd.: London, UK, 2002.
9. Thelandersson, S.; Larsen, H.J. *Timber Engineering*; John Wiley & Sons Ltd.: Chichester, UK, 2003; Volume 16.
10. Ilharco, T.; Lechner, T.; Nowak, T. Assessment of timber floors by means of non-destructive testing methods. *Constr. Build. Mater.* **2015**, *101*, 1206–1214. [CrossRef]
11. Parisi, M.A.; Piazza, M. Seismic strengthening of traditional carpentry joints. In Proceedings of the 14th World Conference on Earthquake Engineering, Beijing, China, 12–17 October 2008.
12. Parisi, M.A.; Piazza, M. Seismic strengthening of traditional timber structures. In Proceedings of the 13th World Conference on Earthquake Engineering, Vancouver, BC, Canada, 1–6 August 2004.
13. Perez, L.P. Design and construction of timber roof structures, built over different structural systems. Cases studium at the Valencia Community. In Proceedings of the First International Congress on Construction History, Madrid, Spain, 20–24 January 2003.
14. Tampone, G.; Semplici, M. *Rescuing the Hidden European Wooden Churches Heritage, an International Methodology for Implementing a Data Base for Restoration Projects*; In Cooperation with Fly Events and Alter Ego Ing Arch, S.r.l.: Città di Castello, Italy, 2006.
15. Aira, J.R.; Arriaga, F.; Íñiguez-González, G.; Guaita, M. Failure modes in halved and tabled tenoned timber scarf joint by tension test. *Constr. Build. Mater.* **2015**, *96*, 360–367. [CrossRef]
16. Mirabella-Roberti, G.; Bondanelli, M. Study and analysis of XIV century timber built-up beams in Verona. *Adv. Mater. Res.* **2013**, *778*, 511–516. [CrossRef]
17. Rug, W.; Linke, G. Study on the load bearing capacity and the load-deferral behavior of wooden composite beams with a teethed joint. In Proceedings of the 3rd International Conference on Structural Health Assessment of Timber Structures—SHATIS'15, Wrocław, Poland, 9–11 September 2015.
18. Rug, W.; Thoms, F.; Grimm, U.; Eichbaum, G.; Abel, S. Untersuchungen zur Biegetragfähigkeit von verzahnten Balken. *Bautechnik* **2012**, *89*, 26–36. [CrossRef]
19. Alberti, L.B. *L'Architettura di Leon Batista Alberti, Tradotta in lingua Fiorentina da Cosimo Bartoli. Con la aggiunta de disegni. Et altri diuersi Trattati del medesimo Auttore*; Appresso Lionardo Torrentino: Nel Monte Regale, Italy, 1565. (In Italian)

20. Ceraldi, C.; Costa, A.; Lippiello, M. Stop-Splayed Scarf-Joint Reinforcement with Timber Pegs Behaviour. In *Structural Analysis of Historical Constructions*; Aguilar, R., Torrealva, D., Moreira, S., Pando, M., Ramos, L.F., Eds.; Springer: Cham, Switzerland, 2019; pp. 360–369.

21. Hirst, E.; Brett, A.; Thomson, A.; Walker, P.; Harris, R. The structural performance of traditional oak tension & scarf joints. In Proceedings of the 10th World Conference on Timber Engineering, Miyazaki, Japan, 2–5 June 2008.

22. Sangree, R.H.; Schafer, B.W. Experimental and numerical analysis of a halved and tabled traditional scarf joint. *Constr. Build. Mater.* **2009**, *23*, 615–624. [CrossRef]

23. Sangree, R.H.; Schafer, B.W. Experimental and numerical analysis of a stop-splayed traditional scarf joint with key. *Constr. Build. Mater.* **2009**, *23*, 376–385. [CrossRef]

24. Arciszewska-Kędzior, A.; Kunecký, J.; Hasníková, H. Mechanical response of a lap scarf joint with inclined faces and wooden dowels under combined loading. *Wiad. Konserw. J. Herit. Conserv.* **2016**, *46*, 80–88.

25. Arciszewska-Kędzior, A.; Kunecký, J.; Hasníková, H.; Sebera, V. Lapped scarf joint with inclined faces and wooden dowels: Experimental and numerical analysis. *Eng. Struct.* **2015**, *94*, 1–8. [CrossRef]

26. Kunecký, J.; Hasníková, H.; Kloiber, M.; Milch, J.; Sebera, V.; Tippner, J. Structural assessment of a lapped scarf joint applied to historical timber constructions in central Europe. *Int. J. Archit. Herit.* **2018**, *12*, 666–682. [CrossRef]

27. Kunecký, J.; Sebera, V.; Hasníková, H.; Arciszewska-Kędzior, A.; Tippner, J.; Kloiber, M. Experimental assessment of full-scale lap scarf timber joint accompanied by a finite element analysis and digital correlation. *Constr. Build. Mater.* **2015**, *76*, 24–33. [CrossRef]

28. Kunecký, J.; Sebera, V.; Tippner, J.; Hasníková, H.; Kloiber, M.; Arciszewska- Kędzior, A.; Milch, J. Mechanical performance and contact zone of timber joint with oblique faces. *Acta Univ. Agric. Et Silvic. Mendel. Brun.* **2015**, *63*, 1153–1159. [CrossRef]

29. Kunecký, J.; Sebera, V.; Tippner, J.; Kloiber, M. Numerical assessment of behavior of a historical central European wooden joint with a dowel subjected to bending. In Proceedings of the 9th International Conference on Structural Analysis of Historical Constructions, Mexico City, Mexico, 15–17 October 2014.

30. Fajman, P.; Máca, J. Stiffness of scarf joints with dowels. *Comput. Struct.* **2018**, *207*, 194–199. [CrossRef]

31. Fajman, P.; Máca, J. The effect of inclination of scarf joints with four pins. *Int. J. Archit. Herit.* **2018**, *12*, 599–606. [CrossRef]

32. Fajman, P. A scarf joint for reconstructions of historical structures. *Adv. Mater. Res.* **2015**, *969*, 9–15. [CrossRef]

33. Fajman, P.; Máca, J. Scarf joints with pins or keys and dovetails. In Proceedings of the 3rd International Conference on Structural Health Assessment of Timber Structures—SHATIS'15, Wrocław, Poland, 9–11 September 2015.

34. Fajman, P.; Máca, J. The effect of key stiffness on forces in a scarf joint. In Proceedings of the 9th International Conference on Engineering Computational Technology, Naples, Italy, 2–5 September 2014; Civil-Comp Press: Stirlingshire, UK, 2014; Volume 40.

35. Šobra, K.; Fajman, P. Utilization of splice skew joint with a key in the reconstruction of historical trusses. *Adv. Mater. Res.* **2013**, *668*, 207–212. [CrossRef]

36. *Structural Timber—Determination of Characteristic Values of Mechanical Properties and Density*; PN-EN 384: 2016-10; PKN: Polish Committee for Standardization: Warsaw, Poland, 2016.

37. *Timber Structures. Structural Timber and Glued Laminated Timber. Determination of Some Physical and Mechanical Properties*; PN-EN 408+A1: 2012; PKN: Polish Committee for Standardization: Warsaw, Poland, 2012.

Article

Effect of PBO–FRCM Reinforcement on Stiffness of Eccentrically Compressed Reinforced Concrete Columns

Tomasz Trapko * and Michał Musiał

Faculty of Civil Engineering, Wroclaw University of Science and Technology, Pl. Grunwaldzki 11, 50-377 Wroclaw, Poland; michal.musial@pwr.edu.pl
* Correspondence: tomasz.trapko@pwr.edu.pl; Tel.: +48-71-320-35-48

Received: 20 January 2020; Accepted: 5 March 2020; Published: 9 March 2020

Abstract: This paper examines the effect of PBO (P-phenylene benzobisoxazole)–FRCM (Fabric Reinforced Cementitious Matrix) reinforcement on the stiffness of eccentrically compressed reinforced concrete columns. Reinforcement with FRCM consists of bonding composite meshes to the concrete substrate by means of mineral mortar. Longitudinal and/or transverse reinforcements made of PBO (P-phenylene benzobisoxazole) mesh were applied to the analyzed column specimens. When assessing the stiffness of the columns, the focus was on the effect of the composite reinforcement itself, the value and eccentricity of the longitudinal force and the decrease in the modulus of elasticity of the concrete with increasing stress intensity in the latter. Dependences between the change in the elasticity modulus of the concrete and the change in the stiffness of the tested specimens were examined. The relevant standards, providing methods of calculating the stiffness of composite columns, were used in the analysis. For columns, which were strengthened only transversely with PBO mesh, reinforcement increases their load capacity, and at the same time, the stiffness of the columns increases due to the confinement of the cross-section. The stiffness depends on the destruction of the concrete core inside its composite jacket. In the case of columns with transverse and longitudinal reinforcement, the presence of longitudinal reinforcement reduces longitudinal deformations. The columns failed at higher stiffness values in the whole range of the eccentricities.

Keywords: column; stiffness; FRCM; PBO mesh; PBO–FRCM

---

1. Introduction

The FRCM (Fabric Reinforced Cementitious Matrix) system, in which a composite mesh is bonded to the substrate with mineral mortar, is becoming the preferred method of choice in increasing the load capacity of concrete elements or the method in repairing them. FRCM reinforcements are applied to strengthen or repair concrete and reinforced concrete elements subjected to bending, shearing and compression. Increasingly more experimental and theoretical investigations into the behavior of elements strengthened with a composite mesh on mineral mortar are reported [1–8].

The FRCM system is characterized by higher resistance to elevated temperatures than the FRP (Fiber Reinforced Polymer) system in which non-metallic composite fibers are embedded in epoxy resin [9,10]. An additional advantage of the FRCM system is its better compatibility with the concrete substrate in comparison with FRP systems, where the composite tends to separate from the substrate. FRCM reinforced structural elements show more plastic behavior than the ones strengthened with FRP. This is attributed to the slip, which occurs in the mortar–fiber interface. In the FRCM system's behavior, one can distinguish two main phases separated by the moment at which the cement mortar cracks. Sometimes an intermediate phase, connected with the initial development of cracks in the matrix, can be distinguished. The behavior of the composite in its prior-to-cracking state depends on both the fibers and the matrix. The behavior of this reinforcement in its post-mortar fail state depends mainly on the fibers [11–14].

The knowledge about the considered subject can be extended through tests and analyses of the confined concrete elements subjected to compression [15–26]. The effectiveness of the composite reinforcement is determined by its stiffness, which affects the ratio of transverse (circumferential) strains to longitudinal strains. The stiffness of the compressed columns determines its ability to deform plastically and redistribute the internal forces in the structure. The concrete core's compressibility is limited, and when the limit is exceeded, the ability of the core to resist the increasing load diminishes, which initiates its failure. As long as longitudinal stresses σ_c in the concrete do not exceed its compressive strength f'_{co}, the strains in the composite remain low. From the instant when $\sigma_c > f'_{co}$, the transverse stresses in the composite increase, so does the PBO mesh action on the concrete core. In order to develop a general method of dimensioning columns reinforced with FRCM, it is necessary to determine the effect of this reinforcement on the change in the stiffness of such columns with increasing stresses.

Tests [17–20] (on which the present analyses are based) carried out by the authors on reinforced concrete columns reinforced with PBO mesh on mineral mortar (PBO–FRCM) show that the reinforced columns are characterized by greater ductility than unreinforced columns. Longitudinal FRCM reinforcement improves the ductility of eccentrically compressed columns. The presence of longitudinal composite reinforcement brings about an increase in the longitudinal stiffness of the columns and consequently affects the ultimate compressive strain value. In reinforced concrete columns longitudinally and transversely reinforced with PBO–FRCM, the ultimate strain value increases with increasing eccentricity and depends on the number of transverse reinforcement layers.

Ombres and Verre [23] carried out tests on reinforced concrete columns reinforced with PBO mesh on mineral mortar. They analyzed the effectiveness of the PBO–FRCM reinforcement in increasing the load capacity of the tested elements, focusing on the effect of eccentric loading and transverse reinforcement intensity on the structural response of the confined (wrapped) columns. In the first series of columns, the load was applied eccentrically to the top of the specimens, whereas the reaction force at their base was eccentrically applied on the other side of the longitudinal axis of the columns. In the second test series, the eccentric load was applied to the top and base of the specimens on the same side of the longitudinal axis of the columns. It was found that the PBO–FRCM confinement (winding) increased the load capacity of the columns by 20–39% relative to the unconfined columns. For comparison, in experimental studies on reinforced concrete columns with only a transverse winding (C_1H and C_2H) carried out by the present authors [18,19], a 5–24% increase in load capacity, where the load capacity values were dependent on the number of reinforcement layers and the eccentricity value, was obtained. Ombres and Verre [23] also showed that the increase in compressive strains is linear until the peak load is reached. This finding is important for the present authors since it corroborates the research results presented [19] and provides the basis for the current analyses of the change in the stiffness of columns reinforced with PBO–FRCM.

1.1. Flexural Stiffness of Compressed Reinforced Concrete Columns

Research into the behavior of compressed columns shows that the stiffness of their cross-sections is not constant. When evaluating the stiffness of such elements, one should take into account the effect of the longitudinal force and its eccentricity and obviously the decrease in the modulus of elasticity of the concrete (E_c) with the increasing load. The value of the modulus of elasticity of the concrete (E_c) in a given stress state is highest at a stress close to zero. As the stress increases, the modulus of elasticity of the concrete decreases. For pure concrete at longitudinal stresses $\sigma_c > 0.5f'_{co}$, the ratio of the concrete's instantaneous modulus of elasticity $E_{c,time}$ to its initial modulus of elasticity E_0 amounts to about 0.77 [27]. At higher stresses, this ratio is difficult to estimate since the concrete enters the plastic phase characteristic related to its grade.

1.2. Standard Analysis of Stiffness of Compressed Reinforced Concrete Columns

Bearing in mind the similarity of PBO–FRCM reinforced columns to composite steel–concrete columns made of steel tubes filled with concrete (CFST—Concrete Filled Steel Tube), let us recall the most important standards providing methods of calculating the stiffness of CFST columns.

According to Eurocode 4 [28], the value of characteristic effective flexural stiffness $(EI)_{\text{eff}}$ of the cross-section of a composite column should be calculated from the formula:

$$(EI)_{eff} = E_a I_a + E_s I_s + K_e E_{cm} I_c,\tag{1}$$

where K_e is a (correction) factor reducing the stiffness component originating from the cross-section of the concrete, amounting to 0.6 according to the standard. Eurocode 4 does not directly specify what this correction factor includes, but certainly it does not include long-term effects. The latter are taken into account through the reduction of the modulus of elasticity of the concrete from E_{cm} to $E_{c,\text{eff}}$.

Eurocode 2 [29], recommends to calculate the stiffness of slender columns with any cross-section from the formula:

$$EI = K_c E_{cd} I_c + K_s E_s I_s,\tag{2}$$

where K_c is a coefficient dependent on the effects of cracking and creep. Moreover, the Eurocode 2 [29], recommends to use $E_{cd,\text{eff}}$ instead of E_{cd} for statically indeterminate columns.

$$E_{cd,eff} = \frac{E_{cd}}{\left(1 + \varphi_{ef}\right)}\tag{3}$$

When, for the purposes of the analyses, one omits the effect of the creep of the concrete, K_c can be calculated from the relation, given in [29]:

$$K_c = k_1 k_2,\tag{4}$$

where:

$$k_1 = \sqrt{\frac{f_{ck}}{20}},\tag{5}$$

$$k_2 = \left(\frac{P}{A_c f_{cd}}\right)\left(\frac{\lambda}{170}\right) \le 0.2.\tag{6}$$

As it is apparent, this coefficient takes into account the strength parameters of the concrete, the slenderness of the column and most importantly, as applied in this paper, the stress intensity of the member, whereas it does not take into account the effect of the eccentric load.

K_s is a factor for contribution of steel reinforcement—$K_s = 1, 0$ when $\rho \ge 0.002$ and $K_s = 0$ when $\rho \ge 0.01$.

2. Test Specimens

In order to determine the effect of the PBO–FRCM reinforcement on the change in the stiffness of columns strengthened in this way, 1500 (height) × 200 × 200 column specimens were subjected to tests, the results of which were presented in more detail in [18,19] (Figure 1). The spacing of the stirrups was concentrated, at the element ends to 1/3 of the spacing, over a section longer than 200 mm (the cross-section size of the column). In order to ensure the parallelism of the holding-down planes and uniform pressure on concrete and reinforcement bars in the columns, front metal plates were made through. The longitudinal concrete steel reinforcement was made of four Ø12 bars (RB500W, f_{yk} = 500 MPa) [29] and the transverse concrete steel reinforcement had the form of Ø6 stirrups (St0S, f_{yk} = 220 MPa) [29]. All of the columns were prefabricated. All of the test columns and concrete specimens were made from a single concrete batch during mixing and vibrating at the concrete prefabrication plant. The concrete-mix design is shown in Table 1.

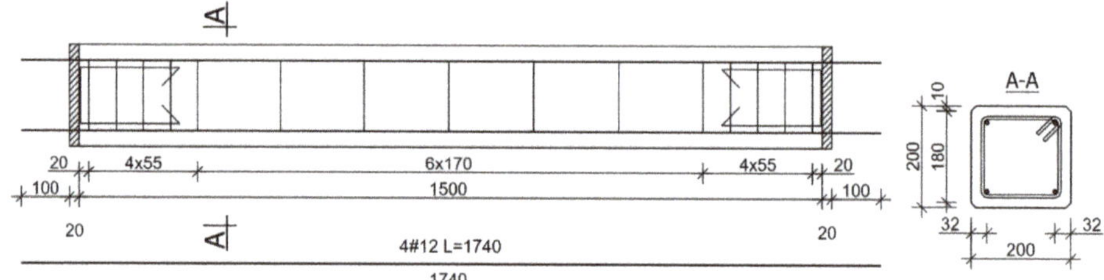

Figure 1. Test specimens.

Table 1. Concrete mixture.

Portland cement of grade 42.5R	455 kg/m^3
Sand 0–2 mm	710 kg/m^3
Aggregate 2–8 mm (river pebble)	710 kg/m^3
Aggregate 8–16 mm (river pebble)	610 kg/m^3
Water	200 kg/m^3
Superplasticizer	0.75% cement mass—3.375 kg/m^3
w/c ratio	0.44

The specimens were made of concrete with mean cubic compressive strength $f_{cm,cube}$ = 55.5 MPa, mean cylinder compressive strength $f_{cm,cyl}$ = 48.7 MPa and mean modulus of elasticity E_{cm} = 33.8 GPa. The mechanical properties of the concrete were determined by standard tests [30,31].

Ruredil X Mesh Gold PBO (P-phenylene benzobisoxazole) mesh (Ruredil, San Donato Milanese, Italy) and mineral mortar Ruredil X Mesh M750 (Ruredil, San Donato Milanese, Italy) were used as the composite reinforcement [32–34]. The mesh is a two-way woven sheet on a matrix, in which there are four times more fibers in the primary direction than in the perpendicular direction. The specifications of the PBO–FRCM strengthening materials are given in Table 2.

Table 2. Mechanical and geometrical characteristic of the PBO–FRCM strengthening materials.

Parameter	Unit	PBO Mesh [32,33]	Cement Based Matrix [32,33]	PBO–FRCM System [34]
Tensile strength	(MPa)	5800	-	1664
Compressive strength	(MPa)	-	29.0	-
Young modulus	(GPa)	270	6.0	128
Nominal thickness	(mm)	0.0455 longitudinal 0.0224 transversal	-	-

Column specimens simultaneously strengthened longitudinally with one layer of the mesh and transversely with one (1H) or two (2H) layers of the mesh were selected for the investigations. The particular layers of this reinforcement were made of a single continuous PBO mesh sheet. The longitudinal composite reinforcement was laid with its fibers running parallel to the column's axis. The columns were wrapped such that the fibers ran horizontally in the primary direction. In the specimens of type C_1V1H and C_1V2H, first, the longitudinal layer was made and then the horizontal layers were laid. The successive layers of mesh were separated from one another with layers of mortar. The length of the finish mesh overlap amounted to 100 mm and the overlap was located on the side perpendicular to the compression plane. The layers of the composite embedded in the binder were topped with a mortar layer closing and leveling the outer surface.

The tests were carried out at the axial force eccentricity within the core of the cross-section amounting to 0, *h*/12 (16 mm) and *h*/6 (32 mm) (Figure 2). The distance between the cylinder axes (rotational axes of the columns) was 1690 mm. For each of the eccentricities, one of the columns was tested as the reference specimen without the C_C reinforcement (Table 3).

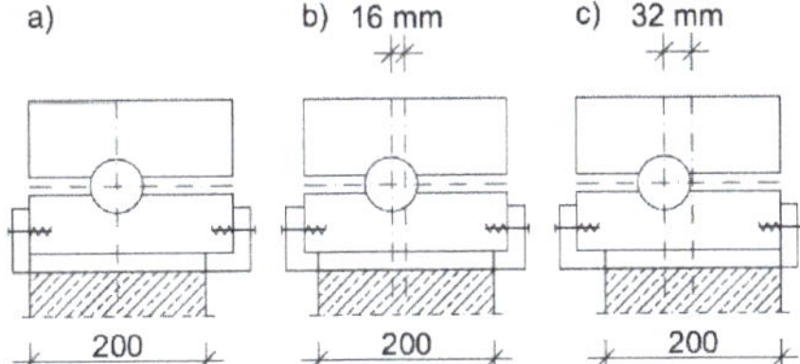

Figure 2. Cylindrical bearing used in the experiment: (**a**) e = 0; (**b**) e = 16 mm; (**c**) e = 32 mm.

Table 3. Configuration of specimens [18,19].

Specimens	Cross-Section	Height	Internal Reinforcement	FRCM Type		Eccentricity
				Horizontal	Vertical	
	(mm)	(mm)	-	-	-	(mm)
C_C_0	200 × 200	1500	4∅12	No	No	0
C_C_16				No	No	16
C_C_32				No	No	32
C_1H_0				1 layer	No	0
C_1H_16				1 layer	No	16
C_1H_32				1 layer	No	32
C_2H_0				2 layers	No	0
C_2H_16				2 layers	No	16
C_2H_32				2 layers	No	32
C_1V1V_0				1 layer	1 layer	0
C_1V1V_16				1 layer	1 layer	16
C_1V1V_32				1 layer	1 layer	32
C_1V2H_0				2 layers	1 layer	0
C_1V2H_16				2 layers	1 layer	16
C_1V2H_32				2 layers	1 layer	32

3. Experimental Results and Analysis

3.1. Changes in Elasticity Modulus of Concrete in Tested Columns

The mean elasticity modulus E_{cm} = 33.8 GPa of the concrete of the columns was determined on five 350 mm cylindrical specimens with a diameter of 113 [31]. After the reference failure load had been determined, six initial load cycles up to the level of $0.5\sigma_{c,max}$, followed by one load cycle up to $0.8\sigma_{c,max}$ and the final cycle until failure, were carried out (Figure 3).

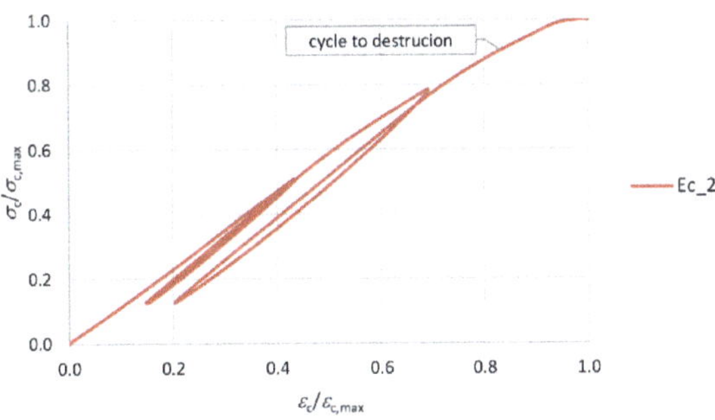

Figure 3. Method of determining modulus of elasticity of concrete.

From the ultimate load cycle dependence, σ_c-ε_c secant elasticity modulus values were determined at every $0.1\sigma_c$ for the three selected specimens. Figure 4 shows how the elastic modulus values change as the stresses in the concrete increase. The relative elasticity modulus $E_{c\_i}/E_{c,max\_i}$ and stress $\sigma_{c\_i}/\sigma_{c,max\_i}$ values in the concrete were determined by relating them to the maximum value for a given specimen (i = 1, 2 and 3).

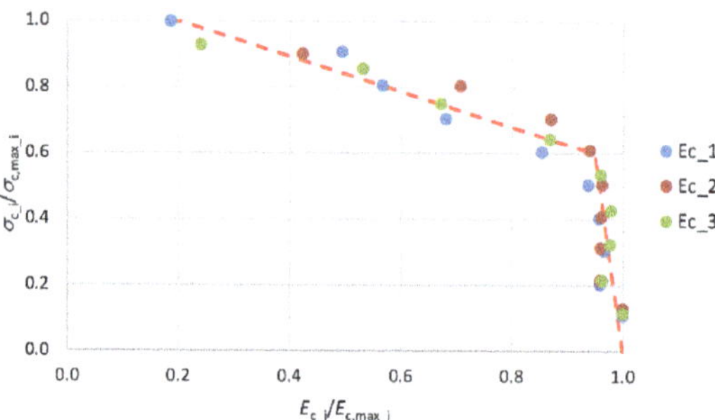

Figure 4. Character of change in secant elasticity modulus values as stresses in concrete increase (i—number of sample; i = 1, i = 2, i = 3).

Figure 4 shows that up to the stress level of about $0.6f_{c,cyl}$ the elasticity modulus values increase only slightly (by about 5%). At higher stress values of $\sigma_c > 0.6f_{c,cyl}$, the elasticity modulus values decrease more sharply until the minimum value of about $0.2E_{c,max}$ is reached immediately before failure. The character of this change can be linearly described, as shown by the broken line in Figure 4.

3.2. Change in Stiffness of Columns with PBO–FRCM Reinforcement

As the columns were being tested, the longitudinal and transverse strains were measured by strain gauges arranged along the circumference of the columns at half of their height. Depending on the column type, different arrangements of strain gauges were adopted. In the reference columns C_C_0, C_C_16 and C_C_32, two vertical strain gauges, V0 and V2, and two horizontal strain gauges, H1 and H3, were used. Strain gauges V0 and H1 were located on the side where the force acted at the eccentricity (the more compressed side) (Figure 5a). Six strain gauges were used in the case of columns C_1H, C_2H, C_1V1H and C_1V2H. Two vertical strain gauges, V0 and V5, were located in the plane of compression. The next four strain gauges, H2, H4, H7 and H9, measured circumferential strains (Figure 5b).

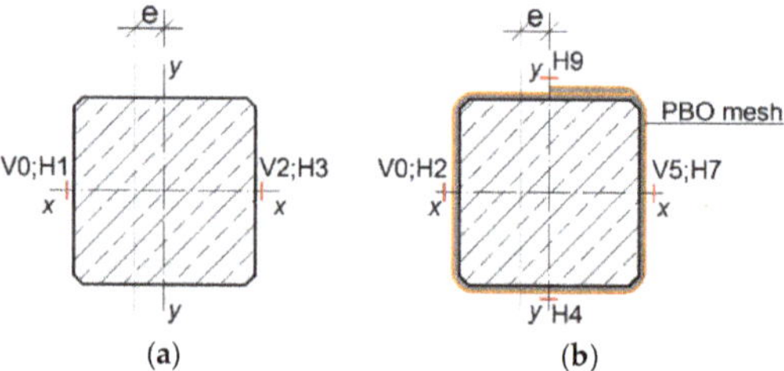

Figure 5. Instrumentation for columns. (**a**) reference columns; (**b**) reinforced columns.

The horizontal columns' displacements (deflections) were measured by Linear Variable Differential Transformers (LVDTs, HBM Masstechnik, Darmstadt, Germany). The measurement span of the transducers was ±10 mm. The LVDTs were mounted on a separate steel frame while the measurement took place at half the height of the columns (Figure 6) [19]. The columns were tested until failure under monotonically increasing displacement. The load, strains and horizontal displacements were acquired with an automatic data acquisition system.

(a)   (b)

Figure 6. C_1V2H_32 specimen on test stand. (**a**) test setup; (**b**) failure of column

The curvatures of the specimens were calculated from Equation (7) on the basis of the measured maximum longitudinal strains $\varepsilon_{v2,lim}$ and $\varepsilon_{v1,lim}$ on the more and less compressed (tensioned) sides of the cross-section, respectively.

$$\frac{1}{r} = \frac{\varepsilon_{v2,lim} - \varepsilon_{v1,lim}}{h},\tag{7}$$

While analyzing the value of longitudinal strains ε_v, in the confined columns with the PBO mesh only, with horizontal layout fibers over the main direction (C_1H and C_2H), failure was observed at a comparable level of strain (Table 4). For the columns in the group C_1H, the limit compression strains amount to 2.736‰–2.962‰; the values in group C_2H amount to 2.827‰–3.200‰. In both columns groups that were loaded at the core limit, C_1H_32 and C_2H_32, at the failure stage, there occurred tension on the side opposite to the action of load. The presence of the longitudinal strengthening reduces the limit strains ε_{v2} of axially compressed columns at which point the destruction of the section occurs, which is fairly unfavorable. For instance, in the element C_1H_0, the strain $\varepsilon_{v2,max} = 2.736‰$, and the additional longitudinal strengthening in the element C_1V1H_0, resulted in a decrease in these strains to $\varepsilon_{v2,max} = 2.392‰$. In contrast, the strains for the elements C_2H_0 and C_1V2H_0 were recorded: $\varepsilon_{v2,max} = 3.200‰$ and $\varepsilon_{v2,max} = 1.734‰$, respectively. The impact of the longitudinal PBO mesh on the limit values of compression strains is evident in the element groups C_1V1H and C_1V2H. It is evident in both groups C_1V1H and C_1V2H that eccentrically compressed elements are capable of transferring considerably higher compression strains on the side of the action of force than axially compressed elements. In addition, the value of these strains rises jointly with the rise in eccentricity.

Table 4. Summary of testing results.

Column	Vertical Strain at Peak Load	
	$\varepsilon_{v2,max}$	$\varepsilon_{v2,max}$
	(‰)	(‰)
C_C_0	−2.121	−1.752
C_C_16	−3.183	−0.052
C_C_32	−3.135	+0.369
C_1H_0	−2.736	−2.459
C_1H_16	−2.845	−0.622
C_1H_32	−2.962	+0.283
C_2H_0	−3.200	−1.489
C_2H_16	−2.907	−0.715
C_2H_32	−2.827	+0.106
C_1V1H_0	−2.392	−1.762
C_1V1H_16	−2.941	−0.115
C_1V1H_32	−3.112	+0.460
C_1V2H_0	−1.734	−1.510
C_1V2H_16	−1.842	−0.903
C_1V2H_32	−2.890	+0.477

The bending moments at the instant of failure (M_{max}) were calculated from (8) on the basis of the maximum deflections w_{max} (Figure 7).

$$M_{max} = P_{max} \cdot (e_0 + w_{max}) \tag{8}$$

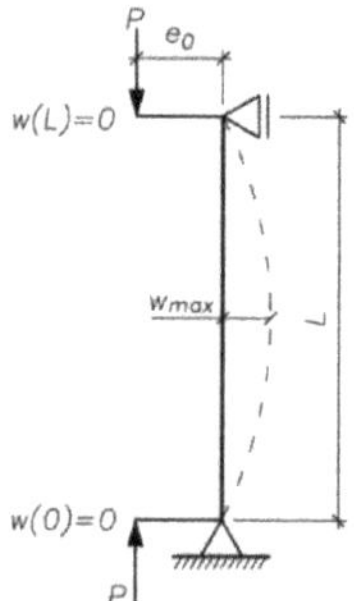

Figure 7. Static diagram of column.

The bending stiffness of columns can be numerically analyzed with the use of Bernoulli's hypothesis with or without eccentric load and additional reinforcements such as fiber materials. The additional longitudinal composite reinforcements contribute to the increasing bending stiffness directly and transverse composite reinforcements give confinement effect to increase stiffness. The axial stiffness should not be evaluated under the combination of axial force and bending moment.

Assuming that Bernoulli's hypothesis is applicable in this case (plane section remains plane) and starting with the general dependence between the curvature of the specimen's deformed axis ($1/r$), bending moment M_{max} and bending stiffness EI (9), the change in stiffness was analyzed depending on the type of strengthening of the column and the stress intensity in the latter.

$$\frac{1}{r} = \frac{M_{max}}{EI}. \tag{9}$$

The next three diagrams (Figures 8–10) show the change (decrease) in the stiffness of the analyzed columns depending on their stress intensity. The horizontal axis represents the ratio of column stiffness at failure *EI* to initial column stiffness $(EI)_{P=0}$ for the load eccentricity of, respectively, 0, 16 and 32 mm. The vertical axis represents the ratio of the ultimate force to the load capacity of the axially compressed column in a given group for the load eccentricity of 0, 16 and 32 mm. The broken line marks the trend in stiffness change.

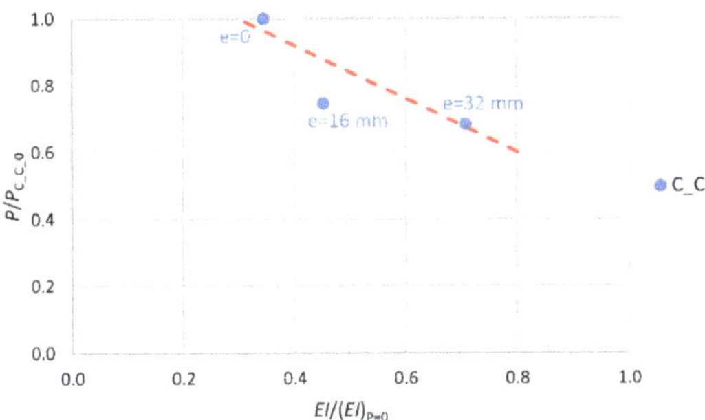

Figure 8. Change in stiffness of reference specimens depending on intensity of their stress.

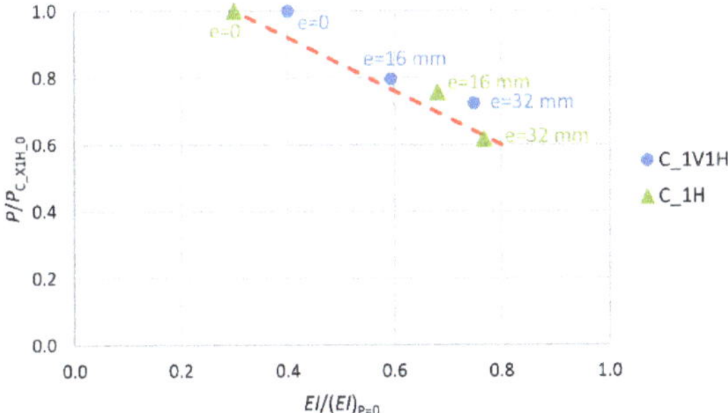

Figure 9. Change in stiffness of specimens with single layer of transverse composite reinforcement (X—number of layers of longitudinal composite reinforcement V: 0 or 1).

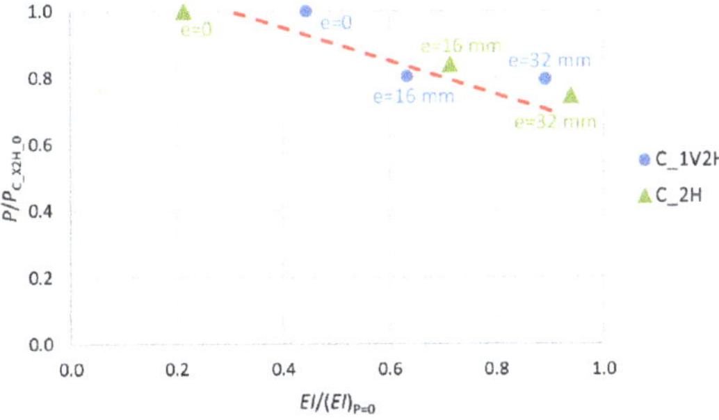

Figure 10. Change in stiffness of specimens with two layers of transverse composite reinforcement (X—number of layers of longitudinal composite reinforcement V: 0 or 1).

In the case of column C_C_0 (most stressed), the stiffness at the point of failure amounts to 35% of the initial value (Figure 8). For the unstrengthened columns loaded at the initial eccentricity of 16 mm and 32 mm (C_C_16 and C_C_32), which were put under less stress, the stiffness at the point of failure amounts to, respectively, 45% and 71% of the initial stiffness value. The elasticity modulus value of the "plain concrete" decreases until about $0.2E_{c,max}$ before failure. The smaller decrease in stiffness of the reinforced concrete columns, than that resulting from the change in the elasticity modulus of the "plain concrete" itself, is evident due to the presence of the longitudinal reinforcement and the shape of the cross-section of the columns.

A similar trend in the change of stiffness is observed in the columns with a single layer (1H) of transverse composite reinforcement (Figure 9). The addition of another layer (2H) of transverse composite reinforcement results in greater stiffness of the composite jacket, and so of the whole cross-section (Figure 10). This is illustrated by the slope of the trend line in the two diagrams.

The stiffness of the composite jacket in these investigations is defined with the equivalent modulus of elasticity of the PBO–FRCM strengthening according to the following formula:

$$E_1 = \frac{t}{R} \cdot E_f, \tag{10}$$

where E_f is given in Table 2 and R is the radius of a circle with the circumference equals the circumference of a considering cross-section. For the considered columns with the square cross-section with the side length a:

$$R = \frac{4 \cdot a}{2 \cdot \pi}. \tag{11}$$

One should note here that in comparison to the reference columns, the decreases in load capacity were observed for columns C_1V2H_0 and C_1V2H_16 (Table 5) [18]. This is not surprising as it was caused by the increase in the stiffness of the columns due to the little-deformable composite jacket. The longitudinal composite reinforcement reduces the longitudinal deformability of the columns, which is rather disadvantageous. Stiffer transverse composite reinforcement reduces the ability of the columns to deform (deflect) in the bending plane. This observation applies particularly to axially compressed columns at a slight eccentricity. An analysis of the diagrams shows that the stiffness of the columns strengthened with PBO mesh on mineral mortar depends on the intensity of stress in the concrete confined by the composite jacket. The stress intensity depends on the eccentricity, which equals the sum of the initial eccentricity and the deflection of the column.

Table 5. Change in stiffness of columns C_1H, C_2H, C_1V1H, C_1V2H as a function of failure load.

Specimens	EI	P_{max}
	(kNm²)	(kN)
C_C_0	1556	2214
C_1V1H_0	1801	2227
C_1V2H_0	1999	2035
C_C_16	2042	1652
C_1V1H_16	2673	1775
C_1V2H_16	2850	1636
C_C_32	3188	1516
C_1V1H_32	3371	1613
C_1V2H_32	4013	1618
C_C_0	1556	2214
C_1H_0	1352	2587
C_2H_0	962	2434
C_C_16	2042	1652
C_1H_16	3062	1957
C_2H_16	3214	2044
C_C_32	3189	1516
C_1H_32	3454	1596
C_2H_32	4226	1812

The next two diagrams (Figures 11 and 12) and Table 5 show the change in the stiffness of the columns as a function of the maximum (ultimate) force registered in the course of the tests. In Figure 11, which illustrates the behavior of the columns strengthened only transversely, one can see that the introduction of one (1H) or two (2H) layers of transverse composite reinforcement results in an increase in the stiffness of the composite jacket. The stiffness of the columns in the state of the ultimate bearing capacity depends on the intensity of stress in the cross-section at the instance of failure. The stress intensity does not increase geometrically with the number of strengthening layers. The columns with longitudinal composite reinforcement behave completely differently (Figure 12). In this case, the columns' stiffness is determined by the presence of the longitudinal composite reinforcement. The lower stress intensity, in comparison with the specimens of type C_1H and C_2H, is accompanied by a reduction in the flexural rigidity of the columns. The application of composite reinforcement along the axis of the columns resulted in an increase in their longitudinal stiffness. Both types of columns: C_1V1H and C_1V2H show considerably greater ductility than the corresponding columns without longitudinal composite reinforcement C_1H and C_2H. This is reflected in the lower value of stiffness at failure at lower stress intensities, in comparison with the columns of type C_1H and C_2H.

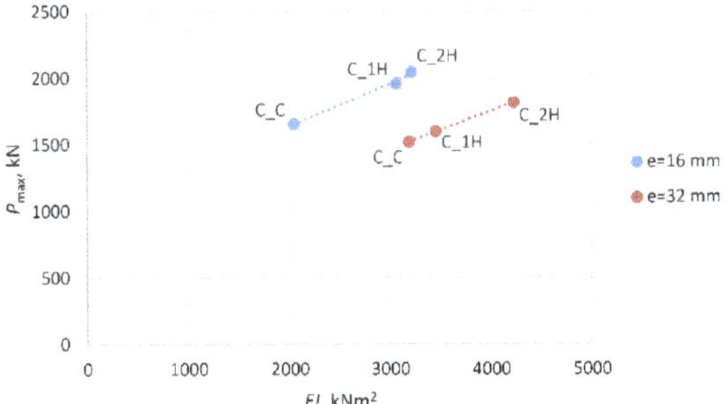

Figure 11. Change in stiffness of columns C_1H and C_2H as a function of failure load.

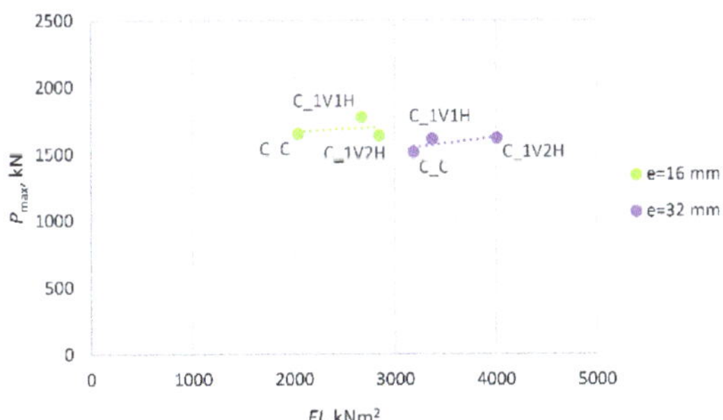

Figure 12. Change in stiffness of columns C_1V1H and C_1V2H as a function of failure load.

The ductility of the columns in these investigations is defined as the ability to horizontally displace the columns, which is induced with the bending moments (eccentric load) what is presented in Figure 13. M_{max} is the first-order moment. The slenderness ratio of the RC columns $\lambda < \lambda_{lim}$ according to [29].

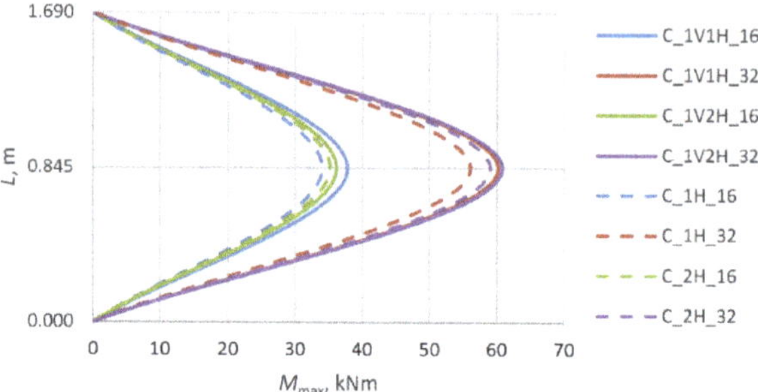

Figure 13. The bending moment of the eccentrically loaded columns.

The effect of the composite jacket in PBO–FRCM columns is closely connected with the variation in the elasticity modulus of the concrete (E_c) due to the stress destruction of the concrete core. Microcracks develop in the concrete beyond the level of stress in the column at which Poisson's ratio v is no longer a liner [35]. As a result of the damage, the load-carrying surface area is reduced and consequently the stiffness of the member decreases. The next graphs (Figures 14–17) show Poisson's ratio versus eccentricity for the analyzed columns. One can see that beyond a certain stress value, Poisson's ratio v quickly increases, which is due to the extensive microcracking of the concrete core. This stress level corresponds to 60–70% of the maximum (ultimate) force P_{max} observed during the tests.

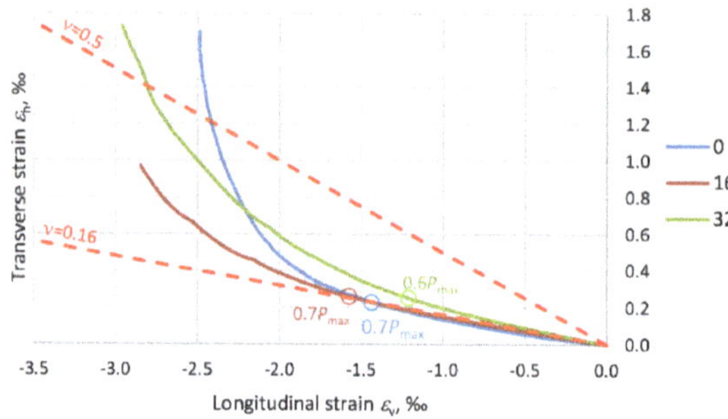

Figure 14. Graphs showing the variation of Poisson's ratio v for columns C_1H.

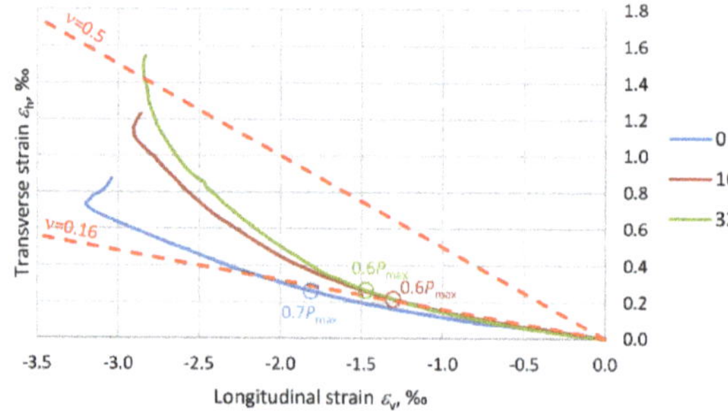

Figure 15. Graphs showing the variation of Poisson's ratio v for columns C_2H.

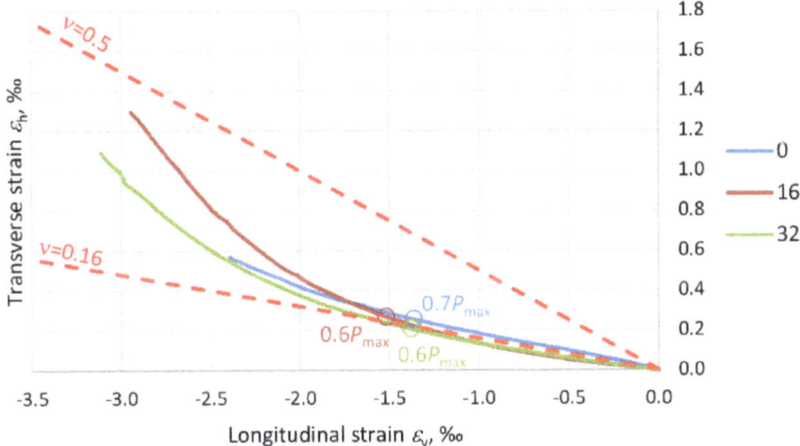

Figure 16. Graphs showing the variation of Poisson's ratio ν for columns C_1V1H.

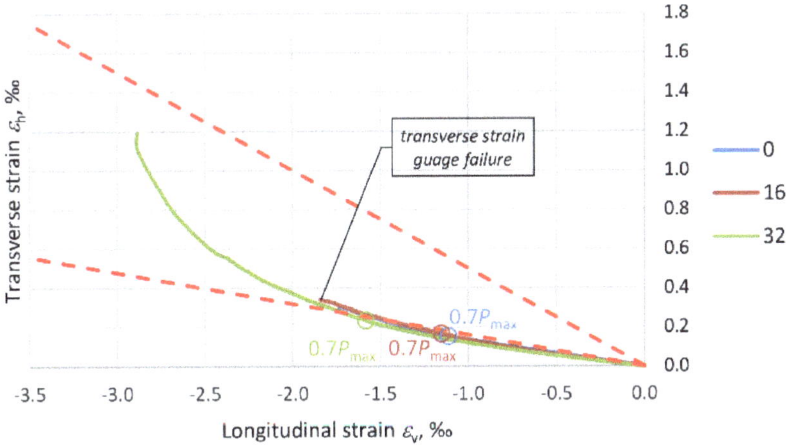

Figure 17. Graphs showing the variation of Poisson's ratio ν for columns C_1V2H.

As the stress further increases, the rate of volumetric changes begins to fall. The concrete is no longer a continuous body, undergoes disintegration and is held only by the external composite jacket. This situation lasts until the reinforcement at the end of the overlap of the PBO mesh starts to delaminate. In the columns strengthened only transversely, i.e., C_1H and C_2H, the Poisson ratio exceeds 0.5, and the volumetric strain assumes negative values. In the case of columns C_1V1H and C_1V2H, the effect of the longitudinal PBO mesh (reducing the compressive stress increment) is clearly visible and ratio $\nu < 0.5$.

With regard to the variability of PBO–FRCM column stiffness, the variation in the ratio of transverse strain to longitudinal strain (Poisson's ration ν), due to the destruction of the concrete inside the composite jacket should be taken into account in the standards.

4. Conclusions

The determination of the stiffness of columns strengthened with composite materials (PBO mesh on mineral mortar in the considered case) is a difficult and complicated task. One cannot adopt, a priori, the standard regulations dedicated to reinforced concrete or composite elements to determine the stiffness of such columns. The aim of the investigations presented in this paper was to assess the effect of the PBO–FRCM reinforcement of eccentrically compressed columns on their stiffness depending on the level of stress intensity.

In the case of unstrengthened columns, the expected results were measured. As eccentricity increased, the longitudinal ultimate force decreased and the deflection of the columns increased. As the curvature of the columns increased, so did the bending moment at half of their height. Each of the columns failed at a correspondingly lower intensity of the stress in the whole cross-section, which was produced by applying longitudinal force on the more compressed side.

In the case of columns C_1H and C_2H, which were strengthened only transversely, our measurements and evidence collected point to the fact that this kind of reinforcement increases their load capacity, but at the same time, the stiffness of the columns increases due to the confinement of the cross-section. In comparison with the reference (unstrengthened) specimens, the curvature would decrease at the initial eccentricities after the first and second PBO layers were laid. This was accompanied by an increase in the ultimate force and a decrease in the horizontal deflection of the columns. At eccentricities of 16 and 32 mm, the columns failed at correspondingly higher stiffness values than the reference members, owing to the use of transverse composite reinforcement. Whereas in the case of columns C_1H_0 and C_2H_0, their load capacity was reached at lower stiffness values, which indicates considerable destruction of the concrete core under axial compression.

In the considered case of quadrangular columns, the tri-axial state of stress induced by the confinement of the concrete is insignificant and the addition of another layer of transverse composite reinforcement only increases the stiffness of the composite jacket.

Unfortunately, in the case of columns C_1V1H and C_1V2H, their higher ductility and greater horizontal deflectability in the bending plane do not directly translate into an increase in their load capacity in comparison with the columns without longitudinal composite reinforcement. As shown in the earlier research, longitudinal PBO–FRCM reinforcement improves the ductility of eccentrically compressed columns. The presence of this reinforcement, however, reduces longitudinal deformations at which the mesh ruptures, which is disadvantageous. Thanks to the presence of longitudinal PBO reinforcement, the columns failed at higher stiffness values in the whole range of the eccentricities: 0, 16 and 32 mm.

This paper investigated the effect of PBO–FRCM reinforcement on the stiffness of eccentrically compressed RC columns. The dependencies between the change in the elasticity modulus of the concrete and the change in the stiffness of the tested specimens were examined. This subject is of great practical relevance. Few such experiments are available in the literature. The analyses presented in this paper can be used to design a wider series of tests of PBO–FRCM columns with various types of reinforcement, which are subjected to eccentric compression. The results of such tests can be used to formulate a standard relation for a column stiffness reduction coefficient depending on the type of reinforcement, the longitudinal force and the cross-section's stress intensity.

Author Contributions: Conceptualization, T.T. and M.M.; methodology, T.T.; formal analysis, T.T. and M.M.; investigation, T.T. and M.M.; writing—original draft preparation, T.T.; writing—review and editing, M.M.; visualization, T.T. and M.M. All authors have read and agreed to the published version of the manuscript.

Funding: This research received no external funding.

Conflicts of Interest: The authors declare no conflicts of interest.

Nomenclature

A_c	Concrete cross-section, m^2
E_a	Modulus of elasticity of structural steel, gpa
E_c	Modulus of elasticity of concrete, gpa
E_{cm}	Secant modulus of elasticity of concrete, gpa
E_{cd}	Design value of modulus of elasticity of concrete, gpa
E_f	Modulus of elasticity of PBO–FRCM strengthening system, gpa
E_s	Modulus of elasticity of reinforcing steel, gpa
I_c	Moment of inertia of concrete, cm^4
I_s	Moment of inertia of reinforcing steel, cm^4
I_a	Moment of inertia of structural steel, cm^4
P	Load, kn
e_0	Initial eccentricity, mm
$f_{c,cyl}$	Cylindrical compressive strength of concrete, mpa
$f_{c,cube}$	Cubic compressive strength of concrete, mpa
f'_{co}	Compressive strength of unconfined concrete, mpa
f_{ck}	Characteristic value of concrete compressive strength, mpa
f_{cd}	Design value of concrete compressive strength, mpa
f_u	Ultimate strength of PBO mesh, mpa
f_y	Yield strength of steel, mpa
f_t	Ultimate tensile strength of steel, mpa
t	Nominal thickness of PBO mesh, mm
w	Column deflection, mm
λ	Column slenderness,
ε_c	Strain in concrete, ‰
ρ	Reinforcement ratio for longitudinal reinforcement
σ_c	Stress in concrete, mpa
ϕ_{ef}	Effective value of the creep coefficient

References

1. Ombres, L. Flexural analysis of reinforced concrete beams strengthened with a cement based high strength composite material. *Compos. Struct.* **2011**, *94*, 143–155. [CrossRef]
2. Ombres, L. The structural performances of PBO-FRCM strengthened RC beams. *Struct. Build.* **2011**, *164*, 265–272. [CrossRef]
3. Ombres, L. Shear Capacity of Concrete Beams Sstrengthened with Cement Based Composite Mmaterials. In Proceedings of the 6th International Conference on FRP Composites in Civil Engineering (CICE 2012), Roma, Italy, 13–15 June 2012.
4. Loreto, G.; Babaeidaarabad, S.; Leardini, L.; Nanni, A. RC beams shear-strengthened with fabric-reinforced-cementitious-matrix (FRCM) composite. *Int. J. Adv. Struct. Eng.* **2015**, *7*, 341–352. [CrossRef]
5. Trapko, T.; Urbańska, D.; Kamiński, M. Shear strengthening of reinforced concrete beams with PBO-FRCM composites. *Compos. Part B* **2015**, *80*, 63–72. [CrossRef]
6. Ombres, L. Structural performances of reinforced concrete beams strengthened in shear with a cement based fibre composite material. *Compos. Struct.* **2015**, *122*, 316–329. [CrossRef]
7. Trapko, T.; Musiał, M. PBO mesh mobilization via different ways of anchoring PBO-FRCM reinforcements. *Compos. Part B* **2017**, *118*, 67–74. [CrossRef]
8. Nerilli, F.; Ferracuti, B. On Tensile Behavior of FRCM Materials: An Overview. In Proceedings of the 8th International Conference on Fibre-Reinforced Polymer (FRP) Composites in Civil Engineering (CICE 2016), Hong Kong, China, 14–16 December 2016.
9. Trapko, T. The influence of temperature on the durability and effectiveness of strengthening of concrete with CFRP composites. *Eng. Build.* **2010**, *10*, 561–564.
10. Trapko, T. The effect of high temperature on the performance of CFRP and FRCM confined concrete elements. *Compos. Part B* **2013**, *54*, 138–145. [CrossRef]

11. D'Antino, T.; Carloni, C.; Sneed, L.H.; Pellegrino, C. Matrix-fibre bond behaviour in PBO FRCM composites: A fracture mechanics approach. *Eng. Fract. Mech.* **2014**, *117*, 94–111. [CrossRef]

12. D'Ambrisi, A.; Feo, L.; Focacci, F. Experimental analysis on bond between PBO-FRCM strengthening materials and concrete. *Compos. Part B* **2013**, *44*, 524–534. [CrossRef]

13. Carozzi, F.G.; Poggi, C. Mechanical properties and debonding strength of Fabric Reinforced Cementitious Matrix (FRCM) systems for masonry strengthening. *Compos. Part B* **2015**, *70*, 215–230. [CrossRef]

14. Ombres, L. Analysis of the bond between Fabric Reinforced Cementitious Mortar (FRCM) strengthening systems and concrete. *Compos. Part B* **2015**, *69*, 418–426. [CrossRef]

15. Ombres, L. Confinement Effectiveness in Concrete Strengthened with Fibre Reinforced Cement Based Composite Jackets. In *Proceedings of the FRPRCS-8: 8th International Symposium on Fibre-Reinforced Polymer Reinforcement for Concrete Structures, Patras, Greece, 16–18 July 2007*; University of Patras, Department of Civil Engineering: Patras, Greece, 2007; Paper No. 17–11.

16. De Caso y Basalo, F.J.; Matta, F.; Nanni, A. Fibre reinforced cement-based composite system for concrete confinement. *Constr. Build. Mater.* **2012**, *32*, 55–65. [CrossRef]

17. Trapko, T. Fibre Reinforced Cementitious Matrix confined concrete elements. *Mater. Des.* **2013**, *44*, 382–391. [CrossRef]

18. Trapko, T. Behaviour of Fibre Reinforced Cementitious Matrix strengthened concrete columns under eccentric compression loading. *Mater. Des.* **2014**, *54*, 947–957. [CrossRef]

19. Trapko, T. Effect of eccentric compression loading on the strains of FRCM confined concrete columns. *Constr. Build. Mater.* **2014**, *61*, 97–105. [CrossRef]

20. Trapko, T. Confined concrete elements with PBO-FRCM composites. *Constr. Build. Mater.* **2014**, *73*, 332–338. [CrossRef]

21. Colajanni, P.; De Domenico, F.; Recupero, A.; Spinella, N. Concrete columns confined with fibre reinforced cementitious mortars: Experimentation and modelling. *Constr. Build. Mater.* **2014**, *52*, 375–384. [CrossRef]

22. Colajanni, P.; Fossetti, M.; Macaluso, G. Effects of confinement level, cross-section shape and corner radius on the cyclic behaviour of CFRCM confined concrete columns. *Constr. Build. Mater.* **2014**, *55*, 379–389. [CrossRef]

23. Ombres, L.; Verre, S. Structural behaviour of fabric reinforced cementitious matrix (FRCM) strengthened concrete columns under eccentric loading. *Compos. Part B* **2015**, *75*, 235–249. [CrossRef]

24. Ombres, L. Concrete confinement with a cement based high strength composite material. *Compos. Struct.* **2017**, *109*, 294–304. [CrossRef]

25. Ombres, L. Structural performances of thermally conditioned PBO FRCM confined concrete cylinders. *Compos. Struct.* **2017**, *176*, 1096–1106. [CrossRef]

26. Donnini, J.; Spagnuolo, S.; Corinaldesi, V. A comparison between the use of FRP, FRCM and HPM for concrete confinement. *Compos. Part B* **2019**, *160*, 586–594. [CrossRef]

27. Hoła, J. Relation of Initiating and Critical Stress to Stress Failure. In *Scientific Papers of the Institute of Building Engineering of the Wroclaw University of Technology*; No. 76, Monographs No. 33; Wroclaw University of Technology Publishing House: Wroclaw, Poland, 2000; p. 182.

28. Polish Standard PN-EN 1994-1-1:2008. *Eurocode 4: Design of Composite Steel and Concrete Structures—Part 1-1: General Rules and Rules for Buildings*; European Committee for Standardization: Brussels, Belgium, 2008.

29. Polish Standard PN-EN 1992-1-1:2008. *Eurocode 2. Design of Concrete Structures—Part 1-1: General Rules and Rules for Buildings*; European Committee for Standardization: Brussels, Belgium, 2008.

30. Polish Standard PN-EN 12390-3:2009. *Testing Hardened Concrete. Part 3: Compressive Strength of Test Specimens*; National Standards Authority of Ireland: Dublin, Ireland, 2009.

31. *Testing Hardened Concrete. Part 13: Determination of Secant Modulus of Elasticity in Compression*; National Standards Authority of Ireland: Dublin, Ireland, 2009.

32. Ruredil, X. *Mesh Gold Data Sheet*; Ruredil SPA: Milan, Italy, 2009.

33. *Technical Approval*; The Road and Bridge Research Institute: Warszawa, Poland, 2004.

34. ACI Committee 549. *Guide to Design and Construction of Externally Bonded Fabric-Reinforced Cementitious Matrix (FRCM) Systems for Repair and Strengthening Concrete and Masonry Structures*; American Concrete Institute Committee: Farmington Hills, MI, USA, 2013.
35. Neville, A.M. *Properties of Concrete*, 5th ed.; Polish Cement Association: Kraków, Poland, 2012; p. 931.

 *materials*

Article

Shear Behavior of Concrete Beams Reinforced with a New Type of Glass Fiber Reinforced Polymer Reinforcement: Experimental Study

Czesław Bywalski *, Michał Drzazga, Maciej Kaźmierowski and Mieczysław Kamiński

Faculty of Civil Engineering, Wroclaw University of Science and Technology, Wybrzeże Wyspiańskiego 27, 50-370 Wrocław, Poland; michal.drzazga@pwr.edu.pl (M.D.); maciej.kazmierowski@pwr.edu.pl (M.K.); mieczyslaw.kaminski@pwr.edu.pl (M.K.)
* Correspondence: czeslaw.bywalski@pwr.edu.pl; Tel.: + 48-71-320-3397

Received: 6 February 2020; Accepted: 3 March 2020; Published: 5 March 2020

Abstract: The article presents experimental tests of a new type of composite bar that has been used as shear reinforcement for concrete beams. In the case of shearing concrete beams reinforced with steel stirrups, according to the theory of plasticity, the plastic deformation of stirrups and stress redistribution in stirrups cut by a diagonal crack are permitted. Tensile composite reinforcement is characterized by linear-elastic behavior throughout the entire strength range. The most popular type of shear reinforcement is closed frame stirrups, and this type of Fiber Reinforced Polymer (FRP) shear reinforcement was the subject of research by other authors. In the case of FRP stirrups, rupture occurs rapidly without the shear reinforcement being able to redistribute stress. An attempt was made to introduce a quasi-plastic character into the mechanisms transferring shear by appropriately shaping the shear reinforcement. Experimental material tests covered the determination of the strength and deformability of straight Glass Fiber Reinforced Polymer (GFRP) bars and GFRP headed bars. Experimental studies of shear reinforced beams with GFRP stirrups and GFRP headed bars were carried out. This allowed a direct comparison of the shear behavior of beams reinforced with standard GFRP stirrups and a new type of shear reinforcement: GFRP headed bars. Experimental studies demonstrated that GFRP headed bars could be used as shear reinforcement in concrete beams. Unlike GFRP stirrups, these bars allow stress redistribution in bars cut by a diagonal crack.

Keywords: GFRP; FRP reinforcement; shear; capacity; reinforced concrete beams

1. Introduction

Corrosion of reinforcing steel is the main factor causing material degradation in reinforced concrete structures, which, if exposed to particularly adverse environmental conditions, cease to meet the requirements of durability and reliability in a facility in a relatively short time [1–4]. The durability of reinforced concrete structures is particularly important for industrial facilities [5–7]. The use of non-metallic reinforcement as the main reinforcement of concrete elements is one way to exclude corrosion and thus extend the service life of the structure. Reinforced concrete elements embedded in structures exposed to particularly adverse weather and operating conditions require time-consuming and cost-intensive renovation. There is growing interest in using non-corrosive Fiber Reinforced Polymer (FRP) reinforcement, especially in engineering structures that will be exposed to harmful effects for a long period of use [2,6,8–10]. Among types of bar-shaped polymer reinforcement, we distinguish bars made of glass fiber (Glass Fiber Reinforced Polymer (GFRP)), carbon fiber (Carbon Glass Fiber Reinforced Polymer (CFRP)), basalt fiber (Basalt Glass Fiber Reinforced Polymer (BFRP)), and aramid fiber (Aramid Glass Fiber Reinforced Polymer (AFRP)). FRP bars are made of continuous fibers impregnated with polymeric resins. Continuous fibers with high stiffness and high strength are embedded in and bonded together by the polymeric matrix with low stiffness. In the FRP composites, the reinforcing fibers are the core, which determine the stiffness and strength of the material in the direction of the fibers. FRP composite bars are marked by good mechanical (high tensile strength) and physical properties (much lower density than reinforcing steel) [11]. FRP bars have been used in concrete elements in facilities particularly exposed to aggressive environments and in facilities whose proper functioning is dependent, among other things, on the electromagnetic neutrality of the construction elements. FRP composite bars are electromagnetically neutral and are therefore used in facilities requiring special precision of operation (no interference with devices operating inside the facility) and in infrastructure facilities (elimination of stray currents causing electro-corrosion) [4,11]. Composite reinforcements can be easily cut, which enables their effective use in temporary elements, e.g., as parts of tunnels [11].

Based on, among others, Fib Bulletin 40 [11] and ACI 440.1R-15 [12], the use of composite bars as longitudinal bending reinforcement has been much better researched than the use of these bars as transverse reinforcement. In order for the entire structural element to be corrosion-resistant and electromagnetically neutral, not only longitudinal reinforcement (flexural reinforcement), but also transverse reinforcement must be made of a suitable material. The most popular type of shear reinforcement is closed frame stirrups surrounding longitudinal reinforcement, put closest to the surface of the concrete element. This is why stirrups in reinforced concrete structures should be classified as the most vulnerable to corrosion. The above-mentioned factors have stimulated researchers to find a solution to the problem of material degradation due to corrosion by introducing transverse reinforcement in the form of frames made of polymer bars. Nagasaka et al. [13], Shehata [14], and Ahmed [15] indicated that direct replacement of steel stirrups with FRP stirrups is not possible due to differences resulting, for example, from mechanical properties. Polymer reinforcement is marked by high tensile strength, a relatively low modulus of elasticity (except CFRP), and linearly, elastic behavior in the entire strength range. Bentz et al. [16] indicated that the above also implies differences in the behavior of support zones reinforced longitudinally with FRP bars. Oller et al. [17] and Kosior-Kazberuk [18] indicated that, in accordance with the superposition principle, the shear load capacity of an element loaded with transverse force is dependent on the following factors: shear reinforcement contribution, load capacity resulting from aggregate interlocking, the load capacity of concrete in the compressed zone, the dowel action of longitudinal reinforcement, and the residual tensile strength of concrete across the crack. Composite reinforcement is characterized by a much lower modulus of elasticity than steel reinforcement. The distance from compressed fibers to the neutral axis in a concrete element reinforced longitudinally with FRP bars is smaller after cracking, than in the case of concrete elements reinforced with steel, the range of the compressed zone of the imported cross-section being smaller. This is due to the lower axial stiffness of FRP reinforcement. The range of the compression zone is smaller, which means that the shear capacity of concrete in the compression zone is also smaller. The crack opening width is larger in the case of FRP reinforcement [6,9–11], and the component related to the aggregate interlocking is reduced. The low transverse stiffness of FRP bars significantly reduces the component resulting from dowel action [12,17,19,20]. Assuming the same longitudinal reinforcement surface, a concrete element reinforced with FRP bars thus has a lower shear capacity than a concrete element reinforced with steel bars [21–23]. Ahmed [15], Kurth [24], and Jumaa et al. [25] indicated that a significant difference between an element reinforced with FRP bars and a concrete element reinforced with steel is a considerable reduction in the tensile strength of the bent FRP bar (stress concentration in the bent zone). This is why it is necessary to control the stresses in the stirrup bars to avoid rupture of the bent fragment. Another aspect is the lack of plasticity of composite reinforcement. Nagasaka et al. [13] and Razaqpur et al. [26] distinguished two failure modes of an FRP bar reinforced element subjected to shearing force: stirrup rupture and crushing of the compressed concrete. The rupture of composite bars is more fragile than crushing concrete and occurs when one of the cut stirrups reaches its tensile strength. Support zones reinforced with steel stirrups are characterized by a lack of stress redistribution ability.

Considering the above and the current state of knowledge reported in the literature, it should be stated that the behavior of support zones reinforced with FRP bars differs from those reinforced with steel. The limited number of studies on the subject does not permit explicit determination of the behavior of support zones reinforced with composite bars, especially support zones with low shear slenderness. Drzazga [27] indicated that while few research results are presented in the relevant literature concerning slender support zones, there are no results whatsoever available of research on short support zones reinforced transversally with GFRP bars, where the failure mechanism would be determined by the failure of transverse reinforcement (the shear–tension failure). Among others, Krall et al. [28], when testing deep beams, obtained only the failure in the form of crushing concrete without the GFRP stirrups' rupture. No papers were found in which GFRP bars with head anchorage were used as shear reinforcement. The pull-out tests [29,30] of this type of anchorage indicated a more plastic nature of failure, if the bar slipped from the anchor head. In view of the shortage of research, this article presents the results of experimental tests of short shear zones reinforced with GFRP stirrups and GFRP headed bars where the failure mechanism is determined by the failure of transverse reinforcement. The most commonly used polymer bar reinforcement is GFRP reinforcement. This is due not only to the good properties of reinforcement made of glass fiber, but also to its relatively low price [11,31]. Therefore, the research program included this type of fiber. The article presents the results of experimental tests of GFRP bars and eight concrete beam supporting zones with differently shaped longitudinal (GFRP or steel) and transverse (GFRP stirrups or GFRP headed bars) reinforcement.

2. Experimental Program

The main purpose of the experimental research was to verify the failure modes of beams reinforced with GFRP stirrups and GFRP headed bars. Four beams with differently shaped longitudinal and transverse reinforcement were tested. All beams were loaded with forces concentrated at a distance of 600 mm from the support (Figure 1). It was possible to verify the shear load capacity and failure modes of beams with low shear slenderness (*a*/*d* ratio ≈ 1.7 < 2.5).

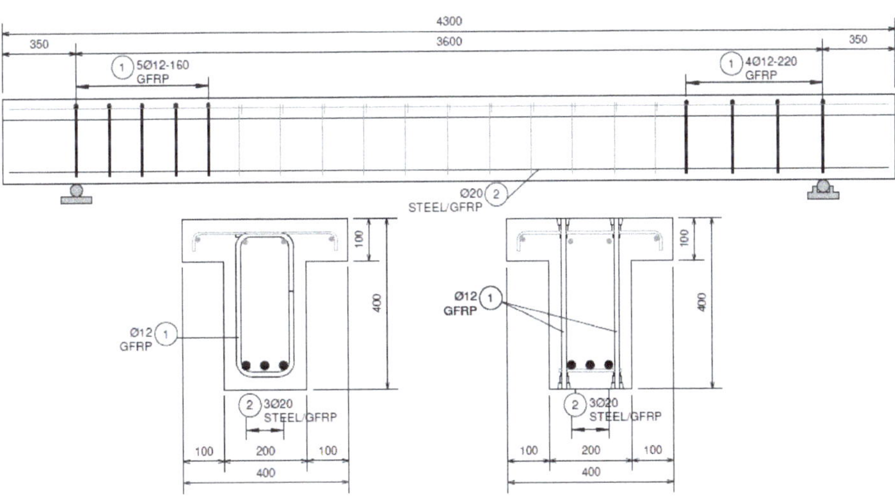

Figure 1. Reinforcement drawing of beams.

2.1. Test Specimens

The beams were designed as T-sections with a 400 mm section height, 100 mm flange height, 400 mm flange width, and 200 mm web width. The total length of the beams was 4300 mm. The elements differed in the type of longitudinal reinforcement and the type of shear reinforcement. Twenty millimeter steel bars and 20 mm GFRP bars were used as longitudinal tensile reinforcement. GFRP stirrups with a diameter of 12 mm and GFRP headed bars with a diameter of 12 mm were used as shear reinforcement. Each beam was designed with two differently reinforced support zones; the same type of reinforcement was used throughout the entire beam, but the spacing of stirrups and headed bars was differentiated. The beams were designed so that the shear zones had the same area of tensioned bars taken into account in dowel action. A high class of concrete was used (concrete compressive strength around 70 MPa), and the T-section was used in order to increase the flexure strength. The method of reinforcing research elements is shown in Table 1 and Figure 1. Figure 2 presents the reinforcement cages of selected beams.

Table 1. Reinforcement details of the support zones of tested beams.

Beam Support Zone	Tensile Longitudinal Reinforcement (Shear Zone)			Shear Reinforcement			
	Diameter, $\varnothing_L$	Reinforcement Type	$\rho_f = \frac{A_{fL}}{b_w d}$	Diameter, $\varnothing_w$	Reinforcement Type	Spacing, s	$\rho_{fw} = \frac{A_w}{b_w s}$
	(mm)	(-)	(%)	(mm)	(-)	(mm)	(%)
B-1_160 B-1_220	20	Steel	1.37	12	GFRP stirrups	160 220	0.71 0.51
B-2_160 B-2_220	20	GFRP	1.37	12	GFRP stirrups	160 220	0.71 0.51
B-3_160 B-3_220	20	GFRP	1.37	12	GFRP headed bars	160 220	0.71 0.51
B-4_160 B-4_220	20	Steel	1.37	12	GFRP headed bars	160 220	0.71 0.51

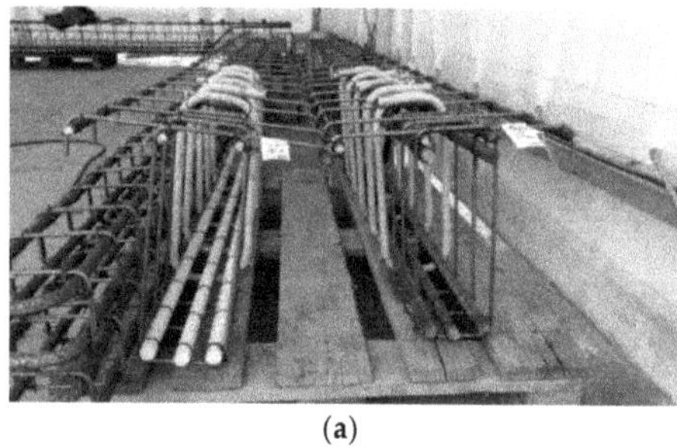
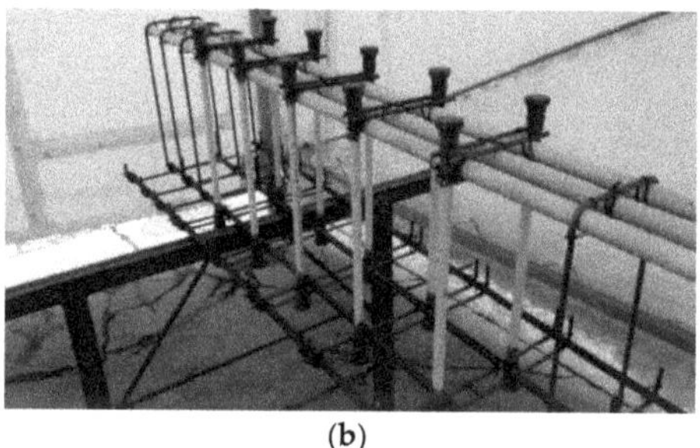

(a)   (b)

Figure 2. Reinforcement details of test specimens based on example beams: (**a**) B-1 and B-2; (**b**) B-3.

2.2. Material Properties

The beam testing was accompanied by the determination of the average compressive strength and the average modulus of elasticity of the concrete. The mechanical properties of the concrete were determined in cylindrical samples with a diameter of 150 mm and a height of 300 mm. The average compressive strength of the concrete for beams B-1, B-2, B-3, and B-4 was 68.94 MPa, 70.13 MPa, 65.47 MPa, and 71.39 MPa, respectively. The average modulus of the concrete elasticity under compression in the case of beams B-1, B-2, B-3, and B-4 was 35.47 GPa, 36.63 GPa, 35.97 GPa, and 38.07 GPa, respectively.

As longitudinal reinforcement, steel bars with a diameter of 20 mm and a characteristic yield strength of 500 MPa and GFRP bars with a diameter of 20 mm and an average tensile strength > 1100 MPa, as declared by the manufacturer, were used. The modulus of elasticity of the steel bars and GFRP bars was 200 GPa and 57 GPa, respectively, as declared by the manufacturer. GFRP stirrups with a diameter of 12 mm and headed bars with a diameter of 12 mm were used as shear reinforcement.

Both stirrups and headed bars were a Schöck product described as Schöck Combar. The bars used were made of E-CR (Electrical/Chemical Resistance) glass fiber composite with a diameter of approximately 20 μm and vinyl ester resin. The head anchorage was a fiber reinforced polymer mortar. The anchor head was 60 mm long, and the maximum diameter was 30 mm (Figure 3a). In order to compare the mode of failure and tensile strength of GFRP bars, five samples of Φ12 headed bars and five samples of Φ12 straight bars were tested. The bar tests were carried out in accordance with Annex A of the American standard ACI 440.3R-04 [32] and the instruction ISO 10406-1:2015 [33]. Test samples were made by embedding the bar ends in Sikadur-330 epoxy resin filling internally screwed steel pipes (Figure 3b). The head anchorage was at one end of the test bar. The end of the head anchoring was embedded in a steel element with a special flange (Figure 3c) to eliminate clamping forces that could distort the test results. For the tensile test, samples with a multiplicity of 10 (measuring base length 120 mm for bars Φ12) were used. The modulus of elasticity was determined on the basis of strain results based on a 60 mm long measurement base.

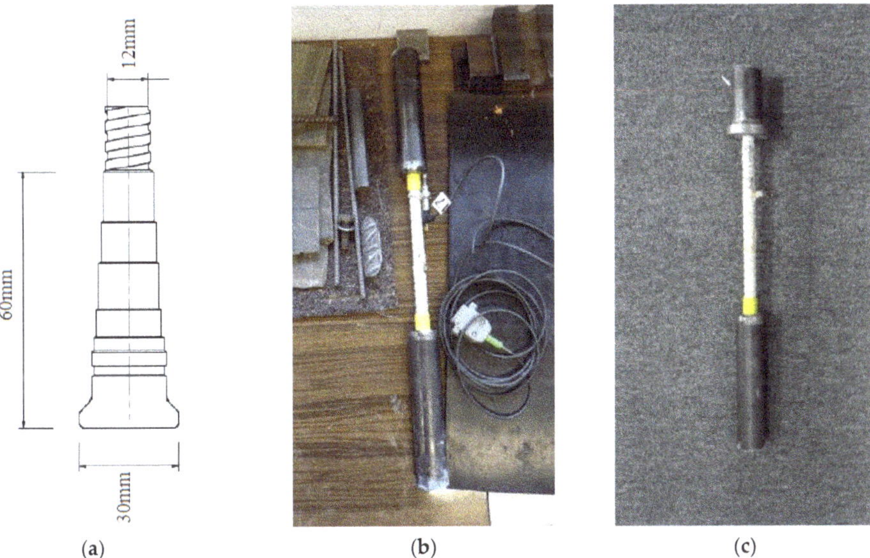

(a)   (b)   (c)

Figure 3. Samples for determining the tensile behavior of GFRP bars: (**a**) dimensions of the head anchorage of GFRP Schöck Combar; (**b**) straight bar sample; (**c**) headed bar sample.

Tensile strength f_u, modulus of elasticity E_f, and ultimate strain ε_u were determined by the procedure in [32] using the following relationships:

$$f_u = \frac{F_u}{A} \tag{1}$$

$$E_f = \frac{F_1 - F_2}{(\varepsilon_1 - \varepsilon_2)A} \tag{2}$$

$$\varepsilon_u = \frac{F_u}{E_f A} \tag{3}$$

where: F_u, the tensile capacity of FRP bar; A, the cross-sectional area of FRP bar; F_1, the tensile load at approximately 50% of the ultimate load capacity; ε_1, the tensile strain at approximately 50% of the ultimate load capacity; F_2, the tensile load at approximately 20% of the ultimate load capacity; ε_2, the tensile strain at approximately 20% of the ultimate load capacity.

Straight bars were stretched in a testing machine using classic jaws used for the tensile test (Figure 4a), steel bars among others. The end in the form of a steel flange was fixed in a special jaw shown in Figure 4b. The use of this type of jaw, with an appropriate wall thickness of the steel tube, eliminated clamping forces that might have distorted the result of the actual anchoring strength.

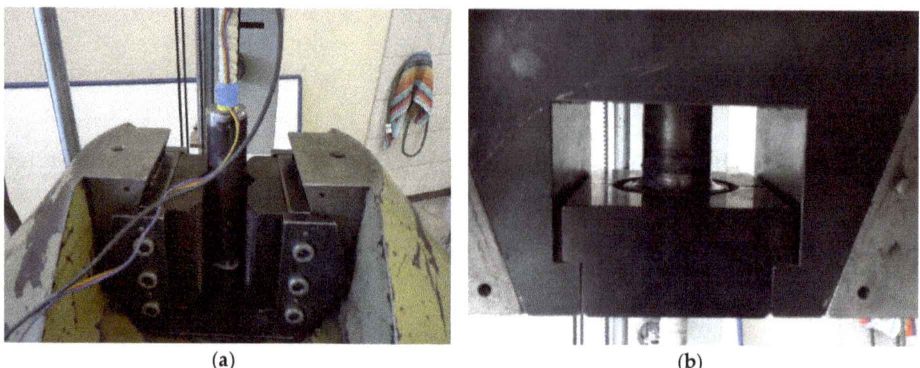

(a)   (b)

Figure 4. The method of fixing the sample in the testing machine: (**a**) fixing the straight steel pipe; (**b**) fixing of a flanged steel pipe.

Based on the data provided by the manufacturer and by Kurth [24], the average tensile strength of the stirrups used was set at f_{bend} = 770 MPa and the mean value of modulus of elasticity at E = 57 GPa. The stirrups used in this study were bent during manufacture, and the bending diameter was seven times the diameter of the bar (84 mm). The stirrups shaped in the same way were the subject of research by the manufacturer and Kurth in [24]. In this study, the stress-strain characteristics of GFRP stirrups were not experimentally determined, and the values of the tensile strength and modulus of elasticity of GFRP stirrups were taken from the manufacturer's data, which were consistent with Kurth's results [24].

2.3. Instrumentation, Test Setup and Procedure

The examination of each beam consisted of two stages. In the first stage, the beam was subjected to a 4-point bending until one of the support zones was damaged (Figure 5). In the second stage, the support at which damage occurred was moved directly under the force, obtaining a 3-point bending scheme.

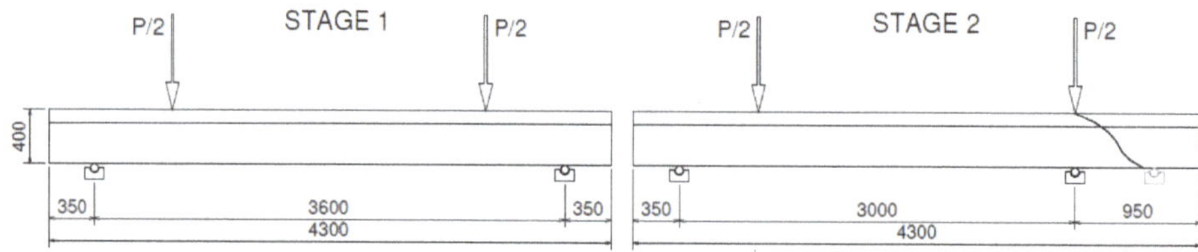

Figure 5. Test setup of two stages. Dimensions in mm.

For the first stage, the load causing the appearance of diagonal and perpendicular cracks was determined. The angle of inclination of the compressed concrete strut was also established for each stage. The element was loaded until the appearance of cracks, and the load was then increased at a speed of about 0.2 kN/s, stopping at ca. every 50 kN; each time the crack pattern and the load at which it occurred was monitored. For each beam, strain measurements of concrete were made in the compression zone and the tension bars. Concrete strains were measured using RL300/50 resistivity strain gauges. Strain gauges were glued onto the concrete around the stirrups (on both sides of the web). Beam deflections were recorded in the middle of the span and under concentrated forces using inductive sensors. Displacement of supports was also measured using inductive sensors located next to the support points. The strains of steel bars and GFRP bars were measured with the use of RL120/20 strain gauges. Strain gauges on bars were placed in the middle of the length of the vertical arms of stirrups/bars with head anchoring. Strain gauges were also glued to longitudinal bars in the middle of the span and in cross-sections under applied forces.

3. Experimental Results and Discussion

3.1. Rebar Test

The failure mode of straight bars was of an explosive nature, and all bars were characterized by linear-elastic behavior over the entire strength range. The average tensile strength of a straight GFRP bar was 1299 MPa with a standard deviation s = 69 MPa. The mean modulus of elasticity was 58.1 GPa, with a standard deviation of s = 3.4 GPa. The mean value of the ultimate strain was 22.4‰ with a standard deviation s = 0.7‰. The damaged sample after the test is shown in Figure 6a, while the static equilibrium paths of the tested bars are shown in Figure 6c.

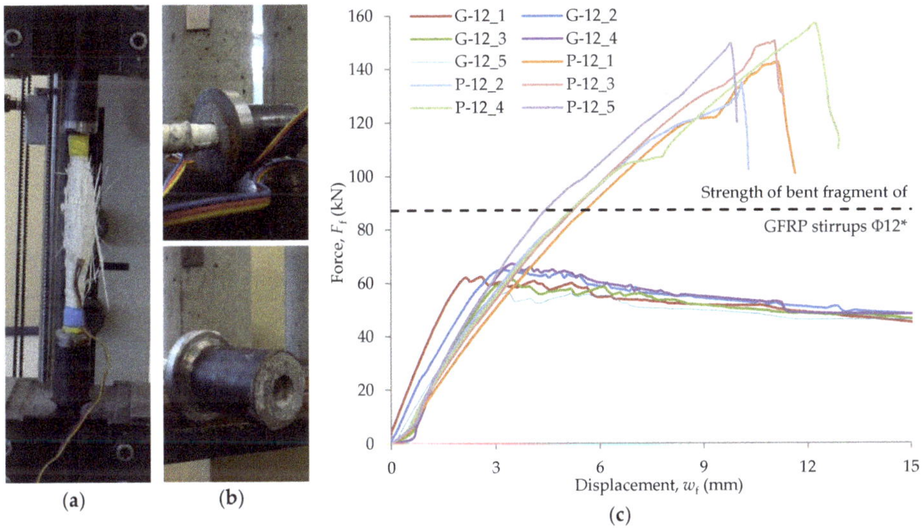

Figure 6. Rebar test results: (**a**) straight bar sample failure; (**b**) headed bar sample failure; (**c**) static equilibrium paths in the tensile test of various types of GFRP bars. * Based on the manufacturer's data and Kurth's tests [24].

Headed bars were characterized by linear-elastic behavior until stress initiating the loss of bar adhesion to the anchor head was reached. The failure mode consisted of the slipping of the bar from the anchor head and was radically less violent than bar rupture. Vint tested the anchoring of GFRP headed bars in concrete and came to similar conclusions [21]. A relatively high value of bar displacement relative to the anchor head was achieved, at which the sample was still able to carry the load. The average value of the tensile strength of the GFRP headed bar was 545 MPa with a standard deviation s = 15 MPa. The average modulus of elasticity was 61.9 GPa, with a standard deviation of s = 3.0 GPa. The damaged sample after the test is shown in Figure 6b, while the static equilibrium paths of the tested bars are shown in Figure 6c.

Before the test, the bar diameter was measured with an accuracy of 0.1 mm. The diameter was measured taking into account the ribs, then the thickness of the ribs was measured, and the core diameter of the bar cross-section $d_{f,i}$ was calculated. Fifteen diameter measurements were made for each bar over a distance of 200 mm. The average value of the diameter d_f is given in Table 2. The tensile strength f_u, modulus of elasticity E_f, and ultimate strain ε_u were determined based on Formulas (1)–(3), and the results are shown in Table 2.

Table 2. Samples details and test results.

Type of Bar	Sample Number	d_f	F_u	f_u	E_f	ε_u
		(mm)	(kN)	(MPa)	(GPa)	(‰)
Straight bars	P-12_1	11.8	141.9	1298	57.8	22.4
	P-12_2	12.0	134.4	1188	53.3	22.3
	P-12_3	12.1	150.1	1305	61.3	21.3
	P-12_4	11.9	156.4	1406	62.4	22.5
	P-12_5	12.1	149.2	1298	55.6	23.3
	Average:	12.0	146.4	1299	58.1	22.4
Headed bars	G-12_1	12.1	62.3	542	65.8	-
	G-12_2	12.2	64.9	555	63.1	-
	G-12_3	12.0	61.2	541	63.5	-
	G-12_4	12.3	67.2	566	57.7	-
	G-12_5	11.8	56.9	520	59.3	-
	Average:	12.1	62.5	545	61.9	-

The short-term tensile strength of headed bars was lower than that of straight bars. The strength of headed bars was about 42% of the tensile strength of straight bars. In addition, based on data from the manufacturer, it was found that the strength of headed bars constituted about 71% of the strength of the bent stirrup section (Figure 6c). The modulus of elasticity for both straight bars and headed bars was similar and amounted to about 60 GPa, which roughly corresponded to the manufacturer's data, including the research by Kurth [24]. Similarly to the paper [24], where GFRP Φ16 headed bars were tested, the modulus of elasticity obtained as a result of testing headed bars was slightly larger than that obtained in the tensile test of bars without anchoring. The average ultimate strain recorded for the tensile test of straight bars was about 22.4‰, which corresponded to an elongation of about 1.3 mm (on a measuring base of 60 mm). In the case of stirrups, based on the assumption of linear-elastic behavior in the tensile test and the value of the average tensile strength and modulus of elasticity declared by the manufacturer, the estimated ultimate strain was about 13.5‰, which corresponded to an elongation of about 0.8 mm (based on measurement equal to 60 mm). The head length of 60 mm introduced the possibility of transferring the force of almost equal tensile strength of the anchorage at an elongation of about 2.5 mm. It could be concluded that a single bar with a head anchorage had a lower tensile strength than a straight bar or stirrup. In the case of a support zone reinforced with headed bars, greater stress redistribution could be expected than in the case of the corresponding reinforcement in the form of frame stirrups.

3.2. Beam Test

Each beam was tested in two stages in order to obtain results from eight support zones. Figure 7 presents an example on the B-1 beam on the test stand during the first and second stages of the test.

(a)   (b)

Figure 7. B-1 beam on the test stand during: (a) first stage of the test; (b) second stage of the test.

The subject of the analysis was both the general behavior of concrete elements reinforced with composite bars and the shear issue, which was the main subject of this study. This article presents the results in terms of bearing capacity and failure mode, static equilibrium paths, and the strains of the reinforcement of the support zone. Complete results of leading and accompanying studies were presented by Drzazga in [27].

3.2.1. Failure Load and Mode of Failure

All beams were designed in order to obtain the failure in the support zone. All the support zones were damaged as a result of the failure of the transverse reinforcement cut through a diagonal crack (shear–tension failure) (Figure 8).

Figure 8. Illustration of the failure mode of a support zone (B-2_160).

The failure of the support zones reinforced with GFRP stirrups was of a violent nature and occurred with the rupture of the stirrup near the bend (Figure 9a). The support zones reinforced with headed bars were marked by a less violent nature of failure, which was associated with the slipping of the bar out of the anchor head (Figure 9b). Figure 10 shows a comparison of the load capacity of individual support zones due to shear reinforcement spacing.

(a)   (b)

Figure 9. Failure mode of support zones: (**a**) reinforced with GFRP stirrups; (**b**) reinforced with GFRP headed bars.

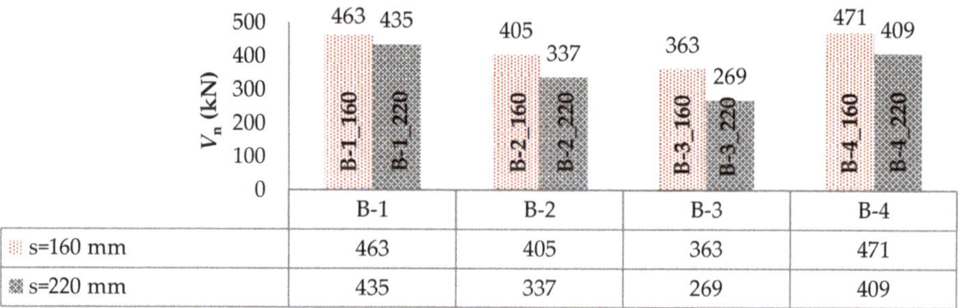

Figure 10. Comparison of the load-bearing capacity of support zones due to shear reinforcement spacing.

The compaction of shear reinforcement, from 220 mm to 160 mm, resulted for the B-1, B-2, B-3, and B-4 beams in increasing the load capacity by approximately 6%, 20%, 35%, and 15%, respectively. The increase in load capacity was clearly greater for beams with longitudinal composite reinforcement. It can also be observed that using a larger number of bars for shear stress transfer, which being cut by a diagonal crack accounted for the contribution of transverse reinforcement in shear capacity, resulted in a greater increase in load capacity for support zones reinforced with headed bars than with stirrups. Along with the increase in load, an increase in the strains of headed bars was recorded up to the initiation of the slipping of the bars (furthest from the top of the crack) out of their anchor heads. Along with the slipping of the bar out of the anchor head, the level of strains on this bar stabilized and the stress was redistributed, which allowed for a more regular distribution of headed bars' strains than in the case of stirrups. Figure 11 shows a comparison of the load capacity of individual support zones by type of longitudinal reinforcement. Support zones longitudinally reinforced with steel bars were marked by a higher load carrying capacity than those longitudinally reinforced with GFRP bars. This confirmed the current conclusions from studies by other authors, including [10–15]. For GFRP stirrups spaced 220 mm and 160 mm, the use of longitudinal steel reinforcement resulted in an increase in load capacity of approximately 29% and 14%, respectively. In the case of GFRP headed bars spaced 220 mm and 160 mm, the use of longitudinal steel reinforcement resulted in an increase in load capacity by approximately 52% and 30%, respectively.

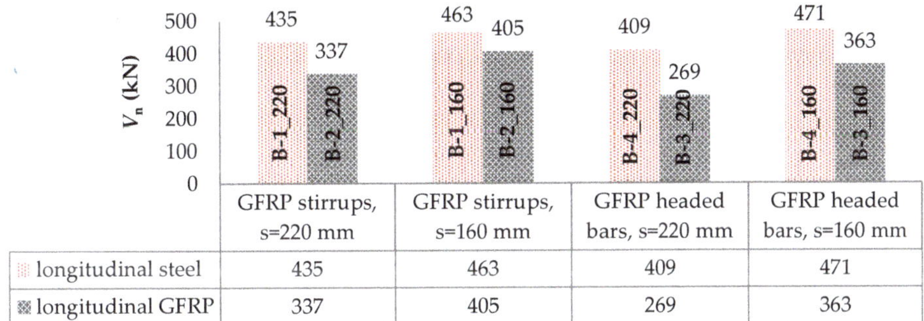

Figure 11. Comparison of the load-bearing capacity of support zones by type of longitudinal reinforcement.

Figure 12 shows a comparison of the load capacity of individual support zones by type of shear reinforcement.

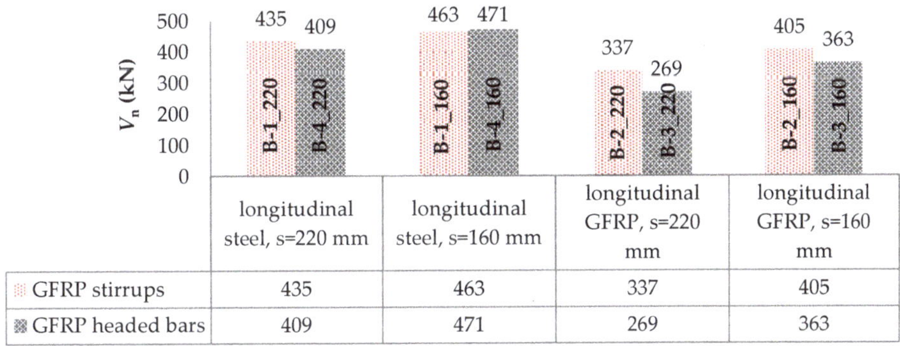

Figure 12. Comparison of the load-bearing capacity of support zones by type of shear reinforcement.

Support zones reinforced with GFRP stirrups were characterized by a higher load carrying capacity than those reinforced with GFRP headed bars. The exception were the support zones B-1_160 and B-4_160, where the support zone with headed bars had a carrying capacity of about 2% higher. This was caused, among other things, by the greater ability of headed bars to redistribute strains (destructive crack cut two pairs of bars). In the case of longitudinal steel reinforcement and shear reinforcement spaced 220 mm, the load capacity of the supporting zone reinforced with stirrups was greater by about 6% than that reinforced with headed bars. Similar values of bearing capacity of support zones longitudinally reinforced with steel bars may result from the high axial stiffness of longitudinal reinforcement. As a consequence, a substantial part of the shear capacity was the bearing capacity of concrete and longitudinal reinforcement. In the case of longitudinal GFRP reinforcement, this increase, for the spacing $s = 220$ mm and $s = 160$ mm, was about 25% and 12%, respectively.

3.2.2. Load Deflection Behavior

The deflection of the research elements was determined on the basis of the results recorded by five induction sensors with an accuracy of 0.001 mm. Two sensors were located at the supporting plates, two directly under the forces and one in the middle of the span. Recording of displacement values took place continuously.

Longitudinally reinforced beams with fiber glass bars (B-2 and B-3) were characterized by higher deflection values at individual load levels than longitudinally reinforced beams with steel bars (B-1 and B-4). This was due to the fact that in the case of beams with longitudinal steel reinforcement, five longitudinal bars were used in the span zone. In addition, the greater axial stiffness of steel reinforcement (as compared to GFRP) resulted in greater bending stiffness of the reinforced concrete element, which was particularly evident in the cracked phase, where the stiffness of an element was largely determined by the modulus of elasticity of longitudinal reinforcement, and this modulus was more than three times smaller for GFRP bars than for steel bars.

In Figure 13a,b, static equilibrium paths for B-1, B-2, B-3, and B-4 beams are shown for Stages 1 and 2, respectively. In the case of Stage 1, the displacement w was recorded in the middle of the span, while for Stage 2, under concentrated force.

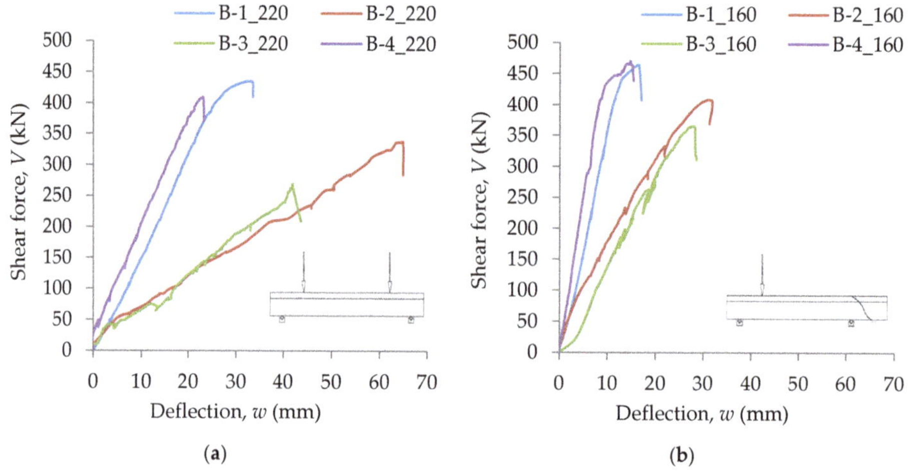

Figure 13. Static equilibrium paths of beams: (**a**) Stage 1; (**b**) Stage 2.

3.2.3. Strains in GFRP Stirrups and Headed Bars

In the course of the tests, the strains of stirrups and headed bars were recorded continuously. Strains were recorded in the central part of the vertical straight segment of stirrups and in the middle of headed bars. The test results were presented in the form of shear reinforcement strains for different load levels together with the crack pattern of the support zone. The strain values given were the average of two vertical straight sections of stirrups or headed bars. Figures 14–17 show the distribution of bars strains for B-1, B-2, B-3, and B-4 beam support zones, for different load levels.

In the case of support zones reinforced with GFRP stirrups, along with the increase in load, the increase in stirrup strains was recorded, almost proportionally, until failure. Strains of headed bars increased almost in proportion until strains initiating the loss of bar adhesion to the anchor head were reached. If one pair of headed bars was cut, failure occurred as a result of the slipping of the bar out of the anchor head. If two pairs of bars were cut, along with an increase in load, an increase in the strains of bars with head anchoring was recorded up to the initiation of the slipping of the bars (furthest from the top of the crack) out of the anchor heads. Along with the slipping of the bar from the anchor head, the level of strains on this bar stabilized, and stress was redistributed, which allowed a more regular distribution of strains in the case of headed bars than in the case of stirrups.

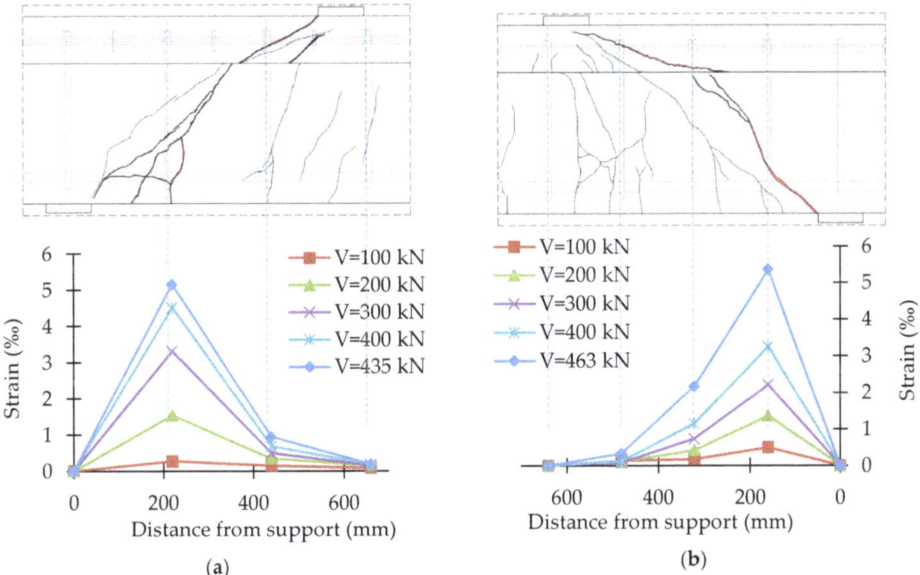

Figure 14. Distribution of strains of the B-1 beam support zone: (**a**) B-1_220; (**b**) B-1_160.

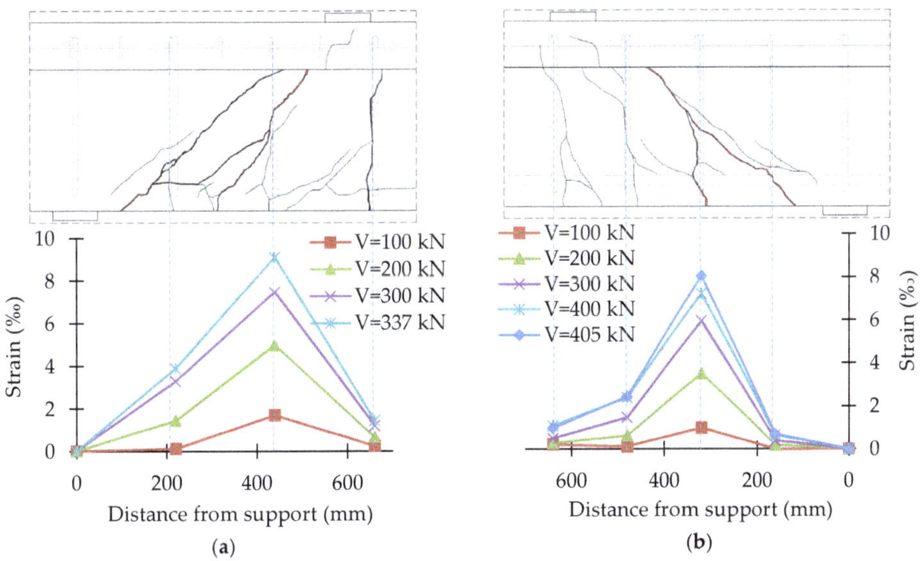

Figure 15. Distribution of strains of the B-2 beam support zone: (**a**) B-2_220; (**b**) B-2_160.

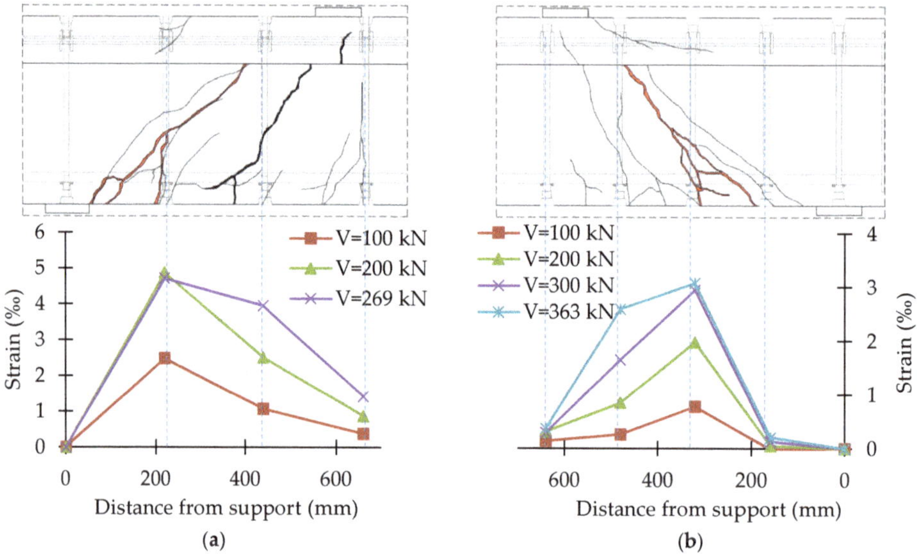

Figure 16. Distribution of strains of the B-3 beam support zone: (**a**) B-3_220; (**b**) B-3_160.

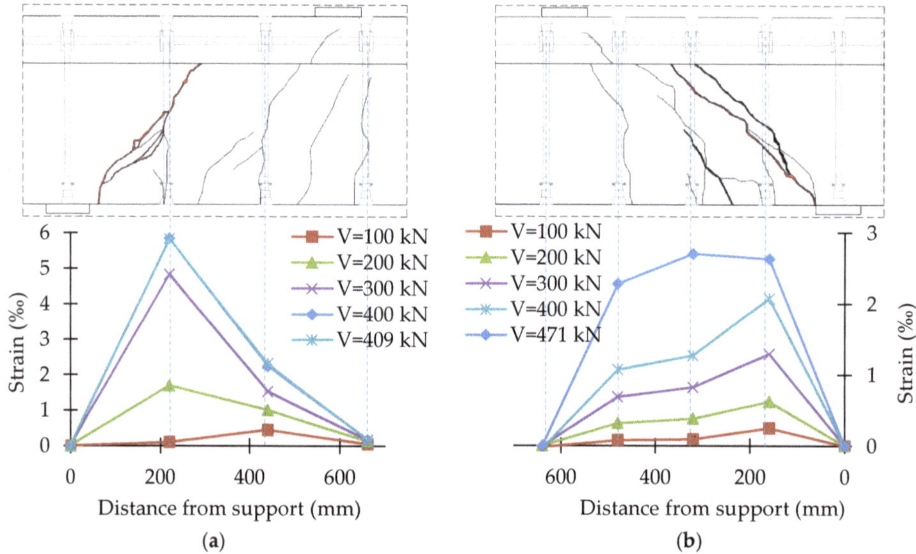

Figure 17. Distribution of strains of the B-4 beam support zone: (**a**) B-4_220; (**b**) B-4_160.

For the B-1 beam, the maximum strains recorded on the straight section of stirrups were 5.15‰ (B-1_220) and 5.37‰ (B-1_160). For the B-2 beam, the maximum strains recorded on the straight section of stirrups were 9.10‰ (B-2_220) and 8.07‰ (B-2_160). The difference in the maximum strains recorded may be due to the fact that the strain gauges were glued to the middle of the vertical sections of stirrups. In the case of B-1_220 and B-1_160, the destructive diagonal crack cut the stirrup slightly above the bend. In the case of B-2_220 and B-2_160, the destructive diagonal crack cut the stirrup much closer to the central part of the vertical sections.

For the B-3 beam, the maximum strains recorded on bars with head anchoring were 4.87‰ (B-3_220) and 3.08‰ (B-3_160). For the B-4 beam, the maximum strains recorded on the headed bars were 5.85‰ (B-4_220) and 2.64‰ (B-4_160). Maximum strains in the case of the more heavily reinforced support zone were much smaller than in the case of the less heavily reinforced support zone, as during the tests on a less heavily reinforced support zone, the bars of a more heavily reinforced support zone were also deformed. As a result of the load in Stage 1, the process of bars slipping from anchor heads was initiated.

4. Conclusions

The article presented experimental tests of a new type of composite bar that was used as shear reinforcement for concrete beams. The results of the experimental studies of bar tests and short support zones reinforced with GFRP stirrups or GFRP headed bars were presented. On the basis of the experimental studies, the following final conclusions were formulated:

(1) The failure mode of straight bars Φ12 GFRP (Schöck Combar) had an explosive character, and all bars were characterized by linear-elastic behavior over the entire strength range. Based on the manufacturer's data and research, including Kurth's research [24], it was found that a similar failure mode accompanied the attempt to stretch the stirrups, where the section in the vicinity of the bend ruptured. GFRP headed bars were characterized by linear-elastic behavior until stresses initiating the loss of bar adhesion to the anchor head were reached. The failure mode consisted of the slipping of the bar from the anchor head and had a radically less violent character than failure by bar rupture. A relatively high value of bar displacement relative to the anchor head was achieved, at which the sample was still able to carry the load.

(2) GFRP headed bars could be used as shear reinforcement. The failure mode of under reinforced support zones consisted in the bar slipping out of the anchor head.

(3) The immediate force initiating the failure of a GFRP headed bar was smaller than the strength of the bent fragment of GFRP stirrups. In the case of a support zone reinforced with headed bars, a more regular distribution of strains could be expected than in the case of analogous reinforcement in the form of frame stirrups, which was confirmed by tests on support zones of beams in a natural scale.

(4) All eight support zones were damaged as a result of the failure of the transverse reinforcement cut by a diagonal crack. The failure of support zones reinforced with GFRP stirrups was violent in its nature and occurred along with the rupture of the stirrup near the bend. Support zones reinforced with headed bars had a less violent mode of failure.

(5) The change of shear reinforcement spacing from 220 mm to 160 mm resulted in an increase in load bearing capacity. The increase in load bearing capacity was clearly greater for beams with longitudinal composite reinforcement. In addition, a greater increase in load bearing capacity was observed for shear zones reinforced with headed bars than for shear zones reinforced with stirrups. Using a larger number of bars cut by a diagonal crack (the contribution of transverse reinforcement in shear capacity) resulted in a greater increase in load bearing capacity for support zones reinforced with headed bars than with stirrups, which indicated a more regular distribution of stress on individual bars cut by a diagonal crack.

(6) The presented research was a new contribution to the experimental study of support zones reinforced with FRP bars. GFRP headed bars could be an alternative for GFRP stirrups and could be used in concrete elements in facilities particularly exposed to aggressive environments and in facilities whose proper functioning is dependent, among other things, on the electromagnetic neutrality of the construction elements. Nevertheless, continuing research in the field of shear behavior of concrete beams reinforced with different type of FRP reinforcement is necessary, both support zones with low and high shear slenderness.

Author Contributions: Conceptualization, C.B. and M.D.; investigation, M.D.; methodology, C.B.; supervision, M.K. (Mieczysław Kamiński); validation, C.B.; visualization, M.D. and M.K. (Maciej Kaźmierowski); writing, original draft, M.D. All authors have read and agreed to the published version of the manuscript.

Materials **2020**, *13*, 1159

Funding: This research received no external funding.

Conflicts of Interest: The authors declare no conflict of interest.

References

1. Li, G.; Zhao, J.; Wang, Z. Fatigue Behavior of Glass Fiber-Reinforced Polymer Bars after Elevated Temperatures Exposure. *Materials* **2018**, *11*, 1028. [CrossRef] [PubMed]
2. Benmokrane, B.; Ali, A.H.; Mohamed, H.M.; El Safty, A.; Manalo, A. Laboratory assessment and durability performance of vinyl-ester, polyester, and epoxy glass-FRP bars for concrete structures. *Compos. Part B Eng.* **2017**, *114*, 163–174. [CrossRef]
3. Kosior-Kazberuk, M.; Wasilczyk, R. Influence of static long-term loads and cyclic freezing/thawing on the behaviour of concrete beams reinforced with BFRP and HFRP bars. In Proceedings of the MATEC Web of Conferences, 3rd Scientific Conference Environmental Challenges in Civil Engineering (ECCE 2018), Opole, Poland, 23–25 April 2018. [CrossRef]
4. Bywalski, C.; Drzazga, M.; Kamiński, M.; Kaźmierowski, M. Analysis of calculation methods for bending concrete elements reinforced with FRP bars. *Arch. Civ. Mech. Eng.* **2016**, *16*, 901–912. [CrossRef]
5. Błaszczyński, T.; Łowińska-Kluge, A. The influence of internal corrosion on the durability of concrete. *Arch. Civ. Mech. Eng.* **2012**, *12*, 219–227. [CrossRef]
6. Mias, C.; Torres, L.; Turon, A.; Barris, C. Experimental study of immediate and time-dependent deflections of GFRP reinforced concrete beams. *Compos. Struct.* **2013**, *96*, 279–285. [CrossRef]
7. Prusiel, J.A.; Rogowski, M.G. Obciążenia termiczne w eksploatowanym żelbetowym kominie przemysłowym (Thermal loads in operated RC industrial chimney). *Mater. Bud.* **2018**, *4*, 74–75. (In Polish) [CrossRef]
8. D'Antino, T.; Pisani, M.A.; Poggi, C. Effect of the environment on the performance of GFRP reinforcing bars. *Compos. Part B Eng.* **2018**, *141*, 123–136. [CrossRef]
9. Kosior-Kazberuk, M. Application of basalt-FRP bars for reinforcing geotechnical concrete structures. In Proceedings of the MATEC Web of Conferences, International Geotechnical Symposium "Geotechnical Construction of Civil Engineering & Transport Structures of the Asian-Pacific Region" (GCCETS 2018), Yuzhno-Sakhalinsk, Russia, 4–7 July 2019. [CrossRef]
10. Urbański, M.; Łapko, A.; Garbacz, A. Investigation on Concrete Beams Reinforced with Basalt Rebars as an Effective Alternative of Conventional R/C Structures. *Procedia Eng.* **2013**, *57*, 1183–1191. [CrossRef]
11. Fib Task Group 9.3. FRP: Reinforcement for Concrete Structures. In *FRP Reinforcement in RC Structures*; Technical Report; Bulletin 40; International Federation for Structural Concrete (fib): Lausanne, Switzerland, 2007; ISBN 978-2-88394-080-2.
12. ACI Committee 440. *ACI 440.1R-15—Guide for the Design and Construction of Structural Concrete Reinforced with Fiber-Reinforced Polymer (FRP) Bars*; American Concrete Institute: Farmington Hills, MI, USA, 2015; ISBN 978-1-942727-10-1.
13. Nagasaka, T.; Fukuyama, H.; Tanigaki, M. Shear performance of concrete beams reinforced with FRP stirrups. In Proceedings of the International Symposium on Fiber-Reinforced-Plastic Reinforcement for Concrete Structures (FRPRCS-1), Vancouver, Canada, 28–31 March 1993; pp. 789–812.
14. Shehata, E.F.G. Fiber Reinforced Polymer (FRP) for Shear Reinforcement in Concrete Structures. Ph.D. Thesis, Department of Civil and Geological Engineering, University of Manitoba, Winnipeg, MB, Canada, March 1999.
15. Ahmed, E.A.M. Shear Behaviour of Concrete Beams Reinforced with Fibre-Reinforced Polymer (FRP) Stirrups. Ph.D. Thesis, Universite de Sherbrooke, Sherbrooke, QC, Canada, June 2009.
16. Bentz, E.C.; Massam, L.; Collins, M.P. Shear strength of large concrete members with FRP reinforcement. *J. Compos. Constr.* **2010**, *14*, 637–646. [CrossRef]
17. Oller, E.; Mari, A.R.; Cladera, A.; Bairan, J.M. Shear design of FRP reinforced concrete beams without transverse reinforcement. *Compos. Part B Eng.* **2014**, *57*, 228–241. [CrossRef]
18. Kosior-Kazberuk, M.; Krassowska, J. Failure mode in shear of steel fiber reinforced concrete beams. In Proceedings of the MATBUD'2018—8th Scientific-Technical Conference on Material Problems in Civil Engineering, Cracow, Poland, 25–27 June 2018. [CrossRef]

19. Acciai, A.; D'Ambrisi, A.; de Stefano, M.; Feo, L.; Focacci, F.; Nudo, R. Experimental response of FRP reinforced members without transverse reinforcement: Failure modes and design issues. *Compos. Part B Eng.* **2016**, *89*, 397–407. [CrossRef]

20. Oller, E.; Mari, A.R.; Cladera, A.; Bairan, J.M. Shear design of reinforced concrete beams with FRP longitudinal and transverse reinforcement. *Compos. Part B Eng.* **2015**, *74*, 104–122. [CrossRef]

21. Lee, S.; Lee, C. Prediction of shear strength of FRP-reinforced concrete flexural members without stirrups using artificial neural networks. *Eng. Struct.* **2014**, *61*, 99–112. [CrossRef]

22. Wegian, F.M.; Abdalla, H.A. Shear capacity of concrete beams reinforced with fiber reinforced polymers. *Compos. Struct.* **2005**, *71*, 130–138. [CrossRef]

23. Yang, K.H.; Mun, J.H. Cyclic Flexural and Shear Performances of Beam Elements with Longitudinal Glass Fiber Reinforced Polymer (GFRP) Bars in Exterior Beam-Column Connections. *Appl. Sci.* **2018**, *8*, 2353. [CrossRef]

24. Kurth, M. Zum Querkrafttragverhalten von Betonbauteilen mit Faserverbundkunststoff-Bewehrung (Shear Behaviour of Concrete Members With Fibre Reinforced Polymers as Internal Reinforcement). Ph.D. Thesis, Rheinisch-Westfälische Technische Hochschule Aachen, Aachen, Germany, December 2012. (In German).

25. Jumaa, G.B.; Yousif, A.R. Size effect on the shear failure of high-strength concrete beams reinforced with basalt FRP bars and stirrups. *Constr. Build. Mater.* **2019**, *209*, 77–94. [CrossRef]

26. Razaqpur, G.; Spadea, S. Shear Strength of FRP Reinforced Concrete Members with Stirrups. *J. Compos. Constr.* **2015**, *19*. [CrossRef]

27. Drzazga, M. Stany Graniczne Betonowych Belek Zbrojonych na Ścinanie Prętami z Włókien Szklanych, Poddanych Działaniu Obciążeń Doraźnych (The Limit States of Concrete Beams Reinforced with Glass Fiber Reinforced Polymer Shear Reinforcement, Subject to Short Term Load). Ph.D. Thesis, Wrocław University of Science and Technology, Wrocław, Poland, July 2019. (In Polish).

28. Krall, M.; Polak, M.A. Concrete beams with different arrangements of GFRP flexural and shear reinforcement. *Eng. Struct.* **2019**, *198*. [CrossRef]

29. Chin, W.J.; Park, Y.H.; Cho, J.R.; Lee, J.Y.; Yoon, Y.S. Flexural Behavior of a Precast Concrete Deck Connected with Headed GFRP Rebars and UHPC. *Materials* **2020**, *13*, 604. [CrossRef] [PubMed]

30. Vint, L.M. Investigation of Bond Properties Of Glass Fibre Reinforced Polymer (GFRP) Bars in Concrete under Direct Tension. Ph.D. Thesis, Department of Civil Engineering, University of Toronto, Toronto, ON, Canada, November 2012.

31. Rduch, A.; Rduch, Ł.; Walentyński, R. Właściwości i zastosowanie kompozytowych prętów zbrojeniowych (Properties and application of composite reinforcement rods). *Przegląd Bud.* **2017**, *88*, 43–46. (In Polish)

32. ACI Committee 440. *ACI 440.3R-04—Guide Test Methods for Fiber-Reinforced Polymers (FRPs) for Reinforcing or Strengthening Concrete Structures*; American Concrete Institute: Farmington Hills, MI, USA, 2004; ISBN 9780870317811.

33. *ISO 10406-1:2015 Fibre-Reinforced Polymer (FRP) Reinforcement of Concrete—Test Methods—Part 1: FRP Bars and Grids*; International Organization for Standardization (ISO): Geneva, Switzerland, 2015.

 *materials*

Article

Stiffness Identification of Foamed Asphalt Mixtures with Cement, Evaluated in Laboratory and In Situ in Road Pavements

Lukasz Skotnicki *, Jarosław Kuźniewski and Antoni Szydlo

Roads and Airports Department, Faculty of Civil Engineering, Wroclaw University of Science and Technology, 50-370 Wroclaw, Poland; jaroslaw.kuzniewski@pwr.edu.pl (J.K.); antoni.szydlo@pwr.edu.pl (A.S.)
* Correspondence: lukasz.skotnicki@pwr.edu.pl; Tel.: +48-71-320-45-38

Received: 7 February 2020; Accepted: 2 March 2020; Published: 3 March 2020

Abstract: The article presents the possibilities of using foamed asphalt in the recycling process to produce the base layer of road pavement constructions in Polish conditions. Foamed asphalt was combined with reclaimed asphalt pavement (RAP) and hydraulic binder (cement). Foamed asphalt mixtures with cement (FAC) were made, based on these ingredients. To reduce stiffness and cracking in the base layer, foamed asphalt (FA) was additionally used in the analyzed mixes containing cement. The laboratory analyzes allowed to estimate the stiffness and fatigue durability of the conglomerate. In the experimental section, measurements of deflections are made, modules of pavement layers are calculated, and their fatigue durability is determined. As a result of the research, new fatigue criteria for FAC mixtures and correlation factors of stiffness modules and fatigue durability in situ with the results of laboratory tests are developed. It is anticipated that FAC recycling technology will provide durable and safe road pavements.

Keywords: recycling; foamed asphalt mixtures with cement (FAC); base layer; reclaimed asphalt pavement (RAP); fatigue durability

---

1. Introduction

Recycling of road pavements makes it possible to reuse road materials that have been refined with binders such as asphalt or cement. The main reason for recycling is the decrease in the availability of stone raw materials, the reduction of aggregate transport costs, and thus the relief of the road and rail network, as well as the liquidation of landfills from damaged road pavements. One of the recycling solutions is the use of foamed asphalt as a binder for recycled aggregates. Foamed asphalt is created by injection through a binder nozzle heated to a temperature of approximately 170 °C with the addition of water, as a result of which, the volume of asphalt increases 15 to 25 times, which in turn allows the smallest grains of the mix to be surrounded. The optimum amount of water for foaming, depending on the type of binder, ranges from 2.0%–3.5% [1–3]. There are also attempts to use ethanol instead of water to foam the asphalt binder [4].

Foamed asphalt (FA) is used in cold, deep recycling technology during the modernization of road pavement construction and in modern technology of warm mix asphalt (WMA), in which the production temperature of asphalt materials is reduced by about 50 °C compared to traditional production technology of hot mix asphalt (HMA).

All over the world, attempts are made to use foamed asphalt in road engineering. Foamed asphalt is used primarily through recycling to make the sub-base layers of road pavement constructions. According to [5], tests have shown that the stiffness modulus of the mixture with foamed asphalt depends on both the stress state and the test temperature. On the basis of triaxial tests, it was found that for mixes without active filler, the hardening of the mixture is generally independent of temperature.

According to the Marshall stability results, the water content of the foam has no significant influence on the performance of the foam asphalt mixture [6]. The results from the study [7] suggest that foamed asphalt cold recycling mixtures have a high modulus and small temperature shrinkage stress, reducing early damage caused by pavement cracks.

In Saudi Arabia, foamed asphalt is mainly used for the production of the sub-base layer and upper base layer, made of reclaimed asphalt pavement (RAP) [1]. In foamed asphalt production processes, the asphalt of high penetration 160/220 is often used [8].

In Indonesia, attempts are being made to use foamed asphalt in asphalt concretes containing only a mineral mixture. Foamed asphalt replaces the regular road binder [9]. These asphalt mixtures based on foamed asphalt can be used in the upper base layer of road pavements.

Mixtures with RAP can be beneficial to the moisture resistance of warm mix asphalt (WMA) and hot mix asphalt (HMA) mixtures. Moisture resistance of asphalt mixtures increases with the increase in RAP content [10]. In addition, the results presented in [11] indicate that the foam processing slightly reduced high-temperature performance and temperature sensitivity while improving the resistance for fatigue cracking.

In the USA, foamed asphalt is commonly used for construction layers of road pavements. An innovative solution is the use of foamed asphalt to stabilize foundations based on ashes [12]. In the USA, in Johnson County, Iowa, RAP temperatures were found to have a significant effect on indirect wet tensile strength, asphalt foam blends produced in the cold recycling site. As the RAP temperature increased, the optimal foam asphalt content decreased—this is due to the activation of the asphalt from waste at a higher temperature and facilitating compaction [13]. Furthermore, the type of asphalt binder contained in the recovered asphalt material has a significant impact on the change in the complex modulus of the recycled mixture [14].

During the 8th Conference on Asphalt Pavements for Southern Africa in Sun City, the authors of [15] and the authors of the research presented in [1] showed an increase in indirect tensile strength (ITS) along with an increase in cement content. Additionally, mixtures containing foamed asphalt showed higher strength values (ITS) than mixtures containing asphalt emulsion. For the same mixtures, a similar relationship was observed in determining the stiffness modulus [16]. Furthermore, the authors' experience show that mixtures containing foamed asphalt have higher durability than mixtures containing asphalt emulsion. This is caused by the different properties of these binders.

The content of asphalt binder has a significant effect on the wet and dry ITS values of materials stabilized with foamed asphalt [17] but a smaller effect on materials stabilized with asphalt emulsion [1,2,15].

Due to the climatic conditions in Central European countries, road pavements should be water- and frost-resistant. On the basis of the research of recycled foamed asphalt pavement, it was found that the use of foamed asphalt improves its tensile stress strength and the mechanical properties of the pavement [18–21]. Moreover, the use of foamed asphalt in mixtures ensures higher water and frost resistance, higher creep stiffness modulus, and higher resistance to plastic strain than when using asphalt emulsion [2,19].

The optimal content of foamed asphalt and hydraulic binder (Portland cement) for mixtures of the base layer gives the desired physical (the air void content) and mechanical parameters (wet-dry ITS) [22–24].

In the test section on the heavily trafficked Greek highway pavement presented in [24], the results of strain and deformation in a layer made of foamed asphalt and recycled material showed that the critical in situ stress in the FA layer was lower than the maximum expected tensile stress threshold. This fact indicates the improvement of fatigue properties of this type of mixture.

Water-based disintegration asphalt emulsions are mainly used for the recycling of asphalt layers in Poland. As a result of mixing reclaimed asphalt pavement (RAP), cement binder and asphalt emulsion, a so-called mineral cement emulsion mixture (MCE) is created. Innovative use of foamed asphalt in mineral-cement mixtures (FAC) in exchange for asphalt emulsion may have a positive effect on the properties of renovated road pavements and the process of building them.

The article presents alternative possibilities for using foamed asphalt to produce the base layer. Foamed asphalt was combined with a mineral mix (reclaimed asphalt pavement (RAP) + possible material for improving gradation) and a hydraulic binder (cement). On the basis of these ingredients, foamed asphalt mixtures with cement were made and marked with the symbol FAC.

2. Materials and Methods

Materials from recycled degraded pavement (test section) were used in the research process. For this purpose, RAP was used from degraded wearing and base course layers as well as crushed granite stone from the base.

Based on control laboratory tests of density, bulk density, Marshall stability, and flow, the technology of production (composition design) of the FAC mixtures was proposed. The durability of the future road pavement is significantly affected by the stiffness and fatigue life of the FAC mixture. Stiffness and fatigue durability in laboratory conditions were determined for FAC-type mixtures. Then, reconstruction of the recycled pavement layers and implementation of the experimental section's pavement layers began. Foamed asphalt, with the addition of cement binder to the base layer, was used. Nevertheless, the presence of too rigid mixtures in the base layers can cause the formation of shrinkage cracks, which copy to the pavement layers of asphalt mixtures in the form of reflected cracks. The use of foamed asphalt in combination with cement allowed the limitation of stiffness and shrinkage cracking of the mixture in terms of its use in the base of the road pavement.

Asphalt layers (asphalt concrete and SMA) were laid on the base layer of the test section made of the FAC mixture. After making the pavement on the experimental section, deflection measurements were made, pavement layer modules were determined, and its fatigue durability was determined. As a result of research, an attempt was made to correlate the stiffness and fatigue life determined in the laboratory with the parameters of FAC mixtures of the test section. New fatigue criteria have been introduced for FAC mixtures used in the base layers. Laboratory tests of the stiffness modulus and fatigue life and developed fatigue criteria can be used to estimate the durability of future road pavement constructions, based on base layers of FAC mixes. The flow chart of the research approach is shown in Figure 1.

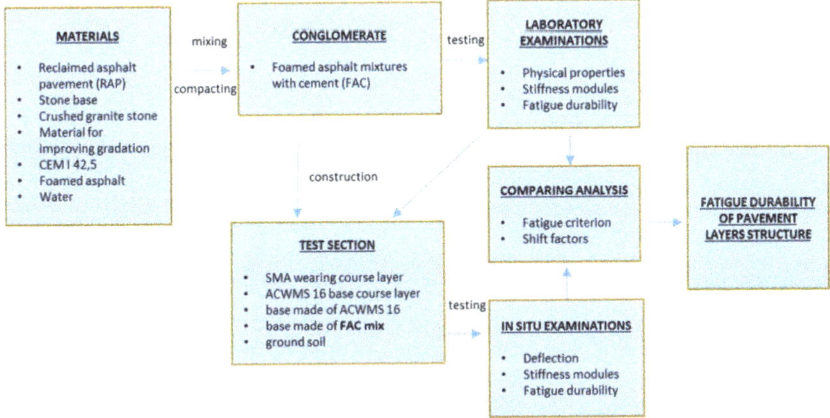

Figure 1. Research methodology flow chart.

The design of the mixture should be correlated with the design of the pavement construction and the organization of works, depending on the method of its implementation. The procedure for designing the composition of the FAC mixture for rebuilding an existing road requires the following steps:

- recognition of pavement structure and layer properties on the basis of samples drilled from it, together with material taken from the base,

    – type and group determination of bearing capacity of the base,
    – determination of the thickness and type of structural layers of the old pavement,
    – recognition of the material forming particular layers,
    – marking of the old binder content in bituminous layers,

- formulation of recycling conditions—method of preparing the mixture, determination of base thickness,
- preparation of analytical samples, performance of tests related to the development of a recipe and determination of the physical and strength characteristics of the designed FAC mixture.

The mineral mixture may consist of the recycled asphalt (RA) material itself obtained directly from milling the pavement or from crushing the lumps from the demolition of the pavement, if it meets the requirements for grading according to [25]. Without this condition, the RA material should be improved with a mineral aggregate.

In the research process, materials from the recycling of the degraded pavement (test section) were used in the form of RAP + stone material from the base + 0/31.5 mm material for improving the mixture gradation of the igneous rock fraction—gabbro. Grading boundary curves for mineral cement emulsion mixtures (MCE) were adapted for research purposes [25]. The recovered mixture (with a content of 5.05% asphalt) did not meet the conditions for gradation, therefore the material for improving gradation was used—0/31.5 mm of igneous rock—gabbro. Table 1 describes the grading of the mineral mixture, taking into consideration the grain size and the percentages of individual components.

Table 1. Grading of the mineral mixture.

Sieve Size	Reclaimed Asphalt (Test Section)	Stone Base (Test Section)	Crushed Granite Stone (Test Section)	Material for Improving Gradation (Aggregate Mining)	Passes	Bottom Curve	Upper Curve
[mm]	[%]	[%]	[%]	[%]	[%]	[%]	[%]
63.0	0.00	0.00	0.00	0.00	100.00	100	100
3.5	0.00	12.05	0.00	0.00	87.95	70	100
25.0	0.00	4.25	1.42	0.00	82.29	65	100
20.0	0.00	4.24	1.82	0.00	76.23	60	100
16.0	0.17	1.57	1.41	0.09	72.99	55	100
12.8	0.11	0.99	1.28	0.94	69.68	49	93
8.0	0.29	1.28	2.21	6.15	59.75	40	84
6.3	0.65	0.37	0.96	3.13	54.64	35	78
4.0	2.60	0.45	1.58	4.90	45.10	25	68
2.0	2.39	0.33	2.78	4.40	35.19	15	50
0.85	2.96	0.41	5.39	4.05	22.38	10	37
0.42	1.74	0.35	3.16	1.56	15.58	8	28
0.30	0.78	0.22	1.59	0.54	12.46	4	18
0.15	0.79	0.37	2.69	0.75	7.86	3	11
0.075	0.44	0.29	1.42	0.79	4.93	3	8
<0.075	2.17	1.15	0.60	1.00	0.00	0	0
Total	15.10	28.30	28.30	28.30	100.00		

The optimal cement addition was estimated on the basis of the compressive strength test at various foamed asphalt contents, according to [26]. FAC mixtures embedded in the base layers should be characterized by susceptibility to deformation on the one hand, and rigidity associated with the transmission of strains from higher layers on the other. The use of cement in typical asphalt mixtures is usually limited to 6% [27], which is why laboratory tests during this work for FAC mixtures were carried out with cement content of 2.38%, 3.38%, and 4.38%.

The optimal water content needed to foam asphalt binder is about 2–3% [1–3,28]. According to [29], in a foaming process, the injection of higher foaming water content (FWC) results in a higher volume expansion but lower stability of foamed asphalt at a certain foaming temperature and air pressure. The amount of water used for foaming the asphalt allows optimal foaming of the binder in the amount of 2.5%.

Due to the cement binder present in the mixture, additional water content was necessary for its proper compacting and setting. The water content of the mineral mix with cement, guaranteeing its maximum compaction, was determined on the basis of optimization according to Proctor methods—method II, based on [30]. The optimum moisture content of the mix was 6.35%.

The FAC mixtures use Nynas Nyfoam 190 of high penetration asphalt with a penetration of 160–220 [31]. Estimating the content of asphalt binder and allowing for the maximum stability of the mixture is possible on the basis of the Marshall test [32]. However, FAC mixtures should not be too susceptible to deformation, but also not too stiff, because of the possibility of cracking. This condition was adopted, due to the need to reduce the shrinkage of the mixture and the formation of cracks in it that could copy into the upper layers of the road pavement. For this stability, the percentage content of foamed asphalt was estimated, which should be added to the mixture. Two levels of asphalt content of 3.5% and 5.5% were used in the analyzed mixtures.

Based on the optimization of foamed asphalt and cement content, a laboratory composition of FAC mixtures was proposed, see Table 2.

Table 2. Composition of FAC mixtures.

No.	Material Name	Share in MM	Share in FAC [%]			
		[%]	Recipe No.			
			C3A3	C4A3	C3A5	C2A3
1	Reclaimed asphalt pavement (RAP)	15.10	13.14	12.98	12.83	13.29
2	Stone base	28.30	24.62	24.34	24.05	24.90
3	Crushed granite stone	28.30	24.62	24.34	24.05	24.90
4	Material for improving gradation	28.30	24.62	24.34	24.05	24.90
5	CEM I 42.5	C	3.38	4.38	3.38	2.38
6	Foamed asphalt	A	3.50	3.50	5.50	3.50
7	Water	W	6.13			

In order to reduce pavement damage and increase its durability, it is important to verify the material parameters of individual pavement layers and carry out the necessary tests, depending on the operating conditions of these materials. As part of the study, FAC mixtures were tested (Table 2), which were applied to the base layers of the pavement.

The qualitative evaluation of the proposed mixtures, collected from recycled old road pavements, consisted of several compatibility tests. Analyzes included in the research program are shown in Table 3.

Table 3. Laboratory research program.

No.	Examination	Temperature [°C]	Curing Period in the Air [days]
1	Density of FAC mix according to [33]	20	7, 14, 28
2	Bulk density of the FAC mixture according to [34]	20	7, 14, 28
3	Air void content according to [35]	20	7, 14, 28
4	Compressive strength according to [26]	25	7, 14, 28
5	Marshall stability and flow according to [32]	60	7, 14, 28

All laboratory samples were compacted using the Marshall method with 75 blows per side [25]. The presented parameters of the FAC mixture are necessary for the correct execution and compaction of the road pavement layer of the test section. The proper load-bearing capacity of the base layer determines the increased durability of the entire road pavement construction.

The main research element were analyzes of stiffness and fatigue life of FAC mixtures. The bending tests of the 4-point beam (4PB-PR) were used to perform them:

- the complex modulus was determined according to [36] at −10 °C, +10 °C, +30 °C, and +55 °C,
- fatigue life according to [37] at +10 °C.

The loading frequency in 4PB-PR tests was 10 Hz. The device for testing stiffness and fatigue life is shown in Figure 2. The fatigue machine allows for simultaneous testing of changes in stiffness of the material being analyzed, determining the so-called complex modulus. The tests use prismatic beams with nominal dimensions: effective length (beam span between supports) L = 357 mm, b = 60 mm, h = 50 mm, as seen in Figure 3.

Figure 2. Beam-Flex apparatus.

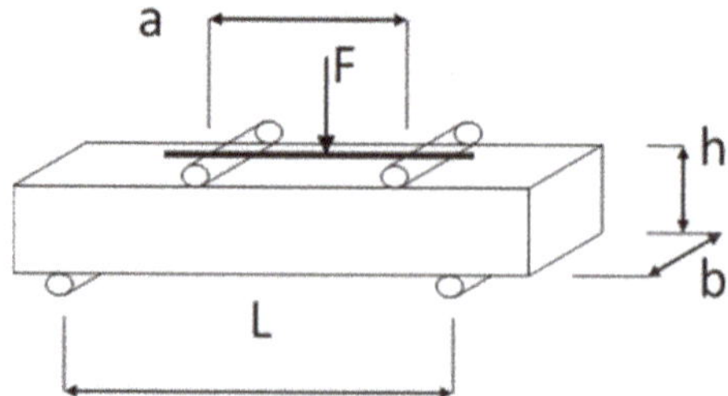

Figure 3. Load diagram for FAC mixture samples in the 4PB test.

The study of the complex modulus of FAC mixture stiffness was carried out by the method of permanent deformation $\varepsilon = 50 \times 10^{-6}$ m/m. During fatigue tests, using the permanent deformation method, 5 load levels were adopted in the form of given strains: $\varepsilon = 500 \times 10^{-6}$ m/m, $\varepsilon = 400 \times 10^{-6}$ m/m, $\varepsilon = 200 \times 10^{-6}$ m/m, $\varepsilon = 100 \times 10^{-6}$ m/m, $\varepsilon = 50 \times 10^{-6}$ m/m. The number of load cycles $N_{f/50}$ was recorded until the complex stiffness modulus dropped to 50% of the initial value—conventional fatigue criterion.

After the laboratory research, the trial field phase was implemented. The FAC mixtures designed in the laboratory were built into the base layer of the road pavement section. The stiffness modulus and fatigue durability of FAC mixtures layer were estimated. Results of laboratory examinations were compared with results of bearing capacity of this test section.

3. Results and Discussion

3.1. Basic Research

The designed FAC mixtures are intended for the lower base layer of the test section pavement. Tests were conducted to determine the basic properties of the mixtures on density, bulk density, air void content, Marshall stability, flow and compressive strength for the FAC mixture used. Base on the optimisation process the mixture C3A3 was choosen for further analysis and for applying to trial field phase. Results from the laboratory tests and analyses carried out for optimal mixture C3A3 are given in Table 4. The results given in Table 4 are mean values calculated from a minimum of three representative samples for each feature.

Table 4. Properties of the FAC mixture (C3A3).

Analyzed Feature	Samples Compacted in the Laboratory		
	Curing Conditions		
	7 Days	14 Days	28 Days
Density [g/cm^3]	2.520	2.522	2.525
Bulk density [g/cm^3]	2.218	2.220	2.226
Air void content [%]	11.75	11.69	11.66
Marshall stability [kN]	10.06	11.33	11.84
Marshall flow [mm]	0.84	0.78	0.76
Compressive strength [MPa]	1.62	2.21	2.45

The compressive strength of samples from the FAC mixture increases as the curing period increases and takes values from 1.5 to 2.5 MPa over 7 to 28 days. This is influenced by the presence of the cement added and its hydration time in the mixture.

The Marshall stability in FAC mixtures has a similar relationship to compressive strength. Its value increases with the length of the sample's "life" in the range of 10.0–12.5 kN. The increase in stability value and decrease in deformation value over time suggests a significant effect of cement presence added to the mixture.

3.2. Stiffness and Fatigue Durability in Laboratory Conditions

On the basis of the analyses, the values of the complex stiffness modulus of the innovative material in a wide temperature range were determine, see Figure 4. The complex stiffness modulus was determined as a mean value form four samples for each applied temperature.

As the temperature increases, the stiffness of FAC mixtures decreases. For a set average annual temperature of +10 °C, the complex stiffness modulus of the FAC mixture is about 2883 MPa.

Temperature changes affect the stiffness of FAC mixtures, but the gradient of changes is smaller compared to conventional asphalt mixtures.

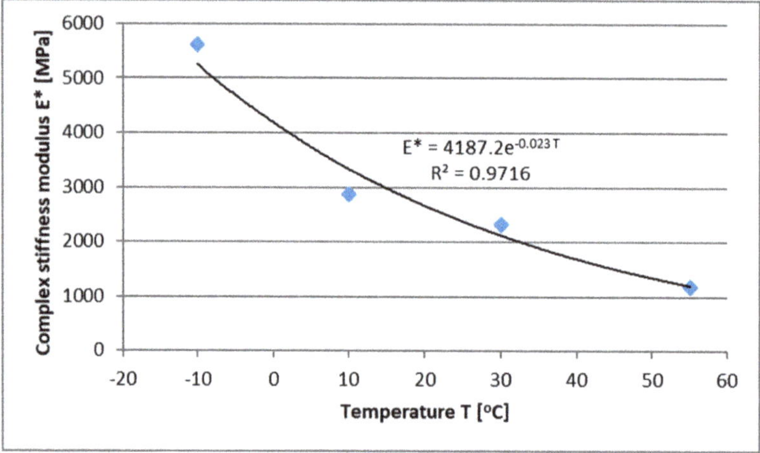

Figure 4. Changes in the FAC mixture stiffness as a function of temperature.

In the fatigue life analysis, the 4-point 4PB bending method was used—the dynamic method at constant deformation. To demonstrate the nature of the work of the FAC recycled material and to determine the fatigue criteria, tests were carried out for FAC mixtures in various environmental conditions.

It was assumed that the number of load cycles $N_{f/50}$, to achieve a decrease in the complex stiffness modulus to 50% of the initial value, is equivalent to the destruction of the sample, then the test was also discontinued. The $N_{f/50}$ criterion applies to mineral-asphalt mixtures according to [38]. After the fatigue tests, data was obtained that allowed the estimation of the FAC mix fatigue curve—Figure 5. The fatigue curve is a relationship of fatigue life (on a logarithmic scale) as a function of the applied load value (strain). Fatigue life was determined as a mean value form six samples for each applied load strain.

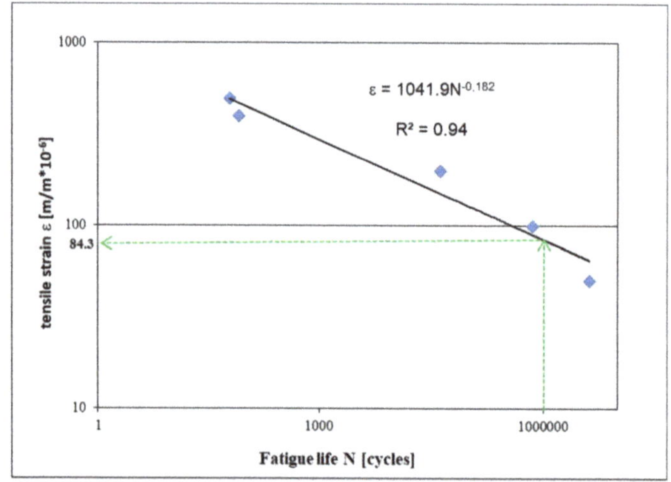

Figure 5. FAC mixture fatigue curve—permanent deformation method.

From the slope of the fatigue curve, it should be concluded that the destructive deformation for 1 million load cycles is equal $\varepsilon_6 = 84.3 \times 10^{-6}$ m/m. The different nature of the decrease in the value of the complex stiffness modulus was found in comparison with typical asphalt mixtures during fatigue tests, see Figure 6. The level of the applied load is marked in green, while the red color indicates the decrease in the value of the complex stiffness modulus.

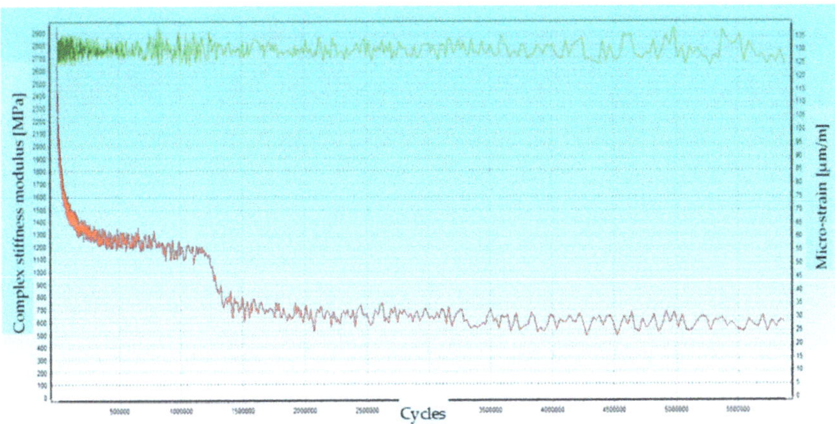

Figure 6. The course of the FAC mixture fatigue test at standard load.

FAC mixtures lose a significant part of the complex stiffness modulus relatively quickly and may be characterized by local decreases in stiffness, but they are still able to carry the given load in the form of strain. This is because microcracks appear in the analyzed material, which do not disqualify it for use in the layers of the road pavement base, in which such cracks are acceptable [39,40]. All tested mixtures had a similar character in the fatigue test.

Due to the presented conditions of FAC mixtures, the allowable decrease in the complex stiffness modulus should be modified to a level of approx. 30%, compared to the initial value. For the modified fatigue criterion $N_{f/30}$, the number of load cycles, to obtain a decrease in the complex stiffness modulus to the level of 30% of the initial value, the level of destructive strain was estimated in the millionth cycle of loading ε_6.

Additional fatigue life tests of FAC mixtures were carried out for this reason, with the modified fatigue criterion $N_{f/30}$. To obtain fatigue curves, fatigue tests were performed using strain levels: $\varepsilon = 200 \times 10^{-6}$ m/m, $\varepsilon = 180 \times 10^{-6}$ m/m, $\varepsilon = 170 \times 10^{-6}$ m/m. On the basis of the fatigue characteristics of the material, destructive strains were estimated in a millionth load cycle ε_6. The results of laboratory tests are shown in Figure 7.

On the basis of the obtained fatigue characteristics (fatigue equation - Figure 7), the permissible level of destructive strain ε_6 in a millionth cycle of loading was estimated at level $\varepsilon_6 = 168.7 \times 10^{-6}$ m/m. Similar analyses were made for all FAC mixtures characterized by a different cement content and different asphalt content. The obtained values of destructive strain ε_6 are presented in Table 5.

Table 5. Fatigue durability of FAC mixtures.

Material Name	Share in MM	Share in FAC			
	[%]	Recipe No.			
		C3A3	C4A3	C3A5	C2A3
CEM I 42.5	C	3.38	4.38	3.38	2.38
Foamed asphalt	A	3.50	3.50	5.50	3.50
Destructive strain $_6$ m/m*10^{-6}		168.7	171.1	187.1	198.6

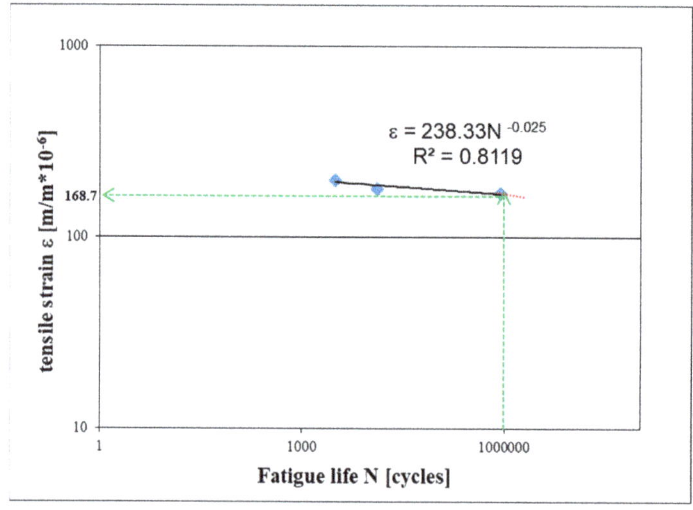

Figure 7. Fatigue curve—FAC.

While analyzing the fatigue lives of tested mixtures it was set that with increasing the amount of asphalt and decreasing the amount of cement the mixtures became more flexible—the destructive strain ε_6 increased.

Using the conducted fatigue tests, the final form of the fatigue equation described by Equation (1), taking into account changes in foamed asphalt and cement content was obtained.

$$\varepsilon = \varepsilon_6 \cdot \left(\frac{E_{10}}{E_T}\right)^{0.78721} \cdot \left(\frac{N_{\frac{f}{30}}}{10^6}\right)^{(-0.57403 \cdot A + 0.64234 \cdot C)}, \tag{1}$$

where:

ε—acceptable tensile strain,

ε_6—tensile strain at which the sample is destroyed after 10^6 load cycles in the following test conditions: bending of a 4-point beam, temperature +10 °C, frequency 10 Hz,

$N_{f/30}$—number of load cycles to achieve a decrease in the complex stiffness modulus to 30% of the initial value [-],

A—"A" asphalt content [%],

C—"C" asphalt content [%],

E_{10}—stiffness modulus of the mixture at +10 °C,

E_T—stiffness modulus of the mixture at temperature T.

Laboratory tests and the developed fatigue equation can be used to predict the fatigue life of structural layers of road pavement, taking into account the shifting factors.

4. Test Section

4.1. Road Pavement Technology

Because the embedded materials in the existing road pavement have lost their bearing capacity and fatigue durability, it was proposed to make this pavement in cold recycling technology with existing materials and to make a FAC mixture on their basis. The C3A3 mixture was built into the base layer of the road pavement. Before finishing the test section, the road had numerous damages, which are shown in Figure 8.

Figure 8. A section of the existing road—west lane.

A mobile deep recycler was used to produce the recycled FAC mixture—Figure 9.

Figure 9. Recycling on-site—crushing and mixing of ingredients.

To verify the compaction of the base layer, the compaction index was checked. This indicator was determined by comparing the bulk density of samples formed from the FAC mixture in the laboratory with the bulk density of samples out of the finished pavement layer. The compaction index was 0.98, which is a satisfactory value for the base layers [25].

After constructing the recycled layers, the surface of the experimental section was finished with asphalt layers: a base layer and a base course layer of asphalt concrete (with a high stiffness modulus (ACWMS) and a wearing course layer of the SMA mixture, see Figure 10.

Figure 10. Wearing course layer, SMA type.

The design of the innovative road pavement structure assumed the following layering:

- SMA wearing course layer, 4 cm thick,
- ACWMS 16 base course layer, 8 cm thick,
- base made of ACWMS 16, 12 cm thick,
- base made of FAC mix, 20 cm thick,
- ground soil.

According to [38], all implemented mineral-asphalt mixtures met the design requirements for the layers of flexible pavement road constructions in Poland. As a result of the applied technology of the road base made of FAC-type mixture, the road durability forecasting was carried out before the road traffic admission. For this purpose, measurements of deflections of the pavement were carried out, along with the identification of layer modules and the subgrade.

4.2. Identification of Layer Modules

The measurements of pavement deflections were done on the street pavement using an FWD (Falling Weight Deflectometer). It is a device that induces a force impulse using a falling weight onto a measuring plate (through a specially designed spring system). The set of displacements determined on a given measuring stand creates the so-called "displacement bowl", which is then used to identify modules of layers and the subrade. Measurements of deflections made by FWD were carried out during the construction of the section on the layers: the FAC layer, the ACWMS layer, and the SMA wearing course layer.

During pavement tests, displacement was measured at the following distances from the load axis: d1 = 0.0; d2 = 0.2; d3 = 0.3; d4 = 0.45; d5 = 0.6; d6 = 0.9; d7 = 1.2; d8 = 1.5; d9 = 1.8 m. The tests were carried out at different temperatures. Figure 11 shows a diagram of deflection testing using an FWD deflectometer. Figure 12 shows a view of the FWD deflectometer during the testing of this pavement.

The results of deflection measurements were used to estimate the layer modules and the subgrade modules of the road pavement construction. The calculation model presented in Figure 13 was adopted for the identification calculations of the modules of the FAC layer. It is an elastic two-layer system, i.e., a layer arranged on the elastic half-space.

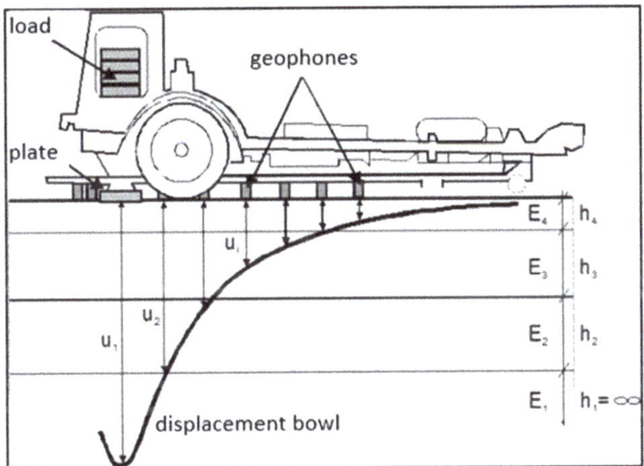

Figure 11. Measurement diagram done with an FWD deflectometer.

Figure 12. View of the FWD device during the test.

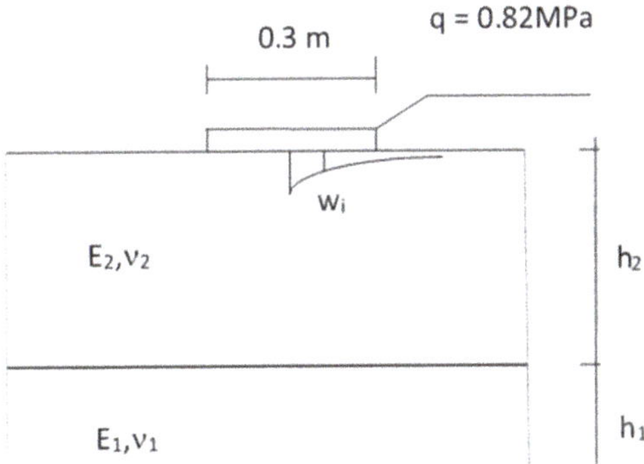

Figure 13. Calculation model of the tested pavement construction.

Particular layers model the layout of the pavement structure. The h_2 layer models FAC, the h_1 layer—an improved subgrade. The FAC layer is described by the E_2 stiffness modulus and Poisson's ratio v_2. The subgrade is described by the modulus of elasticity E_1 and Poisson's ratio v_1. The thicknesses were assumed by the in-depth identification $h_2 = 0.20$ m. It was assumed that Poisson's ratio did not have a significant impact on the state of stress and strains and was assumed to be constant, i.e., $v_2 = 0.3$ and the factor $v_1 = 0.35$.

The essence of identification is to minimize the objective function described by Equation (2):

$$\Delta = \frac{\sqrt{\frac{F}{k}}}{\frac{\sum_{j=1}^{k} w_j}{k}} \tag{2}$$

where:

$$F = \sum_{j=1}^{k} (w_j - u_j)^2 \tag{3}$$

w_j—theoretical deflections calculated in the model,

u_j—measured deflections,

k—the number of deflections measured at one point, forming the deflection bowl.

Of course, the number of layers n should be smaller than the number of k points forming the deflection bowl. Calculations were made on the basis of the CZUG program [41].

As a result of identification, the following values of modules (Ei) of the FAC layer and subgrade were obtained. Measurements of deflections on the FAC layer were made for different temperatures: -2 °C, $+10$ °C, $+25$ °C, $+32$ °C.

The obtained modulus values for the subgrade and the FAC layer are summarized in Table 6 for a 95% level of confidence.

Table 6. List of identified module values.

Temperature during the Test [°C]	E1—Subgrade Modulus [MPa]	E2—FAC Layer Modulus [MPa]
−2	132	5450
+10	135	3490
+25	140	3120
+32	128	2120

After the FAC layer was laid and FWD tests were done, mma layers were laid, after which the deflection bowl measurements were again carried out using an FWD deflectometer.

Figure 14 shows a model of the pavement structure after laying layers of mma. It is a three-layer system. Two layers are arranged on a half-space. The h_3 layer is a layer with mma described by the E_3 stiffness modulus and Poisson's ratio v_3. The h_2 layer is the FAC layer described by the E_2 stiffness modulus and Poisson's ratio v_2. The subgrade is described by the E_1 modulus and Poisson's ratio v_1. It was assumed that $h_3 = 0.24$ m, $h_2 = 0.20$ m, $v_3 = v_2 = 0.35$, and $v_1 = 0.3$.

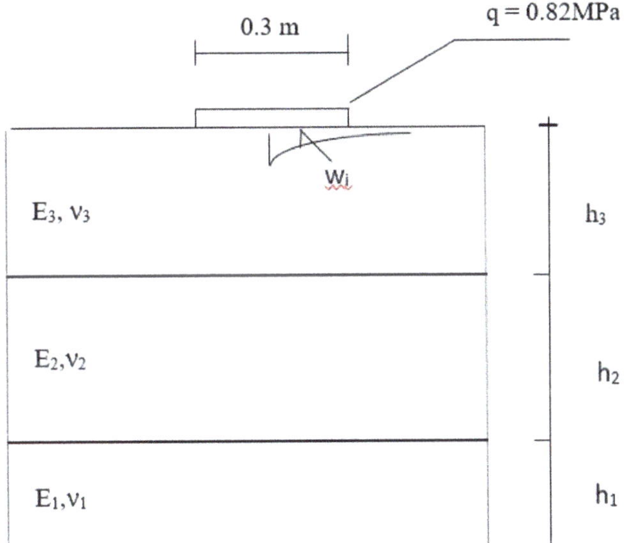

Figure 14. Calculation model of the tested pavement structure.

The module values were calculated on the basis of the deflection bowl measurements using the FWD deflectometer and optimization calculations. The results are summarized in Table 7. Measurements were taken at the approx. temperature +10 °C.

Table 7. List of identified modules for the temperature +10 °C.

E1—Subgrade Modulus [MPa]	E2—FAC Modulus [MPa]	E3—Mma Modulus [MPa]
138	3620	16,250

Figure 15 summarizes the test results and compares the FAC mixture modules obtained in the laboratory and in situ layers.

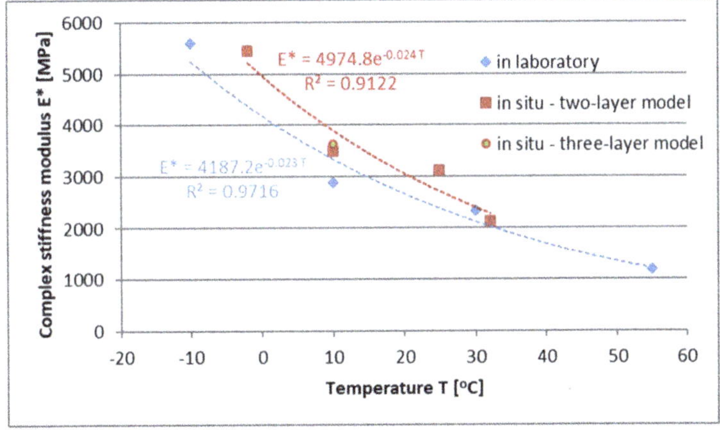

Figure 15. Values of modules in the laboratory and in situ.

When analyzing the results from Tables 6 and 7 and Figure 15 the stiffness modulus value determined in the laboratory is comparable with the modules of the material used in situ. As a result of the analyzes, a good correlation between field and laboratory tests was obtained in the analyzed range of temperatures. The conversion factor of the modules "k" determines the relationship (4), which describes the ratio of the stiffness modules defined in the laboratory to the values of the field modules, as a function of temperature.

$$k = 0.8417 \cdot e^{0.001 \cdot T},$$

(4)

where:

k—module conversion factor [-],
T—FAC layer temperature [°C].

The "k" conversion factor takes values from 0.84–0.87 depending on the temperature—Table 8.

Table 8. List of conversion factor "k" values.

Temperature T [°C]	Factor k [-]
−2	0.84
10	0.85
25	0.86
32	0.87

4.3. Fatigue Durability of Pavement Layers Structure

Using the obtained module values and the model presented in Figure 14, the fatigue life of the structure was calculated for the designed thicknesses. Equation (5), developed by the authors, describes the criterion for the FAC mixture:

$$\varepsilon = \varepsilon_6 \cdot (k)^{0.78721} \cdot \left(\frac{N_{f/30}}{10^6}\right)^{(-0.57403 \cdot A + 0.64234 \cdot C)} \cdot f_1 \cdot f_2 \cdot f_3$$

(5)

where:

ε—strain in the FAC layer,
ε_6—strain at millionth load cycle, 0.000168 was adopted,
$N_{f/30}$—number of load cycles to achieve a decrease in the complex stiffness modulus to 30% of the initial value,
A—percentage of new asphalt in the FAC layer, 4.16% was adopted,
C—percentage of cement in the FAC layer, 3.38% was adopted,
k—a conversion factor of the modules defined in the laboratory to the modules in situ, for a temperature of 10 °C,
f_1—a shift factor dependent on stiffness of FAC mixture, range between 0.8–1.0, was adopted 0.81,
f_2—a shift factor dependent on bearing capacity of subgrade, range between 0.8–1.0, was adopted 0.83,
f_3—a shift factor dependent on heterogeneity of FAC mixture, range between 0.8–0.95, was adopted 0.92.

For the identified modules and model from Figure 14, strains at the bottom of the FAC layer were calculated. $\varepsilon = 0.0000412$ was obtained. Using Equation (5), N = 29,500,000 axles of 115 kN were calculated (fatigue life). The required minimal number of load axles for the road pavement is equal 12,000,000 axles 115 kN.

5. Conclusions

The conducted analyses for FAC mixtures (foamed asphalt mixtures with cement) showed that:

- the proposed conglomerates can be used for incorporation into road base layers.
- Identification tests of embedded layers confirmed the results of laboratory tests. The presented results indicate that the innovative technology used allows for the use of recycled materials, which significantly speeds up repairs.
- During the research work, a technology for modernizing degraded road pavements was developed. The cold recycling technology based on a FAC mixtures was implemented to provide adequate load-bearing capacity and fatigue durability of new road pavement construction.
- The results of in-situ tests in the field of module evaluation correlated with the results of laboratory tests.
- A new fatigue criterion for FAC mixtures and a correlation factor for stiffness modules determined in the laboratory and in situ were developed.
- Stiffness and bearing capacity tests showed that the pavement construction made with innovative technology, i.e., recycled material bonded with foamed asphalt and cement, has sufficient bearing capacity and fatigue life.
- As a result of the bearing capacity analyses, it was found that the layers of the test section meet the requirements for safe exploitation and their durability is satisfactory. Thus, the pavement on the test section could be put into exploitation.
- Laboratory tests of the stiffness modulus and fatigue durability developed fatigue criteria, and correlation factors can be used to estimate the durability of future road pavement structures based on the base layers of FAC mixtures.

Author Contributions: Methodology, L.S., J.K. and A.S.; formal analysis, L.S., J.K. and A.S.; investigation, L.S., J.K. and A.S.; writing—original draft preparation, L.S., J.K. and A.S.; writing—review and editing, L.S. and J.K.; funding acquisition A.S. All authors have read and agreed to the published version of the manuscript.

Funding: This research was funded the project entitled "The innovative technology used the binding agent optimization that provides the long service life of the recycled base layer" (TECHMATSTRATEG1/349326/9/NCBR/2017) within the scientific undertaking of Strategic Research and Development Program entitled "Modern Materials Technology" (TECHMATSTRATEG I), which is financed by the National Centre for Research and Development (Polish NCBR).

References

1. Al-Abdul Wahhab, H.I.; Baig, M.G.; Mahmoud, I.A.; Kattan, H.M. Study of road bases construction in Saudi Arabia using foam asphalt. *Constr. Build. Mater.* **2012**, *26*, 113–121. [CrossRef]
2. Mrugała, J. Soil stabilization with foamed bitumen. In *Structure & Environment*; Kielce University of Technology: Kielce, Poland, 2011; Volume 3, pp. 40–44, ISSN 2081-1500.
3. Road and Bridge Research Institute. *Pavement Technology Division—Report. Subject TN-236 (stage I) Implementing the Foamed Asphalt to Cold Recycling Technology*; Road and Bridge Research Institute: Warsaw, Poland, 2004.
4. Hasan, M.R.M.; You, Z. Comparative study of ethanol foamed asphalt binders and mixtures prepared via manual injection and laboratory foaming device. *J. Traffic Transp. Eng.* **2019**, *6*, 383–395. [CrossRef]
5. Fu, P.; Harvey, J.T. Temperature sensitivity of foamed asphalt mix stiffness: Field and lab study. *Int. J. Pavement Eng.* **2007**, *8*, 137–145. [CrossRef]
6. Hailesilassie, B.W.; Hugener, M.; Partl, M.N. Influence of foaming water content on foam asphalt mixtures. *Constr. Build. Mater.* **2015**, *85*, 65–77. [CrossRef]
7. Zhang, Z.; Cong, C.; Xi, W.; Li, S. Application research on the performances of pavement structure with foamed asphalt cold recycling mixture. *Constr. Build. Mater.* **2018**, *169*, 396–402. [CrossRef]
8. Nguyen, H.V. Effects of mixing procedures and rap sizes on stiffness distribution of hot recycled asphalt mixtures. *Constr. Build. Mater.* **2013**, *47*, 728–742. [CrossRef]

9. Sunarjono, S. Performance of foamed asphalt under repeated load axial test. *Procedia Eng.* **2013**, *54*, 698–710. [CrossRef]

10. Shu, X.; Huang, B.; Shrum, E.D.; Jia, X. Laboratory evaluation of moisture susceptibility of foamed warm mix asphaltcontaining high percentages of RAP. *Constr. Build. Mater.* **2012**, *35*, 125–130. [CrossRef]

11. Yu, X.; Liu, S.; Dong, F. Comparative assessment of rheological property characteristics forunfoamed and foamed asphalt binder. *Constr. Build. Mater.* **2019**, *205*, 186–195. [CrossRef]

12. Mallick, R.B.; Hendrix, G., Jr. Use of foamed asphalt in recycling incinerator ash for construction of stabilized base course. *Resour. Conserv. Recycl.* **2004**, *42*, 239–248. [CrossRef]

13. Kim, Y.; Lee, H.D. Influence of reclaimed asphalt pavement temperature on mix design process of cold in-place recycling using foamed asphalt. *J. Mater. Civ. Eng.* **2011**, *23*, 961–968. [CrossRef]

14. Buczyński, P.; Iwański, M. Complex modulus change within the linear viscoelastic region of the mineral-cement mixture with foamed bitumen. *Constr. Build. Mater.* **2018**, *172*, 52–62. [CrossRef]

15. Hodgkinson, A.; Visser, A.T. The role of fillers and cementitious binders when recycling with foamed bitumen or bitumen emulsion. In Proceedings of the 8th Conference on Asphalt Pavements for Southern Africa (CAPSA 2004), Sun City, South Africa, 12–16 September 2004.

16. Yan, J.; Ni, F.; Yang, M.; Li, J. An experimental study on fatigue properties of emulsion and foam cold recycled mixes. *Constr. Build. Mater.* **2010**, *24*, 2151–2156. [CrossRef]

17. Gui-Ping, H.; Wing-Gun, W. Effects of moisture on strength and permanent deformation of foamed asphalt mix incorporating RAP materials. *Constr. Build. Mater.* **2008**, *22*, 125–130. [CrossRef]

18. Gui-ping, H.; Wing-gun, W. Laboratory study on permanent deformation of foamed asphalt mix incorporating reclaimed asphalt pavement materials. *Constr. Build. Mater.* **2007**, *21*, 1809–1819.

19. Iwański, M.; Chomicz-Kowalska, A. Resistance of the pavement to water and frost in the cold recycling technology. In *Structure and Environment*, 1st ed.; Kielce University of Technology: Kielce, Poland, 2010; pp. 9–17.

20. Iwański, M.; Chomicz-Kowalska, A. Water and frost resistance of the recycled base rehabilitated with the foamed bitumen technology. In Proceedings of the 10th International Conference "Modern Building Materials, structures and Techniques", Vilnius, Lithuania, 19–21 May 2010; pp. 99–105.

21. Jamshidi, A.; White, G.; Hosseinpour, M.; Kurumisawa, K.; Hamzah, M.O. Characterization of effects of reclaimed asphalt pavement (RAP) source and content on dynamic modulus of hot mix asphalt concrete. *Constr. Build. Mater.* **2019**, *217*, 487–497. [CrossRef]

22. Iwański, M.; Chomicz-Kowalska, A. Laboratory study on mechanical parameters of foamed bitumen mixtures in the cold recycling technology. *Procedia Eng.* **2013**, *57*, 433–442. [CrossRef]

23. Li, Z.; Hao, P.; Liu, H.; Xu, J. Effect of cement on the strength and microcosmic characteristics of cold recycled mixtures using foamed asphalt. *J. Clean. Prod.* **2019**, *230*, 956–965. [CrossRef]

24. Papavasiliou, V.; Loizos, A. Field performance and fatigue characteristics of recycled pavement materials treated with foamed asphalt. *Constr. Build. Mater.* **2013**, *48*, 677–684. [CrossRef]

25. Road and Bridge Research Institute. *Technical Conditions for Performing Base Layers Built from ACM Mixture, I-61*; Road and Bridge Research Institute: Warsaw, Poland, 1999.

26. PN-S-96012. *Base and Soil-Cement Sub-Grade*; Polish Committee for Standardization: Warsaw, Poland, 1997.

27. Guha, A.H.; Assaf, G.J. Effect of Portland cement as a filler in hot-mix asphalt in hot regions. *J. Build. Eng.* **2020**, *287*, 101036. [CrossRef]

28. Iwanski, M.; Mazurek, G.; Buczyński, P. Bitumen Foaming Optimization Process on the Basis of Rheological Properties. *Materials* **2018**, *11*, 1854. [CrossRef] [PubMed]

29. Bairgi, B.K.; Mannan, U.A.; Tarefder, R.A. Influence of foaming on tribological and rheological characteristics of foamed asphalt. *Constr. Build. Mater.* **2019**, *205*, 186–195. [CrossRef]

30. PN-EN 13286-2. *Unbound and Hydraulically Bound Mixtures-Part 2: Test Methods for Laboratory Reference Density and Water Content-Proctor Compaction*; Polish Committee for Standardization: Warsaw, Poland, 2010.

31. Nynas UK-PdS Nyfoam 190 Specifications-30/12/2010. Available online: https://notes.nynas.com/Apps/1112.nsf/wpds/GB_EN_Nyfoam_190/$File/Nyfoam_190_GB_EN_PDS.pdf (accessed on 3 March 2020).

32. PN-EN 12697-34. *Bituminous mixtures-Test Methods for Hot Mix Asphalt–Part 34: Marshall Test*; Polish Committee for Standardization: Warsaw, Poland, 2012.

33. PN-EN 12697-5. *Bituminous Mixtures-Test Methods for Hot Mix Asphalt–Part 5: Determination of the Maximum Density*; Polish Committee for Standardization: Warsaw, Poland, 2019.

34. PN-EN 12697-6. *Bituminous Mixtures-Test Methods for Hot Mix Asphalt–Part 6: Determination of Bulk Density of Bituminous Specimens*; Polish Committee for Standardization: Warsaw, Poland, 2012.

35. PN-EN 12697-8. *Bituminous Mixtures–Test Methods for Hot Mix Asphalt–Part 8: Determination of Void Characteristics of Bituminous Specimens*; Polish Committee for Standardization: Warsaw, Poland, 2019.

36. PN-EN 12697-26. *Bituminous Mixtures-Test Methods for Hot Mix Asphalt–Part 26: Stiffness*; Polish Committee for Standardization: Warsaw, Poland, 2018.

37. PN-EN 12697-24. *Bituminous Mixtures-Test Methods for Hot Mix Asphalt–Part 24: Resistance to Fatigue*; Polish Committee for Standardization: Warsaw, Poland, 2018.

38. General Directorate for National Roads and Motorways-Asphalt Pavements on National Roads. *WT-2 2014–Part I: Bituminous Mixtures*; Technical requirements; General Directorate for National Roads and Motorways: Warsaw, Poland, 2014.

39. Kuźniewski, J.; Skotnicki, L.; Szydło, A. Fatigue durability of asphalt-cement mixtures. *Bull. Pol. Acad. Sci. Tech. Sci.* **2015**, *63*, 107–111. [CrossRef]

40. Szydło, A.; Kuźniewski, J.; Skotnicki, Ł. *Recycling of Existing Road Pavements, Reuse of Recycled Materials-Durability Forecasting-Report of Institute of Civil*; Engineering on Wroclaw University of Science and Technology, SPR 19; WUoSaT: Wroclaw, Poland, 2014.

41. Szydło, A. *Parameters Statistical Identification of Airfield Pavements Models–Report of Institute of Civil*; Engineering on Wroclaw University of Science and Technology No. 45/95; WUoSaT: Wroclaw, Poland, 1995.

Article

Sustainable Test Methods for Construction Materials and Elements

Ewa Szewczak [1,*], Agnieszka Winkler-Skalna [2] and Lech Czarnecki [3]

[1] Group of Testing Laboratories, Instytut Techniki Budowlanej, Filtrowa 1, 00-611 Warsaw, Poland
[2] Thermal Physics, Acoustics and Environment Department, Instytut Techniki Budowlanej, Filtrowa 1,
    00-611 Warsaw, Poland; a.winkler-skalna@itb.pl
[3] Scientific Secretary, Instytut Techniki Budowlanej, Filtrowa 1, 00-611 Warsaw, Poland; l.czarnecki@itb.pl
*   Correspondence: e.szewczak@itb.pl

Received: 20 December 2019; Accepted: 27 January 2020; Published: 29 January 2020

Abstract: The laboratory testing of the construction materials and elements is a subset of activities inherent in sustainable building materials engineering. Two questions arise regarding test methods used: the relation between test results and material behavior in actual conditions on the one hand, and the variability of results related to uncertainty on the other. The paper presents the analysis of the results and uncertainties of the simple two independent test examples (bond strength and tensile strength) in order to demonstrate discrepancies related to the ambiguous methods of estimating uncertainty and the consequences of using test methods when method suitability for conformity assessment has not been properly verified. Examples are the basis for opening discussion on the test methods development direction, which makes possible to consider them as 'sustainable'. The authors address the negative impact of the lack of a complete test models taking into account proceeding with an uncertainty on erroneous assessment risks. Adverse effects can be minimized by creating test methods appropriate for the test's purpose (e.g., initial or routine tests) and handling with uncontrolled uncertainty components. Sustainable test methods should ensure a balance between widely defined tests and evaluation costs and the material's or building's safety, reliability, and stability.

Keywords: testing of building materials; test uncertainty; validation of test methods; sustainable test methods

1. Introduction

The construction sector operates a huge number of test methods. Some of them are used to reveal the truth (in a technical sense) about a material, element, or building, while others are simple conformity assessments of products placed on the market. Single laboratory test may give the impression of an irrelevant issue against the complexity of civil engineering. This makes it hard to consider test result uncertainty in this context. However, taking into account all activities related to construction from the moment of its design through all its life cycle, the tests are present at all stages (Figure 1).

The matrix of building science, shown by Czarnecki and van Gembert [1], its multi-faceted nature and complexity in defining building properties implies a colossal number of research and laboratory test methods used in this field including tests of physical and mechanical properties of building materials, chemical, biological, electromagnetic, electric and electronic tests, acoustic and fire tests of buildings and building elements, tests concerning environmental engineering, tests of radiation of building materials, structural tests of materials and others. On the other hand, there is a centuries-old tradition of following the rules of thumb in this area. For many centuries, it gave quite good results. So what is the point of saying about development of precise and modern laboratory test methods? The reason for the need to improve test methods is the accelerated development of materials and technologies, which forces the development of new ways of assessing their suitability. Currently, construction engineering ceased to operate only in the meters and reached the nanoscale [2].

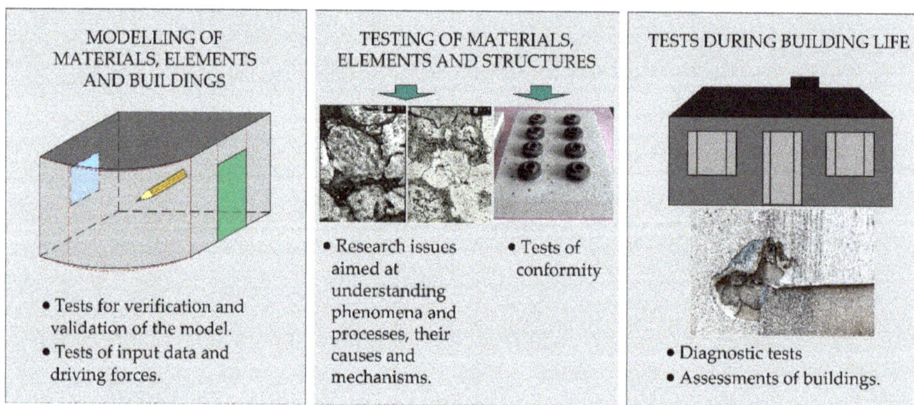

Figure 1. Presence of tests at all stages of the creation and life of buildings.

The type of test depends on its purpose. Some of them are individual and specially designed, while others are standardized. If we consider the tests methods according to the scheme in Figure 1, most of them, and above all methods used for validation of models or for the purpose of solving scientific issues are rather individual and developed for a specific need. Examples of such individual tests are on macro-scale the experimental and numerical methods used to study the mechanical behavior of objects at a natural scale (e.g., [3,4]), fire tests of walls or doors [5]. Micro-scale tests are among others scientific methods based on advanced material research, such as X-ray diffraction (XRD), energy dispersive spectroscopy (EDS), scanning electron microscopy (SEM) (e.g., [6–8]), and other advanced methods used to understand the mechanisms of phenomena occurring from the macro to the nano-scale. There is also a wide range of individual test and assessment methods used during the service life. Most of them are diagnostic methods such as chemical, ultrasonic, electrical, etc. [9,10] used for building safety inspections. Methods and systems of internal environment testing and sustainability of buildings assessment are rapidly developing (e.g., [11–13]). All test methods in all groups shown in Figure 1 are associated with uncertainty and potential dispersion of the results (the dispersion of some results, especially those from individual methods, is not observed because they are performed only once). The issue of uncertainty and precision of the method is, however, particularly important for laboratory tests used to assess the conformity of construction products. These methods should be standardized and ensure an adequate level of compatibility of test results. In this paper, the laboratory tests are generally divided into the research scientific oriented and routine tests of conformity.

The diagram of this article is shown in Figure 2. The focus is on laboratory tests for conformity assessment because, based on them, decisions are made regarding the possibility of using the material or element in the building. Whenever there is a decision involved, regardless of human activity, there is an uncertainty aspect [14]. Test methods are inseparable from result uncertainty. Realizing the role of uncertainty of the building materials tests results can be easy if we consider that the next, higher assessment level of a complete structure takes into account the sum of partial assessment results, which causes further uncertainty amplification. Thus, the bottom-up procedure is a cluster of sub-models for subsequent assessment models. Result uncertainty may multiply by subsequent uncertainty accumulation at higher assessment levels.

Wherever a decision is related to compliance or non-compliance with specific criteria, this uncertainty involves the risk of committing either a type I error, i.e., rejection of an element which meets the requirements; or a type II error, namely accepting an element which does not meet the requirements. There is also the risk, however, of committing a broadly defined type III error [15] related to asking a wrong question (using the wrong null hypothesis).

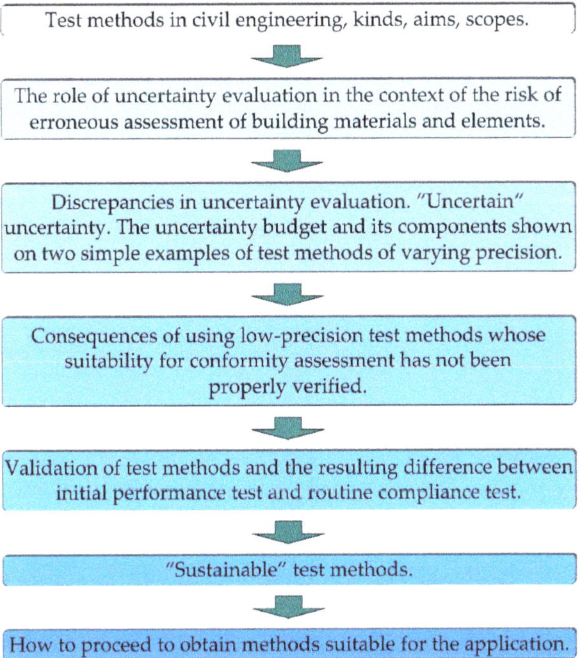

Figure 2. Graphic diagram of the article.

Consider the type I risk error in relation to construction materials. This error may incur costs related to the need to create materials or element of a higher performance level than required. This is done in order to avoid erroneous rejection as the result of test uncertainty. The second error type risk is associated with a building's use-phase and ranges from reduced durability and premature loss of properties through construction disaster extreme cases. In view of these analyses, the type III error risk may mean adoption of incorrect assessment criteria, inadequate testing methods, or incorrect result uncertainty estimation, which may result in the case of wrong material assessment. Therefore, each risk is associated with both: cost and environmental impact (waste, materials with higher content of specific ingredients etc.). In an 'era of sustainability' the scientific and research community should strive to minimize such risks at all levels of decisions [16].

Sustainable development in building material engineering is multifaceted and multidimensional [17]. "Sustainable" test method development seems to be part of the progress—such methods help to determine material suitability for specific applications in a credible way, i.e., reducing the risk of threat to life and human health yet avoiding the risk of unnecessary cost generation, excessive use of raw materials, and unnecessary industrial waste generation.

The way uncertainty is taken into account in determining compliance may vary, as outlined in the document of the Joint Committee for Guides in Metrology (JCGM) [18]. When the safe acceptance principle is applied, a test result together with uncertainty must be within the specified limits outlined by permitted tolerance in order to declare a material compliant with the requirements. This material evaluation and material suitability for a specific application in construction involves a typical dilemma: increased cost or increased evaluation safety. This has been discussed in relation to measurements by, e.g., Pendrill [19]. The cost increase encompasses both production costs, including material composition changes in order to be on the 'safe side', as well as costs of precise tests. Evaluation error risk reduction can be achieved through measuring device accuracy improvement and increasing the number of samples.

Experience of laboratory test results indicates that, apart from structural materials for the construction industry, uniform and correct rules for uncertainty estimation for other construction material testing are not available. The uncertainty of measurement paradigm evolution has resulted in divergence in uncertainty treatment and estimation ([20–24]). Today's scientific milieus are inclined to argue that measurement uncertainty is the researcher's statement on the level of knowledge on a measured quantity. What is worth noting is that uncertainty related issues only pertain to measurements as such. Uncertainty, however, is ignored in scientific discussions in relation to test results. There are few guidance documents [25], which recommend treating tests and measurements alike, but this seems unrealistic, especially in the context of testing building materials. The difference between measurement and test, lies in the fact that in the case of measurements there is a reference value (RV), such as a standard, which is the yardstick to verify trueness. A measurement results should be independent of the measurement method because measurand is physical quantity (like temperature or mass). Uncertainties associated with different measurement methods may be different. A test method is usually specific procedure used to determine the characteristic and there is no reference value for tests. A test quantity definition [26], i.e., a test process model, may be based on several measurements and assessments. A test may offer quantitative results or qualitative results. Quantitative results may not always be a function of the measured values. No uncertainty is attributed to qualitative results at the current knowledge stage. Knowledge obtained from the repeatability and reproducibility of experiments is recommended when estimating the uncertainty of test results. Such experiments may be very expensive and labor-intensive and, thus, should not be the responsibility of a laboratory alone, which has utilized a specific test method. Experiments should be the responsibility of organizations, which develop test methods—these bodies are required to validate methods for intended use. The EA-4/16 guidelines indicate that 'laboratories cannot in general be expected to initiate scientific research to assess the uncertainties associated with their measurements and tests' [25].

The ambiguities regarding the basic uncertainty estimation level of test results may result in numerous discrepancies between laboratories. These include model equation discrepancies, considered interactions, estimation of uncertainty components, which stem from result dispersion based on historical data or current results, comparative test results inclusion and different coverage factors. Szewczak and Piekarczuk [27] showed discrepancies in uncertainty estimation and their consequences for final material assessment.

Laboratories differ in the level of their research equipment and staff competencies. We will consider the ideal situation, i.e., laboratories with proven competences (e.g., accredited), having the same level of equipment and eliminating accidental and subjective errors through detailed analysis and participation in proficiency testing. There will still be discrepancies in results due to a number of factors related to the test method. At the same time, even laboratories with the highest level of competence, based on general guidelines for estimating uncertainty, can receive significantly different values of uncertainty, which then affects the assessment of the material tested.

The uncertainties indicated by laboratories may create a wrong impression on the knowledge level of the possible result variability. In some cases, uncertainty is underestimated—this may result in a material obtaining positive assessment results, when such assessment should, in fact, be negative. An overestimated uncertainty used in assessment may, in turn, result in cost increase, which results from the need to improve material properties. It should be noted, however, that both underestimated and overestimated uncertainty values are based on blurred, non-unified prerequisites. Therefore, general principles for considering uncertainty in conformity assessment [18] become irrational if the uncertainty is burdened with even greater uncertainty resulting from the non-uniformity of methods for uncertainty estimation.

The aim of the paper is to show that, when developing test methods, it is not always appropriate to combine the purpose of revealing the truth with the purpose of conformity assessment. Test methods that simulate real conditions may have too many non-controllable parameters that increase the variability of results. This can lead to irrational decisions.

This paper uses simple two independent and contrast test examples, such as bond strength and tensile strength, in order to demonstrate the consequences of using test methods when method suitability for conformity assessment has not been properly verified (i.e., a method has not been properly validated). Uncertainty concept ambiguity in relation to such methods undermines the assessment's rationality. Although in this article the authors considered only examples of mechanical tests, they believe that the conclusions resulting from simple examples are more universal and form the basis for opening a discussion on the test methods development direction, which makes it possible for the methods to be considered 'sustainable'.

2. Methods

2.1. Bond Strength Tests of Adhesives

The tests were carried out on 15 different polymer-cement adhesives, four samples for each adhesive. The adhesives compositions are cement mortars modified with various polymers [28]. In all cases, same-property foamed polystyrene was used as the substrate. The tests were carried out by three different research teams, which used various equipment (reproducibility conditions). They used the pull-off method according to ÖNORM B 6100:1998 10 01 [29]. A layer of polymer-cement adhesive was applied onto the substrate. Once cured in standardized conditions, 50 mm diameter round steel stamps were glued on. A force perpendicular to the surface was applied to the stamps until they pulled off.

Additional tests were also carried out in accordance with the ETAG 004 method [30]—square-shaped stamps of 50 mm were used in this case. The stamp shape impact on the test results is discussed in Section 4.

This can be a basic equation which describes the test model:

$$\sigma = \frac{4F}{\pi D^2} \tag{1}$$

where:

σ—strength result, MPa

F—breaking force, N

D—diameter of the stamp in mm.

It should be noted that the pull-off test is commonly used in different variants, according to different standards and procedures. Configuration of tests used in this article is presented in Figure 3. In addition to adhesives, a pull-off test is used for a wide spectrum of construction materials, including all types of coatings, varnishes, materials for the protection and repair of concrete structures, etc.

Figure 3. Pull-off test configuration. (**a**) Test scheme; (**b**) photographic documentation of an example test.

2.2. Tensile Strength Tests of Plastics

The object of the tensile strength tests were two plastics: ABS-GF and PC. The plastics tensile strength tests example uses the results obtained in laboratory proficiency tests carried out in 69 laboratories. This exercise was organized by the Deutsches Referenzbüro für Lebensmittel-Ringversuche und Referenzmaterialien GmbH (Kempten, Germany). The tests were carried out in accordance with the method described in ISO 527-1/-2 [31] standard. Reproducibility values were published in the organizer's report [32].

This can be a basic equation, which describes the test model:

$$\sigma = \frac{F}{ab} \qquad (2)$$

where:
    σ—strength result, MPa
    F—force, N
    a—sample thickness, in mm
    b—sample width, in mm.

3. Estimation of Uncertainty

3.1. Bond Strength Tests and Their Influence on the Results

The general adhesion testing principle is based on stress determination, which results in the tested material detachment from the substrate—the substrate is a laboratory equivalent of the substrate on which the material is actually used. Although the test is intended to simulate the actual conditions of use, it differs in a number of aspects, including:

- The test substrate is usually standardised, thus adhesion to some other substrate in actual conditions may significantly vary.
- The sample is subjected to perpendicular force during the test, whilst the system of forces, e.g., on a façade, may differ in real conditions.
- The material is pulled off with a stamp glued to the sample with epoxy glue, which may affect the sample properties.

Therefore, it is difficult to translate the test result directly into the material's behavior when installed in the actual building; nevertheless, a test is a uniformed material assessment method and should reflect the differences between particular materials.

The most important factor affecting adhesion results is the adhesive composition. However, in conformity laboratory tests (like in our case) the laboratory only knows the use of the material and the resulting assessment criterion. The criterion must be met regardless of the composition.

The factors which can affect the test results (their uncertainty) can be divided into three groups:

1. Measurements which directly affect the result. According to Equation (1), these are the force measurement at pull-off and the stamp diameter attached to the sample. These measurement uncertainties can be determined in two ways—either in relation to the calibration certificates or in relation to the confirmation of compliance with conditions specified in the standards as uncertainty. For example, if the force ought to be measured with a Class 1 device, it means that the total uncertainty of measurement should not exceed 1% of the measured value.
2. Measurements with uncertainties which do not affect the result according to a known relationship; however, they are measured in order to maintain compliance with the tolerance limits given in the standard: ambient temperature, air humidity measurement, force application speed measurement.
3. Interactions whose level is not measured: force application direction (some non-perpendicular components of stress) which should be controlled by the construction of the measurement device, but there are always some imperfections, the substrate and stamp surface difference, the epoxy

glue chemical and physical influence on the adhesive, repeatability, and reproducibility of sample preparation (the sample preparation process for testing is multi-stage and involves a number of interactions, such as layer thickness, clamping force, conditioning variability, and others that may affect the test result), the material and sample heterogeneity.

It is noteworthy that in the case of all tensile strength tests there are usually some non-perpendicular components of stress. In some cases, first of all, related to scientific oriented research, emerging non-perpendicular components are the subject of analysis aimed at understanding material behavior and enriching knowledge about the material, such as in rock-engineering performance where visible discontinuities created by nature and adjacent areas (on a mm-scale) differing significantly in mechanical properties in tested material cause variability of tangent modules [33]. In the case of bond strength conformity tests, however, they will constitute an undesirable factor resulting from the imperfections of the measurement system and preparation of the sample. Ultimately, it reveals itself in a dispersion of results along with other uncontrolled factors. In this example, one can see a significant difference between compliance and scientific tests. The same influencing factor is, in one case, a source of information, and in the other, an obstacle to the precise assessment.

The bond strength test, like most tests used to assess construction materials, is of a destructive nature; thus, it is not possible to repeat the action sequence on the same sample. This situation makes it impossible to isolate the impact of individual components from groups 2 and 3 on the result. Thus, from a practical point of view, a laboratory which performs a specific test obtains a limited test result pool (3 to 5 results) and the knowledge of the lab encompasses the device measurement characteristics and the final dispersion of the results.

If the tolerance limits are given in the description of the test method, it is assumed by default that observance of these limits ensures that group 2 interactions have a negligible impact on the test results. The variations related to material heterogeneity, sample preparation, and the test itself are manifested in the final dispersion of the results. Given the fact that a laboratory normally tests 3–5 samples, the listed components must be considered in conjunction.

Thus, in this situation, we can determine the components determined by measurement of force and diameter. These components can be included in the model equation. The rest of the components are contained in the dispersion of results; however, their impact remains unknown. It would be necessary to perform numerous inter-laboratory experiments in order to separate components related to the test from the components which result from the tested material or samples heterogeneity.

Such experiments are justified during the method validation period but few specifications indicate the method precision (standard deviation of repeatability and reproducibility), which could be treated as the uncertainty component.

3.1.1. Precision of Bond Strength Tests

Documents [29,30], which describe and recommend the use of bond strength tests, include no data on test precision. Numerous outcomes based on both repeatability and reproducibility tests indicate that a very large dispersion of results is characteristic for bond strength tests.

Table 1 shows the bond strength test results (average values of four lab tests for each adhesive in each laboratory) in the three research teams. The tests used 15 different polymer-cement adhesive samples of 0.02 to 0.11 MPa adhesion range. The table also includes the result variability for individual adhesives in individual laboratories, as well as variations resulting from repeatability and reproducibility calculated in accordance with the ISO 5725 [34] based on all laboratories' results.

All variations in Table 1 are expressed as a v coefficient of variation for better presentation:

$$v = \frac{s}{\bar{\sigma}}\ 100\% \tag{3}$$

where:
    s—standard deviation of repeatability or reproducibility
    $\bar{\sigma}$—mean value of bond strength in the result set with standard deviation calculated.

Table 1. Bond strength values, standard deviations of bond strength results for 15 different adhesives (marked as a–p in the first row) and three laboratories (marked as 1, 2, 3 in the first column). The individual laboratory variations, repeatability and reproducibility variations are given as the coefficient of variation v.

Laboratory	a	b	c	d	e	f	g	h	i	k	l	m	n	o	p
						Bond Strength Results, MPa									
1	0.022	0.026	0.045	0.043	0.042	0.040	0.048	0.047	0.046	0.060	0.044	0.066	0.062	0.084	0.108
2	0.022	0.054	0.046	0.045	0.044	0.057	0.038	0.047	0.059	0.045	0.071	0.093	0.085	0.083	0.099
3	0.034	0.040	0.051	0.056	0.061	0.054	0.066	0.072	0.070	0.079	0.080	0.089	0.103	0.112	0.113
					Average Bond Strength for Three Laboratories for Particular Adhesives										
	0.026	0.040	0.047	0.048	0.049	0.050	0.051	0.056	0.058	0.061	0.065	0.083	0.084	0.093	0.106
		The v Coefficient of Variation of Bond Strength Results for 15 Different Adhesives and Three Laboratories in %													
1	48.7	24.5	15.6	17.8	16.6	28.4	8.9	17.1	33.1	16.9	9.1	15.4	33.4 [1]	18.0	11.5
2	64.6	14.6	53.5	12.7	55.6 [2]	6.8	9.1	12.2	4.5	54.3	12.4	6.8	4.4	8.3	10.8
3	18.5	23.2	35.6	19.4	11.0	10.9	8.6	8.3	16.1	11.9	5.5	9.9	5.6	16.2	8.9
			Repeatability Expressed as Coefficient of Variation v for Each Adhesive in %												
	41.9	19.8	38.4	17.4	30.9	15.3	9.0	12.1	18.9	26.5	9.5	10.4	15.1	15.3	10.4
			Reproducibility Expressed as v Coefficient of Variation of Each Adhesive in %												
	45.3	39.0	42.3	21.4	34.5	22.5	28.5	28.4	26.6	35.7	29.6	19.8	27.7	22.1	11.4

[1] Outlier variance (outlier variance significant at the 99% level of confidence in the Cochran test); [2] Questionable variance (outlier variance significant at the 95% level of confidence in the Cochran test).

The data was subjected to the Grubbs test for outliers within the result set for a given adhesive. The test showed no outliers. The individual standard deviations for the three laboratories and each adhesive were subjected to the Cochran test. A questionable standard deviation value of 0.024 MPa (v = 55.6%) was obtained for the 'e' sample and Laboratory 2. One standard deviation outlier of 0.021 MPa (v = 33.4%) was obtained for sample 'n' and Laboratory 1. In each case, an increased standard deviation was recorded in a different research team; thus, the questionable and outlier variance results were not rejected in the overall variance of repeatability and reproducibility calculations.

There is no statistically significant correlation between the standard deviation of repeatability and reproducibility values and the adhesion force value (respectively r = 0.13 and r = 0.38 for the critical value with the α = 0.05 coefficient r_{cr} = 0.514). There is no statistically significant correlation between intra-laboratory and inter-laboratory standard deviation either (r = 0.16) or between standard deviations in particular laboratories (r lies between 0.18–0.28).

The dispersion values are random and do not indicate that any of the laboratories involved in the experiment has demonstrated standard deviations, which were too high or too low. Minimum and maximum v values for repeatability are 9% and 42%, and for reproducibility 11% and 45% respectively. These values are very high and, thus, indicate that the test method is of low precision. Given that the adhesive assessment criterion is: σ > 0.08 MPa, the result variability may suggest that this adhesive assessment test method involves a high risk.

3.1.2. Estimation of Bond Strength Results Uncertainty

The uncertainty estimation was carried out with the use of methods recommended in JCGM documents [20,21] and EA [25].

Considering the interaction impact assessment on the test results, the model equation, which makes it possible to determine the individual components influence on the total uncertainty of the adhesion test results, is

$$\sigma = \frac{4F}{\pi D^2} + A_\sigma \tag{4}$$

where:

$A\sigma$—unknown value interaction impacts, which contribute to the result dispersion (includes material heterogeneity). The value $A\sigma$ = 0—the error value which could be corrected is unknown; however, $A\sigma$ uncertainty contributes to σ uncertainty. This uncertainty is revealed in a random dispersion of results expressed as the standard deviation of the σ value.

Other elements as in Equation (1).

The total value of standard uncertainty estimation σ is usually based on equation:

$$u_c^2(y) = \sum \left(\frac{\partial f}{\partial x_i}\right)^2 u^2(x_i) \tag{5}$$

hence

$$u_\sigma^2 = \left(\frac{\partial \sigma}{\partial D}\right)^2 u_D^2 + \left(\frac{\partial \sigma}{\partial F}\right)^2 u_F^2 + \left(\frac{\partial \sigma}{\partial A_\sigma}\right)^2 u_{A_\sigma}^2 \tag{6}$$

where:

u_D—standard uncertainty of D

u_F—standard uncertainty of F

u_{A_σ}—dispersion of σ results.

Expanded uncertainty is

$$U_\sigma = k u_\sigma^2 \tag{7}$$

where k is the coverage factor corresponding to 95 percent coverage interval.

Owing to the fact that there are no uniform uncertainty estimation principles and that, according to recommendations, uncertainty estimation should be based on the state of knowledge, laboratories may make different assumptions about the value of u_D, u_F, a, and k, although they use the same equations as the starting point. For example, when estimating type B uncertainty related to force values, either measurement tolerances or specific uncertainty from the calibration certificate might be taken into consideration. Although u_{A_σ} laboratories mostly rely on historical data of the repeatability test on a larger number of samples in a given laboratory, the current series standard deviation might be used in many instances. Reproducibility standard deviation obtained from inter-laboratory experiments should be used, as recommended [25]. This, however, is a rare case.

$k = 2$ is sometimes taken as the coverage factor, although this assumption is not always justified (approximate normal distribution value corresponding with 95 percent coverage interval). k obtained from Student's t-distribution is also used, based on the effective number of degrees of freedom calculated from the Welch—Satterthwaite equation. The current approach outlined in the JCGM document [21] consists in the probability density distribution (PDF) propagation rather than the uncertainty propagation—this may also produce different uncertainty results.

Table 2 presents some examples of how uncertainty components can be estimated and how can be used with the model equation presented herein. All of them have been created by the authors in accordance with the possibilities presented in the JCGM documents [20,21].

Table 2. Four example-approaches to uncertainty estimation.

Approach	Force F, N Diameter D, mm	Random Dispersion of σ Results	Coverage Factor, k
I approach		No evaluation. Approach based on the assumption that uncertainty relates to the accuracy of the direct measurements only (F and D)	$k = 2$ (sometimes inappropriate but often used).
II approach	Type B evaluation, based on the standard requirements regarding accuracy: F-force measurement accuracy 1%. Rectangular PDF [1]	Type A evaluation based on the current test results. 4 results, $v = 3$ degrees of freedom.	$k = t_{0.95}(v_{\text{eff}})$, where v_{eff}—value of t_p from t-distribution with an effective degree of freedom v_{eff} obtained from the Welch-Satterthwaite formula. p-fraction of the distribution
III approach	D-accuracy of the sample preparation 1 mm. Rectangular PDF	Type B evaluation, based on historical results (e.g., inter-laboratory tests–reproducibility). Normal PDF.	$k = 2$, based on the assumption that the sR component with normal distribution is dominant.
IV approach		Type A evaluation based on the current test results. Normal PDF	No k-factor, 95% coverage intervals obtained by the Monte Carlo method [35]

[1] Probability density function.

Figure 4a,b show the test results and coverage intervals resulting from the use of different approaches towards uncertainty estimation (I, II, and IV). The uncertainties estimated with the use of different approaches differ significantly, which seems particularly important. The detailed results are shown in Table 3.

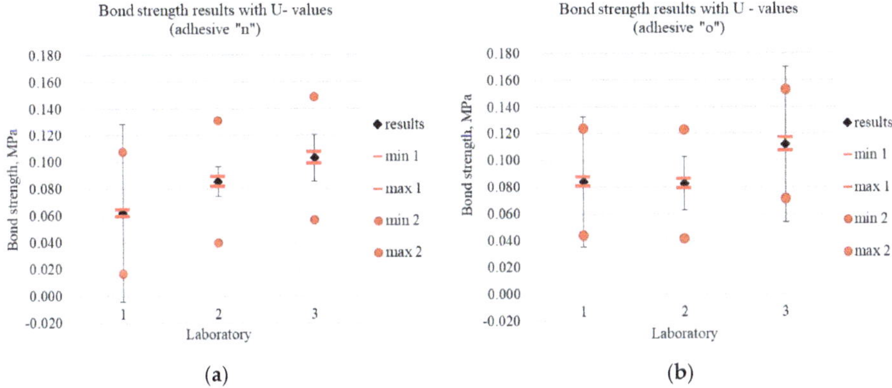

Figure 4. Bond strength results with U value for three laboratories (1–3) and two materials (**a**) adhesive 'n'; (**b**) adhesive 'o'. Black error bars are based on actual standard deviation (approach II). Points min 1 and max 1 are 95% coverage intervals obtained in approach I (only B-uncertainty of F and D values), points min 2 and max 2 are 95% coverage intervals obtained by the Monte Carlo method in approach IV, where reproducibility standard deviation obtained from inter-laboratory comparisons was used.

Table 3. Bond strength and U uncertainty values assessed with the use of I–IV approaches for three laboratories (1–3) and two materials: adhesive 'n' and adhesive 'o'.

Approach	Adhesive 'n'			Adhesive 'o'		
	Lab. 1	**Lab. 2**	**Lab. 3**	**Lab. 1**	**Lab. 2**	**Lab. 3**
Results, MPa	0.062	0.085	0.103	0.084	0.083	0.112
Uncertainties, MPa						
I approach	0.003	0.004	0.005	0.004	0.004	0.005
II approach	0.066	0.011	0.017	0.048	0.020	0.058
III approach	0.074	0.074	0.074	0.065	0.065	0.066
IV approach	0.046	0.047	0.047	0.039	0.041	0.041

The result differences in particular laboratories when testing the same adhesive are significant enough to affect the final conformity assessment against the requirements. The uncertainty value differences obtained by different laboratories using different approaches, and uncertainty values which exceed the test result value (e.g., approaches II and III for Lab 1, adhesive 'n') weaken the significance of such "uncertain" uncertainty when assessing the risk of incorrect assessment.

ISO 527 [31] based tests examine the tensile stress of plastics. Unlike bond strength testing, application conditions simulation is not the test objective; rather, the test results describe the material properties. A designer outlines the material conditions and, thus, determines material suitability for specific construction use.

3.2. Tests of Tensile Strength and Their Influence on the Results

The factors which can affect the test results can be divided into three groups, as with bond strength:

1. Measurements which directly affect the result. According to Equation (2), these are the force, thickness and width of the sample measurements.
2. Measurements with uncertainties which do not affect the result according to a known relationship; however, they are measured: ambient temperature, sample geometry, and stress increase pace.
3. Interactions whose level is not measured yet affects the result, include, but are not limited to, the jaw design which may result in excessive strain exerted on the sample or may cause the sample to slip, heterogeneous sample thickness, and material heterogeneity. The method assumes

perpendicular stress, however, the design of the equipment and inaccurate clamping of the sample may imply the formation of tangential and torsional components that may affect the result.

3.2.1. Precision of Tensile Strength Tests

Tensile strength test precision is described in the ISO 527 standard applicable for various types of plastics. For PC repeatability and reproducibility standard deviations: sr = 0.18 MPa, sR = 0.89 MPa, for ABS sr = 0.18 MPa, sR = 1.93 MPa respectively.

As part of the laboratory proficiency testing organized by the Deutsches Referenzbüro für Lebensmittel-Ringversuche und Referenzmaterialien GmbH (DRRR), the standard reproducibility deviation values were determined: for PC sr = 0.26, sR = 0.73 MPa, for ABS-GF sr = 0.51 MPa, sR = 1.35 MPa. DRRR inter-laboratory studies involved 69 laboratories, thus statistical data is very extensive.

The experiment results make it possible to conclude that, in contrast to the bond strength tests, according to ISO 527-1,2 [31] a random dispersion of results in the tensile strength test expressed as the reproducibility standard deviation is relatively small.

Variation coefficient v defined by Equation (3)—PC and ABS-GF respectively—repeatability: 0.4% and 0.9%, and reproducibility: 1.3 and 2.3%. Compared to the v value determined in bond strength tests, these values are significantly lower.

3.2.2. Estimation of Tensile Strength Results Uncertainty

The model equation which makes it possible to determine the individual components influence on the total uncertainty of tensile strength test results can be presented in the same way as was the case with bond strength tests (Equation (4)):

$$\sigma = \frac{F}{ab} + A_\sigma \tag{8}$$

where:

A_σ—unknown value interaction impacts which contribute to the results dispersion (includes material heterogeneity). Other elements as in Equation (2).

Standard uncertainty of σ is:

$$u_\sigma^2 = \left(\frac{\partial \sigma}{\partial a}\right)^2 u_a^2 + \left(\frac{\partial \sigma}{\partial b}\right)^2 u_b^2 + \left(\frac{\partial \sigma}{\partial F}\right)^2 u_F^2 + \left(\frac{\partial \sigma}{\partial A_\sigma}\right)^2 u_{A_\sigma}^2 \tag{9}$$

where:

$u_{a,b,F}$—standard uncertainty of a, b, F
u_{A_σ}—dispersion of σ results.

Uncertainties were estimated based on ISO 527 standard assumptions, i.e., force measurement with the use of a class 1 device, requirements for the sample shape: a = 10.0 ± 0.2 mm, b = 4.0 ± 0.2 mm.

Figure 5a,b show the test results and coverage intervals which result from the use of different approaches towards uncertainty estimation (I, II, and IV). The different approaches to uncertainty estimation differ slightly, especially when compared to the differences in the uncertainty values in the bond strength test. Therefore, the estimated uncertainties appear to provide the basis to claim that the risk of incorrect assessment can be determined in a credible manner. The data referring to the method precision given in the standard may be the basis for a uniform uncertainty estimation.

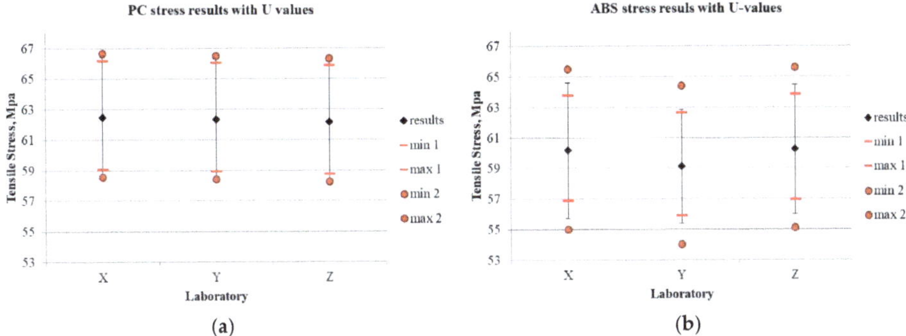

Figure 5. Tensile strength results with U value for three laboratories (X, Y, Z) and two materials (**a**) PC, (**b**) ABS. Black error bars are based on actual standard deviation (approach II). Points min 1 and max 1 are 95% coverage intervals obtained in approach I (only B-uncertainty of F and D values), points min 2 and max 2 are 95% coverage intervals obtained by the Monte Carlo method in approach IV, where reproducibility standard deviation is obtained from ISO 527-2 standard.

4. Discussion

4.1. Comparison of Methods in the Context of Uncertainty

The comparison of similar model-based test methods (Equations (1) and (2)) indicates a large disproportion in reliability of the values' uncertainty determination.

What is important is that the majority of uncertainty components in tests in accordance with ISO 527 [31] can be controlled, as opposed to the bond strength test. This can be observed by comparing Figures 4 and 5. Points min1–max1 indicate the uncertainty extent, which results from the accuracy of the measuring devices used (measurement of sample strength and geometry). This is the measurement uncertainty and is epistemic in its very nature (according to the Walker matrix [14]). The measurement uncertainties described in the test method requirements outlined by the author of the method are normally used (as is the case with this paper). This is the easiest way and provides the most consistent uncertainty data. In some cases, this uncertainty can be overestimated, as the actual measuring device used in a given laboratory may be more accurate than the one referred to in the standard. Thus, more accurate devices can also be used in order to reduce uncertainty.

The component $A\sigma$ presented in this paper is the uncertainty of aleatory nature associated with external and internal interactions (the test model inherent), which remain outside the scope of control. If introduced, this component determines the limits of the min 2 and max 2 interval in Figures 4 and 5. The impact of individual components on uncertainty is shown in Figure 6. While the uncontrolled component ($s^* = \frac{\partial \sigma}{\partial A\sigma} u_{A\sigma}$) in the test carried out according to ISO 527 [31] is of the same order as the other components, in the case of the bond strength test, the uncontrolled component is many times higher, although there are also some results with a randomly smaller dispersion.

The presence of large-scale uncontrolled components of variable values in bond strength tests may undermine confidence in both the test results and the reported uncertainty. The results' confidence is expressed and demonstrated by the error risk assessment. Type I errors are directly related to the production costs increase. Type II errors may affect the material users' safety. In the above case, there may be a high risk of type III errors related to the test model, which sets the assessment criteria incorrectly.

Compared to the bond strength test, the tensile stress test method according to ISO 527 demonstrates much better precision (repeatability and reproducibility). Uncertainty components that result from uncontrolled interactions contribute less towards total uncertainty. Thus, it can be concluded that the tested value definition in the tensile strength minimizes uncontrolled interactions, in contrast to this definition of bond strength. This is related to the fact that the bond strength test is more complex, and involves more operations and more factors affecting variability.

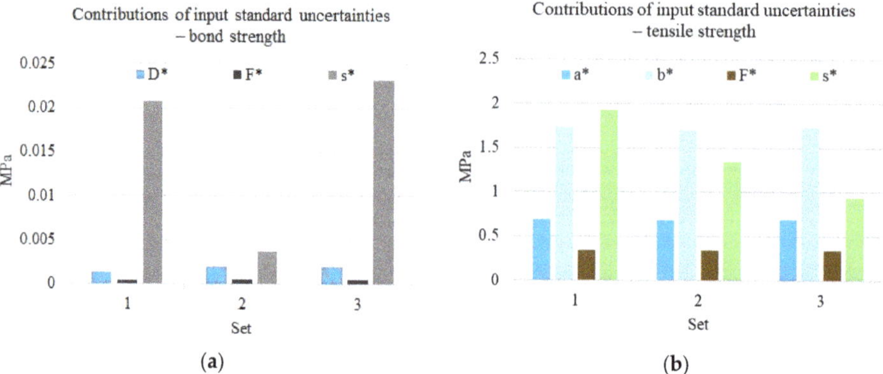

Figure 6. Contributions of input standard uncertainties: (**a**) D, F and A_σ contributions to bond strength combined uncertainty. Sets of bars 1 and 2 have been obtained for two laboratories with different actual standard deviation (II approach), set 3 was obtained for laboratory 2 which used reproducibility standard deviation. D^* is the $\frac{\partial \sigma}{\partial D} u_D$, F^* is $\frac{\partial \sigma}{\partial F} u_{FB}$, and A^* is $\frac{\partial \sigma}{\partial A_\sigma} u_{A_\sigma}$. (**b**) a, b, F, and A_σ contributions to ABS tensile strength combined uncertainty. Set 1 is based on reproducibility standard deviation from ISO 527-2; set 2 is based on reproducibility standard deviation obtained in DRRR inter-laboratory experiment; set 3 is the actual standard deviation of an exemplary laboratory. a^* is the $\frac{\partial \sigma}{\partial a} u_a$, b^* is the $\frac{\partial \sigma}{\partial b} u_b$, F^* is $\frac{\partial \sigma}{\partial F} u_{FB}$, and s^* is $\frac{\partial \sigma}{\partial A_\sigma} u_{A_\sigma}$.

4.2. Effects Related to Material Assessment

According to current JCGM guidelines [18], there are three main ways to account for uncertainty in conformity assessment test results with the tolerance limits delineated in the material requirement:

(a) Guarded acceptance (as shown in Figure 7, a lower acceptance limit $A_L = T_L + U$, where T_L—lower tolerance limit and U—expanded uncertainty)
(b) Simple acceptance (a lower acceptance limit $A_L = T_L$)
(c) Guarded rejection (lower acceptance limit $A_L = T_L - U$)

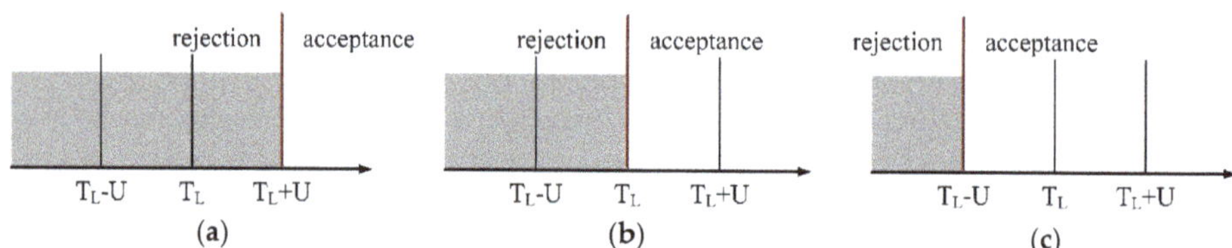

Figure 7. Methods of conformity assessment with the use of uncertainty values, based on the example of the lower tolerance limit T_L. (**a**)—guarded acceptance; (**b**)—simple acceptance; (**c**)—guarded rejection. U—expanded uncertainty value.

In the case of applying the principle of guarded rejection, sometimes used in factory production quality assessment, the lower acceptance limit might be $(0.08 - U)$ (MPa). For uncertainty estimated using approach II by laboratories 1, 2, and 3 it is: 0.014, 0.069, and 0.063 MPa respectively (adhesive "n"). Note that the first value is more than five times lower than the tolerance limit (0.08 MPa).

If guarded acceptance is used, a sample tested by any of 1–3 laboratory which estimates uncertainty according to Approach III, would have to find an adhesion of 0.15 MPa and, thus, almost double the required value (0.08 MPa). This would involve a significant change in the adhesive composition—in polymer cement adhesives this would most likely result in a significant increase of polymer amount in relation to cement. Table 4 presents decisions about rejection or acceptance of adhesive used as insulation material bonding to a wall, depending on the laboratory approach used for uncertainty evaluation and the way of using uncertainty in the assessment.

Table 4. Decisions about rejection or acceptance of the adhesive used for bonding of insulation material to a wall, depending on the laboratory approach used for uncertainty evaluation and the way of using uncertainty in assessment (based on the example of the "n" adhesive).

Way of Assessment	Approach for Uncertainty Evaluation	Laboratory 1	Laboratory 2	Laboratory 3
Guarded acceptance	I	rejection	acceptance	acceptance
	II	rejection	rejection	acceptance
	III	rejection	rejection	rejection
	IV	rejection	rejection	rejection
Simple acceptance	I, II, III, IV	rejection	acceptance	acceptance
Guarded rejection	I	rejection	acceptance	acceptance
	II	acceptance	acceptance	acceptance
	III	acceptance	acceptance	acceptance
	IV	acceptance	acceptance	acceptance

4.3. Difference between Initial Performance Tests and Routine Compliance Tests on the Background of Validation of Test Methods

Validation before method implementation is normally used in order to confirm the test method suitability for a particular application. The first validation stage is setting the requirements a test method should meet. If the method is supposed to provide the material assessment against established criteria, e.g., tolerance limits (T_L—lower limit, T_U—upper limit, tolerance range width $T = T_U - T_L$) assigned to a given material class, this parameter is described as

$$\alpha = \frac{U}{T} \tag{10}$$

the measurement accuracy factor. The factor should be of a certain value in order to make it possible to evaluate whether the criteria based on the results of the measurement are met. Metrology, in a customary manner, assumes that this factor is not greater than 0.1 for greater responsibility measurements and 0.2 for measurements of smaller responsibility. In the bond strength test, where $T = 0.08$ MPa, the uncertainty should be around 0.008 MPa (or 0.016 MPa). Uncertainty at the level of $0.1T$ would only be achievable when approach I is used, i.e., the budget includes component uncertainties arising only from requirements regarding the accuracy of measuring devices. The uncertainty of $0.2T$ is sometimes achievable when the dispersion of results budget resulting from repeatability is included. Given the large differences in the standard deviations obtained in the tests carried out by different laboratories for different materials; however, the uncertainty estimated seems unreliable.

The examples above clearly indicate the test method's deficiencies when used for material evaluation; these deficits are manifested by result differences. From a mathematical point of view, a rational assessment becomes impossible. In addition, assessment is unreliable because there are no unified rules for uncertainty estimation and no arbitrary assessment principles adopted for all parties (Table 4).

As already mentioned, the reasons for the test method deficiencies are mainly related to the large number of interactions which impact the result. This stems from the attempts to develop a method similar to natural conditions.

The substrate to which the tested adhesive is applied in order to assess adhesion is one of these interactions. The following cases can be considered:

1. The test is to be carried out with the use of the substrate which the adhesive will be applied to in regular conditions of use. This solution may prove very costly for adhesives which can be used for a variety of materials.
2. The test is to be carried out with the use of a reference substrate (the same for all laboratories which use a particular method, irrespective of subsequent use of the material). This increases the method precision; however, it makes the method-based test results diverge from the actual conditions of use, which in turn increases the type III risk associated with the use of an incorrect assessment criterion.

In the test methods cited in this paper, the second solution is used. This means that the result variability related to the impact of the substrate is limited. The assessment of how the adhesive will behave in practical application on the actual substrate; however, remains outside the scope of this model.

If R denotes an abstract difference value between the material performance in actual conditions (Yu) and the test result (Yt), the use of the actual substrate for tests results in R reduction and, thus, increases the testing costs (many substrates for different material applications). It may also increase uncertainty. This stems from the fact that there is no specification regarding the substrate type which may cause discrepancies between laboratories' choice regarding test substrate types which approximate the substrate of intended use.

The shape of the adhesive layer on the substrate is another interaction example where result variability has been limited. For instance, ETAG 004 [30] based tests require the use of a 50 mm side square stamp in the adhesive removal test. For adhesive property tests carried out according to previous recommendations in Poland, the method described in point 2 was used—a round stamp of 50 mm diameter. Theoretically, this should not affect adhesion which, by definition, is the force related to a surface unit. The Building Research Institute's practical tests, however, demonstrated different results for the same adhesives and substrates (Table 5).

Table 5. Medium bond strengths obtained according to ÖNORM (round stamp) and ETAG 004 (square stamp).

Material	Method 1, Round Stamp, $D = 50$ mm, Medium Bond Strength, MPa	Method 2, Square Stamp $a = 50$ mm Medium Bond Strength, MPa
Adhesive 1	0.046	0.068
Adhesive 2	0.066	0.097
Adhesive 3	0.051	0.085

The surprisingly large average value differences shown in Table 5 may also be attributed to the poor repeatability of the method, as discussed in Section 3.1.1.

The test method requirements impose specific stamp dimensions and shapes. The adhesive layer at the actual construction site is applied to several places on a foamed polystyrene board. The adhesive layer shapes and surfaces are uncontrollable (or controlled within wide tolerances, as the instructions are not very precise and are related to workers' practical experience). Considering the above variable, there is no rational possibility to establish the relation between material performance in actual conditions (Y_U) and test results (Y_t). In the case of the stamp, the imposed shape may only reduce the dispersion of test results.

It is apparent that whenever the test method does not perfectly duplicate the actual conditions of use, the variability increase related to the test method itself also increases the risk, when the test results are transferred to the conditions of use. In-test interactions and actual conditions interactions do not compensate, because they are random. Therefore, the final properties variation in actual conditions of use must include the test result variability resulting from the test interactions, as well as the variability resulting from the actual conditions of use. This can be expressed symbolically with equation:

$$D^2(R) = D^2(Y_u) + D^2(Y_t) \tag{11}$$

where D^2 means variance.

Hence, the greater the variability of the test method itself, the greater the variability when the method is transposed to the actual conditions of use and, thus, the greater is the risk of erroneous assessment.

It should be noted, however, that both Y_u and $U(Y_u)$ are virtually unknowable and, thus, can be only the subject of modeling, burdened with further uncertainties of the model and its implementation. As for the test result and its uncertainty, the examples presented in this paper also demonstrate the important property of tests methods consisting in the 'unknowability' of 'true' results. The issue of unknowability in relation to measurements and measurement uncertainty are extensively discussed by Grégis [36]. Unlike measurement methods, the level of unknowability of test results is even greater—there is no 'true' value. There is no reference value (RV) either. If one is capable of imagining a 'real' length value (although this 'real' is still in the sphere of abstraction) as there are reference values to which the measurement results' trueness can be referred to, it is very difficult to imagine RV for test methods, especially for those used for construction materials. The bond strength tests seem to be one of the simpler examples where it is not completely unimaginable to have a test standard (although it would be destroyed with each test). It would be extremely difficult, however, to create a test standard of a curtain wall resistance to heavy body impact or of a fire door resistance. Measurements are included in each of the tests mentioned above (e.g., force measurement, the weight of the impact body, temperature), which may have RVs; however, these are only components whose influence on the final test result is unknown in most cases due to the very large number of other interactions.

Considering the above, the unknowability of properties in the actual conditions of use (Y_u), the test result (Y_t) and the variance of these values undermine the possibility of material assessment by means of laboratory tests. There are laboratory tests, however, which are the basis for material assessment. Material property criteria for actual use are, thus, developed. This is the domain of reality modeling. Attempts to achieve the best convergence between the model and reality are universal. This should also apply to test methods. The actual purpose of the test, however, should be considered.

As already mentioned, bringing the test model closer to actual conditions reduces the R value (Figure 8a); however, it may increase test result uncertainty with the increased number of interactions that impact on the result. The ideal situation we seem to be striving for is both small test result uncertainty and maximum approximation to actual conditions. As shown in Figure 8b, this is practically impossible, because as R and $D^2(Y_t)$ are concurrently reduced, the point is reached where the common part of fields (Y_u, $D^2(Y_u)$) and (Y_t, $D^2(Y_t)$) is also reduced.

Thus, this optimization issue can only be an objective function (Figure 8c). If the comparability of the test results of individual materials carried out by different laboratories is the objective, the minimization attempt should characterise the $D^2(Y_t)$ test method. If the method is of a cognitive nature, where reality modeling and 'revealing the truth' is the objective [37], it is necessary to minimize the R value. Following this logic, the initial material test in the design phase should use a different method from the routine test method used in the assessment and verification of constancy of performance (AVCP) [38] or any conformity assessment with the standards.

The initial material and elements test should be a model of reality and should simulate actual conditions. The individual application of such methods reduces uncertainty mainly to aspects related to modeling. Verification and validation of models is a slightly different issue than the validation of simple, repeatable test methods. The uncertainty inclusion in model verification is described in the publications of Scheiber et al. [39] and Oberkampf et al. [40].

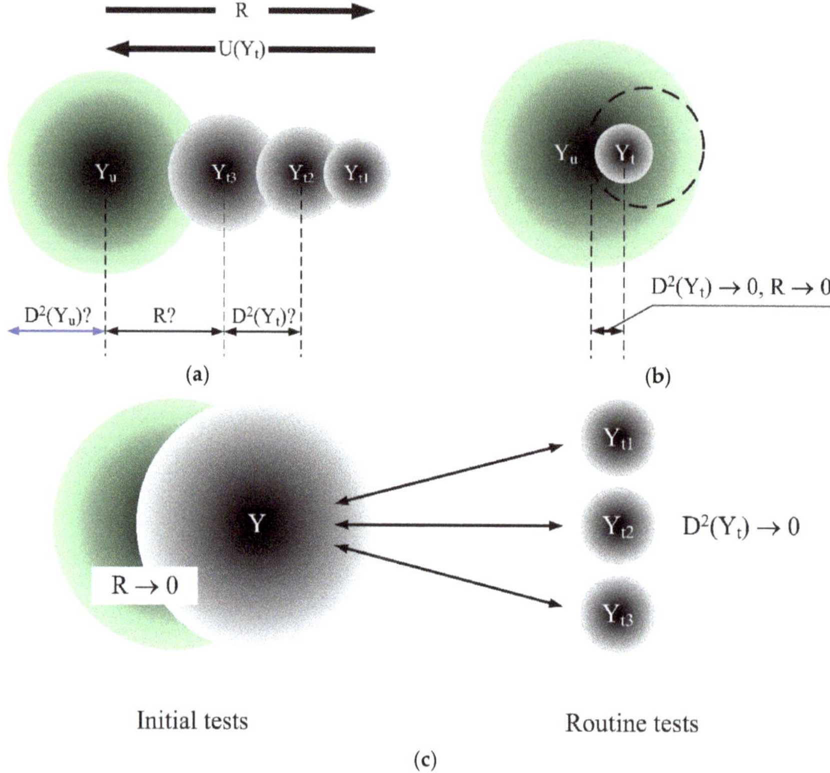

Figure 8. Illustration of the discrepancy between the material's properties in actual conditions of use (Y_u) and the test result (Y_t), $R = Y_u - Y_t$. $D^2(Y_t)$ —variability of test results. $D^2(Y_u)$—symbolic variability of actual conditions. (**a**) Test results uncertainty reduction which results in a difference increase when related to actual conditions. (**b**) Simultaneous R and $D^2(Y_t)$ minimization. (**c**) Tests modeling depending on the purpose.

A routine test should only be a model of a particular aspect of reality simple enough to deliver predictable results. In the conformity assessment tests in AVCP processes, the test development phase should include the incorrect assessment risk estimation resulting from the simplified test model, as well as the incorrect assessment risk resulting from the method precision. The rational level of required uncertainty U determination should be the responsibility of the organization which develops the method.

The test method principles should therefore be formulated as:

$$u_c^2(y) = \sum \left(\frac{\partial f}{\partial x_i}\right)^2 u^2(x_i) + s^2 < (\alpha T)^2 \tag{12}$$

where:

α—targeted accuracy factor related to standard uncertainty and tolerance limits
T—material properties tolerance range
f—function which describes the measurement model.

All components related to the test devices used and other factors that are controlled can be included.

$$u_k^2 = \sum \left(\frac{\partial f}{\partial x_i}\right)^2 u^2(x_i) \tag{13}$$

The method suitability condition can, therefore, be represented by equation

$$s^2 < (\alpha T)^2 - u_k^2 \tag{14}$$

which would limit the dispersion of results not attributed to the specific, known components of uncertainty described by Equation (13).

This validation method can be relevant for the organization which develops the method in order to ensure that the reproducibility variance in the final validation experiment does not exceed value s^2. This is also important for laboratories which would have to show a sufficiently small dispersion of results with the use of the particular method. Therefore, in specific tests the laboratories would reject results whose deviations would demonstrate a higher variance, as high probability assumption means that the variance is related to material properties rather than the test. As can be seen from the above, material assessment criteria on the basis of tests should not only relate to the values but also to their variance.

The presented approach to improving conformity test models by supplementing them with conditions regarding the required variability of results offers benefits. The ambiguities associated with uncertainty estimation are eliminated. Instead of presenting uncertainty associated with a high risk of error, as has been shown, the laboratories would have to demonstrate the test method's individual requirements fulfilment. This, in consequence, would lead to uniform uncertainty levels and uniform material assessments in different assessment bodies.

4.4. Sustainable Test Methods

Test method development, both the initial material assessment in the design phase as well as assessment routine, is based on the balance between economy (research costs) and safety (error risk). Both aspects relate to energy, raw materials consumption, waste generation, and environmental and social costs in the event of an accident.

Sustainable test methods ensure a balance between widely defined tests and evaluation costs and the material's or building's safety, reliability and stability.

Considering the general risk assessment principle:

$$risk = (probability\ of\ undesirable\ outcome) \times (effects\ of\ undesirable\ outcome) \tag{15}$$

where effects can be broadly defined. One should consider probability and effects (economic, environmental, social, and other costs) associated with unnecessary spending on the development of a well-validated assessment method: the increase of the number of samples and equipment accuracy on the one hand and risk failure as a result of incorrect assessment (product of the failure probability and failure effects) on the other. Such comparison differs among material types and applications. This explains the high expenditure and great attention paid to construction materials load capacity assessment methods and the limited attention paid to test methods of finishing materials. In the first case, the failure effects impact human safety.

If we consider the effects associated with building durability, raw materials, energy consumption, and repair costs, however, well-justified and validated testing and assessment methods should also be applied from a sustainability point of view in relation to materials which are less important.

Test method optimization to ensure their 'sustainability', as shown in Figure 8, can be obtained by separating test models designed to learn about material or element behavior in actual conditions from method precision problems. Thus, test methods should be divided into initial and routine.

4.5. Final Remarks

A model only approximates reality. There is a common goal, however, to bring a model as close to reality as possible. This is also science's objective. Given what Czarnecki et al. [37] noted, a picture of reality is dependent on the reference system used. Construction materials engineering uses reality models and may use different reference systems. The latter should be appropriate for the purpose the model has been created. A test method used to assess future material behavior in actual conditions is one of the sub-models, which constitute the final model of material properties. Real conditions, however, are associated with a full range of influences. Some of the interactions are controlled by the model, others may be partially known, yet not included in the model structure. This knowledge has not yet been researched. There is also a group of interactions outside the research scope when the model was created. The creation of test methods, which simulate the actual conditions of use, often leads to the following situation: too many aspects are outside the study model's control in real conditions. An attempt to clarify these aspects increases method precision but, at the same time, makes the method actually diverge from real conditions. This can be optimized by the development of a test method set used depending on the purpose: as an initial test when designing a material or as a routine test when assessing its legal conformity.

Considering the risks and the costs associated with the use of certain types of assessment methods based on tests, increased uncertainty of test results used for assessment increases the likelihood of making a wrong decision, and results in more adverse effects associated with that wrong decision. The associated tipping points are as follows:

- a non-compliant material or element is put on the market, which implies broadly defined costs for the producer related to the batch's withdrawal
- installation of non-compliant materials—this may result in failures, repair mitigation costs, as well as immeasurable costs of impact on health and safety.

Considering erroneous assessments, type III error risk cannot be ignored, as it has been demonstrated in the paper, because of non-validated test methods, lack of unambiguous findings regarding uncertainty estimation in relation to given tests, and lack of uniform rules of material assessment which account for uncertainty. All adverse effects of erroneous assessments also impact the environment through raw materials, energy consumption, and waste generation. Inadequate construction material durability falls short of future building users' expectations and results in higher future renovations costs.

The bond strength study example shows that inappropriate test methods, which do not ensure adequate results reliability, can increase the assessment risk area so much that the assessment becomes irrational. It should also be noted that the simple examples shown in this paper constitute only a fraction of the uncertainty problem. The study considers methods, which produce quantitative results. There is a plethora of methods, however, which produce results on a nominal scale or offer qualitative results (such as ageing methods for durability assessment, resistance tests, etc.) where uncertainty, and hence the risk of incorrect assessment, is extremely difficult to estimate.

Given the importance of sustainability in all areas of human activity, it also seems reasonable to refer to this aspect in test methods in the construction sector. As shown in the paper, a sustainable test method is a method which does not generate undesirable economic, environmental and social effects because of incorrect assessments. In order to consider a method as sustainable, it must be first of all fit for the intended purpose. Whenever necessary (initial tests of innovative products), it ought to present the actual conditions of use. In other cases (methods used routinely in assessments), the method should offer sufficiently good precision for the assessments of materials and elements results to be unambiguous. The development of methods and understanding the relationship between initial and routine assessment and actual conditions of use are scientific issues.

The applicable and reasonable basis for the uncertainty level is an important aspect of test methods, primarily routine ones. This uncertainty level should be the same for all laboratories involved in AVCP and any conformity assessment, established on assessment criteria and methods which take previously established uncertainty into account, and available for all stakeholders. Therefore, the sustainable test methods model should also include the uncertainty estimation procedure, which ensures a uniformed code of conduct for all those who use the method. Universal methods of estimating uncertainty can be interpreted in many ways, as have been shown in point 3, therefore, the only solution is to specify a specific and strict uncertainty assessment procedure addressed to a specific method. This procedure should be developed by the institution that develops the test method during its validation (which is currently very rare) and used by the method users. The level of controlled and uncontrolled interactions should be determined. Controlled interactions should be presented as requirements (e.g., related to the metrological properties of the equipment, the test environment, etc.). Uncontrolled interactions and its acceptable levels within a single test series under repeatability conditions, as well as within reproducibility should be evaluated during the validation experiment's. Laboratories, which use the test method, should be able to prove their capacity to carry out the test by showing a dispersion of results not greater than that given in the procedure. Then, in the case of results with a larger dispersion, a material can be rejected.

In the construction sector, there is a number of test methods aimed at construction product performance assessment; however, there is a limited number of methods which meet the conditions described above, although the method development costs (largely the costs of validation experiments) may be incomparably smaller compared to the adverse effects resulting from incorrect assessment. Development in this area should include better characterization (model improvement) of traditionally used test methods, as well as the development of new ones.

5. Conclusions

Among the huge number of research and test methods employed in the construction sector, two groups can be distinguished: methods used to reveal the technical truth about materials, elements or building and methods used to simple conformity assessments of products placed on the market. Combining these two purposes (which is often found in standard tests) does not always give the desired results.

Conformity test methods which do not ensure adequate results precision, would be able to increase the assessment risk area so much that the assessment becomes irrational. Use of such methods can also generate undesirable economic, environmental, and social effects because of incorrect assessments. Sustainability in the context of test methods means minimizing such effects.

Development of test methods should be preceded the determination of the aim. The initial material test should be a model of reality while a routine test should only be a model of a particular aspect of reality simple enough to deliver predictable results.

Trueness of the test results in the context of the actual conditions of the building working life does not go hand in hand with the precision of the tests, so these two aspects should be split between initial (trueness) and routine (precision) tests.

Author Contributions: Conceptualization, E.S., A.W.-S., and L.C.; Methodology, E.S.; Formal analysis, E.S.; Investigation, E.S.; Writing—Original Draft Preparation, E.S.; Writing—Review and Editing, E.S., A.W.-S., and L.C.; Visualization, E.S. and A.W.-S.; Supervision, L.C. All authors have read and agree to the published version of the manuscript.

Funding: This research was funded by Ministerstwo Nauki i Szkolnictwa Wyższego (LN-2/2019).

Acknowledgments: Special thanks for Artur Piekarczuk for valuable advices.

Conflicts of Interest: The authors declare no conflict of interest. The funders had no role in the design of the study; in the collection, analyses, or interpretation of data; in the writing of the manuscript, or in the decision to publish the results.

References

1. Czarnecki, L.; Van Gemert, D. Scientific basis and rules of thumb in civil engineering: conflict or harmony? *Bull. Polish Acad. Sci. Tech. Sci.* **2016**, *64*, 665–673. [CrossRef]
2. Czarnecki, L. Sustainable Concrete; Is Nanotechnology the Future of Concrete Polymer Composites? *Adv. Mater. Res.* **2013**, *687*, 3–11. [CrossRef]
3. Malesa, M.; Malowany, K.; Pawlicki, J.; Kujawinska, M.; Skrzypczak, P.; Piekarczuk, A.; Lusa, T.; Zagorski, A. Non-destructive testing of industrial structures with the use of multi-camera Digital Image Correlation method. *Eng. Fail. Anal.* **2016**, *69*, 122–134. [CrossRef]
4. Piekarczuk, A. Experimental and numerical studies of double corrugated steel arch panels. *Thin-Walled Struct.* **2019**, *140*, 60–73. [CrossRef]
5. Węgrzyński, W.; Turkowski, P.; Roszkowski, P. The discrepancies in energy balance in furnace testing, a bug or a feature? *Fire Mater.* **2019**. [CrossRef]
6. Rajczakowska, M.; Habermehl-Cwirzen, K.; Hedlund, H.; Cwirzen, A. The Effect of Exposure on the Autogenous Self-Healing of Ordinary Portland Cement Mortars. *Materials* **2019**, *12*, 3926. [CrossRef]
7. Ukrainczyk, N.; Muthu, M.; Vogt, O.; Koenders, E. Geopolymer, Calcium Aluminate, and Portland Cement-Based Mortars: Comparing Degradation Using Acetic Acid. *Materials* **2019**, *12*, 3115. [CrossRef]
8. Gorzelańczyk, T.; Schabowicz, K. Effect of Freeze–Thaw Cycling on the Failure of Fibre-Cement Boards, Assessed Using Acoustic Emission Method and Artificial Neural Network. *Materials* **2019**, *12*, 2181. [CrossRef]
9. Kwan, A.K.H.; Ng, P.L. Building Diagnostic Techniques and Building Diagnosis: The Way Forward. In *Engineering Asset Management-Systems, Professional Practices and Certification*; Tse, P., Mathew, J., Wong, K., Lam, R., Ko, C.N., Eds.; Springer: Berlin, Germany, 2015.
10. Schabowicz, K. Non-Destructive Testing of Materials in Civil Engineering. *Materials* **2019**, *12*, 3237. [CrossRef]
11. Piasecki, M. Practical Implementation of the Indoor Environmental Quality Model for the Assessment of Nearly Zero Energy Single-Family Building. *Buildings* **2019**, *9*, 214. [CrossRef]
12. Piasecki, M.; Kostyrko, K.; Pykacz, S. Indoor environmental quality assessment: Part 1: Choice of the indoor environmental quality sub-component models. *J. Build. Phys.* **2017**, *41*, 264–289. [CrossRef]
13. Bragança, L.; Mateus, R.; Koukkari, H. Building Sustainability Assessment. *Sustainability* **2010**, *2*, 2010–2023. [CrossRef]
14. Walker, W.E.; Harremoës, P.; Rotmans, J.; van der Sluijs, J.P.; van Asselt, M.B.A.; Janssen, P.; Krayer von Krauss, M.P. Defining Uncertainty: A Conceptual Basis for Uncertainty Management in Model-Based Decision Support. *Integr. Assess.* **2003**, *4*, 5–17. [CrossRef]
15. Trzpiot, G. Some remarks of type iii error for directional two-tailed test. *Stud. Èkon.* **2015**, *219*, 5–16.
16. Conte, E. The Era of Sustainability: Promises, Pitfalls and Prospects for Sustainable Buildings and the Built Environment. *Sustainability* **2018**, *10*, 2092. [CrossRef]
17. Czarnecki, L.; Van Gemert, D. Innovation in construction materials engineering versus sustainable development. *Bull. Polish Acad. Sci. Tech. Sci.* **2017**, *65*, 765–771. [CrossRef]
18. *JCGM 106:2012 Evaluation of Measurement Data—The Role of Measurement Uncertainty in Conformity Assessment*; JCGM: Sèvres, France, 2012.
19. Pendrill, L.R. Using measurement uncertainty in decision-making and conformity assessment. *Metrologia* **2014**, *51*, S206–S218. [CrossRef]
20. *JCGM 100:2008 Evaluation of Measurement Data—Guide to the Expression of Uncertainty in Measurement*; JCGM: Sèvres, France, 2008.
21. *JCGM 101:2008 Evaluation of Measurement data—Supplement 1 to the "Guide to the Expression of Uncertainty in Measurement"—Propagation of Distributions using a Monte Carlo Method*; JCGM: Sèvres, France, 2008.
22. Bich, W. From Errors to Probability Density Functions. Evolution of the Concept of Measurement Uncertainty. *IEEE Trans. Instrum. Meas.* **2012**, *61*, 2153–2159. [CrossRef]
23. Kacker, R.N. Measurement uncertainty and its connection with true value in the GUM versus JCGM documents. *Measurement* **2018**, *127*, 525–532. [CrossRef]
24. Grégis, F. On the meaning of measurement uncertainty. *Measurement* **2019**, *133*, 41–46. [CrossRef]
25. *EA-4/16 G:2003 EA Guidelines on the Expression of Uncertainty in Quantitative Testing*; EA: Paris, France, 2003.
26. *JCGM 200:2012 International Vocabulary of Metrology–Basic and General Concepts and Associated Terms (VIM)*; JCGM: Sèvres, France, 2012.

27. Szewczak, E.; Piekarczuk, A. Performance evaluation of the construction products as a research challenge. Small error-Big difference in assessment? *Bull. Polish Acad. Sci. Tech. Sci.* **2016**, *64*, 675–686. [CrossRef]

28. Czarnecki, L.; Łukowski, P. Polymer-cement concretes. *Cem. Lime Concr.* **2010**, *5*, 243–258.

29. *ÖNORM B 6100:1998 10 01 Außenwand-Wärmedämmverbundsysteme-Prüfverfahren*; Austrian Standards: Vienna, Austria, 1998.

30. *ETAG 004 External Thermal Insulation Composite Systems with Rendering*; EOTA: Brussels, Belgium, 2013.

31. *ISO 527-1,2:2012 Plastics—Determination of Tensile Properties—Part 1: General Principles, Part 2: Test Conditions for Moulding and Extrusion Plastics*; ISO: Geneva, Switzerland, 2012.

32. Helbig, T.; Beyer, S. DRRR-Proficiency Testing RVEP 16151 Tensile Test (strength/elongation) ISO 527-1/-2. Unpublished work, 2016.

33. Shang, J.; Hencher, S.R.; West, L.J. Tensile Strength of Geological Discontinuities Including Incipient Bedding, Rock Joints and Mineral Veins. *Rock Mech. Rock Eng.* **2016**, *49*, 4213–4225. [CrossRef]

34. *ISO 5725-2:1994 Accuracy (Trueness and Precision) of Measurement Methods and Results—Part 2: Basic Method for the Determination of Repeatability and Reproducibility of a Standard Measurement Method*; ISO: Geneva, Switzerland, 1994.

35. NIST Uncertainty Machine. Available online: https://uncertainty.nist.gov/ (accessed on 20 December 2019).

36. Grégis, F. Assessing accuracy in measurement: The dilemma of safety versus precision in the adjustment of the fundamental physical constants. *Stud. Hist. Philos. Sci. Part A* **2019**, *74*, 42–55. [CrossRef] [PubMed]

37. Czarnecki, L.; Sokołowska, J.J. Material model and revealing the truth. *Bull. Polish Acad. Sci. Tech. Sci.* **2015**, *63*, 7–14. [CrossRef]

38. *Laying Down Harmonised Conditions for the Marketing of Construction Products and Repealing Council Directive 89/106/EEC*; European Union: Brussels, Belgium, 2011.

39. Scheiber, F.; Motra, H.B.; Legatiuk, D.; Werner, F. Uncertainty-based evaluation and coupling of mathematical and physical models. *Probabilistic Eng. Mech.* **2016**, *45*, 52–60. [CrossRef]

40. Oberkampf, W.L.; Trucano, T.G.; Hirsch, C. Verification, validation, and predictive capability in computational engineering and physics. *Appl. Mech. Rev.* **2004**, *57*, 345–384. [CrossRef]

Article

Monitoring of Thermal and Moisture Processes in Various Types of External Historical Walls

Dariusz Bajno [1], Lukasz Bednarz [2,*], Zygmunt Matkowski [2] and Krzysztof Raszczuk [2]

[1]   Faculty of Civil and Environmental Engineering and Architecture, UTP University of Science and Technology, Al. Prof. S. Kaliskiego 7, 85-796 Bydgoszcz, Poland; dariusz.bajno@utp.edu.pl

[2]   Faculty of Civil Engineering, Wroclaw University of Science and Technology, Wybrzeze Wyspianskiego 27, 50-370 Wroclaw, Poland; zygmunt.matkowski@pwr.edu.pl (Z.M.); krzysztof.raszczuk@pwr.edu.pl (K.R.)

*   Correspondence: lukasz.bednarz@pwr.edu.pl; Tel.: +48-71-320-22-63

Received: 7 December 2019; Accepted: 15 January 2020; Published: 21 January 2020

Abstract: In order to create and make available the following: Design guidelines, recommendations for energy audits, data for analysis and simulation of the condition of masonry walls susceptible to biological corrosion, deterioration of comfort parameters in rooms, or deterioration of thermal resistance, the article analyzes various types of masonry wall structures occurring in and commonly used in historical buildings over the last 200 years. The summary is a list of results of particular types of masonry walls and their mutual comparison. On this basis, a procedure path has been proposed which is useful for monitoring heat loss, monitoring the moisture content of building partitions, and improving the hygrothermal comfort of rooms. The durability of such constructions has also been estimated and the impact on the condition of the buildings that have been preserved and are still in use today was assessed.

Keywords: monitoring; historical masonry wall; hygrothermal processes; internal insulation

---

1. Introduction

Problems related to heat loss and the dampening of building partitions have always accompanied construction and in the most recent literature can be found in many works on this subject, e.g., regarding internal thermal insulation [1–4], thermal insulation of historical buildings [5–8], and even innovative insulation materials, 100% plant-based, allowing for the reduction of CO^2 emissions in construction [9].

The preservation of buildings against soil moisture and condensation was and still is one of the main priorities of their builders, despite significant technological progress. As presented in [10–13], they are the most frequent causes of degradation of entire objects and low utility comfort.

The paper analyzes the technical condition of the external masonry walls of historical buildings with constructions known to be [14–16], constantly exposed to the influence of the external environment (UV radiation, insolation, precipitation, variable temperature, relative air humidity), as well as internal one (relative humidity of rooms and limited ventilation).

While monitoring hygrothermal processes, the measurement possibilities and the test results were analyzed, as presented in [17–23]. The conclusions of the work [24], in which the results of research on historical bricks were presented, were also used. Experimental data was compared with the results of standard procedures used to assess the thermo-physical behavior of historical masonry walls. The requirements set by the standard [25] were also taken into account.

When considering the subject of the additional thermal insulation of historical masonry walls from the inside aspects of this issue, presented in [26–31] have been taken into account. The problem of internal thermal insulation using a capillary active mineral plate or block systems is barely described in the literature, especially when it comes to the protection of external surfaces against moisture.

The paper discusses only part of the issues related to historical buildings, which are quite often ignored by people responsible for their technical condition, as they may seem to be of low importance. They concern dampness of building walls with moisture coming from outside and inside, resulting from physical processes taking place in them, which have direct influence on safety and durability of whole buildings. Not only the loss of load-bearing parameters of structures may lead to their destruction and even construction disasters, but also slow degradation due to processes taking place inside the walls. There are no two identical structures, therefore there can be no standard and universal methods of their maintenance and rescue.

Nowadays, quite often there is a need to carry out thermo-modernization of the buildings, but due to the historical value of the elevation it will not always be possible to do it on the external side of walls. Then, the other method, e.g., insulation from the inside, remains to be chosen. In this case, the method should be thoroughly verified by a calculation model and an "in situ" study, in terms of the possibility of negative effects that may appear even after several years of use, after its implementation. A very important element of such activities is the assessment of their impact on the durability, safety of the structure, functional safety of the facility, and the internal microclimate.

The authors of the paper have focused on selected research techniques and computational simulations in the field of moisture transfer (coming from various sources) inside building walls. These measures should lead to the development of reliable assessments of the technical condition of historical buildings, with particular attention paid to the elements and structures directly exposed, i.e., vertical walls. At the same time, they should confirm or not, the effectiveness of the designed solutions and their subsequent implementations. Simulation calculation models should cover longer, several-year periods (it is recommended that these periods should be at least 10 years) of prognosis of the impact of the implementation of the above methods on the safety and durability of these buildings and their components. In the further part of the paper, the authors have limited their attention to selected problems of appearance and migration of moisture in the walls of historical buildings.

The research methodology was proposed below, in Figure 1 describing the monitoring and intervention procedure. Research was based on several selected representative types of historical masonry walls (being a part of public buildings, located in southern Poland) for which material characteristics of particular masonry elements were presented.

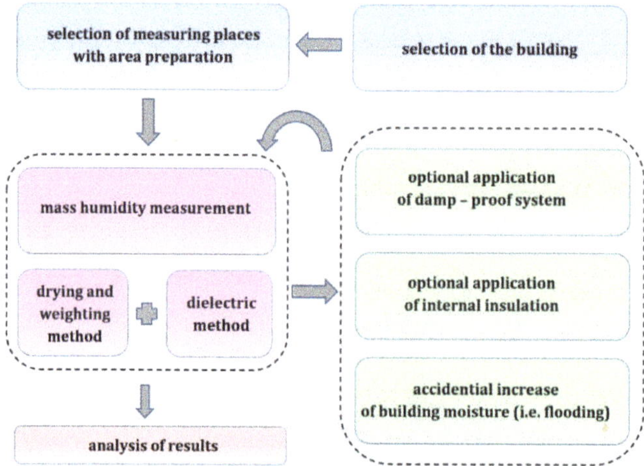

Figure 1. Monitoring and intervention procedure.

The results of moisture measurements (made in several masonry walls of this type) at the turn of the last dozen or so years were presented. In the next step, simulation calculations were made on the basis of which the exploitation evaluation of external masonry walls, made in different technologies and in a different period of time, was made. The possibility of interference in order to improve the hygrothermal comfort of rooms was also proposed.

2. Materials and Monitoring Methods

Nowadays, there are still many unique old constructions, with the unusual shaping of external partitions of buildings and structures in relation to current requirements. A present-day engineer, who has a lot of possibilities of computer-aided design, would highly appreciate them. The structures of historic buildings are distinguished by their age, technology depending on the level of technical knowledge available at the time of construction and the type of building materials used. One of the basic building materials used in constructions in the past was the material easiest to obtain in a given area. With the passage of time, as a result of the development of trade, the applied technologies and building materials were disseminated to other regions. Some of the first building materials used were clay, wood, and stone. These were followed by ceramics (dried and fired bricks, terracotta, etc.), air and hydraulic binders, cast iron, iron (steel), concrete, ferroconcrete, composites.

The main component of masonry structures was and still is stone and brick. While stone was the cheapest material obtained in the immediate vicinity of the works, a brick was already a masonry element produced artificially, through various clay processing technologies, i.e., its drying and later firing. Stone, as a masonry element, had irregular shapes (even when it was carved up), while bricks from the very beginning were given into the shape of a cuboid, or wedge-shaped pieces with appearance and size adapted to the needs of execution and the shape of the structure. Bricks were formed from clay, lime, and sand, and in times closer to ours, bricks were produced and still are produced on the basis of clay. Both stone and bricks were used to make the foundations and walls of buildings, vaults, tanks, chimneys, aqueducts, viaducts, etc., and in the nineteenth century, they also became one of the two basic components of ceramic-steel ceilings of the sectional type and reinforced flat slabs such as the Klein type.

Elements of stone, brick and mixed brick, and stone masonry walls with the following material parameters were used in the research:

- Limestone, unit density $\rho = 2.68$ g/cm^3, open porosity $p = 12.80\%$,
- Solid brick, unit density $\rho = 1.90$ g/cm^3, open porosity $p = 22.30\%$,
- Hollow brick, unit density $\rho = 1.40$ g/cm^3, open porosity $p = 22.30\%$.

The physical properties of the bricks and stones were adopted on the basis of numerous tests performed by the authors on the basis of current technical standards (taking into account the quantity and quality of samples, testing methods, and the method of calculating the value). The presented values are average values obtained from laboratory (destructive) and in situ (NDT) measurements performed in these and other historical objects, including the region (southern Poland). The obtained measurement results were each time confronted with other results available in the literature.

To monitor the mass moisture of masonry walls, the traditional drying and weighting method is used (also in the version using an automatic moisture analyzer) and indirect methods consisting in measuring the selected physical or chemical properties of the material. The division of these methods is presented in Table 1.

After the analysis of methods used for monitoring humidity processes in the walls presented in [23,32–36] and in Table 1, it was decided that the measurements of the state and scope of humidity of wall elements would be studied by one of the most accurate methods, i.e., the drying and weighting method. To verify the results, the nondestructive method based on the measurements of the dielectric properties of the material was used. The drying and weighting method was used to determine the distribution of moisture on the thickness of the wall and to scale electric meters. Samples for mass moisture determination were taken by drilling. The taken samples were tested on an automatic moisture analyzer. The mass moisture content was determined from Equation (1):

$$u_m = \frac{m_w - m_s}{m_s} \times 100\%$$

(1)

where:

u_m—mass humidity (%),

m_w—wet sample mass (g),

m_s—dry sample mass (g).

Table 1. A classification of nondestructive methods determining moisture content in building materials.

Group of Methods		Name of Method	Measured Parameter
Chemical methods	1.	indicator paper method	Change in indicator paper colour under the influence of damp material
	2.	carbide method (CM)	Pressure of acetylene (formed from reaction of carbide with water) in a hermetic container
Physical electrical methods	1.	electric resistance method	Change in material electric resistance as a result of change in dampness
	2.	dielectric method	Change in material dielectric constant as a result of change in dampness
	3.	microwave method	Attenuation of microwaves as they pass through the damp material
Physical nuclear methods	1.	neutron method	Number of neutrons slowed down by collisions with hydrogen atoms
	2.	gammascopy method	Change in γ radiation after it passes through investigated material

Considering that gauge reading–weight moisture content correlations depend on the material's other properties such as its chemical composition, porosity, porosity structure and the kind and concentration of the salts, the commonly used gauges require calibration.

The tests of mass humidity were carried out on four different masonry walls made of solid brick (Figure 2).

In building "A" the first research on the moisture content of walls was carried out in 1997. Next, horizontal damp-proof insulation was made (by chemical barrier–mineral injection). The measurements were repeated in 2004 and then after flooding the cellar interior with water (as a result of a failure of the sewage system) in 2018 (more than 20 years after the first measurement). In building "B" humidity measurements were carried out in 2007. Next, moisture horizontal insulations (by chemical barrier–mineral injection) were made. The measurements were repeated in 2018. In building "C" the measurements were made in 2009. The measurements were repeated in 2019. Anti-humidity and waterproof insulation were not performed in this building. In building "D" the measurements were made in 2016. Next, horizontal damp-proof insulation (by chemical barrier–mineral injection) and internal insulation with a thickness of 5 cm were performed. The measurements were repeated in 2019.

Figure 2. Testing areas and measurement techniques (nondestructive and destructive).

The results of mass moisture tests for these four load-bearing masonry walls are shown in Figure 3. In all cases in which the anti-humidity insulation was made, a significant (about 10%) decrease in mass moisture of the walls was observed. In the absence of such insulation, the mass humidity increased slightly during the 10 years of measurements.

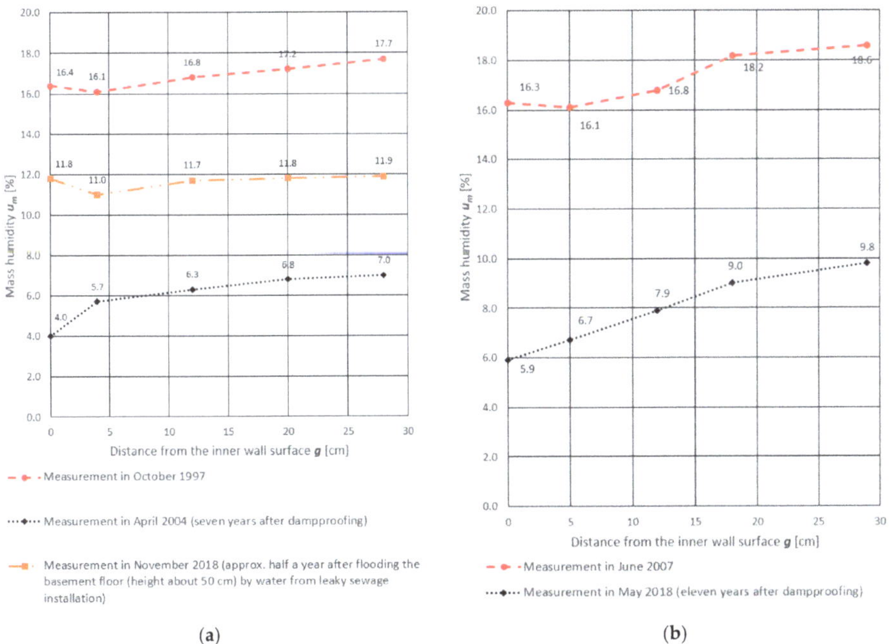

(**a**) (**b**)

Figure 3. *Cont.*

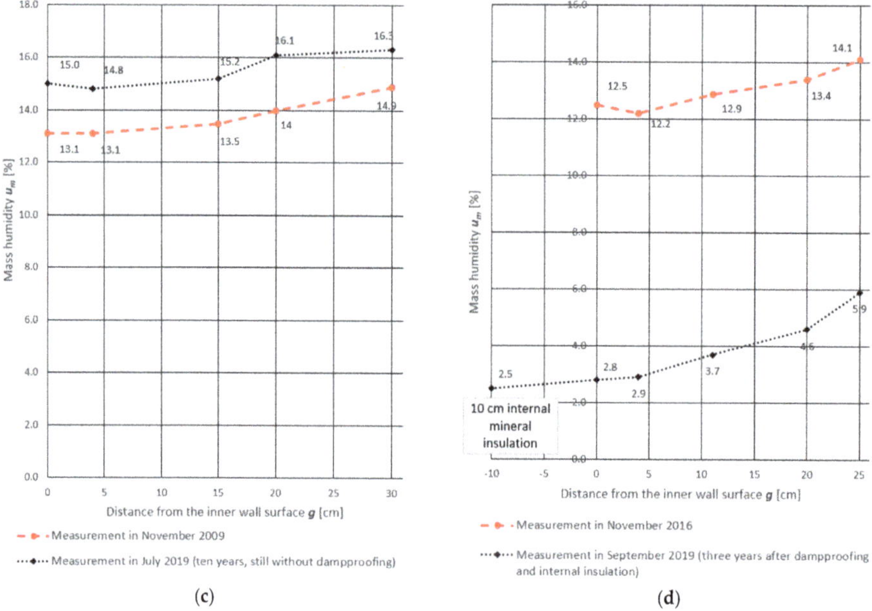

Figure 3. Results of mass moisture measurements on partition thickness in four different masonry walls made of solid brick: (**a**) Wall of building "A"; (**b**) wall of building "B"; (**c**) wall of building "C"; (**d**) wall of building "D".

3. Hygrothermal Calculations

Old historical buildings did not have any moisture or waterproof insulation. Nowadays, such objects are often waterproofed. They are revitalized and thermo-modernized. Revitalization and thermo-modernization are elements of the construction process, which should take into account the conditions of subsequent exploitation of the facilities. Each practical contact with historical buildings and monuments enriches the experience, but does not allow creating a universal model that comprehensively covers all the problems related to their safe maintenance or ensuring their durability. In order to be able to develop guidelines for this type of activities, thermal and humidity calculations were made with the assumption of the required amount of air exchange in the interior of buildings according to [36]. As can be seen, when analysing the data from Table 2, if this condition is not met, the results of the calculations in Table 3 will not even be close to reality. Calculations described in more detail in [4] were carried out for a homogeneous 38 cm thick wall (Figure 3, wall No. 4). It can be clearly observed here that in the case of a decrease in the level of air exchange in a room, the amount of moisture remaining permanently in the partition significantly increases.

Table 2. Dependence of the degree of barrier moisture on the air exchange rate.

No.	Air Exchange Rate per Hour	Moisture Level Increase during Central Heating Months	Increase of Moisture Retained within the Barrier
1	2	1	1
2	1	×2	×3
3	0.5	×3	×5

Table 3. Wall cross sections with moisture level, distribution model, etc.

No.	Wall Cross Section	Moisture Level during Central Heating Months and Retained Moisture kg/m^3	Moisture Distribution Model (Winter)	Coefficient of Heat Transfer U W/m$^2\cdot$K	Temperature Factor f_{Rsi}	The Possibility of Surface Condensation X—yes V—no
1		45.3/21.7		1.551	0.962	X
2		22.0/2.8		1.359	0.824	X
3		8.5/0.8		1.755	0.772	V
4		15.1/11.6		1.567	0.796	X
5		56.7/32.1		1.392	0.819	X
6		50.0/29.2		1.258	0.852	X
7		41.0/22.0		0.857	0.915	V
8		50.2/46.7		0.927	0.768	V

Models of each analysed type of masonry wall occurring in southern Poland (including those in areas belonging to Germany before World War II) are presented in Figure 4.

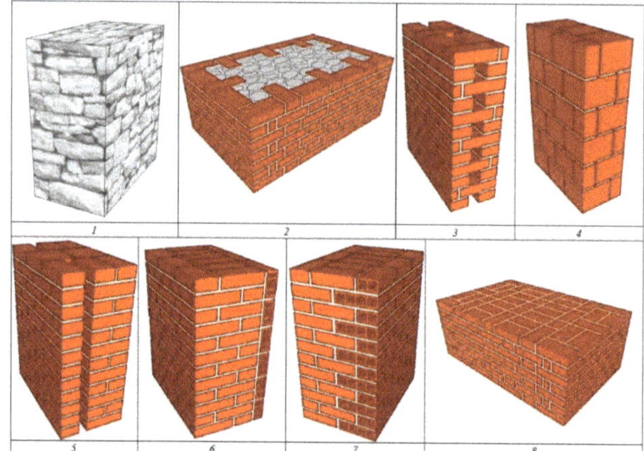

Figure 4. Models of the analyzed types of load-bearing and/or curtain masonry walls.

An attempt was made to make calculations for as many representative types of walls as could currently be found in historical buildings, as well as those erected in the last 200 years. The following types of walls were adopted:

1. Stone wall 80 cm thick. Walls of this type were made of field or lime stone (not treated or cut up), merged with lime mortar or lime mortar with the cement addition (from the late nineteenth century). A field stone is obtained from the fields (irregular, unworked, without clear petrographic features). In southern Poland it is usually granite left by retreating ice or sandstone.

2. Brick and stone or stone layered wall 62 cm thick. A wall such as wall No. 1, but very often the face of these walls was made of bricks.

3. Layered brick wall with air void (with vertical lining) 62 cm thick. A rarely occurring wall structure, due to its low strength and low thermal parameters. It consisted of two external facing walls, mainly 12 cm thick each, tied with stiffeners of the same thickness perpendicular to them, i.e., 12 cm (at the entire height of the joined layers).

4. Solid brick wall 38 cm thick. A very often used type of wall in residential and public buildings. The walls were made mainly of solid bricks with lime mortar, cement, and lime mortar, less frequently cement bricks.

5. Solid brick wall 12 + 25 cm thick with a 6 cm gap. A similar model of a wall to the one described above (No. 4), separated by an internal air gap.

6. Solid brick wall with a single layer of perforated brick on the inside. A solid brick wall finished on the inside with a single layer of perforated brick to limit heat loss and to allow easier removal of moisture accumulated in it.

7. Solid brick wall with double layer of hollow bricks on the inside. A solid brick wall finished on the inside with a double layer of hollow brick to reduce heat loss and to allow for easier removal of moisture accumulated in it.

8. Solid brick wall approx. 70 cm thick. A very often used type of wall in the ground level of large objects. The walls were made mainly of solid bricks of lime mortar, cement, and lime mortar.

The current regulations [25], in Annex 2, let us assume the required critical value of the temperature factor $f_{Rsi} = 0.72$ in rooms heated to the temperature of at least 20 °C, in residential buildings, collective residence, assuming that the average monthly relative humidity of indoor air is equal to $\phi = 50\%$.

The accepted calculation parameters and results are presented in Table 3.

Practically all partitions listed in Table 3 do not meet the standard [25], (results in column coefficient of heat transfer U), while detailed calculations performed with WUFI 2D v.3.4 [37] showed that surface moisture condensation will occur on the inner surfaces of some of the partitions (the possibility of surface condensation). Table 4 presents diagrams describing the amount of moisture in partitions in a 12-month cycle, in a 10-year period, counted from 2017.

Table 4. Dependence of the degree of barrier moisture on the air exchange rate.

No.	Graphs Showing Moisture Content in the Whole Barrier Cross Section over 10 Years in kg/m^3
1	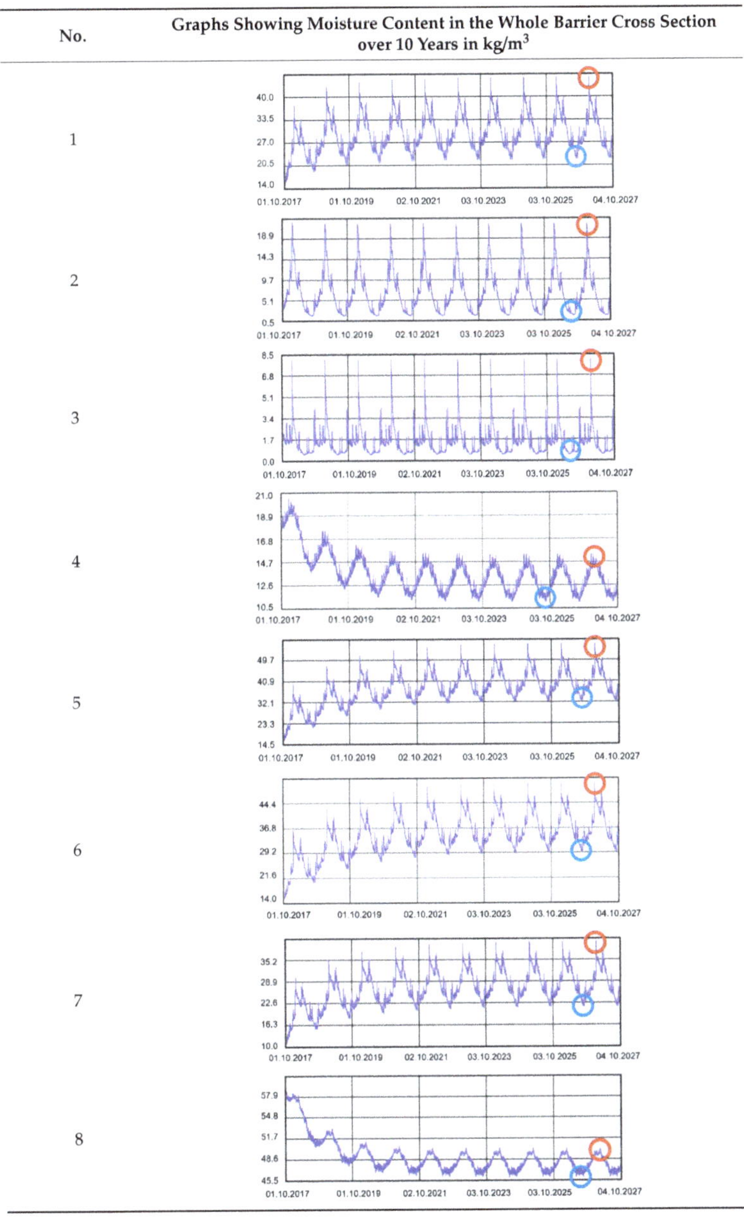
2	
3	
4	
5	
6	
7	
8	

In each of the analyzed partitions, the moisture level stabilizes after about 4–5 years of operation, assuming proper replacement of used and damp indoor air. The service life of partitions 5, 6, 7 during the first 4–5 years results in the increase of moisture inside them. None of the partitions meet the current requirements in terms of thermal insulation.

4. Monitoring and Hygrothermal Calculations of Masonry Adapted to New Conditions of Use

In order to meet modern guidelines, especially concerning the heat consumption of masonry walls in historical buildings, one of the best technical solutions seems to be the use of internal thermal insulation systems using capillary active mineral plaster, plasterboards, or blocks. Thanks to their very porous structure and very low density ($\rho < 150$ kg/m^3), these systems achieve high thermal insulation parameters. The thermal conductivity coefficient ($\lambda \leq 0.05$ W/m·K) and the diffusion resistance coefficient of water vapour (μ = approximately 5) are very low.

For a detailed analysis of the problem, one type of wall, a solid brick masonry wall of approximately 70 cm thickness (Table 3, No. 8), in which a horizontal damp-proof insulation was made and an additional insulation was applied from the inside with blocks of capillary active mineral layer 10 cm thickness. For bricklaying the blocks from the capillary active mineral layer was used as a systemic lightweight capillary active mortar with heat conduction coefficient λ equal to 0.2 W/m·K. Moisture monitoring was performed by means of 2 FP probes of the FOM2 system (Figure 5a), which were placed in the mortar layer between the original wall and the internal insulation layer (Figure 5b). The research was conducted continuously from November 2016 to September 2019.

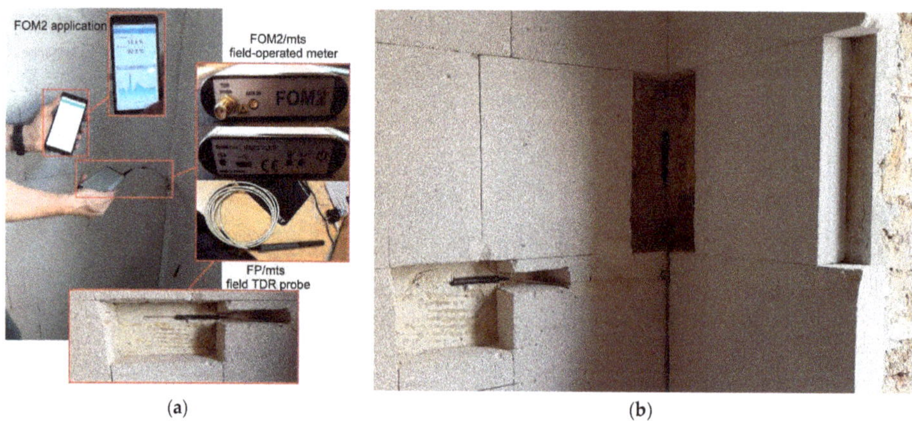

(a)   (b)

Figure 5. FOM2 measuring system: (**a**) Elements of the system; (**b**) probe No. 1 and probe No. 2 installed between the wall and the internal thermal insulation.

The FOM2 system used for measurements is a set consisting of FOM2/mts measuring device and FP/mts probes operating on the basis of TDR (time domain reflectometry) and electrical conductivity. The measurement results can be sent to mobile devices thanks to the use of a Bluetooth connection. Measurement parameters of the FOM2 device are presented in Table 5.

Table 5. FOM2 specification according to [38].

Range of Volumetric Moisture %	Range of Temperature °C	Range of Electrical Conductivity S/m	Moisture Absolute Error %	Temperature Absolute Error °C	Electrical Conductivity Relative Error %
0–100	(−20)–(+50)	0.000–1	±2	±0.5	±10

Probe No. 1 was placed on the flat surface of the wall and probe No. 2 in the corner at a height of approximately 1.2 m above the floor level. Both probes were connected with the measuring device FOM2/mts. Before the probe was placed in the wall, it was calibrated using information on the length of cable connections and two dielectric standards of reference: Air and water. After calibration in the air, the probe should be placed in water (preferably pure and deionized at a temperature of about 20 °C) and recalibrated. Only by carrying out the calibration process can you be sure of the results obtained.

5. Results and Discussion

The results of the humidity measurements together with a diagram of the probe location in the tested masonry wall are presented in Figure 6. The results of temperature measurement (probe No. 1, probe No. 2, indoor temperature, outdoor temperature) are presented in Figure 7.

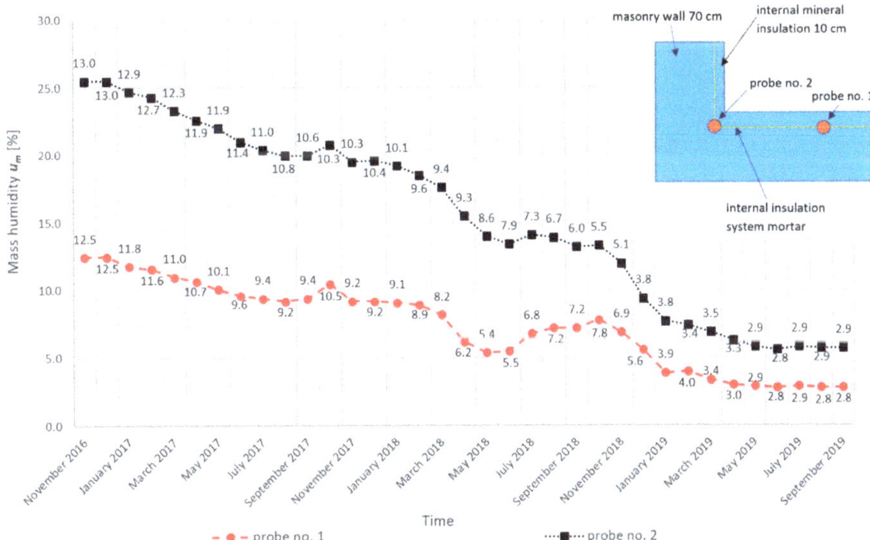

Figure 6. Results of humidity measurements together with a diagram of probes' location in the tested wall.

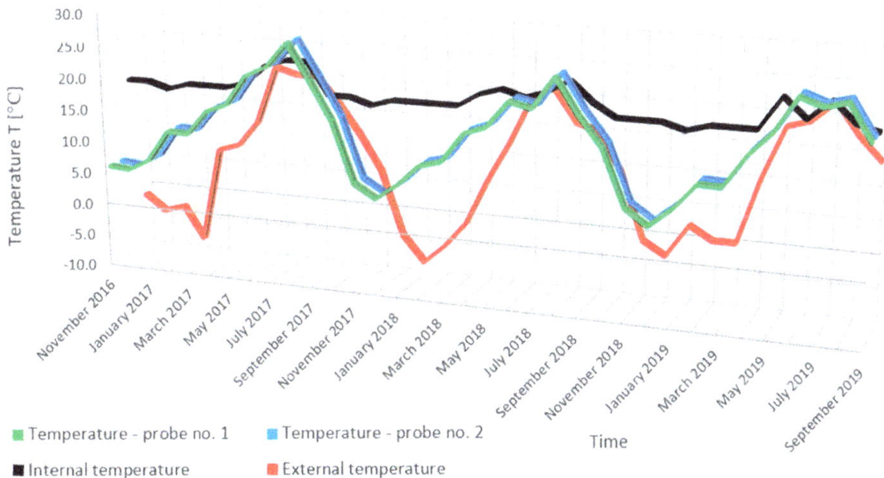

Figure 7. Temperature measurement results (probe No. 1, probe No. 2, indoor temperature, outdoor temperature).

Monitoring of hygrothermal parameters was performed by analyzing the data obtained from humidity and temperature measurements and material parameters using WUFI 2D v.3.4 software [37]. When analyzing the state and distribution of temperature isotherms and the adiabate of heat streams from the last 18 months (Figure 8) in the case of the corner of an uninsulated masonry wall and capillarily insulated from the inside with a capillary active mineral layer of 10 cm thickness, it can be stated that a significant decrease in the cooling of the corner is visible. A significant decrease in the water content of the insulated wall was also recorded (Figure 9).

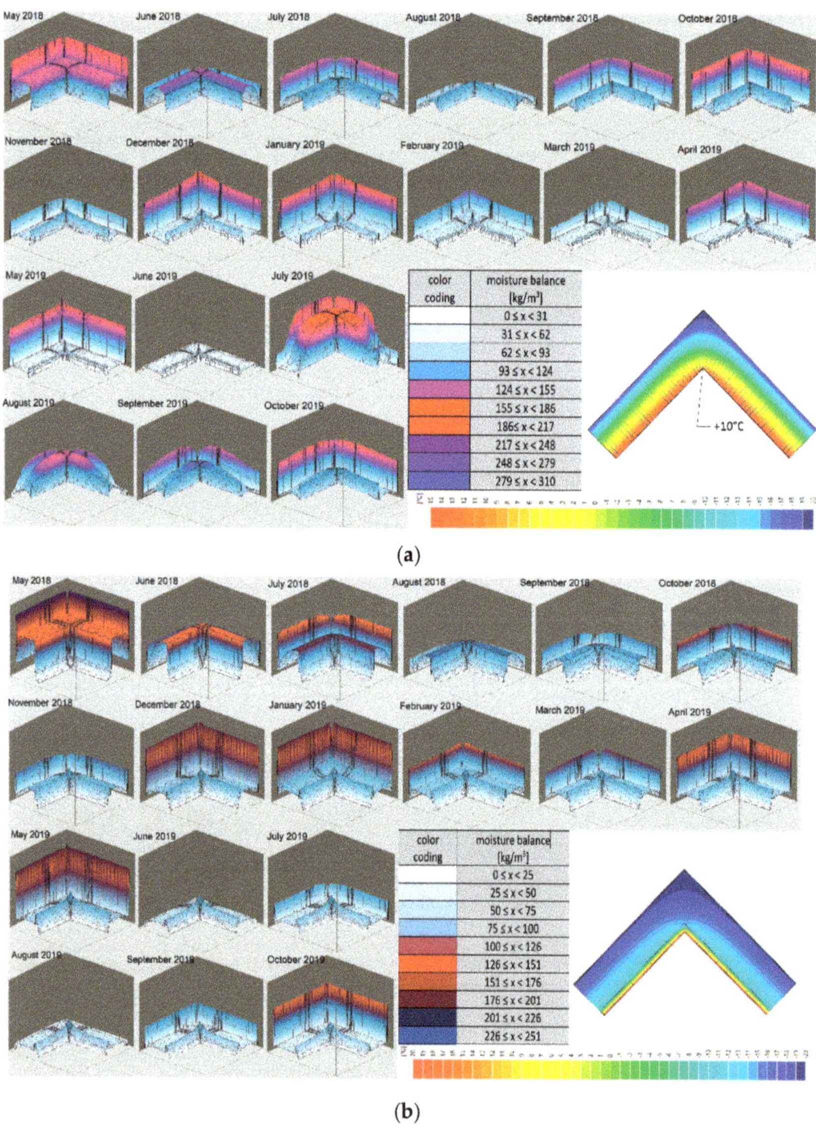

(**a**)

(**b**)

Figure 8. Isothermal distribution of temperature in the corner of the wall: (**a**) Uninsulated from the inside; (**b**) insulated from the inside.

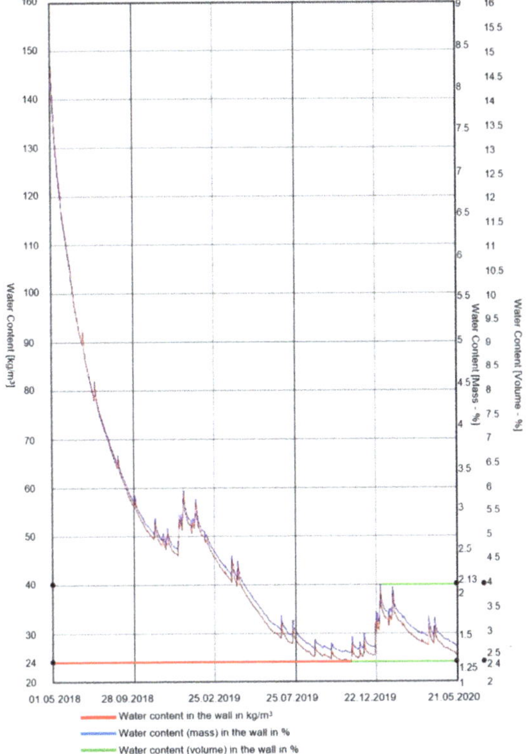

Figure 9. Water content in the insulated wall.

Similar numerical measurements were made for other types of the most popular historical external masonry walls presented in Figure 3. All of them were insulated from the inside with 10–15 cm mineral material (depending on demand). In all cases the coefficient of heat transfer U reached the values acceptable for the current standards of heat and moisture comfort of rooms. The lack of possibility of surface condensation was also observed.

The results obtained, apart from confirming the usefulness of internal thermal insulations performed with mineral systems, indicate one very important problem. In the case of insulation of walls from the inside, it is important to remember the proper protection of external wall surfaces. The paper [39] presents the results of the research which clearly show that additional insulation of historical masonry walls from the inside cannot be performed in walls without additional protection of their external surfaces against precipitation. In particular, if the masonry walls are exposed to acid rain corrosion worsens the mechanical properties of brick masonry. Otherwise, despite the internal insulation, it is possible to destroy the facade and see mould develop on the surfaces behind the internal insulation. Measurements and observations presented in [39–41] confirm these theses.

Thermal insulation of external partitions of buildings from the inside each time requires detailed hygrothermal analyses, because this method should not be treated as a universal one. Improper application of this method may lead to a faster technical wear and tear of partitions or even to their total degradation.

6. Conclusions

Walls are vertical partitions of buildings, which have been improved for many years not only to provide them with the required load-bearing capacity, but also to provide them with adequate protection against the external environment. The article, giving the example of a few selected external partitions of buildings, recalls the solutions applied over the last 200 years and a short history of their evolution to modern requirements. From a physical point of view, attention is drawn not only to the permanent problem of heat loss accompanying the building envelope, but also to the condensation and accumulation of moisture within it. The use of newer technologies did not solve this problem at all, and even showed how little is needed to damage such a structure the more complicated the structure is. New materials have improved the thermal insulation of partitions, but at the same time have made them more sensitive to both external and internal moisture, including condensation.

Author Contributions: Conceptualization, D.B. and L.B.; methodology, L.B.; software, D.B.; validation, D.B., L.B., Z.M. and K.R.; formal analysis, L.B.; investigation, D.B., L.B. and Z.M.; resources, D.B., L.B. and Z.M.; data curation, L.B. and D.B.; writing—original draft preparation, L.B. and D.B.; writing—review and editing, L.B.; visualization, L.B. and K.R.; supervision, D.B. and L.B.; project administration, L.B.; funding acquisition, Z.M. All authors have read and agreed to the published version of the manuscript.

Funding: This research received no external funding.

Conflicts of Interest: The authors declare no conflict of interest.

References

1. Amirzadeh, A.; Strand, R.K.; Hammann, R.E.; Bhandari, M.S. Determination and assessment of optimum internal thermal insulation for masonry walls in historic multifamily buildings. *J. Archit. Eng.* **2018**, *24*, 04018016. [CrossRef]
2. Al-Homoud, M.S. Performance characteristics and practical applications of common building thermal insulation materials. *Build. Environ.* **2005**, *40*, 353–366. [CrossRef]
3. Brauers, W.K.M.; Kracka, M.; Zavadskas, E.K. Lithuanian case study of masonry buildings from the Soviet period. *J. Civ. Eng. Manag.* **2012**, *18*, 444–456. [CrossRef]
4. Bajno, D.; Bednarz, Ł.; Grzybowska, A. Selected issues concerning thermal efficiency improvement of buildings. *Mater. Bud.* **2017**, *11*, 122–125.
5. De Berardinis, P.; Rotilio, M.; Marchionni, C.; Friedman, A. Improving the energy-efficiency of historic masonry buildings. A case study: A minor centre in the Abruzzo region, Italy. *Energy Build.* **2014**, *80*, 415–423. [CrossRef]
6. Ascione, F.; Cheche, N.; De Masi, R.F.; Minichiello, F.; Vanoli, G.P. Design the refurbishment of historic buildings with the cost-optimal methodology: The case study of a XV century Italian building. *Energy Build.* **2015**, *99*, 162–176. [CrossRef]
7. Hensley, J.E.; Aguilar, A. *Improving Energy Efficiency in Historic Buildings*; Government Printing Office: Washington, DC, USA, 2012.
8. Grytli, E.; Kværness, L.; Rokseth, L.S.; Ygre, K.F. The impact of energy improvement measures on heritage buildings. *J. Archit. Conserv.* **2012**, *18*, 89–106. [CrossRef]
9. Karaky, H.; Maalouf, C.; Bliard, C.; Moussa, T.; El Wakil, N.; Lachi, M.; Polidori, G. Hygrothermal and Acoustical Performance of Starch-Beet Pulp Composites for Building Thermal Insulation. *Materials* **2018**, *11*, 1622. [CrossRef]
10. Park, S.C. *Holding the Line: Controlling Unwanted Moisture in Historic Buildings*; Government Printing Office: Washington, DC, USA, 1996.
11. Park, S.C. *Moisture in Historic Buildings and Preservation Guidance*; ASTM International: West Conshohocken, PA, USA, 2009.
12. Smith, B.M. *Moisture Problems in Historic Masonry Walls: Diagnosis and Treatment*; U.S. Department of the Interior, National Park Service, Preservation Assistance Division: Washington, DC, USA, 1984.
13. D'Agostino, D. Moisture dynamics in an historical masonry structure: The Cathedral of Lecce (South Italy). *Build. Environ.* **2013**, *63*, 122–133. [CrossRef]

14. Menzel, C.A. *Der Praktische Maurer: Ein Handbuch für Architekten, Bauhandwerker und Bauschüler*; Verlag von J. J. Arnd.: Leipzig, Germany, 1900.

15. Warth, O.; Breymann, G.A. *Allgemeine Bau-Constructions-Lehre mit Besonderer Beziehung auf das Hochbauwesen. Band 1: Die Konstruktionen in Stein*; Gebhardt: Leipzig, Germany, 1896.

16. Żenczykowski, W. *General Construction. Volume II—Structures and Construction of Walls and Vaults*; Arkady: Warszawa, Poland, 1965.

17. Agliata, R.; Greco, R.; Mollo, L. Moisture measurements in heritage masonries: A review of current techniques. *Mater. Eval.* **2018**, *76*, 1478–1487.

18. Ferrara, C.; Barone, P.M. Detecting moisture damage in archaeology and cultural heritage sites using the GPR technique: A brief introduction. *Int. J. Archaeol.* **2015**, *3*, 57–61. [CrossRef]

19. Ocaña, S.M.; Guerrero, I.C.; Requena, I.G. Thermographic survey of two rural buildings in Spain. *Energy Build.* **2004**, *36*, 515–523. [CrossRef]

20. Lourenço, P.B.; Luso, E.; Almeida, M.G. Defects and moisture problems in buildings from historical city centres: A case study in Portugal. *Build. Environ.* **2006**, *41*, 223–234. [CrossRef]

21. Stojanović, G.; Radovanović, M.; Malešev, M.; Radonjanin, V. Monitoring of water content in building materials using a wireless passive sensor. *Sensors* **2010**, *10*, 4270–4280. [CrossRef] [PubMed]

22. Hoła, A.; Matkowski, Z. Nondestructive Testing of Damp Vault Brickwork in a Gothic-Renaissance City Hall. In Proceedings of the 11th European Conference on Non-Destructive Testing, Prague, Czech Republic, 6–11 October 2014; pp. 1–7.

23. Jasieńko, J.; Matkowski, Z. Salinity and moisture of brick walls in historic buildings-diagnostics, research methodology, rehabilitation techniques. *J. Herit. Conserv.* **2003**, *14*, 43–48.

24. Lucchi, E. Thermal transmittance of historical brick masonries: A comparison among standard data, analytical calculation procedures, and in situ heat flow meter measurements. *Energy Build.* **2017**, *134*, 171–184. [CrossRef]

25. *PN-EN ISO 13788:2012 Hygrothermal Performance of Building Components and Building Elements-Internal Surface Temperature to Avoid Critical Surface Humidity and Interstitial Condensation-Calculation Methods*; ISO: Geneva, Switzerland, 2012.

26. Brachaczek, W. Insulation of salt containing and moist walls of historic buildings using thermal insulation boards with sorption properties. *Constr. Rev.* **2017**, *12*, 56.

27. Zagorskas, J.; Zavadskas, E.K.; Turskis, Z.; Burinskienė, M.; Blumberga, A.; Blumberga, D. Thermal insulation alternatives of historic brick buildings in Baltic Sea Region. *Energy Build.* **2014**, *78*, 35–42. [CrossRef]

28. Walker, R.; Pavía, S. Thermal performance of a selection of insulation materials suitable for historic buildings. *Build. Environ.* **2015**, *94*, 155–165. [CrossRef]

29. Roberti, F.; Oberegger, U.F.; Lucchi, E.; Troi, A. Energy retrofit and conservation of a historic building using multi-objective optimization and an analytic hierarchy process. *Energy Build.* **2017**, *138*, 1–10. [CrossRef]

30. Toman, J.; Vimmrova, A.; Černý, R. Long-term on-site assessment of hygrothermal performance of interior thermal insulation system without water vapour barrier. *Energy Build.* **2009**, *41*, 51–55. [CrossRef]

31. Pavlík, Z.; Pokorný, J.; Pavlíková, M.; Zemanová, L.; Záleská, M.; Vyšvařil, M.; Žižlavský, T. Mortars with Crushed Lava Granulate for Repair of Damp Historical Buildings. *Materials* **2019**, *12*, 3557. [CrossRef]

32. Pender, R.J. The behaviour of water in porous building materials and structures. *Stud. Conserv.* **2004**, *49*, 49–62. [CrossRef]

33. Oxley, T.A.; Gobert, E.G. *Dampness in Buildings: Diagnosis, Treatments, Instruments*; Butterworth-Heinemann: Oxford, UK, 1994.

34. Hoła, A.; Matkowski, Z.; Hoła, J. Analysis of the moisture content of masonry walls in historical buildings using the basement of a medieval town hall as an example. *Procedia Eng.* **2017**, *172*, 363–368. [CrossRef]

35. Binda, L.; Cardani, G.; Zanzi, L. Nondestructive testing evaluation of drying process in flooded full-scale masonry walls. *J. Perform. Constr. Facil* **2010**, *24*, 473–483. [CrossRef]

36. Available online: https://sklep.pkn.pl/pn-b-03430-1983-az3-2000p.html (accessed on 2 December 2019).

37. Available online: https://wufi.de/en/wp-content/uploads/sites/9/2014/09/WUFI2D-3_Manual.pdf (accessed on 2 December 2019).

38. Available online: https://www.e-test.eu/uploads/5/4/4/3/54435037/fom2mpts-v1.2.pdf (accessed on 3 December 2019).

Materials **2020**, *13*, 505

39. Finken, G.R.; Bjarløv, S.P.; Peuhkuri, R.H. Effect of façade impregnation on feasibility of capillary active thermal internal insulation for a historic dormitory–A hygrothermal simulation study. *Constr. Build. Mater.* **2016**, *113*, 202–214. [CrossRef]
40. Zheng, S.; Niu, L.; Pei, P.; Dong, J. Mechanical Behavior of Brick Masonry in an Acidic Atmospheric Environment. *Materials* **2019**, *12*, 2694. [CrossRef] [PubMed]
41. Stryszewska, T.; Kańka, S. Forms of Damage of Bricks Subjected to Cyclic Freezing and Thawing in Actual Conditions. *Materials* **2019**, *12*, 1165. [CrossRef] [PubMed]

Article

Non-Destructive Evaluation of Rock Bolt Grouting Quality by Analysis of Its Natural Frequencies

Mario Bačić *, Meho Saša Kovačević and Danijela Jurić Kaćunić

Faculty of Civil Engineering, University of Zagreb, 10000 Zagreb, Croatia; msk@grad.hr (M.S.K.); djk@grad.hr (D.J.K.)
* Correspondence: mbacic@grad.hr; Tel.: +385-1-4639-636

Received: 5 November 2019; Accepted: 6 January 2020; Published: 8 January 2020

Abstract: Grouted rock bolts represent one of the most used elements for rock mass stabilization and reinforcement and the grouting quality has a crucial role in the load transfer mechanism. At the same time, the grouting quality as well as the grouting procedures are the least controlled in practice. This paper deals with the non-destructive investigation of grouting percentage through an analysis of the rock bolt's natural frequencies after applying an artificial longitudinal impulse to its head by using a soft-steel hammer as a generator. A series of laboratory models, with different positions and percentages of the grouted section, simulating grouting defects, were tested. A comprehensive statistical analysis was conducted and a high correlation between the grouting percentage and the first three natural frequencies of rock bolt models has been established. After validation of FEM numerical models based on experimentally obtained values, a further analysis includes consideration of grout stiffness variation and its impact on natural frequencies of rock bolt.

Keywords: rock bolt; grouting quality; dynamic response; natural frequency; finite element method

---

1. Introduction

Ensuring the stability of rock mass excavation is a challenging task for all persons involved in a construction process. Any kind of instability, which is a result of an inadequate support system can lead to undesirable temporal, financial and, what is most important, safety consequences. For decades, there has been a practice of using cement-based grouted rock bolts for rock mass reinforcement and much experience, both positive and negative, has been gained through their use. Rock bolts are an inevitable part of a large number of projects, such as rock slopes (Figure 1), where different levels of rock bolt design are used, ranging from empirical, analytical, and numerical, the latter having the most expanding applications.

Regardless of the design method used and the level of detail of the rock bolt parameter assignment, the usual practice is to take into consideration fully grouted rock bolts, without any grout defects along their length. Although some authors [1] have stated that grouted rock bolts present the optimal reinforcement method for strengthening weak or fractured rocks, a number of authors point to the fact that such rock masses may cause issues with grouting quality. For example, application of grouted rock bolts in karstic rock mass, characterized by the presence of voids and discontinuities, can lead to the severe grout loss [2,3], limiting their applicability in these types of rock masses [4,5]. The presence of karstic phenomena can lead to poor grouting, as shown in Figure 2. However, most of rock mass discontinuities are non-persistent, even though their visible traces can lead to conclusions as if they were open fractures. Assumption of full persistence gives incorrect predictions of hydraulic conductivity, as well as of grout flow. Since the visible trace length of a discontinuity can be a poor indicator of true persistence, some innovative methods, such as FERM given by Shang et al. [6] may help to detect the internal connections of rock bridges that hinder the flow of grout.

Figure 1. An example of the installed rock bolt system.

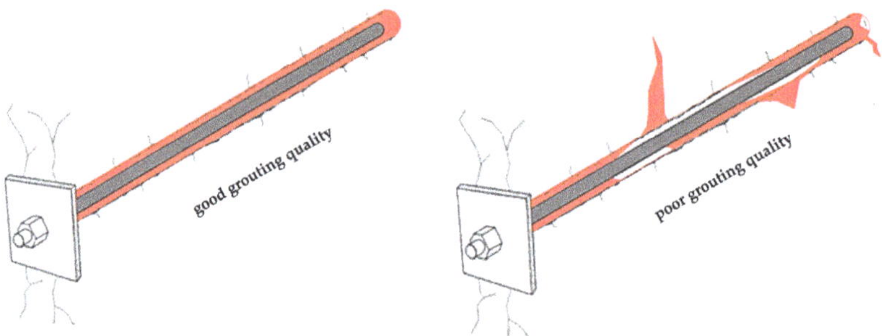

Figure 2. Good and poor grouting quality of a rock bolt.

Although the grouted rock bolts can produce a higher degree of load transfer in comparison to the other types of bolts, the grouting procedures and characteristics of the bonds between the bar and the grout and with the rock mass are the least controlled in practice [7], while the rock mass grouting procedures remain a predominantly empirical practice [8]. The importance of grouting quality can be seen through the potential rock bolt failure mechanisms, where most significant mechanisms include pull-out failures at bar–grout (B–G) contact, as well on grout–rock mass contact (G–RM). The main role of the grout is to provide a mechanism for transferring the load between the rock mass and the reinforcement element, and grouting quality has an important role in this task, along with the interface strength between bolt, grout, and rock, influenced by the adhesion, friction, and mechanical interlocking. Ren et al. [9] conclude that, based on many laboratory and in situ tests, the most common failure type of fully grouted rock bolts is at the bar–grout or grout–rock mass interface. It can be understood that if the grout and rock have similar strengths and if the required grouted rock bolt length is inadequate, then failure could occur at the bolt–grout interface [10]. If the surrounding rock mass is softer, as is the often case with karst, then the failure could happen at the grout–rock interface. There are other pull-out mechanisms which are not directly linked with grouting quality but depend on rock mass characteristics. However, these are out of scope of this paper. The relevant equations for the rock bolt pull-out capacities are:

$$R_{B\text{-}G} = d_{BR}{\cdot}\pi{\cdot}l{\cdot}\mu{\cdot}\tau_{B\text{-}G} \tag{1}$$

$$R_{G\text{-}RM} = d_{BH}{\cdot}\pi{\cdot}l{\cdot}\mu{\cdot}\tau_{G\text{-}RM} \tag{2}$$

The product of bar diameter d_{BR} (m) or borehole diameter d_{BH} (m), installed rock bolt length l (m), π and unit shear strength of bar–grout contact $\tau_{B\text{-}G}$ or unit shear strength of grout–rock mass contact $\tau_{G\text{-}RM}$ (kN/m^2), theoretically give the values of pull-out capacity at bar–grout contact $R_{B\text{-}G}$ (kN) and pull-out capacity at grout–rock mass contact $R_{G\text{-}RM}$ (kN), respectively. Since the pull-out resistance can be achieved only on the grouted portion of a rock bolt, it is of importance to properly determine the actual grouting percentage of the installed rock bolt, μ (—). It is a common designer's assumption of a fully grouted rock bolt ($\mu = 1$), but due to the above-mentioned reasons, a $\mu < 1$ value is more appropriate for the rock mass characterized by the presence of voids and discontinuities.

Much research has dealt, with greater or lesser success, with problems of determining the rock bolt grouting quality taking in consideration its importance for the pull-out capacity. Bačić et al. [11] and Song et al. [12] give the overview of the testing methods. Among many acoustic methods, the first one developed for determining grouting quality is the Boltometer method, which has several limitations when applied to soft and highly fractured rock masses. By using this method, the assumption of a lower sound velocity in the grout in comparison to the velocities in the bar and rock mass is made [13], but this may not be the case with karstic rock masses characterized by low stiffness [14]. If the rock mass has similar stiffness as the grout, a large portion of the energy will dissipate in the rock mass, making the reflections more difficult to recognize [15]. Some progress in this field was made through analysis of ultrasound guided waves [16], where the upgraded version of the Boltometer is currently in development under the name of rock bolt tester [17]. When it comes to vibration-based methods, the GRANIT system [18] stands out and it is getting frequently implemented in practice. This non-destructive system is however dominantly oriented towards determining the force in an anchor. Still, many findings of the GRANIT development process have been significant for the implementation of vibration based non-destructive testing of these types of structural elements. Kovačević et al. [3] had the basic objective of determining the dominant frequency, determined from the power spectrum, of rock bolt models and developing a correlation between the dominant frequency and the grouting percentage. A laboratory testing was conducted on 16 models representing different grouting defects and constructed of 25 mm bar embedded in square concrete blocks of 25 cm side length, having the total length of 3 m. Additionally, 30 field models of 3 m length were tested on location of a tunnel. Overall, the research was not continued and no clear link between the dynamic response and the grouting quality was established, but the authors stress that the procedure is very attractive for further research. Other methods based on an analysis of the frequency response are in development, but need further improvement. The potential of using electrical (such as time-domain reflectometry, electrical resistance, or the Mise-a-la-Masse method) and electromagnetic methods (such as ground penetrating radar) for determining grouting quality have been considered by many authors [19–21], but these methods have not found wider application in this domain.

This paper presents a research on the possibility of determining grouting percentage of rock bolts based on an analysis of the dynamical response by considering its natural frequencies after the generation of an artificial impulse to its head. By considering three natural frequencies instead of analyzing only the dominant frequency in the spectrum, as was done by Kovačević et al. [3], a more concrete correlation could be established so that the value of grouting percentage factor (μ) used for verification of designed grouting assumptions is determined in a more reliable manner.

2. Theoretical Background

Vibration based inspection (VBI) is a research domain, which has found its place not just in rock engineering but also in other branches of engineering. Much research has dealt with direct or inverse solutions [22], that is, with the assessment of the effect of structural damage on its parameters as well as with the problem of detecting, locating, and quantifying the extent of the problem. In order to understand the behavior of longitudinal wave propagation in a bar, a certain explanation must be given. Figure 3 shows a segment of a bar with cross section A, material density r, and elasticity modulus E, where the infinitesimal element of δx is in equilibrium. If the bar is loaded with dynamical force F(x,t) in the direction of its axis, it will yield a displacement u(x,t).

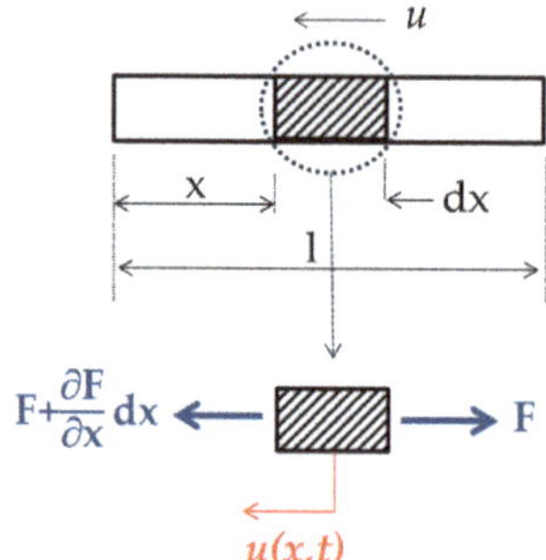

Figure 3. A segment of a bar with position of infinitesimal element.

By solving the equation:

$$\frac{\partial F}{\partial x} = \rho A \frac{\partial^2 u}{\partial t^2} \tag{3}$$

the natural frequencies of the bar can be obtained.

By rearrangement of the equation, while employing the expression for the velocity $v = \sqrt{\frac{E}{\rho}}$ and considering a solution in the form:

$$u(x,t) = U(x) \cdot (A \cos \omega t + B \sin \omega t) \tag{4}$$

where A and B are constants depending on the initial conditions and ω is the angular frequency, Equation (3) becomes:

$$\frac{\partial^2 U}{\partial x^2} + \frac{\omega^2}{v^2} U = 0 \tag{5}$$

and it has a solution of the form:

$$U = M \cos \frac{\omega x}{v} + N \sin \frac{\omega x}{v} \tag{6}$$

where M and N are constants depending on the bar's boundary conditions.

In the present paper, as will be shown later, it is of interest to determine the natural frequencies of a rock bolt model, where a system with both ends fixed, as well as a system with one end fixed and the other free, is considered. Due to the infinite number of solutions there are an infinite number of natural frequencies, one for each $n = 1, 2, 3, \dots$, which can be determined by solving Equation (3):

$$\omega_n = \frac{n\pi}{L} \sqrt{\frac{E}{\rho}} \text{ (for both fixed ends)} \tag{7}$$

$$\omega_n = \frac{(2n-1)\pi}{2L} \sqrt{\frac{E}{\rho}} \text{ (for one end fixed, the other free)} \tag{8}$$

3. Experimental Testing of Rock Bolt Models

3.1. Rock Bolt Models

For the purpose of this research, 51 physical models of grouted bars were made and tested at 1:1 scale. The models were designed to simulate different cases of grouting from the aspect of different grouting percentages and in regard to the position of the grout. Since steel rods of 2100 mm (2000 m bar with 50 mm thread on each side of a bar) were used and the grouting sections had a resolution of 10% of the total length of the rod, the grouted and non-grouted sections were 200 mm long. Therefore, a grouting section is represented with a 200 × 100 × 100 mm element (length × width × height) and the steel bar of 25 mm diameter is centered within this section. Figure 4 shows the schemes of model combinations. The reason for many models is to cover as large as possible a range of grouting percentages and defect positions. Based on the schemes and considering the possibility of generating impulses on both sides of the steel rod (50 mm bar threads are made on each side), a total of 94 testing combinations in respect to different grouting percentages and grout positions were made. In addition, the natural frequencies of the bar alone, i.e., 0% grouted percentage, were determined.

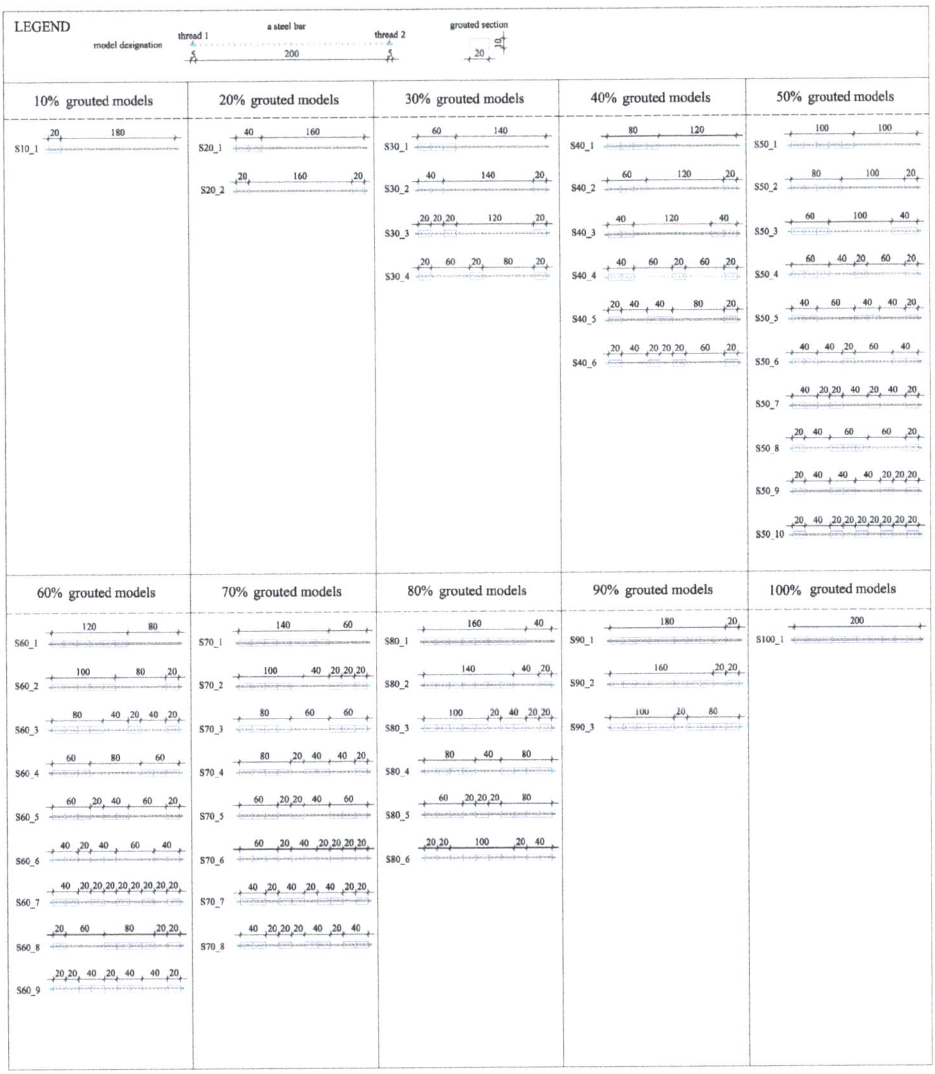

Figure 4. Complete scheme of rock bolt laboratory models.

After the wooden framework was filled with a grout based on the grouting scheme from Figure 4, and after smoothing of the upper surface, the framework was removed, and models were left to reach a 28-day strength before being subjected to the tests. Figure 5a shows, as an example, ten models that have a regular increase in the length of the grouted section. Figure 5b refers to, as an example, ten randomly selected rock bolts from each group of different grouting percentages. These two groups will be later used for numerical modelling.

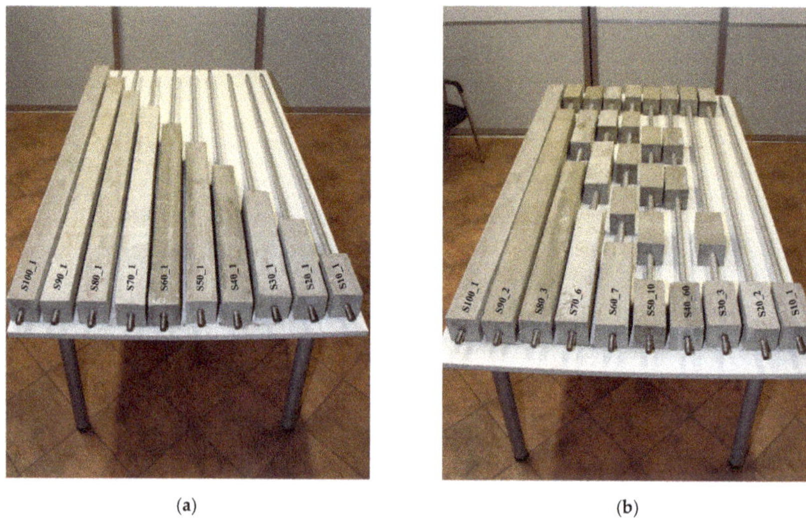

(a)   (b)

Figure 5. Rock bolt laboratory models: (**a**) Group I and (**b**) Group II.

3.2. Characteristics of the Models' Materials

For the construction of the physical models, the reinforcing steel bars B500B with 25 mm diameter were used. Since the parameters of the steel bar can be considered as 'reliable', in the sense that these steel bars were produced under strictly controlled manufacturing conditions, no additional tests for determination of their geometrical and physical and mechanical parameters were performed. In the present study, a cement-based mixture was used as the grout, made from pure Portland cement, water, and filling while no additives were used. During the preparation of the grout, samples were continuously taken in order to determine the physical–mechanical and chemical characteristics. A w/c ratio of 0.42 was used for all models. The laboratory tests to which all samples, were subjected after reaching the 28-day strength are:

1. Determination of the sample length (L), diameter (d), and mass (m) in order to determine its density (r).
2. Ultrasonic test, Figure 6, used to determine the elastic wave propagation velocity through samples. Based on this, the small strain stiffness could be determined for each sample.

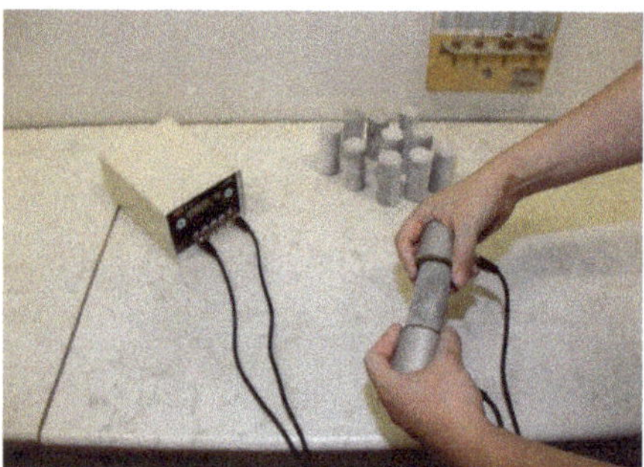

Figure 6. Conduction of ultrasound testing on grout samples.

All tests were documented, and the results were used as an input for subsequent numerical simulations. Figure 7 shows a wave velocity determined by the ultrasonic tests (a), density (b), and stiffness at a small strain (c). In particular, the figure shows the results for samples taken for the purpose of preparing a rock bolt model with 30% of grouted section, as an example. Since numerical analyses require only one value input, the testing results are averaged. This procedure has been carried out for samples of all rock bolt models.

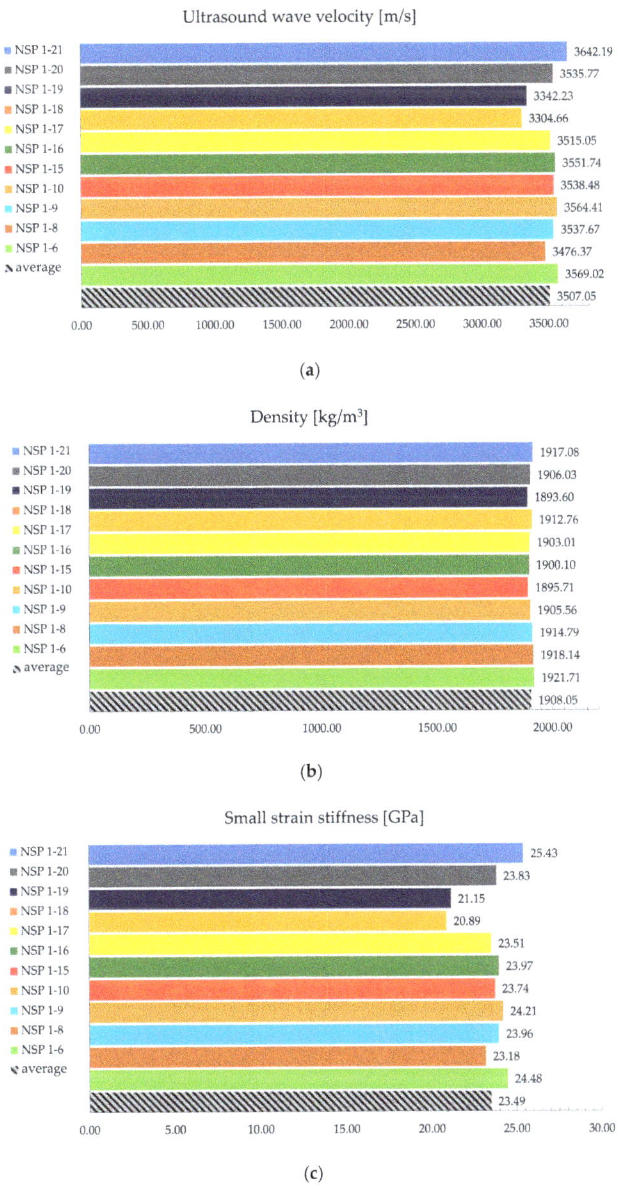

Figure 7. Laboratory test results for 30% grouted model: (a) ultrasound wave velocity, (b) density, and (c) small strain stiffness.

It could be seen from Figure 7 that the difference in wave velocities for all samples is up to 10%, while the density difference is up to 1.5%. A small strain stiffness difference is up to 18% which is expected taking into consideration that the square of wave velocity value is implemented into the equation. These exemplary results are in line with laboratory test results for all samples taken for preparation of all rock bolt models.

3.3. Acquisition Equipment

The equipment for acquisition of the rock bolt's dynamical response consisted of several elements, Figure 8, including a custom made nut with accelerometer, a hammer for generation of impulse and a laptop with developed state machine and a vibration input module.

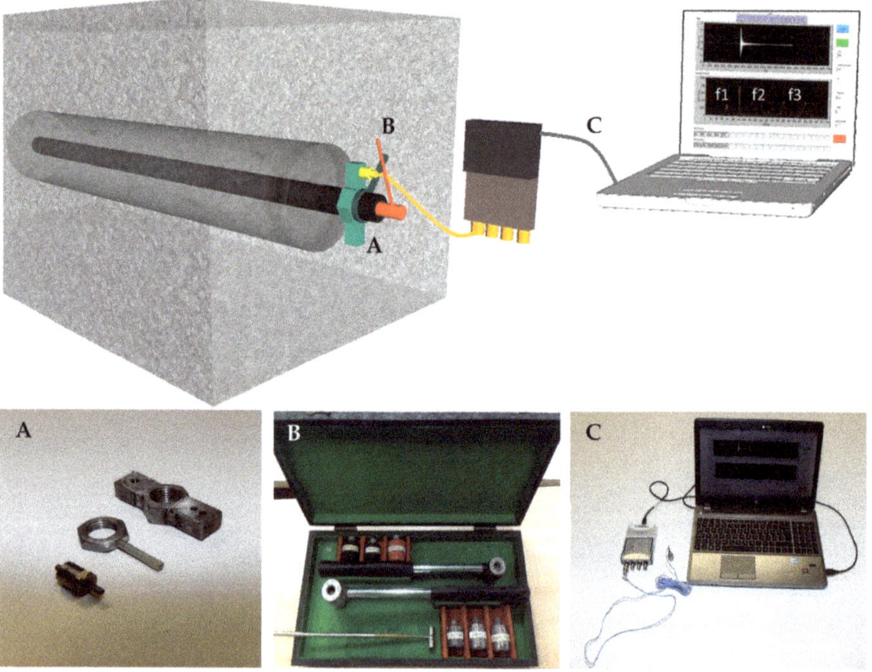

Figure 8. Elements of acquisition equipment: (**A**) a custom made nut with accelerometer; (**B**) a hammer for generation of impulse and (**C**) a laptop with developed state machine and a vibration input module.

At one end of the model, a nut is fitted on which an accelerometer for detection of longitudinal waves is mounted. The sensitivity of the accelerometer is 10.2 mV/(m/s^2), with a frequency range of 5 to 15,000 Hz, which has proved to be sufficient for this research. To generate an impulse at the top of the rock bolt, a soft-steel hammer was used. The impulse generated by a hammer is not an ideal delta function but has a certain time duration, which is, along with the shape of the frequency function, determined by the mass and rigidity of the hammer and the steel bar. Therefore, a soft-steel head proved to be the optimal form among an array of materials (hard steel, lead, rubber, wood, plastic). The registered signal goes through a vibration input module, which has built-in anti-aliasing filters that automatically adjust to define the sample rate. After arriving at the developed state machine, the signal is transformed from the time to the frequency domain by using the Fourier transformation.

Considering that even the accelerometer positioning nut itself influences the frequency response, it is necessary to determine such undesirable frequencies in order to eliminate them from the frequency spectrum. Therefore, a calibration system for the direct positioning of the accelerometer on the rock bolt head using a magnet was developed. Even though, in this case, the impact of the positioning equipment is minimized, such positioning cannot be considered as acceptable in real case conditions, because it would be impossible to generate an impact on the rock bolt head. By testing laboratory models, a magnet can be positioned at one end of the bar and the impulse imposed on the other end of the bar. Thus, the frequency response can be compared with the frequency response of the system using the nut as an integral part of the equipment. By overlapping these two spectra, conclusions as to the nut's 'contribution' from the aspect of the additional, undesirable, frequencies in the spectrum can be deduced.

It was found that the nut, after generating an impact on the rock bolt head, vibrates at higher frequencies than those of interest in this research. By overlapping the frequency spectrums, it can be concluded with a high level of reliability that the first three frequencies from the spectrum correlate relatively well, whereas at higher frequencies, especially at those greater than 4000 Hz, there is a significant difference in the frequency response. If only first three natural frequencies are considered, the proposed mode of accelerometer positioning using a nut can be considered acceptable. This was confirmed for all 94 tested combinations.

3.4. The State Machine

A state machine, developed within this research, provides the ability to collect data in the time domain and their real-time transformation into the frequency domain, thus gaining insight into the frequency spectrum already during the investigation. As a programming tool, the LabVIEW platform was used, where the concept of so-called 'state machines' is implemented. This concept relies on three elements necessary for proper functioning [23]: States, events, and actions. Several states were developed including 'acquire' state, 'analyses' state which includes implementation of the fast Fourier transform (FFT), 'save' state, 'wait' state, and 'stop' state. All these states provide a faster and more efficient data collection and analysis. After the user has defined all the states, the user interface is visible, where the user has, at any time, access to representations of the signal in the time domain and in the frequency domain. Additionally, a stacking procedure involving averaging three frequency logs for each tested configuration was conducted, Figure 9.

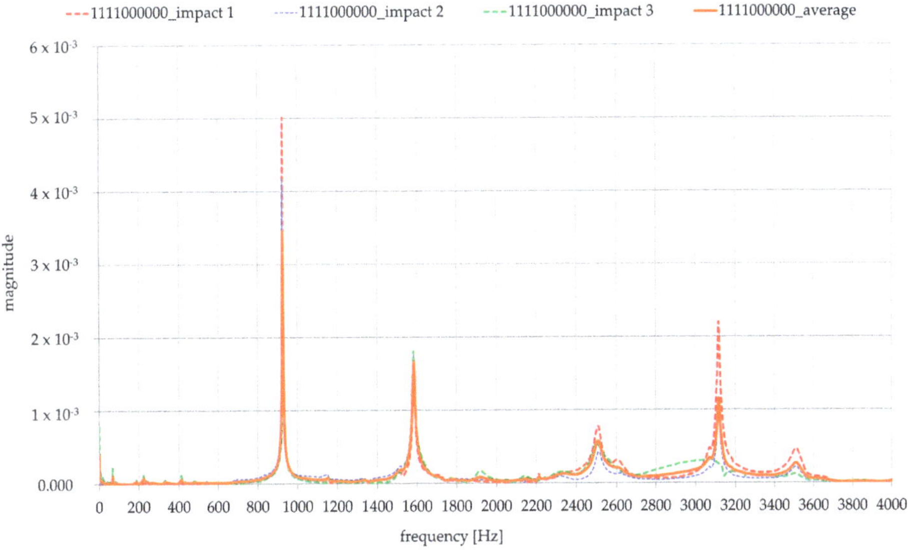

Figure 9. Frequency spectrum averaging procedure.

3.5. The Relevant Frequency Range

As a basic point for investigation of dynamical response, a frequency range needs to be determined. Taking into consideration Equations (7) and (8) and the relation between angular and ordinary frequency $\omega = 2\pi f$, a following arises:

$$f_n = n\frac{v}{2L}, \text{ where n } = 1, 2, 3\dots \tag{9}$$

The wave velocity in a steel bar can be calculated from the ratio of the elasticity and the density of the steel. It is 5048 m/s. In this case, the first three vibration frequencies of the 2100 mm steel rod are 1202, 2404, and 3060 Hz, as shown in Table 1. Comparing these first three calculated frequencies with the measured values in the laboratory shows only slight differences (up to 2.70%). Furthermore, the average wave velocity of a 2000 m long grout (length of bar minus the threads) is 3824 m/s (determined by conducted ultrasound laboratory tests) so the first three frequencies, using Equation (9), have values of 956, 1912, and 2868 Hz.

Table 1. Comparison of calculated and measured first three natural frequencies of steel bar and fully grouted model.

	Steel Bar (No Grout)		
	Calculated [Hz]	Measured [Hz]	Difference [%]
f_1	1202	1170	2.70
f_2	2404	2353	2.10
f_3	3606	3559	1.30
	Fully Grouted Model		
	Calculated [Hz]	Measured [Hz]	Difference [%]
f_1	544	603	9.78
f_2	1088	1143	4.81
f_3	1632	1676	2.63

In the case of the fully grouted model (fg_model), comprising a 2100 mm long bar and 2000 long grout section, both bar and grout have an influence on the natural frequencies of the model. Therefore, these elements can be considered as parallel coupled resistors which provide a resistance to the motion of the wave through the model. In this case, the wave velocity can be calculated as follows:

$$v_{fg\_model} = \frac{v_{steel} \cdot v_{grout}}{v_{steel} + v_{grout}} \tag{10}$$

The wave velocity for fully grouted model is 2176 m/s. The first three frequencies are then 544, 1088, and 1632 Hz, as shown in Table 1. Given the many factors that are the result of the expected error of the experimental model relative to the above calculations, it can be concluded that the matching of the first three frequencies is relatively good and a relevant frequency range could be established. The fact that the presence of grout causes a decrease of the wave velocity in rock bolts was stated in the literature [24].

Further, calculation of natural frequencies of the partially grouted models (pg_models) can be analytically calculated by considering the grouted sections as the parallel coupled resistors which are connected in series with the non-grouted sections. A coefficient of grouting percentage (μ) has to be implemented into equation, which has a form:

$$v_{pg\_model} = \left(\frac{v_{steel} \cdot v_{grout}}{v_{steel} + v_{grout}}\right) \cdot \mu + v_{steel} \cdot (1 - \mu) \tag{11}$$

The properties of a grout are implemented in Equation (11) through the velocity, based on Young's elasticity modulus and density obtained from laboratory tests conducted on grout samples taken during construction of models. For example, in case of a 70% grouted rock bolt model ($\mu = 0.7$), wave velocity is 3038 m/s, which gives the values of first three frequencies 723, 1447, and 2170 Hz. It is worth noting that Equation (11) determines the wave velocity based on grouting percentage and grout properties. The position of the defects in grouting are not covered by this equation.

3.6. Regression Analysis

Figure 10 shows the first three frequencies of each testing combination, set in the corresponding diagrams. Here, the distribution scatter plots show the pairs of values of natural frequency vs. grouting percentage. The presented scatter plots are a starting point in the correlation and regression analysis for establishing a connection between the grouting percentage and the first three natural frequencies for the whole dataset for each frequency (average trendline marked as 'A'). As an optimal regression function, the second order polynomial function was applied. The fitting uses the least squares method to minimize the squares of the residual deviations. The value of the coefficient of correlation implies no correlation or minor correlation when $0 \leq R < 0.2$, mild correlation for $0.2 \leq R < 0.4$, significant correlation for $0.4 \leq R < 0.7$, and high or very high correlation for $0.7 \leq R \leq 1.0$. In this case, the correlation coefficients R of the regression polynomial functions describing the overall dependence of the natural frequencies on the grouting percentage have the values of 0.79 (first natural frequency), 0.83 (second natural frequency), and 0.86 (third natural frequency). These values of correlation coefficients clearly demonstrate benefits of analyzing first, second, and third natural frequency for determination of the grouting percentage along the rock bolt. Therefore, better understanding of rock bolt dynamic response was achieved when compared to the Kovačević et al. [3] study where it was shown that there is no clear correlation when only dominant frequency is observed. The diagrams on Figure 10 also show the theoretical function (T curves), obtained by the means of Equations (10) and (11). The trend and values of a T-curve from Figure 10, are influenced by the grouting percentage and grout characteristics. The overall trend of natural frequency increase with the decrease of grouting percentage is evident for both theoretical and experimental curves. However, certain deviations of experimental results are noticed in comparison to theoretical values and these can be attributed to additional influence of grout position on natural frequencies values. Additionally, minimum and maximum boundaries for each natural frequency are stressed out.

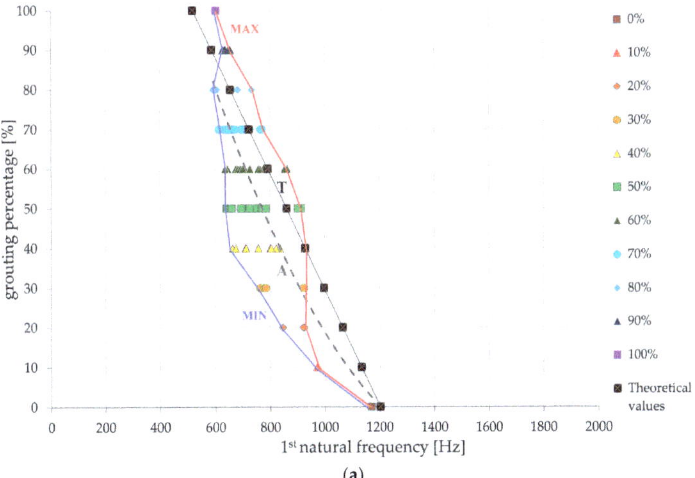

(a)

Figure 10. *Cont.*

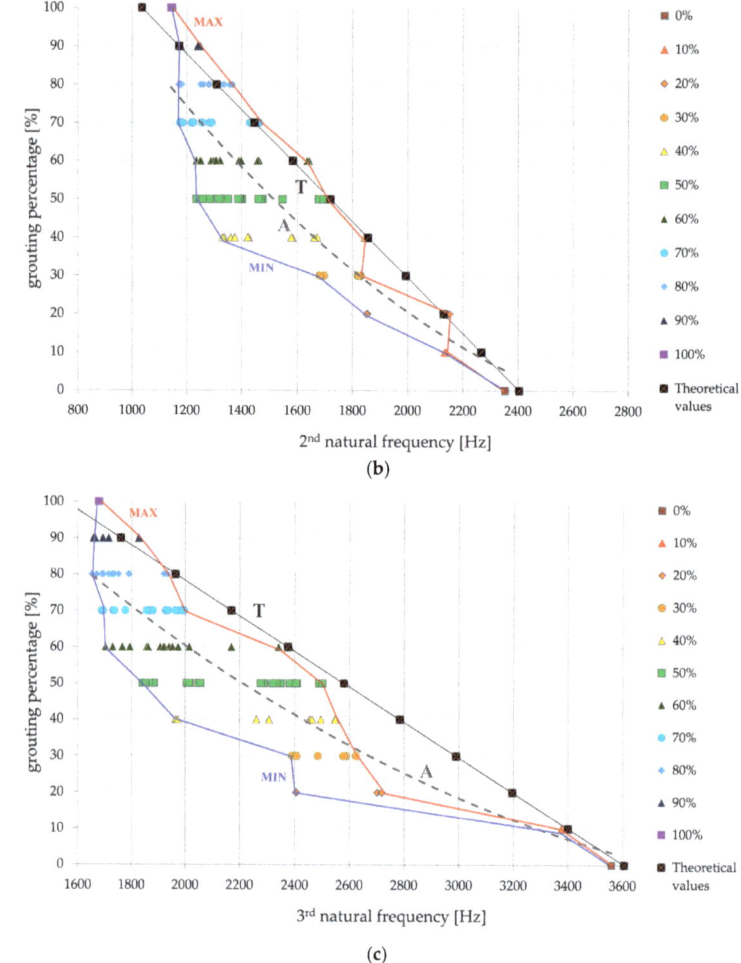

Figure 10. Regression functions for natural frequencies vs. grouting percentage for: (**a**) first, (**b**) second and (**c**) third natural frequency.

4. Numerical Modelling of Grouting Defects

In order to verify the experimental results and to further analyze the applicability for rock bolts installed in rock mass, a numerical modelling was conducted. Within the scope of this paper, two groups of rock bolt models were selected among the experimental models, as shown in Figure 5:

(A)   Group I refers to the models that have a regular increase in the length of the grouted section.

(B)   Group II refers to randomly selected rock bolts from each group of different grouting percentages.

4.1. Determination of Numerical Input Parameters

The laboratory test presented in Section 3.2 were conducted for all samples taken for the preparation of all the rock bolt models. An overview of material characteristics used as input for all numerical simulations is given in Table 2.

Table 2. Material properties of steel bar and grout mixtures.

Steel Bar, B500B	
Elasticity Modulus [Pa]	2×10^{11}
Density [kg/m^3]	7850
Poisson's Coef. [—]	0.30
Grouting Mixture, w/c = 0.42	
Elasticity Modulus [Pa]	2.2×10^{10} to 2.7×10^{10}
Density [kg/m^3]	1800 to 2000
Poisson's Coef. [—]	0.18

4.2. Composition of Numerical Models

Numerical analyses were carried out using a finite element method Ansys Workbench computer program, which is widely accepted in a wide range of human activities and is very often applied for scientific research purposes. Of the many modules offered by Ansys Workbench, in this case, the Harmonic Response module was used, since it enables the acquisition of a full frequency spectrum for the result of the force applied to the rock bolt head.

To carry out the analysis, the following steps were followed:

1. Design of geometric models.
2. Selection of a constituent model and bonds.
3. Input of the physical characteristics of the individual parts of the model.
4. Load input.
5. Discretization of models with appropriate elements.
6. Numerical calculation using the FEM method.
7. Overview and interpretation of results.

Geometrical models were created using the Autocad 3D 2019 computer program, which is a sophisticated universal utility that supports three-dimensional modelling of complex objects. The advantage of its application is its simplicity and compatibility with the selected Ansys Workbench program. For the purposes of the analysis, the two groups of rock bolts that were analyzed through the experimental setup, Group I and Group II, were modelled. They are shown in Figure 11.

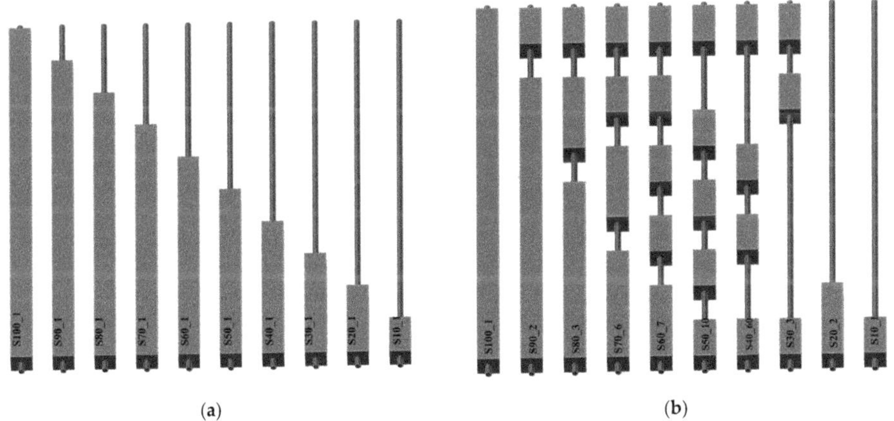

(a)   (b)

Figure 11. Rock bolt geometrical models: (a) Group I and (b) Group II.

Since the numerical analyses in the presented research are dynamic analyses, which imply small strains, the constituent model chosen for all materials is linear-elastic, which assumes that the stresses are directly proportional to deformation. Once the constitutive model and the material characteristics of the material are defined, it is necessary to define the bonds between the individual materials. Since this numerical modelling employs a linear dynamic analysis, it is recommended to avoid nonlinear contacts. Therefore, the "no separation" or "bonded" contacts from a range of types of contact between the steel and the grout could be chosen. The "no separation" contact configuration can be applied to the sides of the region or to its edges, and by its application the separation of the sides that are in contact is not allowed, but a little slip of the sides may appear without friction. However, in case the so-called "bonded" type of contact is chosen, the 3D contact surface–surface element (contact 174) is selected for simulation of contacts between the target surfaces (steel–grout). A bonded contact, used in this research, can be applied to all regions that are in contact and in that case sliding or separation is not allowed, and the bodies tied to this contact act as mutually glued. As a method for simulating contacts, the so-called MPC (multi-point constraint) method is chosen where the contact is directly and efficiently formulated due to the internal addition of edge shift equations to match the displacement between the contact points. Since the magnitude of the force applied to the rock bolt head was not considered in experimental testing, it did not have a significant impact on the numerical modelling results. Namely, by varying the force magnitude, frequency spectra with equal values of natural frequencies were obtained which, depending on the magnitude of the force applied, have different amplitudes and these are of no interest for the subject of this research. Since the load magnitude has no effect on the value of the natural frequencies but only on their amplitudes, an anchorage load of 1 N has been imposed on the rock bolt head in all numerical analyses, in the direction of the anchor axis. The model discretization was performed in such a way that the steel bar was divided into 50 mm length elements while the grout was divided into 50 × 50 × 50 mm bodies. With mesh refinement, the numerical results are closer to the experimental, but the calculation time is prolonged. Figure 12a shows the discretization of the fully grouted model (S100_1), while Figure 12b shows the discretization of the rock bolt model S60_7 from Group II.

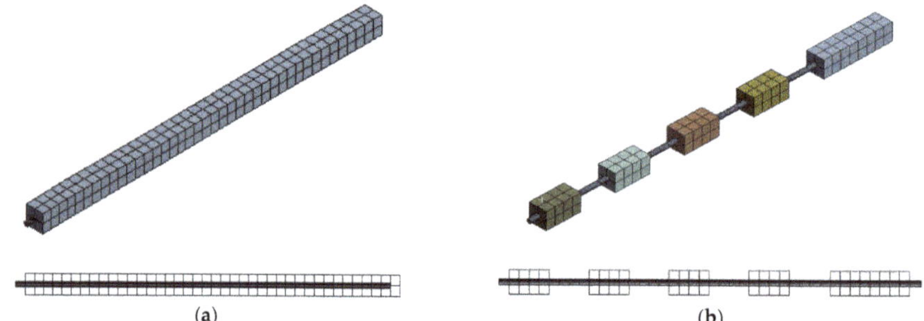

(a)   (b)

Figure 12. Discretization of numerical models: (**a**) S100_1 and (**b**) S60_7.

4.3. Comparison of Numerical Modelling and Experimental Results

The results of the numerical analysis are presented as a comparison of the first three natural frequencies with the results of the laboratory model testing of models from Group I and Group II. Figure 13 shows the comparison between calculated and measured values of natural frequencies for the Group I and Group II, respectively.

The difference between the numerically calculated first three frequencies and the experimentally measured first three frequencies for both groups of selected groups of models, Group I and Group II, are up to 12%, which can be considered acceptable taking into account possible irregularities in the production of laboratory models (despite the strictly controlled conditions), as well as variations in the measuring environment. Accepting the above, the further numerical calculations include the consideration of grout stiffness variation influence on the dynamic response of models.

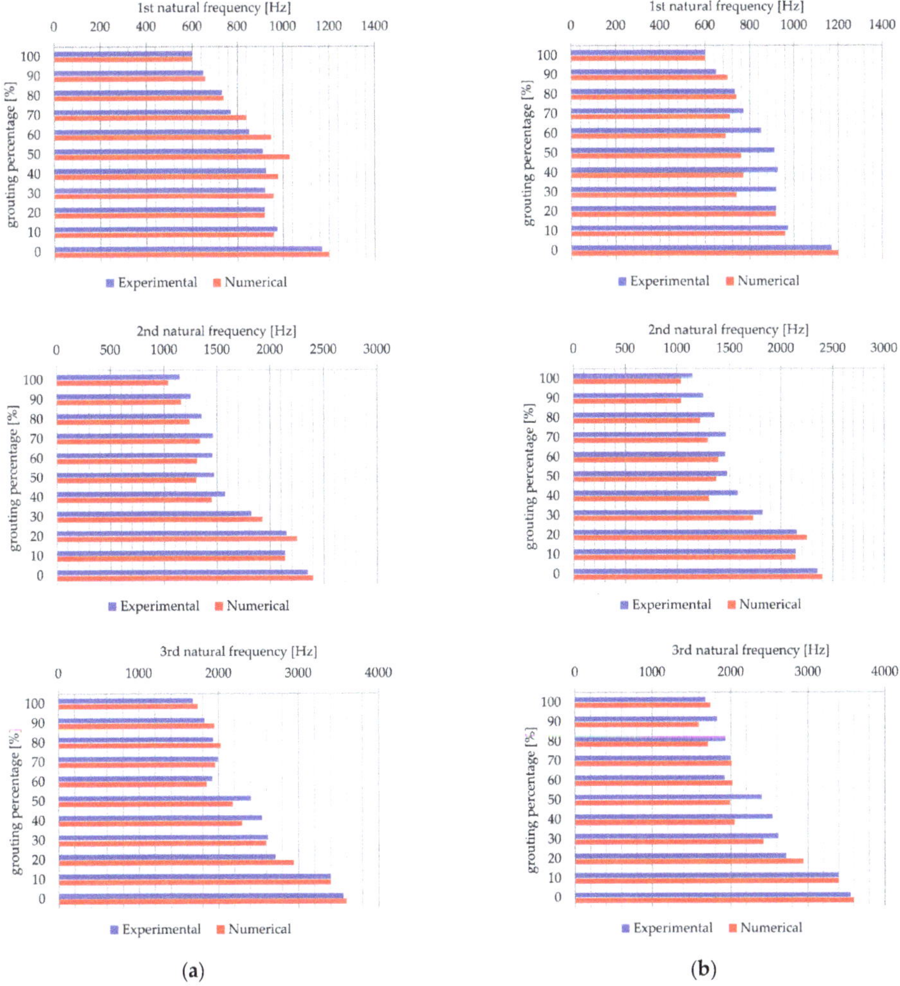

Figure 13. Comparison of the values of first, second, and third calculated natural frequency and experimentally obtained values for: (**a**) Group I and (**b**) Group II.

4.4. Analysis of Grout Stiffness Influence

Following the comparison of numerical and experimental results, the next phase of numerical modelling implies variation of small strain stiffness of grouts for all models within Groups I and II, in order to evaluate the influence of stiffness on the natural frequencies. Considering that the numerical input, besides values of elasticity modulus, requires a density value input, it is selected in all analysis as constant reference value of 2000 kg/m^3. Used values of elasticity modulus, density, and (calculated) wave velocities are shown in Table 3.

A range of elasticity modulus from 20 to 40 GPa is varied since these values represent commonly achieved values for rock bolt systems. Figure 14 shows the influence of the modulus of elasticity of the grouting mixture on the values of the natural frequencies for Group I, while in Figure 15 shows the same for Group II. A clear reduction of the natural frequency values is apparent from Figures 14 and 15 with a decrease of the modulus of elasticity of the grout. This reduction is least pronounced for the first natural frequency and is most clearly expressed for the third natural frequency. In addition, the reduction of the value of the natural frequencies is generally more significant in the models with a higher percentage of grouting, although locally there are values that do not meet the specified. The variation of the elastic modulus does not, however, affect the general trend of increasing natural frequency with the reduction in grouting percentage, whereby the shape of the natural frequency—the grouting percentage curve is independent of the value of modulus of elasticity.

Table 3. Parameters for analysis of grout stiffness influence.

Analysis No.	Elasticity Modulus [Pa]	Density [kg/m^3]	Wave Velocity [m/s]
1	4.0×10^{10}	2000	4472
2	3.5×10^{10}	2000	4183
3	3.0×10^{10}	2000	3873
4	2.5×10^{10}	2000	3536
5	2.0×10^{10}	2000	3162

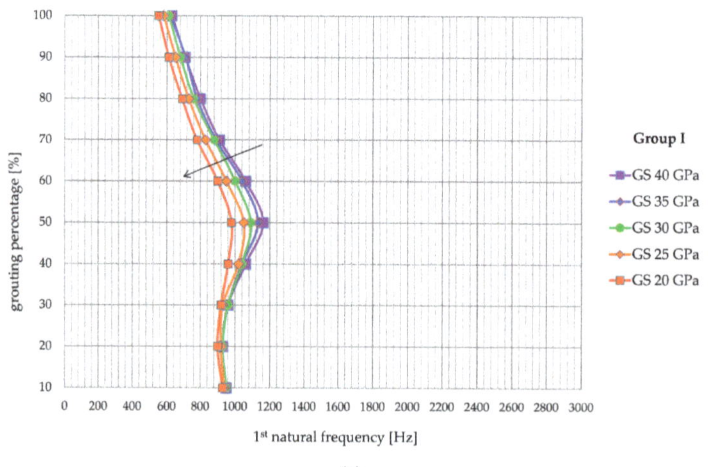

(a)

Figure 14. *Cont.*

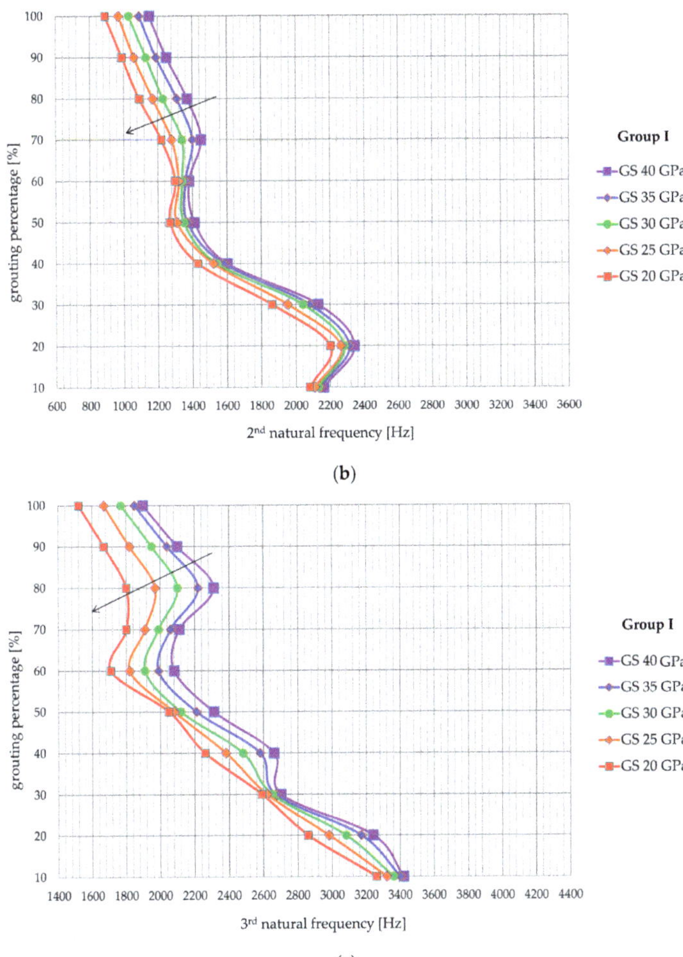

(**b**)

(**c**)

Figure 14. Influence of grout stiffness on (**a**) first, (**b**) second, and (**c**) third natural frequency of Group I models.

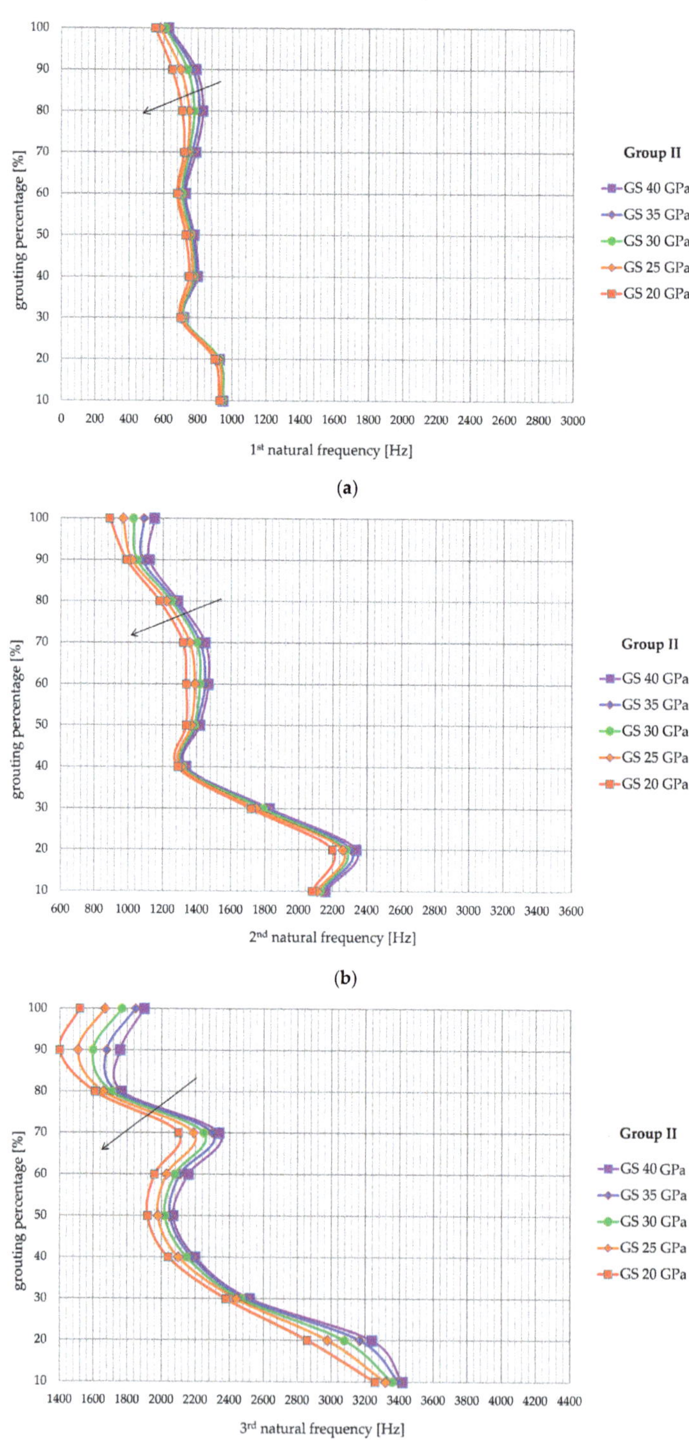

Figure 15. Influence of grout stiffness on (**a**) first, (**b**) second, and (**c**) third natural frequency of Group II models.

5. Conclusions

An experimental study including determination of a correlation between grouting percentage and first three natural frequencies of a laboratory rock models was conducted. In total, 94 different combinations of different grouting percentages and positions were tested. Prior to testing, a data acquisition setup was established consisting of a custom-made nut for positioning of the accelerometer, a hammer made of soft steel for the generation of a dynamical impulse, a vibration input module, and a developed software program. By analyzing the statistical distribution of the data and by establishing a regression function (second order polynomial functions were employed) and regression coefficients, it was shown that a high correlation exists between the natural frequencies and the grouting percentage. This clearly demonstrates the benefits of analyzing first, second, and third natural frequency for determination of grouting percentage along the rock bolt and provides better understanding of rock bolt dynamic response when compared to previous studies which analyzed only dominant frequency. Following experimental testing, the objectives of the numerical simulation included comparison with experimental results, as well as the analysis of the effect of the grout stiffness on the frequency response of rock bolts. When comparing the results of numerical analyses with the results of experiments, a similar general increasing trend of the first three natural frequencies with a reduction in grouting percentage is clearly visible. Moreover, the values of the natural frequencies obtained by numerical analysis have approximately the same values as the natural frequencies obtained by experimental testing. Numerical analyses have also shown that the reduction of the modulus of elasticity of the grouting mixture, leads to the reduction of the values of the natural frequencies while the variation of the elastic modulus does not affect the general trend of increasing natural frequencies with the reduction in grouting percentage, whereby the shape of the natural frequency—the grouting percentage curve is independent of the modulus of elasticity. The potential of using numerical methods, which are properly calibrated, for analysis of more complex grouting defects is therefore evident. Overall findings of this study are in establishing a clear correlation between first, second, and third natural frequency of rock bolt and the percentage of grouting along the rock bolt. After the proposed grouting evaluation method is validated in situ condition, it can be used as a control quality measure to check whether the installation procedures conform with the design requirements. However, because of the change of natural frequencies with rock bolt load increase, the method is applicable only to non-loaded rock bolts, tested immediately after their installation.

Author Contributions: Conceptualization, M.B. and M.S.K.; Data curation, M.B.; Formal analysis, M.B. and M.S.K.; Investigation, M.B. and D.J.K.; Methodology, M.B. and M.S.K.; Validation, M.B. and D.J.K.; Visualization, M.B. and D.J.K.; Writing—original draft, M.B. and M.S.K.; Writing—review & editing, D.J.K. All authors have read and agreed to the published version of the manuscript.

Funding: This research was funded by the European Commission Innovation and Networks Executive Agency, project DESTinationRAIL (Decision Support Tool for Rail Infrastructure Managers), grant number. 636285.

References

1. Peng, S.; Tang, D. Roof Bolting in Underground Mining: State of the Art Review. *Int. J. Min. Eng.* **1984**, *2*, 1–42. [CrossRef]
2. Rustan, A.; Cunningham, C.; Fourney, W.; Spathis, A.; Simha, K.R.Y. *Mining and Rock Construction Technology Desk Reference: Rock Mechanics, Drilling & Blasting*; Taylor and Francis Group, CRC Press: Boca Raton, FL, USA, 2010.
3. Kovačević, M.S.; Vrkljan, I.; Szavits-Nossan, A. Nondestructive procedure for testing grouting quality of rock bolt anchor. In Proceedings of the 9th International Congress on Rock Mechanics, Paris, France, 25–28 August 1999; Vouille, G., Berest, P., Eds.; pp. 1475–1478.
4. Yassien, A.M. 2-D Numerical Simulation and Design of Fully Grouted Bolts for Underground Coal Mines. Ph.D. Thesis, West Virginia University, Morgantown, WV, USA, 2003.

5. Luo, J.A. New Rock Bolt Design Criterion and Knowledge-Based Expert System for Stratified Roof. Ph.D. Thesis, Virginia Polytechnic Institute and State University, Blacksburg, VA, USA, 1999.

6. Shang, J.; Hencher, S.R.; West, L.J.; Handley, K. Forensic Excavation of Rock Masses: A Technique to Investigate Discontinuity Persistence. *Rock Mech. Rock Eng.* **2017**, *50*, 2911–2928. [CrossRef]

7. Arbanas, Ž.; Grošić, M.; Jurić-Kaćunić, D. Influence of grouting and grout mass properties on reinforced rock mass behavior. In Proceedings of the 4th symposium of Croatian Geotechnical Society, Opatija, Croatia, 4–7 October 2006; Szavits-Nossan, V., Kovačević, M.S., Eds.; pp. 55–64.

8. Kim, J.-W.; Chong, S.-H.; Cho, G.-C. Experimental Characterization of Stress- and Strain-Dependent Stiffness in Grouted Rock Masses. *Materials* **2018**, *11*, 524. [CrossRef] [PubMed]

9. Ren, F.F.; Yang, Z.J.; Chen, J.F.; Chen, W.W. An analytical analysis of the full-range behavior of grouted rockbolts based on a tri-linear bond-slip model. *Constr. Build. Mater.* **2010**, *24*, 361–370. [CrossRef]

10. Jalalifar, H. A New Approach in Determining the Load Transfer Mechanism in Fully Grouted Bolts. Ph.D. Thesis, School of Civil, Mining and Environmental Engineering, University of Wollongong, Wollongong, Australia, 2006.

11. Bačić, M.; Gavin, K.; Kovačević, M.S. Trends in non-destructive testing of rock bolts. *Građevinar* **2019**, *71*, 823–831. [CrossRef]

12. Song, G.; Li, W.; Wang, B.; Ho, S.C.M. A Review of Rock Bolt Monitoring Using Smart Sensors. *Sensors* **2017**, *17*, 776. [CrossRef] [PubMed]

13. Agnew, G.D. *An Investigation of Methods for Producing a Non-Destructive Grouted Tendon Tester. Consultancy Report Produced for COMRO*; Department of Electrical Engineering, University of Witwatersrand: Johannesburg, South Africa, 1990.

14. Jurić Kaćunić, D.; Arapov, I.; Kovačević, M.S. New approach to the determination of stiffness of carbonate rocks in Croatian karst. *Građevinar* **2011**, *61*, 177–185.

15. Thurner, H.F. Boltometer-Instrument for non-destructive testing of grouted rock bolts. In Proceedings of the 2nd International Symposium on Field Measurements in Geomechanics, Kobe, Japan, 6–9 April 1987; Sakurai, S., Ed.; pp. 135–143.

16. Beard, M.D.; Lowe, M.J.S. Non-destructive testing of rock bolts using guided ultrasonic waves. *Int. J. Rock Mech. Min. Sci.* **2003**, *40*, 527–536. [CrossRef]

17. Stepinski, T.; Matsson, K.J. Rock bolt inspection by means of RBT instrument. In Proceedings of the 19th World Conference on Non-Destructive Testing, Munich, Germany, 13–17 June 2016.

18. Ivanović, A.; Starkey, A.; Neilson, R.D.; Rodger, A.A. The influence of load on the frequency response of rock bolt anchorage. *Adv. Eng. Softw.* **2003**, *34*, 697–705. [CrossRef]

19. Cheung, W.M.; Lo, D.O.K. *Interim Report on Non-Destructive Tests for Checking the Integrity of Cement Grout Sleeve of Installed Soil Nails Non-Destructive Tests for Checking the Integrity of Cement. Report No. 176*; The Government of the Hong Kong Special Administrative Region: Hong Kong, China, 2005.

20. Lee, C.F. *Review of Use of Non-Destructive Testing in Quality Control in Soil Nailing Works Non-Destructive Testing in Quality Control. Report No. 219*; The Government of the Hong Kong Special Administrative Region: Hong Kong, China, 2007.

21. Kelly, A.M.; Jager, A.J. *Critically Evaluate Techniques for the In Situ Testing of Steel Tendon Grouting Effectiveness as a Basis for Reducing Fall of Ground Injuries and Fatalities*; CSIR (The Council for Scientific and Industrial Research): Pretoria, South Africa, 1996.

22. Ruotolo, R.; Surace, C. Natural frequencies of a bar with multiple cracks. *J. Sound Vib.* **2004**, *272*, 301–316. [CrossRef]

23. Bitter, R.; Mohiuddin, T.; Nawrocki, M. *LabView: Advanced Programming Techniques*, 2nd ed.; Taylor and Francis Group, CRC Press: Boca Raton, FL, USA, 2006.

24. Milne, G.D. Condition Monitoring & Integrity Assessment of Rock Anchorages. Ph.D. Thesis, University of Aberdeen, Aberdeen, UK, 1999.

MDPI
St. Alban-Anlage 66
4052 Basel
Switzerland
Tel. +41 61 683 77 34
Fax +41 61 302 89 18
www.mdpi.com

Materials Editorial Office
E-mail: materials@mdpi.com
www.mdpi.com/journal/materials

www.ingramcontent.com/pod-product-compliance
Lightning Source LLC
LaVergne TN
LVHW071629170726
843515LV00009B/2521